IET ENERGY ENGINEERING SERIES 254

Fuzzy Logic Control in Energy Systems, 2nd Edition

Other volumes in this series:

Volume 1	**Power Circuit Breaker Theory and Design** C.H. Flurscheim (Editor)
Volume 4	**Industrial Microwave Heating** A.C. Metaxas and R.J. Meredith
Volume 7	**Insulators for High Voltages** J.S.T. Looms
Volume 8	**Variable Frequency AC Motor Drive Systems** D. Finney
Volume 10	**SF_6 Switchgear** H.M. Ryan and G.R. Jones
Volume 11	**Conduction and Induction Heating** E.J. Davies
Volume 13	**Statistical Techniques for High Voltage Engineering** W. Hauschild and W. Mosch
Volume 14	**Uninterruptible Power Supplies** J. Platts and J.D. St Aubyn (Editors)
Volume 15	**Digital Protection for Power Systems** A.T. Johns and S.K. Salman
Volume 16	**Electricity Economics and Planning** T.W. Berrie
Volume 18	**Vacuum Switchgear** A. Greenwood
Volume 19	**Electrical Safety: A guide to causes and prevention of hazards** J. Maxwell Adams
Volume 21	**Electricity Distribution Network Design, 2nd Edition** E. Lakervi and E.J. Holmes
Volume 22	**Artificial Intelligence Techniques in Power Systems** K. Warwick, A.O. Ekwue and R. Aggarwal (Editors)
Volume 24	**Power System Commissioning and Maintenance Practice** K. Harker
Volume 25	**Engineers' Handbook of Industrial Microwave Heating** R.J. Meredith
Volume 26	**Small Electric Motors** H. Moczala et al.
Volume 27	**AC–DC Power System Analysis** J. Arrillaga and B.C. Smith
Volume 29	**High Voltage Direct Current Transmission, 2nd Edition** J. Arrillaga
Volume 30	**Flexible AC Transmission Systems (FACTS)** Y.-H. Song (Editor)
Volume 31	**Embedded Generation** N. Jenkins *et al.*
Volume 32	**High Voltage Engineering and Testing, 2nd Edition** H.M. Ryan (Editor)
Volume 33	**Overvoltage Protection of Low-Voltage Systems, Revised Edition** P. Hasse
Volume 36	**Voltage Quality in Electrical Power Systems** J. Schlabbach *et al.*
Volume 37	**Electrical Steels for Rotating Machines** P. Beckley
Volume 38	**The Electric Car: Development and future of battery, hybrid and fuel-cell cars** M. Westbrook
Volume 39	**Power Systems Electromagnetic Transients Simulation** J. Arrillaga and N. Watson
Volume 40	**Advances in High Voltage Engineering** M. Haddad and D. Warne
Volume 41	**Electrical Operation of Electrostatic Precipitators** K. Parker
Volume 43	**Thermal Power Plant Simulation and Control** D. Flynn
Volume 44	**Economic Evaluation of Projects in the Electricity Supply Industry** H. Khatib
Volume 45	**Propulsion Systems for Hybrid Vehicles** J. Miller
Volume 46	**Distribution Switchgear** S. Stewart
Volume 47	**Protection of Electricity Distribution Networks, 2nd Edition** J. Gers and E. Holmes
Volume 48	**Wood Pole Overhead Lines** B. Wareing
Volume 49	**Electric Fuses, 3rd Edition** A. Wright and G. Newbery
Volume 50	**Wind Power Integration: Connection and system operational aspects** B. Fox *et al.*
Volume 51	**Short Circuit Currents** J. Schlabbach
Volume 52	**Nuclear Power** J. Wood
Volume 53	**Condition Assessment of High Voltage Insulation in Power System Equipment** R.E. James and Q. Su
Volume 55	**Local Energy: Distributed generation of heat and power** J. Wood
Volume 56	**Condition Monitoring of Rotating Electrical Machines** P. Tavner, L. Ran, J. Penman and H. Sedding
Volume 57	**The Control Techniques Drives and Controls Handbook, 2nd Edition** B. Drury
Volume 58	**Lightning Protection** V. Cooray (Editor)
Volume 59	**Ultracapacitor Applications** J.M. Miller
Volume 62	**Lightning Electromagnetics** V. Cooray

Volume 63 **Energy Storage for Power Systems, 2nd Edition** A. Ter-Gazarian
Volume 65 **Protection of Electricity Distribution Networks, 3rd Edition** J. Gers
Volume 66 **High Voltage Engineering Testing, 3rd Edition** H. Ryan (Editor)
Volume 67 **Multicore Simulation of Power System Transients** F.M. Uriate
Volume 68 **Distribution System Analysis and Automation** J. Gers
Volume 69 **The Lightening Flash, 2nd Edition** V. Cooray (Editor)
Volume 70 **Economic Evaluation of Projects in the Electricity Supply Industry, 3rd Edition** H. Khatib
Volume 72 **Control Circuits in Power Electronics: Practical issues in design and implementation** M. Castilla (Editor)
Volume 73 **Wide Area Monitoring, Protection and Control Systems: The enabler for smarter grids** A. Vaccaro and A. Zobaa (Editors)
Volume 74 **Power Electronic Converters and Systems: Frontiers and applications** A.M. Trzynadlowski (Editor)
Volume 75 **Power Distribution Automation** B. Das (Editor)
Volume 76 **Power System Stability: Modelling, analysis and control** A.A. Sallam and B. Om P. Malik
Volume 78 **Numerical Analysis of Power System Transients and Dynamics** A. Ametani (Editor)
Volume 79 **Vehicle-to-Grid: Linking electric vehicles to the smart grid** J. Lu and J. Hossain (Editors)
Volume 81 **Cyber-Physical-Social Systems and Constructs in Electric Power Engineering** S. Suryanarayanan, R. Roche and T.M. Hansen (Editors)
Volume 82 **Periodic Control of Power Electronic Converters** F. Blaabjerg, K. Zhou, D. Wang and Y. Yang
Volume 86 **Advances in Power System Modelling, Control and Stability Analysis** F. Milano (Editor)
Volume 87 **Cogeneration: Technologies, optimisation and implementation** C.A. Frangopoulos (Editor)
Volume 88 **Smarter Energy: From smart metering to the smart grid** H. Sun, N. Hatziargyriou, H.V. Poor, L. Carpanini and M.A. Sánchez Fornié (Editors)
Volume 89 **Hydrogen Production, Separation and Purification for Energy** A. Basile, F. Dalena, J. Tong and T.N. Veziroğlu (Editors)
Volume 90 **Clean Energy Microgrids** S. Obara and J. Morel (Editors)
Volume 91 **Fuzzy Logic Control in Energy Systems with Design Applications in MATLAB®/Simulink®** İ.H. Altaş
Volume 92 **Power Quality in Future Electrical Power Systems** A.F. Zobaa and S.H.E.A. Aleem (Editors)
Volume 93 **Cogeneration and District Energy Systems: Modelling, analysis and optimization** M.A. Rosen and S. Koohi-Fayegh
Volume 94 **Introduction to the Smart Grid: Concepts, technologies and evolution** S.K. Salman
Volume 95 **Communication, Control and Security Challenges for the Smart Grid** S.M. Muyeen and S. Rahman (Editors)
Volume 96 **Industrial Power Systems with Distributed and Embedded Generation** R. Belu
Volume 97 **Synchronized Phasor Measurements for Smart Grids** M.J.B. Reddy and D.K. Mohanta (Editors)
Volume 98 **Large Scale Grid Integration of Renewable Energy Sources** A. Moreno-Munoz (Editor)
Volume 100 **Modeling and Dynamic Behaviour of Hydropower Plants** N. Kishor and J. Fraile-Ardanuy (Editors)
Volume 101 **Methane and Hydrogen for Energy Storage** R. Carriveau and D.S-K. Ting
Volume 104 **Power Transformer Condition Monitoring and Diagnosis** A. Abu-Siada (Editor)
Volume 106 **Surface Passivation of Industrial Crystalline Silicon Solar Cells** J. John (Editor)
Volume 107 **Bifacial Photovoltaics: Technology, applications and economics** J. Libal and R. Kopecek (Editors)

Volume 108 **Fault Diagnosis of Induction Motors** J. Faiz, V. Ghorbanian and G. Joksimović
Volume 109 **Cooling of Rotating Electrical Machines: Fundamentals, modelling, testing and design** D. Staton, E. Chong, S. Pickering and A. Boglietti
Volume 110 **High Voltage Power Network Construction** K. Harker
Volume 111 **Energy Storage at Different Voltage Levels: Technology, integration, and market aspects** A.F. Zobaa, P.F. Ribeiro, S.H.A. Aleem and S.N. Afifi (Editors)
Volume 112 **Wireless Power Transfer: Theory, technology and application** N. Shinohara
Volume 114 **Lightning-Induced Effects in Electrical and Telecommunication Systems** Y. Baba and V.A. Rakov
Volume 115 **DC Distribution Systems and Microgrids** T. Dragičević, F. Blaabjerg and P. Wheeler
Volume 116 **Modelling and Simulation of HVDC Transmission** M. Han (Editor)
Volume 117 **Structural Control and Fault Detection of Wind Turbine Systems** H.R. Karimi
Volume 118 **Modelling and Simulation of Complex Power Systems** A. Monti and A. Benigni
Volume 119 **Thermal Power Plant Control and Instrumentation: The control of boilers and HRSGs, 2nd Edition** D. Lindsley, J. Grist and D. Parker
Volume 120 **Fault Diagnosis for Robust Inverter Power Drives** A. Ginart (Editor)
Volume 121 **Monitoring and Control Using Synchrophasors in Power Systems with Renewables** I. Kamwa and C. Lu (Editors)
Volume 123 **Power Systems Electromagnetic Transients Simulation, 2nd Edition** N. Watson and J. Arrillaga
Volume 124 **Power Market Transformation** B. Murray
Volume 125 **Wind Energy Modeling and Simulation, Volume 1: Atmosphere and plant** P. Veers (Editor)
Volume 126 **Diagnosis and Fault Tolerance of Electrical Machines, Power electronics and drives** A.J.M. Cardoso
Volume 127 **Lightning Electromagnetics, Volume 1: Return stroke modelling and electromagnetic radiation, 2nd Edition** V. Cooray, F. Rachidi and M. Rubinstein (Editors)
Volume 127 **Lightning Electromagnetics, Volume 2: Electrical processes and effects, 2nd Edition** V. Cooray, F. Rachidi and M. Rubinstein (Editors)
Volume 128 **Characterization of Wide Bandgap Power Semiconductor Devices** F. Wang, Z. Zhang and E.A. Jones
Volume 129 **Renewable Energy from the Oceans: From wave, tidal and gradient systems to offshore wind and solar** D. Coiro and T. Sant (Editors)
Volume 130 **Wind and Solar Based Energy Systems for Communities** R. Carriveau and D.S-K. Ting (Editors)
Volume 131 **Metaheuristic Optimization in Power Engineering** J. Radosavljević
Volume 132 **Power Line Communication Systems for Smart Grids** I.R.S. Casella and A. Anpalagan
Volume 134 **Hydrogen Passivation and Laser Doping for Silicon Solar Cells** B. Hallam and C. Chan (Editors)
Volume 139 **Variability, Scalability and Stability of Microgrids** S.M. Muyeen, S.M. Islam and F. Blaabjerg (Editors)
Volume 142A **Wind Turbine System Design, Volume 1: Nacelles, drive trains and verification** J. Wenske (Editor)
Volume 142B **Wind Turbine System Design, Volume 2: Electrical systems, grid integration, control and monitoring** J. Wenske (Editor)
Volume 143 **Medium Voltage DC System Architectures** B. Grainger and R.D. Doncker (Editors)
Volume 145 **Condition Monitoring of Rotating Electrical Machines** P. Tavner, L. Ran and C. Crabtree
Volume 146 **Energy Storage for Power Systems, 3rd Edition** A.G. Ter-Gazarian
Volume 147 **Distribution Systems Analysis and Automation, 2nd Edition** J. Gers
Volume 151 **SiC Power Module Design: Performance, robustness and reliability** A. Castellazzi and A. Irace (Editors)

Volume 208 **Nanogrids and Picogrids and Their Integration with Electric Vehicles** S. Chattopadhyay
Volume 210 **Superconducting Magnetic Energy Storage in Power Grids** M.H. Ali
Volume 211 **Blockchain Technology for Smart Grids: Implementation, management and security** H.L. Gururaj, K.V. Ravi, F. Flammini, H. Lin, B. Goutham, K.B.R. Sunil and C. Sivapragash
Volume 212 **Battery State Estimation: Methods and models** S. Wang
Volume 215 **Industrial Demand Response: Methods, best practices, case studies, and applications** H.H. Alhelou, A. Moreno-Muñoz and P. Siano (Editors)
Volume 213 **Wide Area Monitoring of Interconnected Power Systems, 2nd Edition** A.R. Messina
Volume 214 **Distributed Energy Storage in Urban Smart Grids** P.F. Ribeiro and R.S. Salles (Editors)
Volume 217 **Advances in Power System Modelling, Control and Stability Analysis, 2nd Edition** F. Milano (Editor)
Volume 222 **Large Scale Grid Integration of Renewable Energy Sources, 2nd Edition: Solutions and technologies** A. Moreno-Munoz
Volume 225 **Fusion–Fission Hybrid Nuclear Reactors: For enhanced nuclear fuel utilization and radioactive waste reduction** W.M. Stacey
Volume 227 **Processing and Manufacturing of Electrodes for Lithium-Ion Batteries** J. Li and C. Jin (Editors)
Volume 228 **Digital Technologies for Solar Photovoltaic Systems: From general to rural and remote installations** S. Motahhir (Editor)
Volume 230 **Self-Organizing Dynamic Agents for the Operation of Decentralized Smart Grids** A. Vaccaro
Volume 235 **Clean Energy Past and Future** P. Tavner
Volume 236 **Green HV Switching Technologies for Modern Power Networks** K. Niayesh (Editor)
Volume 237 **High Voltage Power Transformers: State of the art and technological innovations** M. Sforna, S. Zunino and F. Ferrari
Volume 238 **AI for Status Monitoring of Utility Scale Batteries** S. Wang, K. Liu, Y. Wang, D. Stroe, C. Fernandez and J.M. Guerrero
Volume 239 **Intelligent Control of Medium and High Power Converters** M. Bendaoud, Y. Maleh and S. Padmanaban (Editors)
Volume 241A **Power Electronic Converters and Systems, 2nd Edition, Volume 1: Converters and machine drives** M. Simões and T. Busarello (Editors)
Volume 241B **Power Electronic Converters and Systems, 2nd Edition, Volume 2: Applications** M. Simões and T. Busarello (Editors)
Volume 242 **AI for Power Electronics and Renewable Energy Systems** F. Blaabjerg, W. Hu, G. Zhang, Z. Zhang and S. Abulanwar
Volume 245 **Agrivoltaics: Technical, ecological, commercial and legal aspects** C. Klyk and S. Schindele (Editors)
Volume 247 **Power System Strength Evaluation Methods, Best Practice, Case Studies, and Applications** H. Alhelou, N. Hosseinzadeh and B. Bahrani (Editors)
Volume 250 **Interactions of Wind Turbines with Aviation Radio and Radar Systems** Alan Collinson
Volume 256 **Wireless Power Transfer Technologies: Theory and technologies, 2nd Edition** N. Shinohara (Editor)
Volume 905 **Power system Protection, 4 Volumes**

Fuzzy Logic Control in Energy Systems, 2nd Edition

With design examples in MATLAB®/Simulink®

İsmail Hakkı Altaş

The Institution of Engineering and Technology

Published by The Institution of Engineering and Technology, London, United Kingdom

The Institution of Engineering and Technology is registered as a Charity in England & Wales (no. 211014) and Scotland (no. SC038698).

First published 2016, second Edition 2024

The Institution of Engineering and Technology
Futures Place
Kings Way, Stevenage
Hertfordshire SG1 2UA, United Kingdom

www.theiet.org

British Library Cataloguing in Publication Data
A catalogue record for this product is available from the British Library

ISBN: 978-1-83953-882-7 (hardback)

ISBN: 978-1-83953-883-4 (PDF)

Typeset in India by MPS Limited

Cover image: jia yu/Moment via Getty Images

This book is dedicated to my be loved family,
my wife Mualla and my beautiful daughters Hilal, Elif, and Candan Sıla,
for their patience and understanding
during the preparation of the manuscript,
and of course to my mom and dad who loved me so much.

Contents

About the author

İsmail Hakkı Altaş is a professor at the Department of Electrical and Electronics Engineering of Karadeniz Technical University, Türkiye. He obtained his PhD degree from the University of New Brunswick, Canada. He is a Senior member of the IEEE and member of the IEEE Power & Energy, Industrial Electronics, Industry Applications, Power Electronics, Control Systems and Systems, Man, and Cybernetics Societies. His research interests include intelligent control of power systems, utilization of renewable energy and electrical microgrids.

Preface to 1st edition

This book is about fuzzy logic controller and its applications in energy systems. It aims to give an insight of clear understanding and design approaches of fuzzy logic controllers in MATLAB® and MATLAB/Simulink® environment. It includes a basic theory of fuzzy sets and fuzzy logic to prepare the reader for a better understanding of fuzzy partitioning, fuzzy relation, and fuzzy decision-making processing, which are required for designing fuzzy logic controllers. A fuzzy unit called fuzzy processor is developed and designed to be used as a fuzzy decision-maker and a fuzzy logic controller depending on the application problem.

Energy system is one of the application areas of fuzzy logic. It is used to manage, control, and operate electrical energy systems. Examples in the book are related to the control, operation, and management of electrical energy utilization. The following examples of fuzzy logic control and fuzzy management are discussed and studied in the scope of the book.

- DC motor speed and torque control
- Excitation and load-frequency control in power systems
- Multi-area load-frequency control in power systems
- Wind energy control systems (WECS)
- Photovoltaic (PV) solar energy control systems
- Maximum power point tracking in PV systems
- Energy management in WECS
- Energy management in PV systems

The book addresses undergraduate and graduate students as well as practicing engineers in electrical power, energy, and control systems. They will be able to get sufficient knowledge of fuzzy logic theory, and then a clear understanding of designing fuzzy decision-makers and controller in MATLAB and Simulink. Those who study the book will be able to develop their own fuzzy processor library and design their own fuzzy logic toolbox for the special problems they study. With the given examples, the readers will also get to know the modeling and simulation of electrical power and energy systems.

A novel fuzzy logic controller (FLC) design approach in both MATLAB and Simulink is given in the book such that the user can see every step of the fuzzy logic processor with the ability to interfere with the code in MATLAB .m files and also in operational Simulink blocks. The FLC design approach will make the readers not just as software users but also as software developers.

Chapters 1–7 can be used as an accompanying textbook for teaching *Fuzzy Logic and Fuzzy Decision-Making* as an undergraduate course. Chapters 1–8 can be used as a textbook for teaching *Fuzzy Logic Control* at undergraduate or graduate levels. Chapters 5–9 can be used in a graduate course about *FLC in Power System Control* assuming that students have a basic knowledge of fuzzy set theory and fuzzy logic. Chapters 5, 6, 7, 10, 11, and 12 can be used as an advanced graduate course about *FLC in Renewable Energy and Distributed Generation*. Chapters 10–12 can also be used in an advanced graduate course to teach Fuzzy Logic Controlled Wind and PV Energy Conversion Systems.

The book is organized into 12 chapters with the following contents.

Chapter 1. Introduction: A brief history of fuzzy set theory and its application areas are summarized in this chapter. The concepts of fuzziness, fuzzy membership functions, and fuzzy subsets are introduced.

Chapter 2. Fuzzy sets: Types and properties of fuzzy sets are studied. Modeling of fuzzy sets in MATLAB and MATLAB/Simulink are shown and MATLAB function files are developed to be used as a part of a user-defined toolbox library. Fuzzy intersection, union, and complement are also studied in this chapter.

Chapter 3. Fuzzy partitioning: Fuzzy subclasses and partitioning of the universes into fuzzy subsets are studied in this chapter. The importance and meaning of the portioning are discussed with examples.

Chapter 4. Fuzzy relation: The concepts of fuzzy relation, two-dimensional fuzzy sets, fuzzy extension principle, fuzzy projection, and binary and *n*-ary fuzzy relations are discussed in this chapter. Representing verbal terms and expressions as fuzzy relations are also introduced in this chapter.

Chapter 5. Fuzzy reasoning and fuzzy decision-making: Approximate reasoning, fuzzy reasoning, and fuzzy decision-making processes are given in this chapter. Single input single rule, single input multiple rules, and multiple input multiple rule base systems are studied and examples are given. The concept of fuzzy reasoning is studied and user-defined MATLAB files are used to support the operational behaviors of fuzzy decision-making.

Chapter 6. Fuzzy processor: Fuzzy reasoning and fuzzy decision-making processes are carried ahead with multiple inputs, multiple rules, and multiple decisions as the fuzzy processor. Known fuzzy reasoning algorithms such as Mamdani Fuzzy Reasoning, Sugeno Fuzzy reasoning, and Tsukomoto Fuzzy reasoning are discussed and steps toward fuzzy logic controllers are given.

Chapter 7. Fuzzy logic controller (FLC): Fuzzy logic controller (FLC) is given in this chapter. Rule development, the way of putting expert ideas into rules, and inference system structure are studied. From crisp input variables to crisp output, all process is discussed and shown. Defuzzification, rule processing, fuzzy reasoning, and crisp output after defuzzification are explained. User-developed FLC examples are given.

Chapter 8. System modeling and control: Mathematical modeling of physical systems is given in this chapter. The methods of obtaining differential equations, simulation diagrams, and state-space models of physical systems are studied. The Runge–Kutta numerical solution method is discussed and user-based MATLAB software is developed to show the meaning of controlling physical systems as one of the application areas of fuzzy logic. The reader will be able to develop his/her own FLC code in MATLAB and in MATLAB/Simulink. Examples of controlling electrical, mechanical, and electromechanical systems will be given.

Chapter 9. FLC in power systems: Application of fuzzy logic controller and decision-maker to excitation control, load-frequency control, and power compensation are discussed in this chapter. Single- and multi-area control of power systems are also studied as examples in the chapter.

Chapter 10. FLC in wind energy systems: Application of fuzzy logic control and decision-making processes to wind energy conversion systems are given in this chapter. After giving the problems and control issues in wind energy conversion systems, the utilization of fuzzy logic in solving these problems is shown.

Chapter 11. FLC in PV solar energy systems: Application of fuzzy logic control and decision-making processes in PV solar systems are given in this chapter. Maximum power point tracking, sun tracking, voltage control, battery charging, and management of the generated power are studied.

Chapter 12. Energy management and fuzzy decision-making: The use of fuzzy decision-making and control processes in energy management systems is studied in this chapter. Energy management in PV solar and wind energy systems are discussed and examples are given.

I would like to express my appreciation and thanks to my PhD supervisor Professor Adel M. Sharaf at the Department of Electrical and Computer Engineering at the University of New Brunswick, Canada. He allowed me to use some of his designs in the book.

I wish to thank my graduate students, Onur Ozdal Mengi, Recep Çakmak, Kenan Yanmaz, Yeşim Aysel Baysal, Serhat Duman, Yahya Danayiyen, and Ömer Aydemir at the Department of Electrical and Electronics Engineering in Karadeniz Technical University, Turkey. They were so patient and understanding during the preparation of the manuscript besides providing me with examples from their studies.

I want to thank IET editorial assistants Jennifer Grace and Olivia Wilkins for their understanding and guidance during the process.

Last, but not least, I thank my mom Firdevs, my wife Mualla, and my children Hilal, Elif, and Candan Sıla, who were great supporters and my energy sources throughout this work. I am also sending my love and sincere thoughts to my dad, Lütfü Altaş, whom I lost when I started writing this book.

İsmail H. Altaş
Karadeniz Technical University
June 2017

Preface to 2nd edition

The technology of electrical distribution systems has been changing too fast based on the developments in computers, digital systems, power electronic components, communication capabilities, and control tools. Therefore, it was inevitable to revise and update the content of the book. As a result of this, the topics about the basics of fuzzy logic and fuzzy set theory have been abbreviated and combined under one chapter while more details are given in fuzzy reasoning, fuzzy decision-making, and fuzzy controller. Besides, physical system modeling and numerical solution methods for digital simulation have been expanded. New application areas of fuzzy logic in power flow, solar PV, and microgrid energy management systems have been added. The new chapters and the topics added to the second edition are listed below:

Chapter 1. Introduction
Chapter 2. Fuzzy sets and fuzzy partitioning
Chapter 3. Fuzzy decision-making

New subsection: Human decision-making process

Chapter 4. Fuzzy logic controller (FLC)
Chapter 5. System modeling and control

New subsection: Numerical solution of model equations

Chapter 6. FLC in power systems

New subsection: FLC in load flow analysis
New subsection: FL-accelerated Gauss–Seidel algorithm
New subsection: FL-accelerated Newton–Raphson algorithm
New subsection: FL-accelerated fast decoupled algorithm

Chapter 7. FLC in wind energy systems
Chapter 8. FLC in PV solar energy systems

New subsection: Shading effects in PV arrays
New subsection: PV arrays with identical PV panels
New subsection: PV arrays with Unidentical PV panels
New subsection: MPPT in PV arrays under shading

Chapter 9. Microgrids and energy management

New subsection: Interfacing distributed sources to common power grid
New subsection: Day ahead forecasting in energy management systems
New subsection: Data using for renewable energy source power and demand forecasting
New subsection: Methods for renewable energy source power and demand forecasting
New subsection: Artificial neural network (ANN)
New subsection: Multi-layer artificial neural network
New subsection: Learning process in a neural network
New subsection: Training an artificial neural network for forecasting
New subsection: Cybersecurity in microgrids
New subsection: Fundamentals of cybersecurity
New subsection: Cybersecurity threats
New subsection: Cybersecurity for power systems
New subsection: Future trends in cybersecurity for power systems

With the revised new edition, the book aims to give an insight into clear understanding and design approaches of fuzzy logic applications in MATLAB and MATLAB/Simulink environment. It includes a basic theory of fuzzy sets and fuzzy logic to prepare the reader for a better understanding of fuzzy partitioning, fuzzy decision-making, fuzzy optimization, and fuzzy logic control, which are required for designing fuzzy logic-based application tools. A fuzzy unit called a fuzzy processor is developed and designed to be used as a fuzzy decision-maker, a fuzzy optimizer, and a fuzzy logic controller depending on the application problem type.

Energy and control systems are the main application areas of fuzzy logic control and decision-making in this book. The FL is used to manage, control, and operate electrical energy systems. Examples in the book are related to the control, optimization, management, and operation of electrical energy utilization.

This book addresses undergraduate and graduate students in engineering as well as researchers and practicing engineers in electrical power, energy, and control systems. They will be able to get sufficient knowledge of fuzzy logic theory, and then a clear understanding of designing fuzzy decision-makers, fuzzy optimizers, fuzzy logic controllers, and fuzzy logic-based management systems in MATLAB and MATLAB/Simulink. Those who study the book will be able to model physical systems develop their own fuzzy processor library and design their own fuzzy logic toolbox for the special problems they study. Except for examples on some special research applications, the MATLAB/Simulink codes and simulation files of most of the examples will be stored by the publisher and will be provided to the readers upon their request. With the given examples, the readers will also get to know the modeling and simulation of electrical power and energy systems.

The book will include a novel fuzzy logic controller design approach in both MATLAB and in MATLAB/Simulink such that the user can see every step of the fuzzy logic processor with the ability to interfere with the code in MATLAB .m files and also in Simulink's operational blocks. The proposed approach will make the readers not just software users but also software developers.

İsmail H. Altaş
Karadeniz Technical University
February 2024

Acknowledgments

I would like to express my appreciation and thanks to my PhD supervisor Professor Adel M. Sharaf at the Department of Electrical and Computer Engineering at The University of New Brunswick, Canada. He allowed me to use some of his designs in the book.

I wish to thank my graduate students Sibel Çevik Bektaş and Büşra Özgenç who made a great contribution to the second edition on the day ahead forecasting of solar power and shading effects in PV modeling, respectively. I also wish to send my appreciation to my colleague Associate Professor Dr. M. Şinasi Ayas at the Department of Electrical and Electronics Engineering at Karadeniz Technical University, Turkiye for his support on cyber security issues.

I want to thank to IET editorial office people for their understanding and guidance during the process.

Last but not least, I thank my family, my wife, and my lovely daughters, who gave me their energy and great support throughout this work. I am also sending my love and sincere gratitude to my dad and mom, whom I lost after I started writing this book.

Chapter 1

Introduction

A brief history of fuzzy set theory and its application areas are summarized in this chapter. Besides, an introductory knowledge of electrical energy systems and novel approaches to the problems arising with the developments of distributed generation and microgrids is given.

1.1 Introduction

Many words we use arbitrarily in our daily life are usually fuzzy in terms of verbal meanings. When expressing or describing a system or an event we use words such as old, young, tall, short, cold, warm, hot, sunny cloudy, fast, slow, etc. that are fuzzy in nature. We, humans, use uncertain, vogue, and muddy words when discussing something or making decisions to perform some actions. Depending on his/her age, we call that person old, middle-aged, young, very old, and very young. We press the gas or brake more or less according to the road condition whether it is dry, slippery, ramp, or flat. If the lights in our study room are low, we increase it a little bit otherwise we decrease it a little bit. All these examples show how our brain acts and makes decisions during situations that are uncertain and fuzzy.

Studies on systems with uncertainty and muddy data have gained a new era with the publication of the article "Fuzzy sets" by Lotfi A. Zadeh [1]. Although this first article was published in 1965, the use of fuzzy logic (FL) increased after the second half of the 1970s when Lotfi A. Zadeh published two more articles [2,3], in which the application of fuzzy set theory to uncertain systems and decision-making is described. FL applications have gained a high speed as the Japanese started using them in commercially available appliances. Nowadays, it is possible to find fuzzy-based applications in almost every area [4]. Some of the utilization areas of FL are listed below.

FL is used in robotics, automation, tracking systems, temperature control, flow control, motion control, commercial products, and many more utilization areas of automatic control systems [4–7]. It is used in information systems as a database tool to store and recall knowledge, uncertain data, expert's ideas, and operational behavior of machines. Image processing, signal aliasing, and human–machine interaction are also some of the application areas of FL [4]. It is possible to find many more FL-based applications in social and medical sciences as well [4]. FL is also used as a mathematical tool in areas such as function optimization, filtering, curve fitting, etc. [4].

Control systems are one of the first application areas of FL. Mamdani *et al.* used FL to control a plant comprising a steam engine and boiler combination [5–7] in the early 1970s. Since then there has been a considerable interest in the subject so a massive amount of work has appeared in the literature [4]. FL control and its applications to power and energy systems are given in this book.

Electrical power systems have many issues from the generation site to the user site. Generation, transmission, distribution, and consumption stages of power systems all have specific problems to be solved. Conventional methods have been used in power systems successfully over the years. However, the integration of renewable energy systems into power distribution networks has changed the conventional structure of power distribution systems yielding microgrid power distribution networks. Microgrids include both generation and distribution in the same operating frame that requires a massive control process besides measurements, communication, and management. Therefore, intelligent methods are developed to overcome the problems that occur in microgrids. That's how the concept of smart grids has become an advanced structured power distribution network. As a part of intelligent methods, fuzzy set theory and FL are used in power and energy systems in different ways. This book addresses the use of fuzzy set theory, fuzzy decision-making, and FL control in the control and energy management of energy systems.

Contrary to classical PID controllers, FL controllers do not require a mathematical model of the systems to be controlled. It adjusts the plant input to get the desired output by just looking at the current output from the system. This simple decision process is nothing but an expert's decision process. That means that machines are programmed to make decisions similar to those of humans by using FL and fuzzy set theory. Due to its adaptivity in nature, FL has gained great importance over the years to control both linear and nonlinear systems. Actually, nothing is linear in real life. They are just linearized using proper approximation methods. However, any linearization process reduces the accuracy of the system model leading to operational errors with PID controllers, which require mathematical models for design. The errors that occur due to linearization do not affect the performance of FL controllers since they do not need mathematical modeling to generate the expert's thoughts. Therefore, FL controllers become an alternative to control nonlinear systems. The nonlinearity of the systems is compensated through rules, membership functions, and decision-making processing in fuzzy systems. The use of FL increases the accuracy and system performance while simplifying and reducing the cost of the system. Selecting rules closer to reality enables the FL controllers to deal with nonlinear systems better than conventional linearized approaches. Most of the control applications have multiple inputs and parameters that are required to be modeled and adjusted, which are time-consuming and bring difficulties to the application. On the other hand, the rules of FL controller include nonlinear properties so that they simplify the application by uniting multiple inputs with the implementation of individual verbal expression as *if* ... *then* ... *else* ... rules. This process is called fuzzy implication and is carried out in terms of the fuzzy relation of the fuzzy subsets used in the rule. Fuzzy set theory, or the application of FL itself for a special system is actually not very different from the application of

Boolean logic or probability logic. The difference in FL comes from its ability to create a more general theory of decision process that is called fuzzy processor, which is a special case of approximate reasoning. Fuzzy processor uses fuzzy set and FL theory in the reasoning process and operates as a fuzzy reasoning algorithm. Approximate reasoning or fuzzy reasoning is carried out by fuzzy words we use frequently in our daily life. For example, expressions such as *a little bit cold*, *almost there*, *somehow successful*, and *more than hot* contain approximate terms *a little bit*, *almost*, *somehow*, and *more than*. These terms are used with fuzzy words such as cold, warm, hot, young, and old in order to change the degree of fuzziness.

In order to understand fuzzy systems and fuzzy processes, first the meaning of fuzziness should be understood well.

1.2 Fuzziness

Fuzzy sets are the core elements of FL and are characterized by membership functions. In fact, these membership functions are nothing but some sort of fuzzy numbers. One should know the meaning of *fuzziness* in order to understand the terms *fuzzy logic*, *membership function*, and *fuzzy number*. For example, two colors defined in the universe of colors are blended and shown in Figure 1.1. First, the color is white, then it is changed to black over a transition region such that it becomes light gray, gray, dark gray, and black as we move from left to right. It is not a clear single color during the transition. It includes shades of both white and black and no one can distinguish one color from the other because the transition part is fuzzy. The colors in the transition area can be distinguished as *white*, *light gray*, *gray*, *dark gray*, and *black* as shown in Figure 1.1.

The concept of fuzziness comes from uncertainty. If data are not crisp and hard to distinguish one data from the other one, they are represented by inclusion degrees in related categories. There are two color categories in Figure 1.1 as WHITE and BLACK, which are called fuzzy subsets or fuzzy membership functions. Fuzzy subset WHITE in Figure 1.1 shows the degrees of colors being white while fuzzy

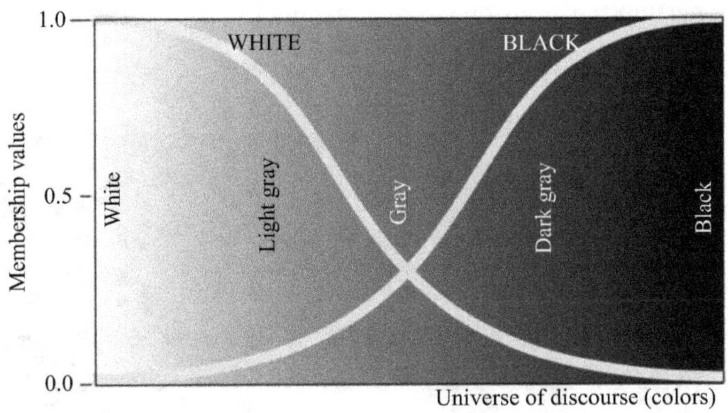

Figure 1.1 Blended colors in the universe of colors

subset BLACK represents the degree of colors being black. The universe of colors starts from white and goes toward black as we move from left to right in Figure 1.1. During this transition, the shades of gray change from lighter to darker while membership degrees become smaller in WHITE and larger in BLACK. Along the transition region, the colors are both white and black. They are neither just white nor black. Therefore, the color spectrum along the transition is fuzzy.

1.3 Fuzzy membership functions

Fuzzy membership functions can be considered as a bridge between uncertain data and fuzzy world. The crisp universe of muddy data is partitioned into subsections and represented by fuzzy membership functions. For example, the crisp universe of gray colors can be partitioned into subsections as *white, light gray, gray, dark gray,* and *black* as shown in Figure 1.2. The shades of gray become darker as we move from left to right or from white to black. The shades of gray color have been categorized into subsections as *white, light gray, gray, dark gray,* and *black* defined by triangular-type fuzzy membership functions.

Fuzzy membership functions are defined in an interval [0, 1]. The lowest value 0 means that there is no inclusion in the related fuzzy set while the highest value 1 means full inclusion. Five triangular-type membership functions representing five fuzzy subsets are used in Figure 1.2 to define the shades of gray. Right triangles are used as boundary membership functions to represent lower (white) and upper (black) limits of the shades of gray. Equilateral triangles are used as membership functions for the other shades that vary from white to black. It is not clear to say where *light gray* ends and *gray* starts. Similarly, one cannot see a clear starting and ending for *gray* and *dark gray*. Therefore, there is a fuzzy transition region from one color to the other. Fuzzy membership functions show how much the color in the universe of discourse belongs to each fuzzy subset, which represents the subcategories of gray. As stated earlier, fuzzy membership functions are used as a tool to represent the fuzzy sets. Various geometrical and pictorial shapes can be used to represent fuzzy sets. However, known mathematical functions are preferred to be used as membership functions due to their simplicity. A detailed analysis of fuzzy membership functions is discussed in Chapter 2.

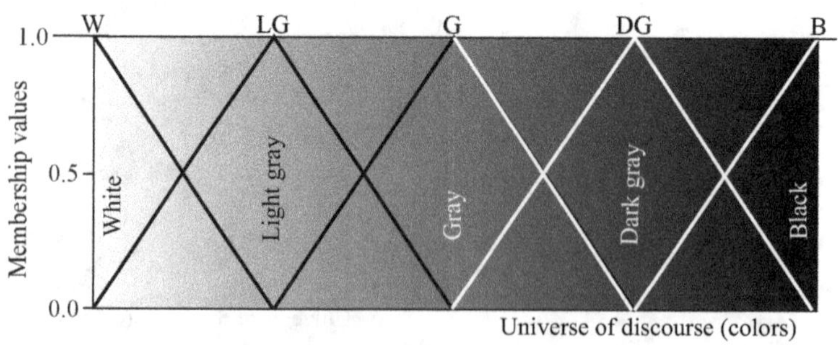

Figure 1.2 Blended colors in the universe of gray colors

1.4 Fuzzy sets

Fuzzy sets are used to categorize the subclasses of uncertainty. Similarly to crisp sets, fuzzy sets represent the inclusion of objects in a categorized group. However, the boundaries of fuzzy sets are not as sharp as they are in crisp sets. As described earlier, an object belongs to a crisp set or not. If an object belongs to a crisp set, its membership degree becomes 1, or 0 otherwise. There is no other choice, just 1 or 0. That is where the difference between crisp and fuzzy sets occurs. The boundaries of fuzzy sets change gradually from 0 to 1 or from 1 to 0 showing how much the object belongs to the set. Therefore, the membership degrees are not just 0 or 1 in fuzzy sets. They are varied between 0 and 1.

1.5 Latest developments in energy systems

Electrical power systems have a very complicated structure including hundreds of generators, transmission lines, bus bars, step-up transformers, step-down transformers, and loads. The generators used in conventional power systems are usually synchronous generators, which require load-frequency and excitation control. The relations both between power and frequency and between excitation and induced voltages are nonlinear and the dynamic model equations are required to be linearized in order to apply PID controllers. Therefore, after the invention of FL, the application of FL controllers in power system control has gained importance since a dynamic model is not required by FL controllers in real-time applications [8].

FL controllers, just like human operators, get information from system output and adjust the input until reaching a desired output. A human operator controls the input variables of the machine by looking at the output variables without knowing or implementing the dynamic model of the machine [5–7]. Since FL can be used to model human thoughts and experienced knowledge, it has been applied in many decision-making processes [9] such as the ones applied to energy management systems [10,11].

Integration of wind and photovoltaic solar-type renewable energy sources into conventional distribution networks has been increasing rapidly [12,13]. As the number of these integrations increases, the structure of electrical distribution networks has changed toward distributed generation networks, microgrids, and smart electrical networks [14,15]. While the power generation units in conventional power networks are far away from the consumption sites, the generation, and consumption are side by side in distributed networks and microgrids. Since generation and consumption are located in the same area with lower tie impedances between the connection points, the dynamic responses of distributed generation networks are different than those of conventional distribution grids [14]. Besides, distributed and microgrids having renewable energy sources carry harmonics, which cause distortions in voltage and current, and additional losses in distributed power. Therefore, additional precautions such as dynamic filters must be employed to clear out the harmonic effects for efficiency and power quality [16]. All new developments in power distribution networks, power electronics, computers, and communication technologies have led intelligent methods such as FL and artificial neural networks to be utilized in solutions to various power system problems [17].

With the integration of renewable energy sources into distributed networks, control, estimation, and management problems have gained more importance in order to operate those systems smoothly. Since measurement, knowledge, information, and data transmission are non-separable parts of microgrids, cybersecurity issues have become another problem in energy systems [18]. Day-ahead estimation of the power to be generated from renewable sources, day-ahead estimation of the power to be consumed on the demand side [19,20], control of generated and consumed power, control of bus voltages and frequencies, generation, and demand side management are all novel topics as well as cybersecurity in microgrids utilizing FL-based solution algorithms [21].

References

[1] Zadeh L.A. 'Fuzzy sets'. *Information and Control.* 1965;8(3):338–353.

[2] Zadeh L.A. 'Outline of a new approach to the analysis of complex systems and decision processes'. *IEEE Transactions on Systems, Man, and Cybernetics.* 1973;3(1):28–44.

[3] Zadeh L.A. 'The concept of linguistic variable and its application to approximate reasoning. Part I, II, and III'. *Information Sciences.* 1975;8 (3):199–249, 8(4):301–357, 9(1):43–80.

[4] Maiers J. and Sherif Y.S. 'Applications of fuzzy set theory'. *IEEE Transactions on Systems, Man, and Cybernetics.* 1985; 15(1):175–189.

[5] Mamdani E.H. and Assilian S. 'An experiment in linguistic synthesis with a fuzzy logic controller'. *International Journal Man-Machine Studies.* 1975;7 (1):1–13.

[6] Mamdani E.H. 'Application of fuzzy algorithms for control of simple dynamic plant'. *IEE Proceedings.* 1974;121(12):1585–1588.

[7] Mamdani E.H. 'Advances in the linguistic synthesis of fuzzy controllers'. *International Journal Man-Machine Studies.* 1976;8(6):669–678.

[8] Silva J.F. and Pinto S.F., 'Linear and nonlinear control of switching power converters', Editor: Rashid M.H., *Power Electronics Handbook* (4th ed.), Oxford: Butterworth-Heinemann, 2018, pp. 1141–1220.

[9] Skalna I., Rębiasz B., Gaweł B., *et al.*, *Advances in Fuzzy Decision Making: Theory and Practice.* Cham: Springer International Publishing, 2015, http://dx.doi.org/10.1007/978-3-319-26494-3.

[10] Mengi O.O. and Altas I.H., 'A fuzzy decision-making energy management system for a PV/wind renewable energy system', *2011 International Symposium on Innovations in Intelligent Systems and Applications*, Istanbul, Turkey, 2011, pp. 436–440, http://dx.doi.org/10.1109/INISTA.2011.5946112.

[11] Ali M., Hossain M.I. and Shafiullah M., 'Fuzzy logic for energy management in hybrid energy storage systems integrated DC microgrid', *2022 International Conference on Power Energy Systems and Applications*

(ICoPESA), Singapore, 2022, pp. 424–429, http://dx.doi.org/10.1109/ICoPESA54515.2022.9754406.

[12] Tareen W.U., Mekhilef S., Seyedmahmoudian M. and Horan B., 'Active power filter (APF) for mitigation of power quality issues in grid integration of wind and photovoltaic energy conversion system', *Renewable and Sustainable Energy Reviews*. 2017;70:635–655.

[13] Rahman S.A., Varma R.K. and Litzenberger W.H., 'Bibliography of FACTS applications for grid integration of wind and PV solar power systems: 1995–2010 IEEE working group report', *2011 IEEE Power and Energy Society General Meeting*, Detroit, MI, USA, 2011, pp. 1–17, http://dx.doi.org/10.1109/PES.2011.6039861.

[14] Lede A.M.R., Molina M.G., Martinez M. and Mercado P.E., 'Microgrid architectures for distributed generation: a brief review', *2017 IEEE PES Innovative Smart Grid Technologies Conference—Latin America (ISGT Latin America)*, Quito, Ecuador, 2017, pp. 1–6, http://dx.doi.org/10.1109/ISGT-LA.2017.8126746.

[15] Shayeghi H. and Alilou M., 'Distributed generation and microgrids', Editor: Ersan K., *Hybrid Renewable Energy Systems and Microgrids*. New York: Academic Press, 2021, pp. 73–102, ISBN 9780128217245, https://doi.org/10.1016/B978-0-12-821724-5.00006-4.

[16] Guerrero J.M., Loh P.C., Lee T.-L. and Chandorkar M., 'Advanced control architectures for intelligent microgrids—Part II: power quality, energy storage, and AC/DC microgrids', *IEEE Transactions on Industrial Electronics*. 2013;60(4):1263–1270, http://dx.doi.org/10.1109/TIE.2012.2196889.

[17] Daryani M.J., Karkevandi A.E. and Usta O., 'A hybrid approach for microgrid protection system based on neural network and fuzzy logic', *2019 IEEE PES Innovative Smart Grid Technologies Europe (ISGT-Europe)*, Bucharest, Romania, 2019, pp. 1–5, http://dx.doi.org/10.1109/ISGTEurope.2019.8905476.

[18] Li Z., Shahidehpour M. and Aminifar F., 'Cybersecurity in distributed power systems', *Proceedings of the IEEE*. 2017;105(7):1367–1388, http://dx.doi.org/10.1109/JPROC.2017.2687865.

[19] Ulutas A., Cakmak R. and Altaş İ.H., 'Hourly solar irradiation prediction by artificial neural network based on similarity analysis of time series', *Innovations in Intelligent Systems and Applications Conference (ASYU)*, Adana, Turkey, 4–6 October 2018, pp. 140–145.

[20] Çevik Bektaş S., Çakmak R. and Altaş İ.H., 'Design of a MATLAB GUI for day ahead forecasting of PV panel power', Akıllı Sistemlerde Yenilikler ve Uygulamaları (ASYU), İzmir, Turkey, 31 October–2 November 2019, pp. 1–6.

[21] Alsuwian T., Shahid Butt A. and Amin A.A. 'Smart grid cyber security enhancement: challenges and solutions—a review'. *Sustainability*. 2022; 14:14226. http://dx.doi.org/10.3390/su142114226.

Chapter 2
Fuzzy sets and fuzzy partitioning

The concept of fuzziness, types, and properties of fuzzy sets, modeling of fuzzy sets in MATLAB® and MATLAB/Simulink® are shown and MATLAB function files are developed to be used as a part of a user-defined toolbox library. Fuzzy intersection, union, and complement have also been studied in this chapter. Besides, fuzzy subclasses and partitioning of the universes into fuzzy subsets are also studied in this chapter.

2.1 Introduction

Membership functions characterizing fuzzy sets and aggregations done using them are the basis of fuzzy sets and fuzzy logic systems. Therefore, understanding fuzzy sets and their aggregations is important in order to know what can be done by using fuzzy sets and fuzzy logic. Therefore, this chapter is reserved to introduce fuzzy sets and analyze their properties from a control application point of view. Well-known membership functions used to represent fuzzy sets will be reviewed one by one and MATLAB functions will be written for each one as a part of developing a user-defined fuzzy logic toolbox.

Fuzzy sets are defined in the interval [0, 1]. Each fuzzy set is characterized by a membership function that assigns membership values between 0 and 1 to each element. This definition of fuzzy sets is made by the founder Lotfi A. Zadeh [1]. If an element is not a member of the fuzzy set, then its membership value is defined as 0. A membership value 1 is assigned as a full membership degree if an element is known to be completely included in that set. The interval [0, 1] is used to assign a membership value between 0 and 1 for an element, which is somehow included in the set with an inclusion degree that may be any number between 0 and 1. On the other hand, crisp sets do not include any element with uncertainty. An element is a member of crisp set or not. Figure 2.1 shows crisp and fuzzy subsets used to categorize the ages of people into subclasses. The center of the figure represents the age of infants as the outer circle corresponds to the age of old people in the universe of ages.

As depicted in Figure 2.1(a), crisp subsets are clearly indicated by their sharp borderlines. For example, those smaller than 2 years of age are members of the crisp subset *Infant*. People between 2 and 15 years old belong to crisp subset *Young*, people between 15 and 30 are included in crisp set *Young*, and people above

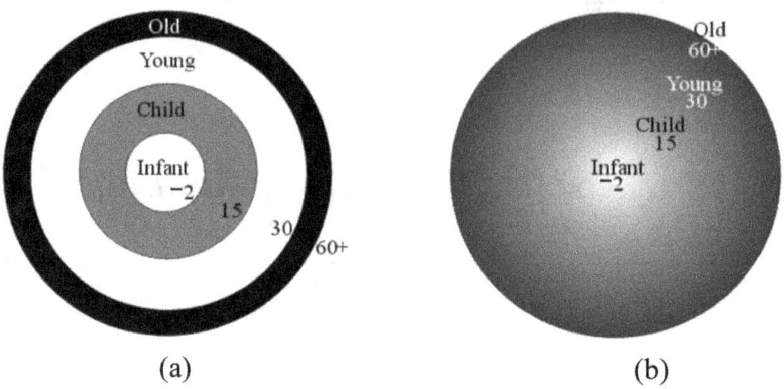

Figure 2.1 Crisp and fuzzy subsets. (a) Crisp subsets. (b). Fuzzy subsets.

Table 2.1 Survey results show the upper limit of fuzzy subset Young

Students	**Upper limit for *Young***
Student 1	35
Student 2	40
Student 3	25
Student 4	33
Student 5	30
Student 6	28
Student 7	35
Student 8	30
Student 9	25
Student 10	32
Average	31.3

30 years old are members of the crisp subset *Old*. According to crisp subsets, a 29-year-old person is considered as young while a 31-year-old person is as same as a person at the age of 60. Figure 2.1(b) shows ages as categorized into subsections, too. However, there is no border between the subsections this time. It is not clear where the subset *Infant* ends and subset *Young* starts. Similarly, it is not easy to say where the subset *Young* ends and the subset *Old* starts. The subclasses of ages in Figure 2.1(b) are defined as fuzzy subsets because the boundaries between these subclasses are not certain. A small survey done with the students attending the Fuzzy Logic course in the Department of Computer Engineering at Karadeniz Technical University, Türkiye, yielded the results given in Table 2.1. The question asked to the students is:

> "What should be the upper limit of a fuzzy set representing young people?"

As the survey shows, the students do not agree on the upper limit of the subset *Young*. The answers to the questions asked vary from 25 to 40 with an average of 31.3. Therefore, age 30 might be assumed to be the upper limit for the subset *Young*. Similarly, the lower and upper limits of all subsets can be defined as given in Figure 2.1 by their pictorial view. In order to work with the subsets, they are represented by their membership functions as shown in Figure 2.2(a) and (b) for crisp and fuzzy subsets, respectively. Since ages are members of crisp subsets or not, they do not have any membership values rather than 0 and 1 as depicted in Figure 2.2(a). However, Membership functions representing the fuzzy subsets have boundary regions with membership values rather than 0 and 1 as given in Figure 2.2(b). Fuzzy membership functions vary between 0 and 1 in an interval [0, 1]. Triangular-type membership functions are used in Figure 2.2(b) to represent fuzzy subsets *Infant*, *Child*, *Young*, and *Old*. Those equal or below 2 years old are considered exactly infants and they are given a full membership degree. Above the age of 2, they start becoming a member of the fuzzy subset *Child* while getting away gradually being the member of *Infant*. At age 15 the membership function representing *Infant* is terminated while the membership function *Child* makes a peak. After the age of 15, the membership function *Child* starts decreasing as *Young* is increasing until it becomes maximum at age 30 with a membership value of 1. Then membership function representing the fuzzy subset

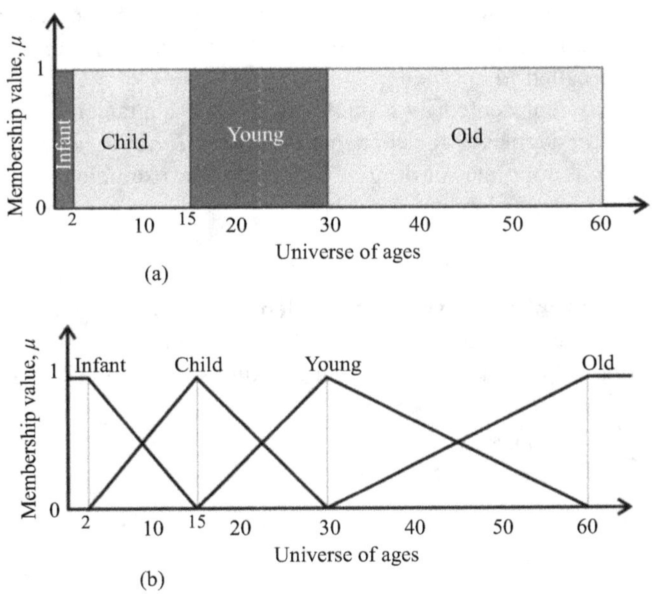

Figure 2.2 Crisp and fuzzy subsets and representation by membership functions. (a) Crisp sets and their membership functions and (b) fuzzy sets and their membership functions.

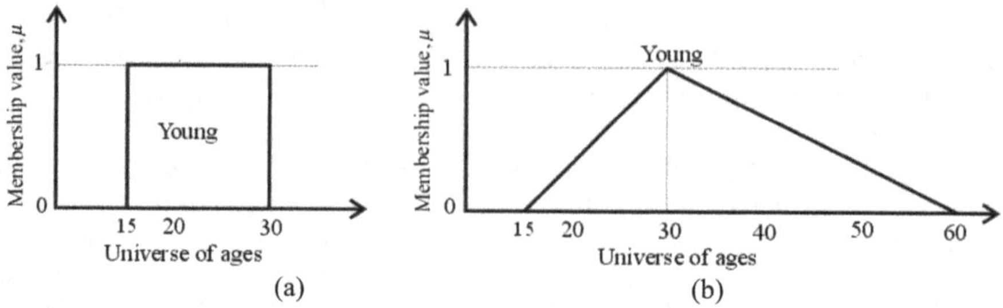

Figure 2.3 Crisp and fuzzy Young subsets and their membership functions. (a) Crisp Young and its crisp membership function and (b) fuzzy Young and its fuzzy membership function.

Young goes to 0 as the one representing *Old* increases to 1 at age 60. Aged 60 and up are assumed to be completely crisp *Old*.

In order to show the differences between crisp and fuzzy subsets and their membership functions, let us analyze crisp and Fuzzy *Young* subsets, which are separated from the others and are given in Figure 2.3 alone.

According to Figure 2.3(a), only the people who are between 15 and 30 years old are exactly *Young* but the others are not. People between 15 and 60 in Figure 2.3(b) are also young. In the latter, people are exactly young at the age of 30 with a full membership value and they are young with some degrees as one gets away from the age of 30. People represented by the membership function *Young* in Figure 2.3(b) are called *fuzzy Young*.

A fuzzy set is represented by a membership or a characteristic function that shows the inclusion degrees of the elements in that set. *Fuzzy membership functions* are defined either as continues or discrete functions. For example, a fuzzy set A can be represented by its membership function as

$$A = \{(x, \mu_{\mathrm{A}}(x)), \text{ Such that } x \in X\} \tag{2.1}$$

where $\mu(x)$ is the membership value of the element x in fuzzy subset A. The crisp variable x is an element of the universe X. Representation of a fuzzy subset A in discrete and continues formats are given in Equations (2.2) and (2.3), respectively.

$$A = \sum_{x_i \in X} \frac{\mu_A(x_i)}{x_i} \text{ (Discrete form)} \tag{2.2}$$

$$A = \int_X \frac{\mu_A(x)}{x} \text{ (Continues from)} \tag{2.3}$$

Expressions (2.1)–(2.3) show how a fuzzy subset is characterized by a membership function. The symbols $\sum$ and $\int$ used in (2.2) and (2.3) do not mean for summation or integration. They are used to represent discreteness and continuity.

Example 2.1

A discrete fuzzy subset C is defined in the crisp universe $X = \{0, 1, 2, 3, 4, 5, 6, 7, 8, 9, 10\}$ and given as

$$C = \sum_{i=1}^{11} \frac{\mu_C(x_i)}{x_i}$$

$$= \frac{0}{0} + \frac{0.2}{1} + \frac{0.4}{2} + \frac{0.6}{3} + \frac{0.8}{4} + \frac{1}{5} + \frac{0.8}{6} + \frac{0.6}{7} + \frac{0.4}{8} + \frac{0.2}{9} + \frac{0}{10}.$$

Plot the membership function $\mu_C(x)$ in a discrete graphical form.

Answer 2.1

A plot of membership function $\mu_C(x)$ in a discrete graphical form is obtained using MATLAB code in Figure E2.1. As depicted in Figure E2.2, a discrete graphic

```
% Example 2.1
clear; clg;
uf=[0 0.2 0.4 0.6 0.8 1.0 0.8 0.6 0.4 0.2 0];
X=0:1:10;
plot(X,uf,'o',0,1.1); axis([-inf inf 0 1.1]);
hold on
for k=1:length(X)
    plot([X(k) X(k)],[0 uf(k)],'-');
end;
hold off; grid
xlabel('Universe of discourse, X');
ylabel('Membership Values')
```

Figure E2.1 MATLAB code for plotting discrete subset C. Discrete plot of fuzzy membership function.

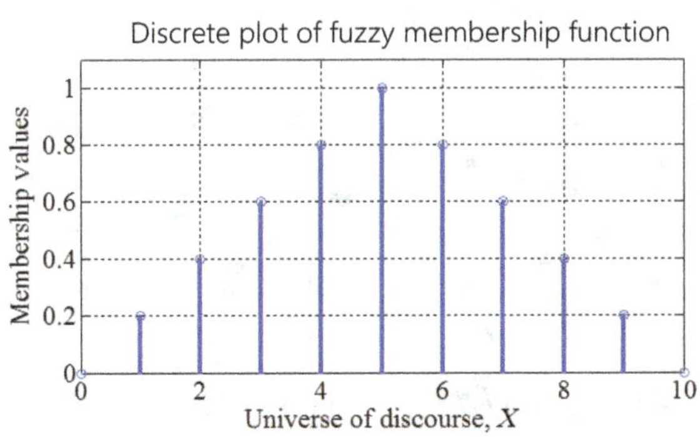

Figure E2.2 MATLAB code for plotting discrete subset C

includes the information only at sampling instants and does not give any ideas during the gaps. Therefore, data points are not connected using straight lines. The discrete form may be useful for the representation of sampled data, which is recorded periodically.

Example 2.2

The membership function representing the fuzzy subset *Young* in Figure 2.3(b) is defined as

$$\mu_{\text{Young}}(x) = \begin{cases} \dfrac{|x| - |a|}{|b| - |a|}, & \text{for } a \le x \le b \\ \dfrac{|c| - |x|}{|c| - |b|}, & \text{for } b \le x \le c \\ 0, & \text{otherwise} \end{cases}$$

where $a = 15$, $b = 30$, $c = 60$, and the universe of discourse X is defined in the interval [0, 70]. Plot the membership function *Young* in continuous form.

Answer 2.2

The continuous membership function representing the fuzzy subset *Young* is plotted and shown in Figure E2.3. This plot is obtained using MATLAB codes given in Figures E2.4 and E2.5. MATLAB functions are useful tools to be used for repeating software codes. A MATLAB function, *triangle00m.m*, given in Figure E2.5 is recalled from the code given in Figure E2.4 in order to plot the triangular-type fuzzy membership function.

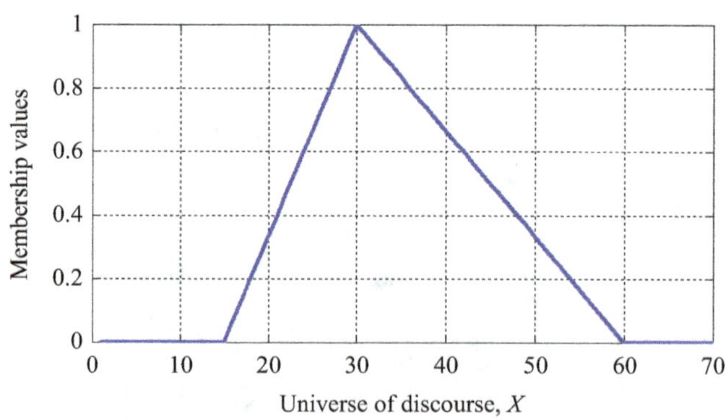

Figure E2.3 A plot of the continuous fuzzy subset in MATLAB

```
% Example 2.2.
% Matlab .m file
% used to plot
% triangular membership function

for i=1:70
   x(i)=i;
   YOUNG(i)=triangle00m(15,30,60,x(i));
end
plot(x,YOUNG)
grid
xlabel('Universe of discourse, X');
ylabel('Membership Values')
```

```
% MATLAB Function triangle00m.m
% Used for triangular function

function mu=triangle00m(x1,x2,x3,x)
if x1==x2
  if x <= x1
    mu=1;
  elseif x <= x3
    mu=(abs(x3)-abs(x))/(abs(x3)-abs(x2));
  else
    mu=0;
  end
elseif x2==x3
  if x <= x1
    mu=0;
  elseif x <= x3
    mu=(abs(x)-abs(x1))/(abs(x2)-abs(x1));
  else
    mu=1;
  end
else
  if x <= x1
    mu=0;
  elseif x <= x2
    mu=(abs(x)-abs(x1))/(abs(x2)-abs(x1));
  elseif x <= x3
    mu=(abs(x3)-abs(x))/(abs(x3)-abs(x2));
  else
    mu=0;
  end
end
```

Figure E2.4 MATLAB code for plotting continuous fuzzy membership function Young

2.2 Fuzzy sets and fuzzy membership functions

Membership functions representing fuzzy sets have different shapes, which are defined by certain types of mathematical formulas. The most used membership function types are listed as triangular, trapezoid, bell, sinusoid, Gaussian, Cauchy, and sigmoid. The properties of these membership functions will be analyzed in the next sections of this chapter. During the analysis, MATLAB functions similar to the one given in Example 2.2 and shown in Figure 2.4 will be developed and used to design toolboxes for decision-making and control applications.

In order to make the operations on fuzzy sets easier, the membership functions are formulated in terms of their parameters [2], which include information about fuzziness and location range in the universe of discourse. The flexibility of adjusting parameters makes membership functions adjustable, too. Due to linearity in their structure, triangular and trapezoid-type membership functions are preferred over the others. However, in some applications, the membership functions are determined and adjusted by learning algorithms that require derivations such as in artificial neural networks. In this case, functions with easy derivatives are preferred.

2.2.1 Triangular membership function

Triangular membership functions are made of lines as shown in Figure 2.4 and realized by the combination of line equations as given in Equation (2.4).

$$\mu_A(x) = \begin{cases} \dfrac{|x| - |x_1|}{|x_2| - |x_1|}, & \text{for } x_1 \leq x \leq x_2 \\ \dfrac{|x_3| - |x|}{|x_3| - |x_2|}, & \text{for } x_2 \leq x \leq x_3 \\ 0, & \text{otherwise} \end{cases} \tag{2.4}$$

where the parameters x_1, x_2, and x_3 give the location of fuzzy membership function A in the universe of X as shown in Figure 2.4. Actually, the parameters x_1, x_2, and x_3 represent the address of the membership function A and tell us where it is located in the corresponding universe. It is sufficient to change the values of parameters in order to define a new membership function with a similar shape or change the location in the universe of discourse. That's why parametrized formulas are important to represent membership functions. Equation (2.4) can be used as a parametrized membership function representing triangular-type fuzzy subsets. However, it seems to be a long equation with conditional statements. Therefore, let us convert it to a simpler form so that it will be more suitable to be used in MATLAB.

Equation (2.4) shows that x_2 is a convergence point and equation

$$\mu_A(x) = \left(\frac{x - x_1}{x_2 - x_1}\right) < \left(\frac{x_3 - x}{x_3 - x_2}\right) \tag{2.5}$$

can be satisfied as long as $x_1 \leq x_2$ and $x \leq x_2$. Similarly,

$$\mu_A(x) = \left(\frac{x_3 - x}{x_3 - x_2}\right) < \left(\frac{x - x_1}{x_2 - x_1}\right) \tag{2.6}$$

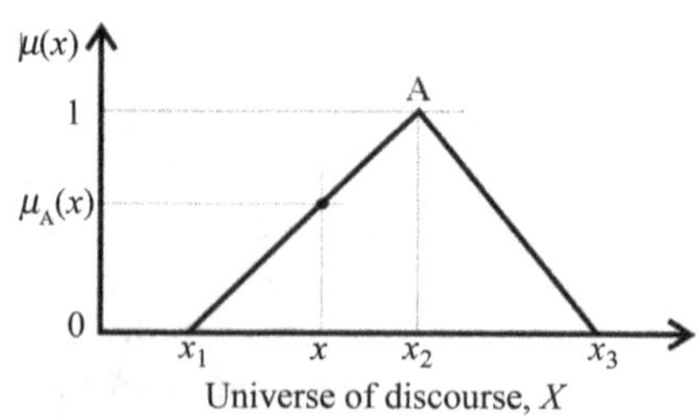

Figure 2.4 A triangular fuzzy membership function

is satisfied as long as for $x \geq x_2$ and $x_2 \leq x_3$. In other words, the output is equal to the smaller part of Equations (2.5) or (2.6). However, these equations give a negative output if $x < x_1$ or $x > x_3$. Since the membership degrees are defined in interval [0, 1], the negative outputs should be changed to 0. Therefore, the maximum value between 0 and the output from Equations (2.5) or (2.6) should be selected. According to these comments, Equation (2.4) can be converted to the one given in (2.7).

$$\mu_A(x) = \max\left(\min\left(\frac{x - x_1}{x_2 - x_1}, \frac{x_3 - x}{x_3 - x_2}\right), 0\right) \tag{2.7}$$

A MATLAB function can be written as an .m file as in Figure 2.5 to execute Equation (2.7). The same function can also be realized in Simulink as shown in Figure 2.6.

```
function mu=triangle01m(x1,x2,x3,x)
a=(x-x1)/(x2-x1);
b=(x3-x)/(x3-x2);
mu=max(min(a,b),0);
```

Figure 2.5 A MATLAB function for simulating triangular fuzzy membership function

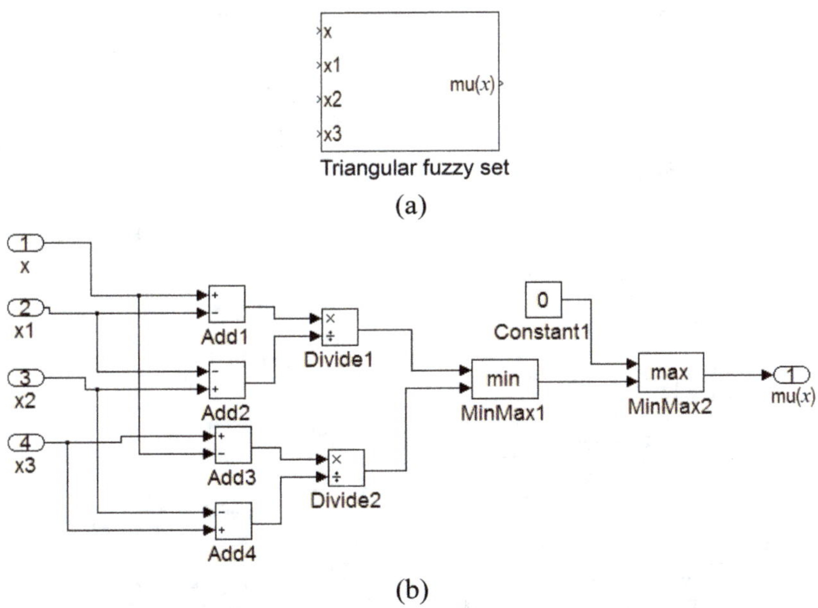

Figure 2.6 A Simulink function for simulating triangular fuzzy membership function. (a) Closed form of the Simulink block and (b) inside view of the block in (a).

Both MATLAB and Simulink functions given in Figures 2.5 and 2.6, respectively, are used to find the membership value of variable x in a fuzzy set triangle defined by its parameters x_1, x_2, and x_3 in the universe of discourse, X. The meaning of these parameters is shown in Figure 2.4. As given in the MATLAB function triangle01m.m in Figure 2.5 and Simulink function in Figure 2.6, the variable x and the parameters x_1, x_2, and x_3 are the inputs to triangular functions.

Example 2.3

Triangles A, B, and C are defined in the universe of $X = \{-5, -4, -3, -2, -1, 0, 1, 2, 3, 4, 5\}$ with their parameters as A $(-4, -2, 0)$, B $(-2, 0, 2)$, and C $(0, 2, 4)$. Use MATLAB and Simulink functions given in Figures 2.5 and 2.6, respectively, and

(a) Plot the triangles A, B, and C in the universe of X
(b) Obtain the membership values of $x = -1.5$ and $x = 1$ on the fuzzy subsets A, B, and C.

Answer 2.3

This example is given to show how MATLAB and Simulink can be used as a tool in order to model and simulate fuzzy subsets. The use of a triangular fuzzy subset modeled in Simulink is given in Figure E2.3(a). In this simulation, a ramp signal with a slope of 1 is added to -5 to generate the universe of X from -5 to 5 as a function of the simulation time. The variable x and triangle parameters x_1, x_2, and x_3 are entered from the input ports of the triangle block. The outputs of triangle blocks are the membership values, which are sent to the workspace so that they can

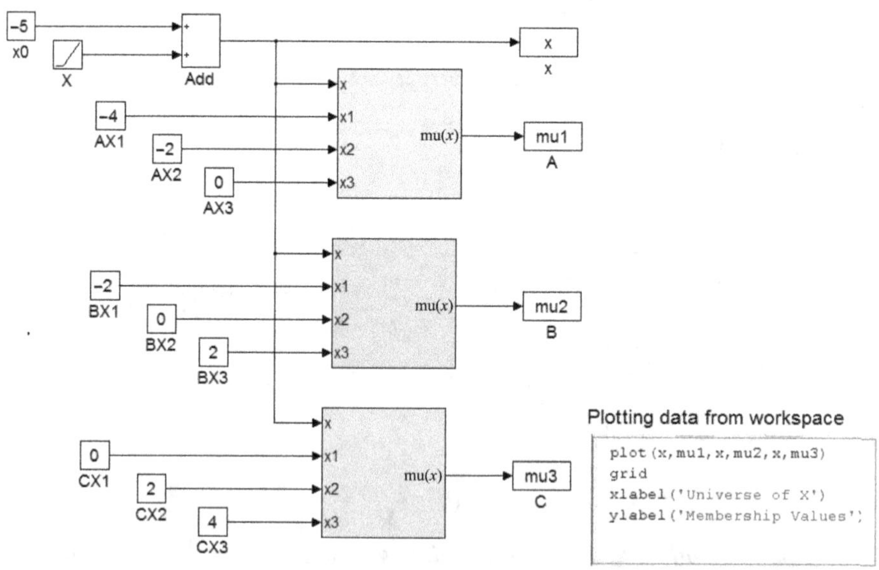

Figure E2.3(a) Modeling and simulating triangular fuzzy subsets in Simulink

be plotted in MATLAB. A small plotting code of the workspace data is also given in Figure E2.3(a). The resultant plots of the fuzzy subsets *A*, *B*, and *C* are shown in Figure E2.3(c). A MATLAB .m file can be used to simulate the same triangular fuzzy subset models, too, as given in Figure 2.3(b). The MATLAB function file *triangle01m.m* defined earlier and given in Figure 2.5 is called as a subprogram to simulate the triangle fuzzy subset.

(a) This example also shows how a function file is recalled from another program. This example will be extended for other applications in the coming chapters. In order to develop our own fuzzy toolbox, we create function files step by step and keep them in a safe place to be used later.

```
% Example 2.3 - Triangle membership function
clear,clc,clf
k=1;
for x=-5:0.1:5
    muA(k)=triangle01m(-4,-2,0,x);
    muB(k)=triangle01m(-2,0,2,x);
    muC(k)=triangle01m(0,2,4,x);
    X(k)=x;
    k=k+1;
end
plot(X,muA,X,muB,X,muC)
grid
xlabel('Universe of X');ylabel('Membership Values')
```

Figure E2.3(b) Modeling and simulating triangular fuzzy subsets in MATLAB

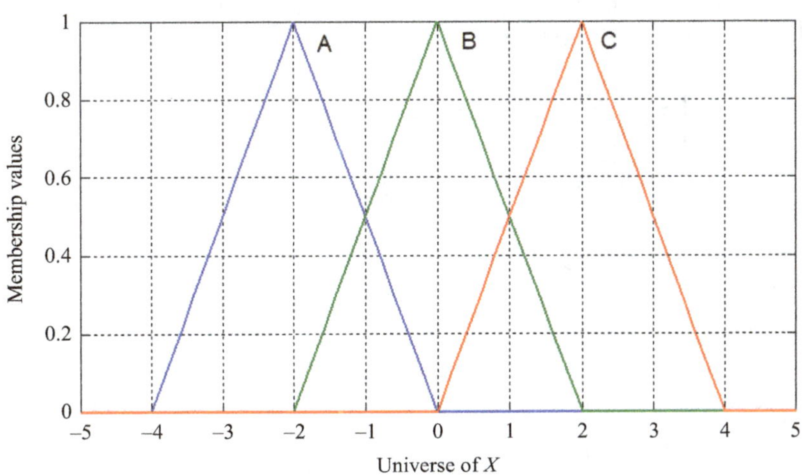

Figure E2.3(c) Triangular fuzzy subsets as a result of simulations given in Figures 2.3(a) and 2.3(b)

Table E2.3 Using function triangle01m.m for the Example 2.3

Function file: *triangle01m.m*	***x* = −1.5**	***x* = 1.0**
μ_A (k) = triangle01m (−4, −2, 0, x);	μ_A $(x_1) = 0.75$	μ_A $(x_2) = 0$
μ_B (k) = triangle01m (−2, 0, 2, x);	μ_B $(x_1) = 0.25$	μ_B $(x_2) = 0.50$
μ_C (k) = triangle01m (0, 2, 4, x);	μ_C $(x_1) = 0$	μ_C $(x_2) = 0.50$

(b) The second part of the example is about using triangular functions modeled in MATLAB and Simulink. If $x = -1.5$ or $x = 1.0$ is entered as input to the model of the triangle in MATLAB or Simulink, the results given in Table E2.3 are obtained.

Due to their linear structures, triangular fuzzy subsets are easy to model and easy to simulate. The sharp peak enables them to react to any changes even if they are very small. Therefore, the sharp peak makes triangle membership functions sensitive to the changes in the crisp variable x.

2.2.2 *Trapezoid membership function*

Trapezoid type membership functions are obtained as the combination of line equations as it is in triangular type. The trapezoidal membership functions differ from the triangular types with its crisp region on the top. Trapezoidal membership functions have a flat top that is not fuzzy while triangular membership functions have a sharp peak.

A parametrized general view of the trapezoid fuzzy membership function is given in Figure 2.7, which is used to draw Equation (2.8). A MATLAB function for trapezoidal fuzzy subset can be developed using Figure 2.7 and Equation (2.8), in which it can easily be seen that, the same line equations used in the triangular membership function can also be used here with the addition of a flat line on the top.

$$\mu_A(x) = \begin{cases} \dfrac{|x| - |x_1|}{|x_2| - |x_1|}, & \text{if } x_1 \le x \le x_2 \\ 1 & \text{if } x_2 \le x \le x_3 \\ \dfrac{|x_4| - |x|}{|x_4| - |x_3|}, & \text{if } x_3 \le x \le x_4 \\ 0 & \text{Otherwise} \end{cases} \tag{2.8}$$

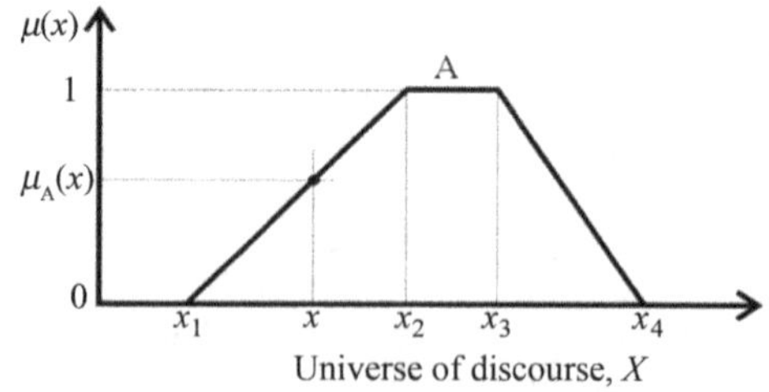

Figure 2.7 A trapezoid fuzzy membership function

where the parameters x_1, x_2, x_3, and x_4 give the location of fuzzy membership function A in the universe of X. Actually, the parameters x_1, x_2, x_3, and x_4 represent the address of the membership function A and tell us where it is located in the corresponding universe. It is sufficient to change the values of parameters in order to define a new membership function with a similar shape or change the location in the universe of discourse. Equation (2.8) may also be used for the triangular membership function by setting $x_2 = x_3$. As it is done for the triangular fuzzy membership function, an analysis of Equation (2.8) shows that there are two convergence points, which are x_2 and x_3. It is clear that

$$\mu_A(x) = \left(\frac{x - x_1}{x_2 - x_1}\right) < 1 \text{ and } \left(\frac{x_4 - x}{x_4 - x_3}\right) \tag{2.9}$$

for $x_1 \leq x_2$ and $x \leq x_2$. And also

$$\mu_A(x) = 1 < \left(\frac{x - x_1}{x_2 - x_1}\right) \text{ and } \left(\frac{x_4 - x}{x_4 - x_3}\right) \tag{2.10}$$

for $x_2 \leq x \leq x_3$. Similarly,

$$\mu_A(x) = \left(\frac{x_4 - x}{x_4 - x_3}\right) < \left(\frac{x - x_1}{x_2 - x_1}\right) \text{ and } 1 \tag{2.11}$$

is obtained for $x \geq x_3$ and $x_3 \leq x_4$. In other words, the output is equal to a smaller part of Equations (2.9), (2.10), or (2.6). However, these equations give a negative output if $x < x_1$ or $x > x_4$. Since the membership degrees are defined in interval [0, 1], the negative outputs should be changed to 0. Therefore, the maximum value between 0 and the output from Equations (2.9), (2.10), or (2.11) should be selected. According to these comments, Equation (2.8) can be converted to the one given in (2.12).

$$\mu_A(x) = \max\left(\min\left(\frac{x - x_1}{x_2 - x_1}, 1, \frac{x_4 - x}{x_4 - x_3}\right), 0\right) \tag{2.12}$$

The MATLAB function given in Figure 2.8 can be used to execute Equation (2.12). The same trapezoid function can also be realized in Simulink as shown in Figure 2.9.

Both MATLAB and Simulink functions given in Figures 2.8 and 2.9, respectively, can be used to simulate the fuzzy set trapezoid defined by its parameters x_1, x_2, x_3, and x_4 in the universe of discourse, X.

```
function mu=trapez01m(x1,x2,x3,x4,x)
a=(x-x1)/(x2-x1);
b=(x4-x)/(x4-x3);
mu=max(min(a,1,b),0);
```

Figure 2.8 Function trapezo1m.m for simulating trapezoidal fuzzy subset

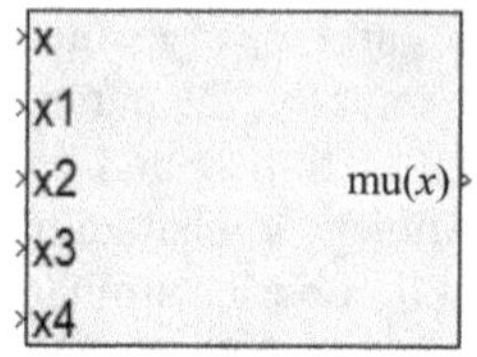

(a)

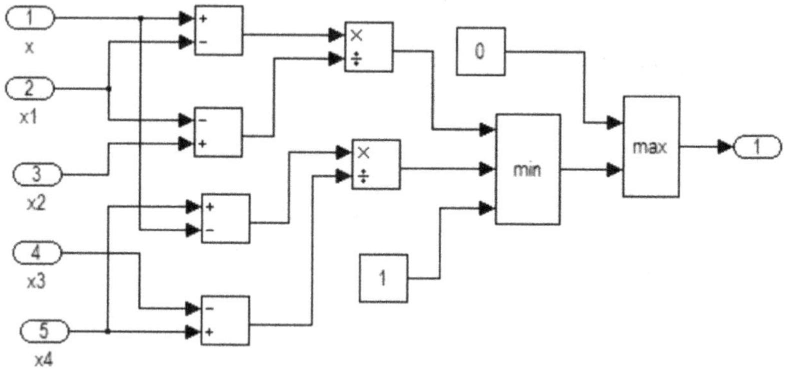

(b)

Figure 2.9 Simulink function for simulating trapezoid fuzzy subset. (a) Closed form of the Simulink block and (b) inside view of the block in (a).

Due to their linear structures, trapezoid fuzzy subsets are also easy to model and easy to simulate. The flat top gives a crisp behavior to trapezoid membership functions so that the sensitivity of the membership function to the changes in the crisp variable is reduced.

Example 2.4

Trapezoid membership functions A, B, and C are defined in the universe of $X = \{-5, -4, -3, -2, -1, 0, 1, 2, 3, 4, 5\}$ with their parameters as A $(-5, -5-4, -1)$, B $(-4, -1, 1, 4)$, and C $(1, 4, 5, 5)$. Use MATLAB and Simulink functions given in Figures 2.8 and 2.9, respectively, and

(a) Plot trapezoid membership functions A, B, and C in the universe of X.
(b) Obtain the membership values of $x = -3.5$, $x = -2$, $x = 2.5$, and $x = 4$ on the fuzzy subsets A, B, and C.

Answer 2.4

(a) Ideal fuzzy subsets modeled in MATLAB and Simulink are given in Figures E2.4(a) and E2.4(b), respectively.

In the MATLAB .m file simulation given in Figure E2.4(a), variable x and trapezoid function parameters x_1, x_2, x_3, and x_4 are entered directly inside the parenthesis where the function is recalled. Then results are plotted as shown in Figure E2.4(c). A ramp source block with a slop of 1 is used to generate the universe of X in an interval [−5, 5] in the simulation model developed in Simulink as shown in Figure E2.4(b). The outputs of trapezoid blocks are the membership values, which are sent to the workspace so that they can be plotted using MATLAB .m files.

```
% Example 2.4 - Trapezoid membership function
clear,clc,clf
k=1;
for x=-5:0.1:5
    muA(k)=trapez01m(-5,-4,-3,-1,x);
    muB(k)=trapez01m(-3,-1,1,3,x);
    muC(k)=trapez01m(1,3,4,5,x);
    X(k)=x;
    k=k+1;
end
plot(X,muA,X,muB,X,muC)
grid
xlabel('Universe of X');ylabel('Membership Values')
```

Figure E2.4(a) Modeling and simulating trapezoid fuzzy subsets in MATLAB

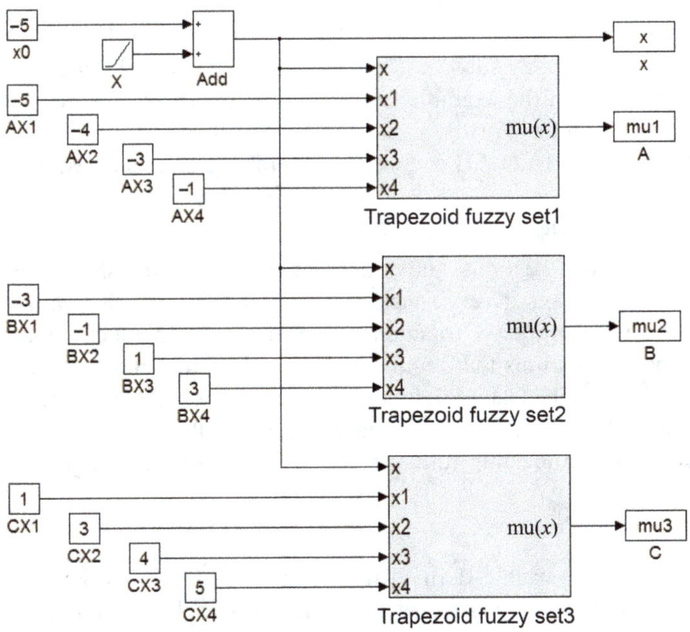

Figure E2.4(b) Modeling and simulating trapezoid fuzzy subsets in Simulink

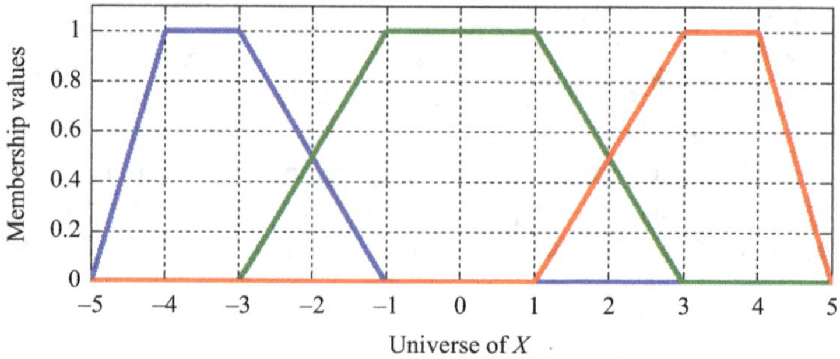

Figure E2.4(c) Trapezoid fuzzy subsets as a result of simulations given in Figures 2.4(a) and 2.4(b)

Table E2.4 Using function trapez01m.m for the Example 2.4

Function file: *trapez01m.m*	**Membership values**			
	$x = -3.5$	$x = -2$	$x = 2.5$	$x = 4$
μ_A = trapez01m (−5, −4, −3, −1, x);	0.5	0.5	0.25	0.0
μ_B = trapez01m (−3, −1, 1, 3, x);	0.0	0.5	0.75	0.0
μ_C = trapez01m (1, 3, 4, 5, x_1);	0.0	0.0	0.0	1.0

A small plotting code of the workspace data is also given in Figure E2.3(a). The resultant plots of the fuzzy subsets *A*, *B*, and *C* are shown in Figure E2.4(c).

The second part of the example is about using trapezoid functions modeled in MATLAB and Simulink. For the values of $x = -3.5$, 2, 2.5, and 4, the model of trapezoid function in MATLAB or Simulink yields the results given in Table E2.4.

2.2.3 Gaussian membership function

As stated in previous sections, the triangular fuzzy membership function has a sharp peak, which makes it very sensitive to the changes in crisp variable. Due to this high sensitivity property, triangular-type membership functions are not preferred in some applications that require soft transition responses to the changes in crisp variables. Since the Gaussian function has a soft peak, it is less sensitive to the changes in crisp variables when compared to triangular-type membership functions. A Gaussian membership function can be defined as in (2.13).

$$\mu_A(x) = e^{-\frac{1}{2}\left(\frac{x-x_P}{w}\right)^2} \tag{2.13}$$

where x is the crisp variable defined in the universe of X. X_P is a crisp number located at the center of the membership function A with the maximum membership degree of 1. The parameter w defines the distribution width of the Gaussian membership function and must be selected carefully. Larger values of w convert the

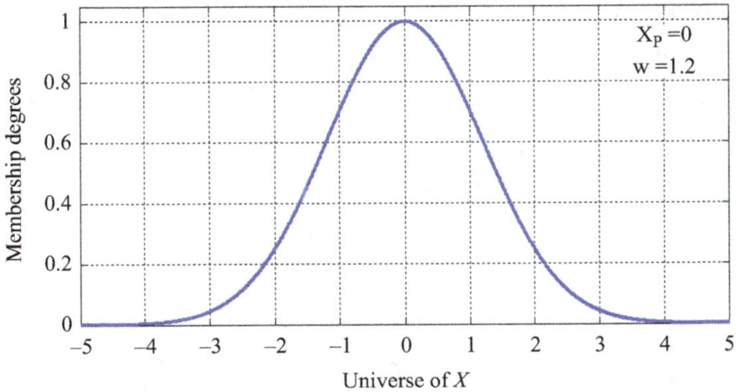

Figure 2.10 Gaussian membership function

Gaussian function into a signal oscillating like a sinewave, whereas smaller values of w convert it to a single line at X_P. A plot of Gaussian function is given in Figure 2.10, which is obtained using Equation (2.13) with the parameters $w = 1.2$ and $X_P = 0$ in the universe of X defined in the interval $[-5, 5]$.

A more generalized Gaussian-type function is given in Equation (2.14), which is used to represent the numbers close to crisp α. This generalized type membership function is also called as fuzzy number due to its normality and convexity. However, the function given by Equation (2.14) is concentrated toward α, which makes it less fuzzy compared to (2.13). The smaller values of β may be decreased to increase the width of the function. If β is selected as $0 < \beta < 1$, the fuzziness of the function is increased along with the width. However, smaller β values cause more increment in the bottom part of the function so that zero crossing points with the crisp axis move toward infinity.

$$\mu_A(x) = \frac{1}{1 + \beta(x - \alpha)^2} \tag{2.14}$$

Figure 2.11 shows the changes due to β in the membership function given in Equation (2.14). It is shown that the fuzziness of the membership function increases as β decreases.

2.2.4 Bell membership function

A bell-type membership function can be described by Equation (2.15), which is a generalized Cauchy distribution used in probability theory. Equation (2.15) yields Figure 2.12 for the parametric values $x_P = 0$, $w = 2$, and $m = 4$.

$$\mu_A(x) = \frac{1}{1 + \left|\frac{x - x_P}{w}\right|^{2m}} \tag{2.15}$$

Parameter m in Equation (2.15) defines the width of the flat top (the crisp section) of the bell function and it is a positive number. Negative values of m yield

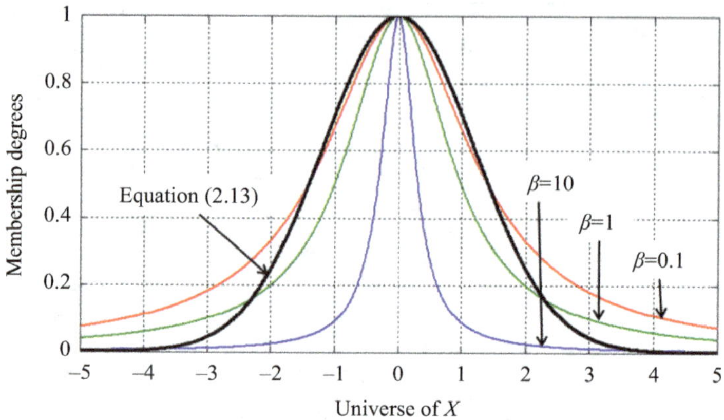

Figure 2.11 A comparison of Equations (2.13) and (2.14)

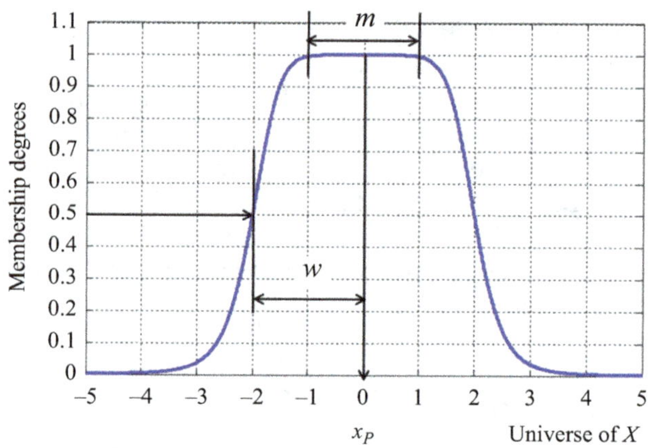

Figure 2.12 Bell membership function

the inverse of the bell function. The bandwidth of the bell function is determined by parameter w as shown in Figure 2.12. In other words, the flat top and bandwidth of the bell function are adjusted by parameters m and w, respectively. The adjustment of the flat top and bandwidth of a fuzzy membership function is nothing but the adjustment of its fuzziness. Larger values of m turn the bell function into a perpendicular or square crisp function while very small values of w change it to a Cauchy curve function, which is described in the next section. The MATLAB code given in Figure 2.13 is an example using the function file *bell01m.m* to plot the bell-type fuzzy subset.

```
%  ********************************
%  BELL Membership Function
%  ********************************
%  Inputs: Crisp number xP at peak,
%  width w, m and variable x
clear;clc;clf
xP=0;  w=2;  m=4; k=1;
for x=-5:0.01:5;
    B(k)=bell01m(xP,w,m,x);
    X(k)=x;
    k=k+1;
end
plot(X,B);
xlabel('Universe of X');
ylabel('Membership Degrees');
grid
```

Figure 2.13 Using the bell function to plot the bell type fuzzy subset

2.2.5 Sinusoid membership function

A sinusoid function is used to simulate this membership function. It is defined along the half cycle of a full period. The function should be formulated carefully so that the properties of a fuzzy membership function are satisfied along the full period. Since a sinewave includes both positive and negative alternates, its absolute value will have two positive half cycles with a doubled frequency. In order to develop a parametrized sinewave function similar to the ones given earlier, let us define a parameter x_P, which is the crisp value corresponding to the peak point of the sinewave function as shown in Figure 2.14. It should be kept in mind that the peak point is the address of a function and is determined by delayed phase angles, which will be addressed later in this section. The length of the universe and the number of fuzzy subsets to be represented are important parameters when designing the sinewave function.

Let us assume that four fuzzy subsets need to be defined in the interval $[x_1, x_2]$ as shown in Figure 2.14. Then the length of the universe X and the parameter α are calculated as

$$X = x_2 - x_1 \tag{2.16}$$

$$\alpha = \frac{X}{n-1} \tag{2.17}$$

where X is the length of the universe, and n is the number of fuzzy subsets. It is obvious from Figure 2.14 that

$$\frac{T}{2} = 2a \rightarrow T = 4\alpha \tag{2.18}$$

The delayed phase distance β can be written as

$$\beta = \frac{X}{2} - a \tag{2.19}$$

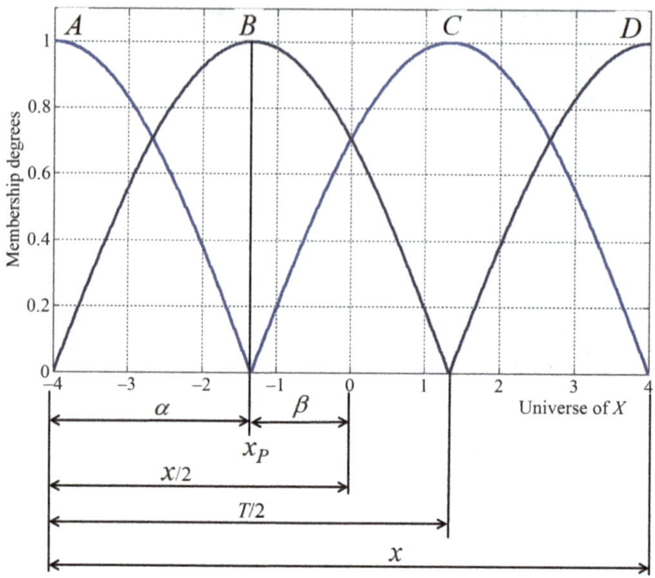

Figure 2.14 Sinusoid membership function defined in X

In order to be used in a sine function, delayed phase distance β should be written in radians. Since the distance α is equal to $\pi/2$ radians as seen in Figure 2.14, a comparison yields the delayed phase angle in radians as

$$\theta_1 = \frac{\beta}{\alpha}\frac{\pi}{2} \tag{2.20}$$

The phase angle θ_1 determines the locations of membership functions A and C in Figure 2.14 as $x = 0$ is the reference point. Therefore, the delated phase angle of the membership functions B and D are written as

$$\theta_2 = \theta_1 + \frac{\pi}{2} \tag{2.21}$$

Thus, the sinusoid membership function is obtained as

$$\mu_A(x) = \left|\sin\left(\frac{\pi}{2\alpha}t + \theta\right)\right| \tag{2.22}$$

where θ is the delayed phase angle in radians, and it is equal to either θ_1 or θ_2. The peak point of each sinusoid membership function can be calculated as

$$x_P(m) = x_1 + (m-1)\alpha, \quad m = 1, 2, \ldots, n \tag{2.23}$$

where n is the number of fuzzy subsets, x_1 is the first element in the universe of X and equal to -4 in Figure 2.14 and, α is given by (2.17). The point x_P corresponds to the peak of fuzzy subsets with the maximum membership degree of 1. This point indicates the location of a fuzzy subset in the related universe.

Sinusoid membership functions are symmetrical due to the characteristic properties of a sinewave. Therefore, they are usually placed symmetrically as 0 line being the reference.

Example 2.5

Sinusoid membership functions *A*, *B*, and *C* are defined in the universe of $X = \{-4, -3, -2, -1, 0, 1, 2, 3, 4\}$. Develop and use MATLAB and Simulink functions, and

(a) Plot sinusoid functions *A*, *B*, and *C* in the universe of *X*.
(b) Add another function *D* and repeat (a) for four membership functions defined in the same universe of *X*.
(c) Obtain the membership values of $x = -3.5$, $x = -2$, $x = 2.5$, and $x = 4$ on the fuzzy subsets *A*, *B*, *C*, and *D*.

Answer 2.5

Let us assume that set *A* and set *C* are one-sided lower and upper boundary sets, respectively, and set *B* is a two-sided center set as shown in Figure E2.5(a), which shows that set *B* is defined on the full universe of *X* from −4 to 4 while set *A* and set *C* are defined on the sections −4 to 0 and from 0 to 4, respectively. It can easily be seen from Figures 2.14 and E2.5(a) that one cycle of the sinewave function is equal to four times the parameter α ($T = 4\alpha$), as in Equation (2.18).

Since the universe of $X = \{-4, -3, -2, -1, 0, 1, 2, 3, 4\}$, the parameters *X*, *T*, and α are obtained as

$$X = x_2 - x_1 = 4 - (-4) = 8$$

$$\alpha = \frac{X}{n-1} = \frac{8}{3-1} = 4$$

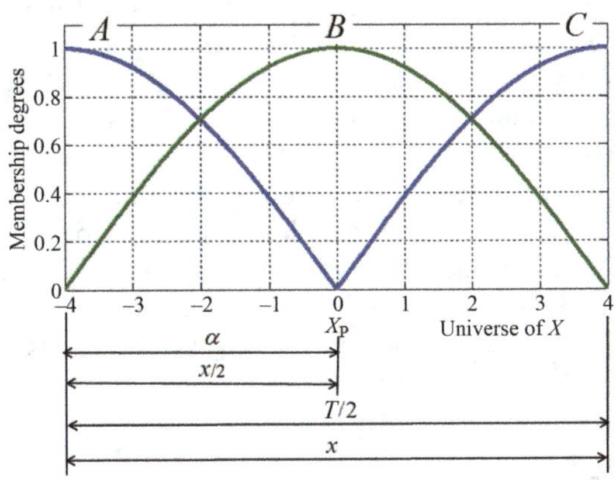

Figure E2.5(a) Sinusoid functions A, B, and C

So, the peak points from (2.23) will be:

$$x_P(m) = -4 + (m-1)4 \quad \text{for } m = 1,2,3$$
$$x_{PA} = -4, \quad x_{PB} = 0, \quad x_{PC} = 4.$$

Then from (2.18) the delayed phase angle is calculated as

$$\beta = \frac{X}{2} - a = 0$$

$$\theta_1 = \frac{\beta}{\alpha}\frac{\pi}{2} = 0$$

which yields that

$$\theta_2 = \theta_1 + \frac{\pi}{2} = \frac{\pi}{2}$$

Thus, the sinewave membership functions are defined as

$$\mu_A(x) = \left|\sin\left(\frac{\pi}{2\alpha}x + \theta_2\right)\right| \mathit{for} -4 \le x \le 0$$

$$\mu_C(x) = \left|\sin\left(\frac{\pi}{2\alpha}x + \theta_2\right)\right| \mathit{for}\ 0 \le x \le 4$$

and

$$\mu_B(x) = \left|\sin\left(\frac{\pi}{2\alpha}x + \theta_1\right)\right| \mathit{for} -4 \le x \le 4$$

(a) Now, let us write a sinusoid membership function and use it in a MATLAB .m file to plot the subsets *A*, *B*, and *C*.
(b) If another function *D* is added, there will be four membership functions in the same universe of $X = \{-4, -3, -2, -1, 0, 1, 2, 3, 4\}$. The input parameters to the MATLAB .m file shown in Figure E2.5(a) become $n = 4$; $x_1 = -4$, and $x_2 = 4$. The rest is just a series of calculations using Equations (2.16)–(2.23) to reach the sinusoid membership functions as shown in Figure E2.5(b).
(c) The membership values of $x = -3.5$, $x = -2$, $x = 2.5$, and $x = 4$ on the fuzzy subsets *A*, *B*, *C*, and *D* are obtained using the sinusoid MATLAB function given in Figure E2.5(c) and listed in Table E2.5. In order to find the membership values for individual x's, it will be useful to convert the MATLAB code given in Figure E2.5(b) to a MATLAB function so that just the value of x is sufficient as input. Such as function file is given in Figure E2.5(c), which is applicable for this example only.

Practice 2.1

The sinusoidal MATLAB function given in Figure E2.5(c) includes repeating lines and processes. Modify the given code and develop a new shorter and simpler-to-use

```
%  **********************************
%   Sinusoid membership function
%  **********************************
%   Inputs  X, n and  x1, x2 and x
clear;  clc;   clf
n=3; x1=-4; x2=-x1; X=x2-x1;
a=X/(n-1); b=(X/2)-a;
for m=1:n
        XP(m)=x1+a*(m-1);
end
XPA=XP(1); XPB=XP(2); XPC=XP(3);
A1=(b*pi)/(2*a);   A2=A1+(pi/2);
k=1;
for x=x1:0.01:x2;
      if x<XPA
           SA(k)=0; SB(k)=0; SC(k)=0;
      elseif x<XPB
           SA(k)=sinus01m(a,A1,x);
           SB(k)=sinus01m(a,A2,x);
           SC(k)=0;
      elseif x<XPC
           SA(k)=0;
           SB(k)=sinus01m(a,A2,x);
           SC(k)=sinus01m(a,A1,x);
      else
           SA(k)=0; SB(k)=0; SC(k)=0;
      end
      X(k)=x;   k=k+1;
end
plot(X,SA,X,SB,X,SC);
xlabel('Universe of X');
ylabel('Membership Degrees');
grid
```

```
%*****************************
% Sinusoid Membership Function
% *****************************
%  Inputs: a,A and t.
%   Calculated: x1, x2 and T.
function mu=sinus01m(a,A,t)
          T=4*a;
          w1=(2*pi/T);
          mu=abs(sin(w1*t+A));
```

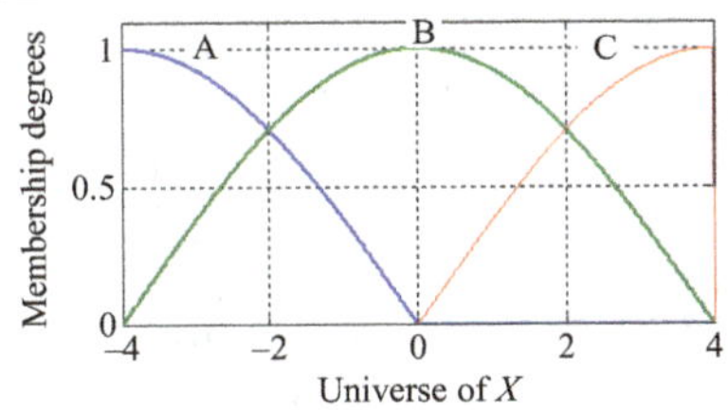

Figure E2.5(b) The use of sinusoid function in MATLAB, for example, 2.5(a)

sinusoid function so that the command $\mu = \text{sinus}\,(n, x_1, x)$ should be sufficient to get the membership values of x in all n fuzzy subsets, which are defined in the universe X in the interval $[x_1, x_2]$.

2.2.6 Sigmoid membership function

When a crisp universe is portioned into subcategories, there are always lower and upper categories. If the subcategories of a crisp universe of X are represented by fuzzy subsets, there will be a starting fuzzy subset at the lowest end and an ending fuzzy subset set at the highest end as shown by fuzzy subsets A and D in Figure 2.14. Right triangle, half trapezoid, and sinusoid type membership functions can be used as lower and upper boundary fuzzy sets. In addition to those mentioned above, the sigmoid functions are the alternatives for the boundary sets used at lower and upper limits. The sigmoid fuzzy membership functions are also called S-curves since they look alike the letter S. The membership values tend to decrease in a sigmoid fuzzy subset if it is used as a starting set at the lower boundary and to increase if it is used at the upper boundary as given in Figure 2.15. Therefore, the sigmoid functions are also called as decreasing sigmoid and increasing sigmoid. Right triangles can be used for decreasing or increasing fuzzy subsets, too.

```
%*******************************
% Sinusoid membership function
%*******************************
%Inputs X, n and  x1, x2 and x
clear;  clc;   clf
n=4; x1=-4; x2=-x1; X=x2-x1;
a=X/(n-1); b=(X/2)-a;
for m=1:n
    XP(m)=x1+a*(m-1);
end
XPA=XP(1); XPB=XP(2);
XPC=XP(3); XPD=XP(4);
A1=(b*pi)/(2*a);  A2=A1+(pi/2);
k=1;
for x=x1:0.01:x2;
    if x<XPA
        SA(k)=0; SB(k)=0;
        SC(k)=0; SD(k)=0;
    elseif x<XPB
        SA(k)=sinus01m(a,A1,x);
        SB(k)=sinus01m(a,A2,x);
        SC(k)=0;  SD(k)=0;
    elseif x<XPC
        SA(k)=0;
        SB(k)=sinus01m(a,A2,x);
        SC(k)=sinus01m(a,A1,x);
        SD(k)=0;
    elseif x<XPD
        SA(k)=0;
        SB(k)=0;
        SC(k)=sinus01m(a,A1,x);
        SD(k)=sinus01m(a,A2,x);
    else
        SA(k)=0; SB(k)=0;
        SC(k)=0; SD(k)=0;
    end
    X(k)=x;  k=k+1;
end
plot(X,SA,X,SB,X,SC,X,SD);
xlabel('Universe of X');
ylabel('Membership Degrees');
grid
```

```
%*****************************
% Sinusoid Membership Function
% ***************************
%  Inputs: a,A and t.
%   Calculated: x1, x2 and T.
function mu=sinus01m(a,A,t)
         T=4*a;
         w1=(2*pi/T);
         mu=abs(sin(w1*t+A));
```

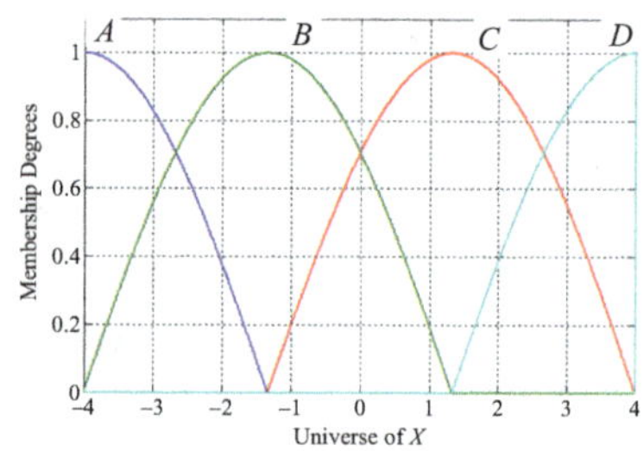

Figure E2.5(c) The use of sinusoid function in MATLAB for example 2.5(b)

However, sigmoid membership functions are usually preferred in applications that require nonlinearity such as learning process in artificial neural network.

A sigmoid function is defined as in (2.24)

$$\mu_A(x) = \frac{1}{1 + e^{-a(x-c)}} \tag{2.24}$$

where the sign of the parameter a indicates the direction of the sigmoid function whether it is decreasing or increasing. If the sing is negative than the sigmoid function shows a decreasing behavior while it has an increasing one for the positive sign. The parameter a defines the slope of the sigmoid curve at $\mu(x) = 0.5$ where $x = c$. Higher absolute values of a increase the slope, whereas lower values decrease it. Such that the function gives a straight line with a zero slope at the membership value of $\mu(x) = 0.5$ for $a = 0$.

```
%   ********************************
%    Sinusoid membership function
%    Example 2.5c
%   ********************************
%    Input x
function mu=sinus02m(x)
n=4; x1=-4; x2=-x1;
X=x2-x1;
a=X/(n-1); b=(X/2)-a;
for m=1:n
     XP(m)=x1+a*(m-1);
end
XPA=XP(1); XPB=XP(2); XPC=XP(3);
XPD=XP(4);
A1=(b*pi)/(2*a);   A2=A1+(pi/2);
if x<XPA
    SA=1; SB=0; SC=0; SD=0;
elseif x<XPB
    SA=sinus01m(a,A1,x);
    SB=sinus01m(a,A2,x);
    SC=0;  SD=0;
elseif x<XPC
    SA=0;
    SB=sinus01m(a,A2,x);
    SC=sinus01m(a,A1,x);
    SD=0;
elseif x<XPD
    SA=0;
    SB=0;
    SC=sinus01m(a,A1,x);
    SD=sinus01m(a,A2,x);
else
   SA=0; SB=0; SC=0; SD=1;
end
mu=[SA SB SC SD];
```

```
Matlab® Command Window

>> M1=sinus02m(-3.5);
>> M2=sinus02m(-2);
>> M3=sinus02m(2.5);
>> M4=sinus02m(4);
>> MM=[M1; M2; M3; M4]

MM =
    0.9569   0.2903   0        0
    0.3827   0.9239   0        0
    0        0        0.7730   0.6344
    0        0        0        1.0000
```

Figure E2.5(d) The use of sinusoid function for specific x's in Example 2.5(c)

The range of sigmoid function in the universe of X is adjusted by the parameters α and c. The parameter c is equal to x when $\mu(x) = 0.5$ and α gives the slope of the curve at $\mu_A(x) = 0.5$. However, these two parameters do not give any information about the crisp numbers (x_1 and x_2) where the membership degree is 0 or 1 such as $\mu_A(x_1) = 0.0$ $\mu_A(x_2) = 1.0$. On the contrary, these two crisp values are required to be known in applications. Therefore, Equation (2.24) should be extended to include these two limits with the membership values of $\mu_A(x_1) = 0.0$, $\mu_A(x_2) = 1.0$ as in (2.25).

$$\mu_A(x) = \begin{cases} 0 & \text{for } x \leq x_1 \\ 2\left[\dfrac{x - x_1}{x_2 - x_1}\right]^2 & \text{for } x_1 \leq x \leq c \\ 1 - 2\left[\dfrac{x - x_2}{x_2 - x_1}\right]^2 & \text{for } c \leq x \leq x_2 \\ 1 & \text{for } x \geq x_2 \end{cases} \tag{2.25}$$

Although Equation (2.25) includes more information, Equation (2.24) is more practical and more preferable.

Table E2.5(c) Using function sinus02m.m for the Example 2.5

Function file: *sinus02m.m*	Membership values			
	$x = -3.5$	$x = -2$	$x = 2.5$	$x = 4$
μ_A = sinus02m (−3.5);	0.95	0.29	0.0	0.0
μ_B = trapez01m (−2);	0.38	0.92	0.0	0.0
μ_C = trapez01m (2.5);	0.0	0.0	0.77	0.63
μ_D = trapez01m (4);	0.0	0.0	0.0	1.0

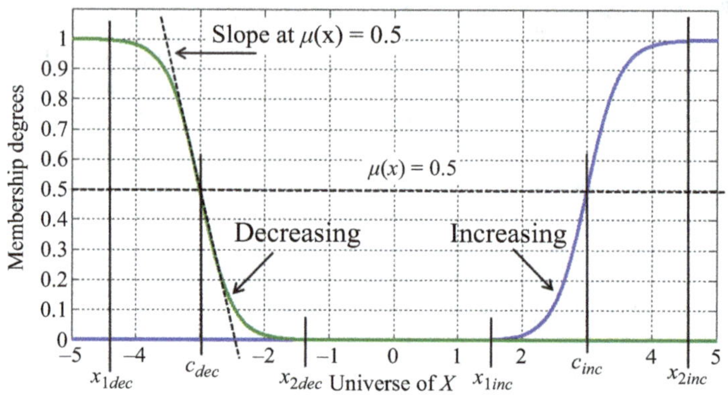

Figure 2.15 Sigmoid membership functions

Example 2.6

Develop a MATLAB function representing sigmoid fuzzy subsets and use it to plot the increasing and decreasing sigmoid functions as shown in Figure 2.15. Assume that the universe of X is defined in the interval [−5, 5] and the parameters are given as $c = 3$ and $a = \pm 4$.

Answer 2.6

The MATLAB function given in Figure E2.6 can be used to plot the sigmoid functions in Figure 2.15.

Example 2.7

The fullness level of the battery charging unit has been defined from 0% to 100% where they mean empty (EM) and full (FU), respectively. Give another definition to the fullness level as half full (HF) and represent them by fuzzy subsets in the universe of fullness (FLNS).

(a) Define the types of starting and ending boundary sets.
(b) Define the type of sets that are applicable to be used as two-sided middle fuzzy subsets.
(c) Plot the defined fuzzy subsets in the universe of fullness.

```
% Sigmoid membership function
clear; clg; clc
ai=4;  ci=3;  % Increasing
ad=-4; cd=-3; % Decreasing
k=1;
for x=-5:0.01:5;
     Si(k)=sigmoid01m(ai,ci,x);
     Sd(k)=sigmoid01m(ad,cd,x);
     X(k)=x;
     k=k+1;
end
plot(X,Si,X,Sd)
xlabel('Universe of X');
ylabel('Membership Degrees');
grid
```

```
%Sigmoid membership funcion
function mu=sigmoid01m(a,c,x)
mu=(1/(1+exp(-a*(x-c))));
```

Figure E2.6 The use of sigmoid function to plot Figure 2.15

Answer 2.7

Since the universe of fullness is defined from 0% to 100%, these two values are the lower and upper limits of the universe and fuzzy subsets will be defined in this universe.

(a) In order to represent the fuzzy sets EM and FU, right triangles, trapezoid, sinusoid, and sigmoid-type membership functions can be used as shown in Figure E2.7.
(b) The fuzzy subset HF is a two-sided middle set and should be represented properly by the sets with similar properties. Two-sided triangular, trapezoid, bell, and Gaussian-type membership functions can be used for HF. Since the sigmoid functions are one-sided, they are not applicable to represent the fuzzy subset HF.
(c) The plot of defined fuzzy subsets E, HF, and F are shown in Figure E2.7(a). This example includes lower and upper boundary fuzzy subsets besides the one in the center. Figures E2.7(a)–E2.7(c) show which types of fuzzy subsets can be used as lower and upper ends and which ones can be used in the middle.

Figure E2.7(a) gives a chance to compare the fuzziness of sinusoid, triangular, and Gaussian membership functions. All three membership functions are symmetrical around the axis passing through their peek points. Gaussian function is concentrated toward 50%, which means that it is less fuzzy than the others. The smoothness of the sinusoid function on the peak point can be seen clearly as the smoothness of the Gaussian function is lost very fast as data moves away from the center. The comparison of sinusoids and triangular-type fuzzy subsets can be made better in Figure E2.7(b) where the peak points are clearly examined. We can conclude that the systems using the triangular membership function respond very fast to the changes in the related data, while the systems using the sinusoid function respond slower. A comparison between trapezoid and bell-type functions is shown

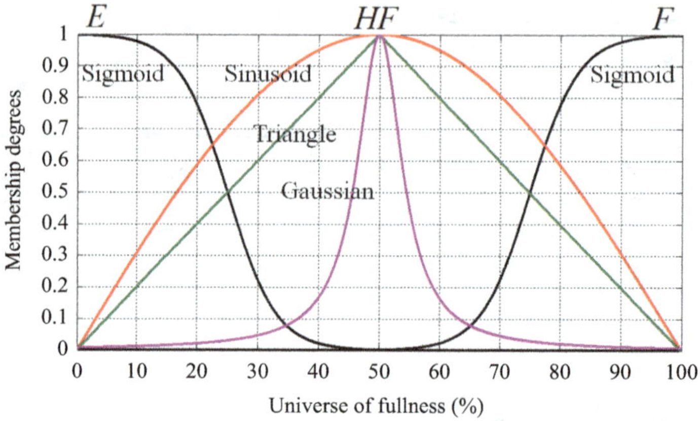

Figure E2.7(a) The use of sigmoid, sinusoid, triangular, and Gaussian functions together

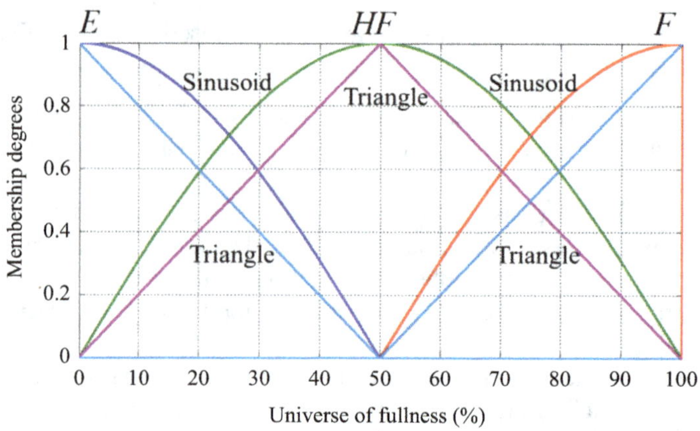

Figure E2.7(b) The use of sinusoid and triangular functions together

in Figure E2.7(c). There is no big difference in the fuzziness for the membership values over 0.5, while the fuzziness is less in the bell function if the membership value is below 0.5.

The MATLAB codes used to simulate the membership functions in Example E2.7 are given in the MATLAB .m-file library, which is available on the publisher site.

Practice 2.2

Develop Simulink-based membership functions and repeat Example 2.7 using Simulink.

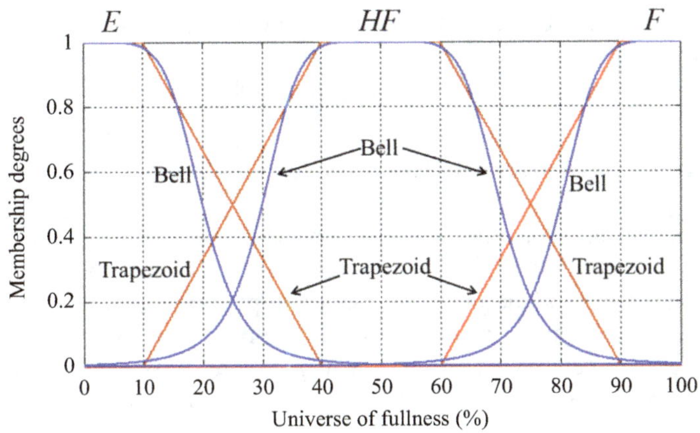

Figure E2.7(c) The use of trapezoid and bell functions together

Practice 2.3

Infant (IN), child (CH), young (YN), middle aged (MA), and old (OL) fuzzy subsets are defined in the universe of age.

(a) Use triangular-type membership functions and plot these fuzzy subsets in the same graph such that the maximum membership values of these functions from infant to old occur at 0, 10, 20, 40, and 60.
(b) Determine the types of fuzzy membership functions that can be used as lower end, upper end, and the ones in between.
(c) Try all possible membership function types to represent the starting and ending fuzzy subsets.
(d) Represent the fuzzy subsets Young and Middle Aged by using triangular, sinusoid, and Gaussian membership functions and comment about the suitability of these membership functions on representing the given fuzzy subsets in terms of fuzziness.
(e) Use trapezoid and bell membership functions to represent the above-mentioned fuzzy subsets in the universe of age and comment on the meaning of the flat tops.
(f) Plot all your solutions to explain them better using MATLAB .m and Simulink.

Practice 2.4

A crisp universe of discourse X is defined in the interval of [0, 10] and partitioned into three fuzzy subsections as SMALL, MEDIUM, and LARGE. Triangular-type fuzzy sets are used to represent these fuzzy subsets.

(a) Draw a software algorithm that gives you the membership value of any x on X in the fuzzy subsets SMALL, MEDIUM, and LARGE.
(b) Write a MATLAB function that simulates the algorithm you developed in part (a).

(c) Write a Simulink function that simulates the algorithm you developed in part (a).
(d) Repeat the previous parts for the sinusoid membership function.

2.3 Properties of fuzzy membership functions

Different types of fuzzy membership functions have been introduced in Section 2.2 focusing on usability and fuzziness of the sets. Besides their shapes, fuzzy membership functions have some other characteristic properties. Typical properties of fuzzy sets and membership functions representing them are covered in this section. The bell-type membership function shown in Figure 2.16 includes most of the characteristic properties of fuzzy subsets and is used as a reference to give the details.

The major properties of fuzzy subsets are summarized in Figure 2.16. Every fuzzy subset may not include all of these properties. Depending upon function type, the properties of the fuzzy subsets may vary. The meaning of these characteristic properties is explained below.

The core of a fuzzy set is a crisp set whose elements have a membership degree of 1.0 in the fuzzy set A. The core of a fuzzy set A is defined as in (2.26).

$$core(A) = \{x, \text{ such that } \mu_A(x) = 1\} \tag{2.26}$$

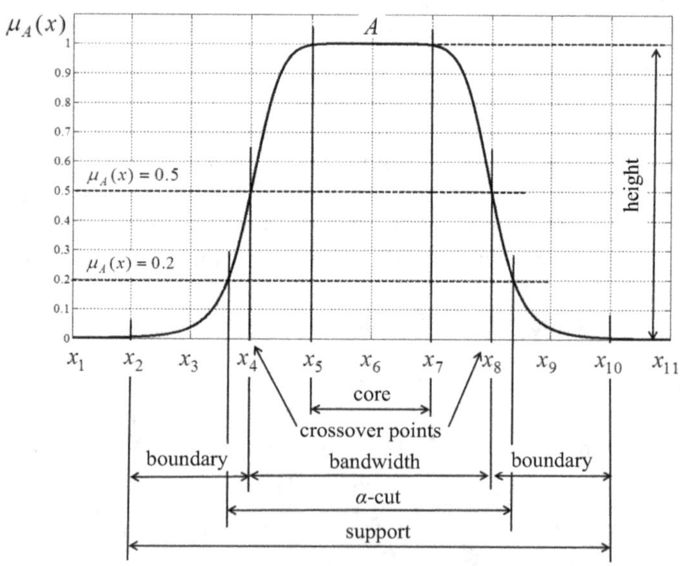

Figure 2.16 Characteristic properties of a fuzzy subset

Crossover points of a fuzzy set A are the crisp elements with membership degrees equal to 0.5 and are defined as

$$crosover(A) = \{x,\ \text{such that}\ \mu_A(x) = 0.5\} \tag{2.27}$$

Bandwidth of a fuzzy set A is described as the crisp elements with the membership values equal to or greater than 0.5 in A.

$$bandwidth(A) = \{x,\ \text{such that}\ \mu_A(x) \geq 0.5\} \tag{2.28}$$

The boundary of a fuzzy set includes the crisp elements whose membership values are between 0 and 1 in set A. The boundary set is defined as

$$boundary(A) = \{x, \text{such that}\ 0 < \mu_A(x) < 1\} \tag{2.29}$$

The support of a fuzzy set consists of the crisp elements of the universe of X such that the membership values of these elements in A are greater than 0.

$$support(A) = \{x\ \text{such that}\ \mu_A(x) > 0\} \tag{2.30}$$

α**-cut** of a fuzzy set A is a crisp set A_α such that the elements of this crisp set are defined in the universe of X and have the membership values equal or greater than α in A. For example, the α-cut of $\alpha = 0.5$ in A is equal to the bandwidth of set A. The α-cut of a fuzzy set A is defined as in (2.31).

$$\alpha - cut(A) = A_\alpha = \{x,\ \text{such that}\ \mu_A(x) \geq \alpha\} \tag{2.31}$$

If the membership degrees of α-cut (A) are greater than α, then it is called **hard** α**-cut** and defined as

$$hard\ \alpha - cut(A) = eA_\alpha = \{x,\ \text{such that}\ \mu_A(x) > \alpha\} \tag{2.32}$$

According to these definitions, $\alpha = 1$ yields the core (A) and $\alpha = 0$ gives the *support* set of A.

$$A_1 = core(A)$$
$$A_0 = support(A)$$

Convexity of a fuzzy set exists if the condition given in (2.33) is satisfied.

$$\mu_A(\lambda x_1 + (1 - \lambda)x_2) \geq \min(\mu_A(x_1), \mu_A(x_1)) \tag{2.33}$$

where x_1 and x_2 are the elements of the universe of X and $\lambda \in [0, 1]$. The convexity can be explained better using Figure 2.17(a) and (b).

Height of a fuzzy set is defined by the membership values of the set. If the membership value of a set is equal to 1.0, which is the highest degree, then this set is called as a normal fuzzy set. If the highest membership value is smaller than 1.0, than the set is called as subnormal fuzzy set. Under normal conditions, the fuzzy subsets are defined as normal sets. However, after some allegations of fuzzy subsets, subnormal sets might be obtained and used in applications. For example, the

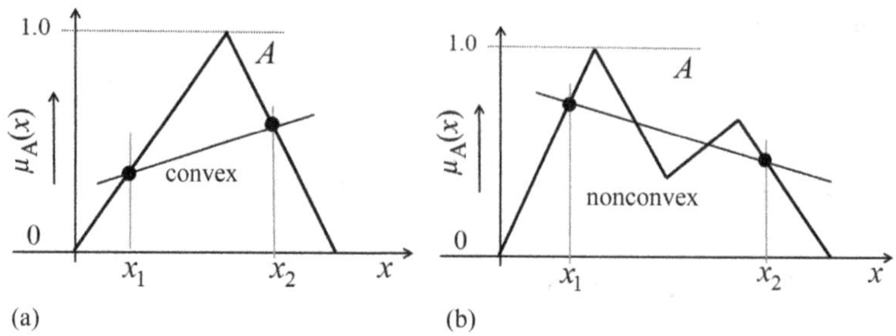

Figure 2.17 Convex and nonconvex fuzzy sets. (a) Convex set, (b) nonconvex set.

intersection of two normal fuzzy set may be a subnormal set. When required, the subnormal fuzzy sets may be normalized to be used.

Cardinality of a fuzzy set is defined as the sum of membership values of the elements belonging to that set and represented as

$$|A| = \sum \mu_A(x) \tag{2.34}$$

where $x \in X$ and X is the universe of discourse. Another term called **relative cardinality** is then defined as

$$\|A\| = \frac{|A|}{|X|} \tag{2.35}$$

where $|A|$ is the cardinality of fuzzy set A and $|X|$ is the cardinality of crisp universe of X. The cardinality of a crisp set is the number of elements in that set.

Example 2.8

Equations (2.13) and (2.22) are used to represent the numbers close to 0 in a universe of discourse X defined in an interval of $[-10, 10]$. Assume that the parameters $x_P = 0$ and $w = 1.5$ for Gaussian and $n = 3$; $x_1 = -6$; $x_2 = 6$; for the sinusoid function. Obtain α-cuts for $\alpha = 0.2$ and 0.6.

(a) Find and comment on the characteristic properties of the sinusoid fuzzy membership function representing the numbers close to 0.
(b) Repeat (a) for the Gaussian function.

Answer 2.8

The plots of the numbers close to 0 are given in Figure E2.8(a) and E2.8(b) for sinusoid and Gaussian functions, respectively. The plots will help to describe the properties. Let us use the definition given in Equation (2.26)–(2.32).

(a) Properties of sinusoid function

The core set is obtained as $core(A) = \{0, \text{ where } \mu_A(x) = 1\}$.

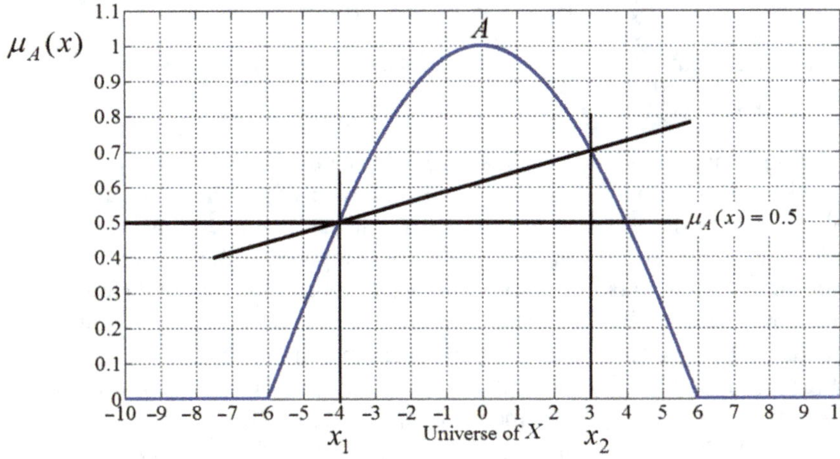

Figure E2.8(a) Properties of sinusoid membership function

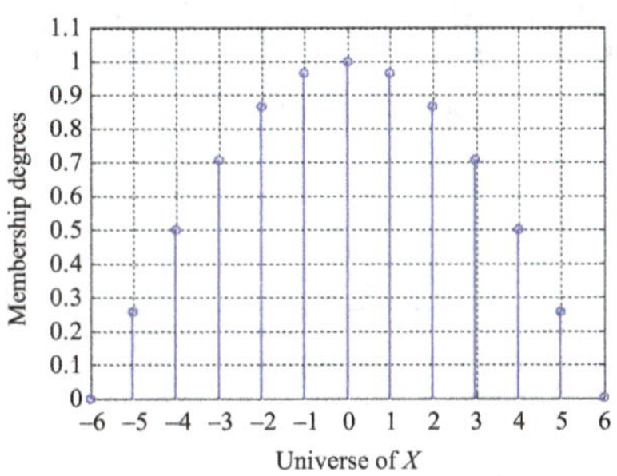


```
clear;  clc;   clf
x1=-6; x2=6; X=x2-x1; n=3;
a=X/(n-1); w=pi/(2*a); x=-6:1:6;
S1=abs(sin((w*x+(pi/2))));
plot(x,S1,'o',0,1); axis([-inf inf 0 1]);
hold on
for k=1:length(x)
plot([x(k) x(k)],[0 S1(k)],'-');
end;  hold off
xlabel('Universe of X');
ylabel('Membership Degrees'); grid
cardS=sum(S1);
cardX=length(x);
rcardA=cardS/cardX;
```

Figure E2.8(b) Cardinality of fuzzy set A

Crossover points are obtained as *crossover* $(A) = \{-4 \text{ and } 4,\ \text{where } \mu_{\mathrm{A}}(x) = 0.5\}$.

Bandwidth set of the fuzzy set A is equal to the crisp interval [−4, 4], which is obtained from $bandwidth(A) = \{[-4, 4] \text{ where } \mu_{\mathrm{A}}(x) \geq 0.5\}$.

The boundary set of a fuzzy set includes the crisp elements as defined by (2.29). Therefore, there are two boundary sets for the sinusoid function used in this example

$$\text{Boundary set1 : boundary}(A) = \{x,\ \text{where } -6 < x < 0 \text{ such that } 0 < \mu_{\mathrm{A}}(x) < 1\}$$

$$\text{Boundary set 2 : boundary}(A) = \{x,\ \text{where } 0 < x < 6 \text{ such that } 0 < \mu_{\mathrm{A}}(x) < 1\}$$

The support set of the fuzzy set A is obtained as

$$\text{support}(A) = \{x,\ \text{where} - < x < 6 \text{ suchthat } 6\mu_{\mathrm{A}}(x) > 0\}$$

α-cuts of the fuzzy set A for $\alpha = 0.2$ and 0.6.

$$\alpha = 0.2 - \text{cut}(A) = A_{0.2} = \{x, \text{ such that } \mu_A(x) \geq 0.2\}$$

$$\alpha = 0.6 - \text{cut}(A) = A_{0.6} = \{x, \text{ such that } \mu_A(x) \geq 0.6\}$$

The numerical results are summarized in Table E2.8(a) where the universe of X is varied from -10 to 10 with an incremental step of 1. Therefore, the steps in membership values are also large and such that the continues function became a discrete one. The first column in Table E2.8(a) represents the universe of X. The second column represents the membership values corresponding to the core set and column three contains the membership values of the crossover points. Similarly, the membership values of the bandwidth are given in column four.

Sinusoid function also satisfies the convexity property, since all of the crisp points between x_1 and x_2 have membership values in fuzzy set A and they are greater than the line connecting $\mu_A(x_1)$ and $\mu_A(x_2)$ each other. The convexity line in Figure E2.8(a) clearly shows the sinusoid membership function is convex.

Cardinality of the fuzzy set A can be obtained by assuming a discrete universe of X in the interval $\{-6, -5, -4, -3, -2, -1, 0, 1, 2, 3, 4, 5, 6\}$ and fuzzy subset A as in Table E2.8(b). The cardinalities of the universe of X and fuzzy subset A are calculated and given in the bottom rows of Table E2.8(b). It should be noted that the cardinality of crisp universe of X is equal to the number of elements in X and the

Table E2.8(a) Properties of sinusoid function in Example 2.8

X	μ_A **(x)**	μ_A **(core)**	μ_A **(co)**	μ_A **(bw)**	μ_A **(bdr)**	μ_A **(sup)**	μ_A **(A_α)**	
							0.2	**0.6**
−10	0							
−9	0							
−8	0							
−7	0							
−6	0							
−5	0.25				0.25	0.25	0.25	
−4	0.50		0.5	0.50	0.50	0.50	0.50	
−3	0.70			0.70	0.70	0.70	0.70	0.70
−2	0.86			0.86	0.86	0.86	0.86	0.86
−1	0.96			0.96	0.96	0.96	0.96	0.96
0	1.00	1.00		1.00	0.00	1.00	1.00	1.00
1	0.96			0.96	0.96	0.96	0.96	0.96
2	0.86			0.86	0.86	0.86	0.86	0.86
3	0.70			0.70	0.70	0.70	0.70	
4	0.50		0.5	0.50	0.50	0.50	0.50	
5	0.25				0.25	0.25	0.25	
6	0							
7	0							
8	0							
9	0							
10	0							

cardinality of fuzzy subset A is the sum of the membership values $\mu_A(x)$ of A. Discrete sinusoid representation of the fuzzy subset A and calculation of its cardinality in MATLAB are given in Figure E2.8(b).

(b) Properties of Gaussian function

The properties of the Gaussian function can be obtained as it is done for the sinusoid function. Similar but the same results may be obtained. The main difference between sinusoid and Gaussian functions is the convexity property. The Gaussian function is not convex as seen in Figure E2.8(c), while sinusoid function is convex as stated earlier.

Table E2.8(b) Cardinality calculation for Example 2.8(a)

Universe X	μ_A **(x)**
−6	0.0000
−5	0.2588
−4	0.5000
−3	0.7071
−2	0.8660
−1	0.9659
0	1.0000
1	0.9659
2	0.8660
3	0.7071
4	0.5000
5	0.2588
6	0.0000
$\|X\| = 13$	$\|A\| = 7.5958$
$\|A\|/\|X\| = 0.5843$	

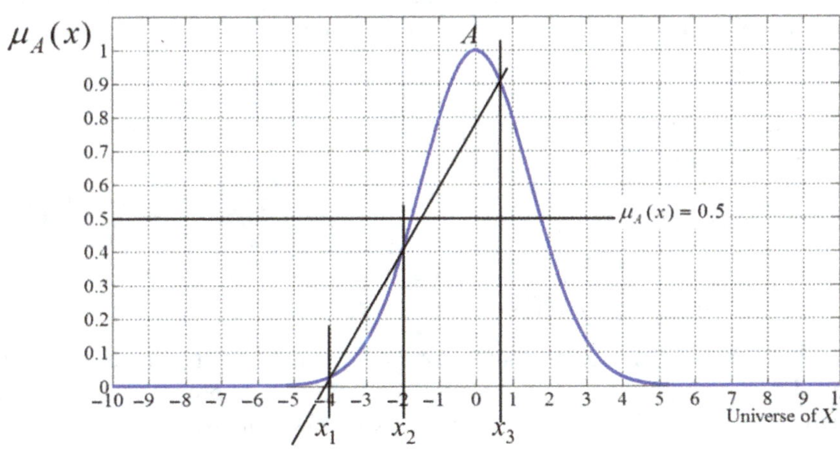

Figure E2.8(c) Properties of Gaussian membership function

As shown in Figure E2.8(c), membership degrees of the crisp points between x_1 and x_2 are smaller than the line connecting $\mu_A(x_1)$ and $\mu_A(x_2)$. The convexity line in Figure E2.8(c) clearly shows that the Gaussian membership function is not convex.

2.4 Fuzzy set operations

Fuzzy sets are characterized by fuzzy membership functions and fuzzy numbers. Operations done using fuzzy membership functions or fuzzy numbers are the basis of fuzzy logic and fuzzy sets. First, these operations must be understood well in order to understand what can be done with fuzzy sets and fuzzy logic. Therefore, the fundamental fuzzy set operations and their properties, especially the properties used in automatic control system applications are studied in this section.

The fuzzy set operations are either done by using the fuzzy sets themselves or membership functions carrying their characteristic properties. As in crisp sets, fuzzy sets also have definitions such as subset, empty set, intersection set, union set, and complement of a set.

Three main fuzzy set operators are given in this part. These operators are

- Intersection
- Union
- Complement

2.4.1 Intersection: t-norm

The intersection of two or more fuzzy subsets is also called t-norm. Intersection corresponds to the operator *and* in Boolean algebra. Therefore, the verbal term *and* in fuzzy systems has the same meaning as intersection and are processed by the intersection operator. Let A and B be two fuzzy subsets and C is the intersection set of A and B, where A, B, and C are all defined in interval [0, 1] and are the elements of universe of X. Then the intersection C is written as

$$C = A \cap B \tag{2.36}$$

where $A\in[0, 1]$, $B\in[0, 1]$, and $C\in[0, 1]$. Equation (2.36) can be written in terms of membership functions, too.

$$\mu_C(x) = \mu_A \cap \mu_B(x) \tag{2.37}$$

$$\mu_C(x) = \mu_A(x) \cap \mu_B(x) \tag{2.38}$$

$$\mu_C(x) = \min(\mu_A(x), \mu_B(x)) \tag{2.39}$$

where $x \in X$. As Equation (2.39) shows, the intersection of two fuzzy subsets can be obtained as the minimum of these subsets. The intersection process using the min operator is called the Zadeh intersection [3]. Researchers on fuzzy sets and systems have developed various equations that can be used to process intersection operators [3]. Due to its simplicity, Zadeh minimum is preferred to be used in this book.

Example 2.9

The fullness percentages of a battery charging unit are defined as empty (EM), full of a quarter (OQ), half full (HF), full of three quarters (TQ), and full (FU), which are represented by triangular fuzzy subsets using their parameters as EM (0%, 0%, 25%), OQ (0%, 25%, 50%), HF (25%, 50%, 75%), TQ (50%, 75%, 100%), and FU (75%, 100%, 100%). These fuzzy subsets are shown in Figure E2.9. The universe of fullness has been partitioned into five sub-sections and each one is represented by a fuzzy membership function. Find the followings

(a) Intersection of the subsets OQ and HF
(b) Intersection of HF and TQ
(c) Intersection of OQ and TQ
(d) Comment on the meanings of the above results.

Answer 2.9

(a) Intersection of the subsets OQ and HF can be obtained using the min operator as given in (2.39). A MATLAB code and the resultant figure for the solution are given in Figure E2.9(a).
(b) Intersection of HF and TQ can be obtained using the min operator as given in (2.39). A MATLAB code and the resultant figure for the solution are given in Figure E2.9(b).
(c) Intersection of OQ and TQ can be obtained using the min operator as given in (2.39). Actually, as can be seen in Figure E2.9, fuzzy subsets OQ and TQ do not have any common area, and therefore their intersection is an EM set.
(d) Comment on the meanings of the above results.

As mentioned earlier, intersection or t-norm operator is used to process the word *and* in Boolean algebra.

Intersection operator in (a) represents the expression "the water tank is full of a quarter (OQ) *and* a half (HF)." This means that the amount of water in the tank is more than a quarter and less than a half so that an intersection exists.

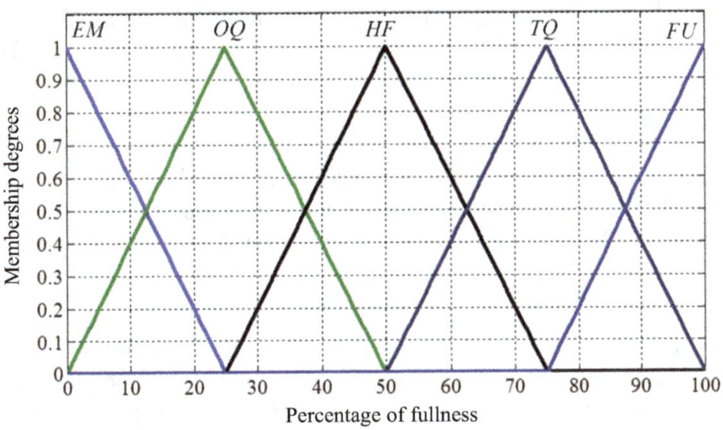

Figure E2.9 Fuzzy subsets represent the percentage of fullness

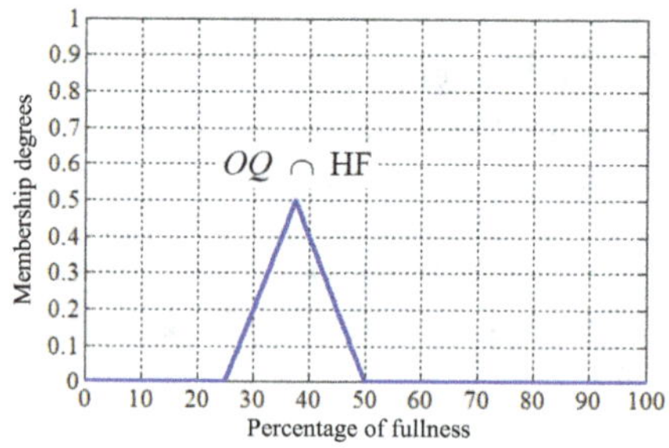


```
% Example 2.9 Figure E2.9a
clear, clc, clf
OQ=[0 25 50]; HF=[25 50 75];
n=1;
for z=0:0.1:100;
  OQZ(n)=triangle02m(OQ,z);
  HFZ(n)=triangle02m(HF,z);
  Z(n)=z; n=n+1;
end
MINOQHF=min(OQZ,HFZ);
plot(Z,MINOQHF)
xlabel('Percentage of Fullness')
ylabel('Membership Degrees')
grid
```

Figure E2.9(a) Intersection of the fuzzy subsets OQ and HF

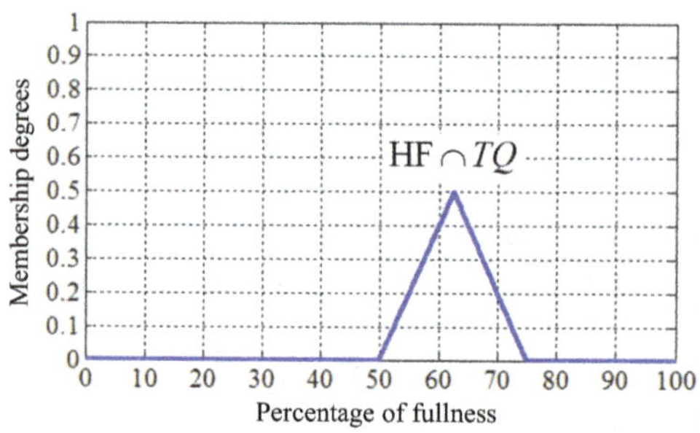


```
% Example 2.9 Figure E2.9b
clear, clc, clf
HF=[25 50 75]; TQ=[50 75 100];
n=1;
for z=0:0.1:100;
  HFZ(n)=triangle02m(HF,z);
  TQZ(n)=triangle02m(TQ,z);
  Z(n)=z; n=n+1;
end
MINHFTQ=min(HFZ,TQZ);
plot(Z,MINHFTQ)
xlabel('Percentage of Fullness')
ylabel('Membership Degrees')
grid
```

Figure E2.9(b) Intersection of the fuzzy subsets HF and TQ

Intersection operator in (b) represents the expression "the water tank is half (HF) *and* three quarter full (TQ)." In this case, the water in the tank is more than half but less than full capacity.

In part (c), the intersection is an EM set since there is no intersection of the sets OQ and TQ. This means that "the water thank is neither full of a quarter nor full of three quarter." In this case, the water thank may be exactly 50% full, which means HF in a crisp manner.

Fuzzy intersection operator has some conditions to be satisfied. Without going into details and proofs of the axioms, these conditions are listed in Table 2.2. Details can be found in [3]. The boundary condition of intersection operator acts as the same as that of a crisp set. The letter *t* in Table 2.2 stands for the operator t-norm.

2.4.2 Union: t-conorm

The union of two or more fuzzy subsets is also called t-conorm or s-norm and corresponds to the operator *or* in Boolean algebra. Therefore, the verbal term *or* in fuzzy systems has the same meaning as union and is processed by the union

Table 2.2 Conditions of intersection operator

Condition	Definition
Boundary	$t(1,1) = 1$ $t(1,0) = 0$ $t(0,1) = 0$ $t(0,0) = 0$
Commutativity	$t(a,b) = t(b,a)$
Associativity	$t(t(a,b),c) = t(a,t(b,c))$
Monotonicity	$\left.\begin{matrix} a_1 \leq a_2 \\ b_1 \leq b_2 \end{matrix}\right\}$ then $t(a_1,b_1) \leq t(a_2,b_2)$
Idempotent	$t(a,a) = a$
Continuity	The intersection (t-norm) continues

operator. Let A and B be two fuzzy subsets and C is the union set of A and B, where A, B, and C are all defined in interval [0, 1] and are the elements of universe of X. Then the union C is written as

$$C = A \cup B \tag{2.40}$$

where $A\in[0, 1]$, $B\in[0, 1]$, and $C\in[0, 1]$. Equation (2.40) can be written in terms of membership functions as

$$\mu_C(x) = \mu_A \cup \mu_B(x) \tag{2.41}$$

$$\mu_C(x) = \mu_A(x) \cup \mu_B(x) \tag{2.42}$$

$$\mu_C(x) = \max(\mu_A(x), \mu_B(x)) \tag{2.43}$$

where $x\in X$. Equation (2.43) shows that the union of two fuzzy subsets can be obtained as the maximum of these subsets. The intersection process using the max operator is called the Zadeh union [3]. Researchers on fuzzy sets and systems have developed various equations that can be used to process the union or t-conorm operator [3]. Due to its simplicity, Zadeh maximum is preferred to be used in this book.

Example 2.10

The fullness percentages of a battery charging unit are defined as empty (EM), full of a quarter (OQ), half full (HF), full of three quarters (TQ), and full (FU), which are represented by triangular fuzzy subsets using their parameters as described in Example 2.9 and shown in Figure E2.9. Find the following:

(a) Union of the subsets OQ and HF
(b) Union of HF and TQ
(c) Union of OQ and TQ
(d) Comment on the meanings of the above results.

Answer 2.10

(a) Union of the subsets OQ and HF can be obtained using the max operator as given in (2.43). The MATLAB code given in Figure E2.9(a) can be used for this example as well by just using the max operator instead of min as MAXOQHF = max (OQZ, HFZ). The resultant union set is shown in Figure E2.10(a).

(b) Union of HF and TQ can be obtained using the max operator as given in (2.43). the MATLAB code in Figure E2.9(b).

(c) can be adapted to this example by just replacing the min operator by max as MAXOQHF = max (HFZ, TQZ). The resultant union set is shown in Figure E2.10(b).

(d) Union of OQ and TQ can be obtained using the max operator as given in (2.43). Actually, as can be seen in Figure E2.9, fuzzy subsets OQ and TQ do not have any common area. However, contrary to the intersection operator, the union of these subsets is not an EM set and includes both of these subsets with FU capacity as shown in Figure E2.10(c).

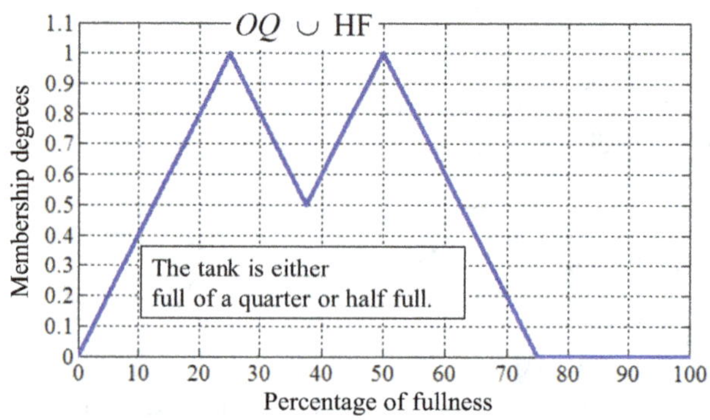

Figure E2.10(a) Union of the fuzzy subsets OQ and HF

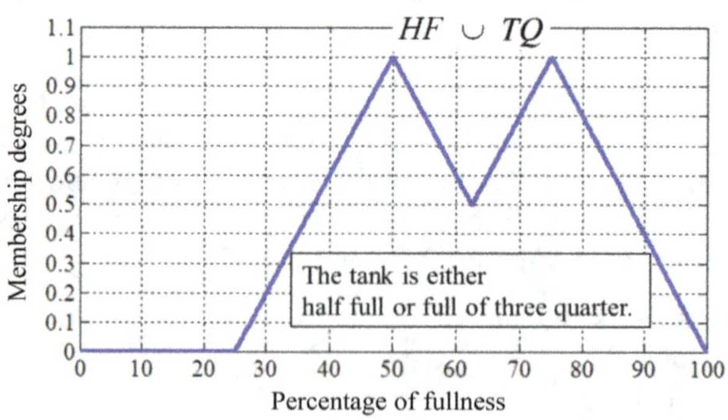

Figure E2.10(b) Union of the fuzzy subsets HF and TQ

(e) Comment on the meanings of the above results.

As mentioned earlier, union or t-conorm operator is used to process the word *or* in Boolean algebra.

Union operator in (a) represents the expression "the battery charging unit is either full of a quarter (OQ) *or* HF." This means that the charging capacity of the battery charging unit falls into both categories with full membership degrees.

Union operator in (b) represents the expression "the battery charging unit is either HF *or* full of three quarter (TQ)." In this case, the charging capacity of the battery charging unit represents both HF and full of three quarters with full membership degrees.

In part (c), the union is not an EM set anymore since the union of the sets OQ and TQ are themselves because they do not have a common area. This means that "the charging capacity of the battery charging unit is either full of a quarter or full of three quarter." In this case, the water thank may be exactly FU of a quarter or exactly FU of three quarters.

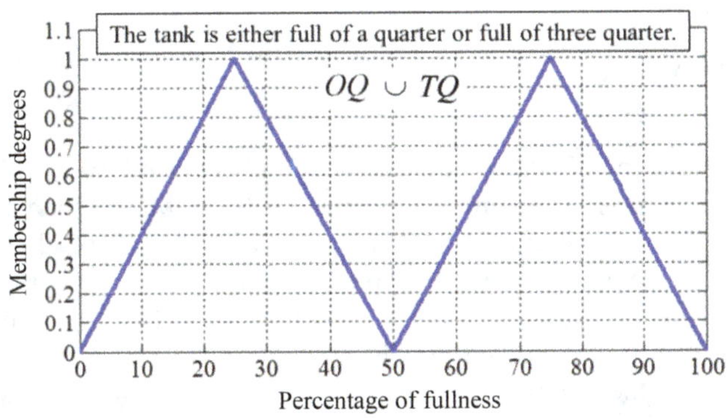

Figure E2.10(c) Union of the fuzzy subsets HF and TQ

Table 2.3 Conditions of union (s-norm) operator

Condition	Definition
Boundary	$s(1,1) = 1$ $s(1,0) = 1$ $s(0,1) = 1$ $s(0,0) = 0$
Commutativity	$s(a,b) = s(b,a)$
Associativity	$s(s(a,b),c) = s(a,s(b,c))$
Monotonicity	$\left.\begin{array}{l} a_1 \leq a_2 \\ b_1 \leq b_2 \end{array}\right\}$ then $s(a_1,b_1) \leq s(a_2,b_2)$
Idempotent	$s(a,a) = a$
Continuity	The union (t-conorm) continues

Fuzzy Union operator has some conditions to be satisfied as in the case of intersection. Without going into details and proofs of the axioms, these conditions are listed in Table 2.3. Details can be found in [3]. The boundary condition of the union operator acts as the same as that of a crisp set. The letter *s* in Table 2.3 stands for the operator t-conorm, which is also called s-norm.

2.4.3 Complement

The complement of fuzzy subsets is also called negation and corresponds to the operator *not* in Boolean algebra. Therefore, the verbal term *not* in fuzzy systems has the same meaning as negation and is processed by the complement operator. Let A be a fuzzy subset and C is the complement of A, where A and C are defined in interval [0, 1] and are the elements of universe of X. Then the union C is written as

$$C = \overline{A} \tag{2.44}$$

where $A \in [0, 1]$ and $C \in [0, 1]$. Equation (2.44) can be written in terms of membership functions as

$$\mu_{\mathrm{C}}(x) = \mu_{\overline{\mathrm{A}}}(x) \tag{2.45}$$

$$\mu_{\mathrm{C}}(x) = 1 - \mu_{\mathrm{A}}(x) \tag{2.46}$$

where $x \in X$. Equation (2.46) shows that a complement of fuzzy subsets can be obtained by subtracting the membership function from the full membership value of 1.0. Since the membership functions are defined in interval [0, 1], the maximum value or the upper boundary of the universe for the membership degrees is 1.0. The complement using the *not* operator is called the Zadeh complement [3]. Researchers on fuzzy sets and systems have developed various equations that can be used to process the complement operator [3]. Due to its simplicity, the Zadeh complement is preferred to be used in this book.

Example 2.11

The fullness percentages of a battery charging unit are defined as empty (EM), full of a quarter (OQ), half full (HF), full of three quarters (TQ), and full (FU), which are represented by triangular fuzzy subsets using their parameters as described in Example 2.9 and shown in Figure E2.9. Find the following:

(a) Complement of the subset EM and FU separately.
(b) Complement of the subset HF and TQ separately.
(c) Complement of the subset OQ ∩ HF and OQ ∪ HF separately.
(d) Complement of the subset OQ ∩ TQ and OQ ∪ TQ separately.
(e) Comment on the meanings of the above results.

Answer 2.11

(a) Complements of the subsets EM and FU can be obtained using the *not* operator as given in (2.46). The MATLAB code given in Figure E2.9(a) can be modified

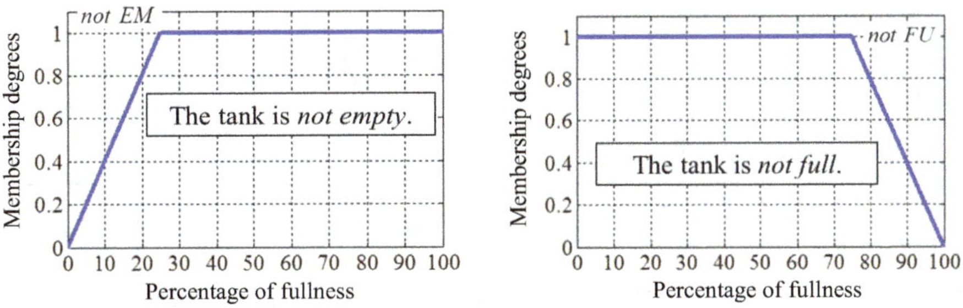

(i) (ii)

Figure E2.11(a) The complements of the sets empty and full. (i) The charging unit is not empty. (ii) The charging unit is not full.

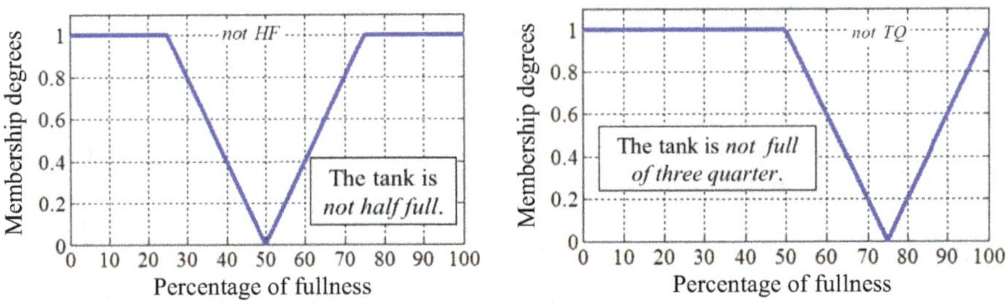

(i) (ii)

Figure E2.11(b) The complements of the sets half full and full of three quarters. (i) The charging unit is not half full. (ii) The charging unit is not full of three quarters.

for this example as well by using *not* operator instead of min as notEM = 1–EM and notFU = 1–FU. The resultant complements are shown in Figure E2.11(a).

(b) Complement of HF and TQ can be obtained using the *not* operator as given in (2.46). the MATLAB code used for part (a) can be used here as notHF = 1−HF and notTQ = 1−TQ. The resultant complement sets are shown in Figure E2.11(b).

(c) Complement of OQ∩HF and OQ∪HF are obtained in two steps. First, the intersection and union of OQ and HF are obtained. Then the complement is applied to the results from the first step to complete the solution. Therefore, the MATLAB code used in previous parts can be modified and used for this part as notOQIHF = 1 – (OQ∩HF) and notOQUHF = 1 – (OQ∪HF).

(d) Complement of OQ∩TQ and OQ∪TQ are obtained in two steps as in part (c). First, the intersection and union of OQ and TQ are obtained. Then the complement is applied to the results from the first step to complete the solution.

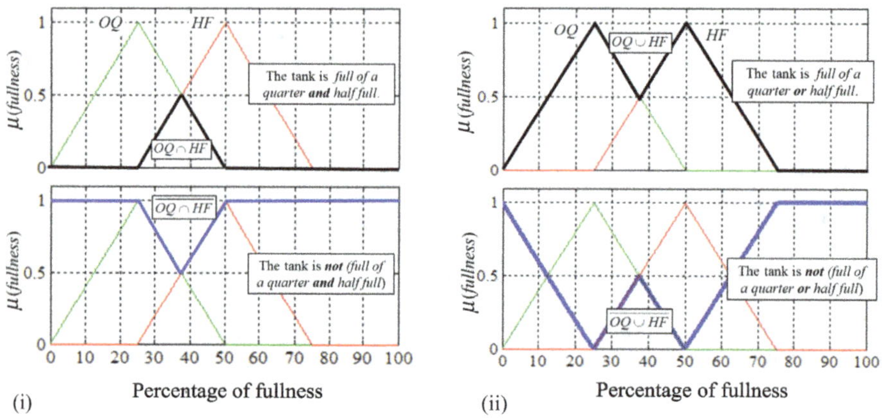

Figure E2.11(c) The complements of the intersection and union of the sets full of a quarter and half full. (i). OQ ∩ HF and its complement. (ii) OQ ∪ HF and its complement.

Therefore, the MATLAB code used in previous parts can be modified and used for this part as notOQITQ = 1 – (OQ∩TQ) and notOQUTQ = 1 – (OQ∪TQ).

(e) The complement operator is used to obtain the negations of the given fuzzy membership functions. The resultant membership functions represent the opposite meaning of the set negated. The use of complement with intersection and union gives additional meaning to these operators. This example shows some cases of using the complement, union, and intersection together.

The complement of the subset EM is given in Figure E2.11(a). The complement of *empty* means *not empty* and obtained as *not empty* = 1-*empty*. Similarly, the complement of the fuzzy subset *full* represents the expression *not full* and is obtained as *not full* = 1-*full* as shown in Figure E2.11(a). The complements of HF and full of three quarters (TQ) are given in Figure E2.11(b). Attention should be given to the complements of intersection and union of the subsets OQ and HF shown in Figure E2.11(c). The complement of the intersection of two neighboring fuzzy subsets looks like the opposite of the intersection set as expected. The complement of the union of two neighboring fuzzy subsets also shows a similar behavior. If the fuzzy subsets do not have any common area, then their intersection will be an EM set as shown in Figure E2.11(d) while the complement of this EM set will be equal to the FU membership value of 1.0. The union of two fuzzy subsets whose intersection is an EM set covers all of these two subsets as given in part (ii) of Figure E2.11(d) as well. The complement of the union of fuzzy subsets OQ and TQ becomes the opposite of the union set.

Fuzzy complement operator has some conditions to be satisfied as in the case of intersection and union. These conditions are listed in Table 2.4. Details can be

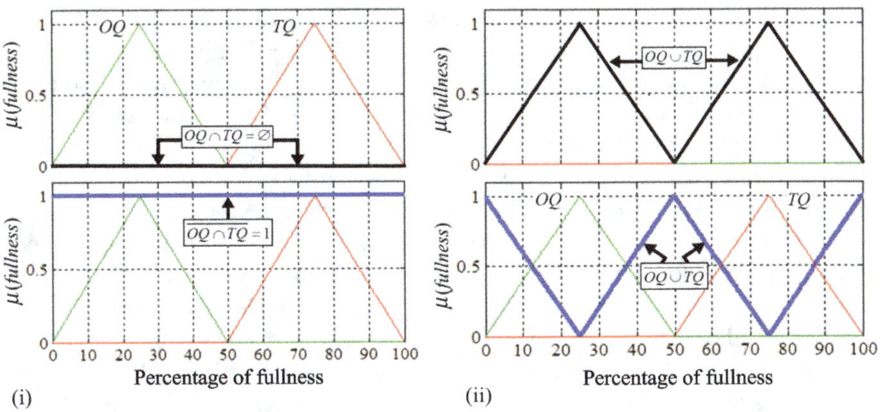

Figure E2.11(d) The complements of the intersection and union of the sets full of a quarter and full of three quarters. (i) OQ∩TQ and its complement. (ii) OQ∪TQ and its complement.

Table 2.4 Conditions of complement operator

Condition	Definition
Boundary	$c(1) = 0$ $c(0) = 1$
Monotonicity	$c(a) \geq c(b)$ if $a < b$
Idempotent	$c(c(a)) = a$
Continuity	The complement continues

where a and $b \in [0, 1]$.

found in [3]. The boundary condition of complement operator acts as the same as that of a crisp set. The letter c in Table 2.4 stands for the operator complement.

2.5 Adjustment of fuzziness

Some fuzzy subset properties such as bandwidth, crossover points, and α-cuts can be changed after some processes on the related set. Changes in these properties affect the fuzziness of the fuzzy subsets. The adjustment of fuzziness is done by using some fuzzy verbal terms that are used frequently in daily life. Many fuzzy verbal expressions adjusting the fuzziness are used unknowingly in daily dialogues. These fuzzy verbal terms and their effects on the fuzziness are discussed in this section.

Verbal terms are the basis of fuzzy modeling techniques because a verbal variable is actually nothing but a fuzzy set name. The verbal expressions *empty*, *full of a quarter*, *half full*, *full of three quarters*, and *full* are all fuzzy words as same as the words *child*, *young*, *middle aged*, *old*, *short*, *medium height*, *tall*, *cold*, *warm*, *hot*, *small*, *normal*, *large*, *big*, *slow*, *fast*, *soft*, *hard*, etc. A fuzzy word is also called as a fuzzy variable. However, they are actually called as *fuzzy variables* if they are used with fuzzy adjectives, which affect the shapes and scopes, i.e., the fuzziness of a fuzzy set. Some examples of fuzzy adjectives can be given as *very*, *few*, *a few*, *more*, *some*, *almost*, *toward*, *close to*, *near*, *far*, *more or less*, etc.

If a fuzzy adjective is added to a fuzzy word or a fuzzy variable, the fuzziness of the fuzzy variable changes. For example, let the fuzzy word *young* be the core fuzzy set of a fuzzy variable. Then many fuzzy variables can be generated by adding fuzzy adjectives as *very young*, *almost young*, *more or less young*, *a little bit young*, *not young*, *not too young*, etc. The use of a fuzzy adjective with fuzzy words generates a subset of the original word. For example, let us define the numbers **close** to 0, which is also called **fuzzy zero**, represented by sinusoid type membership function as:

$$\mu_A(x) = \left|\sin\left(\frac{\pi}{2\alpha}x + 90^\circ\right)\right| \text{ for } -5 \le x \le 5 \tag{2.47}$$

where $\alpha = 2$. Now let us define the numbers **very close** to 0 as the square of the numbers close to 0.

$$\mu_B(x) = [\mu_A(x)]^2 = \left(\left|\sin\left(\frac{\pi}{2\alpha}x + 90^\circ\right)\right|\right)^2 \text{ for } -5 \le x \le 5 \tag{2.48}$$

In order to show the effect of the fuzzy word *very*, both of them are plotted in the same graph as given in Figure 2.18. As it can be understood clearly from Figure 2.18, the fuzzy subset representing the word *close* is more fuzzy than the word *very close*. The adjective *very* has caused the word *close* to shrink toward the core of the original set.

Since the membership values are from 0 to 1, the square of a membership function shrinks toward the core and bandwidth becomes narrower. Except for the

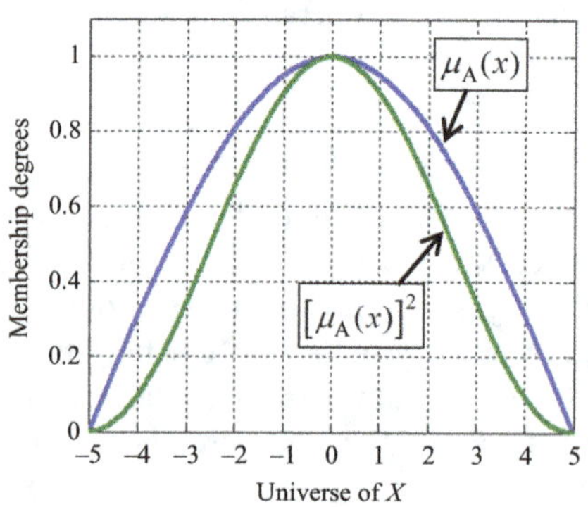


```
clear;  clc;   clf
X=10; n=3;
a=X/(n-1);
w=pi/(2*a);
x=-5:0.01:5;
S1=abs(sin((w*x+(pi/2))));
S2=S1.^2;
plot(x,S1,x,S2);
xlabel('Universe of X');
ylabel('Membership Degrees');
grid
```

Figure 2.18 Fuzzy subsets represent the numbers close to zero and very close to zero

maximum membership value of 1, the other membership values of a fuzzy subset *A* will be smaller and get closer to its core set when it is squared as shown in Figure 2.18. Lowering the membership value means lowering the fuzziness. Therefore, subset *A* is fuzzier than the square of subset *A*. The fuzzy adjective *very* makes a fuzzy subset less fuzzy. Following these descriptions, the following are written:

$$\text{good} = \mu_{\text{A}}(x) \tag{2.49}$$

$$\text{verygood} = [\mu_{\text{A}}(x)]^2 \tag{2.50}$$

$$\text{veryverygood} = \left[[\mu_{\text{A}}(x)]^2\right]^2 \tag{2.51}$$

Having the opposite meaning of *very*, the word *less* is also used as an adjective to affect the meaning of the fuzzy words used *with*. The word *less* increases the fuzziness as *very* decreases it. Therefore, the root square of the membership function is used to implement the fuzzy adjective *less*.

$$\text{lessgood} = [\mu_{\text{A}}(x)]^{1/2} \tag{2.52}$$

The root square of the fuzzy membership function $\mu_{\text{A}}(x)$ increases the bandwidth and membership values between 0 and 1. With larger membership values the fuzzy subset becomes fuzzier, which means the fuzziness is increased. As it is understood from ongoing analysis, the fuzzy word *very* acts as a concentrator and the process is called *concentration*. Similarly, the word *less* expands the fuzzy subset as a *dilation* operator. Based on Equations (2.51) and (2.52), the concentration and dilation are defined as

$$\text{con(A)} = A^n = \int_X \frac{(\mu_{\text{A}}(x))^n}{x} \tag{2.53}$$

$$\text{dil(A)} = A^m = \int_X \frac{(\mu_{\text{A}}(x))^m}{x} \tag{2.54}$$

where n is a positive integer and m is defined as $0 < m < 1$. Another function used for the adjustment of fuzzification is the combination of concentration and dilation operators called *contrast intensification*. This function increases the fuzziness of the bandwidth set while decreasing fuzziness outside the bandwidth as defined in (2.55).

$$\text{ci(A)} = \begin{cases} 2A^2, \ 0 \le \mu_{\text{A}}(x) \le \ 0.5 \\ \overline{2(\overline{A})^2)}, 0.5 \le \mu_{\text{A}}(x) \le 1 \end{cases} \tag{2.55}$$

The effects and differences of concentration, dilation, and contrast intensification are given in Figure 2.19.

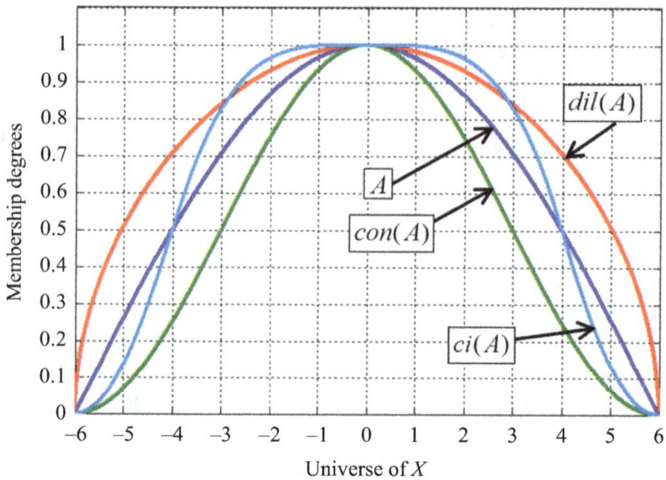

Figure 2.19 Adjustment of fuzziness by the operators con, dil, and ci

2.6 Fuzzy partitioning

Fuzzy sets are defined in universes of discourse to represent groups of objects that are hard to clearly distinguish one from another. They show the inclusion levels of the objects in the group and are characterized by membership functions with various types, which have been discussed in previous sections. Fuzzy membership functions represent linguistic variables that are used to distinguish subgroups from each other. That is the point where the problems arise. When defining and ordering the fuzzy subsets in a universe of discourse, the boundaries, support, and core sets of the fuzzy membership functions gain great importance, because fuzzy linguistic variables do not have a certain or measurable support set. For example, let us partition the universe of age into subcategories as *infant*, *child*, *young*, *middle aged*, and *old* represented by triangular-type fuzzy membership functions. Then let us define the support sets for *infant*, *child*, *young*, *middle aged*, and *old*. Where does the support set of fuzzy set *young* start and end? Where is the starting point of the support set of fuzzy set *middle aged*? The same questions can be asked for all of the fuzzy sets mentioned above. Who do we call as *infant*, *child*, *young*, *middle aged*, and *old*? We do not have a unique answer for each one of these age categories. The support and the core sets of the linguistic variables representing these five groups of ages are not unique and may vary from person to person. 56 undergraduate students with ages 20 to 30 were asked to write down the ending age for *infant*, the starting age for *old*, and the starting and ending ages for *child*, *young*, and *middle-aged*. The results of this small survey come up with different ideas about the age groups as depicted in Figures 2.20–2.24.

As the survey shows, the people do not have the same idea about starting and ending ages for the subcategories *infant*, *child*, *young*, *middle aged*, and *old*.

The boundaries of each subcategory are mixed with the boundaries of the neighboring category. Therefore, partitioning a universe of discourse into subcategories is not an easily distinguishable task. It is uncertainty in transition parts from one category to the other and fuzzy partitioning is required to represent the uncertainty during this transition.

The support set of each subcategory can be defined similarly to (2.30) as

$$\text{support}(A) = \{x \text{ such that } \mu_{\text{A}}(x) > 0\} \tag{2.56}$$

Then the membership function can be defined in terms of the related support set. For example, let us use triangular-type fuzzy membership functions to represent the subcategories given in Figures 2.20–2.24 as

$$\mu_{\text{A}}(x) = \text{triangle}(x_{\text{L}}, x_{\text{P}}, x_{\text{R}}) \tag{2.57}$$

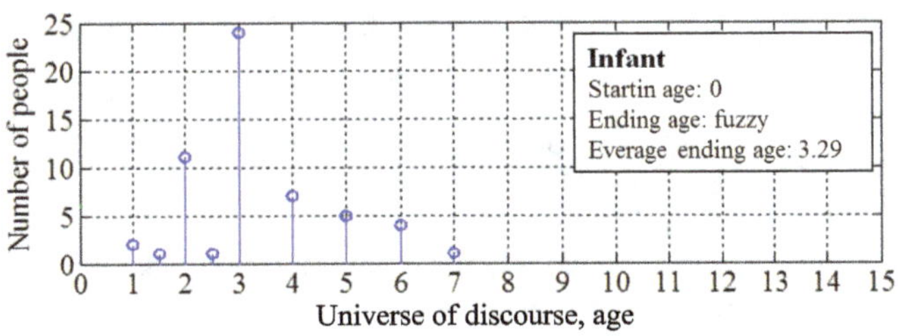

Figure 2.20 Survey results of ending age for infant

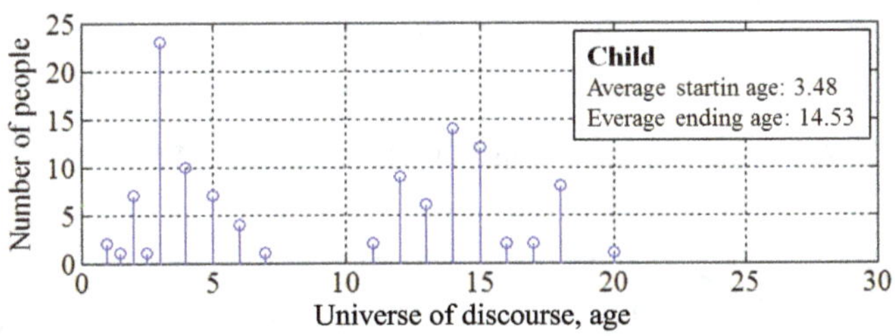

Figure 2.21 Survey results of starting and ending ages for child

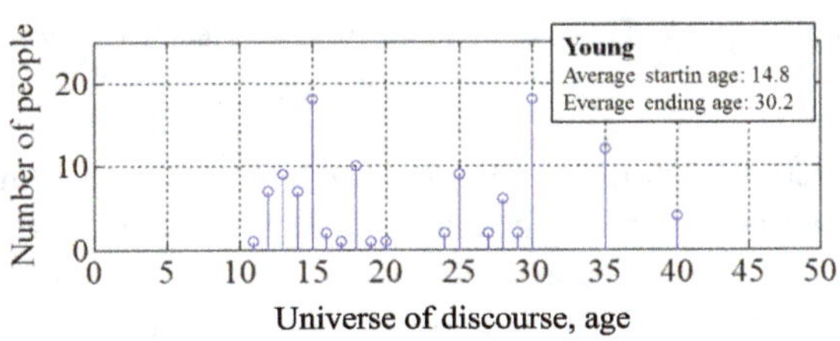

Figure 2.22 Survey results of starting and ending ages for young

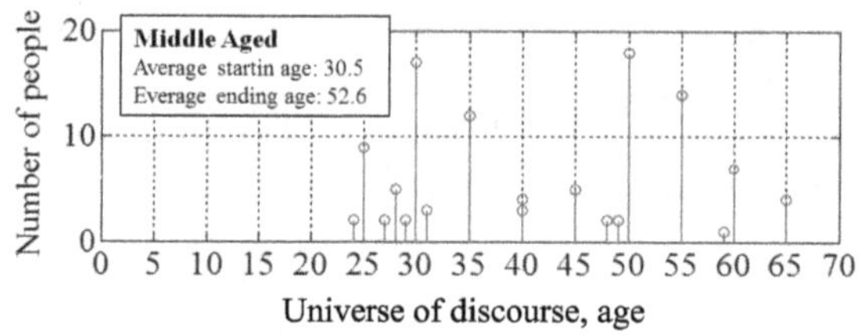

Figure 2.23 Survey results of starting and ending ages for middle aged

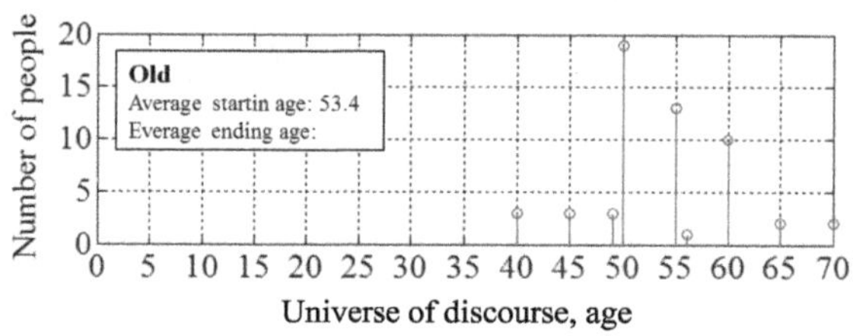

Figure 2.24 Survey results of starting age for old

where x_L, x_P, and x_R are left, peak, and right parameters of triangular membership functions, and are obtained from the survey data as

$$x_L = \min(\text{support}(A)) \tag{2.58}$$

$$x_R = \max(\text{support}(A)) \tag{2.59}$$

$$x_P = x_L + \frac{x_R - x_L}{2} \tag{2.60}$$

Similar approach can be used to define the subcategories by the other types of membership functions

Example 2.12

Use the lowest age value for the starting and highest age value for the ending points from the survey data given in Figures 2.20–2.24 as the parameters of triangular-type fuzzy sets and represent the subcategories by triangular fuzzy subsets. Assume the middle points of the lowest starting and the highest ending points as the parameters for a triangular peak. Besides, use right triangles for *infant* and *old*.

Answer 2.12

As a first step, let us determine the parameters for triangular fuzzy subsets according to the definitions made in question part and put them in a table form. Since a triangular-type fuzzy subset is represented by a membership function in terms of its parameters x_1, x_2, and x_3 as in Equations (2.4) and (2.7), let us obtain these parameters from the survey data shown in Figures 2.20–2.24.

As a second step, let us plot them in the defined universe of discourse, age.

Table E2.12 Parameters of triangular membership functions from the survey data

Fuzzy subset	x_L	x_P	x_R
Infant	0	0	7
Child	1	10	20
Young	11	26	40
Middle aged	24	44	65
Old	40	70	70

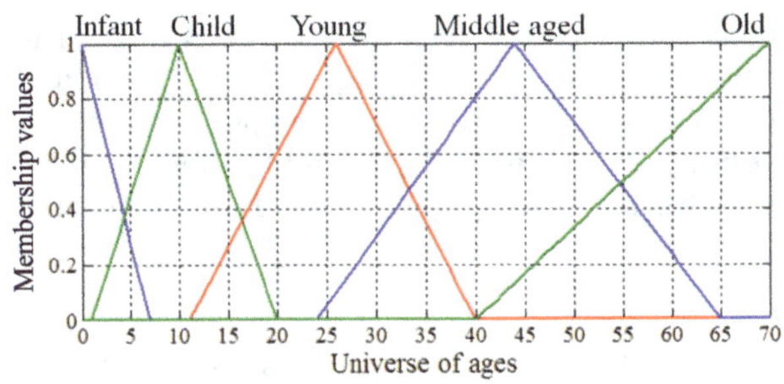

Figure E2.12 A partition of the universe of ages into fuzzy subsets

2.6.1 Theoretical approaches for fuzzy partitioning

Fuzzy sets are used to represent uncertain, vogue, and muddy data in the processing of reasoning, decision-making, optimization, and problem-solving. Fuzzy sets are also used with fuzzy logic in order to model and process the thoughts, ideas, tricks, and knowledge in a way similar to those of a human. When we make a decision, we put our thoughts, ideas, and knowledge to work without processing any math functions such as differentiation, integration, trigonometric functions, etc. Fuzzy logic has the ability to be used to model human decision-making processes. That is one of the main reasons fuzzy logic has become attractive and preferable for problem-solving in the areas of social sciences medicine and engineering [4]. Therefore, it is not intended to draw the readers into heavy math of fuzzy sets and fuzzy logic. Although it is impossible to ignore the importance of math, a standardized method for fuzzy partition does not exist. Usually, expert ideas or trial-and-error methods are applied to establish an initial partition space. Then some algorithms are used for tuning the fuzzy subsets [5–11].

Let X be the finite crisp universe to be partitioned into subclasses and A_1, A_2, …, A_n are the normal fuzzy subsets that represent these subclasses. For the fuzzy subsets and A_1, A_2, …, A_n to be finite fuzzy partition sets of the universe X, the following condition should be satisfied for all $x \in X$

$$\sum_{i=1}^{n} A_i(x) = 1 \tag{2.61}$$

where each one of the fuzzy subsets A_i is normal with the highest membership value of 1.0 for some x_i. Each x in the universe of X has a nonzero membership value for some A_i and zero for others.

More theoretical approaches for fuzzy partitioning can be found in [12–17] as additional information. Fuzzy partition of the crisp universes in this book is based on knowledge and information obtained from experiments, simulations, and experts on the topic. Therefore, the definition of the universe of discourse and partition by fuzzy subsets are straightforward processes for the energy systems.

2.6.2 Fuzzy partition examples in energy systems

Some universes and partitions are given in this section in order to start thinking fuzzy logic with the energy systems problems. Examples given in this section will guide the readers on how to formulate the basic problems in power and energy systems by using fuzzy sets and fuzzy logic. Problem formulation by fuzzy sets at this stage will be the basis for more generalized and hard project topics in the latter chapters.

Example 2.13

The power generated by photovoltaic (PV) solar cells changes as a function of solar irradiation level and operating temperature. So, the generated power increases as the solar irradiation level increases, but decreases as the operating temperature increases. Therefore, meteorologic data about solar irradiation, temperature, wind, and waves are measured daily, monthly, and annual basis to be used in renewable energy projects. These measurements show that 50 mW/cm^2 is low and 200 mW/cm^2 is well over average and assumed to be good. Similarly, the ambient temperature varies between −20°C and 40°C around the world where humans live. Based on these comments, a crisp universe of solar irradiation level (*S*) can be defined in an interval as {50, 200} mW/cm^2, and a crisp universe of temperature (*T*) can be defined in the interval {−20, 40}°C.

(a) Partition the universe of S into fuzzy subsets Low (L), Sufficient (Su), Good (G), and High (Hi), which are represented by evenly distributed triangular-type membership functions.
(b) Partition the universe of *T* into fuzzy subsets Freezing (F), Cold (C), Warm (W), and Hot (Ho), which are represented by evenly distributed triangular-type membership functions.

Answer 2.13

The upper and lower boundary sets in both universes are represented by right triangles assuming that solar irradiation level below 50 mW/cm^2 is crisp low and above 200 mW/cm^2 is crisp high. Similarly, the temperature below −20°C is assumed to be crisp freezing and above 40°C is crisp hot. The other fuzzy sets will be two-sided and placed evenly in the corresponding universe such that the boundaries of all the fuzzy subsets will have the same length.

(a) The universe of S can be portioned into the following fuzzy subsets:

L(50,50,100)
Su(50,100,150)
G(100,150,200)
Hi(150,200,200)

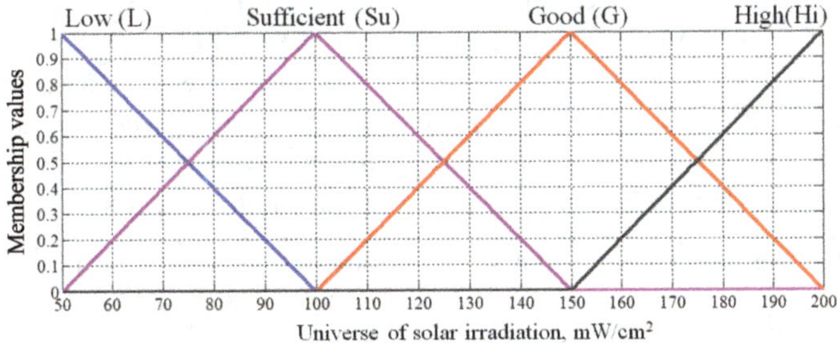

Figure E2.13(a) A partition of the universe of solar irradiation into fuzzy subsets

(b) The universe of T can be portioned into the following fuzzy subsets:

F(–20,–20,0)
C(–20,0,20)
W(0,20,40)
Ho(20,40,40)

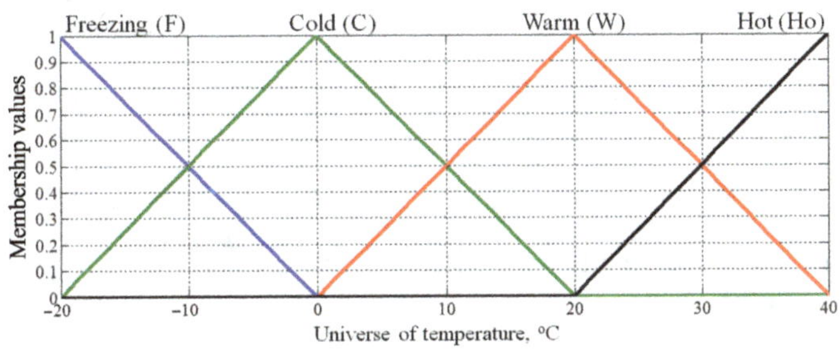

Figure E2.13(b) A partition of the universe of temperature into fuzzy subsets

Example 2.14

The mechanical power generated by a wind turbine is determined by wind speed and the area swept by the rotor blade of the turbine. Such that the generated power increases at higher wind speeds and greater rotor diameters. Studies show that a wind speed of around 15 m/s is a good average speed for wind turbines to be operated. Therefore, a universe of wind speed can be defined in an interval of {0, 20}. The rotor diameter has been increased from 15 m to over 120 m from 1980 to 2013 [18,19] and keeps increasing. Using this information from literature, a universe of rotor diameter is defined in the interval of {0, 130} meters.

(a) Create five fuzzy subsets in the universe of wind speed as Zero, Slow, Medium, Fast, and Faster, and place them evenly in the universe.
(b) Create three fuzzy subsets in the universe of rotor diameter as Small, Medium, and Big, and place them evenly in the universe.

Answer 2.14

The upper and lower boundary sets in both universes are represented by right triangles assuming that there is no wind speed below 0 m/s and wind speed above 20 m/s is not usable. Similarly, the lowest diameter is crisp 0 m and the highest is crisp 130 m. The other fuzzy sets will be placed evenly in the corresponding universe such that the boundaries of all the fuzzy subsets will have the same length because they have the distance between the peak points.

(a) The universe of wind speed is portioned into the following five fuzzy subsets:

Ze(0,0,5)
Sl(0,5,10)
Me(5,10,15)
Fa(10,15,20)
Fr(15,20,20)

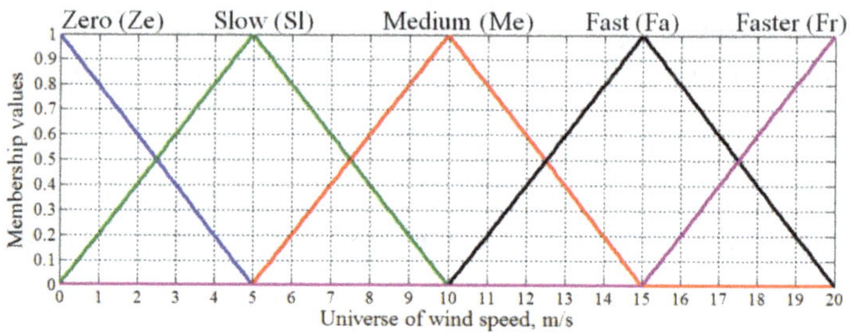

Figure E2.14(a) A partition of the universe of wind speed into fuzzy subsets

(b) The universe of rotor diameter is portioned into the following three fuzzy subsets:

S(0,0,65)
M(0,65,130)
B(65,130,130)

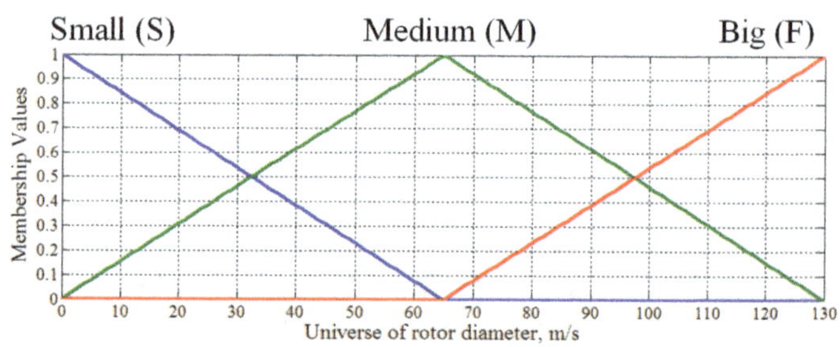

Figure E2.14(b) A partition of the universe of rotor diameter into three subsets

Example 2.15

The power in electrical energy systems is a complex quantity with real and imaginary components. The real component is called active power, which is converted to usable energy by electrical machines and devices such as motors and heaters. The imaginary component is called reactive power and is used by electrical machines and devices such as motors, transformers, and reactors.

The active power component is a factor of the complex power and is determined by a term called a power factor (PF), which is defined as $\cos\phi$ in the interval [0, 1]. In order to have minimum power distribution losses, the PF is kept close to 1, such as 0.95 or higher. In order to keep $\cos\phi = 0.95$, capacitors are connected at the end of the distribution line. However, connecting more capacitors than the required amount results in a voltage increment at the end of the line that is not desired.

The expert operator at the end of the power distribution line, switches the capacitors on or off by looking at the values of PF and bus voltage. The operator uses the following decision rules while deciding the amount of the capacitors to be switched on or off.

If PF is *high* (H_{PF}) and bus voltage is *low* (L_V) then there is *no change* (NC).
If PF is *high* (H_{PF}) and bus voltage is *normal* (N_V) then *decrease* (D) the capacitors.
If PF is *high* (H_{PF}) and bus voltage is *high* (H_V) then *decrease* (D) the capacitors.
If PF is *normal* (N_{PF}) and bus voltage is *low* (L_V) then *increase* (I) the capacitors.
If PF is *normal* (N_{PF}) and bus voltage is *normal* (N_V) then there is *no change* (NC).
If PF is *normal* (N_{PF}) and bus voltage is *high* (H_V) then *decrease* (D) the capacitors.
If PF is *low* (L_{PF}) and bus voltage is *low* (L_V) then *increase* (I) the capacitors.
If PF is *low* (L_{PF}) and bus voltage is *normal* (N_V) then *increase* (I) the capacitors.
If PF is *low* (L_{PF}) and bus voltage is *high* (H_V) then there is *no change* (NC).

This decision process is actually a collection of conditional sentences including verbal terms that all are fuzzy variables. There are three universes in this decision process. Two of them are input universes and one is the output. The verbal terms are defined in these universes depending on the intervals, which may vary from one to another.

(a) Define the universe of discourses for each universe.
(b) Show the fuzzy verbal terms *high*, *normal*, and *low* in the universe of PF assuming a PF of 0.95 is normal.
(c) Assume zero voltage drop as normal and show the fuzzy verbal terms *high*, *normal*, and *low* in the universe of voltage.
(d) Assume no change stands for zero and show the fuzzy verbal terms *increase*, *decrease*, and *no change* in the universe of capacitor change.

Answer 2.15

(a) Definitions of the universes.

Considering the meaning of PF explained above, an interval of {0.8, 1} as in Figure E2.15(a) can be defined as its universe of discourse.

The maximum voltage drop allowed in power distribution lines is specified by standards and published in related guidelines. According to the guidelines, the voltage drop may vary from 0% to 5% for short lines or 10% for long lines. Therefore, the bus voltage can be 10% lower or 10% higher than the normal value. If the rated bus voltage is assumed to be the base voltage, then the bus

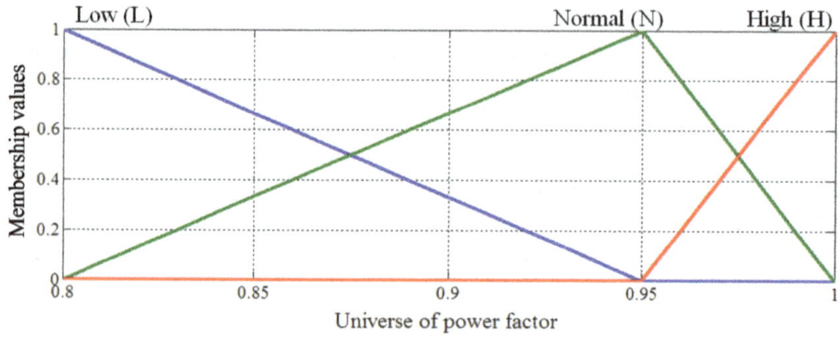

Figure E2.15(a) A partition of the universe of power factor into three subsets

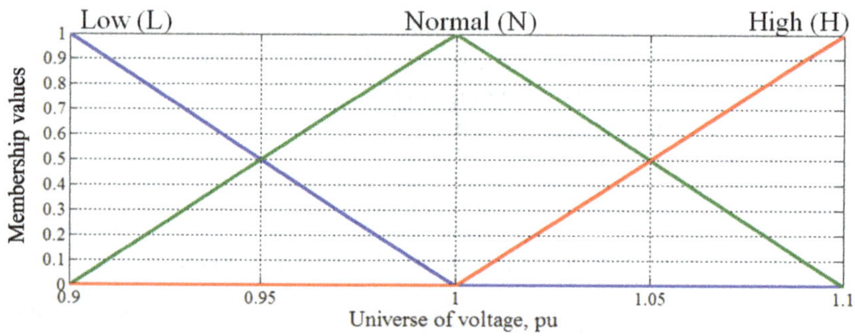

Figure E2.15(b) A partition of the universe of voltage into three subsets

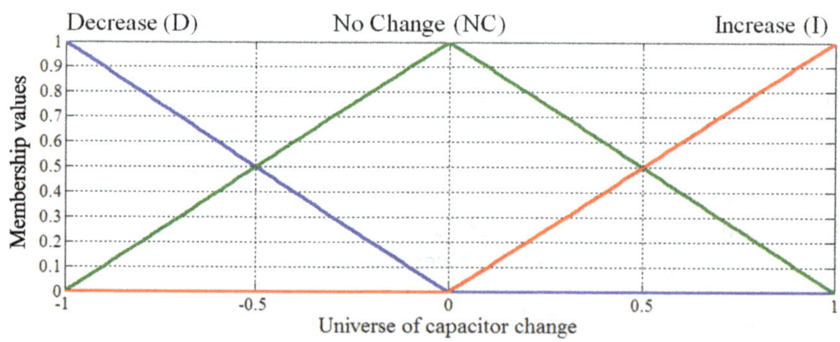

Figure E2.15(c) A partition of the universe of capacitor's changes into three subsets

voltage becomes 1.0 per unit (pu) when it is equal to its rated value, which is called *normal*. Values of the bus voltage up to 10% above the normal value are called *high* and 10% down from the normal value is called *low*. Considering these definitions of low, normal, and high, the universe of bus voltage can be defined in the interval {0.9, 1.1} in pu as shown in Figure E2.15(b).

The increment or decrement in the number of capacitors is determined by the operator's switching. Assuming that a capacitor group is added to the network when the switch is turned on and a capacitor group is disconnected when the switch is turned off. Therefore, +1 can be used to represent the increment while −1 can be used to represent the decrement. As a result, zero,0, will be used for no change. These comments lead us to define a universe of discourse for capacitor change in the interval of $\{-1, 1\}$ as given in Figure E2.15(c).

(b) Let us use triangular membership functions to represent fuzzy verbal terms *high*, *normal*, and *low* in the universe of PF assuming a PF of 0.95 is normal. As depicted in Figure E2.15(a), the universe of PF is not partitioned evenly among the fuzzy subsets. Since the normal value of PF is assumed to be 0.95, which is closer to the upper boundary of 1.0, fuzzy subset density is more than it is in the lower boundary.

(c) Let us place the nominal voltage 1.0 pu, which corresponds to the verbal term *normal*, at the center of the universe. Then *low* representing the bus voltages below the *normal* value will be placed to left and *high* representing the bus voltages above the *normal* value to the right as shown in Figure E2.15(b).

(d) The universe of discourse for the capacitor change is defined as the interval $\{-1, 1\}$. Since the steps of changes in capacitor values are either positive for increment, negative for decrement, or zero for no change, the universe can be partitioned into three subsets called *increase*, *decrease*, and *no change* as shown in Figure E2.15(c).

Example 2.16

The PF described in Example 2.15, is partitioned into six subcategories in its universe, which is the interval $\{0, 1\}$. Instead of three subcategories used in Example 2.15(a), this time six subcategories are defined in the universe of PF. Triangular-type membership functions given below in terms of their parametrized functions are used to represent these subcategories.

Minimum (0.50, 0.50, 0.70)
Lower (0.50, 0.70, 0.80)
Low (0.70, 0.80, 0.90)
Good (0.80, 0.90, 0.95)
Better (0.90, 0.95, 1.0)
Maximum (0.95, 1.0, 1.0)

(a) Show these subcategories in a figure in the universe of PF.
(b) Obtain the membership degrees of $\cos\phi = 0.88$ in the fuzzy subsets described in (a).

Answer 2.16

The MATLAB codes given in previous sections of this chapter for the triangular membership function can be used to plot the fuzzy subsets described in this problem. An updated code used for this problem is given in Figure E2.16(a) where the

MATLAB function code of *triangle01m.m* is given in Figure 2.5. The other types of fuzzy membership functions discussed in in previous sections of this chapter can also be used to represent the subsets given in this example. Trapezoid, Gaussian, bell, sinusoid, and sigmoid-type membership functions all can be used in this example. However, triangular-type membership function is preferred due to its simplicity and linearity.

Besides the partitioning of the universe of PF, this example has an interpretation of a crisp element PF = $\cos\phi = 0.88$ with the fuzzy subsets. In part (b), the inclusion of the crisp PF value in fuzzy subsets is asked to be determined.

(a) Figure E2.16(a) depicts the MATLAB code used to plot six subcategories of PF as described in the question part of this example. The resultant plot is given in Figure E2.16(b).
(b) If PF has a crisp value of PF = $\cos\phi = 0.88$, then it will have nonzero membership values in fuzzy subsets *Low* and *Good*. Figure E2.16(b) shows the

```
% Example 2.16a - Triangle membership function
clear,clc,clf
k=1;
for x=0.5:0.01:1
    Min(k)=triangle01m(0.50,0.50,0.70,x);
    Lower(k)=triangle01m(0.50,0.70,0.80,x);
    Low(k)=triangle01m(0.70,0.80,0.90,x);
    Good(k)=triangle01m(0.80,0.90,0.95,x);
    Bet(k)=triangle01m(0.90,0.95,1.0,x);
    Max(k)=triangle01m(0.95,1.0,1.0,x);
    X(k)=x;
    k=k+1;
end
plot(X,Min,X,Lower,X,Low,X,Good,X,Bet,X,Max)
grid
xlabel('Universe of power factor');
ylabel('Membership Values')
```

Figure E2.16(a) MATLAB code used to plot the fuzzy subsets in Example 2.16

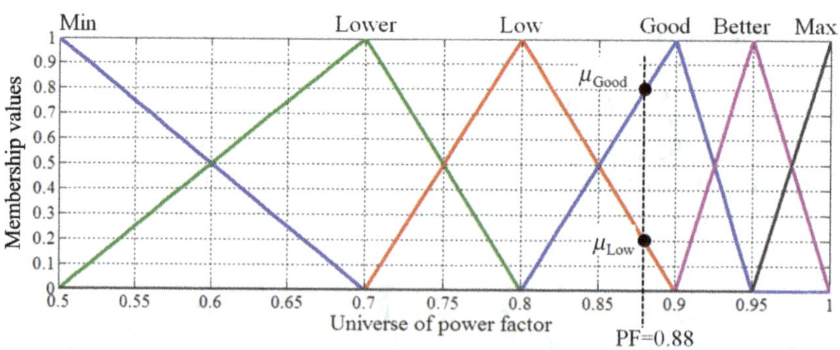

Figure E2.16(b) Fuzzy subsets used to partition the universe of power factors

interpretation of PF = 0.88 with the fuzzy subsets. Using MATLAB function *triangle01m* (x_L, x_P, x_R, PF) with the parameters of the fuzzy sets Low and Good, the membership values of PF = 0.88 are obtained as

$$\mu_{\text{Low}}(\text{PF}) = 0.2$$

$$\mu_{\text{Good}}(\text{PF}) = 0.8$$

Example 2.17

In electrical power transmission and distribution systems, the voltages may include additional frequency components besides the rated frequency. These additional frequency components are called harmonics, which are not wanted at all. Especially devices that cause discontinuity in current are the main harmonic sources such as switched-power electronic devices. The harmonics in power systems cause losses and misoperation of the system. Therefore, they are eliminated or minimized to reduce the unwanted effects by designing proper filters. Since the harmonics are not able to be eliminated completely, standards are developed to guide the users so that the total harmonic distortion (THD) is kept below the standard values. IEEE Std 519-2014, "Recommended Practices and Requirements for Harmonic Control in Electrical Power Systems" provides suggested harmonic values for power systems [20] as shown in Table E2.17.

(a) Partition the universe of bus voltage into the subcategories as low voltage (LV), medium voltage (MV), high voltage (HV), and extra high voltage (EHV), and show them in a graphic.
(b) Partition the universe of individual harmonics into the subcategories as extra low harmonics (ELH), low harmonics (LH), medium harmonics (MH), high harmonics (HH), and extra high harmonics (EHH) and show them in a graphic.
(c) Partition the universe of total harmonic distortion into the subcategories as ELH, LH, MH, HH, and EHH and show them in a graphic.

Answer 2.17

(a) Partition of the universe of bus voltage into four subcategories is shown in Figure E2.17(a1). Four subcategories are defined using parametrized triangular fuzzy membership functions as below.

LV (k) = triangle01m (0, 0, 1, v)

Table E2.17 Voltage distortion limits [20]

Bus voltage at PCC	**Individual harmonic (%)**	**Total harmonic distortion (THD) (%)**
$V \leq 1.0$ kV	5.0	8.0
1 kV $< V \leq$ 69 kV	3.0	5.0
69 kV $< V \leq$ 161 kV	1.5	2.5
161 kV $< V$	1.0	1.5

MV (k) = triangle01m (0, 1, 69, v)
HV (k) = triangle01m (1, 69, 161, v)
EHV (k) = triangle01m (69, 161, 161

where *triangle01m.m* is a MATLAB function used for the examples in this book. LV, MV, HV, and EHV are described in question.

Since the parameters of the fuzzy subset low voltage (LV) are much smaller than the parameters of MV, HV, and EHV, the plot of LV is not visible in Figure E2.17(a1). A zoom-in to the voltage range 0–1 kV shows the partition area of LV and MV as given in Figure E2.17(a2).

(b) Partition of the universe of individual harmonics into the subcategories as extra low harmonics (ELH), low harmonics (LH), medium harmonics (MH), high harmonics (HH), and extra high harmonics (EHH) are shown in Figure E2.17(b). The parameters of triangular membership functions are selected as

ELH (k) = triangle01m (0, 0, 1, h)
LH (k) = triangle01m (0, 1, 1.5, h)
MH (k) = triangle01m (1, 1.5, 3, h)
HH (k) = triangle01m (1.5, 3, 5, h)
EHH (k) = triangle01m (3, 5, 5, h)

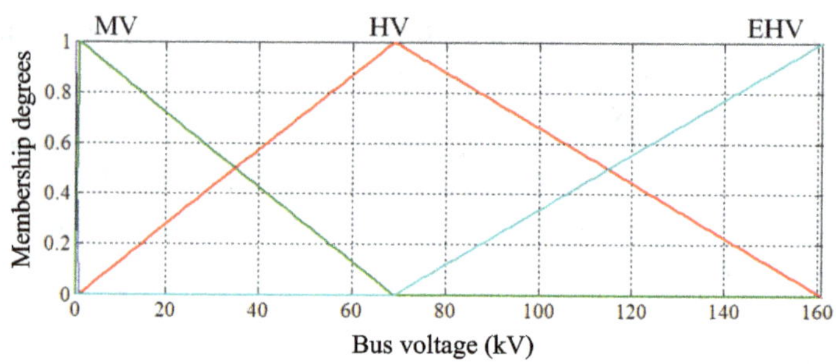

Figure E2.17(a1) Partition of power system bus voltages into fuzzy subsets

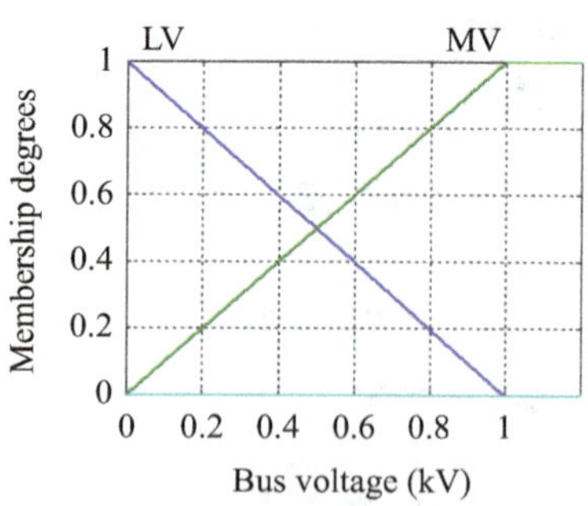

Figure E2.17(a2) Partition of low voltage section into fuzzy subsets LV and MV

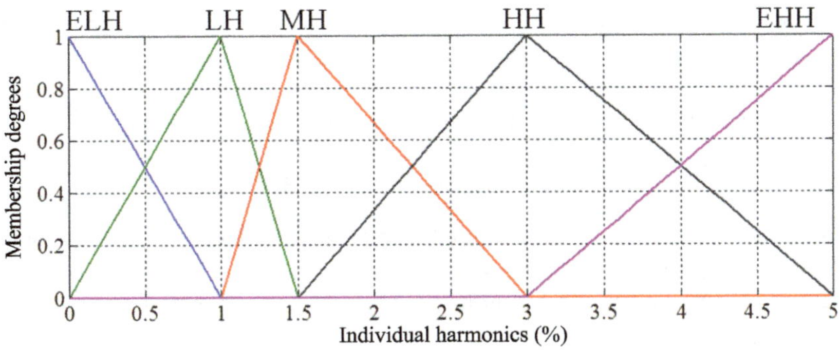

Figure E2.17(b) Partition of individual harmonics into fuzzy subsets

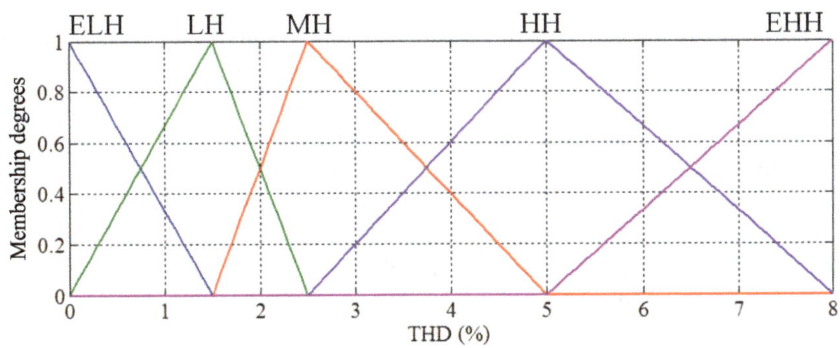

Figure E2.17(c) Partition of total harmonic distortion into fuzzy subsets

(c) Partition of the universe of total harmonic distortion into the subcategories as extra low harmonics (ELH), low harmonics (LH), medium harmonics (MH), high harmonics (HH), and extra high harmonics (EHH) are shown in Figure E2.17(c). The parameters of triangular membership functions are selected as

ELHT (k) = triangle01m (0, 0, 1.5, th)
LHT (k) = triangle01m (0, 1.5, 2.5, th)
MHT (k) = triangle01m (1.5, 2.5, 5, th)
HHT (k) = triangle01m (2.5, 5, 8, th)
EHHT (k) = triangle01m (5, 8, 8, th)

Example 2.18

Oscillatory transients in electrical power systems are described as sudden, non-power frequency changes in the steady-state condition of voltage, current, or both that have both positive and negative polarity values (bidirectional). In other words, the instantaneous voltage or current value of an oscillatory transient varies its polarity quickly. It is described by its spectral content or predominant frequency, magnitude, and duration.

The oscillatory transients are categorized into three subclasses as low-, medium-, and high-frequency oscillations [21]. These frequency ranges correspond to common types of power system oscillatory transient phenomena in terms of frequency, duration, and voltage magnitude as depicted in Table E2.18.

(a) Partition the universe of frequency into fuzzy subsets representing low, medium, and high frequencies.
(b) Partition the universe of duration into fuzzy subsets representing short, medium, and long durations.
(c) Partition the universe of voltage magnitude into fuzzy subsets representing low and high magnitudes in pu.

Answer 2.18

(a) Let us partition the universe of transient frequency into fuzzy subcategories as low frequency (LF), medium frequency (MF), and high frequency (HF). A plot of the portioned universe and fuzzy subsets is shown in Figure E2.18(a).
(b) Let us partition the universes of transient duration into fuzzy subcategories as short duration (SD), medium duration (MD), and long duration (LD). A plot of the portioned universe and fuzzy subsets is given in Figure E2.18(b).
(c) Let us partition the universes of oscillating voltage magnitude into fuzzy subcategories as low voltage (LV) and high voltage (HV). Figure E2.18(c) shows a plot of the portioned universe and fuzzy subsets.

Table E2.18 Oscillatory voltage classification [21]

Frequency categories	Typical spectral content	Typical duration	Typical voltage magnitude
Low	<5 kHz	0.3–50 ms	0–4 pu
Medium	5–500 kHz	20 μs	0–8 pu
High	0.5–5 MHz	5 μs	0–4 pu

For the universe of frequency (kHz):
Low(LF)=triangle(5,5,500)
Medium(MF)=triangle(5,500,5000)
High(HF)=triangle(500,5000,5000)

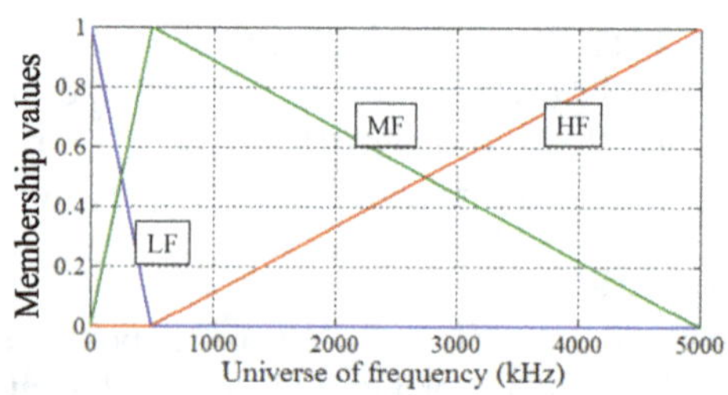

Figure E2.18(a) Frequency categorization for oscillatory transients in power systems

For the Universe of duration (ms):
Short(SD)=triangle(0.005,0.02,50)
Medium(MD)=triangle(0.02,50,300)
Long(LD)=triangle(50,300,300)

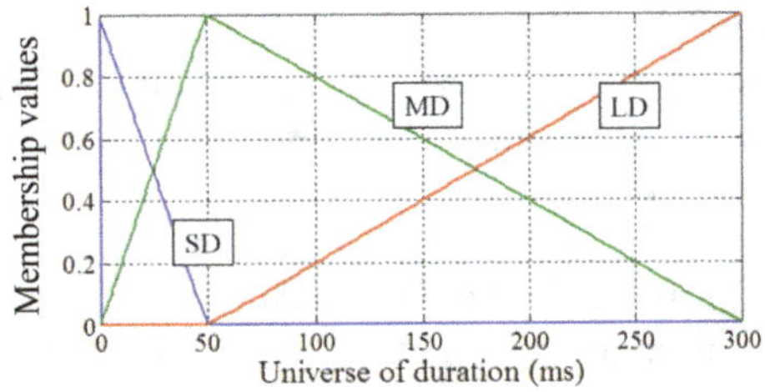

Figure E2.18(b) Subcategories of oscillatory transient durations in power systems

For the universe of voltage (pu):
Low(LV)=triangle(0,0,4)
High(HV)=triangle(0,4,4)

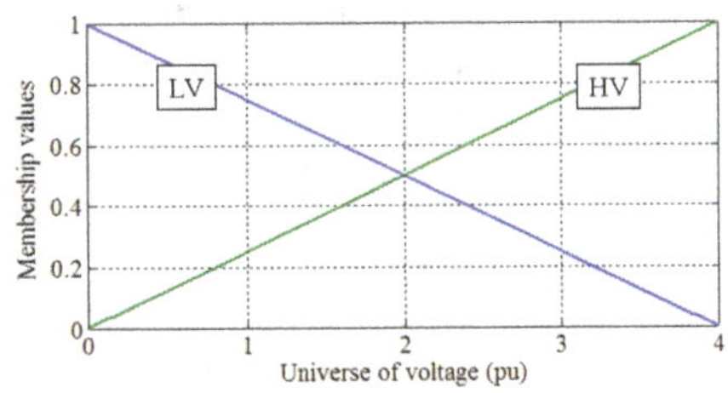

Figure E2.18(c) Subcategories of oscillatory voltage magnitude in power systems

Example 2.19

The active power transferred between two buses is expressed as

$$P_e = \frac{V_S V_R}{X_L} \sin \delta$$

where V_S and V_R are the sending and receiving bus voltages, respectively, and X_L is the reactance of the line connecting the buses to each other. δ is the phase difference between the voltages V_S and V_R and called power angle. The bus voltage magnitudes V_S and V_R are controlled and kept constant in steady-state operation. Since the voltage frequency and dimensional properties of the power line are constant, the line reactance X_L is also constant. Therefore, the electrical power transmission between two buses is written as a function of the power angle δ.

In an interconnected power transmission system, the line connecting two buses has a reactance of 0.5 pu. The bus voltages are measured to be equal and 1.0 pu. If the power angle δ is defined between 0 and π rad, obtain the following:

(a) Partition the universe of power angle δ into subcategories as zero (ZD), quarter (QD), half (HD) three quarter (TQD), and π (PD).
(b) Partition the power Pe into subcategories as zero (ZP), small (SP), medium (MP), high (HP), and maximum (MXP).

Answer 2.19

The equation given for the transmitted active power between two busses is a sinusoid function defined over half a period as shown in Figure E2.19 using the given data.

Both power angle, δ, and active power, P_e, transmitted between the busses are two separate universes and can be partitioned into subcategories similar to the ones given in question part. Now, let us partition the crisp universes δ and P_e given in Figure E2.19.

(a) The partition of the universe of power angle δ into subcategories as ZD, QD, HD TQD, and π (PD) is shown in Figure E2.19(a). It should be noted that these subcategories are not the only choices. Less or more subcategories can be defined as well. Sinusoid membership functions are used to represent the subsets in the partitioned space.

(b) The partition of the power P_e into subcategories as ZP, SP, MP, HP, and MXP are given in Figure E2.19(b). As it was written in part (a), less or more subcategories can be defined for the power P_e as well. Sinusoid membership

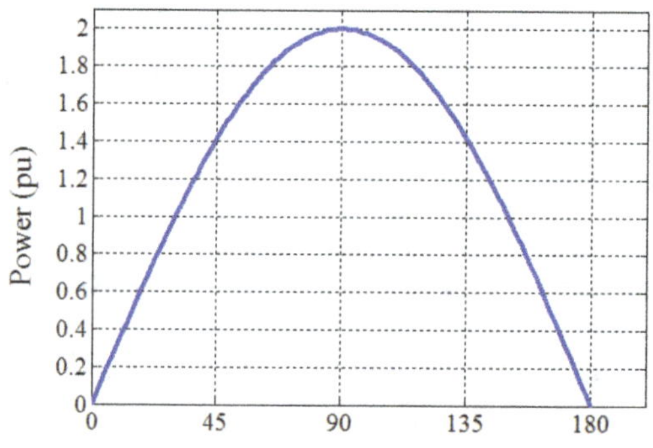

Figure E2.19 Active power transmission between two buses

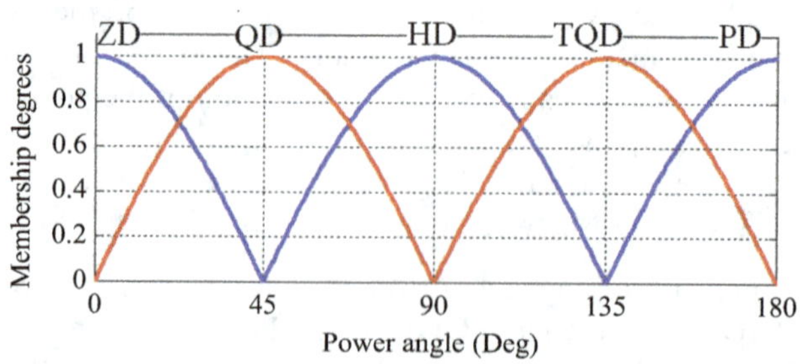

Figure E2.19(a) Fuzzy subcategories of power angle δ

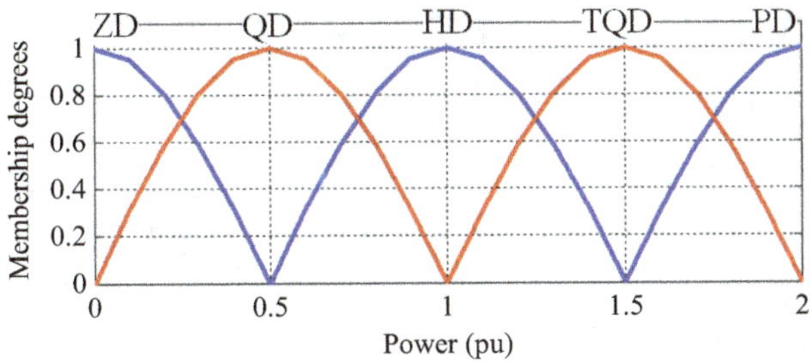

Figure E2.19(b) Fuzzy subcategories of electrical power P_e

functions are used to represent the subsets in the partitioned crisp space of the transmitted electrical power.

Example 2.20

PV solar cells are modeled as a current source such that the generated current is a function of photocurrent, which depends on sunlight or solar irradiation level. Since the PV cell is a sort of P–N diode, a reverse diode saturation current and losses due to heating are subtracted and the remaining current becomes the cell output current, which is the load current. The mathematical model of a PV cell is written for the output voltage as [22–24]

$$V_C = K\frac{BT_C}{q}\ln\left(\frac{I_{ph} + I_0 - I_C}{I_0}\right) - R_S I_C$$

where the curve fitting factor $K = 62$, Boltzmann constant $B = 37.674 \times 10^{-23}$ J/°C, electron charge $q = 1.106 \times 10^{-19}$ C. Assume that the above equation is adapted to a PV array (PVA) as the photocurrent $I_{ph} = 5$ A for solar irradiation level $S_C = 100$ mW/cm^2 and ambient temperature $T_C = 20$°C. Diode saturation current $I_0 = 0.01$ A and the series resistance R_S representing the losses is equal to 0.002 Ohms.

(a) Use the given data in the above equation and obtain current–voltage (I–V) and power–voltage (P–V) curves, assuming that the PVA output current is defined in the interval {0, 5} Amperes. Use per-unit (pu) values of current, voltage, and power in the plots assuming that the maximum power is the base value of power. The current and voltage values corresponding to the maximum power are the base values for current and voltage.

(b) Partition the universes of PVA voltage in pu, PVA current in pu, and PVA power in pu into fuzzy subcategories. Use at least four subcategories for each universe.

Answer 2.20

The MATLAB .m file code given in Figure E2.20 is used to plot the I–V and P–V curves. The values of power, current, and voltage are converted to pu using the base

```
% Example 2.20 - PVA characterisitics
clear, clc, clf
TC=20;              SC=100;
RS=0.002;           I0=0.01;
ICS=5;              IPH=ICS;
q=1.106e-19;        B=273*0.138e-23;
A =62;              PM0=0;
k=1;
for I=0:0.01:IPH
    AA=A*B*TC/q;
    BB=(IPH + I0 - I)/I0;
    VPV=AA*log(BB)-RS*I;
    Vc(k)=limiter(0,VPV,VPV);
    Ic(k)=I;
    Pc(k)=Vc(k)*Ic(k);
    if Pc(k) > PM0
        PB=Pc(k); VB=Vc(k); IB=Ic(k);
        PM0=PB;
    end
    k=k+1;
end
plot(Vc/VB,Ic/IB,Vc/VB,Pc/PB); grid
xlabel('PVA Voltage (pu)');
ylabel('PVA Current and Power (pu)')
```

Figure E2.20 The .m file for PV characteristics

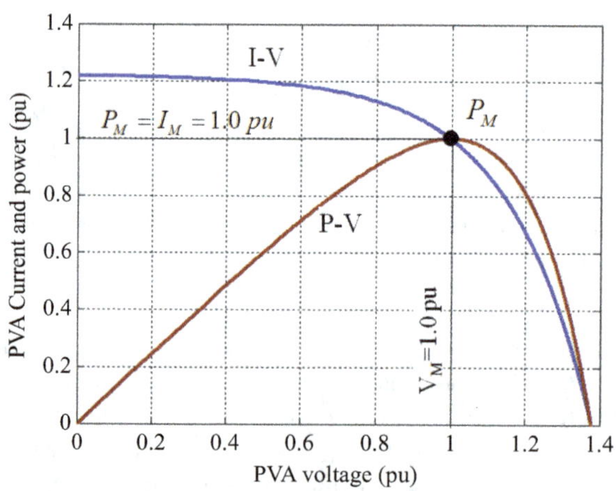

Figure E2.20(a) I–V and P–V characteristics of a PV array.

values as described in the question. This normalization is done to fit both curves into the same graph.

(a) Current–voltage (*I–V*) and power–voltage (*P–V*) characteristics of a PV array.
(b) Partition of the universes of PVA voltage in pu, PVA current in pu, and PVA power in pu into fuzzy subcategories are given in Figures E2.20(b), E2.20(c),

In the universe of PVA voltage:
Zero (ZV)= triangle(0,0,0.4)
Small (SV)= triangle(0,0.4,0.8)
Medium (MV)= triangle(0.4,0.8,1)
Optimum (OV)= triangle(0.8,1,1.2)
Open circuit (OCV)= triangle(1,1.2,1.2)

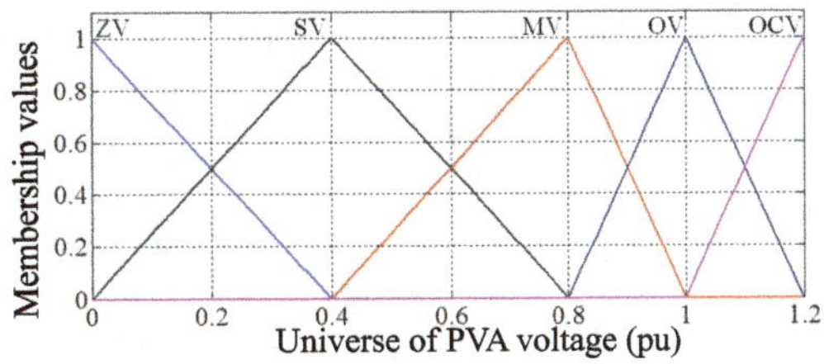

Figure E2.20(b) Fuzzy partition of PVA voltage

In the universe of PVA current:
Zero (ZI)= triangle(0,0,0.4)
Small (SI)= triangle(0,0.4,0.8)
Medium (MI)= triangle(0.4,0.8,1)
Optimum (OI)= triangle(0.8,1,1.2)
Short circuit (SCI)= triangle(1,1.2,1.2)

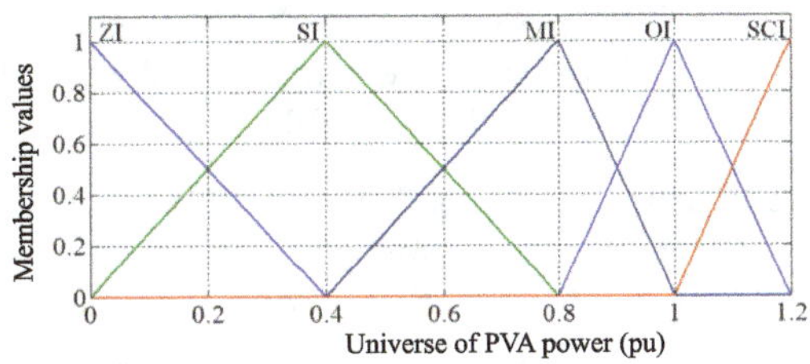

Figure E2.20(c) Fuzzy partition of PVA current

In the universe of PVA power:
Zero (ZP)= triangle(0,0,0.4)
Small (SP)= triangle(0,0.4,0.8)
High (HP)= triangle(0.4,0.8,1)
Maximum (MP)= triangle(0.8,1,1)

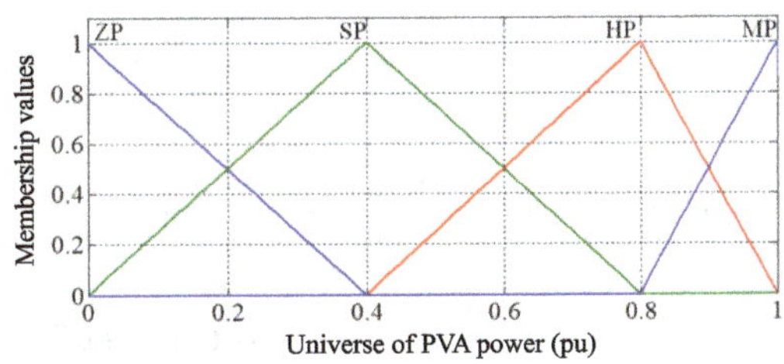

Figure E2.20(d) Fuzzy partition of PVA power

and E2.20(d). The boundaries of the voltage, current, and power are depicted in Figure E2.20(a). These values can be used to define the universes and their fuzzy subcategories represented by triangular membership functions as follows.

2.7 Problems

P2.1 Define three fuzzy sets *A*, *B*, and *C* in the universe of *X* and explain why these sets are fuzzy.

P2.2 Define three fuzzy subsets in each universe of discourses given as *education*, *age*, *height*, and *weight*.

P2.3 List at least five utilization areas where the fuzzy logic is applied.

P2.4 Fuzzy variables A, B, C, and D are defined in the universes $X1$, $X2$, $X3$, and $X4$, respectively, and given as

$$A = \frac{0}{0} + \frac{0.5}{1} + \frac{1}{2} + \frac{0.5}{3} + \frac{0}{4} \quad B = \frac{0}{2} + \frac{0.5}{3} + \frac{1}{4} + \frac{0.5}{5} + \frac{0}{6}$$

$$C = \frac{0}{6} + \frac{0.5}{7} + \frac{1}{8} + \frac{0.5}{9} + \frac{0}{10} \quad C = \frac{0}{11} + \frac{0.5}{12} + \frac{1}{13} + \frac{0.5}{14} + \frac{0}{15}$$

(a) Represent these fuzzy subsets using proper fuzzy membership functions and show them in the corresponding universe of discourses.
(b) Define a common normalized universe of discourse for the given fuzzy subsets such that just three fuzzy subsets will be sufficient in order to represent all four of them.

P2.5 Fuzzy variables A, B, C, and D are defined in the universes $X1$, $X2$, $X3$, and $X4$, respectively, and given as
(a) Two fuzzy subsets A and B are defined in the universe of $X = \{-5, 5\}$.
(b) Obtain $\alpha = 0$ and $\alpha = 0.5$ α-cuts of these fuzzy subsets.
(c) Obtain support sets of these fuzzy subsets.
(d) Comment on the convexity of these fuzzy subsets.

$$A = \frac{0.0}{-5} + \frac{0.2}{-4} + \frac{0.4}{-3} + \frac{0.6}{-2} + \frac{0.8}{-1} + \frac{1.0}{0} + \frac{0.8}{1} + \frac{0.6}{2} + \frac{0.4}{3} + \frac{0.2}{4} + \frac{0.0}{5}$$

$$B = \frac{0.0}{-5} + \frac{0.2}{-4} + \frac{0.4}{-3} + \frac{0.6}{-2} + \frac{1.0}{-1} + \frac{1.0}{0} + \frac{1.0}{1} + \frac{0.6}{2} + \frac{0.4}{3} + \frac{0.2}{4} + \frac{0.0}{5}$$

P2.6 The fuzzy subsets A, B, and C are defined in the universe $X = \{-4, 8\}$ and given as

$$A = \text{Zero} = \frac{0.0}{-2} + \frac{0.5}{-1} + \frac{1.0}{0} + \frac{0.5}{1} + \frac{0.0}{2}$$

$$B = \text{Positive small} = \frac{0.0}{0} + \frac{0.5}{1} + \frac{1.0}{2} + \frac{0.5}{3} + \frac{0.0}{4}$$

$$B = \text{Positive medium} = \frac{0.0}{2} + \frac{0.5}{3} + \frac{1.0}{4} + \frac{0.5}{5} + \frac{0.0}{6}$$

(a) Show the fuzzy subsets A, B, and C on the same figure in the common universe.
(b) Represent the fuzzy subsets A, B, and C by using triangular membership functions.
(c) Repeat (b) for Gaussian and sinusoid membership functions.

P2.7 Cold and Hot fuzzy words are defined in the universe of temperature in °C as given below.

$$\text{Cold} = \frac{1}{-30} + \frac{1}{-25} + \frac{0.9}{-20} + \frac{0.8}{-15} + \frac{0.7}{-10} + \frac{0.6}{-5} + \frac{0.5}{0} + \frac{0.4}{5} + \frac{0.3}{10} + \frac{0.2}{15} + \frac{0.1}{20} + \frac{0}{25} + \frac{0}{30}$$

$$\text{Hot} = \frac{0}{-30} + \frac{0}{-25} + \frac{0.1}{-20} + \frac{0.2}{-15} + \frac{0.3}{-10} + \frac{0.4}{-5} + \frac{0.5}{0} + \frac{0.6}{5} + \frac{0.7}{10} + \frac{0.8}{15} + \frac{0.9}{20} + \frac{1}{25} + \frac{1}{30}$$

Answer the following by generating a fuzzy subset for each question.

(a) The weather was neither cold nor hot yesterday. How was the weather yesterday?
(b) The weather was either cold or hot yesterday. How was the weather yesterday?
(c) The weather in Trabzon is both cold and hot in October. How is the weather in Trabzon in October?
(d) Which ones of the resultant fuzzy subsets in parts (a), (b), and (c) are convex?
(e) Which ones of the resultant fuzzy subsets in parts (a), (b), and (c) are also a fuzzy number?

P2.8 Use the trapezoid fuzzy membership function defined in (2.12) with the parameters given as $x_1 = -1.5$, $x_2 = -0.5$, $x_3 = +0.5$, and $x_4 = +1.5$ in the universe of $X = \{-2, 2\}$.

(a) Develop a MATLAB .m file to plot this membership function.
(b) Develop a Simulink model to plot this membership function.
(c) Develop a Simulink model for the trapezoidal membership function such that the membership value of x is calculated when x, x_1, x_2, x_3, and x_4 are entered as the inputs.

P2.9 A normal fuzzy subset is defined in the universe of $X = \{0, 20\}$. The height of the fuzzy subset corresponds to the crisp 10. Develop a MATLAB code to obtain core, crossover points, bandwidth, boundaries, cardinality, support, height, and α-cuts for $\alpha = 0.2$, $\alpha = 0.5$, and $\alpha = 0.8$.

P2.10 Fuzzy subsets A, B, and C are defined in a universe of $X = \{-4, 4\}$ and are given as

$$A = \frac{0}{-4} + \frac{0.5}{-3} + \frac{1}{-2} + \frac{0.5}{-1} + \frac{0}{0}$$

$$B = \frac{0}{-2} + \frac{0.5}{-1} + \frac{1}{0} + \frac{0.5}{1} + \frac{0}{2}$$

$$C = \frac{0}{0} + \frac{0.5}{1} + \frac{1}{2} + \frac{0.5}{3} + \frac{0}{0}$$

(a) Plot these three fuzzy subsets in the same graph.
(b) Show α-cuts on the figure you obtained in (a) for $\alpha = 0$ and $\alpha = 0.5$.
(c) Perform the following operations on the given fuzzy subsets.

(i) $(\overline{A \cap B})$ or $(\overline{B \cup C})$
(ii) $(\overline{A} \cup \overline{B})$ or $(\overline{B} \cap \overline{C})$

P2.11 A fuzzy subset A is given in Figure P2.11. Find the core set, crossover points, boundaries, support, height, and α-cuts for $\alpha = 0.2$, 0.6, and 0.8. Comment on the normality and convexity of the fuzzy subsets A.

P2.12 Fuzzy subsets *cold, warm*, and *hot* are defined in the universe of temperature $T = \{0, 40\}$ in °C. Triangular fuzzy membership functions defined in (2.4) in terms of the parameters x_1, x_2, and x_3 are to be used to represent these fuzzy subsets *cold*, *warm*, and *hot*.

(a) Show the fuzzy subsets *cold, warm*, and *hot* in the universe of temperature such that the parameters $cold_2$, $warm_2$, and hot_2 corresponding to the maximum membership values of each subset are evenly placed in the universe.
(b) Obtain the membership values of temperatures $T_1 = 5$°C and $T_2 = 25$°C in the fuzzy subsets *cold, warm*, and *hot*.
(c) Write a MATLAB .m file code to solve parts (a) and (b).

P2.13 Fuzzy subsets A and B are defined in the universe of $X = \{-5, 5\}$ and given as

$$A = \left\{\frac{0}{-2} + \frac{0.5}{-1} + \frac{1}{0} + \frac{0.5}{1} + \frac{0}{2}\right\}, B = \left\{\frac{0}{0} + \frac{0.5}{1} + \frac{1}{2} + \frac{0.5}{3} + \frac{0}{4}\right\}$$

(a) Find $C = A$ or B.
(b) Find $D = A$ and B.
(c) Find the cardinality of $E = \overline{A \cup B}$ and $F = \overline{A \cap B}$.
(d) Find the cardinality of $G = \overline{A} \cap \overline{B}$ and $H = \overline{A} \cup \overline{B}$.

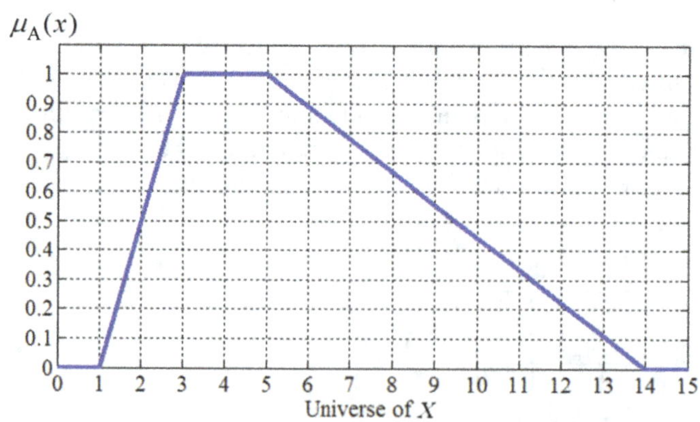

Figure P2.11 Fuzzy membership function A for Problem P2.11

P2.14 A crisp universe of discourse X is defined in the interval of [0, 10] and partitioned into three fuzzy subsections as SMALL, MEDIUM, and LARGE. Triangular-type fuzzy sets are used to represent these fuzzy subsets. Draw a software algorithm that gives you the membership value of any x on X in the fuzzy subsets SMALL, MEDIUM, and LARGE.

P2.15 In electrical power systems, a term called *power factor* is defined as $\cos\phi$ in the interval [0, 1]. In order to reduce the power losses in power distribution networks, the PF is kept close to 1.0 such as 0.5 by connecting power capacitors at the end of the distribution line. However, connecting more capacitors than the required amount causes the bus voltage to be increased and decreases the electrical loads being fed.

The expert, operating the PF unit at the end of the line, determines the right amount and adjusts the capacitors to be connected by looking at PF and line voltage values. The human operator uses a decision process like the one given in Table P2.15(a).

where H, N, and L stand for high, normal, and low, respectively. NC, I, and D mean no change, increase, and decrease. The subscriptions PF, V, and C represent PF, voltage, and capacitor change, respectively.

Crisp numerical values of the verbal terms given in Table P2.15(a) are given in Tables P2.15(b) and P2.15(c). Represent the verbal terms *low*,

Table P2.15(a) Decision table for Problem P2.15

	Line voltage		
Power factor	L_V	N_V	H_V
H_{PF}	NC_C	D_C	D_C
N_{PF}	I_C	NC_C	D_C
L_{PF}	I_C	I_C	NC_C

Table P2.15(b) Crisp data for Problem P2.15

Input space	Low (L)	Normal (N)	High (H)
Power factor (PF)	0.80	0.95	1.0
Live voltage (V)	−10%	0.0	10%

Table P2.15(c) Crisp data for the output space in Problem P2.15

Output space	Decrease (D_C)	No change (NC_C)	Increase (I_C)
Capacitor change (C)	−1.0	0.0	1.0

Table P2.16 Crisp data for the output space in Problem P2.15

Minimum (MIN)	Very low (VL)	Low (L)	Good (G)	Very good (VG)	Maximum (G)
0.50	0.70	0.80	0.90	0.95	1.0

normal, high, decrease, no change, and *increase* as fuzzy subsets in their crisp universes using the crisp values given in Tables P2.15(b) and P2.15(c).

P2.16 The values of PF described in Problem P2.15 are separated into subcategories in the universe of [0, 1] as given in Table P2.16.

(a) Represent these categories as fuzzy subsets and show them in a figure.
(b) Develop a MATLAB and a Simulink model to represent the fuzzy subsets in part (a) by triangular-type membership functions.
(c) Repeat part (b) for Gaussian, sinusoid, bell, and trapezoidal membership functions.
(d) If the PF is measured as $\cos\phi = 0.85$, what would be the membership degrees of PF in the fuzzy subsets developed in parts (b) and (c)?

P2.17 In order to supply electricity to a region with urban and rural areas, three types of electrical power stations are constructed. These power stations are categorized as large, medium, and small size power generating stations. Power stations generating up to 5 MW are called *small*, up to 50 MW are called *medium*, and up to 100 MW and over are called *large* power stations.

Represent these power stations in the universe of power in MW using triangular, trapezoid, bell, Gaussian, sinusoid, and sigmoid membership functions.

P2.18 The management of the power stations described in P2.17 is done according to dissipated power by the loads. Such that the large size power stations are always in service supplying a certain amount of power to the utility. Medium- and small-size power stations are not operated every time. Depending on power demand, medium- and small-size power stations are connected or disconnected from the utility so that generation-dissipation is kept in balance.

During this process, the power generation is increased or decreased by steps. In an intelligent power management system, the step size of power increment or decrement is determined according to power demand in terms of verbal fuzzy variables no change (NC), low Change (LC), medium change (MC), and high change (HC).

Create a partitioned universe of power change considering increments and decrements with the step sizes mentioned above in order to maintain generation-dissipation equality. Triangular fuzzy membership functions can be used to represent the fuzzy variables.

P2.19 In electrical utility systems, the loads are classified according to their importance for energy management. Such that the most important loads such as the ones related to health are always operated with priority. Some loads may have the option to be delayed for a short time, some longer and some may be canceled. In order to use the generated power more efficiently for energy saving and sustainability, energy management algorithms are applied.

(a) The load supplied by the power utility is classified as residential, industrial, street lights, health centers, transportation, and entertainment. Represent these subclasses of loads by fuzzy subsets in the universe of load importance.
(b) Categorize the loads as *can be canceled*, *can be delayed longer*, *can be delayed shorter*, and *must be used* in the universe of load importance.
(c) Categorize the loads as *not important*, *somehow important*, *important*, or *highly important* in the universe of load importance.

P2.20 An industrial plant gets its required electrical power from the utility grid besides its own generation from a wind energy conversion (WEC) system, a PV solar system, and a fuel cell (FC) system. In order to reduce the payments to the utility, the power demand of the industrial plant is managed to be supplied from its own sources by reducing the power drawn from the utility. A fuzzy reasoning process using the rules below is employed in this management system to decide which source is better to use for the moment.

R1: If PV is *sufficient* and WEC is *insufficient* and FC is *empty* then use PV *only*.
R2: If PV is *sufficient* and WEC is *insufficient* and FC is *full* then use PV *and* FC only.
R3: If PV is *insufficient* and WEC is *insufficient* and FC is *empty* then use Utility only.
R4: If PV is *insufficient* and WEC is *sufficient* and FC is *empty* then use WEC only.
R5: If PV is *insufficient* and WEC is *sufficient* and FC is *full* then use WEC only.
R6: If PV is *insufficient* and WEC is *insufficient* and FC is *full* then use FC and Utility.
R7: If PV is *sufficient* and WEC is *sufficient* and FC is *empty* then use WEC and PV.

(a) Determine the verbal fuzzy variables and show them in their universes.
(b) Determine the input and output spaces in this decision process.

P2.21 The windings of a three-phase transformer are diagnosed by a thermal camera. Depending on thermal data, the operator is able to understand whether the transformer is operating normally, under heavy load, light load, or under internal or other faulted conditions. Thermal photographs are obtained and stored from initial tests performed for no load, light load, nominal load, heavy load, extra heavy load, other faulted condition, and internal faulted condition. Then, thermal photographs from real-time operating conditions are taken periodically and compared with the original data to observe the operating conditions and changes. As a result of thermal data analysis, required precautions are taken on time.

A thermal photograph of a faulted point in transformer windings is given in Figure P2.20. In this figure, the color changes from white on the sides to red in the center. It means that the temperature is higher in the center. Assume that the highest density of red in the center represents an unexpected condition such as a faulted case. The white color on the other sides of the image represents lower temperatures that may be considered as cool and healthy conditions.

(a) Use the color spectrum from white to red and create subcategories that represent *no load* (*NoL*), *light load* (*LL*), *nominal load* (*NomL*), *heavy load* (*HL*), *extra heavy load* (*EHL*), and *faulted* (*F*) conditions. Triangular fuzzy membership functions can be used to represent the subcategories.

(b) Define the universe of discourse. What it should be?

P2.22 A crisp color spectrum is shown in Figure P2.21. Each color represents an operating condition as *no load* (*NoL*), *light load* (*LL*), *nominal load* (*NomL*), *heavy load* (*HL*), *extra heavy load* (*EHL*), and *faulted* (*F*) condition.

Convert the crisp color spectrum given in Figure P2.21 into a fuzzy color spectrum representing the fuzzy operating conditions.

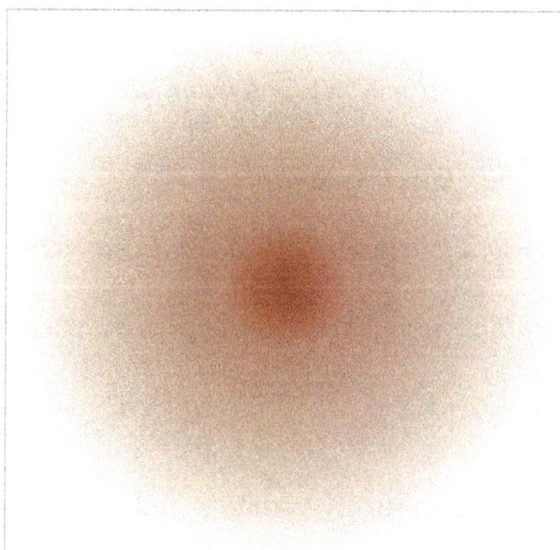

Figure P2.20 An image of a faulted point in transformer windings

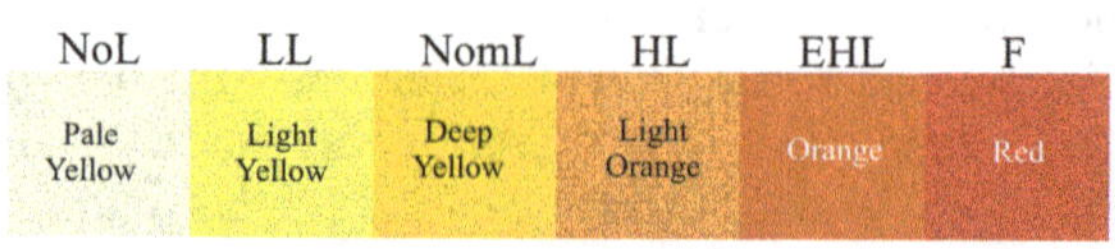

Figure P2.21 Crisp color spectrum representing crisp operating conditions

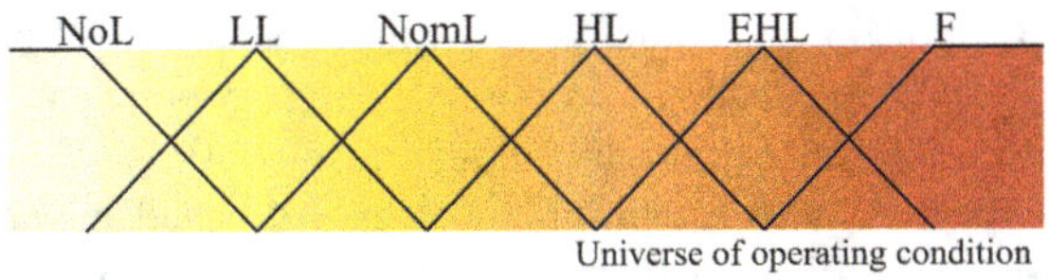

Figure P2.22(a) Color spectrum representing fuzzy loading conditions

Figure P2.22(b) Thermal pictures of a power transmission line under different loadings

P2.23 The fuzzy color spectrum shown in Figure P2.22a is partitioned into six subcategories, which represent the thermal colors of loading conditions in a power line. The subcategories are defined as *no load* (*NoL*), *light load* (*LL*), *nominal load* (*NomL*), *heavy load* (*HL*), *extra heavy load* (*EHL*), and *faulted* (*F*) conditions.

Thermal pictures of a power transmission line are taken at different instants under various operating conditions and given in Figure P2.22b. According to the partition space given in Figure P2.22a, determine the operating condition of each case.

References

[1] Zadeh L.A. 'Fuzzy sets'. *Information and Control*. 1965;8(3):338–353.

[2] Jang J.S.R., Sun C.T. and Mizutani E. *Neuro-Fuzzy and Soft Computing: A Computational Approach to Learning and Machine Intelligence*. Englewood Cliffs, NJ: Prentice-Hall; 1996.

[3] Klir G. and Folger T. *Fuzzy Sets, Uncertainty, and Information*. Englewood Cliffs, NJ: Prentice-Hall; 1988.

[4] Maiers J. and Sherif Y.S. 'Applications of fuzzy set theory'. *IEEE Transactions on Systems, Man, and Cybernetics*. 1985;15(1):175–189.

[5] Tang K.S., Man K.F., Liu Z.F. and Kwong S. 'Minimal fuzzy memberships and rules using hierarchical genetic algorithms'. *IEEE Transactions on Industrial Electronics*. 1998;45(1):162–169.

[6] Ishibuchi H., Nozaki K., Yamamoto N. and Tanaka H. 'Selecting fuzzy if-then rules for classification problems using genetic algorithms'. *IEEE Transactions on Fuzzy Systems*. 1995;3(3):260–270.

[7] Park D., Kandel A. and Langholz G. 'Genetic-based new fuzzy reasoning models with application to fuzzy control'. *IEEE Transactions on Systems, Man, and Cybernetics*. 1994;24(1):39–47.

[8] Chen J.Q., Xi Y.G. and Zhang Z.J. 'A clustering algorithm for fuzzy model identification'. *Fuzzy Sets and Systems*.1998;98(3):319–329.

[9] Seng T.L., Khalid M.B. and Yusof R. 'Tuning of a neuro-fuzzy controller by genetic algorithm'. *IEEE Transactions on Systems, Man, and Cybernetics, Part B (Cybernetics)*. 1999;29(2):226–236.

[10] Nauck D. and Kruse R. 'A neuro-fuzzy method to learn fuzzy classification rules from data'. *Fuzzy Sets and Systems*.1997;89(3):277–288.

[11] Shi Y. and Mizumoto M. 'A new approach of neuro-fuzzy learning algorithm for tuning fuzzy rules'. *Fuzzy Sets and Systems*. 2000;112(1): 99–116.

[12] Nguyen H.T., Prasad N.R., Walker C.L. and Walker E.A. *A First Course in Fuzzy and Neural Control*. London: Chapman & Hall/CRC; 2003.

[13] Lee-Kwang H. and Lee K.M. 'Fuzzy hypergraph and fuzzy partition'. *IEEE Transactions on Systems, Man, and Cybernetics*. 1995;25(1):196–201.

[14] Murata T., Ishibuchi H. and Gen M. 'Adjusting fuzzy partitions by genetic algorithms and histograms for pattern classification problems'. *IEEE International Conference on Evolutionary Computation*; Anchorage, AK, May 1998. pp. 9–14.

[15] Okuda Y., Yamashita H. and Inaida J. 'Fuzzy reasoning using fuzzy partition and its application'. *Conference of the North American Fuzzy Information Processing Society (NAFIPS)*; Pensacola Beach, FL, Aug 1998. pp. 324–328.

[16] Hoppner F. and Klawonn F. 'A new approach to fuzzy partitioning'. *Joint 9th IFSA World Congress and 20th NAFIPS International Conference*; Vancouver, BC, July 2001, vol. 3, pp. 1419–1424.

[17] Chuang C.C. 'Fuzzy weighted support vector regression with a fuzzy partition'. *IEEE Transactions on Systems, Man, and Cybernetics, Part B (Cybernetics)*. 2007;37(3):630–640.

[18] Thresher R.W. and Dodge D.M. 'Trends in the evolution of wind turbine generator configurations and systems'. *Wind Energy*. 1998;1(1):70–86.

[19] Quilter J. *Gamesa progress on 5MW offshore turbine* [online]. Spain: WindPower; 2013. Available from http://www.windpowermonthly.com/article/1184003/gamesa-progress-5mw-offshoreturbine [Accessed 19 Aug 2016].

[20] IEEE Recommended Practices and Requirements for Harmonic Control in Electrical Power Systems. IEEE Std 519-2014 (Revision of IEEE Std 519-1992); New York, NY: IEEE.

[21] IEEE Recommended Practice for Monitoring Electric Power Quality. IEEE Std 1159-2009 (Revision of IEEE Std 1159-1995); 2009. pp. c1–81.

[22] Altas I.H. Control strategies for maximum power tracking and energy utilization of a standalone photovoltaic energy system. PhD Thesis, Faculty of Engineering, Department of Electrical Engineering, University of New Brunswick, Fredericton, NB, Canada, 1993.

[23] Altas I.H. and Sharaf A.M. 'A fuzzy logic power tracking controller for a photovoltaic energy conversion scheme'. *Electric Power Systems Research Journal*. 1992;25(3):227–238.

[24] Altas I.H. and Sharaf A.M. 'Novel maximum power fuzzy logic controller for photovoltaic solar energy systems'. *Renewable Energy*. 2008;33(3): 388–399.

Chapter 3
Fuzzy decision-making

General information on *fuzzy relation* and *the compositional rule of inference* have been given in this chapter. Then, approximate reasoning, fuzzy reasoning, and fuzzy decision-making processes are given. Single input, multiple input, and multiple rule base systems are studied and examples are given. The concept of fuzzy reasoning and fuzzy decision makers is studied and user-defined MATLAB® files are used to support the operational behaviors of fuzzy decision-making. Well-known fuzzy reasoning algorithms such as Mamdani fuzzy reasoning, Sugeno fuzzy reasoning, and Tsukamoto fuzzy reasoning have also been studied here.

3.1 Introduction

Relation is an indication of the connectivity, reflexivity, and similarity between two or more sets. These properties either exist or do not in a crisp relation. The existence of these properties in a fuzzy relation, on the other hand, has some certain degrees. The relation of fuzzy sets is also fuzzy and called fuzzy relation. Therefore, the relation between the elements of fuzzy sets has some membership degrees in the interval [0, 1], not just 0 or 1. Elements of two fuzzy sets may not have any relation as they may have full relation. However, there might be some cases in which one may not be sure to decide whether a relation exists or not. Due to the gradual inclusion of the elements in a fuzzy relation, it is not easy to say certainly that "there is a relation" or "there is not a relation." The gradual inclusion represents the fuzziness of the relation between the elements of the fuzzy sets. Therefore, these types of relations are called fuzzy relations.

The fuzzy relation between two fuzzy sets leads us to define the fuzzy implications, which are the basis of fuzzy rules. A fuzzy relation is nothing but the effective weightings of fuzzy rules, which are also used to develop fuzzy inference systems and fuzzy decision-makers.

This chapter covers the concept of fuzzy reasoning and fuzzy decision-makers besides a brief introduction to fuzzy relation. User-defined MATLAB files are used to support the operational behaviors of fuzzy decision-making in this chapter. Well-known fuzzy reasoning algorithms such as Mamdani fuzzy reasoning, Sugeno fuzzy reasoning, and Tsukamoto fuzzy reasoning have also been explained with examples here.

3.2 Fuzzy relation

Let $A_1, A_2, \ldots, A_n$ be fuzzy sets defined in the universes $Y_1, Y_2 \ldots, Y_n$. The relation of these sets is defined by their cartesian product R as

$$R = A_1 \times A_2 \times A_3 \times \cdots \times A_n \tag{3.1}$$

This cartesian product is a subset of the cartesian product $Y_1 \times Y_2 \times \cdots \times Y_n$. The relation R is represented by the membership values of fuzzy sets $A_1, A_2, \ldots, A_n$, and defined as in (3.2).

$$R = A_1 \times A_2 \times \cdots \times A_n = \int_{Y_1 \times Y_2 \times \cdots \times Y_n} \frac{\mu_{A1}(y_1) \cap \mu_{A2}(y_2) \cap \cdots \mu_{\mathrm{An}}(y_n)}{y_1, y_2, \ldots, y_n} \tag{3.2}$$

where the operator $\cap$ stands for intersection, which is a min operator. Therefore, (3.2) can be written as

$$R = A_1 \times A_2 \times \cdots \times A_n = \int_{Y_1 \times Y_2 \times \cdots \times Y_n} \frac{\min(\mu_{A1}(y_1), \mu_{A2}(y_2), \ldots, \mu_{\mathrm{An}}(y_n))}{y_1, y_2, \ldots, y_n} \tag{3.3}$$

Example 3.1

P–N junction temperature of monocrystalline Si photovoltaic (PV) solar cells increases almost linearly with the ambient temperature [1,2]. Fuzzy subsets High (H_A) for ambient temperature and High (H_J) for the junction temperature are both defined in the same universe of temperature (T) with a universe defined in {0, 100}.

$$H_A = \frac{0}{50} + \frac{0.2}{55} + \frac{0.4}{60} + \frac{0.6}{65} + \frac{0.8}{70} + \frac{1}{75}$$

$$H_J = \frac{0}{75} + \frac{0.2}{80} + \frac{0.4}{85} + \frac{0.6}{90} + \frac{0.8}{95} + \frac{1}{100}$$

(a) Show the fuzzy subsets H_A and H_J as graphics in the universe of temperature.
(b) Obtain the cartesian product $R = H_A \times H_J$ as a relation between high ambient and high junction temperatures in monocrystalline Si PV solar cells.
(c) Obtain the relation R in a matrix format.
(d) Write a MATLAB .m file code to calculate the relation R.
(e) Plot the relation matrix R as a three-dimensional surface in the universe of $T_A \times T_J$.

Answer 3.1

(a) Representing the fuzzy subsets H_A and H_J as graphics in the universe of temperature.

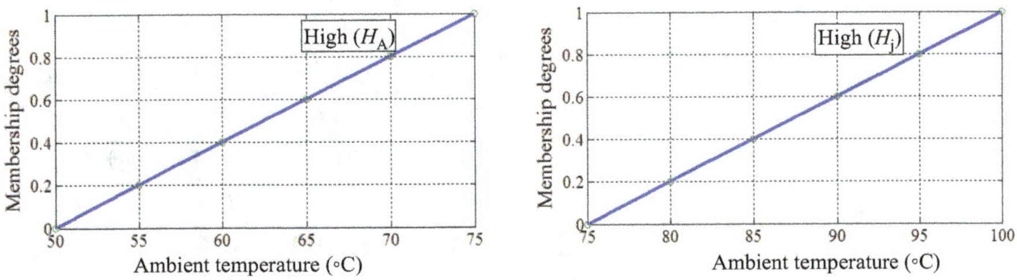

(i) (ii)

Figure E3.1(a) High fuzzy subsets of ambient and junction temperatures. (i) Ambient temperature and (ii) junction temperature.

(b) The cartesian product $R = H_A \times H_J$ is obtained as a relation between high ambient and high junction temperatures in monocrystalline Si PV solar cells as follows.

$$R = H_A \times H_J = \int_{T\times T} \frac{\min(\mu_{\text{HA}}(T), \mu_{\text{HJ}}(T))}{T,T} = \sum_{m=1}^{6} \frac{\mu_{\text{TA}}(T_m) \cap \mu_{\text{TJ}}(T_n)}{(x_m, x_n)}, \; n = 1.$$

Or the cartesian product $R = H_A \times H_J$ can be written as

$$R = H_A \times H_J$$

$$= \left(\frac{0}{50} + \frac{0.2}{55} + \frac{0.4}{60} + \frac{0.6}{65} + \frac{0.8}{70} + \frac{1}{75}\right)$$

$$\times \left(\frac{0}{75} + \frac{0.2}{80} + \frac{0.4}{85} + \frac{0.6}{90} + \frac{0.8}{95} + \frac{1}{100}\right)$$

$$R = H_A \times H_J = \frac{0 \cap 0}{50, 75} + \frac{0 \cap 0.2}{50, 80} + \frac{0 \cap 0.4}{50, 85} + \frac{0 \cap 0.6}{50, 90} + \frac{0 \cap 0.8}{50, 95} + \frac{0 \cap 1}{50, 100}$$

$$+ \frac{0.2 \cap 0}{50, 75} + \frac{0.2 \cap 0.2}{50, 80} + \frac{0.2 \cap 0.4}{50, 85} + \frac{0.2 \cap 0.6}{50, 90} + \frac{0.2 \cap 0.8}{50, 95} + \frac{0.2 \cap 1}{50, 100}$$

$$+ \frac{0.4 \cap 0}{50, 75} + \frac{0.4 \cap 0.2}{50, 80} + \frac{0.4 \cap 0.4}{50, 85} + \frac{0.4 \cap 0.6}{50, 90} + \frac{0.4 \cap 0.8}{50, 95} + \frac{0.4 \cap 1}{50, 100}$$

$$+ \frac{0.6 \cap 0}{50, 75} + \frac{0.6 \cap 0.2}{50, 80} + \frac{0.6 \cap 0.4}{50, 85} + \frac{0.6 \cap 0.6}{50, 90} + \frac{0.6 \cap 0.8}{50, 95} + \frac{0.6 \cap 1}{50, 100}$$

$$+ \frac{0.8 \cap 0}{50, 75} + \frac{0.8 \cap 0.2}{50, 80} + \frac{0.8 \cap 0.4}{50, 85} + \frac{0.8 \cap 0.6}{50, 90} + \frac{0.8 \cap 0.8}{50, 95} + \frac{0.8 \cap 1}{50, 100}$$

$$+ \frac{1 \cap 0}{50, 75} + \frac{1 \cap 0.2}{50, 80} + \frac{1 \cap 0.4}{50, 85} + \frac{1 \cap 0.6}{50, 90} + \frac{1 \cap 0.8}{50, 95} + \frac{1 \cap 1}{50, 100}$$

$$R = H_A \times H_J = \frac{0}{50,75} + \frac{0}{50,80} + \frac{0}{50,85} + \frac{0}{50,90} + \frac{0}{50,95} + \frac{0}{50,100}$$
$$+ \frac{0}{50,75} + \frac{0.2}{50,80} + \frac{0.2}{50,85} + \frac{0.2}{50,90} + \frac{0.2}{50,95} + \frac{0.2}{50,100}$$
$$+ \frac{0}{50,75} + \frac{0.2}{50,80} + \frac{0.4}{50,85} + \frac{0.4}{50,90} + \frac{0.4}{50,95} + \frac{0.4}{50,100}$$
$$+ \frac{0}{50,75} + \frac{0.2}{50,80} + \frac{0.4}{50,85} + \frac{0.6}{50,90} + \frac{0.6}{50,95} + \frac{0.6}{50,100}$$
$$+ \frac{0}{50,75} + \frac{0.2}{50,80} + \frac{0.4}{50,85} + \frac{0.6}{50,90} + \frac{0.8}{50,95} + \frac{0.8}{50,100}$$
$$+ \frac{0}{50,75} + \frac{0.2}{50,80} + \frac{0.4}{50,85} + \frac{0.6}{50,90} + \frac{0.8}{50,95} + \frac{1 \cap 1}{50,100}$$

As seen in Example 3.1, he cartesian product $H_A \times H_J$ is a subset of the cartesian product $T_A \times T_J$, where T_A and T_J are the temperature values in the universe of temperature and they represent ambient and junction temperatures, respectively. The symbols Σ and $\int$ are used to represent the fuzzy subsets in discrete and continues forms, respectively.

(c) The relation R in a matrix format

The result of fuzzy relation can be represented by using the "+" sign between each element in a line format. However, some other formats such as tables, membership matrices, suggested graphs, and basic graphs can be used for clarity and understandability. These formats will be introduced later in this chapter. Since the matrix format gives an overlook view of the membership values of the relation, the matrix format is preferred to be used. The relation matrix for Example 3.1 is given in Table E3.1(a).

As it can be easily understood from the matrix format, the fuzzy relation R is a two-dimensional fuzzy variable while the fuzzy subsets H_A and H_J are one-dimensional fuzzy variables.

(d) A MATLAB .m file code is given in Figure E3.1(b).

Table E3.1(a) Membership matrix of the relation in Example 3.1

		$\frac{0}{75}$	$\frac{0.2}{80}$	$\frac{0.4}{85}$	$\frac{0.6}{90}$	$\frac{0.8}{95}$	$\frac{1}{100}$
$R = H_A \times H_J =$	$\frac{0}{50}$	$\frac{0}{50,75}$	$\frac{0}{50,80}$	$\frac{0}{50,85}$	$\frac{0}{50,90}$	$\frac{0}{50,95}$	$\frac{0}{50,100}$
	$\frac{0.2}{55}$	$\frac{0}{50,75}$	$\frac{0.2}{50,80}$	$\frac{0.2}{50,85}$	$\frac{0.2}{50,90}$	$\frac{0.2}{50,95}$	$\frac{0.2}{50,100}$
	$\frac{0.4}{60}$	$\frac{0}{50,75}$	$\frac{0.2}{50,80}$	$\frac{0.4}{50,85}$	$\frac{0.4}{50,90}$	$\frac{0.4}{50,95}$	$\frac{0.4}{50,100}$
	$\frac{0.6}{65}$	$\frac{0}{50,75}$	$\frac{0.2}{50,80}$	$\frac{0.4}{50,85}$	$\frac{0.6}{50,90}$	$\frac{0.6}{50,95}$	$\frac{0.6}{50,100}$
	$\frac{0.8}{70}$	$\frac{0}{50,75}$	$\frac{0.2}{50,80}$	$\frac{0.4}{50,85}$	$\frac{0.6}{50,90}$	$\frac{0.8}{50,95}$	$\frac{0.8}{50,100}$
	$\frac{1}{75}$	$\frac{0}{50,75}$	$\frac{0.2}{50,80}$	$\frac{0.4}{50,85}$	$\frac{0.6}{50,90}$	$\frac{0.8}{50,95}$	$\frac{1 \cap 1}{50,100}$

Table E3.1(b) Membership values of the relation matrix in Example 3.1

		$\frac{0}{75}$	$\frac{0.2}{80}$	$\frac{0.4}{85}$	$\frac{0.6}{90}$	$\frac{0.8}{95}$	$\frac{1}{100}$
	$\frac{0}{50}$	0	0	0	0	0	0
	$\frac{0.2}{55}$	0	0.2	0.2	0.2	0.2	0.2
$R = H_A \times H_J =$	$\frac{0.4}{60}$	0	0.2	0.4	0.4	0.4	0.4
	$\frac{0.6}{65}$	0	0.2	0.4	0.6	0.6	0.6
	$\frac{0.8}{70}$	0	0.2	0.4	0.6	0.8	0.8
	$\frac{1}{75}$	0	0.2	0.4	0.6	0.8	1

(e) A three-dimensional surface plot of the fuzzy relation R is shown in Figure E3.1(c). The surface in Figure E3.1(c) represents the relation between *high* ambient temperature and *high* junction temperature.

Example 3.1 shows that the Cartesian product $H_A \times H_J$ is a subset of the ordered pairs $(T_A \times T_J)$. The membership values of the elements of the relation R in the cartesian product $T_A \times T_J$ are determined by the membership values of the fuzzy subsets H_A and H_J.

```
% Example 3.1 - Fuzzy relation
%  İ. H. ALTAŞ, KTÜ,  April 1999
clear
HA=[50 0.0;55 0.2;60 0.4;65 0.6;70 0.8;75 1.0];
HJ=[75 0.0;80 0.2;85 0.4;90 0.6;95 0.8;100 1.0];
ha=HA(:,2); hj=HJ(:,2);  TA=HA(:,1); TJ=HJ(:,1);
MM=size(HA); NN=size(HJ);
for m=1:MM(1)
    for n=1:NN(1)
        R(m,n)=min(ha(m),hj(n));
    end
end
plot(TA,ha,TA,ha,'o');
xlabel('Ambiant temperature (Celcius)')
ylabel('Membership degrees');grid
disp('Hit <enter> to continue')
pause
plot(TJ,hj,TJ,hj,'o');
xlabel('Junction temperature (Celcius)'),
ylabel('Membership degrees');grid
disp('Hit <enter> to continue')
pause
mesh(R)
```

Figure E3.1(b) MATLAB .m file code for Example 3.1

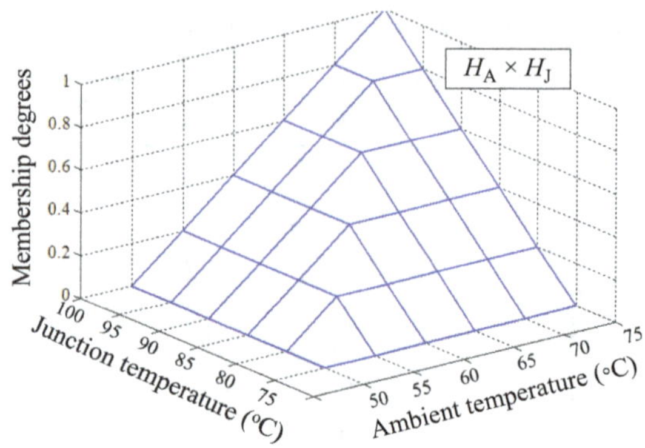

Figure E3.1(c) The relation matrix R as a three-dimensional surface

The fuzzy relation R in Example 3.1 is a relation between two fuzzy subsets. The relation may also exist among more than two fuzzy subsets. A relation between two fuzzy subsets is called as a binary fuzzy relation. If the relation is among three fuzzy subsets, then it is called a ternary fuzzy relation. The names of the relations that happen between two or more fuzzy subsets are listed as

- Binary relation occurs between two fuzzy subsets
- Ternary relation occurs among three fuzzy subsets
- Quaternary relation occurs among four fuzzy subsets
- Quinary relation occurs among five fuzzy subsets
- n-ary relation occurs among n-fuzzy subsets

This book focuses on binary fuzzy relations. Therefore, no more information is going to be given about higher-order relations.

Example 3.2

A description of oscillatory transients in electrical power systems was given in Example 2.18 in Chapter 2. The oscillatory transients are categorized into three subclasses low-, medium-, and high-frequency oscillations [3]. These frequency ranges correspond to common types of power system oscillatory transient phenomena in terms of frequency, duration, and voltage magnitude and are represented by fuzzy subsets in their related universes as depicted in Figures E3.2(a), E3.2(b), and E3.2(c) using triangular-type fuzzy membership functions.

It can be concluded from [3] that the oscillation transients of the voltage magnitude are

(i) Low (LV) and long duration (LD) at low frequencies (LF).
(ii) High (HV) and medium duration (MD) at medium frequencies (MF)
(iii) Low (LV) and short duration (SD) at high frequencies (HF)

where the terms LV, HV, LD, MD, SD, LF, MF, and HF are fuzzy subsets and are given in Figures E3.2(a), E3.2(b), and E3.2(c).

For the universe of frequency (kHz):
Low(LF)=triangle(5,5,500)
Medium(MF)=triangle(5,500,5000)
High(HF)=triangle(500,5000,5000)

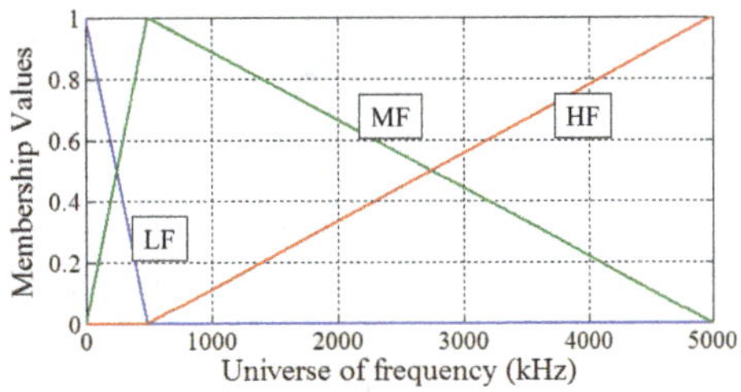

Figure E3.2(a) *Frequency categorization for oscillatory transients in power systems*

For the Universe of duration (ms):
Short(SD)=triangle(0.005,0.02,50)
Medium(MD)=triangle(0.02,50,300)
Long(LD)=triangle(50,300,300)

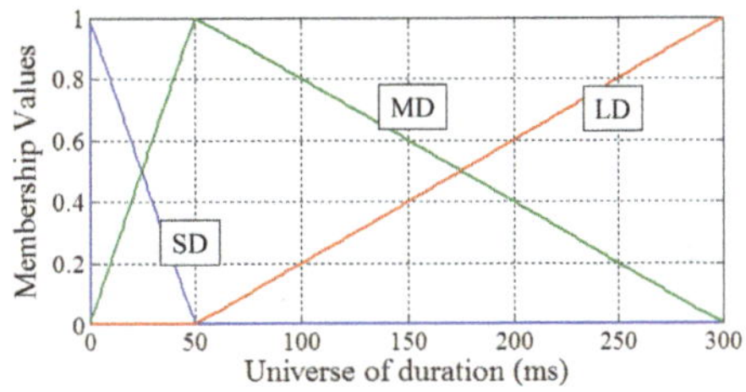

Figure E3.2(b) *Subcategories of oscillatory transient durations in power systems*

For the universe of voltage (pu):
Low(LV)=triangle(0,0,4)
High(HV)=triangle(0,4,4)

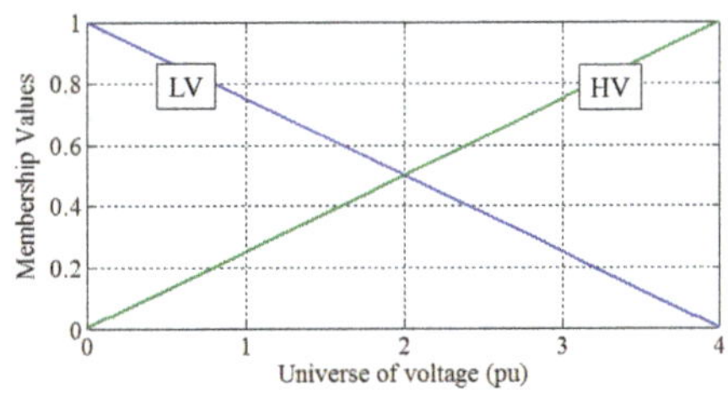

Figure E3.2(c) *Subcategories of oscillatory voltage magnitude in power systems*

(a) Relate the voltage magnitude with oscillation duration at low frequencies.
(b) Relate the voltage magnitude with oscillation duration at medium frequencies.
(c) Relate the voltage magnitude with oscillation duration at high frequencies.

Answer 3.2

(a) The relation between the voltage magnitude and oscillation duration at low frequencies can be expressed as

$$R_1 = \mathrm{LV} \times \mathrm{LD}$$

$$\mu_{R1}(V, D) = \int_{V \times D} \frac{\min(\mu_{\mathrm{LV}}, \mu_{\mathrm{LD}})}{\mathrm{LV}, \mathrm{LD}}$$

where LV and LD stand for low voltage and long duration, respectively. V represents the universe of voltage and D represents the universe of duration. LV and LD are fuzzy subsets defined in the universes of V and D, respectively. The relation surface of $R_1 = LV \times LD$ is shown in Figure E3.2(d), which is obtained using the MATLAB code given in Figure E3.2(e). Instead of a surface plot, the relation could have been shown by its numerical data using the matrix format. However, in order to see the relation surface with higher resolution, the number of numerical data is kept high, which is too much to but in a numerical matrix format. Therefore, the relation surface and the MATLAB code used for plotting are given instead.

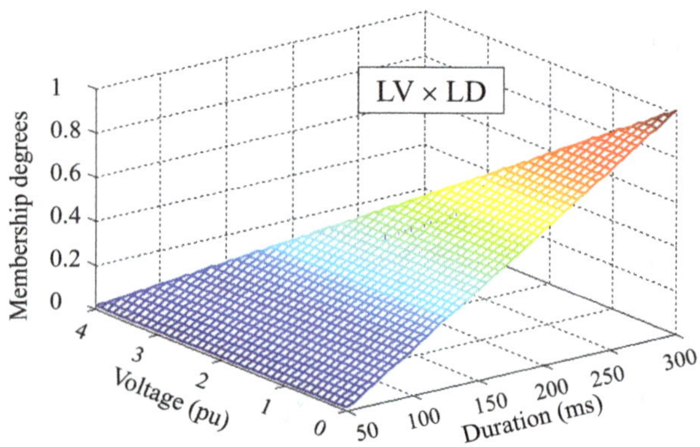

Figure E3.2(d) The relation between low voltage magnitude and long duration

```
clear; clc;
k1=1; k2=1;
for x=0:0.1:4
    LV(k1)=triangle01m(0,0,4,x);
    X1(k1)=x;   k1=k1+1;
end
for x=50:5:300
    LD(k2)=triangle01m(50,300,300,x);
    X2(k2)=x;   k2=k2+1;
end
MM=size(LV); NN=size(LD);
for m=1:MM(2)
    for n=1:NN(2)
        R(m,n)=min(LV(m),LD(n));
    end
end
mesh(R)
```

Figure E3.2(e) The MATLAB code for the relation between voltage magnitude and duration

(b) The relation between the voltage magnitude and oscillation duration at medium frequencies can be expressed as

$$R_2 = \mathrm{HV} \times \mathrm{MD}$$

$$\mu_{R1}(V, D) = \int_{V \times D} \frac{\min(\mu_{\mathrm{HV}}, \mu_{\mathrm{MD}})}{\mathrm{HV}, \mathrm{MD}}$$

where *HV* and *MD* stand for high voltage and medium duration, respectively. *V* represents the universe of voltage, and *D* represents the universe of duration. *HV* and *MD* are fuzzy subsets defined in the universes of *V* and *D*, respectively.

The surface plot of the relation $R_2 = \mathrm{HV} \times \mathrm{MD}$ can be obtained by using a MATLAB code similar to the one given in Figure E3.2(e). The line starting with LV(*k*) in Figure E3.2(e) is replaced by

$$\mathrm{LV}(k1) = \mathrm{triangle01}m(0, 4, 4, x);$$

and the line starting with LD(*k*) is replaced by

$$\mathrm{MD}(k1) = \mathrm{triangle01}m(0, 50, 300, x);$$

The resultant surface plot of the relation $R_2 = \mathrm{HV} \times \mathrm{MD}$ is shown in Figure E3.2(f).

Since the number of the numerical data is small enough as 5×7, we can use the matrix format to show the relation R_2 as in Table E3.2(b).

(c) Since the voltage magnitude is low and the duration is short at high frequencies, the relation between the voltage magnitude and oscillation duration

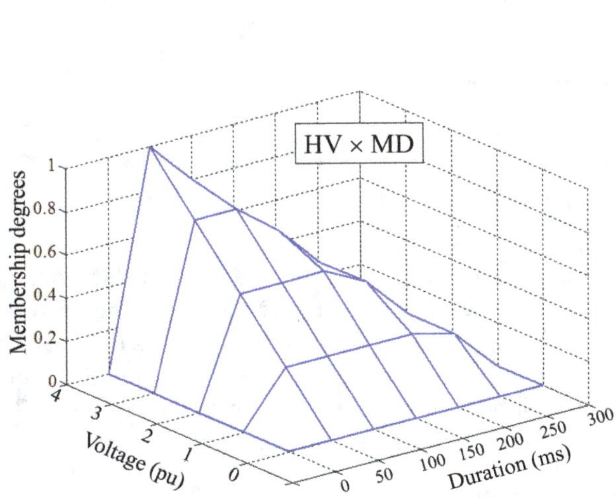

Figure E3.2(f) The relation between high voltage magnitude and medium duration

Table E3.2(b) Membership values of the relation matrix in Example 3.2(b)

		Medium oscillation duration (ms)						
		0	**50**	**100**	**150**	**200**	**250**	**300**
	0	0	0	0	0	0	0	0
	1	0	0.25	0.25	0.25	0.25	0.20	0
High voltage (pu)	**2**	0	0.50	0.50	0.50	0.40	0.20	0
	3	0	0.75	0.75	0.60	0.40	0.20	0
	4	0	1.00	0.80	0.60	0.40	0.20	0

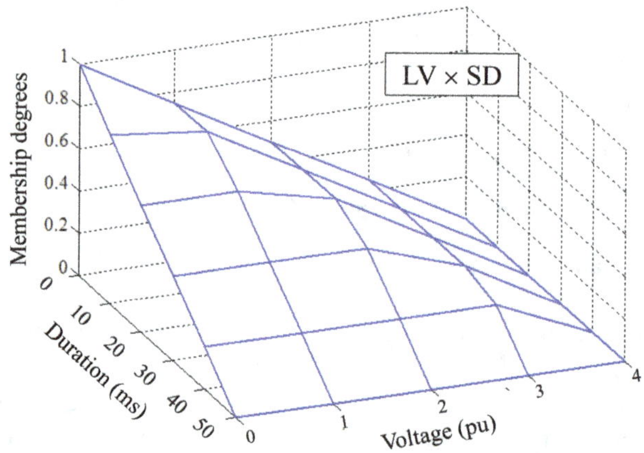

Figure E3.2(g) The relation between low voltage magnitude and short duration

at high frequencies can be expressed as

$$R_3 = \mathrm{LV} \times \mathrm{SD}$$

$$\mu_{R3}(V, D) = \int_{V \times D} \frac{\min(\mu_{\mathrm{LV}}, \mu_{\mathrm{SD}})}{\mathrm{LV}, \mathrm{SD}}$$

where *LV* and *SD* stand for low voltage and short duration, respectively. *V* represents the universe of voltage and *D* represents the universe of duration. *LV* and *SD* are fuzzy subsets defined in the universes of *V* and *D*, respectively.

The surface plot of the relation $R_3 = \mathrm{LV} \times \mathrm{SD}$ can be obtained by using a MATLAB code similar to the one given in Figure E3.2(e). The line starting with the line starting with LD(k) is replaced by

$$\mathrm{SD}(k) = \mathrm{triangle01}m(0, 5, 50, x);$$

The resultant surface plot of the relation $R_3 = \mathrm{LV} \times \mathrm{SD}$ is shown in Figure E3.2(g) and the relation matrix is obtained in Table E3.2(c).

Table E3.2(c) Membership values of the relation matrix in Example 3.2(c)

		Short oscillation duration (ms)					
		0	**10**	**20**	**30**	**40**	**50**
	0	1	0.8	0.6	0.4	0.2	0
	1	0.75	0.75	0.6	0.4	0.2	0
Low voltage (pu)	**2**	0.5	0.5	0.5	0.4	0.2	0
	3	0.25	0.25	0.25	0.25	0.2	0
	4	0	0	0	0	0	0

3.3 Operation with fuzzy relations

Operations done with fuzzy relations are similar to those of fuzzy sets. Since the fuzzy relations are obtained using fuzzy sets, the operators applied to fuzzy sets can be applied to fuzzy relations, as well [4]. The most used operators with fuzzy relations in this book are

- Intersection
- Union
- Negation (Complement)

3.3.1 Intersection of two fuzzy relations

Let R_1 and R_2 be two fuzzy relations defined in universe $X \times Y$. The intersection of two fuzzy relations is also a fuzzy relation and is expressed as

$$R_3 = R_1 \cap R_2 \tag{3.4}$$

In terms of membership functions:

$$\mu_{R_3}(x,y) = \mu_{R_1 \cap R_2}(x,y) = \mu_{R_1}(x,y) \wedge \mu_{R_2}(x,y) = \min\left[\mu_{R_1}(x,y), \mu_{R_2}(x,y)\right] \tag{3.5}$$

where the symbol $\wedge$ stands for intersection and $x \in X$ and $y \in Y$.

3.3.2 Union of two fuzzy relations

The union of two fuzzy relations R_1 and R_2 is written as

$$R_3 = R_1 \cup R_2 \tag{3.6}$$

The union of two fuzzy relations can be obtained using membership functions as

$$\mu_{R_1 V\ R_2}(x,y) = \mu_{R_1}(x,y) \vee \mu_{R_2}(x,y) = \max[\mu_{R_1}(x,y)\ ,\ \mu_{R_2}(x,y)] \tag{3.7}$$

where the symbol $\vee$ is used for union operator.

3.3.3 Negation of a fuzzy relation

The negation or complement of a fuzzy relation is similar to not operator applied in fuzzy sets. Therefore, it can be written as

$$\mu_{\overline{R}}(x, y) = 1 - \mu_R(x, y) \tag{3.8}$$

where $\mu_R(x, y)$ is a fuzzy relation defined on $X \times Y$ such that $x \in X$ and $y \in Y$.

The fuzzy relation operator's intersection, union, and negation are used as they are used in fuzzy sets. However, fuzzy relations are more usable in the representation of verbal expressions.

Example 3.3

In the problem described in Example 3.2, the relation between high voltage and medium duration is obtained and given below as R_1. Similarly, the relation between low voltage and short duration is given as R_2.

$$R_1 = \begin{array}{c} \\ 0 \\ 1 \\ 2 \\ 3 \\ 4 \end{array} \begin{array}{c} \begin{array}{ccccccc} 0 & 50 & 100 & 150 & 200 & 250 & 300 \end{array} \\ \begin{bmatrix} 0 & 0 & 0 & 0 & 0 & 0 & 0 \\ 0 & 0.25 & 0.25 & 0.25 & 0.25 & 0.20 & 0 \\ 0 & 0.50 & 0.50 & 0.50 & 0.40 & 0.20 & 0 \\ 0 & 0.75 & 0.75 & 0.60 & 0.40 & 0.20 & 0 \\ 0 & 1.0 & 0.80 & 0.60 & 0.40 & 0.20 & 0 \end{bmatrix} \end{array}$$

$$R_2 = \begin{array}{c} \\ 0 \\ 1 \\ 2 \\ 3 \\ 4 \end{array} \begin{array}{c} \begin{array}{cccccc} 0 & 10 & 20 & 30 & 40 & 50 \end{array} \\ \begin{bmatrix} 1 & 0.8 & 0.6 & 0.4 & 0.2 & 0 \\ 0.75 & 0.75 & 0.6 & 0.4 & 0.2 & 0 \\ 0.5 & 0.5 & 0.5 & 0.4 & 0.2 & 0 \\ 0.25 & 0.25 & 0.25 & 0.25 & 0.2 & 0 \\ 0 & 0 & 0 & 0 & 0 & 0 \end{bmatrix} \end{array}$$

(a) Obtain the intersection of the relations R_1 and R_2.
(b) Obtain the union of the relations R_1 and R_2.
(c) Obtain the negations of R_1 and R_2.

Answer 3.3

(a) The intersection of the relations R_1 and R_2.
Using (3.4) or (3.5) as

$$\mu_{R_3}(V, D) = \min\left[\mu_{R_1}(V, D), \mu_{R_2}(V, D)\right]$$

$$R_3 = \begin{bmatrix} 0 & 0 & 0 & 0 & 0 & 0 & 0 \\ 0 & 0.25 & 0.25 & 0.25 & 0.25 & 0.20 & 0 \\ 0 & 0.50 & 0.50 & 0.50 & 0.40 & 0.20 & 0 \\ 0 & 0.75 & 0.75 & 0.60 & 0.40 & 0.20 & 0 \\ 0 & 1.0 & 0.80 & 0.60 & 0.40 & 0.20 & 0 \end{bmatrix} \cap \begin{bmatrix} 1 & 0.8 & 0.6 & 0.4 & 0.2 & 0 \\ 0.75 & 0.75 & 0.6 & 0.4 & 0.2 & 0 \\ 0.5 & 0.5 & 0.5 & 0.4 & 0.2 & 0 \\ 0.25 & 0.25 & 0.25 & 0.25 & 0.2 & 0 \\ 0 & 0 & 0 & 0 & 0 & 0 \end{bmatrix}$$

$$R_3 = R_1 \cap R_2 = \begin{bmatrix} 0 & 0 & 0 & 0 & 0 & 0 & 0 \\ 0 & 0.25 & 0.25 & 0.25 & 0.20 & 0 & 0 \\ 0 & 0.50 & 0.50 & 0.40 & 0.20 & 0 & 0 \\ 0 & 0.25 & 0.25 & 0.25 & 0.20 & 0 & 0 \\ 0 & 0 & 0 & 0 & 0 & 0 & 0 \end{bmatrix}$$

(b) The union of the relations R_1 and R_2.
Using (3.6) or (3.7) as

$$\mu_{R_3}(V,D) = \max\left[\mu_{R_1}(V,D), \mu_{R_2}(V,D)\right]$$

$$R_3 = \begin{bmatrix} 0 & 0 & 0 & 0 & 0 & 0 & 0 \\ 0 & 0.25 & 0.25 & 0.25 & 0.25 & 0.20 & 0 \\ 0 & 0.50 & 0.50 & 0.50 & 0.40 & 0.20 & 0 \\ 0 & 0.75 & 0.75 & 0.60 & 0.40 & 0.20 & 0 \\ 0 & 1.0 & 0.80 & 0.60 & 0.40 & 0.20 & 0 \end{bmatrix} \cup \begin{bmatrix} 1 & 0.8 & 0.6 & 0.4 & 0.2 & 0 \\ 0.75 & 0.75 & 0.6 & 0.4 & 0.2 & 0 \\ 0.5 & 0.5 & 0.5 & 0.4 & 0.2 & 0 \\ 0.25 & 0.25 & 0.25 & 0.25 & 0.2 & 0 \\ 0 & 0 & 0 & 0 & 0 & 0 \end{bmatrix}$$

$$R_3 = R_1 \cup R_2 = \begin{bmatrix} 1 & 0.80 & 0.60 & 0.40 & 0.20 & 0 & 0 \\ 0.75 & 0.75 & 0.60 & 0.4 & 0.25 & 0.20 & 0 \\ 0.50 & 0.50 & 0.50 & 0.50 & 0.40 & 0.20 & 0 \\ 0.25 & 0.75 & 0.75 & 0.60 & 0.40 & 0.20 & 0 \\ 0 & 1.0 & 0.80 & 0.60 & 0.40 & 0.20 & 0 \end{bmatrix}$$

(c) The negation of R_1
Using (3.8) as

$$\mu_{\overline{R1}}(V,D) = 1 - \mu_{R1}(V,D)$$

$$\mu_{\overline{R1}}(V,D) = \begin{bmatrix} 1 & 1 & 1 & 1 & 1 & 1 & 1 \\ 1 & 0.75 & 0.75 & 0.75 & 0.75 & 0.80 & 1 \\ 1 & 0.50 & 0.50 & 0.50 & 0.60 & 0.80 & 1 \\ 1 & 0.25 & 0.25 & 0.40 & 0.60 & 0.80 & 1 \\ 1 & 0.0 & 0.20 & 0.40 & 0.60 & 0.80 & 1 \end{bmatrix}$$

The negation of R_2.

$$\mu_{\overline{R2}}(V,D) = \begin{bmatrix} 0.0 & 0.2 & 0.4 & 0.6 & 0.8 & 1 \\ 0.25 & 0.25 & 0.4 & 0.6 & 0.8 & 1 \\ 0.5 & 0.5 & 0.5 & 0.6 & 0.8 & 1 \\ 0.75 & 0.75 & 0.75 & 0.75 & 0.8 & 1 \\ 1 & 1 & 1 & 1 & 1 & 1 \end{bmatrix}$$

3.3.4 Inverse of a fuzzy relation

The inverse of a fuzzy relation is represented by R^{-1} (x, y) or $\mu_{R^{-1}}(x,y)$ as a membership function.

$$R^{-1}(x,y) = \mu_{R^{-1}}(x,y) \tag{3.9}$$

The inverse of a fuzzy relation R is also equal to its transpose and is defined as

$$\mu_{R^{-1}}(x,y) = \mu_{R^T}(x,y) \tag{3.10}$$

where $\mu_R(x, y)$ is a fuzzy relation defined on $X \times Y$ such that $x \in X$ and $y \in Y$. It can also be shown that

$$\left(R^{-1}(x,y)\right)^{-1} = R(x,y) \tag{3.11}$$

Example 3.4

Let $X = \{x, y, z\}$ and $Y = \{a, b\}$ be two universes of discourse and $R(X, Y)$ be a fuzzy relation defined in $X \times Y$ as

$$\mu_R(x,y) = \begin{array}{c} \\ x \\ y \\ z \end{array} \begin{array}{c} \begin{array}{cc} a & b \end{array} \\ \begin{bmatrix} 0.2 & 0.1 \\ 0.3 & 0.5 \\ 0.8 & 1.0 \end{bmatrix} \end{array}$$

Obtain the inverse of the fuzzy relation R.

Answer 3.4

$$\mu_{R^{-1}}(x,y) = \mu_{R^T}(x,y) = \begin{array}{c} \\ a \\ b \end{array} \begin{array}{c} \begin{array}{ccc} x & y & z \end{array} \\ \begin{bmatrix} 0.2 & 0.3 & 0.8 \\ 0.1 & 0.5 & 1.0 \end{bmatrix} \end{array}$$

3.3.5 *Composition of fuzzy relations*

Let $A(X, Y)$ be a crisp relation from X to Y defined in $X \times Y$ and $B(Y, Z)$ another crisp relation from Y to Z defined in $Y \times Z$. Both of these relations use the universe Y as a common universe to transfer the relations A and B onto $X \times Z$. The process of obtaining a third relation from two existing ones sharing a common universe is called *composition* and is defined by

$$C(X,Z) = A(X,Y) \circ B(Y,Z) \tag{3.12}$$

where C is a crisp relation from X to Z. Note that the composition operator generates a third relation from two existing relations, which have a common universe. That means that a common universe must exist for the composition.

The composition operator for binary relations has the following properties:

$$A \circ B \neq B \circ A \tag{3.13}$$

$$(A \circ B)^{-1} = B^{-1} \circ A \circ (B \circ C) \tag{3.14}$$

$$(A \circ B) \circ C = A \circ (B \circ C) \tag{3.15}$$

Recall that various forms of fuzzy intersection and union operators are driven from generalized crisp intersection and union. Similarly, the fuzzy composition is also defined by generalizing the crisp composition. As it is done for the fuzzy intersection and union the composition operator of binary relations can be realized by the membership matrices.

Let us define A, B, and C as three fuzzy relations such that the fuzzy relation A is defined in $X \times Y$ and has a membership matrix M_A

$$M_A = [a_{\text{ik}}] \tag{3.16}$$

where a_{ik} is the membership values of the membership matrix. Similarly, let us define fuzzy relation B in the universe $Y \times Z$. Then the membership matrix M_B can be defined as

$$M_B = [b_{\text{kj}}] \tag{3.17}$$

Then let the fuzzy relation C be a composition of A and B as $C = AB$. The membership matrix for C is then written as

$$M_C = [c_{\text{ij}}] \tag{3.18}$$

The composition $C = AB$ can also be written in terms of membership matrices

$$[c_{\text{ij}}] = [a_{\text{ik}}] \circ [b_{\text{kj}}] \tag{3.19}$$

where

$$c_{\text{ij}} = \max \min(a_{\text{ik}}, b_{\text{kj}}) \tag{3.20}$$

The membership values of the composition matrix are calculated in a similar way as in matrix multiplication. The only difference comes from multiplication and summation, which are replaced by min and max operators, respectively. Due to max and min operators, the composition is called *max–min composition* and is defined as in (4.21).

$$\mu_C(x, z) = \mu_{A \circ B}(x, z) = \max[\min(\mu_A(x, y), \mu_B(y, z))] \tag{3.21}$$

Sometimes the min operator is replaced by arithmetic product and the composition is called *max–product composition*, which is defined as

$$C(X, Z) = A(X, Y) \bullet B(Y, Z) \tag{3.22}$$

$$\mu_C(x, z) = \mu_{A \bullet B}(x, z) = \max[\mu_A(x, y) \bullet \mu_B(y, z)] \tag{3.23}$$

Example 3.5

$X = \{x, y, z\}$, $Y = \{a, b\}$, and $Z = \{\alpha, \beta\}$ are three universes and fuzzy relations $A(X, Y)$ and $B(Y, Z)$ are defined as

$$\mu_A(x,y) = \begin{array}{c} \\ x \\ y \\ z \end{array}\begin{array}{c} \begin{array}{cc} a & b \end{array} \\ \begin{bmatrix} 0.2 & 0.1 \\ 0.3 & 0.5 \\ 0.8 & 1.0 \end{bmatrix} \end{array}, \quad \mu_B(y,z) = \begin{array}{c} \\ a \\ b \end{array}\begin{array}{c} \begin{array}{cc} \alpha & \beta \end{array} \\ \begin{bmatrix} 0.2 & 0.7 \\ 0.1 & 0.8 \end{bmatrix} \end{array}$$

(a) Obtain the fuzzy relation *C*(*X*, *Z*) using *max–min* composition.
(b) Obtain the fuzzy relation *C*(*X*, *Z*) using *max–product* composition.

Answer 3.5

(a) The fuzzy relation *C*(*X*, *Z*) using *max–min* composition.

$$\mu_C(x,z) = \begin{bmatrix} 0.2 & 0.1 \\ 0.3 & 0.5 \\ 0.8 & 1.0 \end{bmatrix} \circ \begin{bmatrix} 0.2 & 0.7 \\ 0.1 & 0.8 \end{bmatrix} = \begin{bmatrix} c_{11} & c_{12} \\ c_{21} & c_{22} \\ c_{31} & c_{32} \end{bmatrix}$$

$$c_{22} = \max[\min(0.3, 0.7), \ \min(0.5, 0.8)] = \max[(0.3), (0.5)] = 0.5$$

all the elements of the composition matrix *C* are obtained in the same way using *max–min composition* and given below.

$$\mu_C(x,z) = \begin{bmatrix} 0.2 & 0.1 \\ 0.3 & 0.5 \\ 0.8 & 1.0 \end{bmatrix} \circ \begin{bmatrix} 0.2 & 0.7 \\ 0.1 & 0.8 \end{bmatrix} = \begin{bmatrix} 0.2 & 0.2 \\ 0.2 & 0.5 \\ 0.2 & 0.8 \end{bmatrix}$$

(b) The fuzzy relation *C*(*X*, *Z*) using *max–product* composition.

$$\mu_C(x,z) = \begin{bmatrix} 0.2 & 0.1 \\ 0.3 & 0.5 \\ 0.8 & 1.0 \end{bmatrix} \bullet \begin{bmatrix} 0.2 & 0.7 \\ 0.1 & 0.8 \end{bmatrix} = \begin{bmatrix} c_{11} & c_{12} \\ c_{21} & c_{22} \\ c_{31} & c_{32} \end{bmatrix}$$

$$c_{22} = \max[(0.3 \bullet 0.7), (0.5 \bullet 0.8)] = \max[(0.21), (0.4)] = 0.4$$

all the elements of the composition matrix *C* are obtained in the same way using *max–product composition* and given below.

$$\mu_C(x,z) = \begin{bmatrix} 0.2 & 0.1 \\ 0.3 & 0.5 \\ 0.8 & 1.0 \end{bmatrix} \bullet \begin{bmatrix} 0.2 & 0.7 \\ 0.1 & 0.8 \end{bmatrix} = \begin{bmatrix} 0.04 & 0.14 \\ 0.06 & 0.4 \\ 0.16 & 0.8 \end{bmatrix}$$

A MATLAB code is given in Figure E3.5(a) for max–min and max–product compositions.

```
% Example 3.5a max-min and max-product composition
clc; clear;
R1=[ 0.2     0.1
     0.3     0.5
     0.8     1.0  ];
R2=[0.2      0.7
    0.1      0.8   ];

row1=size(R1,1);
column1=size(R1,2);
row2=size(R2,1);
column2=size(R2,2);
if column1~=row2
    fprintf('Error in matrix dimension: column size
of R1 must be equal to row size of R2');
end
%%%%%%%%%%% Max-Min Composition Method %%%%%%%%%%
z=1;
for i=1:row1
    for j=1:column2
        for k=1:row2
            mins(1,z)=min(R1(i,k),R2(k,j));
            z=z+1;
        end
        R3(i,j)=max(mins(1,:));
        z=1;
    end
end
fprintf('Max-Min Composition Result ');
R3
%%%%%%%%%%% Max-Product Composition Method %%%%%%
z=1;
for i=1:row1
    for j=1:column2
        for k=1:row2
            products(1,z)=R1(i,k)*R2(k,j);
            z=z+1;
        end
        R3(i,j)=max(products(1,:));
        z=1;
    end
end
fprintf('Max-Product Composition Result ');
R3
```

Figure E3.5 A MATLAB code for max–min and max–product compositions

Example 3.6

The fuzzy relations R_1 and R_2 representing the verbal expressions "*x is greater than y*" and "*y is approximately equal to z*," respectively, are given below. The parameters $x\in X$, $y\in Y$, and $z\in Z$ and the universes X, Y, and Z are given as $X = Y = Z = \{1, 2, 3, 4\}$. Obtain the relation $R_3 = R_1 o R_2$ and comment about the verbal meaning of the relation R_3.

$$\mu_{R1}(x,y) = x\begin{array}{c|c} & y\rightarrow \begin{array}{cccc}1 & 2 & 3 & 4\end{array} \\ \begin{array}{c}1\\2\\3\\4\end{array} & \begin{bmatrix} 0 & 0 & 0 & 0 \\ 0.2 & 0 & 0 & 0 \\ 0.5 & 0.1 & 0 & 0 \\ 0.8 & 0.3 & 0.1 & 0 \end{bmatrix}\end{array} \qquad \mu_{R2}(y,z) = y\begin{array}{c|c} & z\rightarrow \begin{array}{cccc}1 & 2 & 3 & 4\end{array} \\ \begin{array}{c}1\\2\\3\\4\end{array} & \begin{bmatrix} 1 & 0.5 & 0 & 0 \\ 0.5 & 1 & 0.5 & 0 \\ 0 & 0.5 & 1 & 0.5 \\ 0 & 0 & 0.5 & 1 \end{bmatrix}\end{array}$$

Answer 3.6

The relation $R_3 = R_1R_2$ can be obtained using (3.20) and (3.21).

$$\mu_{R3}(x,z) = \mu_{R1\circ R2}(x,z) = \begin{bmatrix} 0 & 0 & 0 & 0 \\ 0.2 & 0 & 0 & 0 \\ 0.5 & 0.1 & 0 & 0 \\ 0.8 & 0.3 & 0.1 & 0 \end{bmatrix} \circ \begin{bmatrix} 1 & 0.5 & 0 & 0 \\ 0.5 & 1 & 0.5 & 0 \\ 0 & 0.5 & 1 & 0.5 \\ 0 & 0 & 0.5 & 1 \end{bmatrix}$$

$$\mu_{R3}(x,z) = \mu_{R1\circ R2}(x,z) = \begin{bmatrix} 0 & 0 & 0 & 0 \\ 0.2 & 0.2 & 0 & 0 \\ 0.5 & 0.5 & 0.1 & 0 \\ 0.8 & 0.5 & 0.3 & 0.1 \end{bmatrix}$$

where for $x = 4$ and $z = 3$.

Each component in relation R_3 is calculated using the max–min composition. In order to show how it is done, the calculation of $R_3(4, 3)$ is given below.

$$\begin{aligned} \mu_{R3}(4,3) &= \max[\min(0.8,0),\ \min(0.5,0),\ \min(0.3,0.5),\ \min(0.1,1.0)] \\ &= 0.3. \end{aligned}$$

Conclusion

R_1 is a relation from X to Y and represents the expression "*x is greater than y*" and R_2 is a relation from Y to Z and represents the expression "*y is approximately equal to z*"

Then, as a relation from X to Z, R_3 should represent the expression "*x is greater than z.*"

This comment is correct as we can see that the relation matrix R_3 is similar to R_1, which is used for *greater*. x is greater than y, y is equal to z, hence x becomes greater than z.

The MATLAB code given in Figure E3.5 can be used to obtain the composition matrix for Example 3.6, too. Just replace the contents of the relations R_1 and R_2 with the input relation matrices given in this example.

3.3.6 The compositional rule of inference

The fuzzy relation R is defined as a relation between two or more fuzzy sets as

$$R = A \times B = \int_{X\times Y} \frac{\mu_A(x) \cap \mu_B(y)}{x,y} \tag{3.24}$$

$$\mu_R(x,y) = \min(\mu_A(x), \mu_B(y)) \tag{3.25}$$

Equations (3.24) and (3.25) show that the relation R is obtained by using the known fuzzy sets A and B. Now, let us assume that the membership matrix $\mu_R(x, y)$ of relation R and the membership function of one of the fuzzy sets, for example, $\mu_A(x)$ of A, used to generate it are known. Then the question is how the membership function $\mu_B(y)$ other fuzzy set B can be calculated. In order to find the fuzzy set B from known fuzzy relation R and fuzzy set A, Equations (3.26) and (3.27) are defined.

$$B = A \circ R \tag{3.26}$$

$$\mu_B(y) = \mu_{A\circ R}(y) = \max \min(\mu_A(x), \mu_R(x, y)) \tag{3.27}$$

where $x \in X$, $y \in Y$, and R is defined in $X \times Y$. Similarly, if the fuzzy set B and the relation R are known, then the other fuzzy set A can be obtained using (3.28) and (3.29).

$$A = B \circ R \tag{3.28}$$

$$\mu_A(x) = \mu_{B\circ R}(x) = \max \min(\mu_B(y), \mu_R(x, y)) \tag{3.29}$$

This form of *the max–min composition* is called as *the compositional rule of inference*.

Example 3.7

A fuzzy subset *Small* defined in universe $X = \{1, 2, 3, 4\}$ and a fuzzy relation R representing verbal expression "*approximately equal*" are given below.

$$\text{Small} = \frac{1}{1} + \frac{0.6}{2} + \frac{0.2}{3} + \frac{0}{4}, \quad R = \begin{bmatrix} 1 & 0.5 & 0 & 0 \\ 0.5 & 1 & 0.5 & 0 \\ 0 & 0.5 & 1 & 0.5 \\ 0 & 0 & 0.5 & 1 \end{bmatrix}$$

If x is known to be small and approximately equal to y, then y should be small, too. Show that y is also small and approximately equal to x.

Answer 3.7

Since x is known to be small, the fuzzy subset *small* given in question can be used to represent x. The fuzzy relation given in question part is a relation that represents the verbal expression "*approximately equal.*" Therefore, relation R can be used for "*x approximately equal to y*," which means that R is a relation from X to Y. Since X is known, then Y can be obtained using the compositional rule of inference.

$$\mu_B(y) = \mu_{A\circ R}(y) = [1 \quad 0.6 \quad 0.2 \quad 0] \circ \begin{bmatrix} 1 & 0.5 & 0 & 0 \\ 0.5 & 1 & 0.5 & 0 \\ 0 & 0.5 & 1 & 0.5 \\ 0 & 0 & 0.5 & 1 \end{bmatrix}$$

The MATLAB code given in Figure E3.5 can be modified to be used for the compositional rule of inference in this example. Actually replacing the relation matrices with A and R will be sufficient for the modification as given in Figure E3.7.

Calculation of *the compositional rule of inference* manually is given below for a clear understanding of the process. Since the result of the composition of a fuzzy

```
% Example 3.7 - Compositional Rule of Inference
clc; clear;
A=[ 1      0.6     0.2     0.1  ];
R=[ 1      0.5     0       0
    0.5    1       0.5     0
    0      0.5     1       0.5
    0      0       0.5     1       ];
row1=size(A,1);    column1=size(A,2);
row2=size(R,1);     column2=size(R,2);
if column1~=row2
    fprintf('Error: column size of A must be equal to
row size of R');
end
%%%%%%%%%%% Max-Min Composition Method %%%%%%%%%%%
z=1;
for i=1:row1
    for j=1:column2
        for k=1:row2
            mins(1,z)=min(A(i,k),R(k,j));
            z=z+1;
        end
        B(i,j)=max(mins(1,:));
        z=1;
    end
end
fprintf('Max-Min Composition Result ');
B
```

Figure E3.7 MATLAB code for the compositional rule of inference

set with a relation yields another fuzzy set with the same set used in the composition, it will look like the one given below.

$$\mu_B(y) = [b_{11} \quad b_{12} \quad b_{13} \quad b_{14}]$$

The components b_{11}, b_{12}, b_{13}, and b_{14} are obtained as follows:

$$\begin{aligned}
b_{11} &= \max[\min(1,1),\ \min(0.6,0.5),\ \min(0.2,0),\ \min(0,0)] = \max(1,0.5,0,0) = 1\\
b_{12} &= \max[\min(1,0.5),\ \min(0.6,1),\ \min(0.2,0.5),\ \min(0,0)] = \max(0.5,0.6,0.2,0) = 0.6\\
b_{13} &= \max[\min(1,0),\ \min(0.6,0.5),\ \min(0.2,1),\ \min(0,0.5)] = \max(0,0.5,0.2,0) = 0.5\\
b_{14} &= \max[\min(1,0),\ \min(0.6,0),\ \min(0.2,0.5),\ \min(0,1)] = \max(0,0,0.2,0) = 0.2\\
\mu_B(y) &= [1 \quad 0.6 \quad 0.5 \quad 0.2]\\
y &= \text{small} = \frac{1}{1} + \frac{0.6}{2} + \frac{0.5}{3} + \frac{0.2}{4}
\end{aligned}$$

3.4 The extension principle

Fuzzy membership functions are usually considered and analyzed as one-dimensional, in other words, as single input sets. However, in some cases, fuzzy membership functions with two input spaces are required to define the problems. Membership functions having two inputs each coming from different universes can only be represented by a two-dimensional membership function, which is defined in both universes such as the membership function of a fuzzy relation. A one-dimensional fuzzy membership function can be extended to a two-dimensional membership function using the *cylindrical extension principle*. On the other hand, a two-dimensional fuzzy membership function can be separated into two one-dimensional membership functions by an operator called *projection*. In some applications, one-dimensional fuzzy membership functions are mapped onto another space rather than the one that is defined. The mapping of a fuzzy membership function from one universe to another is done through a math function, which is used for the fuzzification of crisp numbers, too.

Let a fuzzy set A be defined in a universe of X. The cylindrical extension of this fuzzy set A in two-dimensional universe $X \times Y$ is obtained as

$$\mathrm{SA}(A) = \int_{X \times Y} \mu_A(x)/(x,y) \tag{3.30}$$

This process, in fact, noting but the extending the fuzzy set A from the universe X along the universe Y. The opposite process of the *extension* is called *projection* and is defined as

$$R_X = \int_X \max_y \mu_R(x,y)/x \tag{3.31}$$

$$R_Y = \int_Y \max_x \mu_R(x,y)/y \tag{3.32}$$

Projection gives the reflection of a two-dimensional fuzzy set on both universes. Equation (3.31) gives *the projection* of $R(X, Y)$ on the universe of X while (3.32) gives it on the universe of Y. The projection reduces the dimension of the membership function. If a two-dimensional fuzzy set or relation can be analytically separated into two one-dimensional fuzzy sets, then it is called composite otherwise noncomposite. If a composite two-dimensional fuzzy set is compared with the projection sets, it can be seen that the projection is a result of the two-dimensional fuzzy set being composite. The fuzzy set operator intersection (AND) and union (OR) can also be used to combine the sets from different universes into one two-dimensional common universe as a two-dimensional set.

Example 3.8

Some information about power systems harmonics is in Example 2.17 in Chapter 2. As stated in Example 2.17, standards are developed to guide power system design engineers to keep the total harmonic distortion below the standard values. IEEE Std 519-2014, "Recommended Practices And Requirements For Harmonic Control In Electrical Power Systems" provides suggested harmonic values for power systems [5] as summarized in Table E2.17, which yields the following statements:

If the voltage is low (LV) then the total harmonic distortion is extra high (EHH).
If the voltage is medium (MV) then the total harmonic distortion is high (HH).

LV(0,0,1)
MV (0,1,69)
HV (1,69,161)
EHV(69,161,161)

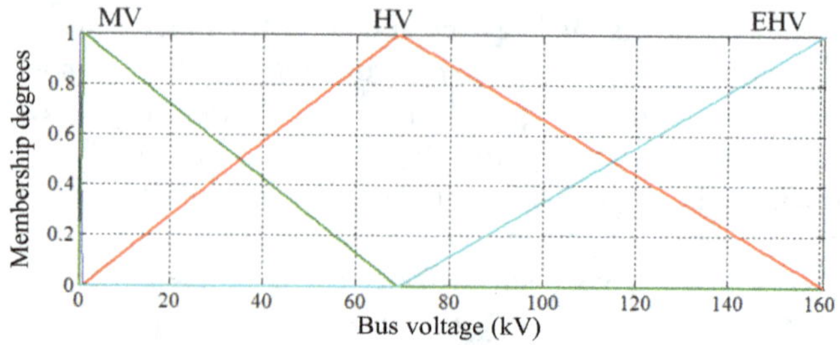

Figure E3.8(a) Partition of power system bus voltages into fuzzy subsets

ELH(0,0,1.5,th)
LH(0,1.5,2.5,th)
MH(1.5,2.5,5,th)
HH(2.5,5,8,th)
EHH(5,8,8,th)

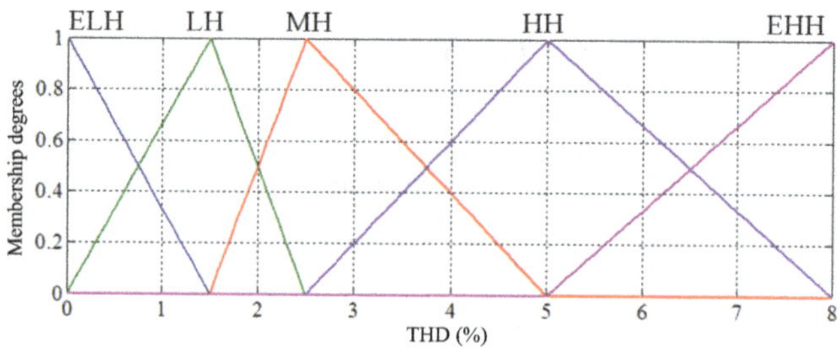

Figure E3.8(b) Partition of total harmonic distortion into fuzzy subsets

If the voltage is high (HV) then the total harmonic distortion is medium (MH).
If the voltage is extra high (EHV) then the total harmonic distortion is low (LH).

where the fuzzy subsets are defined as in Figures E3.8(a) and E3.8(b).

Assume that the statement "If voltage is high (HV) then the total harmonic distortion is medium (MH)" is a relation R = HV × MH. Find the following:

(a) Plot the fuzzy high voltage (HV) in the universe of bus voltage.
(b) Plot the fuzzy medium harmonic (MH) in the universe of total harmonic distortion.
(c) Plot the membership matrix $\mu_{R_1}(V,H) = \min[\mu_{\mathrm{HV}}(v), \mu_{\mathrm{MH}}(h)]$ of relation R

Answer 3.8

(a) The plot of the fuzzy high voltage (HV) in the universe of bus voltage is shown in Figure E3.8(c).
(b) The plot of the fuzzy medium harmonic (MH) in the universe of total harmonic distortion is given in Figure E3.8(d).
(c) The plot of the membership matrix $\mu_{R_1}(V,H) = \min[\mu_{\mathrm{HV}}(v), \mu_{\mathrm{MH}}(h)]$ of relation R is obtained as in Figure E3.8(e).

The MATLAB code given in Figure E3.1(d) can be modified to plot the fuzzy sets and relation HV × MH in this example.

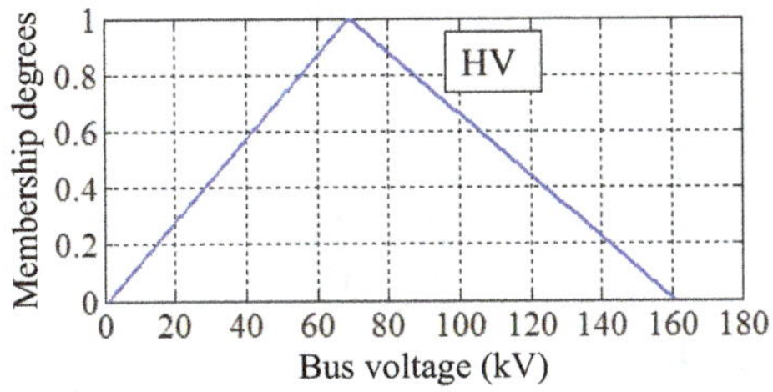

Figure E3.8(c) Fuzzy subset representing fuzzy high voltage (HV)

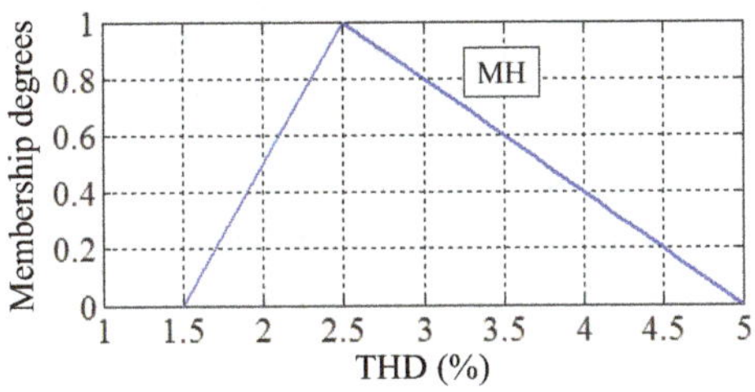

Figure E3.8(d) Fuzzy subset representing fuzzy medium harmonic (MH)

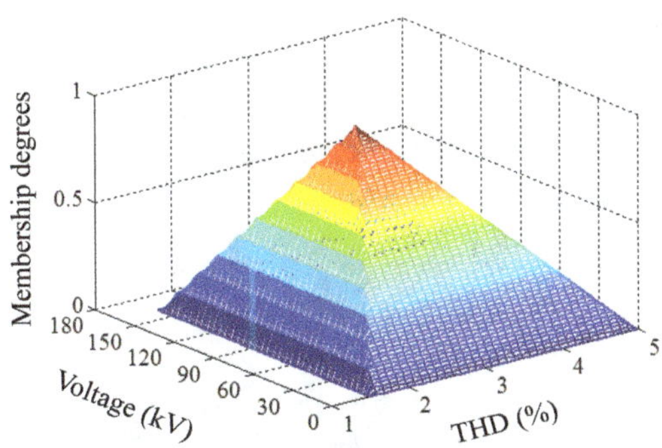

Figure E3.8(e) Fuzzy relation representing HV × MH

Example 3.9

Assume that the fuzzy membership functions HV and MH described in Example E3.8 are represented using Gaussian functions to yield the three-dimensional membership surface $\mu_R(V, H) = \min[\mu_{\text{HV}}(v), \mu_{\text{MH}}(h)]$ as shown in Figure E3.9.

(a) Use the extension principle and extend the fuzzy set HV along the universe of V.
(b) Use the extension principle and extend the fuzzy set MH along the universe of H.

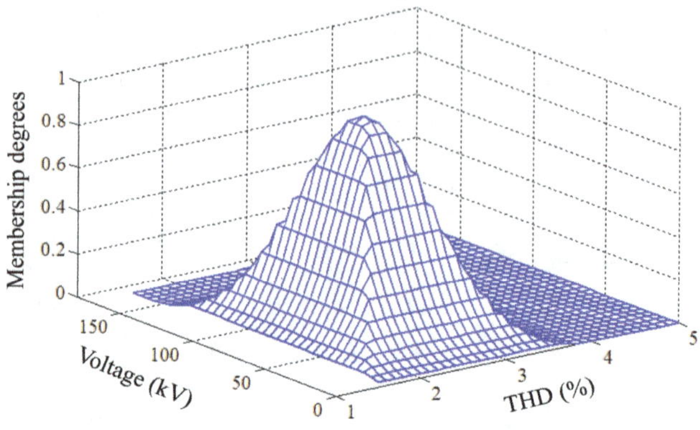

Figure E3.9 Fuzzy relation representing HV × MH

Answer 3.9

(a) Fuzzy membership function representing the high voltage HV in the universe of voltage and its projection from the two-dimensional plot of $\mu_R(V, H)$ are shown in Figures E3.9(a-a) and E3.9(a-b).

(b) Fuzzy membership function representing the medium harmonics MH in the universe of harmonics (H) and its projection from the two-dimensional plot of $\mu_R(V, H)$ are shown in Figures E3.9(b-a) and E3.9(b-b).

Similarly, the statement "If voltage is low (LV) then the total harmonic distortion is extra high (EHH)" is another relation $R_2 = \text{LV} \times \text{EHH}$, and the expression "If voltage is extra high (EHV) then the total harmonic distortion is low (LH)" is also a relation as $R_3 = \text{EHV} \times \text{LH}$. The extension principle can be applied to these

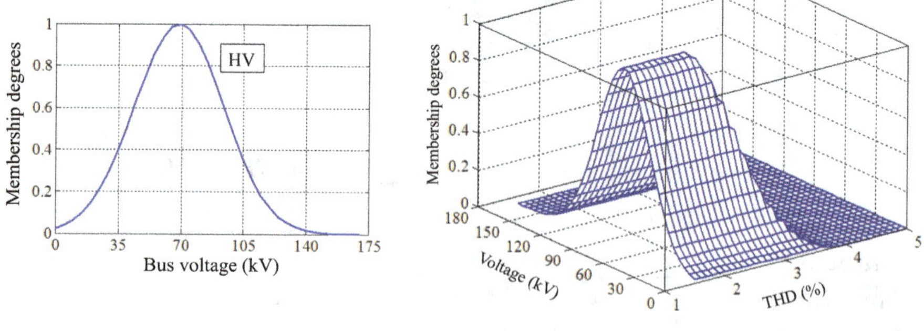

(a) (b)

Figure E3.9(a) Projection of a two-dimensional fuzzy membership function $\mu_R(V, H)$ on the universe of V as a fuzzy set HV. (a) Fuzzy set HV in the universe V and (b) projection of $\mu_R(V, H)$ onto the universe of V.

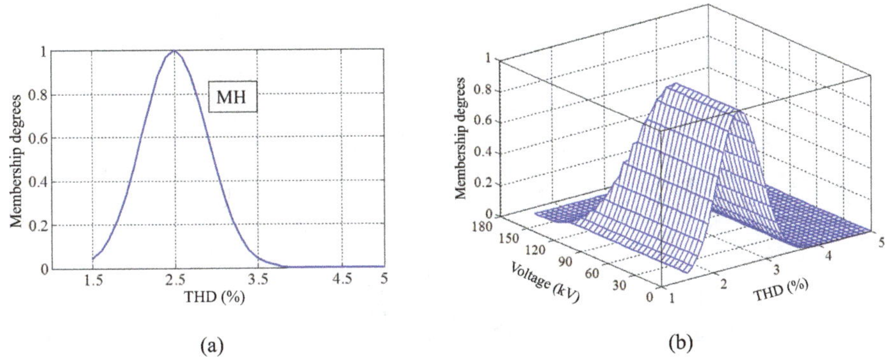

Figure E3.9(b) Projection of a two-dimensional fuzzy membership function $\mu_R(V, H)$ on the universe of H as a fuzzy set MH. (a) Fuzzy set MH in the universe H and (b) projection of $\mu_R(V, H)$ on the universe of H.

statements to obtain the two-dimensional representation of the verbal definitions in graphical formats. The projection process can be applied as well. In Example 3.9, a two-dimensional fuzzy membership function $\mu_R(V, H)$ is separated into two individual fuzzy subsets. Therefore, it is composite.

3.5 Fuzzy mapping

Since Zadeh introduced the concept in 1975 [6–8], the extension principle has become one of the important tools in fuzzy set theory. The extension principle enables the crisp math functions to be carried into a fuzzy environment. Therefore, it is used extensively in fuzzy set theory, especially in engineering for problem definition and solution. A fuzzy subset A defined in the universe of X can be mapped onto the universe of Y as another fuzzy subset B through a math function $f(\cdot)$ using the extension principle.

Let A be a fuzzy subset defined in $X = \{x_1, x_2, \ldots, x_n\}$ and $f(x)$ be the math function to be used in mapping the set A onto universe $Y = \{y_1, y_2, \ldots, y_n\}$ as fuzzy subset B. The fuzzy subset A is defined as in (3.33)

$$A = \frac{\mu_A(x_1)}{x_1} + \frac{\mu_A(x_2)}{x_2} + \cdots + \frac{\mu_A(x_n)}{x_n} \tag{3.33}$$

The mapping of fuzzy set A as a fuzzy set B in the universe of Y is obtained as

$$B = f(A) = \frac{\mu_B(y_1)}{y_1} + \frac{\mu_B(y_2)}{y_2} + \cdots + \frac{\mu_B(y_n)}{y_n} \tag{3.34}$$

where the elements $y_i \in Y$ for $i = 1, 2, \ldots, n$, are obtained as a function of $x_i \in X$ for $i = 1, 2, \ldots, n$ as depicted in (3.35).

$$y_i = f(x_i) \tag{3.35}$$

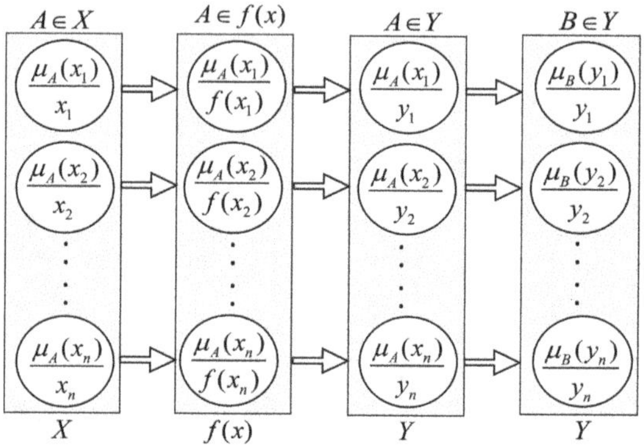

Figure 3.1 One-to-one mapping

The elements x_i of the universe of X are mapped onto the universe of Y as y_i over a math function $f(\cdot)$. According to (3.35), each element x_i may result in an element of y_i. This is called one-to-one mapping. Equations (3.34) and (3.35) show that the membership values $\mu_B(y_i)$ of fuzzy set B defined in the universe of Y are determined by the membership values $\mu_A(x_i)$ of fuzzy set A defined in the universe of X. In one-to-one mapping, there is only one x_i for one y_i, and one membership value of $\mu_B(y_i)$ is obtained from just one membership value of $\mu_A(x_i)$. One-to-one mapping algorithm is shown in Figure 3.1.

Depending on the function used in mapping, two or more elements $x_1, x_2, \ldots, x_n \in X$ may result in the same point $y_i \in Y$, then this is called many-to-one mapping and defined as

$$y_i = f(x_m) = f(x_n), \text{ where } m \neq n \tag{3.36}$$

In many-to-one mapping, there are two or more x_i, where $i = 1, 2, \ldots, n$, resulting in just one y_i, whose membership value, $\mu_B(y_i)$, is obtained as the maximum of the membership values of $\mu_A(x_i)$, where $i = 1, 2, \ldots, n$, as given in (3.37).

$$\mu_B(y) = \max_{i=1,n} \mu_A(x_i) \tag{3.37}$$

Many-to-one mapping algorithm is shown in Figure 3.2 where many elements x in the universe of X yield the same y in the universe of Y. Since there are many membership values for many values of x, the membership value of y must be selected among the membership values of x. According to the definition of the extension principle, the max operator is used for the final decision as depicted in Figure 3.2.

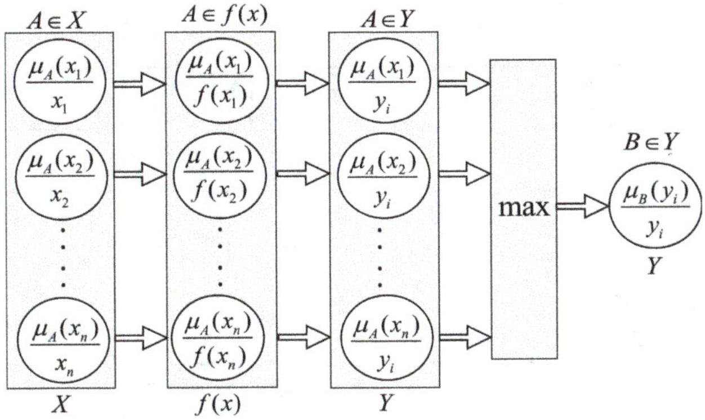

Figure 3.2 Many-to-one mapping

Example 3.10

Recall that current–voltage relationship of a constant resistor is defined by Ohm's law as

$$i_R(t) = \frac{v_R(t)}{R}$$

where there will be an $i_R(t)$ for every $v_R(t)$ indicating a one-to-one mapping. Assume that a 12 Ohm resistor is connected across a voltage source whose output voltage can be adjusted from zero to 12 V. If the source voltage is partitioned into fuzzy subsets Zero (ZV), Medium (MV), and High (HV) in the universe of voltage in volts as shown in Figure E3.10, obtain the fuzzy subsets defined in the universe of current in Amperes.

Answer 3.10

In this example, the membership functions ZV, MV, and HV defined in the universe of voltage are to be mapped onto the universe of current as ZI, MI, and HI using the equation defined by Ohm's law. The calculation of $i_i = f(v_i)$ and then transferring it to $v_i = f(v_i)$ are summarized in Table E3.10 and shown in Figure E3.10(a).

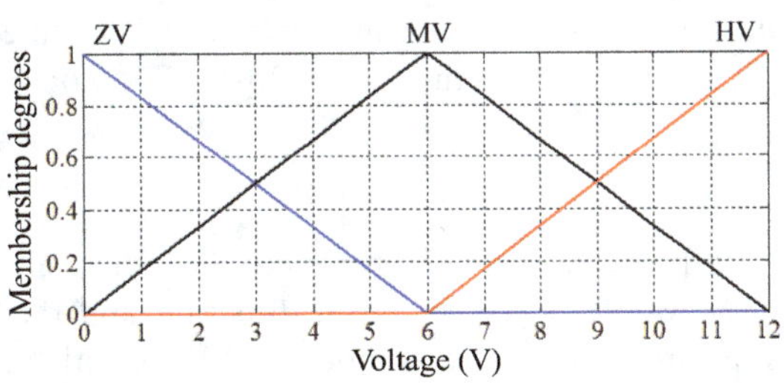

Figure E3.10 Fuzzy partition of a 12 V variable DC voltage source

Table E3.10 Example for one-to-one mapping

$v_i \in V$	$\frac{\mu(v_i)}{v_i}$			$f(v_i) = v_i/R$ $Ri_i = v_i/R$	$\frac{\mu(i_i)}{i_i}$		
	ZV	**MV**	**HV**		**ZI**	**MI**	**HI**
0	1	0	0	0	1	0	0
3	0.5	0.5	0.0	0.25	0.5	0.5	0.0
6	0	1	0	0.5	0	1	0
9	0	0.5	0.5	0. 75	0	0.5	0.5
12	0	0	1	1.0	0	0	1

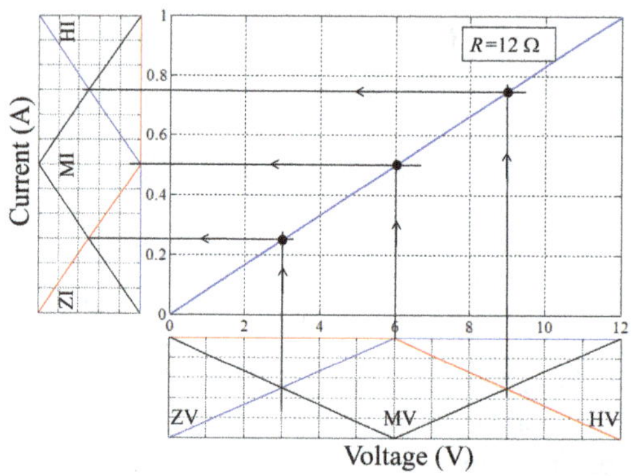

Figure E3.10(a) Graphical representation of one-to-one mapping

Example 3.11

The active power transferred between two buses is expressed as $P_e = P_m \sin \delta$, where $P_m = V_S V_R / X_L$ is the maximum power that can be transmitted between two busses and is usually assumed to be constant. A numerical example of the power transfer between two busses was given in Example 2.19. The partitions of the universes of transmitted power and the power angle into fuzzy subcategories are also given in Example 2.19. The fuzzy subset half (HD) defined in the universe of power angle is shown in Figure E3.11.

Use the formula given as $P_e = 2 \cdot \sin \delta$ where δ is defined in the interval $\{0,\pi\}$ radians or $\{0°, 180°\}$ and map the fuzzy subset HD$\in\delta$ onto a fuzzy subset in the universe of P_e.

Answer 3.11

Figure E3.11 shows both power–power angle curve, Pe $= f(\delta)$, and the membership function half (HD). Since Pe $= f(\delta)$ is a sine function and the fuzzy set HD is represented by a sinusoid membership function and they both have their peaks at 90°, the points over the curves become symmetrical according to the line passing through 90°. As given in Figure E3.11 and in Table E3.11, for five values of δ there

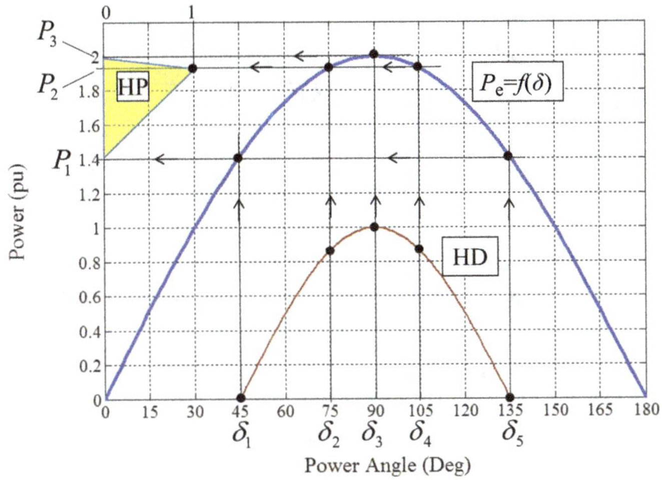

Figure E3.11 Graphical representation of many-to-one mapping

Table E3.11 Example for many-to-one mapping

$\delta_i \in$Deg	$\frac{\mu_{HD}(\delta_i)}{\delta_i}$	$P_{ei} = P_m \sin \delta_i$	P_{ei}	$\frac{\mu(P_{ei})}{\delta_i}$	$\frac{\mu(P_{ei})}{\delta_i}$
45	0.0	1.4	1.4	0.0 ∨ 0.0	0.0
75	0.86	1.9	1.9	0.86 ∨ 0. 86	0.86
90	1.0	2.0	2.0	1.0	1.0
105	0. 86	1.9			
135	0.0	1.4			

are three values of power Pe. Therefore, this is a many-to-one mapping process. The resulting mapped fuzzy membership function is also shown in Figure E3.11 as HP, which stands for high in the universe of active power.

Example 3.12

Photovoltaic array (PVA) characteristics and partitioning of the universes of voltage, current, and power were given in Example 2.20. Use the information from the Example and map the fuzzy subset optimum voltage (OV) from the universe of PVA voltage onto the universe of PVA power. The math function for the mapping is defined by *P–V* curve shown in Figure E2.20a and given below in Figure E3.12. The PV cell operating power as a function of the cell current and voltage can be expressed by the following equation [9].

$$P_C = V_C I_C, \quad \text{where} \quad V_C = K\frac{BT_C}{q}\ln\left(\frac{I_{ph} + I_0 - I_C}{I_0}\right) - R_S I_C$$

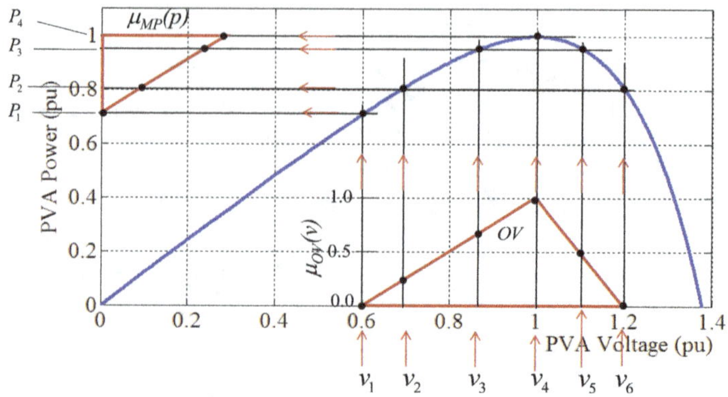

Figure E3.12 Graphical representation of many-to-one mapping

Table E3.12 Example for many-to-one mapping

$v_i \in V$	$\frac{\mu_{OV}(v_i)}{v_i}$	$P_i = V_C I_C$	P_i	$\frac{\mu(P_i)}{v_i}$	$\frac{\mu(P_i)}{P_i}$
0.6	0.0	0.71	0.71	0.0	0.0
0.68	0.2	0.80	0.80	0.2 ∨ 0. 0	0.2
0.88	0.7	0.96	0.96	0.7 ∨ 0.5	0.7
1.0	1.0	1.0	1.0	1.0	1.0
1.1	0.5	0.96			
1.2	0.0	0.80			

Answer 3.12

The *P–V* curve and redefined fuzzy subset OV from Example 2.20 are replotted and shown in Figure E3.12.

3.6 Fuzzy implications

The verbal expressions in fuzzy propositions are represented by fuzzy sets and processed by fuzzy relations. Let (3.38) and (3.39) be fuzzy propositions.

$$\text{if } A \text{ than } B \tag{3.38}$$

$$\text{if } A \text{ then } B \text{ else } C \tag{3.39}$$

where A, B, and C are fuzzy subsets that may be defined in different universes X, Y, and Z, respectively, and they represent some verbal fuzzy variables. The verbal expressions (3.38) and (3.39) can be written in terms of cartesian products as

$$\text{if } A \text{ then } B \text{ else } C = A \times B + \overline{A} \times C \tag{3.40}$$

where the signs + and × represent union (max) and intersection (min) operators, respectively. The propositions given in (3.38) and (3.39) are also considered as fuzzy implications since they have at least one input and one output universe. In order to obtain a generalized form of fuzzy implication, let us define fuzzy subsets $A_1, A_2, \ldots, A_n$ in the universe of X and fuzzy subsets $B_1, B_2, \ldots, B_n$ in the universe of Y and write a series of implications.

$$
\begin{aligned}
&\text{if } A_1 \text{ then } B_1\\
&\qquad \text{else if } A_2 \text{ then } B_2\\
&\qquad \text{else if } A_3 \text{ then } B_3\\
&\qquad \cdot\\
&\qquad \cdot\\
&\qquad \text{else if } A_{n-1} \text{ then } B_{n-1}\\
&\text{else}\\
&\qquad A_n \text{ then } B_n
\end{aligned}
\tag{3.41}
$$

These implications can be written in a compact form as cartesian products

$$R_T = A_1 \times B_1 + A_2 \times B_2 + \cdots + A_n \times B_n \tag{3.42}$$

where each implication is a cartesian product and each cartesian product is a fuzzy relation. Now, let us write each implication as a relation

$$
\begin{aligned}
R_1 &= A_1 \times B_1 &&\Rightarrow \quad \mu_{R1}(x,y) = \min(\mu_{A1}(x), \mu_{B1}(y))\\
R_2 &= A_2 \times B_2 &&\Rightarrow \quad \mu_{R2}(x,y) = \min(\mu_{A2}(x), \mu_{B2}(y))\\
&\cdot \quad \cdot && \qquad\qquad \cdot\\
&\cdot \quad \cdot && \qquad\qquad \cdot\\
R_n &= A_n \times B_n &&\Rightarrow \quad \mu_{Rn}(x,y) = \min(\mu_{An}(x), \mu_{Bn}(y))
\end{aligned}
\tag{3.43}
$$

Since the term *else* used to connect the implications in (3.41) has a similar meaning with *or*, Then the total relation becomes the union of all relations and is written as in (3.44) and (3.45).

$$R_T = R_1 \cup R_2 \cup \cdots \cup R_n \tag{3.44}$$

$$\mu_{RT}(x,y) = \max(R_1, R_2, \ldots, R_n) \tag{3.45}$$

Then (3.45) can be written as a max–min process as in *the compositional rule of inference*.

$$\mu_{RT}(x,y) = \max[\min(\mu_{A1}, \mu_{B1}), \min(\mu_{A2}, \mu_{B2}), \ldots, \min(\mu_{An}, \mu_{Bn})] \tag{3.46}$$

The implications or the relations given in (3.39), (3.40), and (3.41) are all single input–single output. There is an input universe X and an output universe Y. The verbal expressions with multiple input and multiple output spaces can also be represented by implications with multiple input/output (I/O) spaces. For example, let us write the implication in (3.47) as combination of the cartesian products as in (3.48).

$$\text{If } A \text{ then (if } B \text{ then } C \text{ else } D\text{) else } E \tag{3.47}$$

$$A \times B \times C + A \times \overline{B} \times D + \overline{A} \times E \tag{3.48}$$

Both (3.47) and (3.48) have the same meaning and are meant to be the same. There are two fuzzy sets A and B as inputs from different input universes and three output sets C, D, and E defined in three different output universes. The verbal expressions used (3.42) and (3.47) appear in many fuzzy logic applications. Especially the expert's knowledge is stored and processed as fuzzy rules in fuzzy logic control applications. The fuzzy rules are represented by verbal expressions similar to the ones given above.

Example 3.13

In order to supply electrical power to large urban areas, usually three types of power stations are used for efficient energy management. The power stations with a generating capacity of 5 MW and below are called small (S) power stations. The stations with generating capacity of up to 50 MW are considered as medium (M) size power stations and stations with generating capacity of 100 MW and above are classified as big (B) power stations. This classification of power stations is represented by fuzzy subsets as shown in Figure E3.13(a).

The management of the power stations described above is done according to power demand by loads. Such that, the big power stations are always kept in service. Depending on power demand, medium- and small-sized power stations are put into service. Whenever there is a decrement in power demand, some units in medium or small size power stations are turned off. During the increment in power demand, the units are turned on. Based on these explanations, the following single input–single output implications are developed.

R1: *If* increment in demand is *low* then increment in generation is *low*.
R2: *If* increment in demand is *medium* then increment in generation is *medium*.
R3: *If* decrement in demand is *low* then decrement in generation is *low*.
R4: *If* decrement in demand is *medium* then decrement in generation is *medium*
R5: *If* change in demand is *zero* then change in generation is *zero*.

where the words *low*, *medium*, and *zero* are defined in the universes of change in power demand and change in power generation representing both increment and decrement cases as shown in Figure E3.13(b).

Obtain the consequences of these implications.

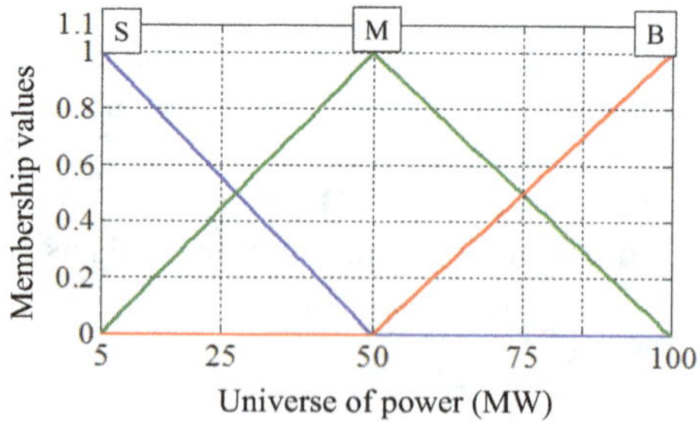

Figure E3.13(a) Fuzzy partition of power stations

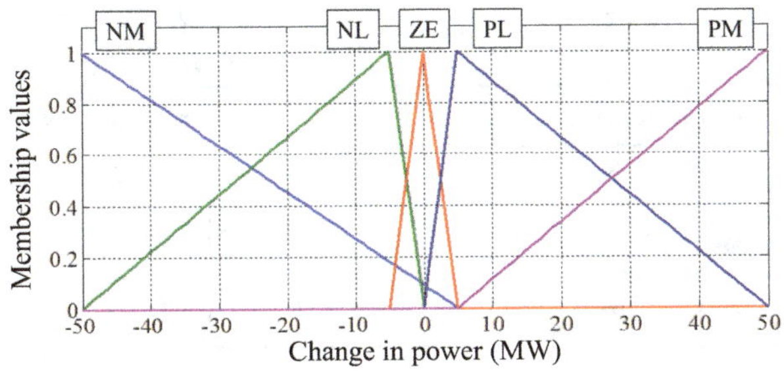

Figure E3.13(b) *Fuzzy partition of the changes in power demand and generation*

Answer 3.13

Let us represent the increments as positive (P) and decrement as negative (N). The letter D can be used with the fuzzy subsets related to demand and the letter G can be used with the fuzzy subsets related to the generation in order to distinguish the fuzzy subsets with the same name. For example positive low in the universe of increment in power demand is represented as PLD and positive low in the universe of increment in power generation is represented as PLG. Then the first implication becomes,

R1: *If* decrement in demand is *medium* (NMD) then decrement in generation is *medium* (NMG). This implication can be abbreviated as *if* PLD and *then* PLG.

Similarly, all implications can be written as

if NMD *then* NMG
If NLD *then* NLG
If ZED *then* ZEG
If PLD *then* PLG
If PMD *then* PMG

These implications can be written as fuzzy relations, which are the consequences. Using the definitions of the fuzzy subsets given in Figure E3.13(b), the following fuzzy relations can be obtained as the consequences of the fuzzy implications expressed in the example.

$$
\begin{aligned}
R_1 &= NMD \times NMG &\Rightarrow\quad \mu_{\mathrm{R1}}(\mathrm{d,g}) &= \min(\mu_{\mathrm{NMD}}(d), \mu_{\mathrm{NMG}}(\mathrm{g})) \\
R_2 &= NLD \times NLG &\Rightarrow\quad \mu_{\mathrm{R2}}(\mathrm{d,g}) &= \min(\mu_{\mathrm{NLD}}(\mathrm{d}), \mu_{\mathrm{NLG}}(\mathrm{g})) \\
R_3 &= ZED \times ZEG &\Rightarrow\quad \mu_{\mathrm{R3}}(\mathrm{d,g}) &= \min(\mu_{\mathrm{ZED}}(\mathrm{d}), \mu_{\mathrm{ZEG}}(\mathrm{g})) \\
R_4 &= PLD \times PLG &\Rightarrow\quad \mu_{\mathrm{R4}}(\mathrm{d,g}) &= \min(\mu_{\mathrm{PLD}}(\mathrm{d}), \mu_{\mathrm{PLG}}(\mathrm{g})) \\
R_5 &= PMD \times PMG &\Rightarrow\quad \mu_{\mathrm{R5}}(\mathrm{d,g}) &= \min(\mu_{\mathrm{PMD}}(\mathrm{d}), \mu_{\mathrm{PMG}}(\mathrm{g}))
\end{aligned}
$$

Since the same fuzzy subsets are used to represent the changes in both power demand and generation, the relation between the same fuzzy subsets yields this set and the implications take form as in Table E3.13. The membership functions of the resultant fuzzy relations are given in Figure E3.13(c). Note that the membership functions given in Figure E3.13(c) are the same as those shown in Figure E3.13(b).

Table E3.13 The relations in Example 3.13

The implication		Resultant rule
If NM *then* NM	:	R1 = NM
If NL *then* NL	:	R2 = NL
If ZE *then* ZE	:	R3 = ZE
If PL *then* PL	:	R4 = PL
If PM *then* PM	:	R5 = PM

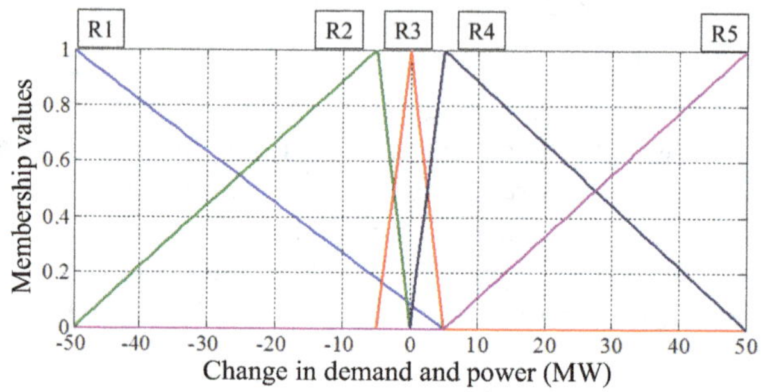

Figure E3.13(c) The membership functions of the resultant fuzzy relations

Example 3.14

The power generated by PV solar cells changes as a function of solar irradiation level and operating temperature. So, the generated power increases as the solar irradiation level increases, but decreases as the operating temperature increases. Some information about PV solar cells were discussed and a partitioning of the universes of temperature and solar irradiation levels were given in Examples 2.13 and 2.20 in Chapter 2. Some knowledge about the effect of junction temperature, *I–V*, and *P–V* characteristics was given in Example 3.1.

From information about the effect of ambient temperature and solar irradiation, the following implications can be written.

If solar irradiation is *good* then the power is *high* else the power is *low*.

where the fuzzy subset *good* (G) in the universe of solar irradiation level is defined as in Figure E3.14(a) and fuzzy subsets *high* (H) and *low* (L) are defined in the universes of power as shown in Figure E3.14(b). Obtain the consequence of this single input–single output fuzzy implication.

Answer 3.14

Let us abbreviate the given sentence as

If Good then High else Low.

This expression can be written in Cartesian product format as

$$\text{Good} \times \text{High} + \overline{\text{Good}} \times \text{Low}$$

There are two fuzzy relations here. They are connected to each other with *else*, which has similar meaning as *or*, i.e., max operator. Therefore, (3.46) can be applied.

$$R = R_1 + R_2$$

$$\mu_R(s,p) = \max[\mu_{R1}(s,p), \mu_{R2}(s,p)]$$

$$\mu_R(s,p) = \max\left[\min(\mu_G(s), \mu_H(p)),\ \min\left(\mu_{\overline{G}}(s), \mu_L(p)\right)\right]$$

The relation R_1 and R_2, then R are obtained using the given fuzzy subsets and shown in Figures E3.14(c), E3.14(d), and E3.14(e).

The surface plot shown in Figure E3.14(e) represents the verbal expression "*If Good* then *High* else *Low*." There are two implications in this sentence. These

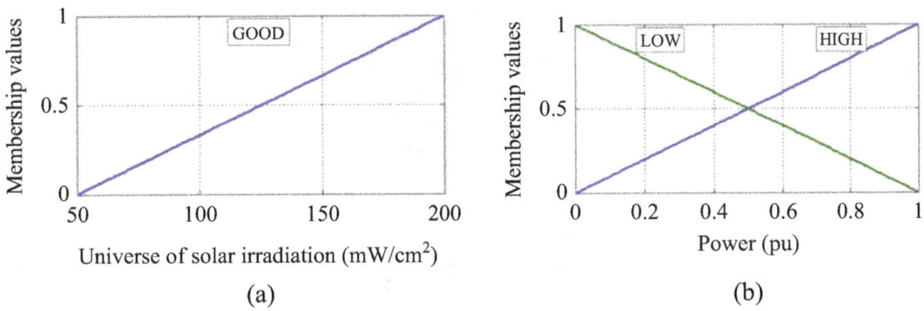

Figure E3.14(a and b) The membership functions for Example 3.14

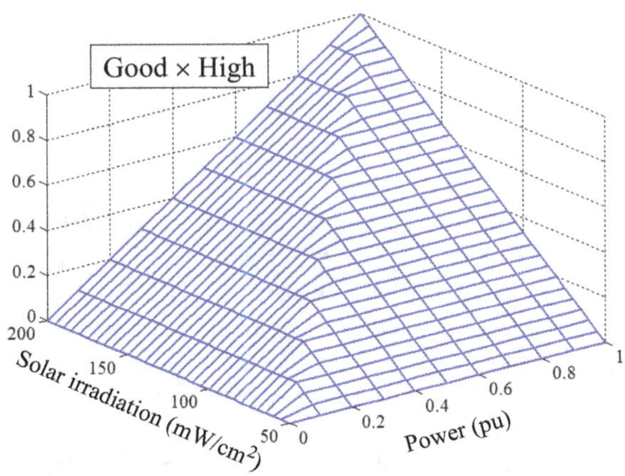

Figure E3.14(c) The relation surface of good × high

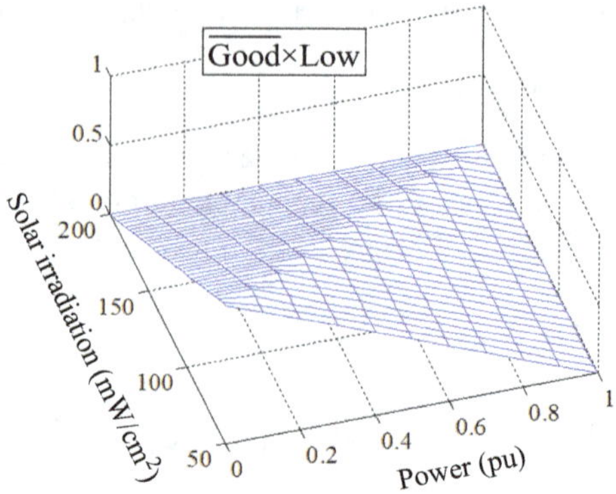

Figure E3.14(d) The relation surface of not good × high

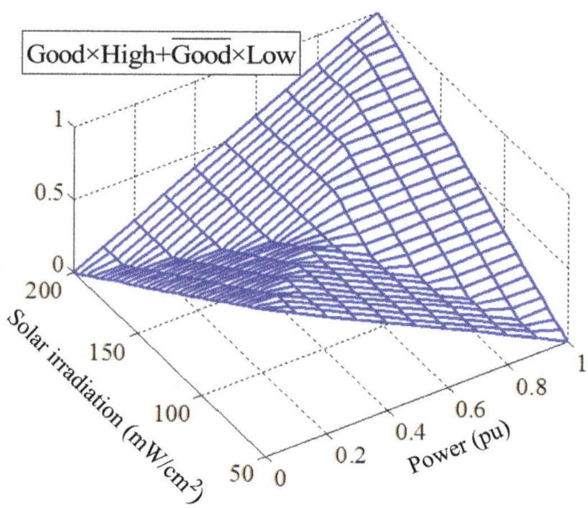

Figure E3.14(e) The relation surface of good × high + not good × high

implications are considered two separate relations and processed separately in Figures E3.14(c) and E3.14(d). The term "…else low." means that "*if it is not good then low.*" Therefore, the relation "*not good then low*" is obtained and shown in Figure E3.14(d). Figure E3.14(e) shows the combination of the rules as $\max(\mu_{R1}, \mu_{R2})$.

Practice 3.1

Consider a 100 km long high voltage power transmission line and write at least five fuzzy implications about the receiving end bus voltage depending on the loading conditions of the line.

3.7 Approximate reasoning

The theory of approximate reasoning was introduced by Zadeh in 1979 [10] as one of the application areas of fuzzy implication to reach a decision from fuzzy data and information. Contrary to the crisp set implication and reasoning, it is possible to reach a decision about the events with incomplete data in fuzzy reasoning. Actually, Zadeh suggested the use of fuzzy sets and fuzzy logic in the decision process in 1973 [7] after the appearance of the concept "fuzzy sets" in literature [11].

In order to see the difference between the crisp and fuzzy reasonings, let us take two crisp numbers $x \in X$ and $y \in Y$. Assume that the relation between the numbers x and y is given by

$$y(x) = 1 - e^{-x} \tag{3.49}$$

It is clear that there is an answer y for every input value of x as depicted in Figure 3.3. The crisp reasoning given by (3.49) and shown in Figure 3.3 has mainly three parts, which are input, calculation formula, and result. In a decision process, the formulation part is called *premise*, which gives a result based on the facts. Therefore, the input is called as *fact* and the output or the result is called as *consequence*. The crisp reasoning algorithm, which is called the inference system is summarized in (3.50).

$$\begin{array}{ll} \textit{Premise} & : \quad y = f(x) \\ \textit{Fact} & : \quad x = x_1 \\ \hline \textit{Consequence} & : \quad y = y_1 \end{array} \tag{3.50}$$

The reasoning algorithm given in (3.50) depends on the function in (3.49) used as the premise. For point-to-point reasoning, the implications in (3.51) can be written. The math functions used as premise may not yield a point-to-point reasoning algorithm every time. For example, there may not be an output y for every input x as in *V–I* characteristics of diodes or photovoltaic cells in electronics. The forward bias current of an ideal diode can be approximated as in (3.52) which yields an *I–V* characteristic similar to the one shown in Figure 3.4.

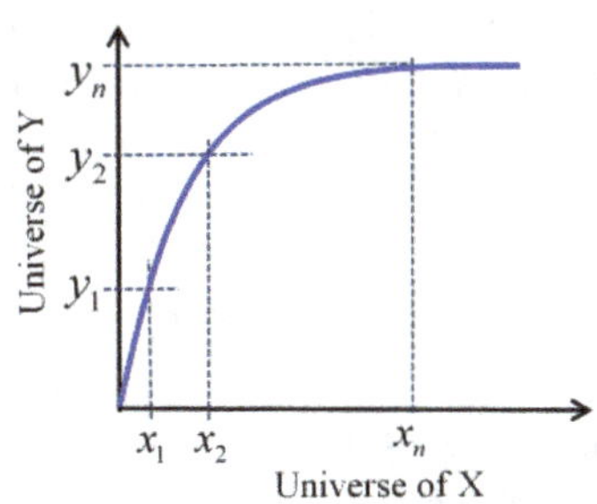

Figure 3.3 A single input–single output crisp reasoning

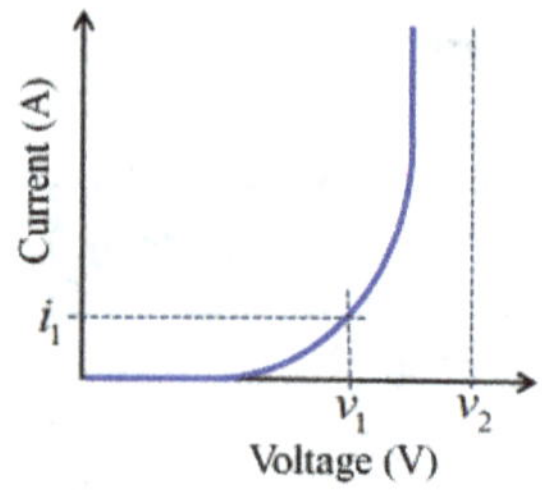

Figure 3.4 I–V characteristics of a diode as a single input–single output system

$$\begin{array}{llll} & \text{if } x \text{ is } x_1 & \text{then } y \text{ is } y_1 \\ \text{also} & \text{if } x \text{ is } x_2 & \text{then } y \text{ is } y_2 \\ \text{also} & \text{if } x \text{ is } x_3 & \text{then } y \text{ is } y_3 \\ \cdots & \cdots & \cdots \\ \cdots & \cdots & \cdots \\ \text{also} & \text{if } x \text{ is } x_n & \text{then } y \text{ is } y_n \end{array} \tag{3.51}$$

$$i = I_0 e^{\mathrm{qv/kT}} \tag{3.52}$$

where q is electron charge, v is bias voltage, k is Boltzmann constant, and T is temperature in Kelvin. Under constant temperature, the current of a diode can be written as just a function of voltage as

$$i = f(v) \tag{3.53}$$

It is clear from Figure 3.4 that every input v does not give a new and different output. Therefore, every input v does not correspond to every output. For example, while v_1 yields a current i_1, the input v_2 does not yield any current at all. In this latter case one cannot give a decision about i_2 and the implication does not have a consequence as in (3.54).

$$\begin{array}{ll} & \text{if } v \text{ is } v_1 \text{ then } i \text{ is } i_1 \\ \text{also} & \text{if } v \text{ is } v_2 \text{ then } i \text{ is?} \end{array} \tag{3.54}$$

Since the function $i = f(v)$ used as the premise here, is a crisp function, each output node must be in pair with an input node. If there is a little uncertainty in data, the output from the crisp reasoning is considered wrong or it is not accepted as an output at all.

Example 3.15

Assume that the word *good* used in Example 3.14 is represented as a crisp set in the universe of solar irradiation level and defined as in Figure E3.15.

A measurement of solar irradiation level on a partially cloudy day shows that the irradiation level is 100 mW/cm^2 at 14:00 and 99 mW/cm^2 at 16:00. Use the

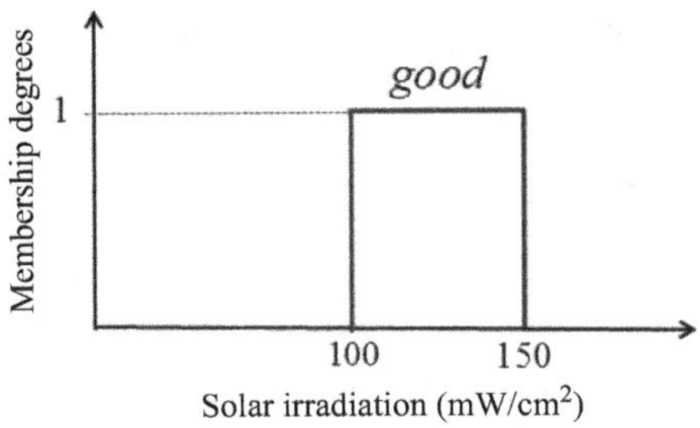

Figure E3.15 Crisp set good

crisp reasoning algorithm given by (3.50) and find the consequences for these two measurements. Are these two solar irradiation levels considered as *good*?

Answer 3.15

The answer can be obtained by using the inference system given in (3.50).

Premise	:	$good = \{100, 150\}$ definition from Figure E 3.15.
Fact	:	$S_1 = 100\ \mathrm{mW/cm^2}$ and $S_2 = 99\ \mathrm{mW/cm^2}$
Consequence	:	S_1 is good S_2 is not good. It is not defined.

3.8 Inference rules of approximate reasoning

The aggregation operators such as intersection, union, and complement discussed in previous chapters, are used with implications, which are called rules in reasoning, to widen the range of approximate reasoning. Zadeh has introduced the concept of approximate reasoning [10] with inference rules that enable the use of verbal expressions together with the implications. Zadeh's inference rules of approximate reasoning can be used to generate different and meaningful inference engines for specific applications. Some of these inference rules can be written as entailment rule, conjunction rule, disjunction rule, negation rule, projection rule, generalized modus ponens (GMP) rule, and compositional rule. Among these inference rules, GMP rule and compositional rule are going to be used and expanded for decision-making and control applications. Therefore, only these two inference rules have been explained in this book.

3.8.1 Generalized modus ponens rule of inference

GMP is the basis of fuzzy inference system. That is why it is one of the most important inference rules of fuzzy implications. The inference algorithm of GMP is given as in (3.55). According to the GMP, the conclusion depends on how the data suits the knowledge. Therefore, the existence of the conclusion is determined by data. If data is true then there will be a conclusion.

Knowledge	:	If *A* then *B*
Data	:	*A*
Conclusion	:	*B*

(3.55)

Example 3.16

Let us use the terms *good* and *high* defined in Example 3.14 for solar irradiation level and power generated by PV panels, respectively.

Knowledge	:	*If* solar irradiation is good *then* solar power is *high*
Data	:	Solar irradiation is *good*.
Conclusion	:	Solar power is high.

3.8.2 Compositional rule of inference

The compositional rule of inference (CRI) was discussed in Section 3.3.6 in this chapter. As stated in Section 3.3.6, this rule is used to extract the previously stored data from the knowledge base. Since the relation $R(X, Y) = A \times B$ includes data from the universe of X as the fuzzy subset A and from the universe of Y as the fuzzy subset B, then a fuzzy subset B' can be extracted from relation R as long as R and A are known. Zadeh has suggested the compositional rule of inference for the fuzzy implications to be used in approximate reasoning [10]. The fuzzy inference used in fuzzy logic control systems is based on the compositional rule of inference, which is written as in (3.56).

Knowledge	:	If x is A then y is B
Data	:	x is A'
Conclusion	:	y is B'

(3.56)

It is clear that the knowledge base here is a fuzzy implication as $A \rightarrow B$. Therefore, B can be obtained as a composition

$$B' = A \circ (A \rightarrow B) \tag{3.57}$$

where $A \rightarrow B$ is a fuzzy implication from A to B.

Example 3.17

Let us use the fuzzy implication R = LV→LD defined in Example 3.2 as the relation between voltage magnitude and transient duration in power systems.

Knowledge	:	*If* voltage is LV then duration is LD
Data	:	Voltage is *LV'*
Conclusion	:	duration is *LD'*.

In fuzzy control, the knowledge base is a collection of fuzzy implications, which are called rules consisting of the required actions to be taken. The input data is the fuzzy subsets indicating the current situation and the conclusion is the resultant action subtracted from the knowledge base or rule base.

3.9 Fuzzy reasoning

The inference system used in fuzzy reasoning is similar to the one used for approximate reasoning, but not exactly the same. Instead of fuzzy relation matrix, a rule base is used as the knowledge base in fuzzy reasoning. Fuzzy inference system used for fuzzy reasoning has fuzzy data as input layer, an *if* ... *then* ... rule processing layer, and a fuzzy conclusion layer. Although these three layers are used in all fuzzy reasoning-based applications, a consensus has not been reached about the name. People are calling the fuzzy inference system with different names as listed below.

- Fuzzy inference systems [12,13]
- Fuzzy rule-based systems [14,15]
- Fuzzy expert systems [16,17]
- Fuzzy model [18,19]
- Fuzzy associative memory [20,21]
- Fuzzy logic controller [22,23]

The processes performed under these names are very similar and all of them use the fuzzy rule-based reasoning algorithms. They have fuzzy input data, fuzzy rule processing, and fuzzy reasoning. Since the inference system is a series of fuzzy processes, it is called a *fuzzy processor* (*FP*) in this book. A fuzzy processor has mainly three parts as shown in Figure 3.5.

In the fuzzy processor, the knowledge, information, or expert's ideas are stored in a rule-based system, which has a processing structure in the form of *if* ... *then* ... *else*. The input data may be fuzzy or crisp. If data is uncertain or unclear, then it can be used as fuzzy input data. The crisp data must be fuzzified first before being used in the rule base. The rule processing unit yields an output for each rule. Then these individual rule outputs are combined together in a reasoning unit to yield a single fuzzy output. The reasoning unit uses fuzzy aggregation operators such as union and intersection. After this interpretation of the fuzzy processor and

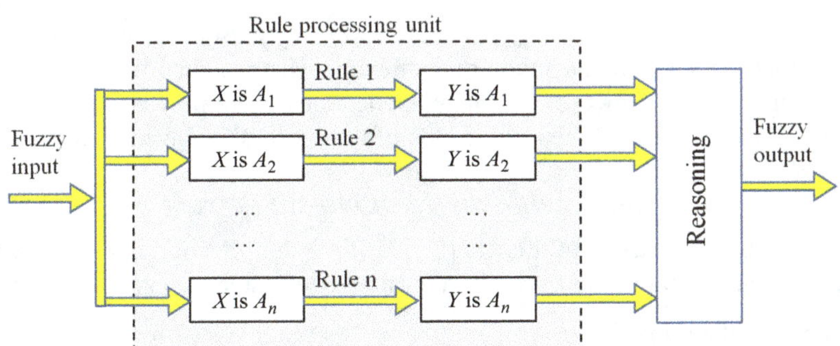

Figure 3.5 The fuzzy processor

Figure 3.5, the compositional rule of inference engine given in (3.56) can be written for the fuzzy processor as in (3.58).

$$\begin{array}{lcl} \textit{Rule base} & : & \text{If } x \text{ is A then y is } B \\ \textit{Data base} & : & x \text{ is } A \\ \hline \textit{Conclusion} & : & y \text{ is } B \end{array} \tag{3.58}$$

The input and the output of the fuzzy processor given in (3.58) and in Figure 3.5, are considered as fuzzy input and fuzzy output. If the input is crisp, then a fuzzifier is added to the input side of fuzzy processor. Sometimes crisp output may be required. This time a defuzzification unit is added to the output terminals of the fuzzy processor in order to convert fuzzy conclusion into a crip number or decision.

The inference algorithm given for the fuzzy processor in (3.58) represents the approximate reasoning more than that of fuzzy. Because for y to be B is possible if and only if x is A. If x is different than A, a connection between y and B cannot be established. It should be noted that the rule base, database, and conclusion triangle are a result of the generalized *modus ponen* inference system. While crisp *modus ponen* inference system gives crisp reasoning, *the compositional rule of inference*, which is suggested by Zadeh [10] as the fuzzy version of *modus ponens*, gives fuzzy reasoning. Therefore, *the fuzzy processor* is nothing but a *fuzzy reasoning* version of *the compositional rule of inference*. In fuzzy processor, for y to be B, x does not have to be A. If x is A' then y may be B' as shown in (3.59).

$$\begin{array}{lcl} \textit{Rule base} & : & \text{If } x \text{ is A then y is } B \\ \textit{Data base} & : & x \text{ is } A' \\ \hline \text{Conclusion} & : & y \text{ is } B' \end{array} \tag{3.59}$$

This forward reasoning inference system is useful, especially in fuzzy logic controllers. A similar but backward inference system can be written from *modus ponen* inference system as in (3.60), which is useful in medical science and expert system applications of fuzzy processors.

$$\begin{array}{lcl} \textit{Rule base} & : & \text{If } x \text{ is A then y is B} \\ \textit{Data base} & : & y \text{ is } B' \\ \hline \textit{Conclusion} & : & x \text{ is } A' \end{array} \tag{3.60}$$

The examples given for the approximate and fuzzy reasoning inference systems were all based on single input–single rule structure. That was fine for explaining the topic and the concept. However, in many applications, there may be more than one input, input space, and multiple rules. Some of these cases are given next.

3.9.1 Inference engine with single input–single rule

The examples given earlier in this chapter for fuzzy implication and approximate reasoning were all single input–single rule inference systems as given in (3.61).

$$\text{If } x \text{ is A then y is } B \tag{3.61}$$

where x is a member of the input space X and y is a member of the output space Y. Fuzzy sets A and B are defined as the input and output sets in X and Y, respectively. The single input–single rule inference engine can be used with both crisp and fuzzy rule-based systems. Some verbal implications can be given as examples:

- *If* ambient temperature is *high* (*HA*) *then* junction temperature is *high* (*HJ*).
- *If* power demand is *medium* (*MD*) *then* generation is *medium* (*MG*).
- *If* solar irradiation is *low* (*LS*) *then* power generation is *low* (*LP*).
- *If* voltage is *medium* (*MV*) *then* the total harmonic distortion is *high* (*HH*).

Example 3.18

Let us refer to Example 3.8 where the relation between voltage magnitude and harmonics in power systems was discussed and some implications were given. Use the following implication as an example here.

If voltage is *medium* (*MV*) *then* the total harmonic distortion is *high* (*HH*).

where the input fuzzy set *MV* is defined in the universe of voltage and the output fuzzy set *HH* is defined in the universe of harmonic distortion. The fuzzy subsets are already given in Example 3.8.

Apply a single input–single rule inference algorithm and obtain the final decision if the input voltage is extra high (*EHV*).

Answer 3.18

The single input–single rule inference engine algorithm is shown in Figure E3.18, which is a graphical representation of (3.59) and (3.61). There is one rule base, one

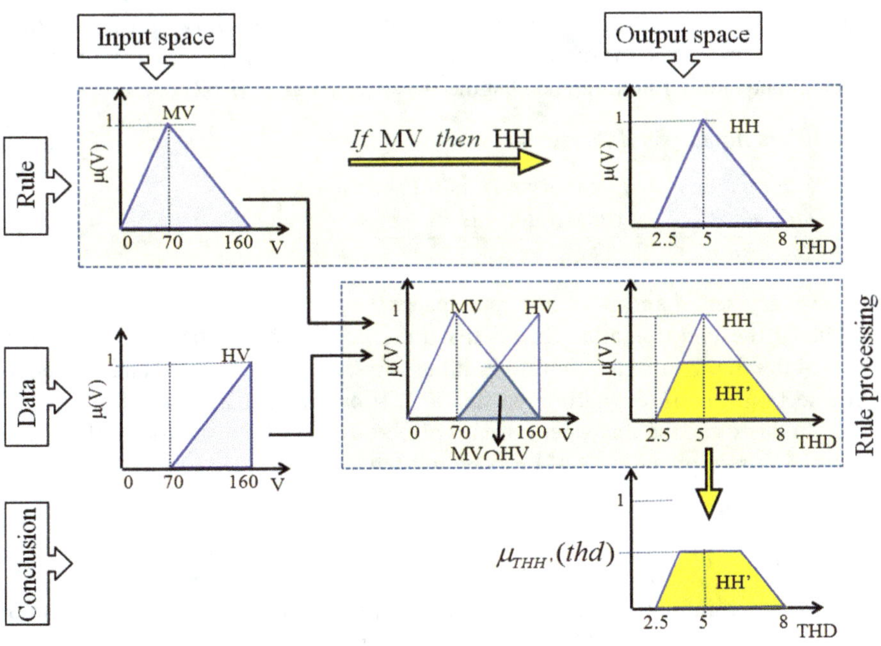

Figure E3.18 Single input–single rule inference engine

input data, and a conclusion part in Figure E3.18, besides the rule processing unit. For this problem, Equation (3.59) can be applied as follows in written format:

$$\begin{array}{lll} \textit{Rule base} & : & \text{If } v \text{ is } MV \text{ then thd is } HH \\ \textit{Data base} & : & v \text{ is } HV \\ \hline \textit{Conclusion} & : & \text{thd is } HH' \end{array}$$

The rule processing is shown graphically in Figure E3.18. The fuzzy set entered as data is not exactly the same as the one used in the rule. However, they are defined in the same universe of V. Since there is just one single rule, then there is just one output fuzzy set. Because a different fuzzy set HV is used as the input data set, the output will not be exactly the same as the one given in the rule. A weighting factor is used to obtain the effect of the output set in the conclusion. This weighting factor is obtained as a membership value using the intersection (min) operator in the same way as it is used for fuzzy relation in (3.43).

$$\mu_{\mathrm{HH'}}(thd) = \min(\mu_{\mathrm{MV}}(v), \mu_{\mathrm{HV}}(v))$$

3.9.2 Inference engine with multiple input–single rule

There may be more than one input space in many applications. If a rule has two or more input spaces, then it is called a multiple input–single rule inference system. The rule base is written by considering all inputs. Since there is one rule, then there will be one single output. The output depends on all input data. In order to yield a conclusion, at least one input should be included in the fuzzy subsets used in the rule base. If the input fuzzy sets do not have any membership value in any subsets of the rule base, the output is not generated. The degree of inclusion of the output fuzzy set as the final decision is obtained from the membership values of the fuzzy sets used in input spaces. Multiple input–single rule inference engine is expressed as in (3.62) and (3.63).

$$\textit{If } x \text{ is } A \textit{ and } y \text{ is } B \textit{ then } z \text{ is } C \tag{3.62}$$

$$\begin{array}{lll} \textit{Rule base} & : & \textit{If } x \text{ is } A \text{ and } y \text{ is } B \textit{ then } z \text{ is } C \\ \textit{Data base} & : & x \text{ is } A_1 \text{ and y is } B_1 \\ \hline \textit{Conclusion} & : & z \text{ is } C_1 \end{array} \tag{3.63}$$

The fuzzy sets A and B in the input spaces X and Y are connected to one input space using the *and* operator. Recall that the *and* operator is used for the *intersection* of the fuzzy subsets. Therefore, it has the same meaning here and is used as an intersection. The rule could be written as *if A or B then C* instead of *if A and B then C*. In this case, the *and* operator is replaced by *or*, which stands for the *union* operator. The rule given in (3.45) can be written as

$$\text{If } A \textit{ and } B \textit{ then } C \tag{3.64}$$

$$(A \cap B) \rightarrow C \tag{3.65}$$

$$\mu_{\mathrm{C}}(\mathrm{z}) = \min[\min(\mu_A(x), \mu_{A1}(x)),\ \min(\mu_B(y), \mu_{B1}(y))] \tag{3.66}$$

where $\mu_C(z)$ is the membership value of the fuzzy set C and represents the weighting effect of the rule at the output universe Z depending on the inclusion of the input fuzzy sets A_1 and B_1 in A and B, respectively. The membership value $\mu_C(z)$ determines the effectiveness of the fuzzy set C at the output as a new fuzzy set C_1 where $C_1 \subseteq C$. It should be noted that arithmetic products can be used instead of the min operator as given by (3.67). However, as stated earlier, the min operator is preferred to be used in this book.

$$\mu_C(z) = [\mu_A(x) \cdot \mu_{A1}(x)] \cdot [\mu_B(y) \cdot \mu_{B1}(y)] \tag{3.67}$$

Example 3.19

The power generated by PV solar cells changes as a function of solar irradiation level and operating temperature. So, the generated maximum power increases as the solar irradiation level increases, but decreases as the operating temperature increases. Fuzzy partitions of solar irradiation (S) and ambient temperature (T) into subcategories are given in Figure E3.19(a) and E3.19(b). Find the conclusion of the fuzzy rule given below if the solar irradiation level is *sufficient* (SI) and the temperature is *warm* (WT) where SI and WT are defined in the universes of S and T, respectively.

If s is good (GI) and t is cold (CT) then p is high (HP)

where $s \in S$ and $t \in T$. The fuzzy sets *good* (GI) and *cold* (CT) are defined in the universes of S and T, respectively. Fuzzy set *high* (HP) is defined in the universe of power (P) as shown in Figure E3.19(c).

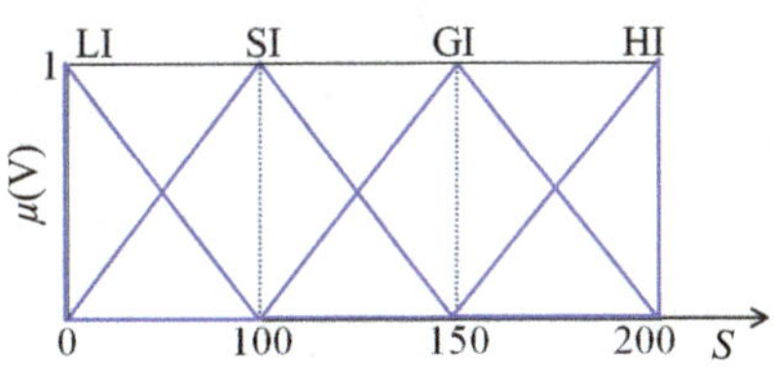

Figure E3.19(a) Subcategories of solar irradiation

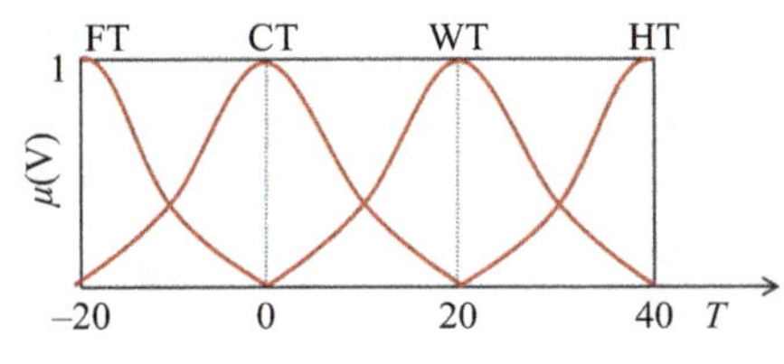

Figure E3.19(b) Subcategories of ambient temperature

Answer 3.19

The rule given in this example has two input spaces and one output space. In order to write a shorter rule let us use some abbreviations such as $s \in SI$ for the solar irradiation level, $t \in T$ for the temperature, and $p \in P$ for the power. Then write the fuzzy inference system as

Rule base	:	*If* s is GI *and* t is CT *then* p is HP
Data base	:	*s* is S *and t* is WT
Conclusion	:	*p* is HP_2

Since there is only one rule, the output will be always *hihg* (HP). However, the effective amount of HP, which is shown as HP_2, is determined from the intersection of the fuzzy sets GI and SI in the universe of S and the intersection of the fuzzy sets WT and CT in the universe of *T*. The interpretation of the inference system given in (3.65) and (3.66) is represented graphically as in Figure E3.19(d).

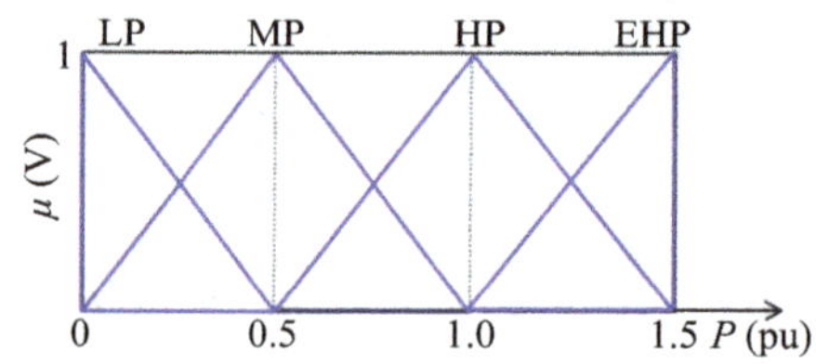

Figure E3.19(c) Subcategories of PV power

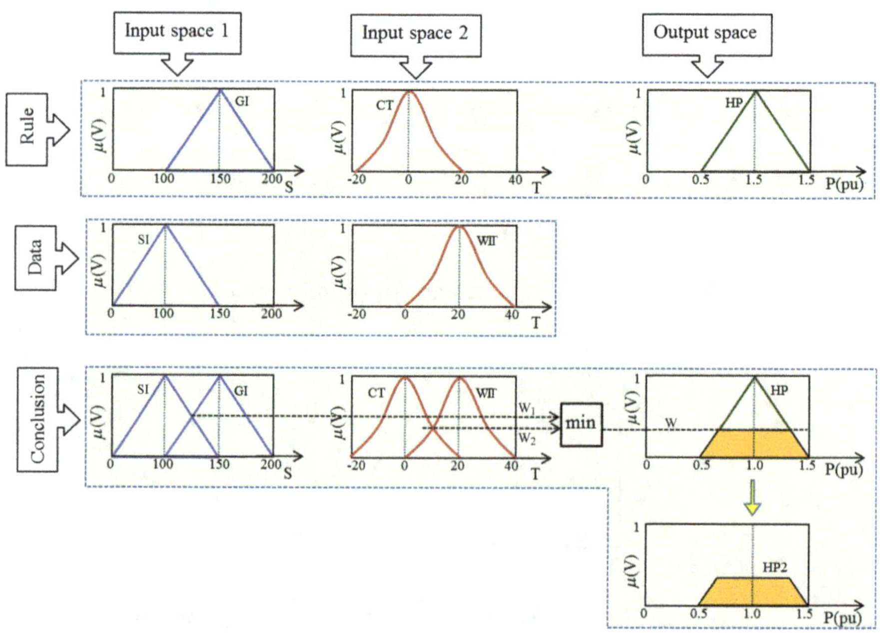

Figure E3.19(d) Double input–single rule inference engine

The final membership value $\mu_{\mathrm{HP2}}(p) = w_2$ of HP_2 can be obtained using (3.66).

$$\mu_{\mathrm{HP2}}(p) = \min[w_1, w_2] = 0.3$$

where

$$w_1(\mathrm{s}) = \min(\mu_{\mathrm{GI}}(s), \mu_{\mathrm{SI}}(s)) = 0.5$$

$$w_2(\mathrm{t}) = \min(\mu_{\mathrm{CT}}(t), \mu_{\mathrm{WT}}(t)) = 0.3$$

3.9.3 *Inference engine with multiple input–multiple rule*

Multiple input–multiple rule inference system is an expansion of multiple input–single rule system. Besides the multiple inputs, this time there are multiple rules and multiple outputs. Multiple inputs may be defined in the same or different universes as multiple outputs may also be defined in the same or different universes. In a multiple rule inference system, there will be one output from each rule yielding multiple outputs. All outputs from the rules may or may not be defined in the same universe.

The number of outputs is determined by the inclusion of input data sets in the fuzzy sets used in the fuzzy rules. In order to generate an output, at least one of the input data sets should have nonzero membership degrees in one of the sets in the rule. Otherwise, no output is generated. Therefore, it can easily be understood that all of the rules in a multiple rule-based system may or may not be active. Some rules may be active while some are not.

A multiple input–multiple rule inference system is expressed as in (3.68).

$$\begin{array}{lll}
\textit{Rule Base} & : & \textit{If } x \text{ is } A_1 \textit{ and } y \text{ is } B_1 \textit{ then } z \text{ is } C_1 \\
 & & \textit{If } x \text{ is } A_2 \textit{ and } y \text{ is } B_2 \textit{ then } z \text{ is } C_2 \\
 & & \quad \cdots \\
 & & \quad \cdots \\
 & & \textit{If } x \text{ is } A_n \textit{ and } y \text{ is } B_n \textit{ then } z \text{ is } C_n \\
\hline
\textit{Data Base} & : & x \text{ is } A' \textit{ and } y \textit{ is } B' \\
\hline
\textit{Conclusion} & : & z \text{ is } C'
\end{array} \tag{3.68}$$

where C_1, C_2, ..., C_n are the outputs from each individual rule and C' is the final effect of all the rules together. The rules in (3.68) are all standing independently from each other. In order to obtain a final single effective output, the rules should be combined. The combination of the rules is done using either union or intersection operators. Therefore, *else if* or *also* are used to connect the rules to each other. Due to the form of their usage, *else if* and *also* are called connectors here.

The fuzzy inference engine uses the connector ***else if*** is given by (3.69).

$$\begin{array}{lll} \textit{Rule Base} & : & \text{If } x \text{ is } A_1 \text{ and y is } B_1 \textit{ then } \text{z is } C_1 \\ & & \textit{else if } x \text{ is } A_2 \text{ and y is } B_2 \textit{ then } \text{z is } C_2 \\ & & \cdots \\ & & \cdots \\ & & \textit{else if } x \text{ is } A_{n-1} \text{ and y is } B_{n-1} \textit{ then } z \text{ is } C_{n-1} \\ & & \text{else x is } A_n \textit{ and } y \text{ is } B_n \textit{ then } \text{z is } C_n \\ \hline \textit{Data Base} & : & x \text{ is A}' \text{ and y is B}' \\ \hline \textit{Conclusion} & : & z \text{ is } C' \end{array} \tag{3.69}$$

The inference engine of multiple input–multiple rule system with the *else* connector given in (3.69) can be written as in (3.70) to (3.75), as well.

$$C' = [(A_1 \cap B_1) \to C_1] \cup [(A_2 \cap B_2) \to C_2] \cup \cdots \cup [(A_n \cap B_n) \to C_n] \tag{3.70}$$

If the input data A' and B' are included, (3.70) becomes,

$$C' = [(A_1 \cap A') \cap (B_1 \cap B')] \cup \cdots \cup [(A_n \cap A') \cap (B_n \cap B')] \tag{3.71}$$

$$C' = C_1{}' \cup C_2{}' \cup C_3{}' \cup \cdots \cup C_n{}' \tag{3.72}$$

where C_i', $i = 1, 2, \ldots, n$ are the weighted outputs from each rule. The last equation is nothing but a max–min composition and can be expressed as

$$\mu_{C'}(z) = \max[\min(w_{A1}(x), w_{B1}(y)), \ldots, \min(w_{\mathrm{An}}(x), w_{\mathrm{Bn}}(y))] \tag{3.73}$$

Equation (3.73) can be written in a shorter form as

$$\begin{aligned} w_{\mathrm{C1}'}(z) &= \min(w_{A1}(x), w_{B1}(y)) \\ w_{\mathrm{C2}'}(z) &= \min(w_{A2}(x), w_{B2}(y)) \\ &\cdots \\ w_{\mathrm{Cn}'}(z) &= \min(w_{\mathrm{An}}(s), w_{\mathrm{Bn}}(y)) \end{aligned} \tag{3:74}$$

where the weightings are obtained as

$$\begin{aligned} w_{A1}(x) &= \min(\mu_{A1}(x), \mu_{A'}(x)), w_{B1}(y) = \min(\mu_{B1}(y), \mu_{B'}(y)) \\ w_{A2}(x) &= \min(\mu_{A2}(x), \mu_{A'}(x)), w_{B2}(y) = \min(\mu_{B2}(y), \mu_{B'}(y)) \\ &\cdots\cdots \\ w_{\mathrm{An}}(x) &= \min(\mu_{\mathrm{An}}(x), \mu_{A'}(x)), w_{\mathrm{Bn}}(y) = \min(\mu_{\mathrm{Bn}}(y), \mu_{B'}(y)) \end{aligned} \tag{3.75}$$

Example 3.20

Consider the PV system described in Example 3.19 and assume that a second rule is added to the existing one as below. Find the conclusion of the fuzzy rules given above if the solar irradiation level is *sufficient* (SI) and temperature is *warm* (WT) where all the fuzzy subsets are given in Figures E3.19(a), E3.19(b), and E3.19(c).

If s is good (GI) *and* t is cold (CT) *then* p is high (HP)
else if s is low (LI) *and* t is hot (HT) *then* p is low (LP)

Answer 3.20

The rules given in this example have two input spaces. Each rule has its own output resulting in two outputs in total. Since the rules are connected using the *else* connector, (3.69) can be applied as the inference engine.

Rule 1 : *If* s is good (GI) and t is cold (CT) *then* p is high (HP)
Rule 2 : *else* if s is low (LI) and t is hot (HT) *then* p is low (LP)

Data : s is low (SI) *and* t is hot (WT)

Conclusion : p is P′

Since there are two rules, each rule will have its own output. The outputs are *high* (HP) and *low* (LP) from Rule 1 and from Rule 2, respectively. The effective amounts of the output set HP and LP are shown as HP' and LP' in Figure E3.20 and

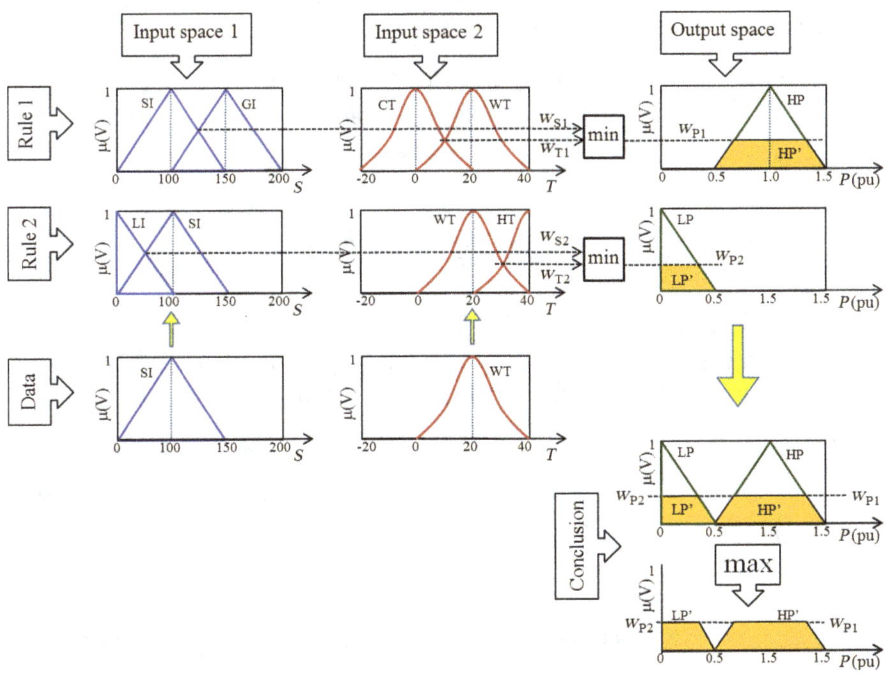

Figure E3.20 Graphical representation of multiple input–multiple rule inference system

are determined by weightings $w_{P1'}$ and $w_{P2'}$ using (3.74) as below.

$$\begin{aligned} w_{P1'}(p) &= \min(w_{S1}(s), w_{T1}(t)) = 0.3 \\ w_{P2'}(p) &= \min(w_{S2}(s), w_{T2}(t)) = 0.3 \end{aligned}$$

where

$$\begin{aligned} w_{S1}(s) &= \min(\mu_{\text{GI}}(s), \mu_{\text{SI}}(s)) = 0.5 \quad & w_{T1}(t) &= \min(\mu_{\text{CT}}(t),\ \mu_{\text{WT}}(t)) = 0.3 \\ w_{S2}(s) &= \min(\mu_{\text{LI}}(s), \mu_{\text{SI}}(s)) = 0.5 \quad & w_{T2}(t) &= \min(\mu_{\text{HT}}(t),\ \mu_{\text{WT}}(t)) = 0.3 \end{aligned}$$

The final conclusion is then obtained as

$$\begin{aligned} \mu_{P'}(p) &= \max[\min(w_{S1}(s), w_{T1}(t)),\ \min(w_{S2}(s), w_{T2}(t))] \\ \mu_{P'}(p) &= \max[\min(0.5, 0.3),\ \min(0.5, 0.3)] \\ \mu_{P'}(p) &= \max[0.3, 0.3] \\ \mu_{P'}(p) &= 0.3 \end{aligned}$$

Note that the calculations done here are done in terms of weightings. Actually, the weighted outputs HP' and LP' are not just single membership values. They are weighted fuzzy subsets and the final output is the union of HP' and LP' as shown in Figure E3.20.

The fuzzy inference engine using the connector ***also*** is given by (3.76) for multiple input–multiple rule systems.

$$\begin{array}{lll} \textit{Rule Base} & : & \textit{If } x \text{ is } A_1 \textit{ and } \text{y is } B_1 \textit{ then } \text{z is } C_1 \\ & & \textit{also } x \text{ is } A_2 \textit{ and } \text{y is } B_2 \textit{ then } \text{z is } C_2 \\ & & \cdots \\ & & \cdots \\ & & \textit{also } x \text{ is } A_{n-1} \text{ and } y \text{ is } B_{n-1} \textit{ then } \text{z is } C_{n-1} \\ & & \textit{also } x \text{ is } A_n \text{ and } y \text{ is } B_n \textit{ then } \text{z is } C_n \\ \hline \textit{Data Base} & : & x \text{ is } A' \textit{ and } \text{y is } B' \\ \hline \textit{Conclusion} & : & z \text{ is } C' \end{array} \tag{3.76}$$

The inference engine of multiple input–multiple rule system with the *also* connector given in (3.70) can be written as in (3.77) to (3.80), as well.

$$C' = [(A_1 \cap B_1) \to C_1] \cap [(A_2 \cap B_2) \to C_2] \cap \cdots \cap [(A_n \cap B_n) \to C_n] \tag{3.77}$$

If the input data A' and B' are included, (3.77) becomes,

$$C' = [(A_1 \cap A') \cap (B_1 \cap B')] \cap \cdots \cap [(A_n \cap A') \cap (B_n \cap B')] \tag{3.78}$$

The last equation is nothing but a min–min composition and can be expressed as

$$\mu_{C'}(z) = \min[\min(w_{A1}(x), w_{B1}(y)), \ldots,\ \min(w_{\text{An}}(x), w_{\text{Bn}}(y))] \tag{3.79}$$

where

$$\begin{aligned} w_{A1}(x) &= \min(\mu_{A1}(x), \mu_{A'}(x)), & w_{A1}(x) &= \min(\mu_{B1}(y), \mu_{B'}(y)) \\ w_{A2}(x) &= \min(\mu_{A2}(x), \mu_{A'}(x)), & w_{A1}(x) &= \min(\mu_{B2}(y), \mu_{B'}(y)) \\ \ldots\ldots & & & \\ w_{\mathrm{An}}(x) &= \min(\mu_{\mathrm{An}}(x), \mu_{A'}(x)), & w_{\mathrm{An}}(x) &= \min(\mu_{\mathrm{Bn}}(y), \mu_{B'}(y)) \end{aligned} \tag{3.80}$$

The only difference between the usage of *else if* and *also* is the final composition, in which the max operator is used for *else if* while the min operator is used for *also*. Replacing the *union* (∪) by *intersection* (∩) and max by min, equations written for *else if* can be converted to equations for *also*.

It should also not be forgotten that arithmetic products can also be used instead of the min operator.

Fuzzy inference engine is explained above with examples. It can be concluded from the explanations and examples given that the fuzzy inference system is nothing but a collection of fuzzy processes. That is why it is suggested to call the inference system as *fuzzy processor*. Fuzzy inference system transfers the facts of input fuzzy data to the output as a fuzzy output. This process can also be called mapping the fuzzy data from input spaces onto the output space as fuzzy output.

In practical applications especially in control systems, the input data is usually measured and crisp. Therefore, it must be converted to fuzzy data before being used in a fuzzy processor. Similarly, the output of the fuzzy processor must be converted to crisp value so that it can be used by devices. The fuzzy processor using crisp input and giving crisp output requires a *fuzzification* unit at the input side and a *defuzzification* unit at the output side. Fuzzy processors with crisp input and crisp outputs have been studied in literature by various researchers. The following are three of the most known fuzzy reasoning algorithms.

- Mamdani fuzzy reasoning
- Sugeno fuzzy reasoning
- Tsukamoto fuzzy reasoning

These three fuzzy reasoning algorithms are explained in the next chapter.

Practice 3.2

In a water station, the tanks are filled manually such that the valve is fully open (100%) when the tank is empty, half open (50%) when the tank is half full, and closed (0%) when the tank is completely full. Representation of the valve opening and water level in the tank by fuzzy membership functions is described in Table P3.2.

(a) Partition the *water level* space into subsets using triangular membership functions such that the peak points of triangular membership functions are equally placed. Show this partitioning in a figure.

(b) Partition the valve position space into subsets using triangular membership functions such that the peak points of triangular membership functions are equally placed. Show this partitioning in a figure.

(c) Write a *single input–single output* fuzzy reasoning algorithm (FRA) and a set of rules for this operation.
(d) Obtain the valve position if the tank is full by 20%.

Table P3.2 Descriptions of fuzzy membership functions for Practice 3.2

Water level in the tank	Valve position	Fuzzy membership function
EW: Empty	FOV: Full open	Right triangle
HW: Half full	HOV: Half open	Triangle
FW: Full	FCV: Fully closed	Right triangle

Example 3.21

Fuzzy set cold (C) in the universe of temperature T is defined by a triangular membership function $C(-20, 0, 20)$ and fuzzy set high (H) representing the PV output voltage is defined as $H(0, 40, 40)$. Obtain a fuzzy relation that represents the fuzzy implication given below.

If the ambient temperature is *cold then* PV output voltage is *high*.

Use the numerical values with membership degrees of 0.0, 0.5, and 1.0 only and show the process in a matrix format.

Answer 3.21

The answer to Example 3.21 is given in Table E3.21.

Table E3.21 Fuzzy relation matrix R for the implication given in Example 3.21

		Temperature, cold (°C)				
		$\frac{0}{-20}$	$\frac{0.5}{-10}$	$\frac{1}{0}$	$\frac{0.5}{10}$	$\frac{0}{20}$
PV voltage, high (V)	$\frac{0}{0}$	$\frac{0}{0,-20}$	$\frac{0}{0,-10}$	$\frac{0}{0,0}$	$\frac{0}{0,10}$	$\frac{0}{0,20}$
	$\frac{0.5}{20}$	$\frac{0}{20,-20}$	$\frac{0.5}{20,-10}$	$\frac{0.5}{20,0}$	$\frac{0.5}{20,10}$	$\frac{0}{20,20}$
	$\frac{1}{40}$	$\frac{0}{40,-20}$	$\frac{0.5}{40,-10}$	$\frac{1}{40,0}$	$\frac{0.5}{40,10}$	$\frac{0}{40,20}$

Example 3.22

A family, consisting of mom, dad, and two children, has gone to a cultural vacation. The satisfaction degree of the vacation is determined by evaluating traveling comfort, accommodation comfort, and the approbation of the visited places. After the vacation, the family members evaluated the vacation as in Table E3.22.

The fuzzy verbal expressions used in Table E3.22 are defined by triangular membership functions as shown in Figure E3.22(a).

Obtain how the vacation was satisfactory for all family members.

Table E3.22 Satisfaction of family members about the vacation in Example 3.22

	Traveling	**Accommodation**	**Places**
Dad	Good	Good	Good
Mom	Fair	Fair	Good
Child HE	Fair	Good	Fair
Child CS	Bad	Good	Bad

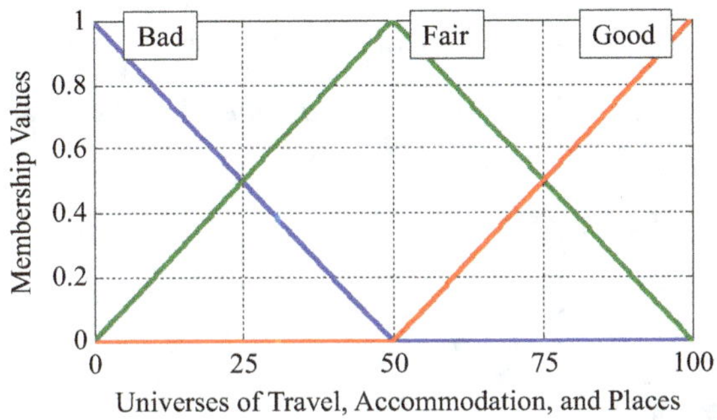

Figure E3.22(a) Fuzzy subsets bad, fair, and good in the related universes

Answer 3.22

Mom thinks that the traveling was *fair and* accommodation was *fair and* the places were *good*. Since the expressions are connected by the term and, the consequence will be the intersection of the fuzzy subsets used in the sentence as shown in Figure E3.22(b).

Mom's thoughts = fair ∩ fair ∩ good

Child HE thinks that the traveling was *fair and* accommodation was *good and* places were *fair*. The consequence will be the intersection of the fuzzy subsets *fair* and *good* as in Figure E3.22(b) given for mom.

Thoughts of child HE = fair ∩ good ∩ fair

According to child CS, the traveling was *bad and* accommodation was *good and* places were *bad*. The consequence will be the intersection of the fuzzy subsets *bad* and *good* as in Figure E3.22(a). The fuzzy subsets *bad* and *good* do not have any common area. Therefore, their intersection becomes an empty set.

Thoughts of child CS = bad ∩ good ∩ bad = ∅

Dad thinks that the traveling was *good and* the accommodation was *good and* places were *good*. Since all the fuzzy subsets are good, the consequent becomes good as shown in Figure E3.22(c).

Dad's thoughts = good ∩ good ∩ good = good

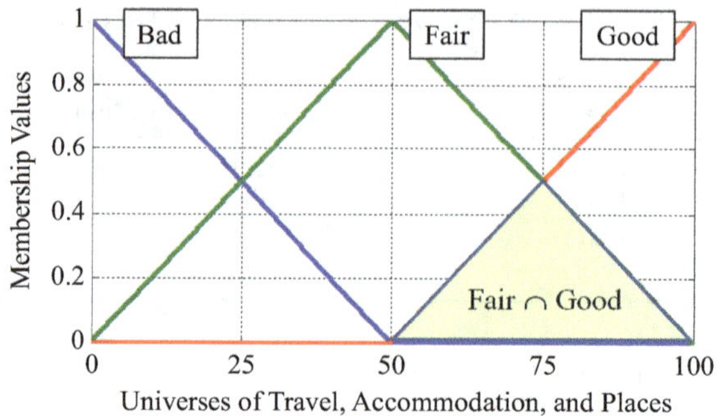

Figure E3.22(b) Mom's satisfaction from the vacation

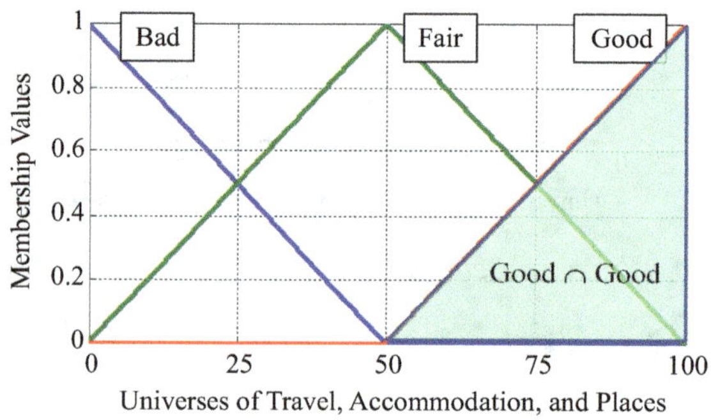

Figure E3.22(c) Dad's satisfaction from the vacation

Example 3.23

The satisfaction degree of the family members in Example 3.22 is represented by a fuzzy subset called *satisfied* in the universe of satisfaction in percentages. A triangular-type membership function is used to represent the fuzzy subset *satisfied* as *fuzzified* = *triangle* (0, 100, 100).

Obtain the satisfaction levels of each family member and determine who is the one who enjoyed the vacation most.

Answer 3.23

It is clear that the output space is the satisfaction in percentages. Since satisfaction is measured in terms of traveling comfort, accommodation comfort, and visited places, these universes are the input spaces. The reasoning process and the consequences are shown in Figure E3.23.

The overall family satisfaction from this vacation can be obtained by getting the weighted average of the consequences from each rule as

$$\text{Family satisfaction} = \frac{100 + 50 + 50 + 0}{4} = \frac{200}{4} = 50\%$$

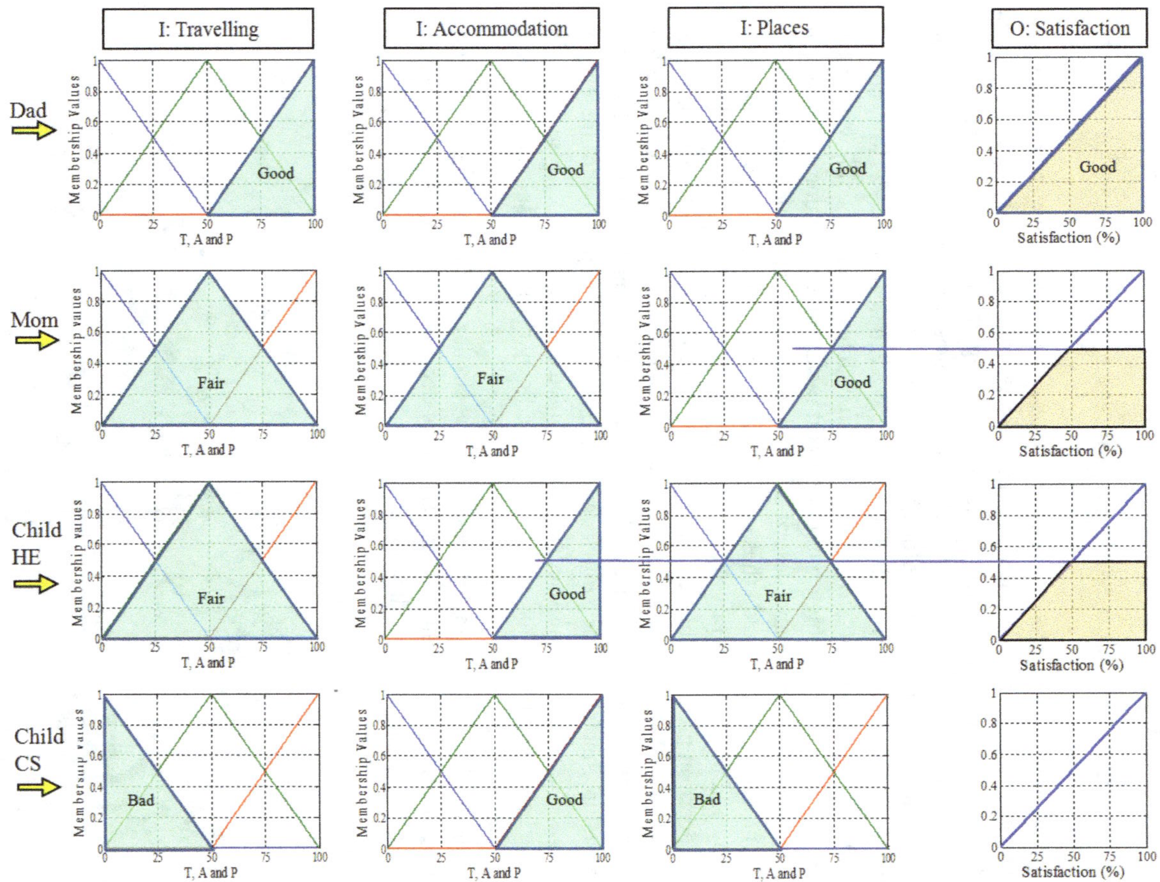

Figure E3.23 The reasoning algorithm for Example E3.23

The individual conclusions from each rule show that dad has 100% satisfaction from the vacation. Therefore, it is clear that dad is completely satisfied. Mom and child HE are pleased with the vacation by 50%. However, Child CS has not been satisfied at all. The overall family satisfaction is found to be 50%.

3.10 Human decision-making process

Fuzzy sets are used to represent uncertain and fuzzy data in the way as human does. We as humans get data from the five sense organs, which are eyes, nose, ears, skin, and tongue. All data one gets from these five sensing organs are fuzzy in nature not crisp. We see around and objects as colored images without knowing the exact length, width, size, and certain color. We evaluate the smell of odor as none, a little, or intolerable without knowing its OUe value. We can hear sounds as low, normal, or high without knowing its dB value. Our skin enables us to feel cold, warm, pain, etc. as low, medium, and high without certain measurements in units. We use our tongue to sense the taste and describe it as sweet, salty, and spicy without measurement units. We can grade these definitions as very salty, very sweet and very spicy by adding another fuzzy word such as adverb "very."

A decision process includes mainly three parts such as *data* (sensing), *knowledge* (rules), and *conclusion* as shown in Figure 3.5. In fuzzy reasoning, the *data* is the current measurement represented by fuzzy subsets, which are also called fuzzy membership functions. The *knowledge* is the ability of experts who are making decisions. Therefore, the knowledge is represented by *if* ... *then* ... *else* type rules in a rule base. And finally the *conclusion* is the fuzzy output as resultant decision to be applied to take action.

Human decision-making process is very similar to that of fuzzy decision. Figure 3.6 depicts how a human driver makes decision in order to keep an ambulance on the road in correct direction.

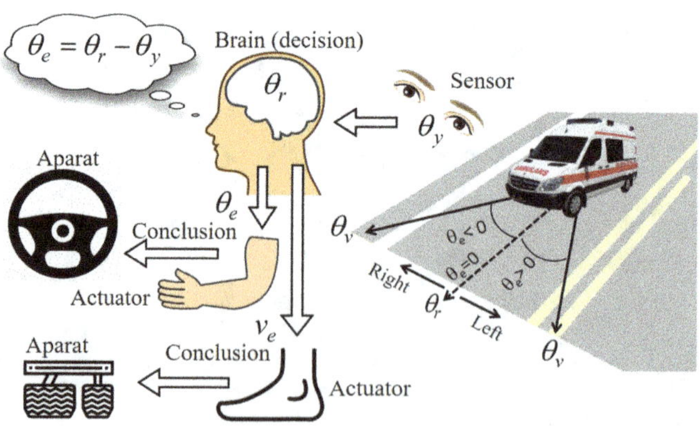

Figure 3.6 Decision process of a human car driver

The driver sees the correct direction, θ_r, by looking at the road and keeping it in mind as the reference direction angle. The reference angle is usually in the same direction as the road and is assumed to be 0°. Then the driver measures the actual direction ambulance as the angle θ_v by using the eyes as sensors. The action to be taken, in order to keep the ambulance in the right direction, depends on the direction angle error, which can be defined as the angular difference between reference and deviation angles as

$$\theta_e(t) = \theta_r(t) - \theta_v(t) \tag{3.80}$$

The driver uses (3.80) in his/her mind and calculates the direction angle error. If the reference direction angle is taken as zero, then (3.80) can be written as

$$\theta_e(t) = -\theta_v(t) \tag{3.81}$$

The aim here is to make the deviation angle (or the direction error) zero. Therefore, the driver reaches a conclusion (gives a decision) whether to turn the steering wheel to right or to left or do nothing. This decision is sent from the brain to the arms, which are the actuators used to perform the action.

In the human decision-making process, the driver sees (measures) the deviation angle as input data, and checks whether it is zero. A non-zero deviation value means that there is a deviation to right or to left. If there is a deviation angle to right or to left, the driver makes a decision in his/her mind to turn the steering wheel toward the opposite direction in order to bring the vehicle to the reference direction. The decision made in the brain is the output of the human decision-making process. This decision is sent to hands through the synapsis of neurons in the arms to perform the decision. The hands of the driver are the actuators in the human case. When human decision-making is required for the speed adjustment of the vehicle, then the actuator would be the driver's foot to press the gas pedal more or less.

The similarities between fuzzy decision-making and human decision-making processes are summarized in Figure 3.6. A human being can understand whether the deviation is to right or to left by just looking at the road and road lines. In the intelligent machine case, the right or left directions of deviation are determined by the sign of deviation angle. Let us assume that the deviation angle, θ_v, is positive if the vehicle is going toward right instead of the reference direction as given in Figure 3.6 and θ_v is negative if the vehicle is going toward left instead of reference direction. According to these assumptions, there would be three options for the direction error, θ_e, as

$$\text{Deviation to right}: \theta_e(t) = 0 - (\theta_v(t)) = -\theta_v(t) < 0 \tag{3.82}$$

$$\text{Deviation to left}: \theta_e(t) = 0 - (-\theta_v(t)) = +\theta_v(t) > 0 \tag{3.83}$$

$$\text{No deviation exist}: \theta_e(t) = 0 - (\theta_v(t) = \theta_r(t)) = 0 \tag{3.84}$$

These three options of the direction error can be evaluated by a person as fuzzy positive (P), fuzzy zero (Z), and fuzzy negative (N) instead of crisp numerical degrees. However, human drivers may think that the deviation to left or right is

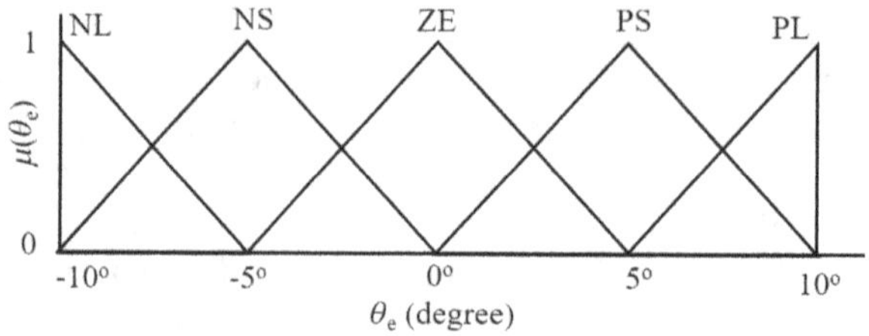

Figure 3.7 Fuzzy subsets for the direction angle error

small or large, and turn the steering wheel toward left or right with small or large changes. Therefore, fuzzy sets P and N should be defined with subcategories as negative large (NL), negative small (NS), positive small (PS), and positive large (PL). As stated in Chapter 1, these definitions of the direction error are the verbal fuzzy variables and are represented by fuzzy subsets NL, NS, ZE, PS, and PL as shown in Figure 3.7.

The boundaries of the direction angle error are set to −10 and +10, which means that any angle error below −10 is assumed to be exactly crisp NL, and any angle error above +10 is assumed to be exactly crisp PL. Any value of the angle error from −10 to −5 is both fuzzy NL and NS. Similarly, any value of the angle error from −5 to 0 would be fuzzy NS and ZE. It is also clear that the value of the angle error above 0 is positive similar to those on negative side.

The driver makes decisions to turn the steering wheel toward right or left in order to bring the direction of the vehicle on the reference and make direction angle error zero. The driver's decisions are based on the state of the deviation angle being N, Z, or P according to the reference direction. From the definitions of (3.82)–(3.84), it can be understood that the vehicle makes an angle of deviation to the right if $\theta_e < 0$, and therefore the driver must turn the steering wheel toward left (LF). If $\theta_e > 0$, then the driver has to turn the steering wheel toward right (RG) because the vehicle is making an angle of deviation to the left. Nothing (NT) is to be done when vehicle is going in the reference direction with a zero deviation angle, $\theta_e = 0$. These statements can be written in a rule-based format as in (3.85). *If* (θ_e is *N*) *then* (turn the steering wheel toward RG).

$$\begin{aligned} &\textit{If } (\theta_e \text{ is } P) \textit{ then } (\text{turn the steering wheel toward RG}) \\ &\textit{else if } (\theta_e \text{ is } N) \textit{ then } (\text{turn the steering wheel toward LF}) \\ &\textit{else } (\text{do nothing (NT)}) \end{aligned} \tag{3.85}$$

where left (LF), right (RG), and nothing (NT) are also fuzzy variables. A driver does not turn the steering wheel suddenly with large steps but turns it gradually toward right or left under normal driving conditions making incremental or decremental changes in rotation angle of the steering wheel. Therefore, the rotation angle steps (changes) of steering wheel should be defined for fuzzy decision-makers in order to have a smoother rotation. The rotation angle changes can be defined as subcategories of LF, and RG as large LF (LLF), small LF (SLF), small RG (SRG), and large RG (LRG), which are given in Figure 3.8.

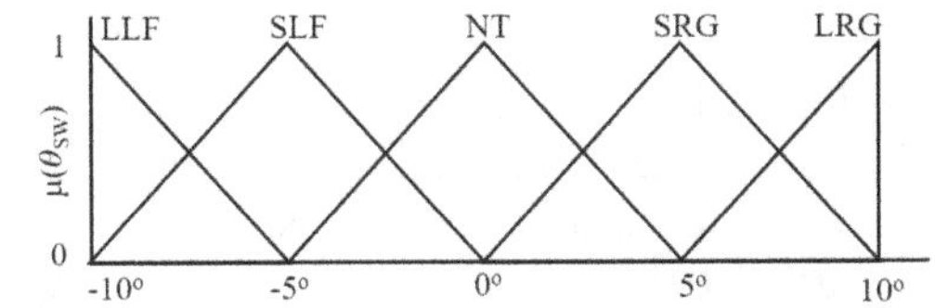

Figure 3.8 Fuzzy subsets for the rotation angle changes of steering wheel

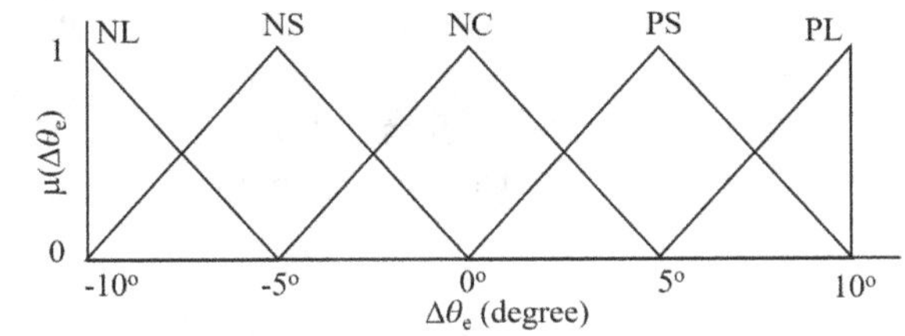

Figure 3.9 Fuzzy subsets for the change in direction angle error

A driver never turns the steering wheel suddenly all the way left or right. He/she observes the change in deviation angle and decides to how much turn the steering wheel. Therefore, to make a healthy conclusion, fuzzy decision-making algorithm needs another input data, which is the amount of change in deviation or in direction angle error, θ_e.

Let us define the change in the error of direction angle as the difference between its two consecutive values as in (3.86).

$$\Delta\theta_e(k) = \theta_e(k) - \theta_e(k-1) \tag{3.86}$$

The amount of change $\Delta\theta_e$ in direction angle error is evaluated as negative large (NL), negative small (NS), no change (NC), positive small (PS), and positive large (PL), which are all fuzzy variables and can be represented by fuzzy membership functions as given in Figure 3.9. Since the change $\Delta\theta$e in direction angle error may increase or decrease, it takes both positive and negative values.

If the amount of change $\Delta\theta_e$ in direction angle error is included in the decision-making process, the rules in (3.85) can be expanded to yield (3.87).

Rule Base	:	*If* θ_e is NL *and* $\Delta\,\theta_e$ is NL *then* $\Delta\,\theta_{SW}$ is LRG
		else if θ_e is NS *and* $\Delta\,\theta_e$ is NL *then* $\Delta\,\theta_{SW}$ is LRG
		else if θ_e is ZE *and* $\Delta\,\theta_e$ is NL *then* $\Delta\,\theta_{SW}$ is SRG
		...
		else if θ_e is PL *and* $\Delta\,\theta_e$ is PS *then* $\Delta\,\theta_{SW}$ is SLF
		else θ_e is PL *and* $\Delta\,\theta_e$ is PL *then* $\Delta\,\theta_{SW}$ is LLF
Data Base	:	θ_e is PS *and* $\Delta\,\theta_e$ is NS
Conclusion	:	$\Delta\theta_{SW}$ is SLF

(3.87)

The rule base in (3.87) is a representation of the expertise and knowledge of human driver. The driving ability of a driver has been stored as the fuzzy rules in the rule base. Therefore, fuzzy decision-maker should have a reasoning algorithm similar to that of a human decision-making process.

Let us consider Figure 3.10, which is a new version of Figure 3.6. Instead of using left (LF) and right (RG) terms, negative (−) and positive (+) are used here. The time response of deviation angle, which is equal to direction angle error, is also depicted in Figure 3.10 in order to understand the movements of the vehicle to stay in the correct direction. The time response of the deviation angle including the states of the change deviation angle is given in Figure 3.11 with more details to make it more understandable.

Now, let us think about what a driver does if deviation angle θ_e and its change $\Delta\theta_e$ is negative, positive, or zero, and list the responses as given in Table 3.1.

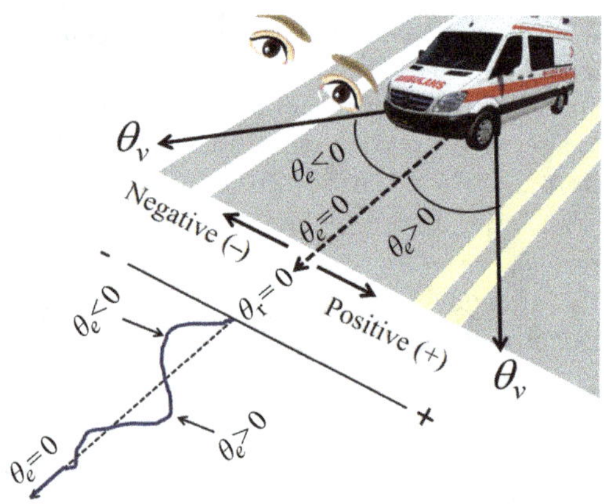

Figure 3.10 Keeping a vehicle on correct route

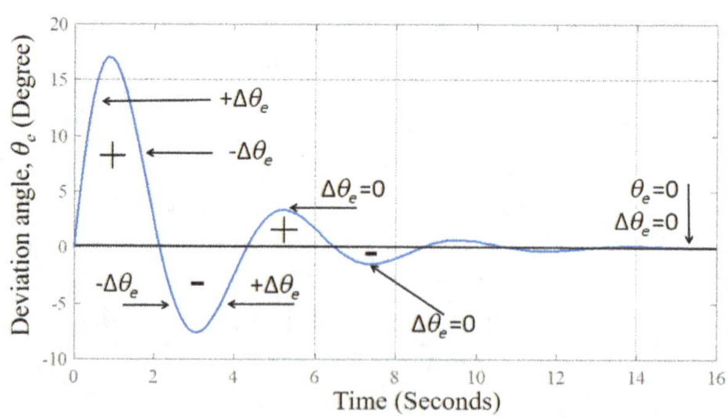

Figure 3.11 Time response of deviation angle

Table 3.1 The states of deviation angle and driver's actions to them

θ_e	+	+	−	−	0	0	+	−	0
$\Delta\theta_e$	+	−	+	−	+	−	0	ö	0
$\Delta\theta_{SW}$	−	−	+	+	−	+	−	+	0

Table 3.2 The states of deviation angle and driver's actions in rule table format

	$\Delta\theta_e<0$	$\Delta\theta_e=0$	$\Delta\theta_e>0$
$\theta_e<0$	$\Delta\theta_{SW}>o$	$\Delta\theta_{SW}>o$	$\Delta\theta_{SW}>o$
$\theta_e=0$	$\Delta\theta_{SW}>o$	$\Delta\theta_{SW}=o$	$\Delta\theta_{SW}<o$
$\theta_e>0$	$\Delta\theta_{SW}<o$	$\Delta\theta_{SW}<o$	$\Delta\theta_{SW}<o$

- If deviation angle θ_e is positive and its change $\Delta\theta_e$ is positive then the driver turns the steering wheel toward right, which is negative side, with step changes.
- If deviation angle θ_e is positive and its change $\Delta\theta_e$ is negative then the driver turns the steering wheel toward right, which is negative side, with step changes.
- If deviation angle θ_e is negative and its change $\Delta\theta_e$ is negative then the driver turns the steering wheel toward left, which is positive side, with step changes.
- If deviation angle θ_e is negative and its change $\Delta\theta_e$ is positive then the driver turns the steering wheel toward left, which is positive side, with step changes.
- If deviation angle θ_e is positive and its change $\Delta\theta_e$ is zero then the driver turns the steering wheel toward right, which is negative side, with step changes.
- If deviation angle θ_e is negative and its change $\Delta\theta_e$ is zero then the driver turns the steering wheel toward left, which is positive side, with step changes.
- If deviation angle θ_e is zero and its change $\Delta\theta_e$ is positive then the driver turns the steering wheel toward left, which is positive side, with step changes.
- If deviation angle θ_e is zero and its change $\Delta\theta_e$ is negative then the driver turns the steering wheel toward right, which is negative side, with step changes.
- If deviation angle θ_e is zero and its change $\Delta\theta_e$ is zero then the driver does nothing.

Table 3.1 can be summarized as follows and as in Table 3.2, which is a rule table representation of driver's actions with positive step changes in upper left corner and negative step changes in lower right corner.

If θ_e is not zero then $\Delta\theta_{SW}$ takes the opposite sign of θ_e

Else $\Delta\theta_{SW}$ takes the opposite sign of $\Delta\theta_e$.

Table 3.2 is a nine-rule table. Three fuzzy verbal variables positive, negative, and zero are used to establish this rule table. If subcategories small and large of negative and positive variables are used, a 25-rule table can be established including the variable zero. A 25-rule table is given in Table 3.3. The content of this 25-rule table is explained in more detail in the next chapter.

Table 3.3 A 25-rule table to stare driver's ability

	$\mathbf{NL}_{\Delta\theta e}$	$\mathbf{NS}_{\Delta\theta e}$	$\mathbf{ZE}_{\Delta\theta e}$	$\mathbf{PS}_{\Delta\theta e}$	$\mathbf{PL}_{\Delta\theta e}$
$\mathbf{NL}_{\theta e}$	$PL_{\Delta\theta SW}$	$PL_{\Delta\theta SW}$	$PS_{\Delta\theta SW}$	$PS_{\Delta\theta SW}$	$ZE_{\Delta\theta SW}$
$\mathbf{NS}_{\theta e}$	$PL_{\Delta\theta SW}$	$PS_{\Delta\theta SW}$	$PS_{\Delta\theta SW}$	$ZE_{\Delta\theta SW}$	$NS_{\Delta\theta SW}$
$\mathbf{ZE}_{\theta e}$	$PS_{\Delta\theta SW}$	$PS_{\Delta\theta SW}$	$ZE_{\Delta\theta SW}$	$NS_{\Delta\theta SW}$	$NS_{\Delta\theta SW}$
$\mathbf{PS}_{\theta e}$	$PS_{\Delta\theta SW}$	$ZE_{\Delta\theta SW}$	$NS_{\Delta\theta SW}$	$NS_{\Delta\theta SW}$	$NL_{\Delta\theta SW}$
$\mathbf{PL}_{\theta e}$	$ZE_{\Delta\theta SW}$	$NS_{\Delta\theta SW}$	$NS_{\Delta\theta SW}$	$NL_{\Delta\theta SW}$	$NL_{\Delta\theta SW}$

Table 3.3 is a symmetrical 25-rule table. The symmetrical structure will be discussed in Chapter 4 after the Mamdani reasoning algorithm is explained. Numerical examples will be given to see how the abilities of an expert are represented by the rule tables and how the decision process is utilized.

3.11 Problems

P3.1 A fuzzy relation R_1 representing the verbal expression "*close to each other*" and another fuzzy relation R_2 representing "*away from each other*" are given below.

$$\mu_{R1}(x,x) = \begin{array}{c|cccc} & 1 & 2 & 3 & 4 \\ \hline 1 & 1 & 0.5 & 0.2 & 0 \\ 2 & 0.5 & 1 & 0.5 & 0.2 \\ 3 & 0.2 & 0.5 & 1 & 0.5 \\ 4 & 0 & 0.2 & 0.5 & 1 \end{array} \qquad \mu_{R2}(y,y) = \begin{array}{c|cccc} & 1 & 2 & 3 & 4 \\ \hline 1 & 0 & 0.2 & 0.5 & 1 \\ 2 & 0.2 & 0 & 0.2 & 0.5 \\ 3 & 0.5 & 0.2 & 0 & 0.2 \\ 4 & 1 & 0.5 & 0.2 & 0 \end{array}$$

Obtain the fuzzy relations expressed as:

R_3 = "The ages of Elif and Hilal are *close to each other* **and** *away from each other*."

R_4 = "The ages of Elif and Hilal are either *close to each other* **or** *away from each other*."

P3.2 A two-dimensional fuzzy set representing "x is close to 6 *and* y is close to 4." is defined in the universe of $R = X \times Y$, where the universes X and Y are given as $X = [0, 12]$ and $Y = [0, 12]$. Trapezoid membership functions are used to represent the fuzzy sets.

(a) Plot the fuzzy set representing the expression "x is close to 6."
(b) Plot the fuzzy set representing the expression "y is close to 4."
(c) Obtain the relation $R = X \times Y$ in terms of fuzzy membership functions.
(d) Develop a MATLAB .m file code and plot this two-dimensional fuzzy set.
(e) Obtain the projections of $R = X \times Y$ in the universes X and Y separately.

P3.3 In an electrical circuit with a voltage source (E) and a resistor (R), it is observed that the current (I) flowing through the resistor is approximately 1.0 A and the resistor is approximately 20 Ohms. The current and resistor are given below. Obtain the source voltage, $E = I \times R$

$$I = \left\{\frac{0}{0} + \frac{0.5}{0.5} + \frac{1}{1} + \frac{0.5}{1.5} + \frac{0}{2}\right\}, \quad R = \left\{\frac{0}{10} + \frac{0.5}{15} + \frac{1}{20} + \frac{0.5}{25} + \frac{0}{30}\right\}$$

P3.4 Two fuzzy numbers zero (ZE) and positive medium (PM) are defined in the universe of $X = \{-5, 5\}$ and are given below.

$$\mathrm{ZE} = \left\{\frac{0}{-2} + \frac{0.5}{-1} + \frac{1}{0} + \frac{0.5}{1} + \frac{0}{2}\right\}, \quad \mathrm{PM} = \left\{\frac{0}{0} + \frac{0.5}{1} + \frac{1}{2} + \frac{0.5}{3} + \frac{0}{4}\right\}$$

(a) Obtain the α-cuts of these two fuzzy subsets for $\alpha = 0$ and $\alpha = 0.5$.
(b) Obtain the support sets of these fuzzy subsets.
(c) Obtain the resultant membership matrix R for the implication "*if* ZE *then* PM."
(d) A third fuzzy subset called positive small (PS) is defined as below. Obtain a fourth fuzzy subset using the compositional rule of inference as $C = \mathrm{PS} \circ R$ and comment on the meaning of fuzzy set C.

$$\mathrm{PS} = \left\{\frac{0}{-1} + \frac{0.5}{0} + \frac{1}{1} + \frac{0.5}{2} + \frac{0}{3}\right\}$$

P3.5 Two series of connected resistors are defined by fuzzy numbers as given below. Obtain total equivalent resistor value $R_T = R_1 + R_2$ in Ohms using the extension principle.

$$R_1 = \left\{\frac{0.5}{3} + \frac{0.8}{4} + \frac{0.6}{5}\right\}\Omega, \mathrm{ve} R_2 = \left\{\frac{0.3}{8} + \frac{1}{9} + \frac{0.4}{10}\right\}\Omega.$$

P3.6 The resistance of conductors varies with temperature. The following equation can be used to define the resistance of a copper conductor as a function of temperature.

$$R_X = R_0(1 + 0.00428 \times T_X)$$

where R_0 is the known resistance when the temperature is 0°C and R_X is the resistance when the temperature is T_X. The resistance value of the cooper conductor at 0°C is approximately 200 Ohms as given below. Obtain the fuzzy resistance value of this conductor if the temperature $T_X = 80$°C.

$$R_0 = \left\{\frac{0.6}{190} + \frac{0.8}{195} + \frac{1}{200} + \frac{0.8}{205} + \frac{0.6}{210}\right\}\Omega$$

P3.7 Fuzzy numbers 8 and 4 are given as

$$A = \left\{\frac{0}{6} + \frac{0.5}{7} + \frac{1}{8} + \frac{0.5}{9} + \frac{0}{10}\right\}, \quad B = \left\{\frac{0}{2} + \frac{0.5}{3} + \frac{1}{4} + \frac{0.5}{5} + \frac{0}{6}\right\}$$

Table P3.9 Parameters of the fuzzy numbers A, B, and C

Fuzzy number	Core set	Support set
A	4	{3, 4, 5}
B	2	{1, 2, 3}
C	3	{2, 3, 4}

If the fuzzy number C is defined as $C = f(A, B) = A/B$, obtain its fuzzy value and comment about the meaning of fuzzy C. Which crisp number is represented by fuzzy C? Can fuzzy C be considered as a fuzzy number?

P3.8 A fuzzy set A is defined in the universe of $X = \{1, 2, 3, 4, 5\}$ as follows. Map the fuzzy set A onto universe of Y using the crisp function $f(x) = 4x-1$.

$$A = \left\{\frac{0.0}{1} + \frac{0.5}{2} + \frac{1}{3} + \frac{0.5}{4} + \frac{0.0}{5}\right\}$$

P3.9 Two fuzzy calculations are given as $y_1 = A + B$ and $y_2 = A \times C$, where A, B, and C are fuzzy numbers and defined in the universes of X, Y, and Z with the parameters given below.

(a) Plot and show the fuzzy numbers A, B, and C in the same plot.
(b) Obtain $y_1 = f(A,B)$ and $y_2 = f(A,C)$ using the extension principle.

P3.10 The fuzzy set given below is defined in the universe of $X = \{-5, -4, -3, -2, -1, 0, 1, 2, 3, 4, 5\}$. Map the fuzzy set $A \in X$ from the universe of X onto universe of Y as a fuzzy set $B \in Y$ using the crisp function $f(x) = 2x^2 - 3x + 1$.

$$A = \frac{0}{-5} + \frac{0.2}{-4} + \frac{0.4}{-3} + \frac{0.6}{-2} + \frac{0.8}{-1} + \frac{1}{0} + \frac{0.8}{1} + \frac{0.6}{2} + \frac{0.4}{3} + \frac{0.2}{4} + \frac{0}{-5}$$

P3.11 The success of a company depends upon mainly two factors, which are the work efficiency and quality of the manufactured goods. The work efficiency and goods quality are given by the fuzzy subsets WE and GQ below. Obtain the fuzzy success of this company.

$$\mathrm{WE} = \frac{0.0}{0} + \frac{0.5}{0.5} + \frac{1.0}{1.0} + \frac{0.5}{1.5} + \frac{0.0}{2.0}$$

$$\mathrm{GQ} = \frac{0.0}{0} + \frac{0.2}{0.2} + \frac{0.4}{0.4} + \frac{0.6}{0.6} + \frac{0.8}{0.8} + \frac{1.0}{1.0}$$

P3.12 The heat control of a room is based on the room temperature and light coming from the windows in order to use the heater during the daytime only.

(a) Partition the universes of heat level, light level, and switching into fuzzy subcategories based on your intuitions.

(b) Write at least three expressions to turn the heater's switch on/off depending on the room temperature and light level.
(c) Write a MATLAB .m file to simulate your expressions in part (b).

P3.13 Membership function of linguistic variable *cold* (C) and membership matrix of relation R representing the expression "*if cold then turn on the heater*" are given below.

$$\text{Cold} = \frac{1}{-5} + \frac{0.6}{0} + \frac{0.4}{5} + \frac{0.0}{10}, \quad R = \begin{bmatrix} 1 & 0.6 & 0.4 & 0.2 \\ 0.6 & 0.6 & 0.4 & 0.1 \\ 0.4 & 0.4 & 0.4 & 0.1 \\ 0.1 & 0.1 & 0.1 & 0.0 \end{bmatrix}$$

(a) How much is the heater *on*?
(b) Obtain the numerical value of the expression "*if cold then on.*"
(c) Obtain a relation matrix to represent the expression "in not cold then not on."

P3.14 The use of a battery is defined by the relation $R = I \times H$ where I is the current in Amperes drawn from the battery and H is the duration of the current drawn from the battery. The relation R represents the total lifetime of the battery. The possibility of fuzziness of each independent factor is about ±20%. Normalized fuzzy sets of these factors are given as

$$I = \left\{\frac{0.0}{0.0} + \frac{0.5}{0.5} + \frac{1}{1} + \frac{0.5}{1.5} + \frac{0.0}{2.0}\right\}, \quad \text{H} = \left\{\frac{0.0}{0.0} + \frac{0.5}{1} + \frac{1}{2} + \frac{0.5}{3} + \frac{0.0}{4}\right\}$$

(a) Obtain the relation $R = I \times H$
(b) The energy stored in the battery is finished after being used fuzzy four hours, which is defined as

$$\text{FF} = \left\{\frac{0.0}{0.0} + \frac{0.5}{2} + \frac{1}{4} + \frac{0.5}{6} + \frac{0.0}{8}\right\}$$

The amount of amper × hour (Ah) of the battery is obtained as a relation between current and usage time as asked in part (a). What can you say about the current drawn from the battery?

P3.15 Two fuzzy sets young (Y) and high school student (HSS) defined in the universe of ages are given below.

$$\textit{Young} = \frac{1}{15} + \frac{0.8}{20} + \frac{0.6}{25} + \frac{0.4}{30} + \frac{0.2}{35} + \frac{0}{40}$$

$$\textit{High School Student} = \frac{0}{12} + \frac{0.5}{14} + \frac{1}{16} + \frac{0.5}{18} + \frac{0}{20}$$

A fuzzy implication is stated as "*If* Hilal is a *high school student then* she is *young*." and obtained by the Cartesian product as given below, which yields Table E3.21.

Table E3.21 The relation matrix for high school student × young

		Young					
		1.0	**0.8**	**0.6**	**0.4**	**0.2**	**0.0**
High school student	**0.0**	0	0	0	0	0	0
	0.5	0.5	0.5	0.5	0.4	0.2	0
	1.0	1	0.8	0.6	0.4	0.2	0
	0.5	0.5	0.5	0.5	0.4	0.2	0
	0.0	0	0	0	0	0	0

$$R = (\text{High School Student}) \times (\text{Young})$$

If Elif's age is defined as below, what can be said about Elif being a high school student.

$$\text{Elif's age} = \frac{1}{5} + \frac{0.8}{8} + \frac{0.6}{10} + \frac{0.4}{12} + \frac{0.2}{14} + \frac{0}{16}$$

P3.16 A fuzzy reasoning process is given as

Rule base	:	*If x is* small *and y* is medium then *z* is big
Data base	:	*x* is very small *and y* is close to medium
Conclusion	:	$z = ?$

The fuzzy subsets small (S), medium (M), and big (B) are defined in the interval {0, 10} as below. Show the fuzzy reasoning process in a graphical format.

$$Small = \frac{1}{0} + \frac{0.5}{2.5} + \frac{0}{5} + \frac{0}{0.75} + \frac{0}{10}$$

$$Medium = \frac{0}{0} + \frac{0.5}{2.5} + \frac{1}{5} + \frac{0.5}{7.5} + \frac{0}{10}$$

$$Big = \frac{0}{0} + \frac{0}{2.5} + \frac{0}{5} + \frac{0.5}{7.5} + \frac{1}{10}$$

P3.17 In the thermal characteristics of conductors show that the conductor resistance increases as it gets heated. Use this information and generate a fuzzy implication by assigning some fuzzy subsets to the changes in heat and resistance of a copper conductor.

P3.18 In a room heating process, the difference between the set value and actual room temperature is called the control error (E), and the difference between two values of control error from two consecutive samples is called the change in error (CE). It is known that the time response of the control error is similar to that of an oscillatory decaying second-order system and can be

Table P3.18 Parameters of the fuzzy membership functions for Problem 3.18

Membership function	**Universe of discourse**	x_1 **as in (2.7)**	x_2 **as in (2.7)**	x_3 **as in (2.7)**
NE	E	−2	−2	0
ZE	E	−2	0	5
PE	E	0	5	5
NCE	CE	−0.2	−0.2	0
ZCE	CE	−0.2	0	0.5
PCE	CE	0	0.5	0.5
NCV	CV	−1	1	0
ZCV	CV	−1	0	1
PCV	CV	0	1	1

made zero by increasing or decreasing the voltage applied to the heater. Let us call these increment and decrement processes the change in voltage (CV).

The universe of error in the control system described above is partitioned into three subcategories as negative error (NE), zero error (ZE), and positive error (PE). Similarly, the universe of the change in error is also partitioned into three subcategories negative change (NCE), zero change (ZCE), and positive change (PCE). In order to make the error to be zero, the applied voltage to the heater is increased or decreased by fuzzy steps such as negative change in voltage (NCV), zero change in voltage (ZCV), and positive change in voltage (PCV). All these fuzzy subsets are represented by parametrized triangular fuzzy membership functions, whose parameters are given in Table P3.18. Note that the error and the change in error are in °C and the maximum change in voltage is 1 V.

Develop a series of control rules by considering the values of error and the change in error to give some increments and decrements to the voltage applied to the heater so that the error will be zero.

P3.19 Two fuzzy numbers defined in the universe $X = \{-5, 5\}$ are given below.

$$A = \text{Zero} = \left\{\frac{0}{-2} + \frac{0.5}{-1} + \frac{1}{0} + \frac{0.5}{1} + \frac{0}{2}\right\}$$

and

$$B = \text{Positive Medium} = \left\{\frac{0}{0} + \frac{0.5}{1} + \frac{1}{2} + \frac{0.5}{3} + \frac{0}{4}\right\}$$

(a) Obtain a fuzzy relation R that represents the implication *If A then B*.

(b) If another fuzzy number C is defined as follows in the same universe of X, obtain the fuzzy number $D = C \circ R$ by using max–min composition.

$$C = \text{Positive Small} = \left\{\frac{0}{-1} + \frac{0.5}{0} + \frac{1}{1} + \frac{0.5}{2} + \frac{0}{3}\right\}$$

P3.20 A city has four main entrance points (E1, E2, E3, and E4) from the surrounding highway. Sensors are placed at each entrance counting the vehicles entering to and going out of the city. These four entrances are also connected to each other by four main streets (S1, S2, S3 and S4). Depending on the traffic on the streets and the entrance points, the vehicles are guided toward the streets and the entrance points with light traffic. The guiding monitors are placed at each entrance and at cross-sections of the streets.

Develop a rule base system such that the traffic conditions at the entrance points and at the streets are used as inputs and the new recommended direction of the traffic is the output.

P3.21 A crisp universe of discourse X is defined in the interval of [0, 10] and partitioned into three fuzzy subsections as *small, medium*, and *large.* Triangular-type fuzzy sets are used to represent these fuzzy subsets. Draw a software algorithm that gives you the membership value of any x on X in the fuzzy subsets *small, medium*, and *large.*

References

[1] Jiang J.A., Wang J.C., Kuo K.C., *et al.* 'Analysis of the junction temperature and thermal characteristics of photovoltaic modules under various operation conditions'. *Energy*. 2012;44(1):292–301.

[2] Jang S.H. and Shin M.W. 'Thermal characterization of junction in solar cell packages'. *IEEE Electron Device Letters*. 2010;31(7):743–745.

[3] *IEEE Recommended Practice for Monitoring Electric Power Quality*. IEEE Std 1159-2009 (Revision of IEEE Std 1159-1995); 2009. pp. c1–81.

[4] Klir G. and Folger T. *Fuzzy Sets, Uncertainty, and Information*. Englewood Cliffs, NJ: Prentice-Hall; 1988.

[5] *IEEE Recommended Practices and Requirements for Harmonic Control in Electrical Power Systems*. IEEE Std 519-2014 (Revision of IEEE Std 519-1992); New York, NY: IEEE.

[6] Zadeh L.A. 'Fuzzy sets as a basis for a theory of possibility'. *Fuzzy Sets and Systems*. 1978;1(1): 3–28.

[7] Zadeh L.A. 'Outline of a new approach to the analysis of complex systems and decision processes'. *IEEE Transactions on Systems, Man, and Cybernetics*. 1973;3(1):28–44.

[8] Zadeh L.A. 'The concept of linguistic variable and its application to approximate reasoning. Part I, II, and III'. *Information Sciences*. 1975;8(3):199–249, 8(4):301–357, 9(1):43–80.

[9] Altas I.H. and Sharaf A.M. 'A fuzzy logic power tracking controller for a photovoltaic energy conversion scheme', *Electric Power Systems Research Journal*. 1992;25(3):227–238.

[10] Zadeh L.A. 'A theory of approximate reasoning' in Hayes J. And Mikulich L. I. (eds.). *Machine Intelligence*. New York: Halstead Press; 1979, vol. 9, pp. 149–194.

[11] Zadeh L.A. 'Fuzzy sets'. *Information and Control*. 1965;8(3):338–353.

[12] Jang J.S.R. 'ANFIS: adaptive-network-based fuzzy inference system', *IEEE Transactions on Systems, Man, and Cybernetics*.1993;23(3):665–685.

[13] Jang J.S.R., Sun C.T. and Mizutani E. *Neuro-Fuzzy and Soft Computing: A Computational Approach to Learning and Machine Intelligence*. Englewood Cliffs, NJ: Prentice-Hall; 1996.

[14] Cordon O., Jesus M.J. and Herrera F. 'A proposal on reasoning methods in fuzzy rule-based classification systems'. *International Journal of Approximate Reasoning*. 1999;20(1):21–45.

[15] Ishibuchi H. and Yamamoto T. 'Rule weight specification in fuzzy rule-based classification systems'. *IEEE Transactions on Fuzzy Systems*. 2015;13 (4):428–435.

[16] Kandel A. (ed.). *Fuzzy Expert Systems*. Boca Raton, FL: CRC Press; 1991.

[17] Siler W. and Buckley J.J. *Fuzzy Expert Systems and Fuzzy Reasoning*. Hoboken, NJ: John Wiley & Sons; 2005.

[18] Takagi T. and Sugeno M. 'Fuzzy identification of systems and its applications to modeling and control'. *IEEE Transactions on Systems, Man, and Cybernetics*. 1985;15(1):116–132.

[19] Hellendoorn H. and Driankov D. (eds.). *Fuzzy Model Identification: Selected Approaches*. London: Springer; 1997.

[20] Kosko B. 'Fuzzy associative memory systems' in Kandel A. (ed.). *Fuzzy Expert Systems*. Boca Raton, FL: CRC Press; 1991. pp. 135–162.

[21] Chung F.L. and Lee T. 'On fuzzy associative memory with multiple-rule storage capacity'. *IEEE Transactions on Fuzzy Systems*. 1996;4(3): 375–384.

[22] Mamdani E.H. 'Application of fuzzy algorithms for control of simple dynamic plant'. *IEE Proceedings*. 1974;121(12):1585–1588.

[23] Lee C.C. 'Fuzzy logic in control systems: fuzzy logic controller. I–II'. *IEEE Transactions on Systems, Man, and Cybernetics*. 1990;20(2): 404–435.

Chapter 4
Fuzzy logic controller (FLC)

Fuzzy logic controller (FLC) theory and design approaches are given in this chapter. Machines operate the human way as if run by expert operators and steps toward FLCs are given. Rule development, the way of putting expert ideas into rules, and inference system structure are studied. From crisp input variables to crisp output, all process is discussed and shown. Defuzzification, rule processing, fuzzy reasoning, and crisp output after defuzzification are explained. Using the FLC as an optimization tool is introduced as a new subsection for this edition. User-developed FLC examples are given.

4.1 Introduction

Theoretical studies in approximate and fuzzy reasoning have increased with the application of fuzzy logic in engineering, medical science, and social science. In engineering, fuzzy logic and reasoning algorithms are used in control, systems identification, modeling, optimization, decision-making, management, and many other similar applications related to industrial developments and management. Zadeh has suggested new concepts on linguistic variables and decision-making processes, Mamdani and his colleagues have applied fuzzy logic to a pressure control system, Takagi and Sugeno used it in system modeling and identification, and Tsukamoto developed a reasoning algorithm. Many other researchers on fuzzy sets and fuzzy logic published very important papers showing their studies in the topic. Among all those studies, the approaches introduced by Mamdani, Takagi–Sugeno, and Tsukamoto have attracted the others. All these approaches use fuzzy processing algorithms that are called by different names as mentioned in the section "Fuzzy Reasoning" in the previous chapter. Since fuzzy reasoning and fuzzy decision include the process of fuzzy data, fuzzy subsets, and fuzzy rules a more generalized name "fuzzy processor" is used to call all of those methods and approaches. In this chapter, the most used fuzzy reasoning approaches by Mamdani, Takagi–Sugeno, and Tsukamoto are discussed and fuzzy processing steps in these algorithms are given.

4.2 Mamdani fuzzy reasoning

After the introduction of the fuzzy logic concept by Lotfi A. Zadeh in 1965 [1], Mamdani and his colleagues developed and used a fuzzy logic controller (FLC) in a

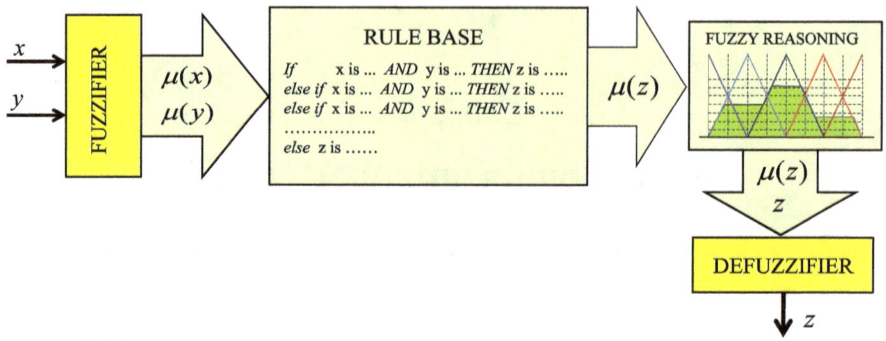

Figure 4.1 Mamdani fuzzy reasoning algorithm

hydraulic pressure system [2–4]. The inference mechanism used by Mamdani is known as Mamdani fuzzy reasoning. This reasoning algorithm uses the *max–min* process explained earlier as the compositional rule of inference. Mamdani reasoning algorithm is similar to the reasoning algorithms given in Section 3.9. However, input data and output in the Mamdani algorithm are not fuzzy but crisp. Due to this property, the Mamdani reasoning algorithm is very suitable to be used in fuzzy control systems. In industrial control, the input and output data are usually measurable and used in the process. The output is measured and feedback to input to generate a new control signal by checking its value and whether an adjustment is necessary. In addition to fuzzy reasoning algorithms given in Section 3.9, there are two additional sections here. The crisp input variables are converted to fuzzy variables before being used in fuzzy reasoning process and fuzzy output variable is converted to crisp output variable to be used in real-time applications. Therefore, besides the *rule base*, *database*, and *conclusions* given in fuzzy reasoning algorithms in Chapter 3, two new sections called *fuzzifier* and *defuzzifier* are included in the Mamdani reasoning algorithm as depicted in Figure 4.1.

The crisp inputs $x \in X$ and $y \in Y$ in Mamdani fuzzy reasoning are mapped from crisp input universes X and Y onto fuzzy input universe in terms of fuzzy membership values $\mu_i(x)$ and $\mu_j(y)$ where i and j are the number of fuzzy sets in which x and y have non zero membership degrees. After the rule processing in the rule base, the fuzzy outputs from each rule are obtained and combined in the reasoning stage of output universe Z. The crisp output is then obtained by applying a process called *defuzzification*, which is usually an averaging math function. *Defuzzification* will be explained later in this chapter.

Example 4.1

The generated maximum power in a PV system changes as a function of solar irradiation level and operating temperature. Some information about the PV systems was described in Examples 3.19 and 3.20. The maximum power points at some solar irradiation and temperature levels are defined as a relation PM.

The crisp universes of solar irradiation, temperature, and maximum power points are partitioned into subcategories as shown in Figures E4.1(a), E4.1(b), and E4.1(c). The relation matrix PM is represented by nine rules, which are written below and depicted in Table E4.1.

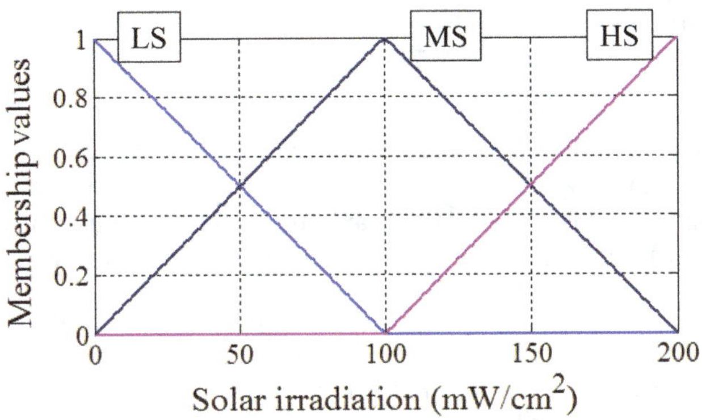

Figure E4.1(a) Subcategories of solar irradiation

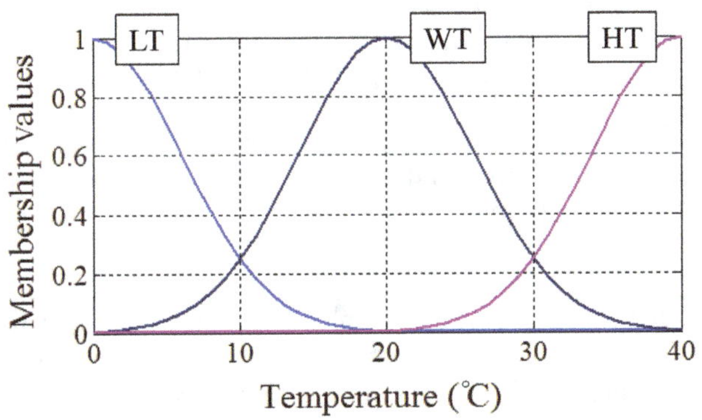

Figure E4.1(b) Subcategories of ambient temperature

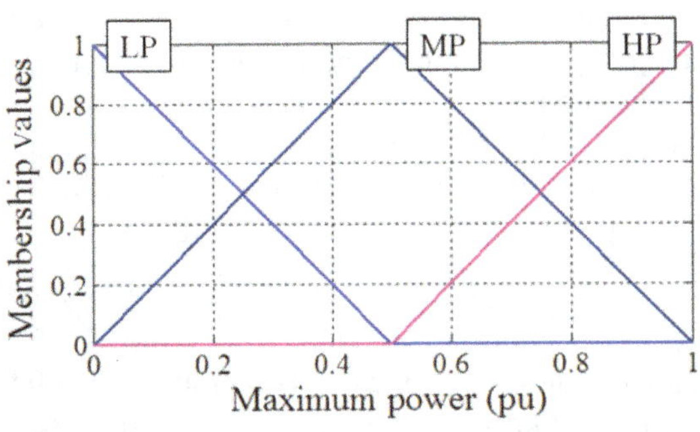

Figure E4.1(c) Subcategories of PV power

Table E4.1 The rule table for Example 4.1

	LT	**WT**	**HT**
LS	LP	LP	LP
MS	MP	MP	MP
HS	HP	HP	HP

Find the power output in pu of this fuzzy rule base system if the solar irradiation level is 125 mW/cm^2 and the temperature is 15°C. Assume that the base power used for pu conversion of the maximum power is equal to the maximum power available from the PV system under the best conditions where it is located.

Rule 1 : *if* s is LS *and* t is LT *then* p is LP
Rule 2 : *else if* s is LS *and* t is WT *then* p is LP
Rule 3 : *else if* s is LS *and* t is HT *then* p is LP
Rule 4 : *else if* s is MS *and* t is LT *then* p is MP
Rule 5 : *else if* s is MS *and* t is WT *then* p is MP
Rule 6 : *else if* s is MS *and* t is HT *then* p is MP
Rule 7 : *else if* s is HS *and* t is LT *then* p is HP
Rule 8 : *else if* s is HS *and* t is WT *then* p is HP
Rule 9 : *else if* s is HS *and* t is HT *then* p is HP

where LS, MS, and HS stand for low, medium, and high solar irradiation. LT, WT, and HT are used for low, warm, and hot temperature. LP, MP, and HP represent low, medium, and high maximum power.

Answer 4.1

The solution to this problem is done graphically and depicted in Figure E4.1(d). The crisp solar irradiation level of 125 mW/cm^2 has non-zero membership degree in fuzzy subsets medium (MS) and high (HS). Since the membership value of solar irradiation is zero in fuzzy subset low (LS), the line starting with LS in Table E4.1 does not have any effects on the output. Therefore, the first three rules are not active and they are not fired. Similarly, the crisp temperature value of 15°C has non-zero membership values in the fuzzy subsets low (LT) and warm (WT), only. Since the membership degree is zero in the fuzzy subset hot (HT), the column of the fuzzy rule matrix has no effect on the output. Therefore, the rules 3, 6, and 9 are not fired. The active rules, which are the fourth, fifth, seventh, and eighth rules, are fired and only these active rules are shown in Figure E4.1(d).

The result of each rule is obtained as follows:

$$\begin{aligned}
\mu_{P4}(p) = \mu_{MP}(p) &= \min(\mu_{MS}(s), \mu_{LT}(t)) = \min(0.75, 0.04) = 0.04 \\
\mu_{P5}(p) = \mu_{MP}(p) &= \min(\mu_{MS}(s), \mu_{LT}(t)) = \min(0.75, 0.7) = 0.7 \\
\mu_{P7}(p) = \mu_{HP}(p) &= \min(\mu_{HS}(s), \mu_{LT}(t)) = \min(0.25, 0.04) = 0.04 \\
\mu_{P8}(p) = \mu_{HP}(p) &= \min(\mu_{HS}(s), \mu_{WT}(t)) = \min(0.25, 0.7) = 0.25
\end{aligned}$$

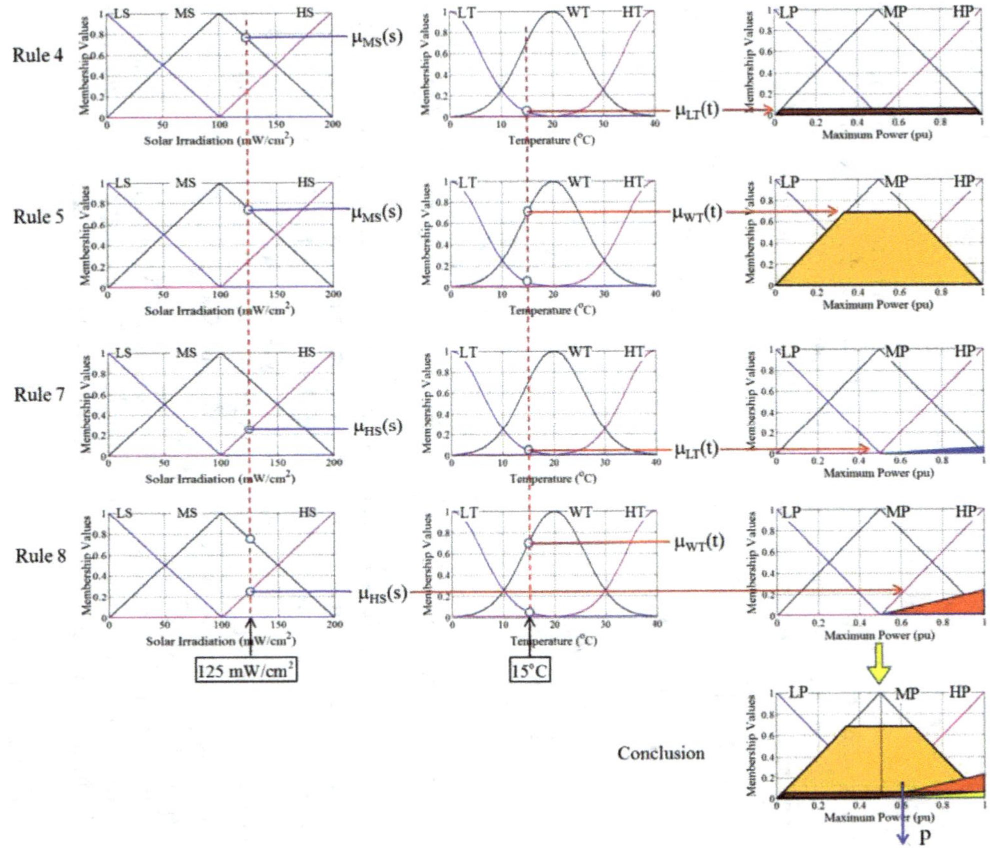

Figure E4.1(d) Mamdani fuzzy reasoning algorithm

Each rule results in a membership value, which is obtained as the minimum of the membership values coming from the input spaces. The resultant membership value from each rule is a weighting factor used to determine the effectiveness of the rule on the output space. For example $\mu_{P4}(p) = 0.04$ is the weighting factor from Rule 4 and it shows how much the fuzzy set *medium* power (MP) is effective as depicted by the shaded area in the upper right corner in Figure E4.1(d). Similarly, $\mu_{P5}(p) = 0.7$ is the weighting factor from Rule 5 and it shows the effectiveness of the fuzzy set *medium* power (MP) by shaded area as seen in the second graph on the upper right corner in Figure E4.1(d). The effects of Rules 7 and 8 are also shown by shaded areas in the same figure.

The final fuzzy output in this problem is the union of the areas obtained from each active rule because the rules are connected to each other by the term *else*. The unified fuzzy output is seen in the lower right corner of Figure E4.1(d). Mamdani fuzzy reasoning has another section called defuzzifier as depicted in Figure 4.1. Defuzzifier is used to convert the fuzzy output to a crisp output by extracting a crisp value from the unified shaded area. There are several defuzzification methods to obtain a crisp output. One of these methods is called the *center of area* (COA) method and is used as follows for this example:

$$p = \frac{\sum_{i=1}^{n} \mu_{Ri}(p) \cdot p_{Ri}}{\sum_{i=1}^{n} \mu_{Ri}(p)}$$

$$p = \frac{(0.04)(0.5) + (0.7)(0.5) + (0.04)(1) + (0.7)(1)}{0.04 + 0.7 + 0.04 + 0.7} = \frac{1.11}{1.48} = 0.75 \text{ pu.}$$

where the subscription Ri is the active rule number and $\mu_{Ri}(p)$ is the weighting factor obtained from the ith Rule. p_{Ri} is the crisp value as $p_{Ri} \in P$ and it is equal to the core of the output fuzzy set for the ith Rule. P is the output universe and it is the maximum power of a photovoltaic (PV) energy system (PVES) as described in Example 4.1. The maximum power that the PVES can generate under the given solar irradiation and temperature levels is 0.75 pu. Note that 1.0 pu is the highest power that can be generated by this PVES.

It was stated earlier and shown in Figure 4.1 that Mamdani fuzzy reasoning algorithm has four stages *fuzzification*, *rule base*, *conclusion*, and *defuzzification* except inputs and outputs. These four stages are explained in detail next.

4.2.1 Fuzzification

The crisp inputs entered into the input spaces are considered as triggering signals. They activate the rules in which they have non-zero membership values in the fuzzy subsets used in the rules. In Example 4.1, the input signal s = 125 mW/cm^2 has non-zero membership degrees in the fuzzy subsets medium (MS) and high (HS) in the universe of solar irradiation level. Therefore, the input s = 125 mW/cm^2 activates the rules that include fuzzy subsets MS and HS. The input signal t = 15°C has non-zero membership degrees in the fuzzy subsets low (LT) and warm (WT) in the universe of temperature and activates the rules that include fuzzy subsets LT and WT. Actually, the fuzzification is nothing but the activation of the rules. The crisp inputs are converted to a fuzzy membership function of the active rules. This process is shown in Figure 4.2 for the active rules in Example 4.1.

The membership values of s = 125 mW/cm^2 in fuzzy subsets MS and HS in one of the input universes are obtained for four active rules and carried out of the fuzzification as $\mu_{\mathrm{MS}}(\mathrm{s})$ from Rules 4 and 5, and as $\mu_{\mathrm{HS}}(\mathrm{s})$ from Rules 7 and 8. The membership values of t = 15°C in fuzzy subsets LT and WT in the other input universe are obtained for four active rules and carried out of the fuzzification as μ_{LT} (t) from Rules 4 and 7, and as $\mu_{\mathrm{WT}}(t)$ from Rules 5 and 8. After the fuzzification, one membership value from each input universe is obtained at the output of the fuzzification. Since there are two input universes in Example 4.1, consequently there are two membership values at the output of the fuzzification from each rule. Therefore, it can be concluded that the number of the membership values from each rule is equal to the number of the input spaces. There are four active rules for the

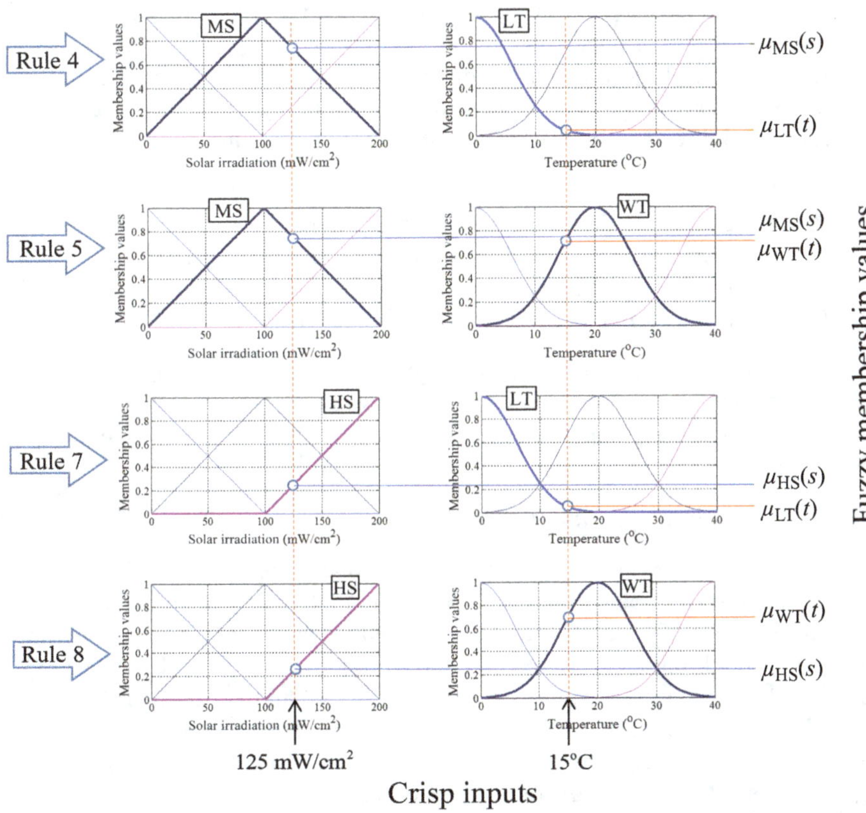

Figure 4.2 Fuzzification

given two crisp inputs in Example 4.1, and therefore there are eight fuzzy membership values at the output of the fuzzifier.

Two membership values from each rule are then used in the reasoning section to obtain the weighting factors in order to determine the effectiveness of the related rule at the output space.

4.2.2 Fuzz rule base

Fuzzy rule base is the brain of fuzzy reasoning algorithms. The rules represent the operational behavior of a system, the decision process, and the thoughts of an expert. Therefore, the construction of a rule base system is very important to get the correct decision. The expert's ideas, knowledge, intuitions, and tricks are included in a decision process, therefore in the rule base. In technical problems, the rules are usually developed based upon input–output characteristics of the systems.

Let us assume that the rules given in Table E4.1 represent an expert's ideas about the maximum power generated by a PVES at various temperature and solar irradiation levels. Although there are nine rules in Table E4.1, only four of them are active for the given input signals. Different rules will be activated for different input signals. Due to the partitioning structure of the universes, maximum of four rules can be activated at once.

The active rule base used in Example 4.1 consists of fuzzification and reasoning parts of Mamdani reasoning algorithm shown in Figure E4.1(d). These parts of the referred figure are a representation of the active rules in (4.1) and are shown in Figure 4.3.

$$\begin{aligned} &\textit{Rule } 4 \ : \textit{else if } s \text{ is MS } \textit{and } t \text{ is LT } \textit{then } p \text{ is MP} \\ &\textit{Rule } 5 \ : \textit{else if } s \text{ is MS } \textit{and } t \text{ is WT } \textit{then } p \text{ is MP} \\ &\textit{Rule } 7 \ : \textit{else if } s \text{ is HS } \textit{and } t \text{ is LT } \textit{then } p \text{ is HP} \\ &\textit{Rule } 8 \ : \textit{else if } s \text{ is HS } \textit{and } t \text{ is WT } \textit{then } p \text{ is HP} \end{aligned} \tag{4.1}$$

After the fuzzification, a fuzzy membership value is obtained from each input universe for each rule corresponding to the crisp input. The membership values obtained from different input universes for each rule are used to obtain the weighting factor of that rule as

$$w_{Ri}(p) = \min(\mu_{\text{AS}}(s), \mu_{\text{BT}}(t)) \tag{4.2}$$

where Ri stands for the ith rule, AS stands for the one of fuzzy subsets LS, MS, and HS in the universe of solar irradiation for Example 4.1. BT stands for one of the fuzzy subsets LT, WT, and HT defined in the universe of temperature. Therefore, $w_{Ri}(p)$ is the weighting factor to be used in the output space of the Rule i. Equation (4.2) is used to obtain the weighting factors w_4, w_5, w_7, and w_8 for the rules R_4, R_5, R_7, and R_8, respectively. The weighting factors are used to find the effectiveness of the corresponding fuzzy subset in output space. The weighting factors give the height of the shaded areas in the output space as shown in Figures 4.3 and 4.4. The shaded areas from each rule are then combined as the final fuzzy conclusion, which is discussed next.

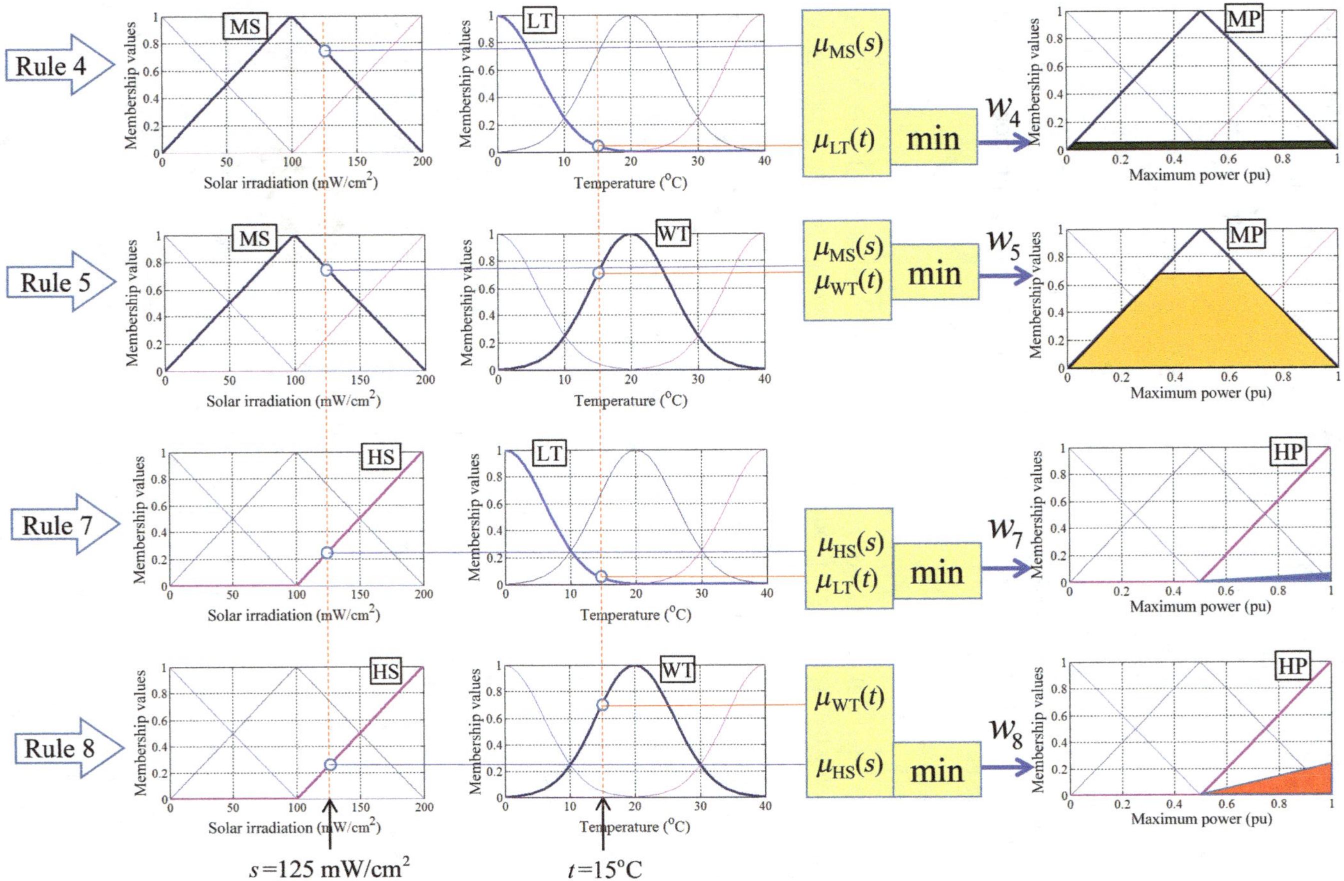

Figure 4.3 Fuzzy rule base

4.2.3 Fuzzy conclusion

The effectiveness of the fuzzy sets for each rule in output space is determined by the weighting factors, which are membership values defined in the interval [0, 1]. The weighting factors show how much of the output fuzzy subset is affective for the related rule. The degree of the effectiveness of the output fuzzy subsets is shown by shaded areas in Figure 4.4.

After the effect of each rule is determined by the weighting factor using (4.2), the shaded areas are obtained using (4.3).

$$\mu_{Ri}(p) = \min[w_{Ri}, \mu_{CP}(p)] \tag{4.3}$$

where $\mu_{Ri}(p)$ is the membership function representing the shaded area of the ith rule. The final conclusion is the union of the result membership functions from all active rules and written as

$$\mu_{RT}(p) = \bigcup_{i=1,N} \mu_{Ri}(p) \tag{4.4}$$

where N is the total number of the rules. The effect of each rule can also be obtained using the arithmetical product as in (4.5) instead of the min operator given in (4.3).

$$\mu_{Ri}(p) = w_{Ri} \cdot \mu_{CP}(p), \text{ for } i = 1, 2, \ldots, N \tag{4.5}$$

where N is the total number of the rules. As stated in the previous chapter, the min operator is preferred to be used in examples of this book.

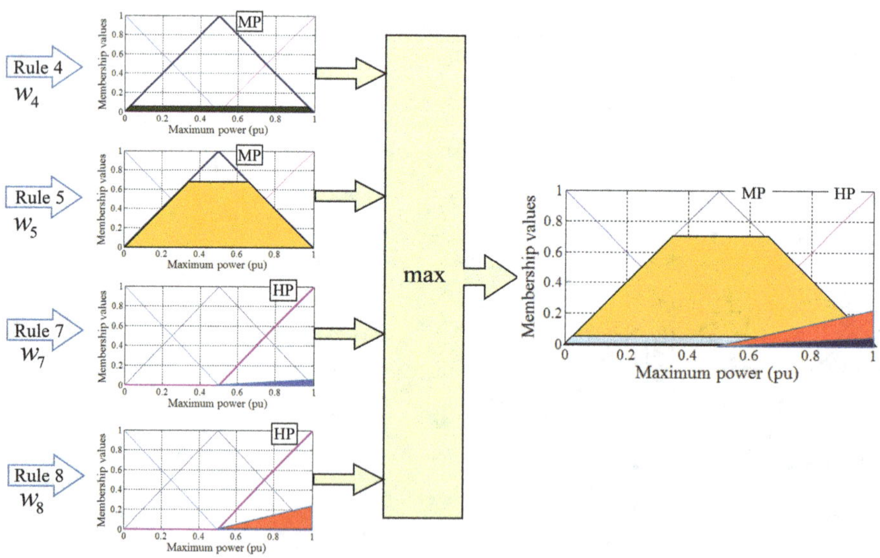

Figure 4.4 Fuzzy conclusion

Practice 4.1

Write a MATLAB® .m file in order to simulate Example 4.1 so that the shaded fuzzy subsets are plotted as the fuzzy conclusion for given crisp input values.

4.2.4 Defuzzification

The union of the conclusions from individual rules is the final conclusion obtained using (4.2), (4.3), and (4.4). The final fuzzy conclusion is usually an area similar to the one shown on the right in Figure 4.4. In order to explain different defuzzification methods a more generalized fuzzy conclusion area given in Figure 4.5 is used in this section. In this figure, $\mu'(z) \leq 1$, and C_T' is the total area representing the final conclusion as the union of all individual rules. Z is the universe of discourse.

Defuzzification is done in order to convert the fuzzy conclusion into a crisp value. In real-time control system applications, the decision is usually a crisp signal used to trigger real-time devices such as electronic switches and relays. Therefore, defuzzification is required in many applications while the fuzzy conclusion may be sufficient for some.

Defuzzification may be explained as the extraction of a single crisp value from an area that represents the fuzzy conclusion of a fuzzy reasoning process. The most known defuzzification methods are listed below.

- Lower maximum
- Upper maximum
- Mean of maxima
- Equal areas
- Center of areas
- Center of gravity

The lower maximum (***LM***) is used as one of the defuzzification methods. It is equal to the first element of the core set of the final fuzzy subset representing the final conclusion as shown in Figure 4.5 and defined in (4.6).

$$z_{\text{LM}}(z) = \{z_i, \text{such that } \mu(z_i) > \mu(z_j) \text{ for } i \neq j \text{ and } z_i < z_k \text{ for } \mu(z_i) = \mu(z_k)\} \tag{4.6}$$

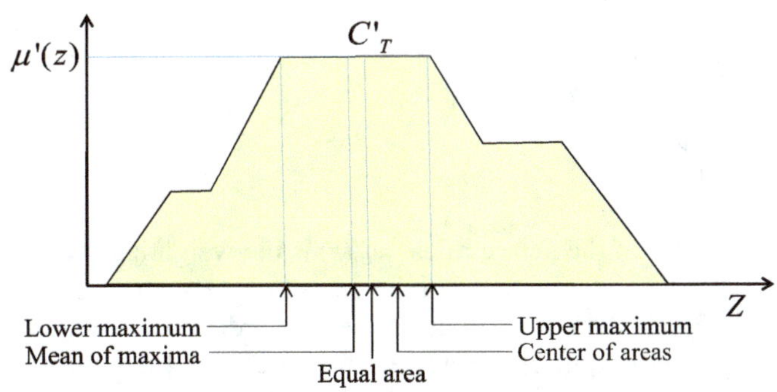

Figure 4.5 Defuzzification methods

The upper maximum (*UM*) is used as another defuzzification method. It is equal to the last element of the core set of the final fuzzy subset representing the final conclusion as shown in Figure 4.5 and defined in (4.7).

$$z_{\mathrm{UM}}(z) = \{z_i, \text{such that}\ \mu(z_i) > \mu(z_j)\ \text{for}\ i \neq j\ \text{and}\ z_i > z_k\ \text{for}\ \mu(z_i) = \mu(z_k)\} \tag{4.7}$$

The mean of maxima (*MOM*) is another defuzzification method. It is equal to the mean of the fuzzy subset or a group of subsets that have maximum membership values in the resulting final fuzzy subset representing the final conclusion as shown in Figure 4.5 and defined in (4.8).

$$z_{\mathrm{MOM}}(z) = \sum_{i=1}^{n} \frac{z_i}{n} \tag{4.8}$$

where the crisp value z_i is the support set of fuzzy memberships $\mu_{C'}(z)$, which have maximum values in the final output set C'_{T}, and n is the number of elements in this support set.

The equal area (***EA***) method is also used for defuzzification. In this method, the final conclusion area is divided into two equal areas. The crisp value corresponding to the point that separates two equal areas is taken as the crisp output and expressed as in (4.9).

$$\int_{z_1}^{Z_{\mathrm{EA}}} \mu_{C'}(z)dz = \int_{Z_{\mathrm{EA}}}^{Z_2} \mu_{C'}(z)dz \tag{4.9}$$

where z_{EA} is the equilibrium point that divides the total output set C'_{T} into two equal areas. The first integration starts from the left starting point of C'_{T} and ends at the point z_{EA}. The equilibrium point z_{EA} is the starting point for the second integral which ends at the right ending point of C'_{T}.

The center of areas (COA) method is the most used one in fuzzy logic control applications [4,5]. In this method, the crisp output is the point that corresponds to the center of areas that are obtained as the union of the areas from individual rules. The defuzzification is done using the formula given in (4.10).

$$z_{\mathrm{COA}} = \frac{\sum_{i=1}^{n} w_{Ri} \cdot z_i}{\sum_{i=1}^{n} w_{Ri}} \tag{4.10}$$

where n is the number of the active rules, $w_{Ri}(z)$ is the weighting factor from the ith rule, and z_i is the crisp value defined in the output universe and corresponds to the center of the core set of the output fuzzy set. In order to understand the use of (4.10), let us use the output space and unified areas of final fuzzy conclusions shown in Figure 4.6.

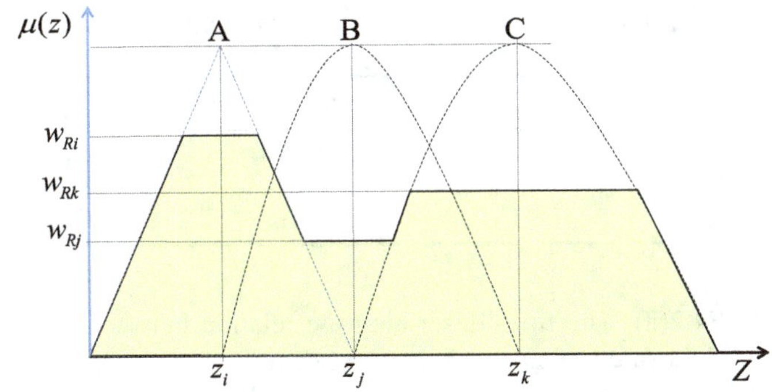

Figure 4.6 A Fuzyy output to be used to explain COA defuzzification method.

The defuzzification result of Figure 4.6 is written as

$$z_{\mathrm{COA}} = \frac{w_{Ri} \cdot z_i + w_{Rj} \cdot z_j + w_{Rk} \cdot z_k}{w_{Ri} + w_{Rj} + w_{Rk}} \tag{4.11}$$

The center of gravity (***COG***) and the center of areas are the same methods. They use the same defuzzification formula given by (4.10). The names are different because some researchers called the method as center of gravity [6,7] instead of center of area.

Example 4.2

In a system, the universe of X is partitioned into three categories A_1, A_2, and A_3. A second universe called Y is also portioned into three subcategories B_1, B_2, and B_3. The fuzzy subcategories are represented by triangular fuzzy membership functions as described in Table E4.2(a).

The relation between the universes X and Y yields another universe Z, which is also partitioned into three subcategories C_1, C_2, and C_3 whose parameters are also

Table E4.2(a) Parameters of the fuzzy membership functions for Example 4.2

Membership function	**Universe of discourse**	x_1 **as in (2.7)**	x_2 **as in (2.7)**	x_3 **as in (2.7)**
A_1	X	−2	−2	0
A_2	X	−2	0	2
A_3	X	0	2	2
B_1	Y	−0.5	−0.5	0
B_2	Y	−0.5	0	0.5
B_3	Y	0	0.5	0.5
C_1	Z	−1	−1	0
C_2	Z	−1	0	1
C_3	Z	0	1	1

Table E4.2(b) The rule table for Example 4.2

	B_1	B_2	B_3
A_1	C_2	C_3	C_3
A_2	C_1	C_2	C_3
A_3	C_1	C_1	C_2

listed in Table E4.2(a). The rules describing the relation between the universes X and Y are given in a rule table as in Table E4.2(b).

Obtain the value of z if $x = -1.8$ and $y = 0.35$. Use the ***Mamdani Reasoning*** approach with the center of area defuzzification method.

Answer 4.2

The solution to this problem will be given step by step in a MATLAB .m file format. *Fuzzyfication* is the first step in Mamdani fuzzy reasoning algorithm and it is used to convert crisp inputs to fuzzy membership values. Therefore, the crisp inputs $x = -1.8$ and $y = 0.35$ should be written in terms of their membership values in the fuzzy subsets defined in the same universe. The MATLAB code given in Figure E4.2 (a) can be used as the *fuzzification*.

The term "triangle01m" is a MATLAB function developed in Chapter 2 for the triangular fuzzy membership function. The code in Figure E4.2(a) gives the membership values of x in the fuzzy subsets as $A_1 = 0$, $A_2 = 0.25$, $A_3 = 0.75$, $B_1 = 0$, $B_2 = 0.6$, and $B_3 = 0.4$.

The second step in Mamdani fuzzy reasoning is the rule base system. Therefore, the rules are fired to yield the weighting factor as a result of each rule. The MATLAB code for the weighting factors is given in Figure E4.2(b).

The numerical values of the weighting factors are given in matrix format for the data used in this example. The matrix format is similar to the rule table and

```
x=1.5; y=0.2;
A1=triangle01m(-2,-2,0,x);
A2=triangle01m(-2,0,2,x);
A3=triangle01m(0,2,2,x);
B1=triangle01m(-0.5,-0.5,0,y);
B2=triangle01m(-0.5,0,0.5,y);
B3=triangle01m(0,0.5,0.5,y);
```

Figure E4.2(a) MATLAB code for fuzzification

```
w1=min(A1,B1);   w2=min(A1,B2);   w3=min(A1,B3);
w4=min(A2,B1);   w5=min(A2,B2);   w6=min(A2,B3);
w7=min(A3,B1);   w8=min(A3,B2);   w9=min(A3,B3);
```

Figure E4.2(b) MATLAB code for weighting factors

shows which rules are active.

$$\mu_R(z) = \begin{matrix} \\ 0 \\ 0.25 \\ 0.75 \end{matrix} \begin{matrix} 0 & 0.6 & 0.4 \\ \left[\begin{matrix} 0 \\ 0 \\ 0 \end{matrix}\right. & \begin{matrix} 0 \\ 0.25 \\ 0.6 \end{matrix} & \left.\begin{matrix} 0 \\ 0,25 \\ 0.4 \end{matrix}\right] \end{matrix}$$

Since $\mu_{A1}(x) = 0$ and $\mu_{B1}(y) = 0$, the first line and column become zero, Therefore, the weighting factors in the first line and column are zero. Only Rules 5, 6, 8, and 9 are active.

At this point let us use the code given in Figure E4.2(c) to plot and see the fuzzy membership functions C_1, C_2, and C_3 in the universe of Z. The resultant plot is shown in Figure E4.2(d).

It was mentioned earlier that (4.5) can be used to determine the effects of the rules on the fuzzy sets in output space. The weighting factors can be obtained by using the code in Figure E4.2(b) and substituted into (4.5) as in Figure E4.2(e) to

```
k=1;
for z=-1:0.01:1
    C1(k)=triangle01m(-1,-1,0,z);
    C2(k)=triangle01m(-1,0,1,z);
    C3(k)=triangle01m(0,1,1,z);
    Z(k)=z;
    k=k+1;
end
plot(Z,C1,Z,C2,Z,C3); grid
xlabel('Universe, Z'); ylabel('Membership Values')
disp('Hit the <enter> to continue'); pause
```

Figure E4.2(c) MATLAB code for plotting the fuzzy subsets in output universe Z

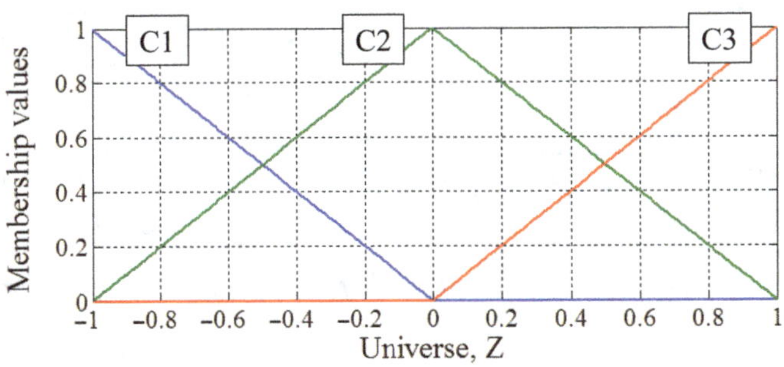

Figure E4.2(d) The fuzzy subsets C_1, C_2, and C_3 in output universe Z

```
R1=min(w1,C2);   R2=min(w2,C3);   R3=min(w3,C3);
R4=min(w4,C1);   R5=min(w5,C2);   R6=min(w6,C3);
R7=min(w7,C1);   R8=min(w8,C1);   R9=min(w9,C2);
```

Figure E4.2(e) The code to find the effect of each rule in output space Z

yield the effective areas of the fuzzy subsets in the output space Z. The resultant effective areas for Rules 1, 2, 3, 4, and 7 become zero because these rules are not fired for the given values of x and y. Therefore, only the effect of the active Rules 5, 6, 8, and 9 in the output space are shown in Figure E4.2(f). The final fuzzy conclusion is then obtained by using (4.4) as given in Figure E4.2(g), which results in the plot shown in Figure E4.2(h).

The final crisp output is obtained by defuzzifying the final fuzzy conclusion shown in Figure E4.2(i). Using the numerical values of the parameters w_i and z_i given in Figure E6.2(f) in (6.10) or (6.11) the crisp output is obtained. A MATLAB code used for defuzzification in this example is given in Figure E6.2(i), besides the

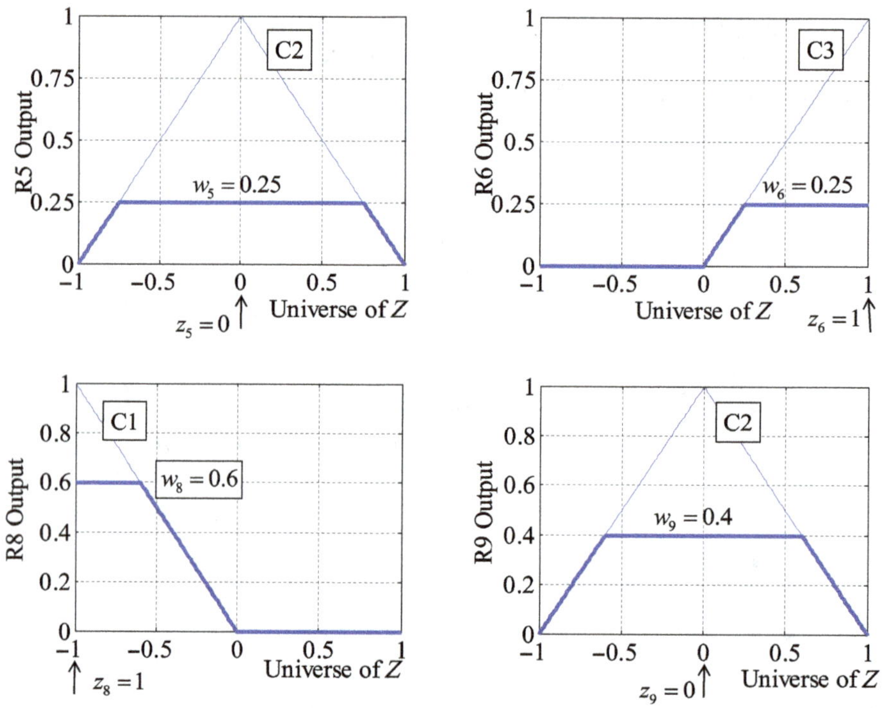

Figure E4.2(f) The effects of the active rules in the output space

```
R=[R1;R2;R3;R4;R5;R6;R7;R8;R9];
RR=max(R);
RRR=plot(Z,R1,Z,R2,Z,R3,Z,R4,Z,R5,Z,R6,Z,R7,Z,R8,Z,R9,Z,RR);
set(RRR(10),'linewidth',2);
set(RRR(10),'Color',[0 0 1])
xlabel('Final Fuzzy Conclusion');
ylabel('Membership values'); grid
```

Figure E4.2(g) The code to find the union of active rule outputs

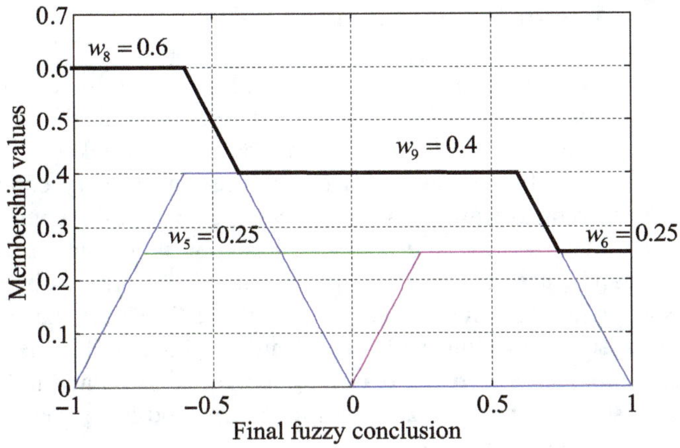

Figure E4.2(h) The unified final fuzzy conclusion for Example E4.2

```
W=[ w1  w2  w3
    w4  w5  w6
    w7  w8  w9  ];
% The rule matrix
RM=[  0  1 1
     -1  0 1
     -1 -1 0  ];
% Defuzzification
WZ=0;
WTOT=0;
for nn=1:3
    for mm=1:3
        WZ=WZ+W(nn,mm)*RM(nn,mm);
        WTOT=WTOT+W(nn,mm);
    end
end
ZOUT=WZ/WTOT
```

Figure E4.2(i) The MATLAB code for defuzzification in Example E4.2

direct solution by hand below.

$$z_{\text{out}} = \frac{w_5 \cdot z_5 + w_6 \cdot z_6 + w_8 \cdot z_8 + w_9 \cdot z_9}{w_5 + w_6 + w_8 + w_9}$$

$$z_{\text{out}} = \frac{(0.25) \cdot (0) + (0.25) \cdot (1) + (0.6) \cdot (-1) + (0.4) \cdot (0)}{0.25 + 0.25 + 0.6 + 0.4}$$

$$z_{\text{out}} = \frac{0.25 - 0.6}{1.5} = -0.23$$

4.3 Takagi–Sugeno fuzzy reasoning

Fuzzy rules used in fuzzy controllers are usually defined by considering the knowledge of human experts in specific areas. Takagi and Sugeno argue that an operator cannot tell linguistically what kind of action he/she takes in a particular situation [8] and they offer model-based control just in ordinary control theory. Mowing from this point forward, they have developed Takagi–Sugeno (TS) fuzzy model for system identification and modeling based on input–output data [8]. TS fuzzy modeling can be used to represent complex nonlinear systems by fuzzy rules that have linear crisp equations as the consequents [9]. This fuzzy approach is widely used in system modeling and identifications [10–14]. The use of the TS fuzzy modeling approach as a fuzzy reasoning method is explained in [15].

The general structure of the fuzzy rules in this method is written as in (4.12).

$$\textit{if } x \text{ is } A \textit{ and } y \text{ is } B \textit{ then } z \text{ is } f(x,y). \tag{4.12}$$

where A and B are fuzzy subsets defined in the universes X and Y, respectively. When compared with the rules in Mamdani fuzzy reasoning method, it is seen that only the output space is different in (4.12). The output space in TS fuzzy reasoning does not include fuzzy subsets. A crisp function $f(x, y)$ is used to find a crisp output for each rule. The crisp output function is usually defined in terms of crisp input variables x and y, and written as

$$z_i = f(x,y) \tag{4.13}$$

where i is the active rule number. After calculating a crisp output z_i for each active rule, the final crisp output is obtained as a weighted sum of the crisp outputs as

$$z = \sum_{i=1}^{N} w_{Ri} \cdot z_i \tag{4.14}$$

where

$$\sum_{i=1}^{N} w_{Ri} \cong 1 \tag{4.15}$$

where N is the total number of the rules or active rules, w_{Ri} is the weighting factor obtained from the ith rule and z_i is given in (4.13). The values of w_{Ri} are obtained as the process of the rule base system using the operator min or arithmetic *product*, which makes the output equal to or smaller than the inputs. Therefore, equation (4.15) is expected to be satisfied, otherwise, if the sum of the weighting factors is much >1, the fuzziness of the process weakens. Since the outputs from individual rules are not fuzzy, a defuzzification is not required in TS fuzzy reasoning, and (4.14) is used. However, in some applications, if (4.15) is not approximately satisfied, then the final crisp output is obtained as a weighted average similar to the center of area method given by (4.10) and repeated in (4.16).

$$z = \frac{\sum_{i=1}^{N} w_{Ri} \cdot z_i}{\sum_{i=1}^{N} w_{Ri}} \tag{4.16}$$

Since the final conclusion is just the weighted sum of the active rules outputs as in (4.14), the process time becomes smaller compared to the use of (4.16). A graphical view of the TS or TS fuzzy reasoning algorithm is shown in Example 4.3.

Example 4.3

Let us repeat Example 4.1 using TS fuzzy reasoning instead of Mamdani fuzzy reasoning for the same input data as solar irradiation level s = 125 mW/cm^2 and temperature t = 15°C. Use the same triangular fuzzy membership functions with the same definitions in the universe of solar irradiation and modify the fuzzy membership functions in the universe of temperature by using sinusoid membership functions instead of Gaussian as shown in Figure E4.3(a). Since the output universe is not fuzzy, then the rule table is modified as in Table E4.3(a).

where

$$\begin{aligned} P_1 &= 1 - e^{-0.1s} - e^{-0.1t} \\ P_2 &= 1 - e^{-0.05s} - e^{-0.06t} \\ P_3 &= 1 - e^{-0.02s} - e^{-0.015t} \end{aligned} \tag{E4.3a}$$

Answer 4.3

The answer to this example is very similar to the answer given for Example 4.1. The difference occurs at the output space since the output space in the TS reasoning algorithm is crisp and defined by the functions P_1, P_2, and P_3. Therefore, the rule table does not contain fuzzy membership functions in the conclusion part. From input variables s and t to the output of the fuzzification are the same as in Example 4.1. Then the weighting actor from each rule is obtained using either min or product operator as the same in Example 4.1 given for Mamdani fuzzy reasoning. Therefore, the difference starts right in the output space, which yields crisp output directly.

The membership values of s = 125 mW/cm^2 in the fuzzy subsets defined in the universe of solar irradiation (S) are obtained using the MATLAB function $\mu(x) = triangle01m(x_1, x_2, x_3, x)$ given in Figure 2.5 as follows:

$$\mu_{MS}(125) = triangle01m(0, 100, 200, 125) = 0.75 \tag{E4.3b}$$

$$\mu_{HS}(125) = triangle01m(100, 200, 200, 125) = 0.25 \tag{E4.3c}$$

The membership values of t = 15°C in the fuzzy subsets defined in the universe of temperature (T) are obtained using the MATLAB function

Table E4.3(a) The rule table for Example 4.3

	LT	WT	HT
LS	P3	P3	P3
MS	P2	P2	P2
HS	P1	P1	P1

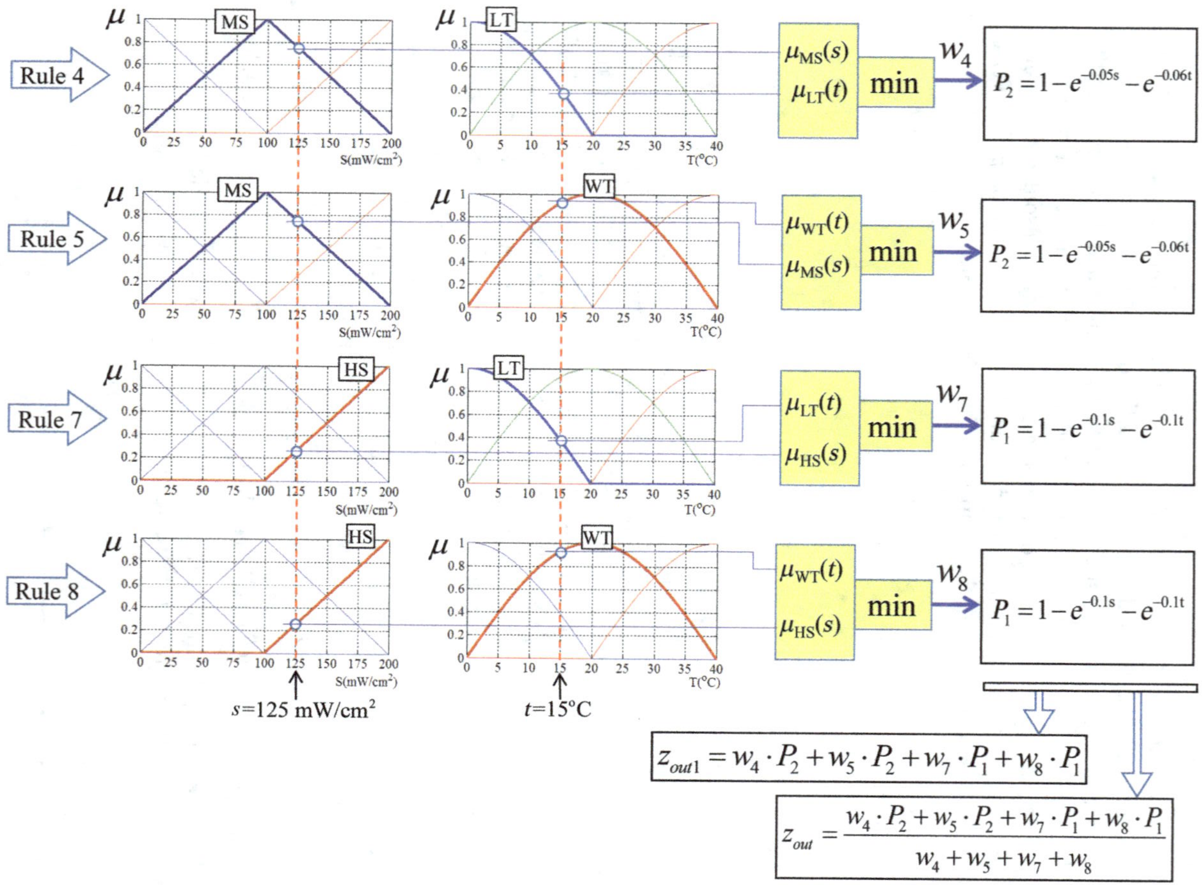

Figure E4.3(a) Takagi–Sugeno fuzzy reasoning algorithm

$\mu(x) = \text{sinus01m}\,(a, A, x)$ given in Figure E2.5(a) as follows:

$$\mu_{\text{LT}}(15) = \text{sinus01m}\,(20, \pi/2, 15) = 0.38 \tag{E4.3d}$$

$$\mu_{\text{WT}}(15) = \text{sinus01m}\,(20, 0, 15) = 0.92 \tag{E4.3e}$$

The given data s = 125 mW/cm^2 has non-zero membership degrees in the fuzzy subsets MS and HS in the universe of S and t = 15°C has non-zero membership degrees in the fuzzy subsets LT and LW in the universe of T. Therefore, the active rules will be related to the fuzzy subsets MS and HS in the universe of S and to the fuzzy subsets LT and WT in the universe of T. The active rules are depicted in Table E4.3(b) by the shaded area, which shows that the active rules are Rule 4, Rule 5, Rule 7, and Rule 8. Rules 4 and 5 both trigger the crisp function P_2 and rules 7 and 8 activate the crisp function P_1. The crisp function P_3 is not activated for the given crisp inputs.

For given data s = 125 mW/cm^2 and t = 15°C the crisp functions in the output space are obtained directly by using the related formulas. Note that P_3 is not activated because the rules using $P3$ are not active.

$$\begin{aligned} P_1 &= 1 - e^{-(0.1)(125)} - e^{-(0.1)(15)} = 0.7769 \text{ pu} \\ P_2 &= 1 - e^{-(0.05)(125)} - e^{-(0.06)(15)} = 0.5915 \text{ pu} \\ P_3 &= 1 - e^{-(0.02)(125)} - e^{-(0.015)(15)} = 0.1194 \text{ pu} \end{aligned} \tag{E4.3f}$$

Let us find the weighting factors for the active rules next.

$$\begin{aligned} w_{R4} &= \min(0.75, 0.38) = 0.38 \\ w_{R5} &= \min(0.75, 0.92) = 0.75 \\ w_{R7} &= \min(0.25, 0.38) = 0.25 \\ w_{R8} &= \min(0.25, 0.92) = 0.25 \end{aligned} \tag{E4.3g}$$

The sum of the weightings becomes

$$\begin{aligned} \sum_{i=1}^{N} w_{Ri} &= w_{R4} + w_{R5} + w_{R7} + w_{R8} \\ \sum_{i=1}^{N} w_{Ri} &= 0.38 + 0.75 + 0.25 + 0.25 = 1.63 \\ \sum_{i=1}^{N} w_{Ri} &= 1.63 > 1 \end{aligned} \tag{E4.3h}$$

Table E4.3(b) The rule table for Example 4.3

	LT	WT	HT
LS	$R_1 = P_3$	$R_2 = P_3$	$R_3 = P_3$
MS	$R_4 = P_2$	$R_5 = P_2$	$R_6 = P_2$
HS	$R_7 = P_1$	$R_8 = P_1$	$R_9 = P_1$

If the sum of the weightings is approximately equal to unity, then the weighted sum of the outputs from individual rules can be considered as the final conclusion for the reasoning algorithm. However, the sum of the weightings is much greater than unity here and the use of weighted sum may lead to the wrong conclusion as given below.

$$\begin{aligned} p_W &= w_4 \cdot P_1 + w_5 \cdot P_1 + w_7 \cdot P_2 + w_8 \cdot P_2 \\ p_W &= 0.38 \cdot 0.7769 + 0.75 \cdot 0.7769 + 0.25 \cdot 0.5915 + 0.25 \cdot 0.5915 \\ p_W &= 0.295 + 0.58 + 0.15 + 0.15 = 1.175 \\ p_W &= 1.175 \end{aligned} \tag{E4.3i}$$

In this problem, the universe of P is the maximum power generated by a specific PV panel under various levels of solar irradiation and temperature. The possible maximum power that can be generated by this panel where it is located is selected as the base power. Therefore, the highest value of the universe of S is given as unity. This means that, p_W cannot be >1 pu. In this case, the weighted average should be used to obtain the final conclusion as follows:

$$\begin{aligned} p_{AV} &= \frac{w_4 \cdot P_1 + w_5 \cdot P_1 + w_7 \cdot P_2 + w_8 \cdot P_2}{w_4 + w_5 + w_7 + w_8} \\ p_{AV} &= \frac{0.38 \cdot 0.7769 + 0.75 \cdot 0.7769 + 0.25 \cdot 0.5915 + 0.25 \cdot 0.5915}{0.38 + 0.75 + 0.25 + 0.25} \\ p_{AV} &= \frac{0.295 + 0.58 + 0.15 + 0.15}{1.63} = \frac{1.175}{1.63} = 0.72 \end{aligned} \tag{E4.3j}$$

The result from the weighted average is close to 0.75, which was obtained in Example 4.1 using Mamdani fuzzy reasoning.

Example 4.4

An industrial plant is fed from wind (WES) and photovoltaic (PES) energy systems besides the utility distribution network (UDN). WES, PES, and the industrial plant belong to the same company. For economical considerations, the company wants to minimize the electrical power from the utility by using WES and PES by utilizing a fuzzy decision process based on the rules given below

R_1: *If* PES is *sufficient* and WES is *insufficient* then the use of UDN = 1-min (μ_{RES}, μ_{GES})
R_2: *If* PES is *sufficient* and WES is *sufficient* then the use of UDN = 1-min (μ_{RES}, μ_{GES})
R_3: *If* PES is *insufficient* and WES is *insufficient* then the use of UDN = 1-min (μ_{RES}, μ_{GES})
R_4: *If* PES is *insufficient* and WES is *sufficient* then the use of UDN = 1-min (μ_{RES}, μ_{GES})

The transition from one rule to the next one is done by using the term *else* and the fuzzy subsets used in the rules are defined as follows:

Insufficient in the universe of PES, (IPES) = triangle (0, 0, 200)
Sufficient in the universe of PES, (SPES) = triangle (0, 200, 200)
Insufficient in the universe of WES, (IWES) = triangle (0, 0, 20)
Sufficient in the universe of WES, (SWES) = triangle (0, 20, 20)

The universe of the use of UDN is defined in interval {0, 1} such that UDN = 0 means that the UDN is not being used, and UDN = 1 means that all the required power (100%) is supplied from UDN. Obtain the percentage of the use of UDN if the solar irradiation level in PES is 80 mW/cm^2 and wind speed in WES is 15 m/s.

Answer 4.4

The graphical representation of the rule processing makes the reasoning algorithm more clear and understandable. Therefore, the solution of this example is done both in graphical and written formats.

Since the input data is crisp, they should be fuzzified first using the MATLAB function $\mu(x) = triangle01m(x_1, x_2, x_3, x)$ given in Figure 2.5 as follows:

$$\begin{aligned}
\mu_{\mathrm{IPES}}(80) &= triangle01m(0, 0, 200, 80) = 0.6 \\
\mu_{\mathrm{SPES}}(80) &= triangle01m(0, 200, 200, 80) = 0.4 \\
\mu_{\mathrm{IWES}}(15) &= triangle01m(0, 0, 20, 15) = 0.25 \\
\mu_{\mathrm{SWES}}(15) &= triangle01m(0, 20, 20, 15) = 0.75
\end{aligned} \tag{E4.4a}$$

Substitution of the numerical values into the rules results in the outputs from each rule.

R1: PES = SPES (0.4) and WES = IWES (0.25) then UDN = 1 − 0.25 = 0.75
R2: PES = SPES (0.4) and WES = SWES (0.75) then UDN = 1 − 0.4 = 0.6
R3: PES = IPES (0.6) and WES = IWES (0.25) then UDN = 1 − 0.25 = 0.75
R4: PES = IPES (0.6) and WES = SWES (0.75) then UDN = 1 − 0.6 = 0.4

The sum of the weightings becomes

$$\sum_{i=1}^{N} w_{Ri} = 0.25 + 0.4 + 0.25 + 0.6 = 1.5 > 1.0 \tag{E4.4b}$$

Since the sum of the weightings is much greater than unity, the weighted average should be used to obtain the final conclusion as follows:

$$\begin{aligned}
p_{\mathrm{UDN}} &= \frac{(0.25)(0.75) + (0.4)(0.6) + (0.25)(0.75) + (0.6)(0.4)}{0.25 + 0.4 + 0.25 + 0.6} \\
p_{\mathrm{UDN}} &= \frac{0.855}{1.5} = 0.57
\end{aligned} \tag{E4.4b}$$

The result from the weighted average is 0.57, which means that 57% of the required power of the industrial plant is supplied from the utility distribution network. The remaining 43% is supplied from PES and WES.

The graphical representation of this solution is shown in Figure E4.4.

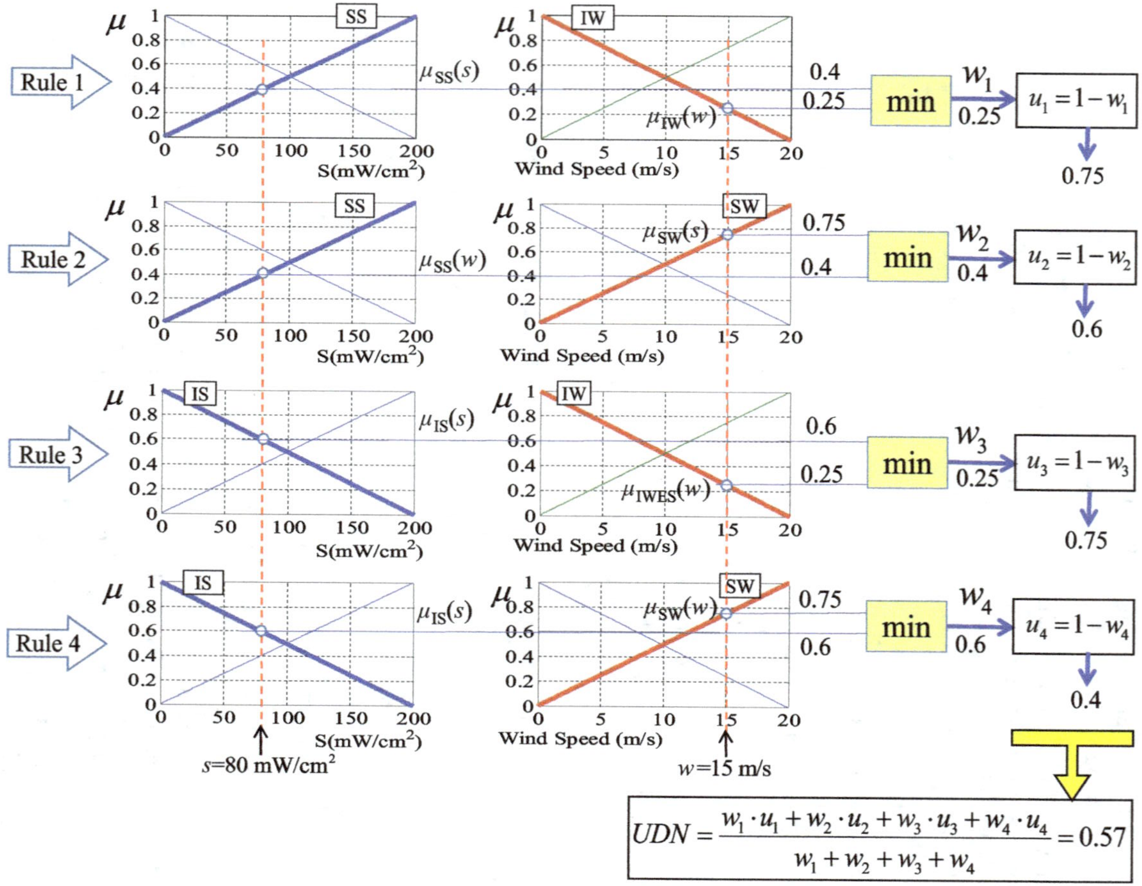

Figure E4.4 Takagi–Sugeno fuzzy reasoning algorithm for Example 4.4

4.4 Tsukamoto fuzzy reasoning

Fuzzification and rule processing stages in Mamdani, TS, and Tsukamoto fuzzy reasoning algorithms are exactly the same. The crisp input data is fuzzified and related rules in the rule base are executed to yield a consequent. The difference between these three reasoning algorithms is seen in this consequent part. The consequent in Mamdani fuzzy reasoning is a weighted fuzzy subset and the overall consequence from all fuzzy rules is the union of the weighted fuzzy sets from all the rules. Therefore, defuzzification is required in Mamdani reasoning in order to yield a crisp output. The consequents of the individual rules in TS are crisp functions and the final overall consequence is a crisp value obtained by using either a weighted sum or a weighted average of the individual consequents. Tsukamoto has suggested another approach to the consequent part of the fuzzy reasoning algorithm [16]. In Tsukamoto fuzzy reasoning algorithm, the consequent is a fuzzy set such that the intercepting of the weighting with this fuzzy set yields a crisp value as the consequent of the related individual rule. Therefore, the effect of a weighting factor on the output of a single rule is written as

$$w_i = C_i(z_i) \tag{4.17}$$

where w_i is the resultant weighting factor from the *i*th rule, C_i is the output fuzzy set of the *i*th rule and $z_i \in Z$ is the crisp consequent of the *i*th rule.

Example 4.5

Assume that the fuzzy rules in Example 4.4 are redefined for the same problem as below

R1: *If* PES is *sufficient* and WES is *insufficient* then the use of UDN is Medium
R2: *If* PES is *sufficient* and WES is *sufficient* then the use of UDN is low
R3: *If* PES is *insufficient* and WES is *insufficient* then the use of UDN is high
R4: *If* PES is *insufficient* and WES is *sufficient* then the use of UDN is Medium

The transition from one rule to the next one is done by using the term *else* and the fuzzy subsets used in the rules are defined as follows:

Insufficient in the universe of sunlight, (IS) = triangle (0, 0, 200)
Sufficient in the universe of sunlight (SS) = triangle (0, 200, 200)
Insufficient in the universe of wind, (IW) = triangle (0, 0, 20)
Sufficient in the universe of wind, (SW) = triangle (0, 20, 20)
Low in the universe of the use of utility (LU) = triangle (0, 0.5, 0.5)
Medium in the universe of the use of utility (MU) = triangle (0.25, 0.75, 0.75)
High in the universe of the use of utility (HU) = triangle (0.5, 1, 1)

Obtain the percentage of the use of UDN if the solar irradiation level in PES is 80 mW/cm^2 and wind speed in WES is 15 m/s.

Answer 4.5

The solution of this example is done both in graphical and written formats. This time the graphical solution is given first as in Figure E4.5.

Since the fuzzification process is the same as in the Mamdani and TS algorithms, there is no need to repeat the same process. Therefore, the weighting factors obtained in Example 4.4 can be used in this example, too.

$$\mu_{\text{IS}}(80) = 0.6,\ \mu_{\text{SS}}(80) = 0.4,\ \mu_{\text{IW}}(15) = 0.25,\ \mu_{\text{SW}}(15) = 0.75 \qquad \text{(E4.5a)}$$

Substitution of the numerical values into the rules results in the outputs from each rule.

R1: PES = SS (0.4) and WES = IW (0.25) then $u1 = 0.25 + w/2 = 0.25 + 0.25/2 = 0.375$.
R2: PES = SPES (0.4) and WES = SWES (0.75) then $u2 = 0 + w/2 = 0.4/2 = 0.2$
R3: PES = IPES (0.6) and WES = IWES (0.25) then $u3 = 0.5 + w/2 = 0.5 + 0.25/2 = 0.625$
R4: PES = IPES (0.6) and WES = SWES (0.75) then $u4 = 0.25 + w/2 = 0.25 + 0.6/2 = 0.55$

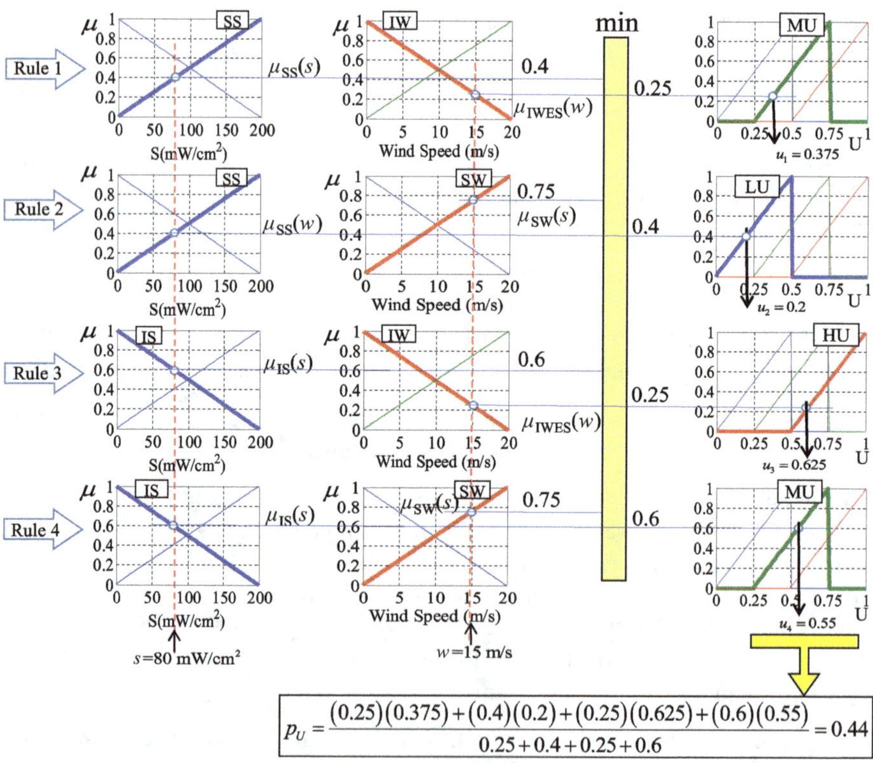

$$p_U = \frac{(0.25)(0.375) + (0.4)(0.2) + (0.25)(0.625) + (0.6)(0.55)}{0.25 + 0.4 + 0.25 + 0.6} = 0.44$$

Figure E4.5 Tsukamoto fuzzy reasoning algorithm for Example 4.5

The final conclusion is obtained using the weighted average as follows:

$$p_U = \frac{(0.25)(0.375) + (0.4)(0.2) + (0.25)(0.625) + (0.6)(0.55)}{0.25 + 0.4 + 0.25 + 0.6}$$
$$p_U = \frac{0.66}{1.5} = 0.44 \tag{E4.5a}$$

The result from the weighted average is 0.44, which means that 44% of the required power of the industrial plant is supplied from the utility distribution network. The remaining 46% is supplied from PES and WES.

Example 4.6

A power management system was introduced in Example 3.13. Three types of power stations were described as small, medium, and large and represented by fuzzy subsets as shown in Figure E3.13(a). It was addressed that the generated power should be equal to the dissipated power demand plus the losses so that the equilibrium of the system is guaranteed. Based on these operating conditions and power system properties, the following single input and single output rules are rewritten for this example.

R1: If change in demand is PLD then change in generation is PLG.
R2: if change in demand is PMD then change in generation is PMG.
R3: if change in demand is NLD then change in generation is NLG.
R4: if change in demand is NMD then change in generation is NMG
R5: if change in demand is ZED then change in generation is ZEG.

where PLD, PMD, NLD, NMD, and ZED stand for positive low, positive medium, negative low, negative medium, and zero changes, respectively, in the universe of power demand. PLG, PMG, NLG, NMG, and ZEG are used for positive low, positive medium, negative low, negative medium, and zero changes, respectively, in the universe of generation. These fuzzy subsets are described in Figure E5.1 for this example, as well. Note that both the changes in power demand and in power generation have the same universes.

During a balanced operating period, while the power system is under equilibrium operating conditions, it is observed that the demand power has decreased about 2 MW. Find the amount of change in generated power and give a decision about the power station where the power change is subject to be applied. Use the center of area defuzzification method where it is necessary.

Answer 6.6

A decrement of 2 MW in power demand means that the change in power demand is −2 MW. Therefore, a power change of −2 MW has membership values in the fuzzy subsets negative low (NL) and zero (ZE) only. Under these conditions, there will be two active rules among the given five rules. The active rules include negative low (NLD) and zero (ZED) changes in power demand.

if NLD *then* NLG
If ZED *then* ZEG

A decrement of 2 MW in power demand yields the following membership values in the fuzzy subsets defined in the universe of change in power demand.

$$\begin{aligned}
\mathrm{NMD} &= \mathit{triangle01m}(-50,-50,-5,-2) = 0.0 \\
\mathrm{NLD} &= \mathit{triangle01m}(-50,-5,0,-2) = 0.4 \\
\mathrm{ZED} &= \mathit{triangle01m}(-5,0,5,-2) = 0.6 \\
\mathrm{PLD} &= \mathit{triangle01m}(0,5,50,-2) = 0 \\
\mathrm{PMD} &= \mathit{triangle01m}(5,50,50,-2) = 0
\end{aligned}$$

where tiangle01m.m is a MATLAB function given in Figure 2.5. Since the fuzzy subsets NLG and ZEG are exactly the same as NLD and ZED, respectively, the membership values in the output space will be the same. Applying the defuzzification method, the center of area, gives the crisp output as the final conclusion.

$$\begin{aligned}
p_{\mathrm{CG}} &= \frac{(0.4)(-5)+(0.6)(0.0)}{0.4+0.6} = \frac{-2}{1} = -2 \\
p_{\mathrm{CG}} &= -2\mathrm{MW}
\end{aligned} \tag{E4.6a}$$

A power reduction −2 MW should be done in generation for the system equilibrium. This reduction can be done by turning off some units in small power stations.

4.5 Introduction to fuzzy logic control

Fuzzy logic (FL) controllers based on fuzzy set theory are used to represent the experience and knowledge of a human operator in terms of linguistic variables that are called fuzzy rules. Since an experienced human operator adjusts the system inputs to get a desired output by just looking at the system output without any knowledge of the system's dynamics and interior parameter variations, the implementation of linguistic fuzzy rules based on the procedures done by human operators does not also require a mathematical model of the system. Therefore, a FLC becomes nonlinear and adaptive in nature having a robust performance under parameter variations with the ability to get desired control actions for complex, uncertain, and nonlinear systems without the requirement of their mathematical models and parameter estimation. FL-based controllers provide a mathematical foundation for approximate reasoning, which has been proven to be very successful in a variety of applications [17]. In modern control techniques, uncertainty and vagueness have a great amount of importance to be dealt with. The use of membership functions quantified from ambiguous terms in fuzzy logic control rules has given a pulse to speed up the control of the systems with uncertainty and vagueness [18–20]. Since the introduction of fuzzy set theory [1] and its application to control systems [3], it has become an important and useful tool in especially controlling nonlinear systems. As computer and chip technologies developed, the applications of fuzzy logic-based controllers have increased tremendously so they have been applied in many different systems [17,21–24].

In this chapter, a simple and direct approach will be addressed in order to construct a fuzzy rule decision table that is suitable for modeling in the MATLAB/Simulink® GUI environment using simple operational blocks. A generalized rule table will be constructed such that also works for different systems without the requirement of a redesign process. Many works based on trial and error [25], artificial neural network (ANN) [26], genetic algorithms (GA)-based [27] algorithms, and clustering methods [28] have appeared in the literature concerning the rule decision table. The trial and error method uses some initial knowledge about how the system works and estimates the rest of the rule table by trying and error until a working table is obtained. The learning algorithms using ANN techniques also start with initial knowledge and fill in the rule decision table by estimating the rest of it using previously trained data. The algorithms using GA methods also consist of a sort of learning system, which learns a set of rules from a set of examples [29]. It has been proven that all these methods work very well. However, it should be noted that they are not just fuzzy systems. They are hybrid systems, which combine other intelligent methods such as neural networks and GA with fuzzy logic. Although the hybrid systems are more powerful and adaptive, they require high-level algorithms with time-consuming processes that are not desirable in control applications. The FLCs that appeared in the literature are mostly modeled for specific applications [30–32] rather than for general cases.

A direct approach to obtain an initial 3×3 rule table, which is then expanded to larger dimensions is discussed in this chapter. The initial rule table is extracted directly from the time response of the system control error and the change in this error over one sampling period. Then the rule table is partitioned into fuzzy sublevels in a similar way for partitioning the spaces of error and error change. So, the rule table is actually nothing but the partitioned version of the output space into fuzzy subsets with a symmetrical structure. Examples representing various applications and cases are given for different control schemes in MATLAB/Simulink environments. Some of these examples may be listed as a permanent magnet DC (PMDC) motor speed control scheme, a PMDC motor-driven position control scheme, a radar tracking control scheme for moving targets, a synchronous generator voltage control scheme, and a power system load–frequency control scheme.

4.6 Physical system behaviors and control

For a physical system operator, one of the most important knowledge is the time response of the system. An operator knows very well how the system responds to a change in reference input or system parameters in order to make necessary adjustments so that the system keeps operating at the reference set point. In classical automatic control systems, an error signal is defined as the difference between reference set value and actual system output.

$$e(t) = r(t) - y(t) \tag{4.18}$$

The output responses of a linear time-invariant (LTI) system for a step-type reference input are shown in Figure 4.7 for different cases of a second-order system. The step response of an LTI stable physical system reaches a constant steady-state value either smoothly or after some oscillations depending on the parameters and order of the system equations. The oscillating response occurs if the mathematical model of the system has second- or higher-order equations. Therefore, a second-order system with a characteristic equation in the s-domain as in (4.19) can be used to analyze and simulate the LTI systems.

$$\phi(s) = s^2 + 2\zeta\omega_n + \omega_n^2 \tag{4.19}$$

where ζ and ω_n are called damping ratio and natural frequency, respectively. If $\zeta = 0$, the system response oscillates like a sinewave and never reaches to a constant steady-state value. If ζ is between 0 and 1 as $0 < \zeta < 1$, then the system response oscillates. The magnitude of the oscillation is called overshoot and it gets smaller as ζ approaches to unity. The response is called damped and no overshoot exists as it is in the first-order systems if $\zeta = 1$. For the values of ζ greater than unity as $\zeta > 1$, the response is called overdamped with a longer settling time than in the case called damped response. The system responses for these four cases of ζ are summarized in Figure 4.7. The natural frequency ω_n determines the oscillation frequency of the response.

The time response of a physical system is important since it reflects the effects of the changes either in reference input or system's interior parameters. The meaning of controlling a physical system is having the system behave or operate as the user wishes. The user or the system operator tells the system what to do and how to operate through the reference input. Then the system is controlled so that the output will be equal or very close to the reference within acceptable error limits. Besides tracking the reference, the operator wants a faster system response without or with minimum overshoot. Therefore, the oscillations should be eliminated or minimized while the time required for the output to reach and settle at the reference

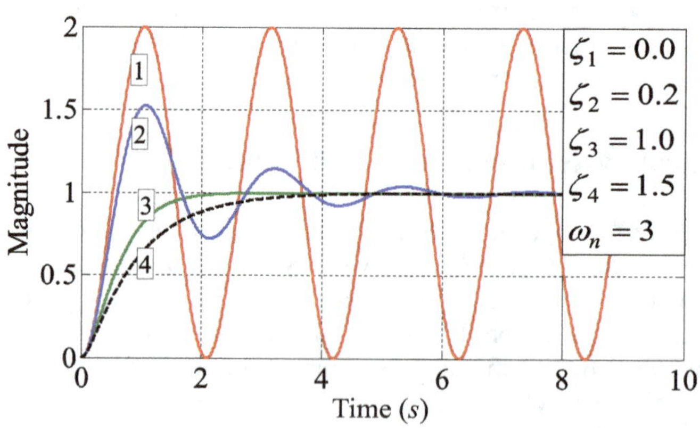

Figure 4.7 Step responses of a second-order system

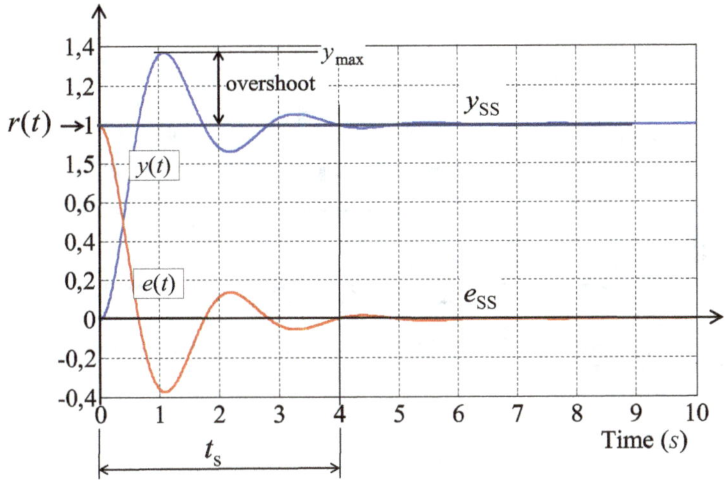

Figure 4.8 Adjustable control variables of a system response

value is kept short or minimum. In order to control a physical system, mainly three parameters, which are overshoot, settling time, and steady-state error as depicted in Figure 4.8 are affected by the controller.

A classical automatic control system uses the error signal in order to generate a control signal to drive the physical system or process so that the actual output tracks the reference with zero or minimum error. The error $e(t)$ is obtained as the difference between reference $r(t)$ and actual system output $y(t)$ as given in (4.18) and shown in Figure 4.8. In a classical control system, the error signal is either amplified, integrated, and/or derivate to yield the required control action. The controllers with these processes are called as proportional (P), integral (I), and derivative (D) controllers and their combination is called as PID controller. The PID controllers have been used successfully in the industry for a long time.

In order to understand and design an FLC, the operation and characteristic properties of the classical PID controller should be understood first. An operational block diagram of a classical feedback control system including the actuator, as well, is given in Figure 4.9.

The system output $y(t)$ is measured and fed back to a summer located at the input side of the controller. The summer acts as a comparator to compare the actual output $y(t)$ with the desired reference input $r(t)$. The difference between the reference and the actual output is the error signal defined as in (4.18) and compensated by the controller so that the output tracks the reference input with a zero or minimum steady-state error. In the human decision process, the error signal is calculated in mind by observing and comparing the output variable with the reference in mind.

The controller generates a control signal $u(t)$, which drives the physical system to be controlled. In a human decision case, the control signal is determined by the human expert using his/her ability and experience on the subject. Depending upon

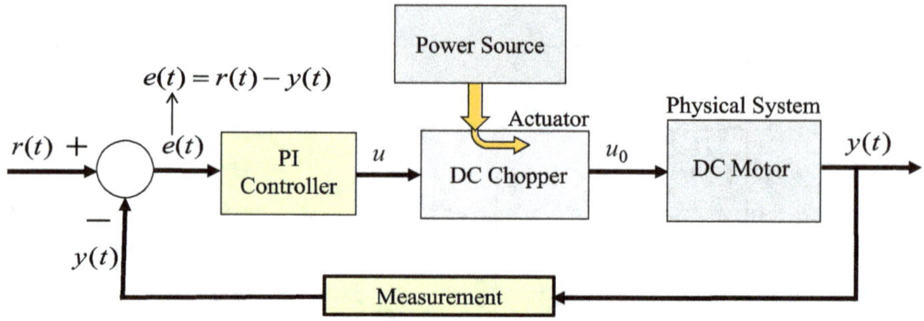

Figure 4.9 Block diagram of a feedback control system

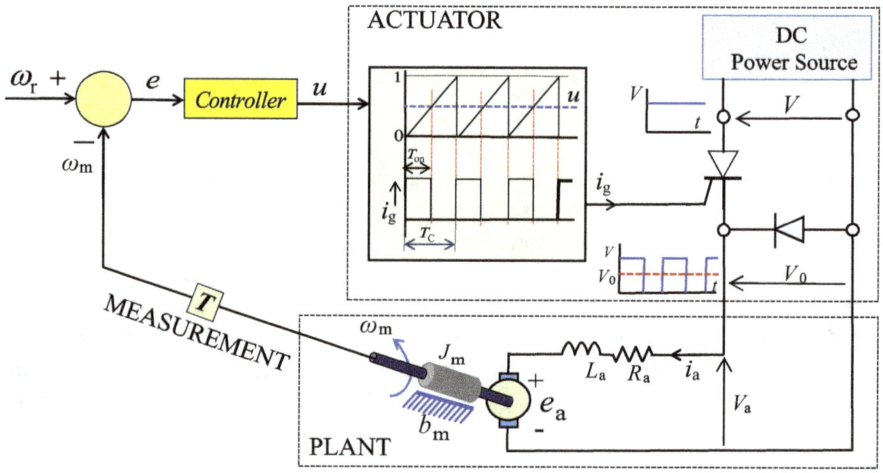

Figure 4.10 A simple detailed PMDC motor control scheme

the physical system the signal $u(t)$ may be used directly to drive the system or it may be used by actuators, which are additional devices or drives used to drive the main physical system, which is called *plant* in control system theory. The actuators, for example, in the human driver's case are the arms, hands, and feet. In most cases, the control signal $u(t)$ is used as a triggering signal to open a gateway for larger quantities to flow through the gate with the amount determined by the control signal $u(t)$. A more detailed representation of the actuator and the plant is given in Figure 4.10. A PMDC motor speed control system is depicted in Figure 4.10. The control signal $u(t)$ is used by the actuator in order to adjust the voltage $v_a(t)$ applied to the PMDC motor. The actuator in Figure 4.10 is a simple controlled DC chopper, which is used to obtain an adjustable DC voltage from the input DC voltage. The DC chopper has an electronic switch such as a thyristor, a GTO, or a MOSFET

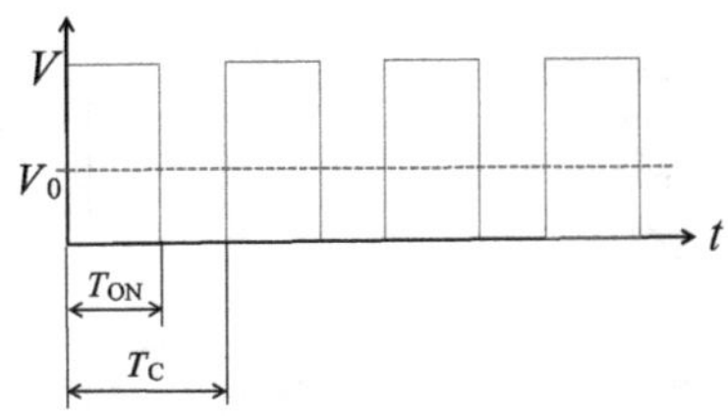

Figure 4.11 Output voltage profile of an ideal DC chopper

device. Assume that this electronic switch is turned on to conducting mode when the gate current $i_g > 0$ and turned off to blocking mode when $i_g = 0$. The gate current is obtained by comparing the control signal $u(t)$ with a reference sawtooth signal $s(t)$ such that a pulse is generated during $u(t) > s(t)$. The width of the pulse becomes longer for larger values of $u(t)$ and shorter for smaller values of $u(t)$.

Since the electronic switch is on conducting mode during i_g pulses, the chopped voltage of an ideal chopper will have the same pulse widths as i_g pulses as given in Figure 4.11. If the voltage from a DC power source has the magnitude V, then the average voltage at the output terminals of the chopper becomes V_0 by getting the integration of the chopped output voltage over a period as in (4.20).

$$V_0 = \frac{1}{T_C}\int_{*}^{T_{ON}} v(t)dt \tag{4.20}$$

$$V_0 = \frac{T_{ON}}{T_C}V \tag{4.21}$$

$$V_0 = CV \tag{4.22}$$

where C is called chopper duty cycle and defined as

$$0 \leq C = \frac{T_{ON}}{T_C} \leq 1 \tag{4.23}$$

which yields the adjustable chopper output voltage

$$0 \leq V_0 \leq V \tag{4.24}$$

Since the duration T_{ON} is determined by $u(t)$, if the value of $u(t)$ is limited between 0 and 1, then (4.22) can be written directly as in (4.25) for simulation purposes.

$$V_0 = u(t)V \tag{4.25}$$

where V is the DC voltage at the input terminals of the chopper. An ideal DC chopper can be simulated using (4.25) and can be represented as a gain block in the block diagram of a closed-loop system. Therefore, Figure 4.9 can be updated as in Figure 4.12.

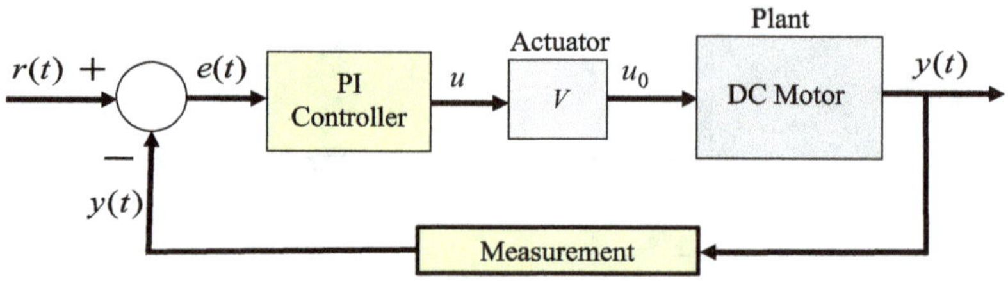

Figure 4.12 A simplified block diagram of feedback control systems

Example 4.7

A PMDC motor is represented by its state-space model equations as given in (E4.7a). Replace the "DC Motor" block in Figure 4.12 with the state-space model equations and simulate the feedback control system shown in Figure 4.12.

$$\begin{bmatrix} \dfrac{di_a}{dt} \\ \dfrac{d\omega_m}{dt} \end{bmatrix} = \begin{bmatrix} -\dfrac{R_a}{L_a} & -\dfrac{K_e}{L_a} \\ \dfrac{K_t}{J_m} & -\dfrac{b_m}{J_m} \end{bmatrix} \begin{bmatrix} i_a \\ \omega_m \end{bmatrix} + \begin{bmatrix} \dfrac{1}{L_a} & 0 \\ 0 & -\dfrac{1}{J_m} \end{bmatrix} \begin{bmatrix} v_a \\ T_L \end{bmatrix} \tag{E4.7a}$$

where the load torque T_L may be a constant or a polynomial equation to represent a fan-type load as

$$T_L = k_0 + k_1\omega_m + k_2\omega_m^2 \tag{E4.7b}$$

The nominal voltage, current, and speed of the PMDC motor are given as $V_a = 36$ V, $I_a = 15$ A, and $N_n = 3400$ rpm, respectively. The no-load current is $I_{NL} = 1.62$ A and the other parameters of PMDC motor and PI controller are given as follows:

Electrical parameters: $R_a = 1.4\ \Omega$, $L_a = 0.008$ H $K_e = 0.095$ V s/rad.
Mechanical parameters: $b_m = 4.5 \times 10^{-3}$ Nm s, $J_m = 1.5 \times 10^{-4}$ kg m^2 and $K_t =$ 0.095 Nm/A.
Load torque: $T_L = 0.5$ Nm.
Controller: $K_P = 0.001$ and $K_I = 0.15$.

Answer 4.7

The simulation algorithm of the PMDC motor speed control system can be constructed from reference input to system output in the same order shown in Figure 4.12. Assume that initially, the DC motor is standing still and the output speed is zero. Therefore, the error as the input to the controller is nonzero and equal to the reference speed input at the starting instant. The controller generates an output $u(t)$, which is multiplied by V in the actuator block to yield he adjusted voltage $u_0(t)$ to be applied to the motor. V in the actuator block is the maximum voltage that can be applied to the motor. The control signal $u(t)$ acts as a scaling factor and changes the value of $u_0(t)$ from 0 to V. Due to starting error signal a nonzero $u_0(t)$ is generated to operate the motor. Once the motor starts rotating, the speed will not be zero anymore and the error signal will be reduced as a result of

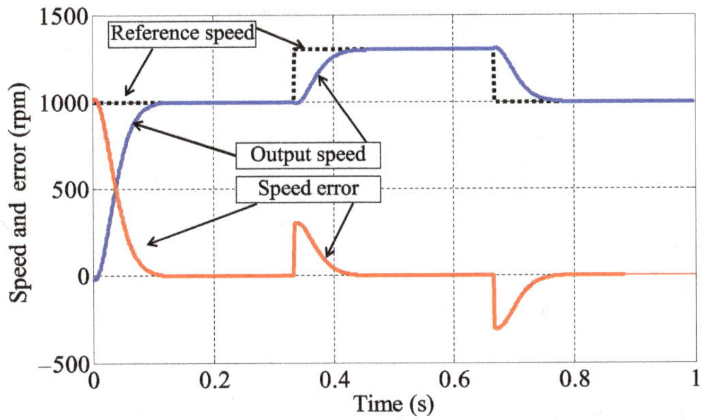

Figure E4.7(a) Controlled speed response of MPDC motor

(4.18). A MATLAB code and resultant speed responses are given in Figures E4.7 (a) and E4.7(b).

The simulation started with a reference speed of 1000 rpm and increased by 30% after some time while the system was operating at a steady state. Then the reference speed was dropped down starting value of 1000 rpm after reaching and operating at a steady state again. These step changes in reference input are done to see the performance of the controller in terms of how the output speed is tracking the reference. The controller shows a good performance so the step input tracking is good.

The same control system can be simulated in Simulink as given in Figure E4.7(c). It should be noted that the block diagram shown in Figure 4.12 can be represented almost with the same view in Simulink. Both MATLAB and Simulink are using rad and rad/s in mathematical calculations. Therefore, the reference and motor speed are converted from rpm to rad/s before the simulation and then converted from rad/s to rpm after the simulation. These conversions are done because the expression of the speed s rpm is more common and understandable in industry and system operators.

The simulation model of the PMDC model to be used in MATLAB and Simulink will be discussed in the next chapter along with the modeling methods of electrical, mechanical, and electromechanical systems.

Example 4.7 is given to understand the dynamic behaviors of a second-order electromechanical system. Let us think that the control signal $u(t)$ is adjusted by a human operator manually so that the operator uses a potentiometer to obtain a voltage from 0 to 1 V, which is used to generate pulses for the chopper drive system. Similar to the automatic controller, the human operator increases and decreases the control signal $u(t)$ manually to increase and decrease the speed such that these increments and decrements may be in large or small steps. If the response is away from the reference the incremental change becomes larger. If the response is closer to the reference then the incremental changes become smaller. Instead of the system output response, similar actions can be done according to the error

```
% Example 4.7 - PMDC Motor speed control
%-----------------------------------------------------------------
clear;   clf;    clc
% ------------------------------------PMDC Motor data
Vn=36;          Wn=3400;      TL=0.5;     In=15;
Ra=1.4;         La=0.008;     Ke=0.095;
Jm=1.5e-4;      Bm=4.5e-3;     Kt=0.095;
% ---------------------------------- State Space model
A=[ -(Ra/La)      -(Ke/La)
      Kt/Jm       -(Bm/Jm)   ];
B=[1/La 0;0  -1/Jm ];  C=[0  1];    D=[0 ] ;
U=[Vn  TL];
% -------------------------------- initials for simulation
 t0=0;        dt=0.001;     tend=1;  k=1;
SA=size(A);  NL=SA(1);     NC=SA(2);
for n=1:NL
    x0(n)=0;
end
% ---------------------------------- Data for controller
R0=1000;          KP=0.001;  KI=0.15;
wr0=2*pi*R0/60;  y0=0;       e10=0;
%----------------------------------- Simulation loop
while t0<tend-dt
   if t0 < tend/3;        wr=wr0;
   elseif t0 < 2*tend/3; wr=wr0+0.30*wr0;
   else;                  wr=wr0;
   end
   e0=wr-y0;                    % Control error
   ekp=KP*e0;                   % P-Control
   e1=e10+dt*KI*e0;  e10=e1;    % I-Control
   U0=(e1+ekp)*U;               % Total control effect
   [x]=runge(A,B,U0,x0,dt);     % Calling runge.m function
   u(k)=U(1); x1(k)=x(1);   x2(k)=x(2);  e(k)=e0;
   r(k)=wr;    y1(k)=x(1);   y2(k)=x(2);  y0=y2(k);
   t(k)=t0+dt;    t0=t(k);
   for n=1:NL
    x0(n)=x(n);    XX(k,n)=x(n);
   end
   k=k+1;
end;
%-------------------------------- Plots
WR=60*r/(2*pi); WM=60*y2/(2*pi);
ER=60*e/(2*pi);
plot(t,WR,t,WM,t,ER );
xlabel('Time (s)'); ylabel('Speed and Error (rpm)');grid
```

Figure E4.7(b) MATLAB .m file code for the speed control of a PMDC motor

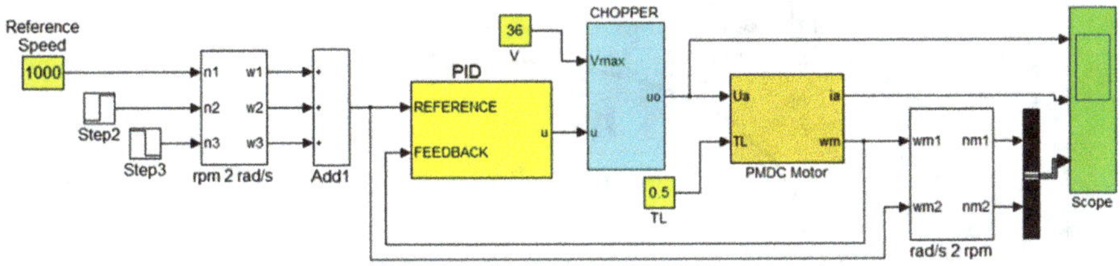

Figure E4.7(c) Simulink model of the speed control of the PMDC motor

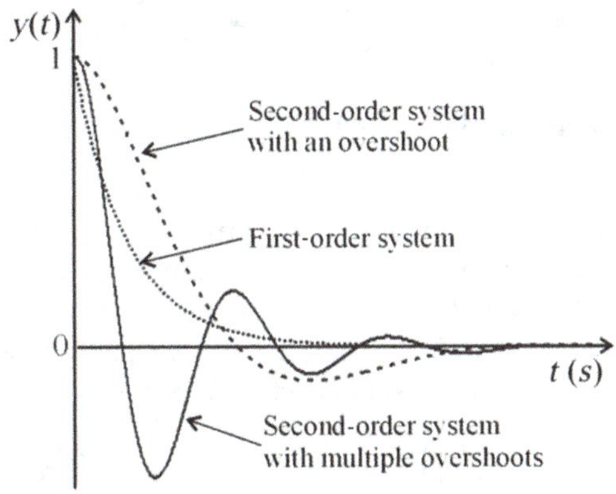

Figure 4.13 Error signals of the first- and second-order systems for a step input

signal. In this case, the incremental changes in the control signal $u(t)$ become larger if the error is large and smaller if the error is getting closer to zero. Therefore, the time response of the control error and its derivation are usually used as two input parameters to carry the decision process and actions of the human operator to the FLC. Since the error response includes the information about system output, it is used as a bridge connecting the system's input to the output over a set of linguistic fuzzy rules.

A linear time-invariant (LTI) system can be represented in different ways such as state-space model, transfer functions, simulation diagrams, signal flow graphs, and differential equations, which are mostly in first, second, or higher order. A more detailed discussion of the mathematical modeling of the physical systems for simulation purposes is given in Chapter 5. However, in order to derive the fuzzy rules from system behaviors, typical dynamic responses in time domain are reviewed in this section. The error response given in Figure 4.8 is redrawn in Figure 4.13 for various responses to the error. In order to make decisions about the size of step changes in the control signal, the deviation of the error is required. For example, the smaller deviations mean that the error is getting closer to zero or its steady-state value. Therefore, smaller steps should be given to increase or decrease

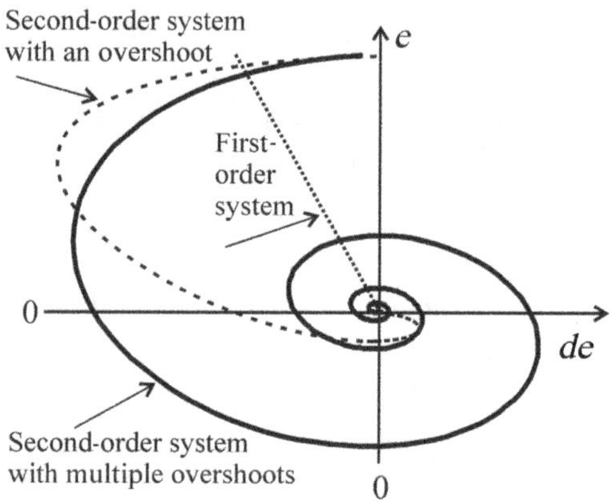

Figure 4.14 e-de space represents the time responses of error and change in error

the control input signal. The deviation of the error signal is used to determine the step size of the control signal while the error signal determines the sign of the change in the control signal. Figure 4.14 depicts a plot of the error versus the change in error in order to see the boundaries and the support sets of both signals. Figure 4.14 represents *e-de* space and is used as a design space for the fuzzy logic system.

The time responses of error signals can be used to represent information related to the system output behavior. As the error signals approach zero the output signals move toward the reference set point. Depending on the performances of the controllers used, the error signals may or may not become zero. The dynamic responses of error signals in a controlled system will be sufficient to derive the control rules since it contains the necessary information about the output. Therefore, the error signals shown in Figure 4.13 are used as the source of information for constructing the rule base systems for FLC. These signals may represent any type of control error with a step-type reference input. In fact, the error signal of a control system with a ramp input will not be much different than that given in Figure 4.13 in spite of the differences between the outputs of the systems with step and ramp inputs.

In a control system, the human operator makes necessary adjustments fast or slowly by looking at the system output. A fast action, a high change in magnitude, is required if the output is away from the target, i.e., the error is large, while a slower action, a small change in magnitude, may be enough for the output closer to the reference target. Therefore, information about the amount of change in error signal over a sampling period is also required. A plot of the time variation of error versus the time variation of its change is given in Figure 4.14, which can be used to obtain this information.

The values of error, *e*, and its change, *de*, start with larger values and terminate at the origin or near the origin in a controlled system. As given in Figure 4.14, the values of *e* and *de* from the first-order system lie only in one-quarter of *e-de* space, while the values of *e* and *de* from a second-order oscillatory system travel all four quarters. Since higher-order LTI systems will also have an oscillating response similar to that of a second-order LTI system, the plot of $e = \mathrm{f}(de)$ on *e-de* space of the second-order system given with the solid line in Figure 4.14 can be used as a general case that valid for both first and higher order oscillating systems.

4.7 Fuzzy processor for control

The studies on FL theory have increased tremendously since its development by Zadeh [1]. The application of FL theory to a pressure control system by Mamdani and his colleges [3] has given a push to the FL real-world applications.

The operation principle of an FLC is similar to the decision process of a human operator doing the same job. It performs the same actions as a human operator does by adjusting the input signal looking at only the system output. The architecture of an FLC is very similar to that of a fuzzy reasoning algorithm. The most used algorithm for FLC is the one given as the Mamdani reasoning algorithm in this chapter earlier. As it was in Mamdani fuzzy reasoning architecture, an FLC also consists of four stages namely *fuzzifier*, *rule base*, *consequent*, and *defuzzifier* as shown in Figure 4.15. The input and output signals of an FLC are crisp quantities, which require fuzzification and defuzzification processes. Therefore, the main signal and its change for each sampling are crisp inputs and converted to fuzzy membership values by the fuzzifier. Then they are used in the rule base system to yield the fuzzy consequences which are combined as fuzzy reasoning and then converted to crisp decisions by the process called defuzzification.

The FL-based controller shown in Figure 4.15 is designed to act as an integral controller, such that the final crisp consequent of the FLC is the change $du(k)$, which is added to the previous value $u(k-1)$ of the control signal $u(k)$ to yield the current output, which is the updated control signal $u(k)$. Recalling the digital solution of an integrator using Euler's integration as

$$u(k) = u(k-1) + du(k) \tag{4.26}$$

In a digital integration, the term $du(k)$ is expressed as

$$du(k) = K_{\mathrm{I}} T e(k) \tag{4.27}$$

where K_{I} is the integral constant, T is the sampling period, and $e(k)$ is the integrated signal. The change $du(k)$ on the output of an integrator becomes zero when the input $e(k)$ is zero. Therefore, the output of an integrator retains the previous value. Hence, (4.26) can be used for simulating an integrator. The difference between an integrator and an FLC is the method that is used to obtain $du(k)$, which is obtained using (4.27) for an integrator, and *fuzzy processor* for the FLC.

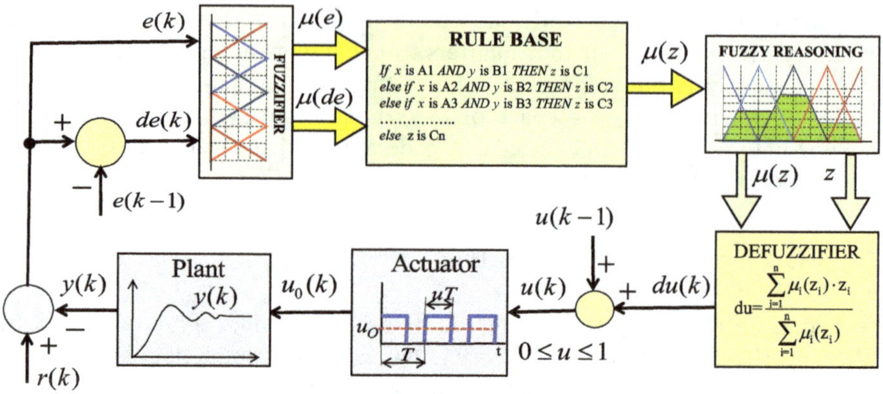

Figure 4.15 The structure of the fuzzy logic controller

The use of fuzzy processor in a feedback control system is shown in Figure 4.15. Since the fuzzy processor is software-based and operates digitally, the crisp inputs $e(k)$ and $de(k)$ are sampled data. Consequently, so is the output $du(k)$. The fuzzy processor uses the error $e(k)$ and its change $de(k)$ as two crisp inputs and converts them to fuzzy membership values in the fuzzifier. The rule base system then activates the related rules to yield a fuzzy consequent from each active rule. The reasoning section of the fuzzy processor combines the individual consequences into a general fuzzy conclusion of the process. Finally, the defuzzifier converts this general fuzzy conclusion to the crisp output $du(k)$ as the final decision for the kth sampling. The fuzzy processor uses the Mamdani fuzzy reasoning method from the inputs $e(k)$ and $de(k)$ to the output $du(k)$. Depending on the magnitude of the control signal $u(k)$, the actuator adjusts the power or energy input to the physical system to be controlled so that the system output tracks the reference. For example, the actuator adjusts the armature voltage of the PMDC motor so that the motor runs at the reference speed. The fuzzy processor, summers, actuator, and the physical system all together are called the fuzzy logic control system, the process results in the control signal $u(t)$ is called the FLC.

The output $du(k)$ of the fuzzy processor may be positive or negative resulting in an increment or a decrement in the control signal $u(k)$ as in (4.26). The decision given by the fuzzy processor is expressed in terms of $du(k)$ either to increase or decrease the control signal, or do nothing if the system output $y(t)$ is tracking the reference $r(k)$ within acceptable limits. In order to get a proper amount of increments and decrements the fuzzy processor should have correct and suitable rules in the rule base. Therefore, the performance of the FLC depends on the rules, which represent the expert's knowledge to operate the system. It is clear that the construction of the rule base is very important for the FLC because it is the modeling of knowledge and thoughts of the human operator. The rules should reflect the knowledge and decision of the operator correctly. There is no direct method to drive the rules. However, some works have been done to complete the rules after having some initial rules. A rule construction process will be given next to show a way of constructing the rules.

4.7.1 Fuzzy rules: the modeling of thoughts and abilities

Human decision-making process is explained in Section 3.10 in the previous chapter. Decision-making process of a driver was modeled in a rule-based system to make an introduction to the concept. This section here addresses the modeling of human thoughts and abilities in more detail by giving examples of the control of physical systems.

As shown in Figure 4.15, there are two inputs to the *fuzzy processor*. One is the control error $e(k)$, which is the difference between the reference signal $r(k)$ and the output signal $y(k)$, the other one is the change in this error $de(k)$. These two inputs, defined as in (4.28) and (4.29), are first fuzzified and converted to fuzzy membership values that are used in the rule base in order to execute the related rules so that an output can be generated.

$$e(k) = r(k) - y(k) \tag{4.28}$$

$$de(k) = e(k) - e(k-1) \tag{4.29}$$

The fuzzy rule base, which may also be called as the fuzzy decision table, is the unit mapping two crisp inputs, $e(k)$ and $de(k)$ to the fuzzy output space defined on the universe of $du(k)$. The time response of the control error $e(k)$ for a step input can be represented by the generalized step response error of a second-order system as shown in Figures 4.8 and 4.13. This error signal may have a damped or an oscillatory response with a decaying exponential component. The latter one is considered for constructing the rule table since it includes overshoot effects leading the rule base to represent more generalized cases. Let us divide the oscillatory error response into 10 regions as shown in Figure 4.16 and look at each region more closely with the eye of a system operator. What does an operator do to operate the

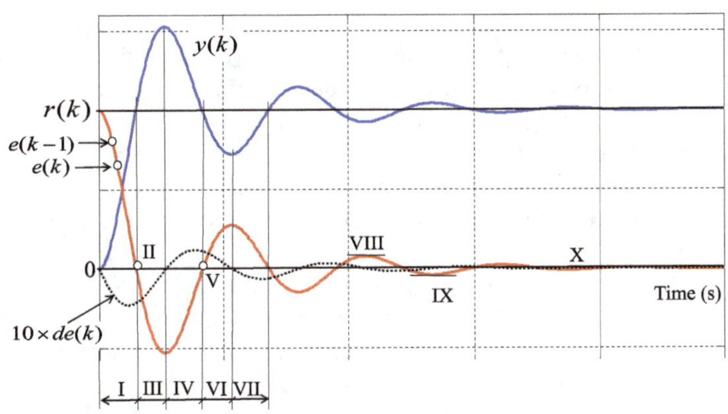

Figure 4.16 Operating regions of error and error change

system with a zero steady-state error? What could be the action taken by the operator at each region of the error response? Does he/she give an increase or a decrease to the control signal depending on the values of $e(k)$ and $de(k)$ in the regions?

The fuzzy rules represent the knowledge and abilities of a human operator who makes necessary adjustments to operate the system with minimum error and fast response. In order to model the actions that a human operator would decide whether the change, $du(k)$, at the output of the fuzzy processor should increase or decrease the control signal according to the error $e(k)$ and its change $de(k)$. It is necessary to observe the behaviors of the error signal $e(k)$ and its change $de(k)$ on different operating regions, as shown by the Roman numbers in Figure 4.16. The output $du(k)$ from the FLC is the change that is required to increase or decrease the overall control action to the controlled system. Therefore, the signs of $e(k)$ and $de(k)$ are used to determine the signs of $du(k)$, which determines whether the overall control signal is to be increased. The sign of $du(k)$ should be positive if $u(k)$ is required to be increased and it should be negative otherwise.

Step input responses of reference $r(k)$, actual system output $y(k)$, control error $e(k)$, and change in error $de(k)$ are given in Figure 4.16 for an oscillatory second-order system. Since the magnitude of the error change $de(k)$ is much smaller than $e(k)$ and the others, it was multiplied by 10 in order to make it visible in the figure. Now, let us look at closely to each region and see what an operator is supposed to do to have a zero steady-state error.

At region I: $e(k)$ is "+" and $de(k)$ is "−." The error is positive and its change is negative, which means that the error is decreasing toward zero. Therefore, the control input $u(k)$ should be increased so that the decrement of error toward zero becomes faster. In order to increase the value of $u(k)$, $du(k)$ should be "+."

At region II: $e(k)$ is zero and $de(k)$ is "−." The error is zero, but its change is negative. This means that the control signal $u(k)$ is higher than the required value and must be reduced. Otherwise, the error will be negative and continue to get away from zero. The error becomes negative when the output $y(k)$ becomes greater than the reference $r(k)$. Therefore, control input $u(k)$ should be reduced by generating a negative "−" $du(k)$.

At region III: Both $e(k)$ and $de(k)$ are negative "−." The error is negative and continues to be more negative. That means that the overshoot is increasing. Something must be done to reduce the control signal $u(k)$ so that the overshoot and error will be smaller or brought to zero. Therefore, $du(k)$ must have a negative "−" sign.

At region IV: $e(k)$ is negative "−" and $de(k)$ is positive "+." The error is still negative, but getting toward zero because of positive $de(k)$. The overshoot is also decreasing. In order to make this decrement faster, the control signal $u(k)$ should be reduced again. Therefore, $du(k)$ is kept negative "−."

At region V: $e(k)$ is zero and $de(k)$ is "+." The error is zero, but its change is positive. This means that the control signal $u(k)$ is smaller than the required value and must be increased. Otherwise, the error will be positive and keep increasing in a positive direction. The error continues to be positive with a larger magnitude and the output $y(k)$ becomes smaller than the reference $r(k)$. Therefore, control input $u(k)$ should be increased by generating a positive "+" $du(k)$.

At region VI: Both $e(k)$ and $de(k)$ are negative "+." The error is positive and increases in a positive direction. That means that the output $y(k)$ is under the reference $r(k)$ and needs to be increased. The control signal $u(k)$ must be increased to increase $y(k)$ so that the error will be smaller or brought to zero. Therefore, $du(k)$ must be positive "+."

At region VII: This is a repeat of region I with smaller magnitudes.

At region XIII: $e(k)$ is positive "+" and $de(k)$ is zero "0." The error is positive and constant since there is no change. This means that the oscillations are over or very small. However, the output is smaller than the reference and the control signal $u(k)$ needs to be increased. Therefore, a positive "+" value for $du(k)$ should be assigned.

At region IX: $e(k)$ is negative "−" and $de(k)$ is zero "0." The error is negative and constant since there is no change. This means that the oscillations are over or very small. However, the output is greater than the reference and the control signal $u(k)$ needs to be reduced. Therefore, a negative "−" value for $du(k)$ should be assigned to decrease the error.

At region X: Both $e(k)$ and $de(k)$ are zero "0." The error is zero and not changing anymore. Therefore, $du(k)$ is set to zero since no change is required for the control signal $u(k)$.

As stated in previous sections, the error $e(k)$ and its change $de(k)$ are two crisp inputs to the fuzzy processor, and $du(k)$ is the crisp output. The sign assignment of $du(k)$ is done using the signs of $e(k)$ and $de(k)$ in all 10 regions. Therefore, the fuzzy rule base system should be designed correctly so that the sign assignments of $du(k)$ are done properly and reflect the signs of $e(k)$ and $de(k)$. The sign assignments of $du(k)$ are summarized in Table 4.1.

An analysis of Table 4.1 shows that the $du(k)$ takes the sign of $e(k)$ as long as $e(k)$ is not zero and it takes the sign of $de(k)$ if $e(k)$ is zero. This conclusion can be expressed as in (4.30).

$$\begin{aligned} \textit{If}\ e(k) \text{ is Zero } \textit{then}\ du(k) &\text{ takes the sign of } de(k) \\ \textit{else}\ du(k) &\text{ takes the sign of } e(k) \end{aligned} \tag{4.30}$$

Table 4.1 shows that each one of $e(k)$, $de(k)$, and $du(k)$ has three different options for the signs to be assigned. They are either positive or negative if not zero.

Table 4.1 The sign assignments of du(k) for basic control actions

	Operating regions									
	I	**II**	**III**	**IV**	**V**	**VI**	**VII**	**VIII**	**IX**	**X**
$e(k)$	+	0	−	−	0	+	+	−	+	0
$de(k)$	−	−	−	+	+	+	−	0	0	0
$du(k)$	+	−	−	−	+	+	+	−	+	0

Table 4.2 An initial nine-rule table

$e(k)$	$de(k)$		
	N_{de}	Z_{de}	P_{de}
P_e	P_{du}	P_{du}	P_{du}
Z_e	N_{du}	Z_{du}	P_{du}
N_e	N_{du}	N_{du}	N_{du}

Therefore, the universes of error, change in error, and change in control signal can be partitioned into three categories negative (N), zero (Z), and positive (P), initially. Keeping in mind these three options, an initial rule decision table with nine rules can be formed as shown in Table 4.2 using the expression (4.30).

Representing the input crisp variables $e(k)$ and $de(k)$ by three fuzzy sets, P, N, and Z means that these input spaces are partitioned into three fuzzy regions each yielding a fuzzy output space with nine rules maximum as given in Table 4.2. A nine-rule fuzzy decision table may be sufficient for some applications. However, many applications require more rules than nine. In order to construct a fuzzy rule decision table with more than nine rules, the input spaces must be partitioned into more than three regions each. For example, the linguistic variable *negative* can be partitioned into subcategories as *negative small* (NS) and *negative big* (NB). Similarly the term *positive* can have subcategories as *positive small* (PS) and *positive big* (PB). Using the subset *zero* (ZE) besides these four subsets, there will be five fuzzy subsets defined in the universes of $e(k)$ and $de(k)$. Therefore, a 5×5 decision space is obtained for $du(k)$ at the output as given in Table 4.3.

The output space $u(k)$ is partitioned into five subcategories as it is done for the input spaces $e(k)$ and $de(k)$. The sign of $du(k)$ is determined from the signs of $e(k)$ and $de(k)$ according to (4.30). Besides the signs, the subsets in the input spaces are also reflected in the output space. For example, if $e(k)$ is PB_e then $du(k)$ is PB_{du}, if $e(k)$ is NS_e, then $du(k)$ is PS_{du}. If $e(k)$ is ZE_e and $de(k)$ is NS_{de}, then $du(k)$ is NS_{du}.

The fuzzy rule table given in Table 4.3 can be expanded to a 49-rule table by adding fuzzy subsets of *negative medium* (NM) and *positive medium* (PM) into negative and positive sections of $e(k)$ and $de(k)$. After obtaining these initial rule tables, they are modified to fit the problem or application where the rule table is going to be used. Too many methods have been proposed and studied in the literature. Some of these studies on fuzzy rule generation and construction can be found in [33–39].

Now let us look at Table 4.3 more closely. The decision space defined in the universe of *du*(*k*) has five fuzzy subsets as NB_{du}, NS_{du}, ZE_{du}, PS_{du}, and PB_{du} similar to the fuzzy subsets defined in the universes *e*(*k*) and *de*(*k*). However, while the fuzzy subset zero stands as a separator between negative and positive fuzzy subsets in the universes *e*(*k*) and *de*(*k*), the negative and positive fuzzy subsets are mixed and border each other without a zero separator between them. Therefore, the initial fuzzy rule table given in Table 4.3 should be modified so that a zero fuzzy subset separate positives and negatives. Table 4.2 shows that the upper right part is mostly positive and the lower left part is mostly negative. Therefore, a fuzzy zero separator can be placed on the main diagonal of Tables 4.2 and 4.3. If zero fuzzy subsets are placed on the main diagonals, these rule tables become as in Tables 4.4 and 4.5.

Table 4.3 An initial 25-rule table

***e*(*k*)**	***de*(*k*)**				
	$\mathbf{NB_{de}}$	$\mathbf{NS_{de}}$	$\mathbf{ZE_{de}}$	$\mathbf{PS_{de}}$	$\mathbf{PB_{de}}$
PB_e	PB_{du}	PB_{du}	PB_{du}	PB_{du}	PB_{du}
PS_e	PS_{du}	PS_{du}	PS_{du}	PS_{du}	PS_{du}
ZE_e	NB_{du}	NS_{du}	ZE_{du}	PS_{du}	PB_{du}
NS_e	NS_{du}	NS_{du}	NS_{du}	NS_{du}	NS_{du}
NB_e	NB_{du}	NB_{du}	NB_{du}	NB_{du}	NB_{du}

Table 4.4 An initial nine-rule table with zero diagonal

***e*(*k*)**	***de*(*k*)**		
	$\mathbf{N_{de}}$	$\mathbf{Z_{de}}$	$\mathbf{P_{de}}$
P_e	Z_{du}	P_{du}	P_{du}
Z_e	N_{du}	Z_{du}	P_{du}
N_e	N_{du}	N_{du}	Z_{du}

Table 4.5 An initial 25-rule table with zero diagonal

***e*(*k*)**	***de*(*k*)**				
	$\mathbf{NB_{de}}$	$\mathbf{NS_{de}}$	$\mathbf{ZE_{de}}$	$\mathbf{PS_{de}}$	$\mathbf{PB_{de}}$
PB_e	ZE_{du}	PB_{du}	PB_{du}	PB_{du}	PB_{du}
PS_e	PS_{du}	ZE_{du}	PS_{du}	PS_{du}	PS_{du}
ZE_e	NB_{du}	NS_{du}	ZE_{du}	PS_{du}	PB_{du}
NS_e	NS_{du}	NS_{du}	NS_{du}	ZE_{du}	NS_{du}
NB_e	NB_{du}	NB_{du}	NB_{du}	NB_{du}	ZE_{du}

The zero diagonal in Table 7.5 is separating positives and negatives from each other. However, two fuzzy subsets, PS_{du} at line 2 column 1, and NS_{du} at line 4 column 5, are breaking this rule. Therefore, these two fuzzy subsets should be changed to proper fuzzy subsets. Since they are next to zero diagonal, just a sign change corrects the table as in Table 4.6.

Macvicar-Whelan has shown the importance of having a symmetrical fuzzy rule table for man–machine interaction [40]. The symmetrical fuzzy rule table can be used as a generalized rule table that works for many applications with small changes. In order to have a generalized fuzzy rule base system let us convert Table 4.6 to a symmetrical one. Since the universes of *e*(*k*) and *de*(*k*) are partitioned into fuzzy subsets in the order of NB, NS, ZE, PS, and PB, let us use the same order in the output space *du*(*k*) and put the fuzzy subsets into the same order from the lower left corner to the upper right corner as shown in Table 4.7, which can be extended to a 49-rule table by including the fuzzy subsets negative medium (NM) and positive medium (PM) as in Table 4.8.

The universes *e*(*k*) and *de*(*k*) are partitioned into the same number of fuzzy subsets in Tables 4.2–4.8. Both universes have three, five, or seven fuzzy subsets,

Table 4.6 A 25-rule table with zero diagonal

e(k)	**de(k)**				
	$\mathbf{NB_{de}}$	$\mathbf{NS_{de}}$	$\mathbf{ZE_{de}}$	$\mathbf{PS_{de}}$	$\mathbf{PB_{de}}$
PB_e	ZE_{du}	PB_{du}	PB_{du}	PB_{du}	PB_{du}
PS_e	NS_{du}	ZE_{du}	PS_{du}	PS_{du}	PS_{du}
ZE_e	NB_{du}	NS_{du}	ZE_{du}	PS_{du}	PB_{du}
NS_e	NS_{du}	NS_{du}	NS_{du}	ZE_{du}	PS_{du}
NB_e	NB_{du}	NB_{du}	NB_{du}	NB_{du}	ZE_{du}

Table 4.7 A generalized symmetrical 25-rule table

e(*k*)	*de*(*k*)				
	NB_{de}	NS_{de}	ZE_{de}	PS_{de}	PB_{de}
PB_e	ZE_{du}	PS_{du}	PS_{du}	PB_{du}	PB_{du}
PS_e	NS_{du}	ZE_{du}	PS_{du}	PS_{du}	PB_{du}
ZE_e	NS_{du}	NS_{du}	ZE_{du}	PS_{du}	PS_{du}
NS_e	NB_{du}	NS_{du}	NS_{du}	ZE_{du}	PS_{du}
NB_e	NB_{du}	NB_{du}	NS_{du}	NS_{du}	ZE_{du}

Table 4.8 A generalized symmetrical 49-rule table

$e(k)$	$de(k)$						
	NB_{de}	NS_{de}	ZE_{de}	PS_{de}	PB_{de}	PB_{de}	PB_{de}
PB_e	ZE_{du}	PS_{du}	PS_{du}	PM_{du}	PM_{du}	PB_{du}	PB_{du}
PS_e	NS_{du}	ZE_{du}	PS_{du}	PS_{du}	PM_{du}	PM_{du}	PB_{du}
ZE_e	NS_{du}	NS_{du}	ZE_{du}	PS_{du}	PS_{du}	PM_{du}	PM_{du}
NS_e	NM_{du}	NS_{du}	NS_{du}	ZE_{du}	PS_{du}	PS_{du}	PM_{du}
NB_e	NM_{du}	NM_{du}	NS_{du}	NS_{du}	ZE_{du}	PS_{du}	PS_{du}
NB_e	NB_{du}	NM_{du}	NM_{du}	NS_{du}	NS_{du}	ZE_{du}	PS_{du}
NB_e	NB_{du}	NB_{du}	NM_{du}	NM_{du}	NS_{du}	NS_{du}	ZE_{du}

resulting in a square output space with nine, 25, or 49 rules. The number of the fuzzy subsets in the universes of $e(k)$ and $de(k)$ do not have to be equal and the output space does not have to be square, neither. Therefore, the output space may be a square or a perpendicular. Whatever the size of the output space is, the number of active rules is small and four in most of the cases. The number of active rules is determined by the interaction of the neighboring fuzzy subsets in the same universe. In other words, the number of active rules is related to the inclusion of the crisp inputs in the fuzzy subsets of the crisp input universes. For example, if $e(k)$ has non-zero membership value in two fuzzy subsets and $de(k)$ has two nonzero membership values in two fuzzy subsets, then there will be four active rules in the output space.

4.7.2 The input–output interaction

A plot of $e(k)$ versus $de(k)$ makes it easy to see the variation space of these variables. Therefore, an FLC designer can see and define the boundaries of the variables. The definitions of the boundaries are very important to select a working area in the input space so that the sparsity is minimized and operations are focused in the active areas of the input universes. Such a plot of $e(k)$ versus $de(k)$ for a second-order oscillatory system is shown in Figure 4.17 where each one of the universes of $e(k)$ and $de(k)$ are partitioned into three regions as negative (N), zero (Z) and positive (P). The usable area in Figure 4.17 is shown by the shaded area. Triangular-type membership functions are used for partitioning the crisp universes into fuzzy subsets. Different membership functions such as Gaussian, trapezoidal, and bell could have also been used. Each one of these membership functions has its own effects on the FLC output [25]. However, triangular membership functions are more convenient for expressing the concept because their

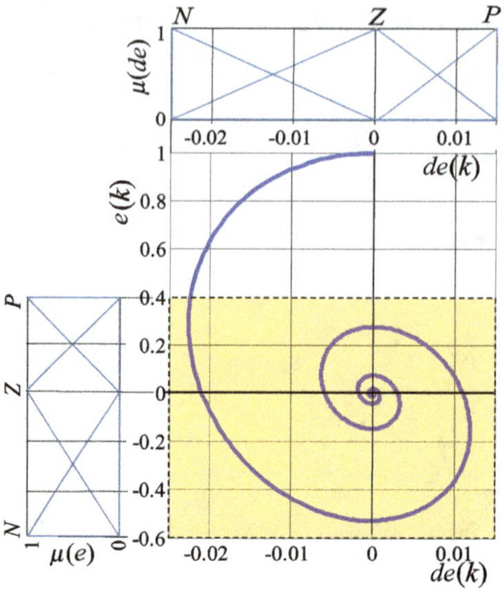

Figure 4.17 Input space of an FLC from the system response

linearity makes it easier to intercept membership degrees from a triangle. Therefore, triangular membership function defined by (2.7) and given in Figure 2.5 is used to represent the fuzzy subsets shown in Figure 4.17.

The boundaries for the universe *e*(*k*) are selected from −0.6 to 0.4, and the boundaries of *de*(*k*) are selected from −0.025 to 0.015 in Figure 4.17. For a more effective FLC, these boundaries can be selected much smaller in order to reduce the sparsity. For example, another selection for *e*(*k*) might be from −0.2 to 0.3 while it might be from −0.01 to 0.01 for *de*(*k*). With the later selection, the working area is focused more closely toward the origin where the data density is higher. These latter boundaries are used in Figure 4.18. However, this time the universes are partitioned into six and five subsets for *e*(*k*) and *de*(*k*), respectively. Depending upon the responses of the physical systems, the fuzzy subsets may or may not be spaced evenly. Since the lower and upper boundaries of the universe *e* (*k*) in Figure 4.18 are not symmetrical according to zero line, the fuzzy subsets might be distributed evenly. Even if they are evenly distributed in Figure 4.18, the fuzzy subsets are not symmetrical. There are two fuzzy subsets in the negative region and three fuzzy subsets in the positive region. The lower and upper boundaries of the universe *de*(*k*) are symmetrical and the fuzzy subsets in *de*(*k*) are distributed evenly. With the reduced sparsity, the fuzzy rules become more effective because any operating point in the input space will have a membership value in the fuzzy subsets defined in both *e*(*k*) and *de*(*k*) as depicted by the point (*e*(*k*), *de*(*k*)) in Figure 4.18.

The input space given in Figure 4.18, enables the FLC design engineers to set the boundaries of the input universes clearly. It should be kept in mind that the

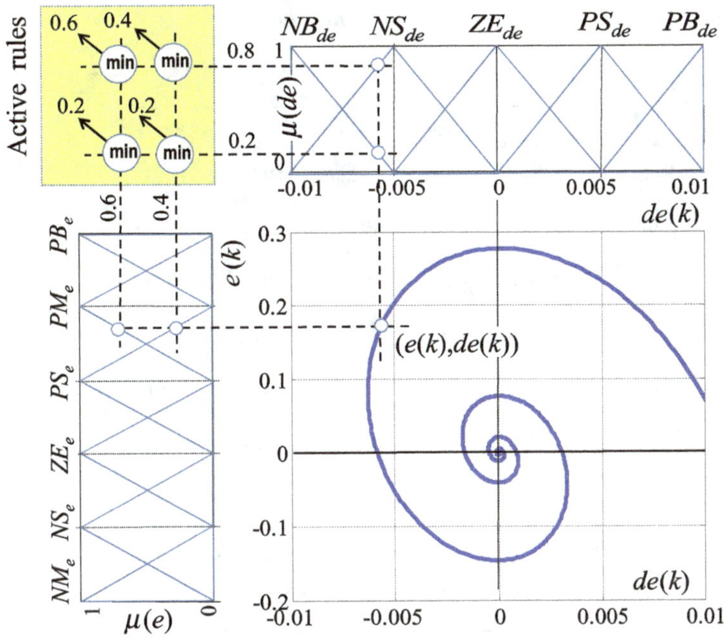

Figure 4.18 Crisp input space, fuzzification, and fuzzy weightings in an FLC

smaller sampling period gives smaller $de(k)$ values. However, the plot of $e(k)$ versus $de(k)$ depicts everything to set the boundaries. A point ($e(k)$, $de(k)$) is marked in the input space in Figure 4.18 for the values of $e(k) = 0.16$ and $de(k) = -0.006$. With this given value, $e(k)$ has nonzero membership degrees in fuzzy subsets PS_e and PM_e, while $de(k)$ has the membership nonzero membership values in the fuzzy subsets NB_{de} and NS_{de}. Using the given parameters of the triangular fuzzy subsets in Figure 2.5, the related membership values are obtained as shown in Figure 4.18 and written as

$$\begin{aligned}
\mu_{\mathrm{PSe}}(e) &= \mathrm{triangle01m}(0, 0.1, 0.2, 0.16) = 0.4 \\
\mu_{\mathrm{PMe}}(e) &= \mathrm{triangle01m}(0.1, 0.2, 0.3, 0.16) = 0.6 \\
\mu_{\mathrm{NBde}}(de) &= \mathrm{triangle01m}(-0.01, -0.01, -0.005, -0.006) = 0.2 \\
\mu_{\mathrm{NSde}}(de) &= \mathrm{triangle01m}(-0.01, -0.005, 0.0, -0.006) = 0.8
\end{aligned} \tag{4.31}$$

These membership values are the outputs of the fuzzification process. They are used in fuzzy rule-based systems to obtain related consequences of the rules as the weighting factors, which are used in the output universe $du(k)$ as rule scaling factors. The resulting weighting factors from each rule are given in the upper left corner of Figure 4.18. Before proceeding more, let us generate the rule base for the system shown in Figure 4.18. Since the number of the subsets in the universes of $e(k)$ and $de(k)$ are six and five, respectively, a rule base with 30 rules can be generated as in Table 4.9.

Table 4.9 A generalized 30-rule table

$e(k)$	$de(k)$				
	NB_{de}	NS_{de}	ZE_{de}	PS_{de}	PB_{de}
PB_e	PS_{du}	PS_{du}	PB_{du}	PB_{du}	PB_{du}
PM_e	ZE_{du}	PS_{du}	PS_{du}	PB_{du}	PB_{du}
PS_e	NS_{du}	ZE_{du}	PS_{du}	PS_{du}	PB_{du}
ZE_e	NS_{du}	NS_{du}	ZE_{du}	PS_{du}	PS_{du}
NS_e	NB_{du}	NS_{du}	NS_{du}	ZE_{du}	PS_{du}
NM_e	NB_{du}	NB_{du}	NS_{du}	NS_{du}	ZE_{du}

Table 4.10 Active rules in Table 4.9 for the data given in Figure 4.18

$e(k)$	$de(k)$	
	NB_{de}	NS_{de}
PM_e	ZE_{du}	PS_{du}
PS_e	NS_{du}	ZE_{du}

where the fuzzy subsets defined in the universe of *du* are represented by triangular fuzzy membership functions as NB_{du} = triangle(-1, -1, -0.5), NS_{du} = triangle(-1, -0.5, 0.0), ZE_{du} = triangle (-0.5, 0, 0.5), PS_{du} = triangle(0, 0.5, 1), PB_{du} = triangle(0.5, 1, 1).

The crisp input $e(k) = 0.16$ has membership values in the fuzzy subsets PS_e and PM_e as $\mu_{PSe}(e) = 0.4$ and $\mu_{PMe}(e) = 0.6$, respectively. Therefore, the active rules will be on the rows corresponding to PS_e and PM_e. The crisp input $de(k) = -0.006$ has membership values in the fuzzy subsets NB_{de} and NS_{de} as $\mu_{NBde}(de) = 0.2$ and $\mu_{NSde}(de) = 0.8$, respectively. Therefore, the active rules will be on the columns corresponding to NB_{de} and NS_{de}. The active rules will be on the cross-section of PS_e and PM_e with NB_{de} and NS_{de} as shown in Table 4.10. There will be four active rules and four resulting weighting factors, which are shown in the upper left corner of Figure 4.18.

The linguistic expressions of the active rules are written as in (4.32), which yields the weighting factors for the consequent fuzzy subsets in the output space as in (4.33).

$$\begin{array}{ll} \textit{If} & e(k) \text{ is } PM_e \textit{ and } de(k) \text{ is } NB_{de} \textit{ then } du(k) \text{ is } ZE_{du} \\ \textit{else if} & e(k) \text{ is } PM_e \textit{ and } de(k) \text{ is } NS_{de} \textit{ then } du(k) \text{ is } PS_{du} \\ \textit{else if} & e(k) \text{ is } PS_e \textit{ and } de(k) \text{ is } NB_{de} \textit{ then } du(k) \text{ is } NS_{du} \\ \textit{else if} & e(k) \text{ is } PS_e \textit{ and } de(k) \text{ is } NS_{de} \textit{ then } du(k) \text{ is } ZE_{du} \\ \textit{else} & \qquad du(k) \text{ is } ZE_{du} \end{array} \tag{4.32}$$

$$\begin{aligned} w_{21} &= \mu_{\text{ZEdu}}(du) = \min[\mu_{\text{PMe}}(e), \mu_{\text{NBde}}(de)] = \min(0.6, 0.8) = 0.6 \\ w_{22} &= \mu_{\text{PSdu}}(du) = \min[\mu_{\text{PMe}}(e), \mu_{\text{NSde}}(de)] = \min(0.6, 0.2) = 0.2 \\ w_{31} &= \mu_{\text{NSdu}}(du) = \min[\mu_{\text{PSe}}(e), \mu_{\text{NBde}}(de)] = \min(0.4, 0.8) = 0.4 \\ w_{32} &= \mu_{\text{ZEdu}}(du) = \min[\mu_{\text{PSe}}(e), \mu_{\text{NSde}}(de)] = \min(0.4, 0.2) = 0.2 \end{aligned} \quad (4.33)$$

Then the final crisp output is obtained after the defuzzification using the center of area method.

$$du(k) = \frac{\sum_{i=1}^{n} w_{nm}(du) \cdot DU_{nm}}{\sum_{i=1}^{n} w_{nm}(du)} \quad (4.34)$$

$$du(k) = \frac{(0.6)(0) + (0.2)(0.5) + (0.4)(-0.5) + (0.2)(0)}{0.4 + 0.2 + 0.4 + 0.2} = \frac{-0.1}{1.2}$$

$$= -0.0833 \text{ pu.}$$

The final conclusion is the change in control input signal *u*(*k*). Therefore, *du*(*k*) is added to the previous value of *u*(*k*) to obtain the new control signal. Since *du*(*k*) is obtained as a negative value, then *du*(*k*) will be reduced by the amount of −0.0833.

4.8 Modeling the FLC in MATLAB

The design and simulation of an FLC are explained in this section. An FLC is designed to control a radar tracking system, which is adapted from [26]. A MATLAB .m file is generated to simulate this FLC-controlled radar tracking system. All steps of the designed FLC and its application to the radar tracking system are given in detail through the codes of a MATLAB .m file.

Example 4.8

The principle diagram of a radar tracking system to be used in this example is shown in Figure E4.8(a). The system consists of a radar, a radar receiver, an amplifier, and a permanent magnet direct current (PMDC) motor. The motor and the radar are coupled together over a gear system. The radar is supposed to track the car passing by. Two angles, θ_{Rd} and θ_{C}, which represent the radar position and the car location axes, respectively, are determined by the radar system. These two inputs from the radar are then used by the controller to generate a control signal which is amplified and applied to the motor.

The difference between the car position axis and the radar axis is defined as the tracking error signal to be compensated and is obtained as

$$e(t) = \theta_{\text{C}}(t) - \theta_{\text{Rd}}(t) \quad \text{(E4.8a)}$$

As long as the error signal *e*(*t*) is not zero, then, the amplified value V_a of the controlled signal *u* is applied to the motor. The motor accelerates until it catches the moving car. If the radar axis overlaps the car axis and they are synchronized, the

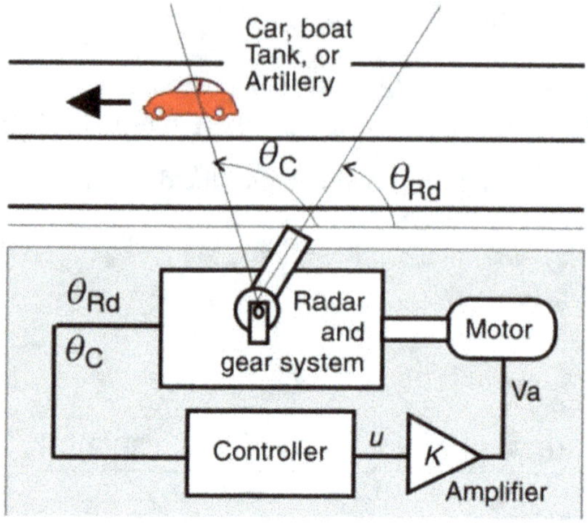

Figure E4.8(a) Schematic diagram of the radar tracking system

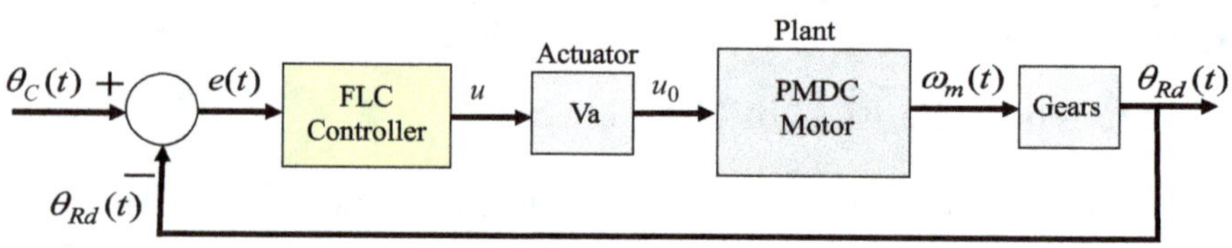

Figure E4.8(b) General block diagram of the control system

error becomes zero. When this condition is satisfied, the changes in both angles must be equal to each other for a zero steady-state error operation.

$$\frac{d\theta_C}{dt} = \frac{d\theta_{Rd}}{dt} \text{ while } \theta_C = \theta_{Rd} \tag{E4.8b}$$

The angle θ_C representing the car position is the reference input to the controller. It is assumed that the radar can be rotated between 0° and 180°. Therefore, the reference angle is also varied from 0° to 180°. The control block diagram of the system studied is given in Figure E4.8(b) where only the blocks of the system components are shown.

Although the mathematical model of the physical system to be controlled is not needed in real-time fuzzy logic-based control systems, one is needed here since the system is to be simulated. Therefore, the simulation diagram of the radar system comprising the PMDC motor and the gears is obtained as in Figure E4.8(c). The methods for obtaining a dynamic model of physical systems are discussed in Chapter 5.

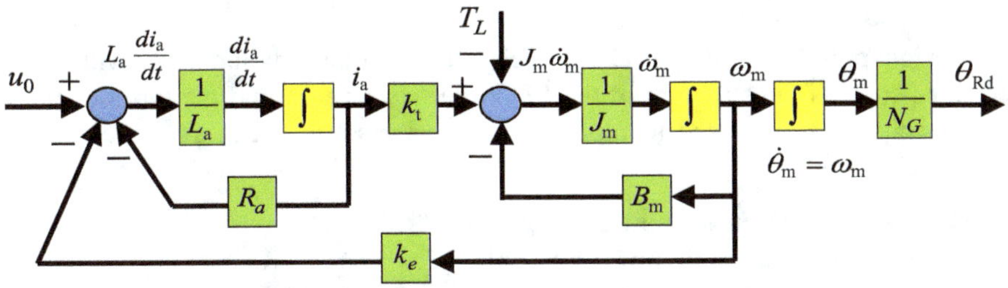

Figure E4.8(c) Simulation diagram of a PMDC motor

The state-space model of the PMDC motor is written below for armature current speed from Figure E4.8(c).

$$\begin{bmatrix} \dot{i}_a \\ \dot{\omega}_m \end{bmatrix} = \begin{bmatrix} -\dfrac{R_a}{L_a} & -\dfrac{k_e}{L_a} \\ \dfrac{k_t}{J_m} & -\dfrac{B_m}{J_m} \end{bmatrix} \begin{bmatrix} i_a \\ \omega_m \end{bmatrix} + \begin{bmatrix} \dfrac{1}{L_a} & 0 \\ 0 & -\dfrac{1}{J_m} \end{bmatrix} \begin{bmatrix} v_a \\ T_L \end{bmatrix} \tag{E4.8c}$$

where ω_m and i_a are the motor speed and the motor armature current, respectively. The rotor position angle θ_m and radar position angle θ_{Rd} are obtained by integrating the speed ω_m and considering the turn ratio N_G of the gear system as

$$\theta_m = \int \omega_m(t)dt \tag{E4.8d}$$

As can be seen in Figure E4.8(b), the signal u is the controlled input to the system, and it is determined by the controller. This signal is multiplied by the nominal voltage V_a of the motor to determine the applied voltage u_0 to the motor.

Assume that the position angle of the moving object varies between 0° and 180° (or 0 and π rad) with respect to the position angle of the radar system, the error between these two position angles may vary from $-\pi$ to $+\pi$ rad. It is also assumed that the car position angle varies from 0 to π rad in 4 s.

Use the following PMDC motor data and turns ratio of the gears and design an FLC for the radar tracking system described above.

R_a = resistance of armature winding = 1.4 Ohm
L_a = inductance of armature winding = 0.0805 H
K_e = voltage constant = 0.095 V s/rad
K_t = torque constant = 0.095 Nm/A.
J_m = moment of inertia = 0.0007432 kg m^2
B_m = viscous constant = 0.000431 Nm s.
V_a = Nominal armature voltage = 36 V
N = Turns ratio of the gears = 2.67

Answer 4.8

The design process of the FLC for the given problem should start with some decisions such as the structure of the FLC, the fuzzy rules, and the types of fuzzy membership functions used to represent the fuzzy subsets. Let us make the following selections.

- Use the FLC structure shown in Figure 4.15.
- Use the fuzzy rule table given by Table 4.7.
- Use sinusoid membership functions defined in (2.23) and given in Figure E2.5(a).

The fuzzy rule table given by Table 4.7 includes 25 rules because the universes of *e* (*k*) and *de*(*k*) are partitioned into five fuzzy subsets as *negative big* (NB), *negative small* (NS), *zero* (ZZ), *positive small* (PS), and *positive big* (PB). These fuzzy subsets are represented by *sine* functions for all universes of *e*, *de*, and *du* in this example using the function in Figure E2.5(a).

Since the position angle of the moving object varies between 0 and π rad with respect to the position angle of the radar system, the error between these two position angles may vary from $-\pi$ to $+\pi$ rad. Therefore, the minimum and the maximum limits of the universe of error, *e*, are taken as $e_{min} = -\pi$ rad and $e_{max} = +\pi$ rad, respectively. After simulating the system without any controllers, it can be observed that the change in error is very small compared to the error depending on the sampling time. Therefore, the minimum and the maximum limits of the error change are assigned as $de_{min} = e_{min}/50$ rad and $de_{max} = e_{max}/50$ rad, in order to make the changes in error more effective in terms of the controller. The minimum and the maximum values of the change in control signal are -1 and $+1$, respectively. Besides, MATLAB uses radians in mathematical calculations. Therefore, data related to position angles should be in radians. The locations of the fuzzy subsets in the universes of *e*(*k*), *de*(*k*), and *du*(*k*) are shown in Figure E4.8(d).

The sinusoid-type membership functions representing the fuzzy subsets mentioned above are evenly spaced in related universes as depicted in Figure E4.8(e).

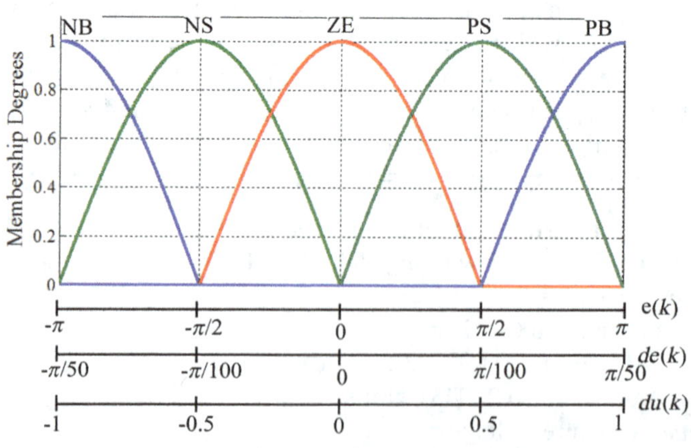

Figure E4.8(d) Fuzzy subsets defined in the universes of e(k), de(k) and du(k)

```
%-------------------------------------------------------
% Example 04.08
clear; clf;  clc;
% ----------------------PMDC Motor and radar data
Vn=36;            Wn=3400;          TL=0;         In=15;
Ra=1.4;           La=0.0805;        Ke=0.095;
Jm=0.0007432;     Bm=0.000431;      Kt=0.095;     NG=2.67;
% ---------------------------  State Space model
A=[ -(Ra/La)      -(Ke/La)
       Kt/Jm      -(Bm/Jm)   ];
B=[1/La     0
     0   -1/Jm ];
C=[0  1];     D=[0 ] ;
Ua=[Vn  TL];
% initials for simulation
Teta0=0;  dt=0.01; t0=0;  tend=4;    k=1;
SA=size(A);  NL=SA(1);    NC=SA(2);
for n=1:NL
    x0(n)=0;
end
```

Figure E4.8(e)   Plant data for the FLC

```
y1=0;    Y1=y1*pi/180;
y2=180; Y2=y2*pi/180;
TETAM0=Y2/2;
TS=dt/100;
```

Figure E4.8(f)   The definition of the universes of car and radar position angles

After this point, the design process of the FLC will be given by MATLAB codes and their explanations. Therefore, let us enter the data first. Figure E4.8(f) shows the input data for the plant, the system to be controlled. In order to generalize the solution, the raw data is converted to state-space model data by constructing *A*, *B*, *C*, and *D* matrices and *U* input matrix. Then starting time and sampling period are given. The iteration counter *k* is set to 1. The initial values of all variables are set to zero.

When the simulation is started, the radar position axis and the car axis might be anywhere in the space defined in the interval $\{y_1, y_2\} = \{0, 180\}$ in degrees or $\{Y_1, Y_2\} = \{0, \pi\}$ in radians. The definition of the operation area is given in Figure E4.8(f). Assume that the car position axis is at 0° and mowing up to 180° in 4 s. The radar position axis might be anywhere between 0° and 180° and assumed to be in the midpoint as being $Y_2/2$, which is converted to radians as TETAM0.

In the next step, the input parameters of the fuzzy subsets used in FLC are entered. Upper and lower boundaries of error $e(k)$, its change $de(k)$, and control input change $du(k)$ are entered as EMAX, EMIN, DEMAX, DEMIN, DUMAX, and DUMIN. Then these three universes are partitioned into five subsets as negative big

(NB), negative small (NS), zero (ZE), positive small (PS), and positive big (PB) for the error *e*, error change *de*, and control input change *du* as given in Figure E4.8(g). In the bottom part of Figure E4.8(g), the initial values of the variables *e*(*k*), *de*(*k*), and *du*(*k*) are entered.

The rule table can be entered as a membership matrix consisting of the values of *du*(*k*) such that

$$\mu_{\text{NBdu}}(du) = \mu_{\text{NSdu}}(du) = \mu_{\text{ZZdu}}(du) = \mu_{\text{PSdu}}(du) == \mu_{\text{PBdu}}(du) = 1 \tag{E4.8e}$$

where the crisp values of NBDU, NSDU, ZZDU, PSDU, and PBDU are defined in Figure E4.8(g). These values correspond to the crisp values *du*(*k*) with the membership value of 1.0. In other words, these crisp values correspond to the peak points of the sinusoid fuzzy subsets and can be considered as their addresses in the universe of *du*. The membership matrix is shown in Figure E4.8(h). The number of

```
% Lower and upper boundaries of the universes e, de and du
EMAX   = Y2;            EMIN   = -EMAX;
DEMAX  = EMAX/50;       DEMIN  = -DEMAX;
DUMAX  = 1;             DUMIN  = -1;
% Partition the universe of e into five fuzzy subsets
TE=2*EMAX;   WE=2*pi/TE;  AE=1;
NBE=-TE/2;   NSE=-TE/4;   ZZE=0;      PSE=TE/4;     PBE=TE/2;
x1E=NBE;     x2E=NSE;     x3E=ZZE;    x4E=PSE;      x5E=PBE;
% Partition the universe of de into five fuzzy subsets
TDE=2*DEMAX;  WDE=2*pi/TDE;  ADE=1;
NBDE=-TDE/2;  NSDE=-TDE/4;   ZZDE=0;     PSDE=TDE/4;
PBDE=TDE/2;
x1DE=NBDE;    x2DE=NSDE;     x3DE=ZZDE;  x4DE=PSDE;   x5DE=PBDE;
% Partition the universe of du into five fuzzy subsets
TDU=2*DUMAX; WDU=2*pi/TDU; ADU=1;
NBDU=-TDU/2;  NSDU=-TDU/4;  ZZDU=0;      PSDU=TDU/4;  PBDU=TDU/2;
x1DU=NBDU;    x2DU=NSDU;    x3DU=ZZDU;   x4DU=PSDU;   x5DU=PBDU;
% Initial values for FLC variables
ee=EMAX;         dee=0;     E=EMAX;     dE=0;
e0=EMAX;         e(1)=0;    e(2)=0;     de=e(2)-e(1);
DUSCALE=0.2;     U00=0;
```

Figure E4.8(g) The definition of the universes of car and radar position angles

```
% ***************         rule matrix
  DU=[  NBDU    NBDU    NSDU    NSDU    ZZDU
        NBDU    NSDU    NSDU    ZZDU    PSDU
        NSDU    NSDU    ZZDU    PSDU    PSDU
        NSDU    ZZDU    PSDU    PSDU    PBDU
        ZZDU    PSDU    PSDU    PBDU    PBDU   ];
DULC=size(DU);   DUR=DULC(1);   DUC=DULC(2);
```

Figure E4.8(h) The rule matrix

rows and columns of the rule matrix is determined as DUR and DUC to be used in the latter parts of the simulation.

Since all definitions, initials, and required data are entered, the simulation loop can be started. The beginning of this loop is given in Figure E4.8(i). The loop will continue to iterative simulation as long as the simulation time t_0 is smaller than the end time *tend*.

The first two lines in the loop are used to limit the values of error $e(k)$ and its change $de(k)$ in order to keep these values between the defined lower and upper boundaries of the related universes. The outputs from the function *limiter* are the values of $e(k)$ and $de(k)$ as E and DE for the current sampling instant k. At this point, the crisp values of two inputs to the fuzzy processor are set and ready to be fuzzified. Therefore, a MATLAB function *sinusoid5.m* is used to fuzzify E and DE. The function *sinusoid5* uses the crisp variable, E or DE as the input along with the parameters of the sinusoid function and fuzzy membership functions, and returns the fuzzified values of E or DE as FSE or FSDE, respectively. The fuzzified value FSE contains the membership values of E in the fuzzy subsets NBE, NSE, ZZE, PSE, and PBE while FSDE includes the membership values of DE in the fuzzy subsets NBDE, NSDE, ZZDE, PSDE, and PBDE.

After the fuzzification, the rules are to be processed. The rule processing stage is shown in Figure E4.8(j). Since the weighting factors of the fuzzy sets in

```
%**************  Iteration loop *************************
while t0 < tend-dt
    E=limiter(EMIN,EMAX,ee);         % --- Limit error
    DE=limiter(DEMIN,DEMAX,dee); % --- limit error change
    % ------------------------------  fuzzification of e
    FSE=sinusoid5(E,WE,AE,NBE,NSE,ZZE,PSE,PBE);
    % ------------------------------ fuzzification of de
    FSDE=sinusoid5(DE,WDE,ADE,NBDE,NSDE,ZZDE,PSDE,PBDE);
```

Figure E4.8(i) Starting the simulation loop

```
% ----------------- simulate the rules
FSDUTOP=0;
DDUTOP=0;
for n=1:DUR
    for m=1:DUC
        FSDU(n,m)=min(FSE(n),FSDE(m));
        DDU(n,m)=FSDU(n,m)*DU(n,m);
        FSDUTOP=FSDUTOP+FSDU(n,m);
        DDUTOP=DDUTOP+DDU(n,m);
    end
end
```

Figure E4.8(j) The rule processing part in the simulation

the universe of u are obtained as the minimum of the intercepting fuzzy membership values of *e*(*k*) and *de*(*k*) in the related rules, the min operator is used for each rule to obtain these weightings as FSDU in Figure E4.8(j). Recall from (4.33) that the numerator of the center of area method is equal to the sum of the product of the weightings and *du* corresponding to the peak of the fuzzy membership function defined as the output decision for the related rule. This multiplication is represented by DDU in Figure E4.8(j). Then the sum of these multiplications is obtained as DDUTOP, which is divided by the sum of the weightings, FSDUTOP, in order to obtain the defuzzified crisp output as given in Figure E4.8(k). The final defuzzified output from the fuzzy processor is *duu*, which corresponds to the change in control signal for the current sampling instant *k*.

Since the boundaries of the universe of *du* are set to as $\{-1, 1\}$, the value *duu* is limited to be in this interval and to yield *du* for the sampling *k*. The change *du* is added to the previous value U00 of the control signal to obtain the new control signal UU. The control signal is defined in the interval $\{0, 1\}$, and a limiter function is used to keep it in this interval to yield the control signal U to be applied to the plant. The actuator, which is assumed to be a chopper is simulated as in (4.25). The remaining part of the process is almost the same as in Example 4.7.

The function *runge.m* is called to solve the PMDC motor equations for the input voltage U0. Then the variables are updated for the next step to continue the simulation until the condition of the while loop is over. The last part of the while loop is given in Figure E4.8(l).

The resultant responses of the moving car and radar positions are depicted in Figure E4.8(m). The radar axis is assumed to be at 90° initially and start following the car axis as soon as the car is in the range of the radar starting from 0°. The radar axis catches and locks the car position axis after a short transient. The error and the change in error become very small and zero during the first 2–3 s as given in Figure E4.8(m).

The speed of the car is assumed to be very fast, so it takes 4 s for the car to be in the range of the radar. If this duration is longer than 4 s, then it will be easier for the radar to catch and track the car.

```
% ------------------ defuzzification
duu=DUSCALE*(DDUTOP/FSDUTOP);
du=limiter(DUMIN,DUMAX,duu);
UU=U00+du;      U00=UU;
U=limiter(0,1,UU);
U0 = U*Ua;          % ---- dc chopper
```

Figure E4.8(k) Defuzzification and actuator

```
% solution of the PMDC motor equations ***
[x]=runge(A,B,U0,x0,dt);
 % PNDC motor current (A) and speed (rad/s)
Ia(k)=x(1);      wm(k)=x(2);
 % Speed rpm) and Torque (Nm)
n(k)=wm(k)*60/(2*pi);     TORG(k)=Kt*Ia(k);
 %  Car and radar position angles
TETAT(k) = (t0*(Y2-Y1) + tend*Y1 )/tend;
TETAM = TETAM0 + TS*wm(k);   TETAM0=TETAM;
TETAR(k)=TETAM/NG;
 %  Position error and its change
e(k)=TETAT(k)-TETAR(k);
de(k)=e(k)-e0;    e0=e(k);
Vmot(k)=U0(2);           duty(k)=U;
 %  Update data for the next iteration
ee=e(k); dee=de(k); t(k)=t0+dt; t0=t(k);
for n=1:NL
    x0(n)=x(n);    XX(k,n)=x(n);
    end
    k=k+1;
end;
```

Figure E4.8(l) The final part of the while loop

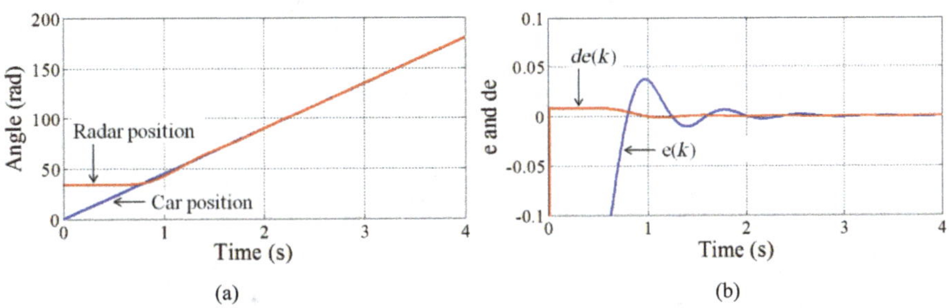

Figure E4.8(m) The simulation results of Example 4.8. (a) The time responses of the car and radar axes and (b) the time responses of the error and the change in error.

The radar tracking example is selected to test the FLC because the reference is a ramp. Using a ramp instead of a step function as the reference to be tracked makes the problem harder. Fuzzy logic-based controllers are used in many different applications. Some of these application areas will be discussed in the latter chapters. Therefore, more examples are left to the application chapters. Example E4.78 is given to address the operational behavior of an FLC and step-by-step algorithm code in a MATLAB .m file. The simulation algorithm given in Example E4.8 is summarized in Figure 4.18.

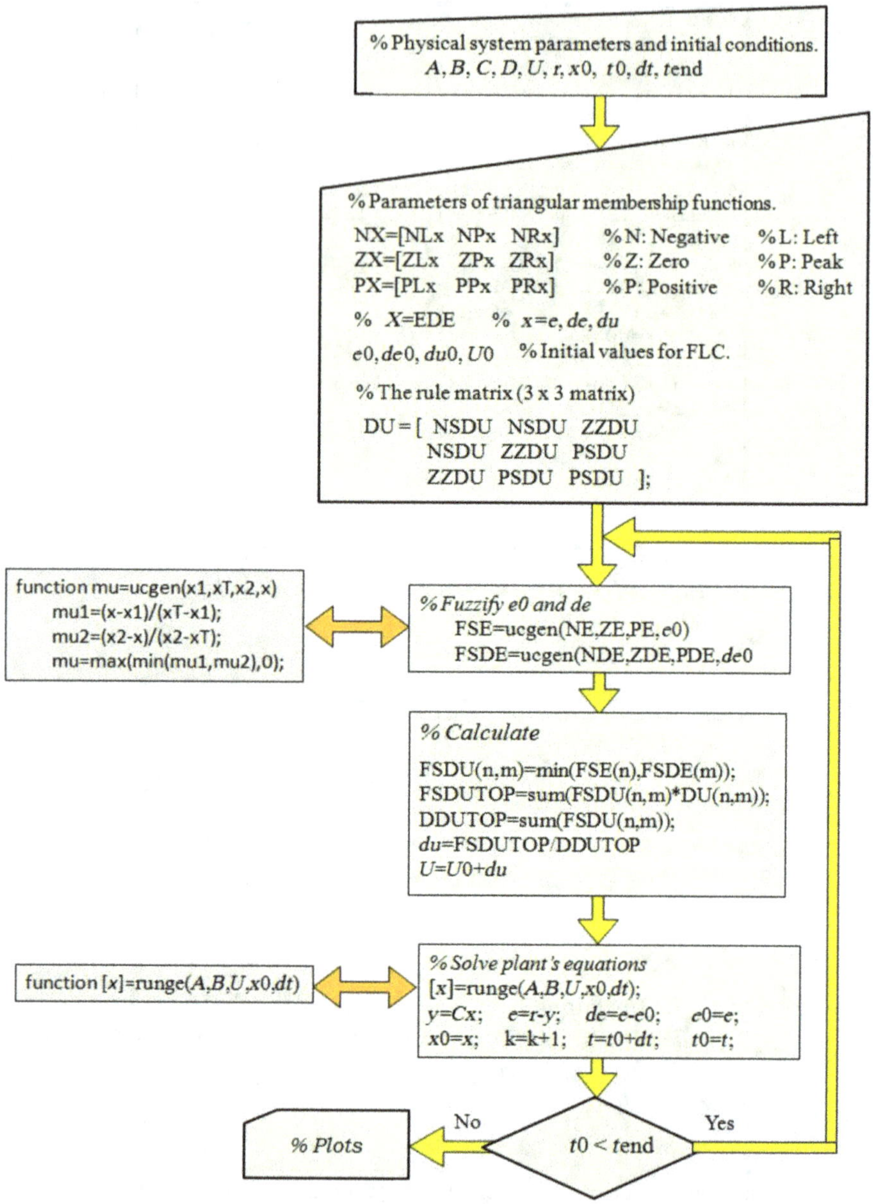

Figure 4.18 Flowchart of fuzzy logic controlled system simulation

4.9 Modeling the FLC in Simulink

The model development of the FLC for the MATLAB/Simulink environment is described in this section using the information given in previous sections and in

[40,41]. The first step here is to generate a fuzzy membership function for partitioning the input and output spaces into fuzzy subsets. As given in Figure 4.15, the process of an FLC can be summarized in four steps: Fuzzification, rule base, fuzzy reasoning, and defuzzification. Due to its simplicity and linearity, triangular fuzzy membership functions are used in both fuzzification and defuzzification stages. The Simulink model of the triangular fuzzy membership function was given in Figure 2.6 in Chapter 2 and is repeated in Figure 4.19, where x_1, x_2, and x_3 are the crisp parameters used to define the location and shape of the triangle as given in Figures 2.4 and 2.5. The input x is the crisp variable whose membership value on this triangle fuzzy subset is the output $\mu(x)$ in Figure 4.19.

Let us work on a fuzzy processor with two inputs and one output universe, which are partitioned into five fuzzy subsets each as shown in Figure 4.20. The fuzzy subsets defined in Figure 4.20 will be used in Simulink modeling of the FLC. Therefore, the partitioning and the membership functions used as fuzzy subsets must be understood well, so that the data points and Simulink blocks can be connected correctly. Assume that the input universes are error, *e*, and error change, *de*, and the output universe is the change *du* in the control signal. The labeling of the fuzzy subset parameters are the definition parameters and represent their locations in the corresponding universe. The universe of *X* in Figure 4.20 may be universe of error *e* or *de*. Therefore, the subscript *x* is replaced by *e* for the universe of *e* and replaced by *de* for the universe of *de*.

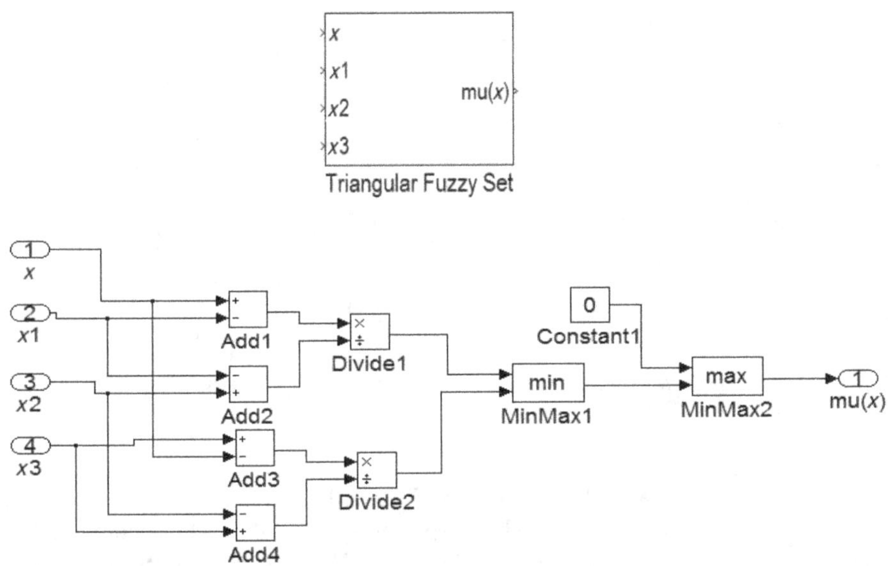

Figure 4.19 A Simulink function for simulating triangular fuzzy membership function. (a) Closed form of the Simulink block and (b) inside view of the block in (a).

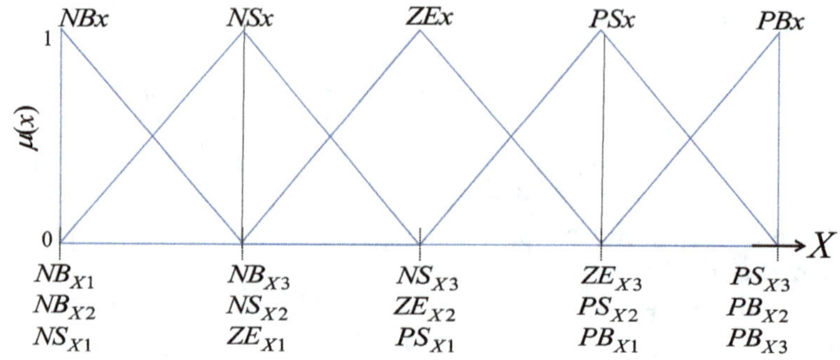

Figure 4.20 Parametric definitions of the fuzzy subsets in a universe of X

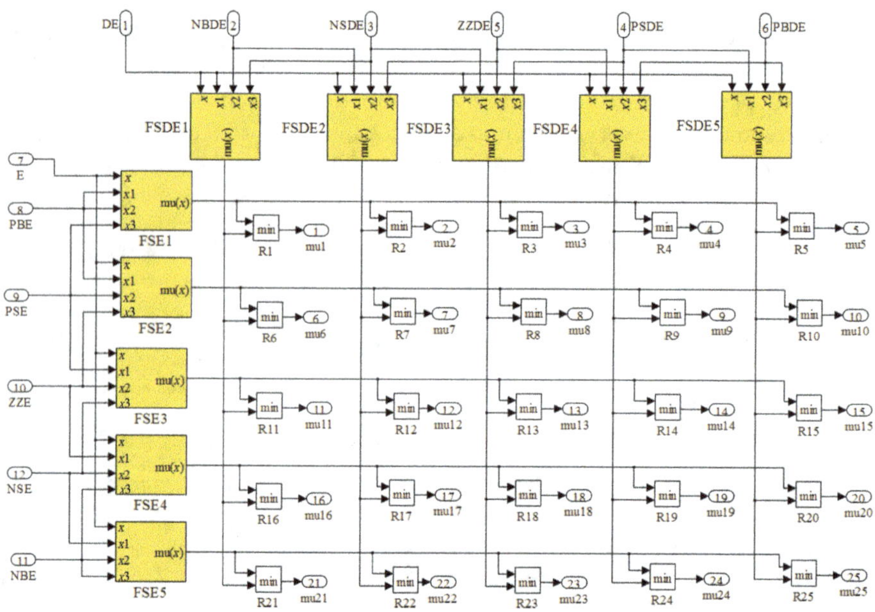

Figure 4.21 The Simulink process of the fuzzification and resulting weightings

Fuzzification stage and the first part of the rule-based fuzzy processing unit of the FLC are given in Figure 4.21. The crisp inputs $e(k)$ and $de(k)$ are converted to fuzzy membership values on the fuzzy subsets NB, NS, ZZ, PS, and PB. Each fuzzy subset (FS) is shown by blocks in Figure 4.19 and represented by a triangular membership function as described in Figure 4.20. The letters E and DE in Figure 4.21 are used to indicate whether the elements belong to error, e, or its change de, respectively.

The Boolean operator "min" is used for the verbal connector "*and*" to simulate the *input space of the rules* that have the structure as in expression (4.35). The input space of the fuzzy rules used here has two inputs and one output as

$$\textbf{\textit{If}}\ e\ is\ A\ \textbf{\textit{and}}\ de\ is\ B\ \textbf{\textit{then}}\ du\ is\ C. \tag{4.35}$$

where A, B, and C in (4.35) represent any one of the fuzzy subsets NB, NS, ZZ, PS, and PB defined in Figure 4.20.

The input spaces in (4.35) are in the part represented by the expression (e is A ***and*** de is B). Therefore, the min operator in Simulink Block Library is used to model the input spaces of 25 rules used by FLC. The outputs of the "min" operators are the weighting factors from each rule and they indicate the strength (membership degree) of the rules in the output space du. The implementation of the rule input universes by the expression (e is A ***and*** de is B) is nothing but the fuzzification of the two crisp inputs e and de for all the rules. The process of fuzzification of the input space with 25 rules is shown in Figure 4.21.

The membership degrees obtained as depicted in Figure 4.21 are multiplied by the crisp values of each corresponding fuzzy subset in the output space du as shown in Figure 4.22. The crisp values of the fuzzy subsets used in this multiplication process are the values that have a maximum membership degree of 1.0 in the corresponding fuzzy subset. In other words, these crisp values indicate the peak locations of the triangular fuzzy subsets. Actually, this multiplication process represents the products in the nominator of *center of area* defuzzification method. Then the sum of these products is divided by the sum of the weighting factors obtained as in Figure 4.22, which has two combined blocks called "Fuzzification" and "Fuzzy Rules and Reasoning." The inside views of these "Fuzzification" and "Fuzzy Rules and Reasoning" blocks are shown in Figures 4.21 and 4.22, respectively.

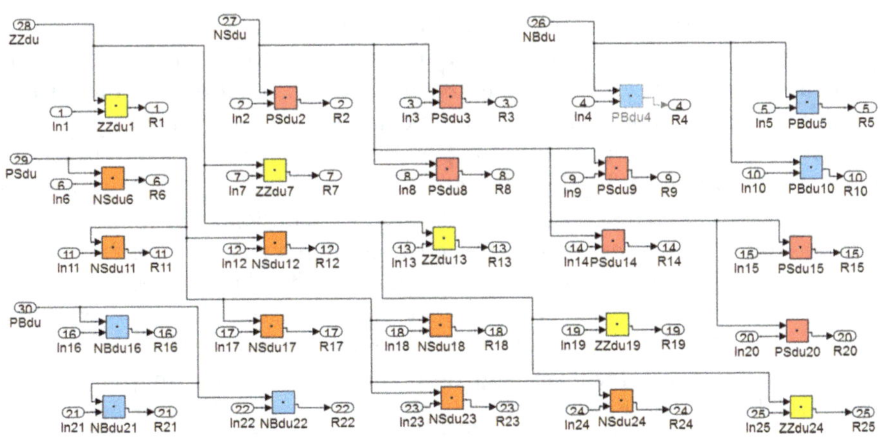

Figure 4.22 The Simulink model of the fuzzy rules

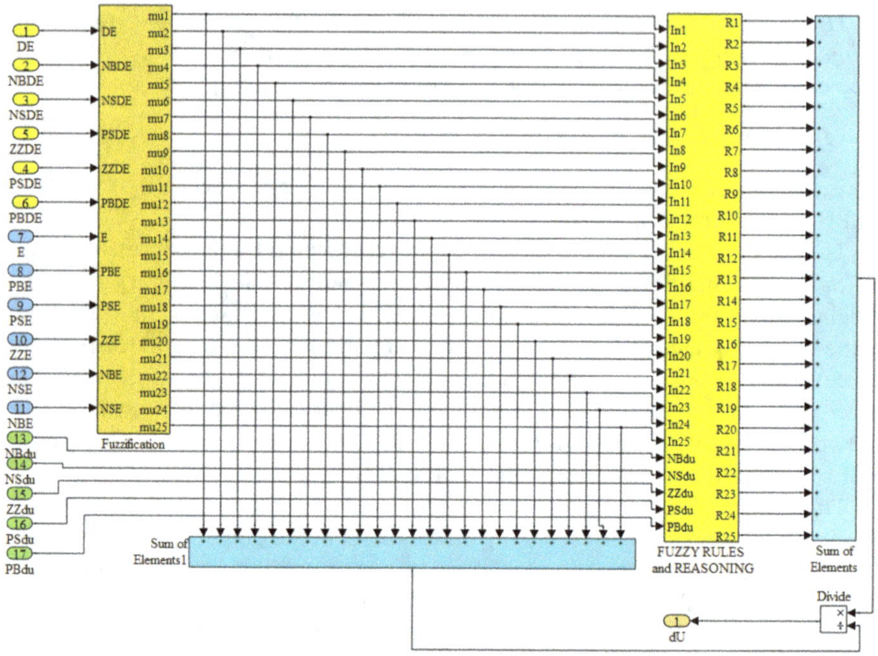

Figure 4.23 Fuzzy reasoning represents the process from fuzzification to defuzzification

Simulation model of the *center of area* method is depicted in Figure 4.23. This is the final stage of the FLC to generate the required change in the control signal for the current *k*th sampling. In order to prevent zero division that causes simulation problems such as delayed simulation time and simulation hanging, a signal route has been provided.

A general overlooked view of the FLC Simulink block diagram is given in Figure 4.24 where the processes from inputs *e* and *de* to output *du* are shown. The input data blocks representing fuzzy membership functions for the error *e*, error change *de*, and the controlled output change *du* are shown in Figure 4.24. The user is able to edit and change the parameters of the membership functions on this stage without going into the details of the FLC. The definition parameters of the fuzzy membership functions are problem-dependable and must be set accordingly. Once the maximum and minimum values of the inputs and output signals are assigned, the other subvalues can be placed in between these limits. If the initial settings are to be used, then some gain blocks can be used to match the signals with the predefined membership parameters. The difference between reference and feedback signals is the error signal *e* and used as the input variable E to the block called "Fuzzy Reasoning," whose inside view is shown in Figure 4.23. The change in error *e* is obtained as the difference between two consecutive values of *e* using a discrete difference block $(z-1/z)$.

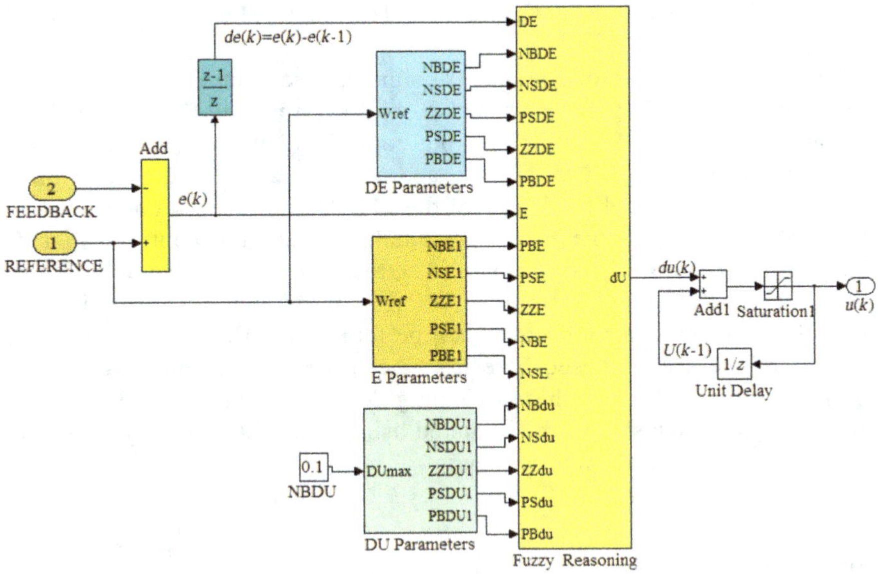

Figure 4.24 The input and output units of the FLC

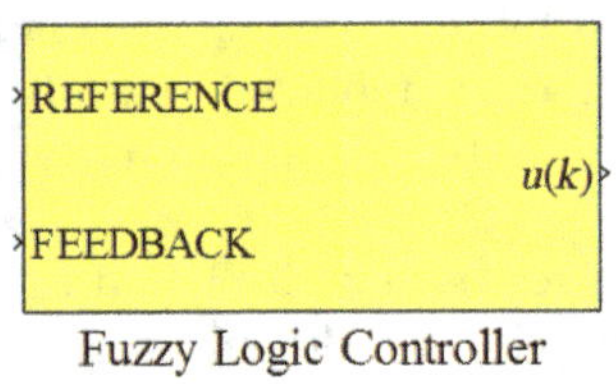

Figure 4.25 A single FLC block

The error change DE is another input variable signal to the "Fuzzy Reasoning" block.

The parameters of the universes E, DE, and DU are entered using the blocks called "E Parameters," "DE Parameters," and "DU Parameters" in Figure 4.24. The reference signal w_{ref} is used to determine the upper and lower boundaries of the input variables E and DE such that $+\ w_{ref}$ is assigned as the upper boundary and $-w_{ref}$ is assigned as the lower boundary. A scaling factor 0.1 is used for the output *du*. Therefore, the upper and lower boundaries of the change in control signal DU are set to +1 and −1, respectively.

The Simulink model of the FLC described and explained above has been represented by a simple block with input and output connection ports so that it can be inserted into different systems as a fully functioning controller. The model blocks in Figure 4.24 are combined as a subsystem to obtain a single FLC block with reference input and feedback signals as the inputs to the block and the control signal u is the output from the controller. The single FLC block is shown in Figure 4.25.

The designed FLC block is used to simulate the control of five different systems namely a PMDC motor speed control system, a PMDC motor-driven position control system, a radar tracking control system, a synchronous generator terminal voltage control system, and a two-area load–frequency control power system. The modeling details of the PMDC motor speed control system are given in [22,25] as the modeling details for position control and radar tracking systems are given in [42] and [43], respectively. The modeling details of the synchronous generator terminal voltage control and two area power system load–frequency control systems can be found in [44]. Since the modeling processes of the systems used here have been done before and given in the references as just indicated, these modeling details are not repeated here. Five different controlled systems are simulated using both FLC and PID controllers for comparison and validation purposes.

Example 4.9

PMDC motor speed control using FLC Simulink model.

MATLAB/Simulink model block diagram of the PMDC motor speed control system is shown in Figure E4.9(a) and the related simulation results are given in Figures E4.9(b) and E4.9(c). For the simulation with a PID controller, the FLC block in Figure E4.9(a) is replaced by a PID controller block with the proportional gain (K_P) set to 0.001, integral gain (K_I) set to 0.01, and derivative gain (K_D) set to 0.

The speed control responses from both FL and PID controllers are plotted on the same graph for better comparison. Although there are some differences during the transient period of speed, both controllers give almost the same settling time and steady-state operation by responding to step changes in reference input as depicted in Figure E4.9(b). However, each controller gives different current responses during the transients. As depicted in Figure E4.9(c), the current has fewer ripple magnitudes during the transients with FLC than it has with the PID controller.

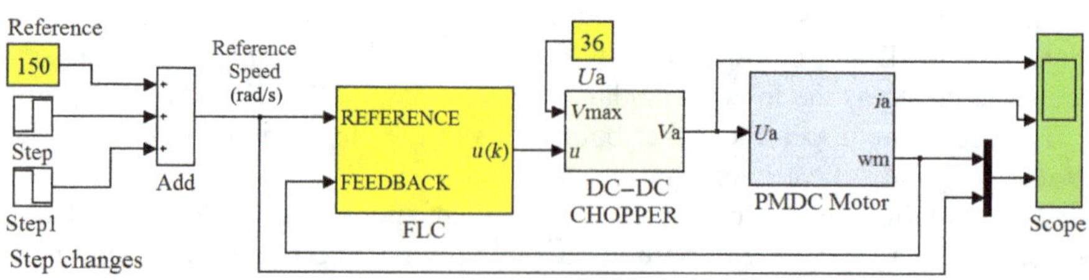

Figure E4.9(a) Using FLC block for PMDC motor speed control

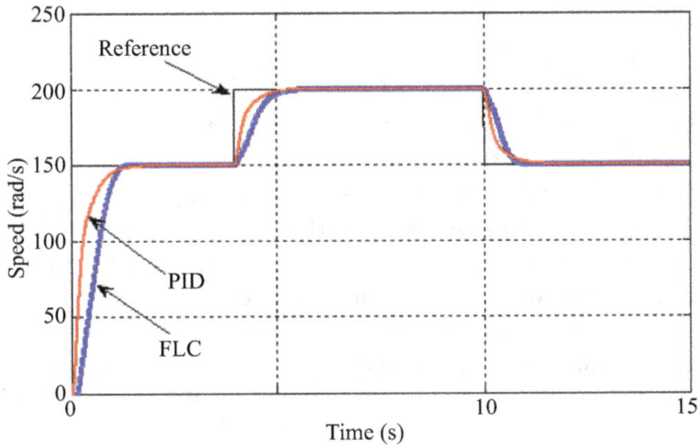

Figure E4.9(b) *PMDC motor speed control responses using FL and PID controllers*

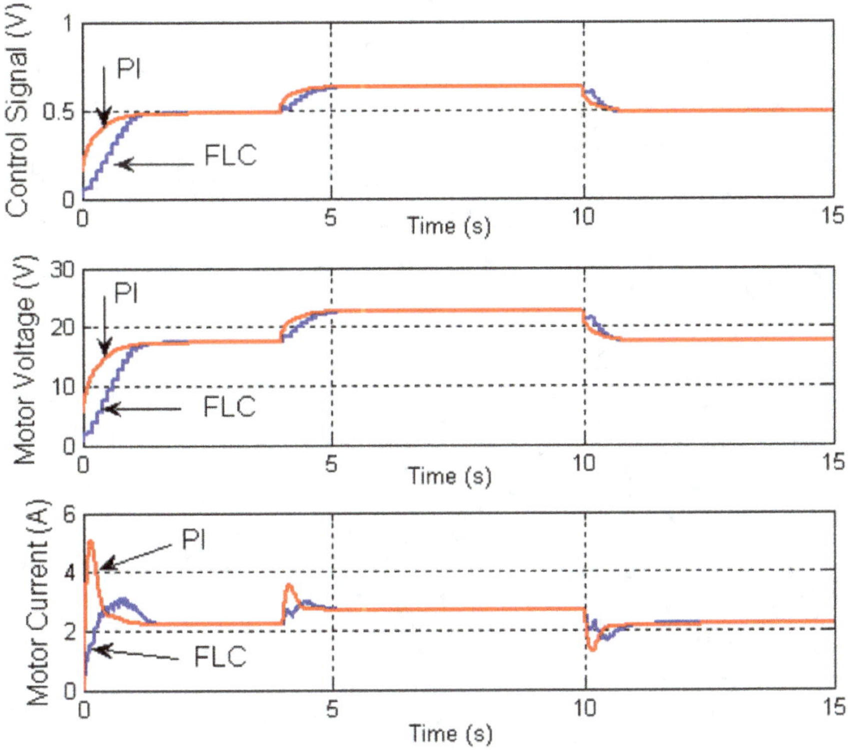

Figure E4.9(c) *Responses of the PMDC motor speed control system using FL and PID controllers*

Example 4.10

Position control using the FLC Simulink model.

The operational block diagram of the position control system using FLC is shown in Figure E4.10(a). The FLC block is replaced by a PID controller block with the parameters set to $K_P = 0.08$, $K_I = 0.001$, and $K_D = 0$ when PID is used.

The simulation results for the position control system are given in Figures E4.10(b) and E4.10(c), where the differences in transient responses of both FLC and PID controllers are shown. The position response from FLC seems to be slower than that obtained with PID when step changes occur in the reference position input level. However, with the FLC controller, the control signal, motor voltage; motor current, and motor speeds have considerably reduced magnitudes during transient periods.

Example 4.11

Radar tracking control using the FLC Simulink model.

The operational Simulink block diagram of the radar tracking system with the FLC is given in Figure E4.11(a). The simulation results of this system are shown in Figure E4.11(b) for both FLC and PID controllers. In the radar tracking system, it is assumed that the radar is tracking moving objects as discussed in Example 4.8, so that

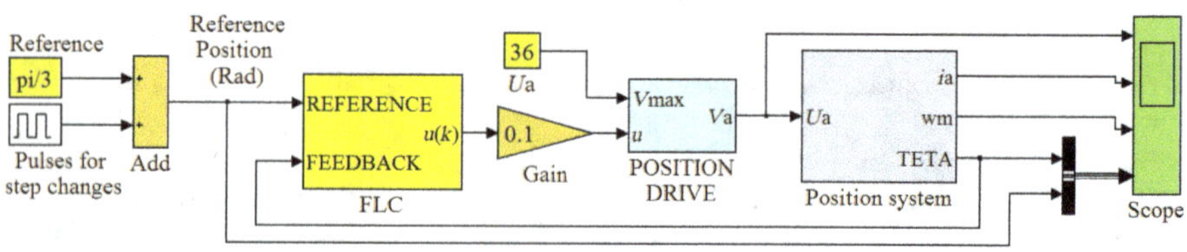

Figure E4.10(a) Using FLC block for PMDC motor-driven position control

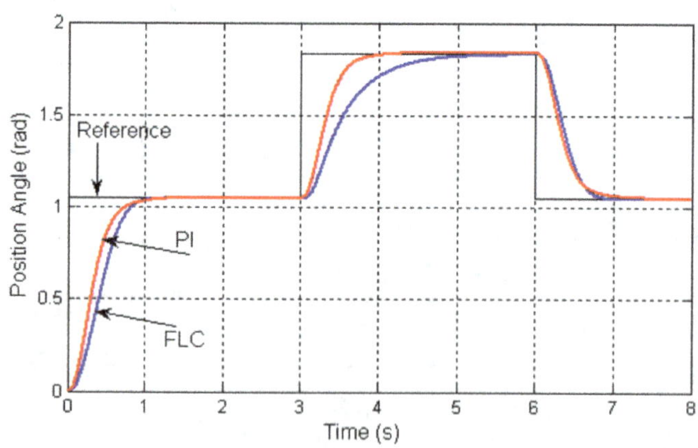

Figure E4.10(b) Position control performances of both FL and PID controllers

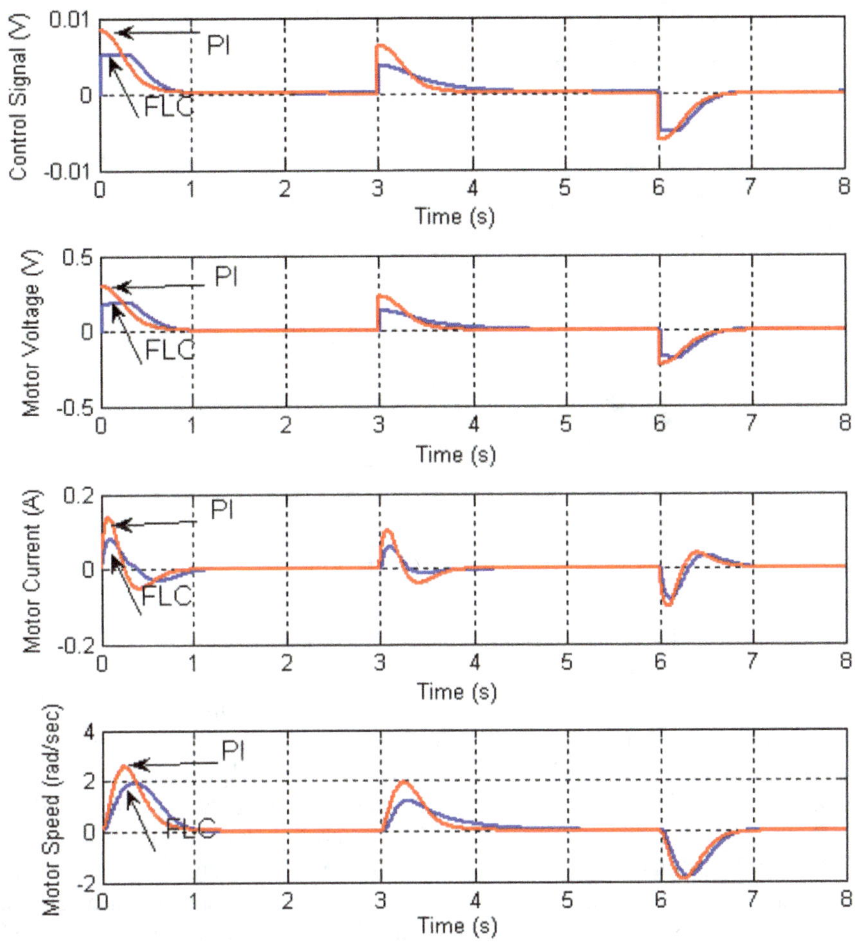

Figure E4.10(c) *Responses from the position control system using FL and PID controllers*

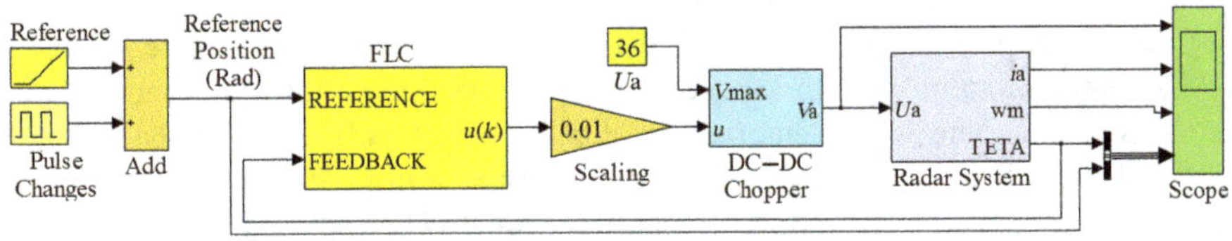

Figure E4.11(a) *Using FLC block for PMDC motor-driven radar control system*

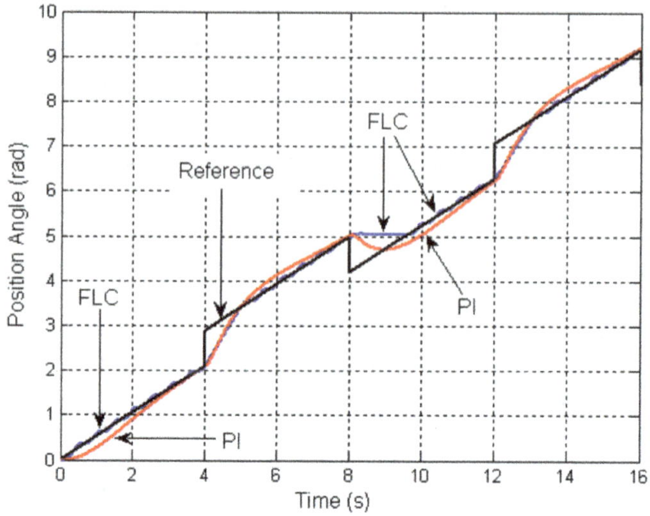

Figure E4.11(b) Radar tracking performances of both FL and PID controllers

after catching the object, the radar then locks and tracks it on a surface with 360° rotating ability. Therefore, the reference input to the radar tracking system is a ramp input as a function of the simulation time. It is also assumed that the moving object slows down and speeds up again during the simulation as shown in Figure E4.11(b), where both FL and PID controllers are tracking the object. However, the FLC has a better and closer tracking performance even if it has some small oscillations. In order to simulate the radar tracking system with the PID controller, the FLC block is replaced by a PID block with the parameters set to $K_P = 1.1$, $K_I = 0.8$, and $K_D = 0$.

The proposed FLC block for the MATLAB/Simulink environment is also validated by simulating two power system examples from [43]. These examples are originally simulated in [43] using PID controllers, which are replaced by the proposed FLC here in order to show the simplicity of using the FLC block as a simple controller that can be inserted in a system model just like it is done with the classical PID controllers.

Example 4.12

Power system voltage control using the FLC Simulink model.

One of the power system control examples is an automatic voltage regulator (AVR) control system used to keep the terminal voltage of the synchronous generators constant. The block diagram of the FL-based voltage control system is given in Figure E4.12(a). The block diagram representing the voltage control system of a synchronous generator is adapted from [44]. The PID block with the parameters, $K_P = 1$, $K_I = 0.25$, and $K_D = 0.28$ in the original block diagram in [44] is replaced by the FLC as shown in Figure E4.12(a). The FLC results in a slower transient response for the system used as shown in Figure E4.12(b). However, both controllers give the same steady-state operating response.

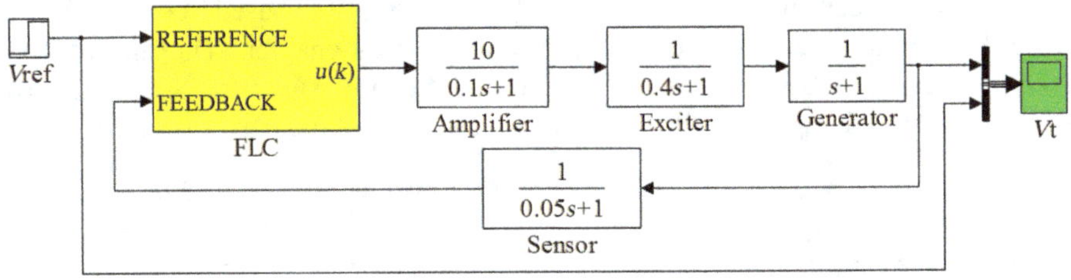

Figure E4.12(a) *Using FLC block as a terminal voltage controller for a synchronous generator*

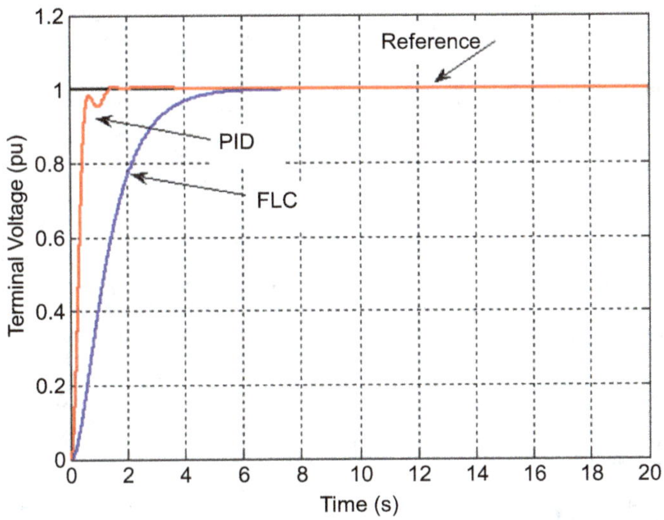

Figure E4.12(b) *Terminal voltage control responses of a synchronous generator using both FL and PID controllers*

Example 4.13

Two-area automatic generation control (AGC) system using FLC Simulink model.

The second power system control example is a two-area AGC based on load–frequency control two areas, which is also taken from [44] for comparison. The PID controller with the parameters $K_P = 0$, $K_I = 0.3$, and $K_D = 0$, is replaced by the FLC as shown in Figure E4.13(a), yielding the simulation results given in Figure E4.13(b) where the FLC responses for both areas have less oscillations and shorter settling time.

Actually, it should be noted that the comparison made here is not given to show which controller is better. The comparison is given to show that the proposed FLC model for MATLAB/Simulink environment works and gives acceptable results without redesigning it for each separate system.

A generalized direct approach for designing FL-based controllers in MATLAB/Simulink environment is presented in this chapter. The approach is based upon using *e-de* space consisting of the plot *e*(*t*) versus *de*(*t*). Since the error response of an undamped second-order system is used as the base system to derive

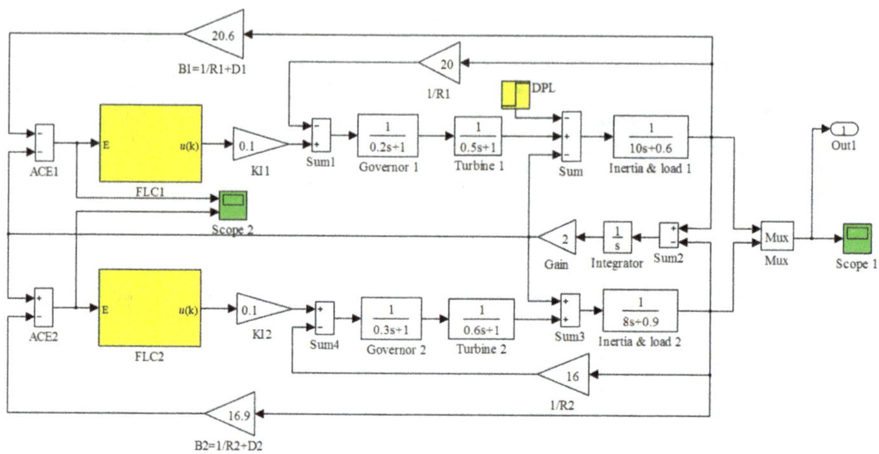

Figure E4.13(a) Using FLC blocks as AGC controllers in power systems

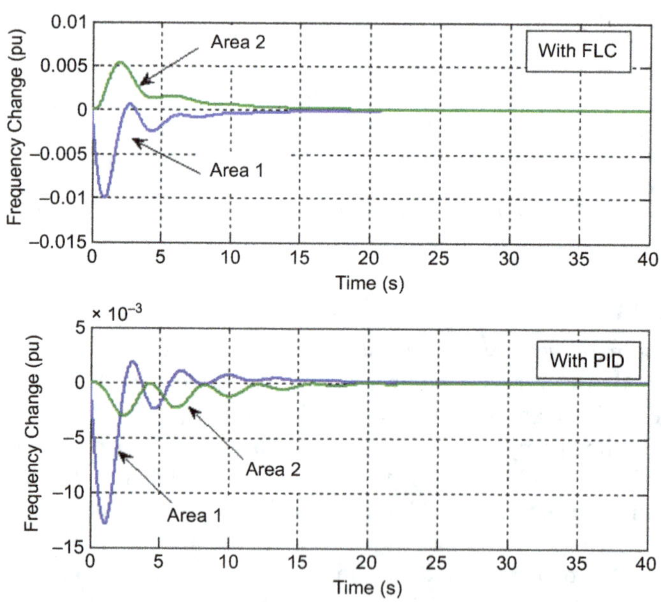

Figure E4.13(b) Load–frequency controller responses in a two-area power system using both FL and PID controllers

the fuzzy rules, the FLC is capable of handling different types of systems with different orders. The rule table is formed in such a way that the table itself becomes fuzzy with a zero main diagonal separating positive and negative fuzzy rules that are also partitioned into sub-fuzzy rules among themselves. The approach is used to generate a generalized FLC Simulink block that can be used in different types of system control schemes by just inserting the block to the model of the system to be controlled. The generated FLC block requires only small gain adjustments to be adapted for different systems. In order to validate the developed FLC model, five different systems, which are a PMDC speed control system, a position control system, a radar tracking system, a voltage regulator system, and a two-area load–frequency control system, are controlled using the proposed FLC and the results are compared with those obtained using PID controllers. The results from five simulations showed good and acceptable performances for the FLC. The purpose here is not to compare the FLC with PID to see which one gives better performance but to observe whether the FLC gives acceptable results as a PID controller does. Therefore, the comparisons are not focused on observing which one gives better response. However, it was observed that the FLC gives lower transient magnitudes in some system variables such as current and input voltage as in position and speed control systems.

4.10 Problems

P4.1 Assume that the fuzzy rules in Example 4.4 are redefined for the same problem and given below in the forms of verbal statements, which are represented by Table P4.1.

Rule 1: If sunlight is low and wind is low then use utility
Rule 2: Else if sunlight is low and wind is medium then use utility
Rule 3: Else if sunlight is low and wind is high then use WES
Rule 4: Else if sunlight is medium and wind is low then use Utility
Rule 5: Else if sunlight is medium and wind is medium then use PES and WES
Rule 6: Else if sunlight is medium and wind is high then use WES
Rule 7: Else if sunlight is high and wind is low then use PES
Rule 8: Else if sunlight is high and wind is medium then use PES
Rule 9: Else if sunlight is high and wind is high then use WES

Table P4.1 The rule table for Problem 4.1

	LW	MW	HW
LPV	U	U	WES
MPV	U	PES + WES	WES
HPV	PES	PES	WEs

The fuzzy subsets low (LPV), medium (MPV) and high (HPV) in the universe of PV, and the fuzzy subsets low (LW), medium (MW), and high (HW) in the universe of wind are represented by triangular fuzzy subsets and defined as follows. If solar irradiation level and wind speed are 80 mW/cm^2 and 15 m/s, respectively, find the decision of this energy management system using a proper fuzzy decision algorithm.

P4.2 Assume that a rule base system with nine rules (3 × 3) is generated in Problem P3.18 in Chapter 3 and the following four rules are active for a sampling instant such that the error and change in error are measured to be $e = -0.5$ and $e = 0.2$, respectively. Obtain the change in the voltage applied to the heater using a proper fuzzy reasoning approach.

$R1$: *If* e is NE *and ce* is ZCE *then cv* is NCV
$R2$: *else if* e is NE *and ce* is PCE *then cv* is NCV
$R3$: *else if* e is ZE *and ce* is ZCE *then cv* is ZCV
$R4$: *else if* e is ZE *and ce* is PCE *then cv* is PCV

P4.3 Two input and one output crisp spaces are defined in the universes X, Y, and Z, respectively, in the interval $\{0, 10\}$ as in Problem P3.16 in Chapter 3. The relation between the input and output spaces is described by four rules, which are given below.

Rule 1: *If* x is *small* and y is *small* then z is *large*
Rule 2: *If* x is *small* and y is *medium* then z is *big*
Rule 3: *If* x is *medium* and y is *small* then z is *small*
Rule 4: *If* x is *medium* and y is *medium* then z is *small*

where the fuzzy subsets small, medium, and big are described in Problem P3.16. If the crisp data from the input spaces are given as $x = 1.5$ and $y = 3$, find the crisp output z by using the following reasoning methods in the fuzzy processor.

(a) Mamdani fuzzy reasoning.
(b) Tsukamoto fuzzy reasoning.

P4.4 A room heating process was described in Problem P3.18 and asked to generate a series of fuzzy rules for the process. Assume that a rule-based system with nine rules is generated and the following four rules are active for the given crisp inputs E = −0.5 and CE = 0.2. Obtain the crisp consequent z by using Mamdani fuzzy reasoning.

Rule 1: *If* E is NE *and* CE is ZCE *then* CV is NCV
Rule 2: *If* E is NE *and* CE is PCE *then* CV is NCV
Rule 3: *If* E is ZE *and* CE is ZCE *then* CV is ZCV
Rule 4: *If* E is ZE *and* CE is PCE *then* CV is PCV

P4.5 The final fuzzy conclusion of a fuzzy processor is shown in Figure P4.5. This conclusion is obtained as the union of the individual consequences from each rule at sampling k. The shaded areas depict the effectiveness of the related rules and output fuzzy subsets, which are defined by the parameters shown in Figure P4.5.

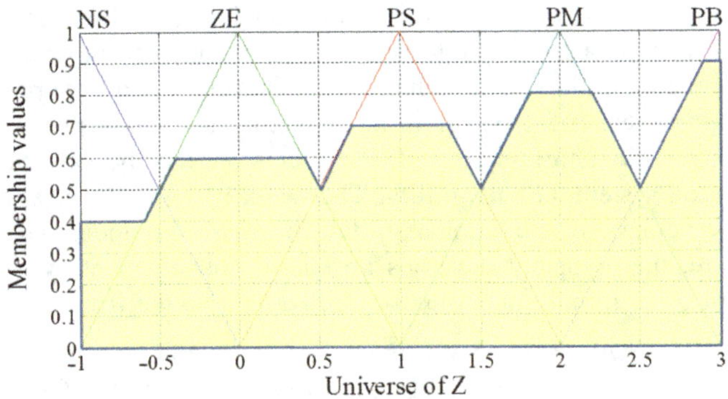

Figure P4.5 A final fuzzy conclusion

Table P4.6 Parameters of the fuzzy membership functions used in P4.6

Membership function	Universe	Function parameters		
		x_1	x_2	x_3
Cold (C)	Temperature (°C)	0	0	20
Warm (W)	Temperature (°C)	0	20	40
Hot (H)	Temperature (°C)	20	40	40
Negative (N)	The interval {0, 1}	−1	−1	0
Zero (Z)	The interval {0, 1}	−1	0	1
Positive (P)	The interval {0, 1}	0	1	1

where NS is used for negative small, ZE is used for zero. PS, PM, and PB stand for positive small, positive medium, and positive big, respectively. Use a defuzzification algorithm and obtain the final conclusion as a crisp output.

P4.6 Two fuzzy rules are described as

Rule 1: if x is cold and y is negative then $z_1 = p_1x + q_1y + r_1$.
Rule 2: if x is hot and y is zero then $z_2 = p_2x + q_2y + r_2$.

where the coefficients are given as $p_1 = -1$, $q_1 = -1.5$, $r_1 = 1$, $p_2 = 1$, $q_2 = -2$, and $r_2 = -1$. The fuzzy subsets cold (C), warm (W), and hot (H) in the universe of temperature and negative (N), zero (Z), and positive (P) in the interval {0, 1} are defined by parametrized triangular fuzzy membership functions as in Figure 2.5 and given in Table P4.6.

The input variables to this system are given as $A(0, 5, 10)$ and $B(0, 0.5, 1)$, where A and B are fuzzy subsets and represented by triangular-type fuzzy membership functions with the parameters given in the parenthesis. Fuzzy set A is defined in the universe of temperature and fuzzy set B is defined in the interval {0, 1}. Obtain the final consequence of these two rule systems using the TS fuzzy reasoning algorithm.

Caution: Since the inputs are fuzzy, the conclusion will not be just a single crisp number but a series of numbers defined in an interval in the output space *Z*. Example 4.3 in [15] will be helpful for the solution of this problem.

P4.7 The required electrical power in an industrial plant is supplied from wind (WE), solar PV, and fuel cells (FC) systems besides the national electricity distribution system, i.e., the utility. The wind, PV, and FC power generating systems belong to the company, which owns the industrial plant, too. Therefore, the company develops an energy management system as given below in order to minimize the use of electrical power from the utility grid.

R1: If PV is *sufficient* and WE is *insufficient* and FC is *empty* then *use* PV.
R2: If PV is *sufficient* and WE is *insufficient* and FC is *full* then *use* PV and FC.
R3: If PV is *insufficient* and WE is *insufficient* and FC is *empty* then *use* utility.
R4: If PV is *insufficient* and WE is *sufficient* and FC is *empty* then *use* WE.
R5: If PV is *insufficient* and WE is *sufficient* and FC is *full* then *use* WE.
R6: If PV is *insufficient* and WE is *insufficient* and FC is *full* then *use* FC and utility.
R7: If PV is *sufficient* and WE is *sufficient* and FC is *empty* then *use* PV and WE.

The power required to charge the FC system is also considered as a part of the load in an industrial plant. The Fuzzy expressions sufficient (SU), insufficient (IS), full (FU), empty (EM), and use (U) used in the rules are defined in their crisp universes as in Figures P4.7(a) and P4.7(b).

If the solar irradiation level for PV system is 80 mW/cm^2, the wind speed is 15 m/s and the fuel cell is 40% full, then how the required power to the industrial plant is supplied?

P4.8 The power in alternating current power systems is a complex quantity called as complex power with real and imaginary components. The real component is called active power, which is the one converted to useful energy by electrical machines and devices. The reactive component is usually required for magnetizing and dissipated by reactors. Since the reactive power can be generated by capacitors, its transmission is not needed. Therefore, the

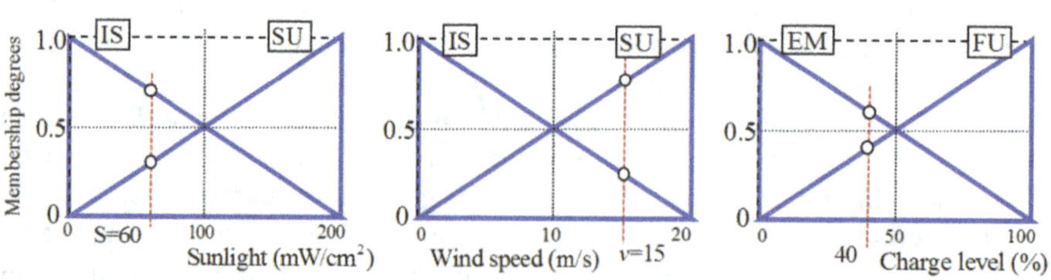

Figure P4.7(a) The input universes and fuzzy subsets defined in these universes

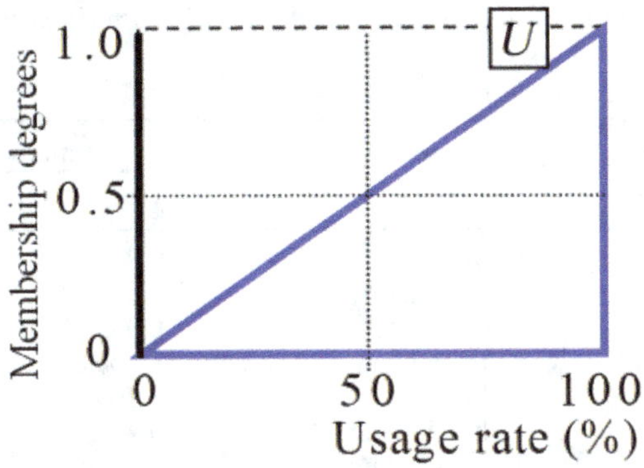

Figure P4.7(b) The output universe and output fuzzy subset use

reactive power is supplied by capacitors near the load and its transmission is minimized in order to reduce the transmission losses. Due to minimized reactive power, the active power component of the transmitted complex power becomes close to the complex power.

The amount of the active power as one of the components of the complex power is determined by a factor called *power factor* and represented by **cosϕ**, which is defined in the interval [0, 1]. In order to minimize the transmission losses in power systems, the power factor is kept closer to 1, such as around 0.95. In order to keep the power factor of the transmitted power around 0.95, capacitors are connected at the end of the power transmission line. However, if the capacitive value of the capacitors connected at the end of the line is greater than the required amount, the voltage at the bus where the capacitors are connected increases resulting in unwanted events. Therefore, the power system operator increases or decreases the number of capacitors according to the values of the bus voltage and the power factor of the transmitted power using a decision process given in Table P4.8(a).

Table P4.8(a) Decision rules for power compensation

Rule table		**Line voltage**		
		Low (LV)	**Normal (NV)**	**High (HV)**
cosϕ	High (HPF)	ZC	NC	NC
	Normal (NPF)	PC	ZC	NC
	Low (LPF)	PC	PC	ZC

Table P4.8(b) Parameters of the fuzzy membership functions used in Problem P4.8

Membership Function	Universe	Function parameters		
		x_1	x_2	x_3
Low (LPF)	Power factor	0.8	0.8	0.95
Normal (NPF)	Power factor	0.8	0.95	1.0
High (HPF)	Power factor	0.95	1.0	1.0
Low (LV)	Voltage change	−10%	−10%	0
Normal (NV)	Voltage change	−10%	0	10%
High (HV)	Voltage change	0	10%	10%
Negative (NC)	The interval {−1, 1}	−1	−1	0
Zero (ZC)	The interval {−1, 1}	−1	0	1
Positive (PC)	The interval {−1, 1}	0	1	1

where the crisp fuzzy subsets used in the rules are represented by triangular fuzzy membership functions with the parameters given in Table P4.8(b).

The abbreviations NC, ZC and PC are used for negative, zero, and positive changes. The measurement devices in the power management center show that the power factor of the transmitted power is $\cos\phi = 0.90$ and the change in bus voltage is −5%.

(a) Determine the active rules in Table P4.8(a).
(b) Defuzzify the crisp inputs and obtain the related membership values.
(c) Determine the weighting factor from each rule.
(d) Determine the final fuzzy conclusion
(e) Determine the final crisp conclusions.
(f) Determine the amount of capacitors to be switched on or switched off.

P4.9 The input data to a fuzzy decision maker is categorized as low (L), normal (N), and high (H), which are used to reach a decision. For example, the fuzzy decision maker (FDM) is used to generate a decision about the amount of detergent in a washing machine depending upon the darkness in the color of output water and washing time.

Assume that the universe of ***water darkness*** and the universe of ***washing time*** in minutes are partitioned into subcategories as follows:

Water darkness: NW (0%, 0%, 50%), LW (0%, 50%, 100%), and HW (50%, 100%, 100%).

The washing time: LT (0, 0, 10), NT (0, 10, 30), and HT (30, 60, 60).

Create a MATLAB .m file in order to perform the following:

(a) Plot the fuzzy subsets on input universes water darkness and washing time.
(b) Assume that the darkness of water color is 60% and the washing time is set to 20 min. Fuzzify these crisp inputs and obtain fuzzy values of them.

P4.10 Assume that the decision maker described in P4.9 is using the rule base system given in Table P4.10(a) in order to determine the amount of detergent to be filled in the detergent box of a washing machine.

Table P4.10(a) Decision rules for Problem P4.10

Rule table		**Washing time (min)**		
		Low (LT)	**Normal (NT)**	**High (HT)**
Water darkness	High (HW)	FLL	TQRT	HLF
	Normal (NW)	ZE	ZE	ZE
	Low (LW)	TQRT	HLF	OQRT

Table P4.10(b) Preset time durations for water darkness in a washing machine

Water darkness (%)														
0			20			40			70			100		
Washing time (min)														
10	30	60	10	30	60	10	30	60	10	30	60	10	30	60

where the fuzzy subsets zero (ZE), one quarter (OQRT), half (HLF), three quarter (TQRT), and full (FLL) are also represented by triangular-type fuzzy subsets with the parameters given as

ZE (0%, 0%, 25%),
OQRT (0%, 25%, 50%),
HLF (25%, 50%, 75%),
TQRT (50%, 75%, 100%), and
FLL (75%, 100%, 100%)

Full means that the detergent box is full. Assume that the darkness of water color is 60% and the washing time is set to 15 min.

(a) Apply Mamdani's Fuzzy Reasoning algorithm to amount of the detergent using hand calculation.
(b) Write a MATLAB .m file to simulate Mamdani's Fuzzy Reasoning algorithm and obtain the crisp decisions for data given in Table P4.10(b).

P4.11 In a water filling station, the tanks are filled manually such that the valve is fully open (100%) when the tank is empty, half open (50%) when the tank is half full, and closed (0%) when the tank is full. Representation of the valve opening and water level in the tank by fuzzy membership functions are described in Table P4.11(a).

(a) Partition the universe of *water level* into subsets using triangular membership functions as described in Table P4.11(a) and plot them in MATLAB.
(b) Partition the universe of *valve opening* into subsets using triangular membership functions as described in Table P4.11(a) and plot them in MATLAB.

Table P4.11(a) Fuzzy subset definitions for Problem P4.11

Water level in the tank	Valve opening	Fuzzy membership function
EM: empty	FO: full open	Triangle (0, 0, 50)
HF: half full	HO: half open	Triangle (0, 0, 100)
FL: full	FC: fully closed	Triangle (50, 100, 100)

Table P4.12 Output response of a physical system as a sampled data

k	1	2	3	4	5	6	7	8	9	10	11	12
$y(k)$	0	0.1	0.3	0.6	0.9	1	1.1	1.05	0.98	1	1	1

(c) Write a single input–single output fuzzy reasoning algorithm (FRA) and a set of rules for the operation of this system based upon the description given in the question part.

(d) Construct a fuzzy processor and obtain the amount of the valve opening if the tank is full about 20%. Use center of area method for defuzzification when needed.

P4.12 An input signal of 100 units is applied to a physical system at the instant of $t = 0$ s and the output is measured with a sampling period of 0.01 s. The measured outputs at each sampling k are listed in Table P4.12.

(a) Obtain a rule-based system with 25 rules using the system behavior given in Table P4.12(a).

(b) Design a fuzzy logic controller for this system.

(c) Obtain the change $du(k)$ as the output from the FLC for $k = 4$ using the rule base you have designed in (a).

Guidance: The following order of processes will be helpful for the solution. Find $e(k)$ and $de(k)$ for every sampling and add them to Table P4.12 as two new rows. Then use the approach described in Section 4.7.1.

P4.13 Fuzzy sets defined in the universes of error e and error change de in a room temperature control system are given below. The room temperature increases linearly with the inversing heater voltage and decreases linearly with decreasing heater voltage. Assume that the error e has a response like the one a second-order system has for $0 < \zeta < 1$.

(a) Develop a $3 \times 3 = 9$ rule base system.

(b) Define the control signal change du in an interval of $\{-1, 1\}$ and partition this universe into three fuzzy subsets.

(c) Assume that the rules given below are the active rules for $e(k) = 0$ and $de(k) = 2$, and obtain the crisp output du of an FLC using the COA method for defuzzification.

Defined fuzzy subsets for *e* and *de*.

$$\text{ZE} = \text{Zero} = \left\{\frac{0}{-2} + \frac{0.5}{-1} + \frac{1}{0} + \frac{0.5}{1} + \frac{0}{2}\right\}$$

$$\text{PS} = \text{PositiveSmall} = \left\{\frac{0}{0} + \frac{0.5}{0.5} + \frac{1}{1} + \frac{0.5}{1.5} + \frac{0}{2}\right\}$$

$$\text{PM} = \text{PositiveMedium} = \left\{\frac{0}{1} + \frac{0.5}{1.5} + \frac{1}{2} + \frac{0.5}{2.5} + \frac{0}{3}\right\}$$

The active rules:
Rule 1: **IF** E is ZE **AND** DE is ZE **THEN** DU is ZE
Rule 2: **IF** E is ZE **AND** DE is PM **THEN** DU is PS
Rule 3: **IF** E is PS **AND** DE is ZE **THEN** DU is PS
Rule 4: **IF** E is PS **AND** DE is PM **THEN** DU is PM

P4.14 In a fuzzy logic controller, the crisp universes of control error *e* error change *de* and control input change *du* are partitioned into fuzzy subsets as shown in Figure P4.14. The responses of the control error *e* (*k*) and its change *de*(*k*) are given in Table P4.14(a) for four sampling instants. Obtain the crisp value of the controller output *u*(*k*) using the rule base given in Table P4.14(b).

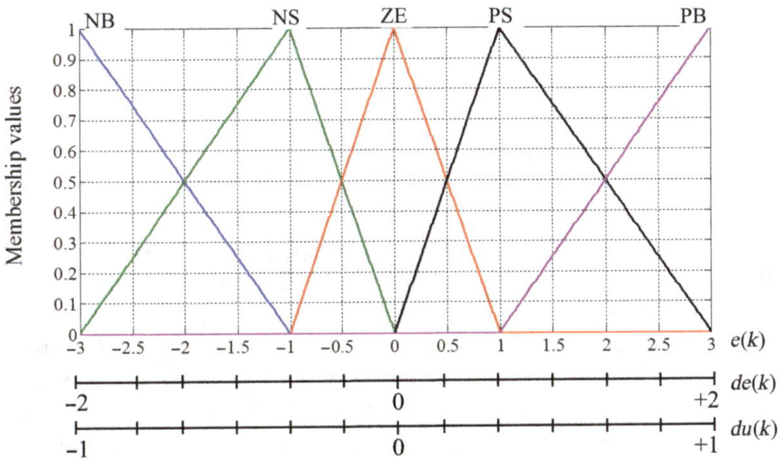

Figure P4.14 Fuzzy subsets defined in e, de, and du

Table P4.14(a) The rule base for Problem P4.14

k	**1**	**2**	**3**	**4**
e(*k*)	−2.4	−1.25	1.1	0
de(*k*)	1.3	0.2	0	0

Table P4.14(b) 25-Rule table for Problem P4.14

e(k)	**de(k)**				
	NB_{de}	**NS_{de}**	**ZE_{de}**	**PS_{de}**	**PB_{de}**
PB_e	ZE_{du}	PS_{du}	PS_{du}	PB_{du}	PB_{du}
PS_e	NS_{du}	ZE_{du}	PS_{du}	PS_{du}	PB_{du}
ZE_e	NS_{du}	NS_{du}	ZE_{du}	PS_{du}	PS_{du}
NS_e	NB_{du}	NS_{du}	NS_{du}	ZE_{du}	PS_{du}
NB_e	NB_{du}	NB_{du}	NS_{du}	NS_{du}	ZE_{du}

P4.15 Develop a fuzzy logic control simulator in MATLAB as an .m file and simulate the physical system given below. Then draw a flowchart representing your FLC and the system to be controlled such that the flowchart must include the following:

- Data input blocks
- FLC processing blocks
- Blocks representing the solution of the physical system
- Logical comparisons
- Loops
- Data updating
- Subfunction recallings
- Subfunctions
- Terminating blocks

State-space model equations of the physical system

$$\begin{bmatrix} \dot{x}_1 \\ \dot{x}_2 \end{bmatrix} = \begin{bmatrix} -3 & 1 \\ -2 & 0 \end{bmatrix} \begin{bmatrix} x_1 \\ x_2 \end{bmatrix} + \begin{bmatrix} 0 & 0 \\ 0 & 1 \end{bmatrix} \begin{bmatrix} u_1 \\ u_2 \end{bmatrix}, \quad [y] = [1 \quad 0] \begin{bmatrix} x_1 \\ x_2 \end{bmatrix}$$

where $u_1 = 0$, $u_2 = 1$ and sampling period $dt = 0.01$.

P4.16 Develop an FLC simulator in Simulink for the physical system given in P4.15.

P4.17 An FLC is to be designed to control the water level $h(t)$ in a water tank shown in Figure P4.17. The input valve is controlled to adjust the amount of water so that the water level is kept constant at the reference level when the water from the output valve is increased or decreased.

The bottom area of the tank is $S = 2$ m^2 and the maximum height of the tank is $H = 3$ m. The maximum amount of water from the input and output valves are $X_i = X_o = 5$ m^3/s. The output flow ratio from the output valve is defined as $0 < K_o < 1$, where $K_o = 1/X_o$ such that if $K_o = 0$, the valve is fully closed and if $K_o = 1$ then the valve is full open.

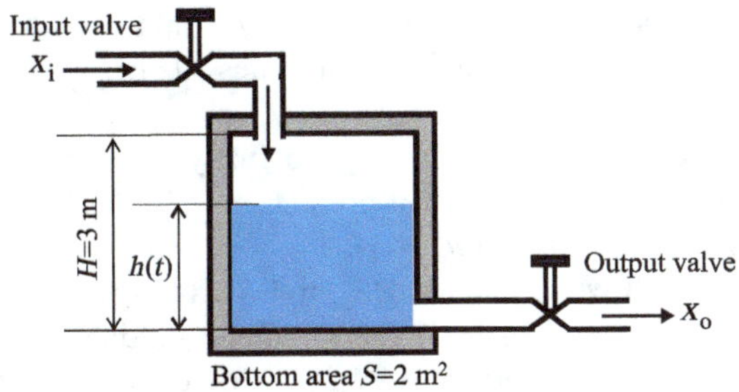

Figure P4.17 Water tank system for Problem P4.17

The mathematical model for the water level is defined as

$$\frac{dh(t)}{dt} = -\frac{K_o}{S}h(t) + \frac{1}{S}X_i$$

References

[1] Zadeh L.A. 'Fuzzy sets', *Information and Control.* 1965;8(3):338–353.

[2] Mamdani E.H. 'Application of fuzzy algorithms for control of simple dynamic plant', *IEE Proceedings*. 1974;121(12):1585–1588.

[3] Mamdani E.H. and Assilian S. 'An experiment in linguistic synthesis with a fuzzy logic controller', *International Journal Man–Machine Studies*. 1975;7 (1): 1–13.

[4] Mamdani E.H. 'Advances in the linguistic synthesis of fuzzy controllers', *International Journal Man–Machine Studies*. 1976;8(6):669–678.

[5] Bernard J.A. 'Use of a rule based system for process control', *IEEE Control Systems Magazine*. 1988;8(5):3–13.

[6] Li Y.F. and Lau C.C. 'Development of fuzzy algorithms for servo systems', *IEEE Control Systems Magazine*. 1989;9(3):65–72.

[7] Takagi T. and Sugeno M. 'Fuzzy identification of systems and its applications to modeling and control', *IEEE Transactions on Systems, Man, and Cybernetics*. 1985;15(1):116–132.

[8] Wong L.K., Leung F.H.F. and Tam P.K.S. 'Design of fuzzy logic controllers for Takagi–Sugeno fuzzy model based system with guaranteed performance', *International Journal of Approximate Reasoning*. 2002;30(1):41–55.

[9] Tanaka K. and Sano M. 'A robust stabilization problem of fuzzy control systems and its application to backing up control of a truck-trailer', *IEEE Transactions on Fuzzy Systems*. 1994;2(2):119–134.

[10] Tanaka K., Ikeda T. and Wang H.O, 'Robust stabilization of a class of uncertain nonlinear systems via fuzzy control: quadratic stabilizability, H1 control theory, and linear matrix inequalities', *IEEE Transactions on Fuzzy Systems*. 1996;4(1):1–13.

[11] Wang H.O., Tanaka K. and Griffin M.F. 'An approach to fuzzy control of nonlinear systems: stability and design issues', *IEEE Transactions on Fuzzy Systems*. 1996;4(1):14–23.

[12] Zeng W., Zhao Y. and Yin Q. 'Sugeno fuzzy inference algorithm and its application in epicentral intensity prediction', *Applied Mathematical Modelling*. 2016;40(13):6501–6508.

[13] Roubos J.A., Mollov S., Babuska R. and Verbruggen H.B. 'Fuzzy model-based predictive control using Takagi–Sugeno models', *International Journal of Approximate Reasoning*. 1999;22(1):3–30.

[14] Jang J.S.R. 'ANFIS: Adaptive-network-based fuzzy inference system', *IEEE Transactions on Systems, Man, and Cybernetics*. 1993;23(3):665–685.

[15] Jang J.S.R., Sun C.T. and Mizutani E. *Neuro-Fuzzy and Soft Computing: A Computational Approach to Learning and Machine Intelligence*. Englewood Cliffs, NJ: Prentice-Hall; 1996.

[16] Tsukamoto Y. 'An approach to fuzzy reasoning method', in Gupta M.M., Ragade R.K. and Yager R.R. (eds.). *Advances in Fuzzy Set Theory and Application*. Amsterdam: North-Holland; 1979. pp. 137–149.

[17] Maiers J. and Sherif Y.S. 'Applications of fuzzy set theory', *IEEE Transactions on Systems, Man, and Cybernetics*. 1985;15 (1):175–189.

[18] Zadeh L.A. 'Outline of a new approach to the analysis of complex systems and decision processes', *IEEE Transactions on Systems, Man, and Cybernetics*. 1973;3(1):28–44.

[19] Lee C.C. 'Fuzzy logic in control systems: fuzzy logic controller I' *IEEE Transactions on Systems, Man, and Cybernetics*. 1990;20(2):404–418.

[20] Lee C.C. 'Fuzzy logic in control systems: fuzzy logic controller II' *IEEE Transactions on Systems, Man, and Cybernetics*. 1990;20(2):419–435.

[21] Altas I. H. and Sharaf A.M. 'A fuzzy logic power tracking controller for a photovoltaic energy conversion scheme', *Electric Power Systems Research Journal*. 1992;25(3):227–238.

[22] Eminoglu I. and Altas I. H. 'A method to form fuzzy logic control rules for a PMDC motor drive system', *Electric Power Systems Research*. 1996;39 (2):81–87.

[23] Altas I.H. 'Stabilization of a two-link robot manipulator using fuzzy logic', *The Proceedings of the 3rd International Mechatronic Design and Modeling Workshop*; Ankara, Turkey, September 1997. pp. 41–51.

[24] Altas I.H. and Sharaf A.M. 'A novel fuzzy logic controller for maximum power extraction from a PV array driving a three-phase induction motor', *7th Mediterranean Electrotechnical Conference (MELECON'94)*; Antalya, Turkey, April 1994, pp. 853–856.

[25] Altas I.H. 'The effects of fuzziness in fuzzy logic controllers', *2nd International Symposium on Intelligent Manufacturing Systems*; Sakarya, Turkey, August 1998, pp. 211–220.

[26] Altas I. H. and Kruse R. 'A Fuzzy Logic Controlled Tracking System For Moving Targets', 12th IEEE International Symposium on Intelligent Control, ISIC'97, July 16–18 1997, Istanbul, Turkey, pp. 43–48.

[27] Herrera F., Lozano M. and Verdegay J. L. *Generating fuzzy rules from examples using genetic algorithms*. Technical Report #DECSAI-93115, Department of Computer Science and Artificial Intelligence, University of Granada, Spain; 1993.

[28] Klawonn F. and Keller A. 'Fuzzy clustering and fuzzy rules', *Proceedings of the 7th International Fuzzy Systems Association World Congress*; Prague, Czech Republic, June 1997, vol.1, pp. 193–198.

[29] Gonzalez A. and Perez R. 'A learning system of fuzzy control rules based on genetic algorithms', in Herrera F. and Verdegay J.L. (eds.). *Genetic Algorithms and Soft Computing*. Heidelberg: Physica-Verlag. 1996;8:202–225.

[30] Hugo E.M. and Du Plessis J.J. 'Automated multi-mode fuzzy logic controller design', *Proceedings of the IEEE International Symposium on Intelligent Control*; Mexico City, September 2001, pp. 376–382.

[31] Lai Y.S. and Lin J.C. 'New hybrid fuzzy controller for direct torque control induction motor drives', *IEEE Transaction Power Electronics*. 2003;18 (5):1211–1219.

[32] Soucy M. and Payeur P. 'Flexible fuzzy logic control for collision-free manipulator operation', *IEEE International Conference on Mechatronics and Automation*; 2005, vol. 2, pp. 723–728.

[33] Ren Y., Chai X., Xing T. and Chen X. 'Fuzzy rule representation and knowledge base construction in expert system', *International Conference on Intelligent Systems and Knowledge Engineering (ISKE)*; Hangzhou, China, November 2010. pp. 106–110.

[34] Taniguchi T., Tanaka K., Ohtake H. and Wang H.O. 'Model construction, rule reduction, and robust compensation for generalized form of Takagi–Sugeno fuzzy systems', *IEEE Transactions on Fuzzy Systems*. 2001;9 (4):525–538.

[35] García D., González A. and Pérez R. 'An iterative strategy for feature construction on a fuzzy rule-based learning algorithm', *11th International Conference on Intelligent Systems Design and Applications (ISDA)*; Cordoba, Spain, November 2011. pp. 1235–1240.

[36] Cervinka O. 'Construction of fuzzy rule base using hinging hyperplanes algorithm from training data', *1996 Biennial Conference of the North American Fuzzy Information Processing Society, NAFIPS*; Berkeley, California, June 1996. pp. 471–475.

[37] Murata T., Ishibuchi H. and Gen M. 'Construction of fuzzy classification systems using multiple fuzzy rule tables', *IEEE International Conference on Systems, Man, and Cybernetics (SMC)*; Tokyo, Japan, October 1999, vol. 4, pp. 875–880.

[38] Juang C.F. 'Construction of dynamic fuzzy if-then rules through genetic reinforcement learning for temporal problems solving', *Joint 9th IFSA World Congress and 20th NAFIPS International Conference*; Vancouver, BC, July 2001, vol. 4, pp. 2341–2346.

[39] Diaz-Robainas R.R., Huang M.Z. and Zilouchian A. 'Fuzzy identification with clustering methods of rules' construction', *Proceedings of IEEE*

Southeastcon'96. Bringing Together Education, Science and Technology; Tampa, FL, April 1996, pp. 462–467.

[40] Macvicar-Whelan P.J. 'Fuzzy sets for man–machine interaction', *International Journal Man–Machine Studies*. 1976;8:687–697.

[41] Altas I.H. and Sharaf, A.M. 'A generalized direct approach for designing fuzzy logic controllers in MATLAB/Simulink GUI environment', *International Journal of Information Technology and Intelligent Computing*. 2007;1(4).

[42] Altas I. H. 'A fuzzy logic based door position control system', *Proceedings of the 3rd International Mechatronic Design and Modeling Workshop*, Middle East Technical University, Ankara, Turkey, 15–18 September 1997, pp. 107–116.

[43] Altas I. H. 'A fuzzy logic controlled tracking system for moving targets', *12th IEEE International Symposium on Intelligent Control, ISIC'97*, Istanbul, Turkey, 16–18 July 1997, pp. 43–48.

[44] Saadat H. *Power System Analysis*. New York: McGraw Hill Book Company, 1999.

Chapter 5
System modeling and control

This chapter starts with mathematical modeling of physical systems. The methods of obtaining differential equations, simulation diagrams, and state-space models of physical systems are studied. Runge–Kutta numerical solution method is discussed and user-based MATLAB® software is developed to show the meaning of controlling physical systems as one of the application areas of fuzzy logic. The reader will be able to develop his/her own FLC code in MATLAB and in MATLAB/Simulink®. Examples of controlling electrical, mechanical, and electromechanical systems are given.

5.1 Introduction

It was stated in the previous chapter that fuzzy logic controllers do not need mathematical models of the systems to be controlled. That is correct. The design process of FLCs does not require mathematical models of the physical systems because the input–output relation is sufficient without knowing the interior structure and mathematical models. However, in order to simulate the fuzzy logic-controlled physical systems, their mathematical models are required. In the simulation of an FLC-controlled system, the mathematical model of the plant is combined with the FLC algorithm. Therefore, the mathematical model of the physical systems is required not for designing FLCs but for simulating the FLC combined physical systems. The mathematical model of a system is the basis for all control system analysis and design methods in classical feedback control systems. A detailed model allows some verification of the performance of the control system through simulation before it is implemented in the actual system.

The system models appear either as physical models or mathematical models. The physical models are usually scaled or analog types. The mathematical models are obtained by using analytical or experimental-based methods and solved by proper numerical solution methods such as Euler integration or Runge–Kutta algorithms given in Section 5.6. During the modeling process the basic physical properties, physical laws and input–output data relations are used. Some acceptable assumptions, approximations, and model reductions are applied to get an accurate enough model. The mathematical modeling of physical systems and solution methods for control purposes is addressed in this chapter. Although there are many software such as MATLAB/Simulink to obtain the solutions of the model equations, Euler's integration and Runge–Kutta algorithm are explained in the final section of this chapter. This numerical solution method of linear time-invariant (LTI) differential equations will be helpful to those who want to develop his/her own solution tool.

5.2 System modeling

As just mentioned above, the mathematical model of a system is the basis for all control system analysis and design methods in classical feedback control systems. A detailed model allows some verification of the performance of the control system through simulation before it is implemented in the actual system. All these comments are also valid for fuzzy logic-controlled system simulation. Computer or software-based simulation tools require the mathematical models of the physical system. A model for a given system depends upon defined system boundaries, the objective of the study, and the required level of approximation A design model includes many assumptions and simplifications in order to have linear, time-invariant models so that analytical methods can be applied easily to solve the model equations numerically using computer simulations.

The mathematical models of the physical systems are driven by experimental data, basic physical properties, input–output data relations, physical laws such as the law of conservation of mass, the law of conservation of energy, the law of conservation of momentum, and theoretical laws such as Newton's law, Kirchhoff's law, Ohm's law, Ampere's law, Faraday's law, and Lenz's law. Depending on the subject studied, many other laws can be added.

The mathematical model of a physical system is based on the objectives of the modeling. Because the input–output variables and the operating boundaries depend upon the defined objective function of the system. The system's function is expressed in terms of measured output variables, which are manipulated through the control input variables. The system's operation is also affected by the disturbance input variables that may be measurable or unmeasurable [1]. Therefore, the model of a physical system includes control and disturbance input variables, output variables, boundaries, and subsystems as shown in Figure 5.1.

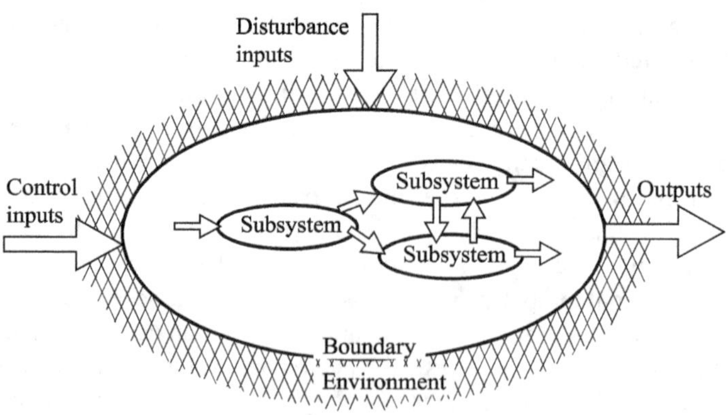

Figure 5.1 An overlook at a physical system

The models of physical systems are represented in different forms such as differential equations, transfer functions, state-space equations, signal flow graphs, simulation diagrams, and block diagrams depending on analysis and design domains, which are either time domain or frequency domain.

The modeling process is summarized as follows:

(I) Define the purpose or objective of the model considering system boundaries, interconnecting variables, and input–output relations.
(II) Identify each individual component of the physical system as the subsystem.
(III) Write the mathematical equations to define the input–output relationship of each subsystem using physical laws or experimental data.
(IV) Represent each component by its input–output equation in a software code or by a functional block with clearly identified input and output connection ports.
(V) Integrate the subsystem models into an overall system model either in a software code or in a simulation block diagram.
(VI) Verify the model validity and accuracy by implementing a simulation of the model equations and comparing the results with experimental data for the same conditions.
(VII) Combine equations and make simplifications to create an approximate model suitable for control design. In order to make this simplification, eliminate ineffective dynamics for model reduction, use linearization to linearize the nonlinear equations, use lumped parameters to convert partial differential equations into ordinary differential equations, and use sequence models in order to convert time-varying equations into time-invariant equations.
(VIII) Simulate the model equations and check the LTI reduced order model accuracy again by comparing the experimental results. The model accuracy must be remained while the model complexity is reduced.

The mathematical model equations might be in three classes, which are continuous differential equations, discrete difference equations, or algebraic equations. Within each class, there are subclasses of equations such as partial differential equations, nonlinear equations, and time-varying equations [2]. However, these subclasses of equations are converted to LTI equations in order to use the solutions methods that are applicable in linear system theory [3].

Let us represent the smallest subsystem by a block consisting of one input and one output port as shown in Figure 5.2. The system with just one input and one output port is called as single input–single output system in control systems.

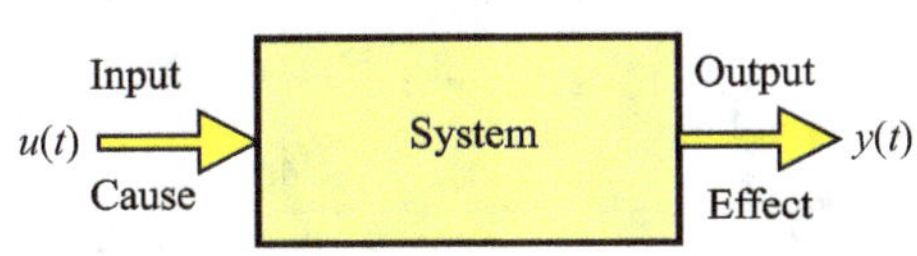

Figure 5.2 A block represents a subsystem or a system

Block diagram representation of mathematical models helps in the realization, synthesis, or fabrication of systems. A simulation of the system may be realized from the block diagram by interconnecting the basic building blocks that represent the smallest subsystems [4]. It also helps in the analysis and the design of systems as it provides a clear picture of the cause-and-effect relationship governing the various signals within the system. A simple block usually includes a linear gain, a nonlinear gain, or an integral gain as in (5.1), (5.2), and (5.3), respectively.

$$y_1 = \alpha u_1 \tag{5.1}$$

$$y_2 = f(u_2) \tag{5.2}$$

$$y_3 = \int u_3(\tau)d\tau \tag{5.3}$$

where a is α constant gain, u is the input and y is the output. Block diagram representations and input–output relations of linear, nonlinear, and integral gains are shown graphically in Figures 5.3–5.5 assuming that $\alpha = 3$, $f(u) = \sin(u)$ and the inputs are assigned as $u_1 = t$ for $t < 3$ and $u_1 = 0$ otherwise, $u_2 = 2\pi/20$ and $u_3 = 1$ for $t < 1$ and $u_3 = 0$ otherwise.

Simulation diagram is one of the methods used to represent the mathematical models of the systems. They give a clear and detailed pictorial view of the model. They are looked over like a map to see the paths, blocks, intersections, joining in and departing out nodes. The simplest elements of the simulation diagrams are the gain blocks, connecting paths, summers, departing nodes, and input–output nodes. The gain blocks are linear gain, nonlinear gain, and integral gain blocks as shown in Figures 5.3–5.5. The paths are unidirectional lines with an arrow head showing the signal flow direction.

Digital simulation of an integrator is done using different approaches such as Euler's method, trapezoid method, and Simpson's rule. Due to its simple algorithm

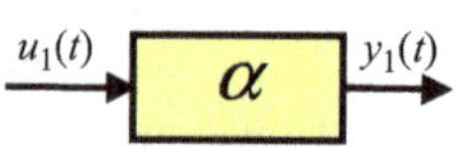


```
% Linear gain
Te=6; T=1/50;
a=3;   t=-2;   k=1;
while t < Te
     if t < (Te-Te/2)
         u1(k)=t;
     else
        u1(k)=0;
     end
     y1(k)=a*u1(k);
     t1(k)=t; t=t+T;
     k=k+1;
end
plot(t1,u1,t1,y1);
xlabel('Time
(sec)');grid
```

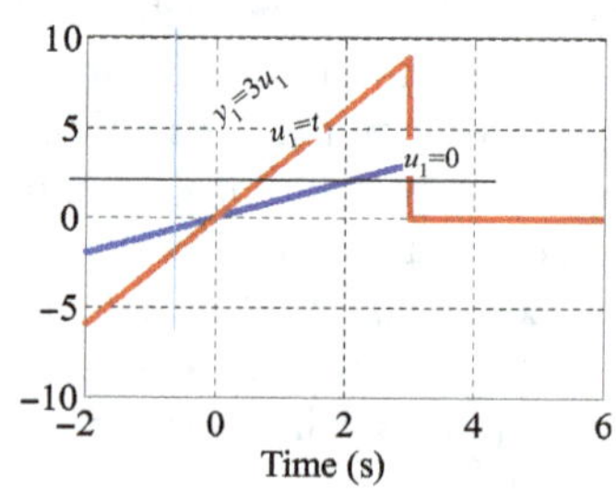

Figure 5.3 Simple block, .m file, and input–output response of linear gain

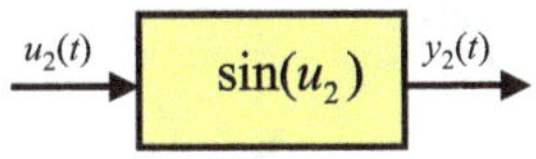


```
% Nonlinear gain
Te=5; T=1/50;t=-5;
w=2*pi/20; k=1;
while t < Te
     u=w*t;
     y2(k)=sin(u);
     t2(k)=t; t=t+T;
     u2(k)=u; k=k+1;
end
plot(u2,y2);
xlabel('Angle (rad)');
grid
```

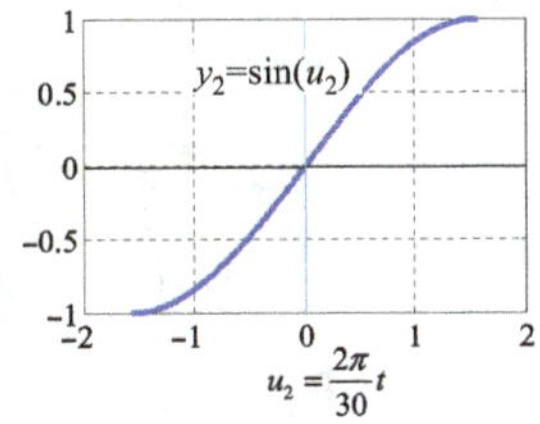

Figure 5.4 Simple block, .m file, and input–output response of nonlinear gain

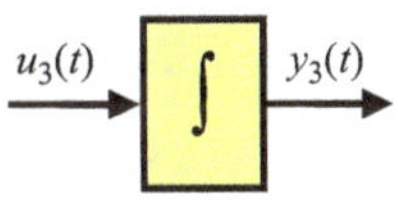


```
T=1/50;Te=2;y30=0;
u30=1;tt=0; k=1;
while tt < Te
  if tt < Te/2)
     u3=u30;
  else
     u3=0;
  end
  y3(k)=y30+T*u3;
  y30=y3(k);    U3(k)=u3;
  t3(k)=tt; tt=tt+T;
  k=k+1;
end
plot(t3,U3,t3,y3);
xlabel('Time (sec)');
grid
```

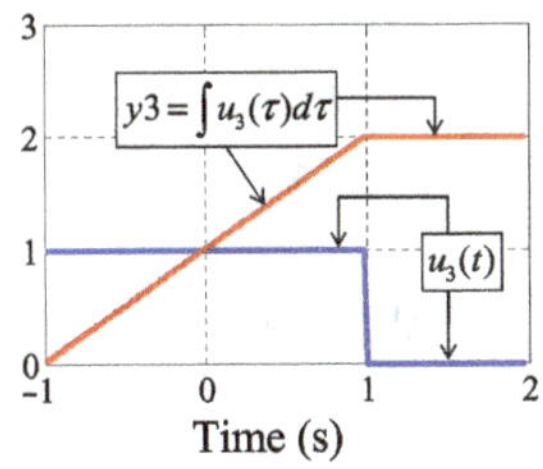

Figure 5.5 Simple block, .m file, and input–output response of integral gain

and faster speed, Euler's method given in (5.4) is preferred for the simulation of control systems.

$$y(k) = y(k-1) + Tu(k) \tag{5.4}$$

where T is the sampling period, $u(k)$ input signal to be integrated, $y(k\text{–}1)$ is the value of output y from the previous sample and it is called the initial value for $k = 1$, and $y(k)$ is the new value of the output y for the kth sample. Actually, an indefinite integrator can be separated into two parts. Assume that the input signal $u(t)$ is to be integrated from any time in the past, such as from to ∞ a finite time t in the future as written in (5.5).

$$y(t) = \int u(\tau)d\tau = \int_{-\infty}^{t} u(\tau)d\tau \tag{5.5}$$

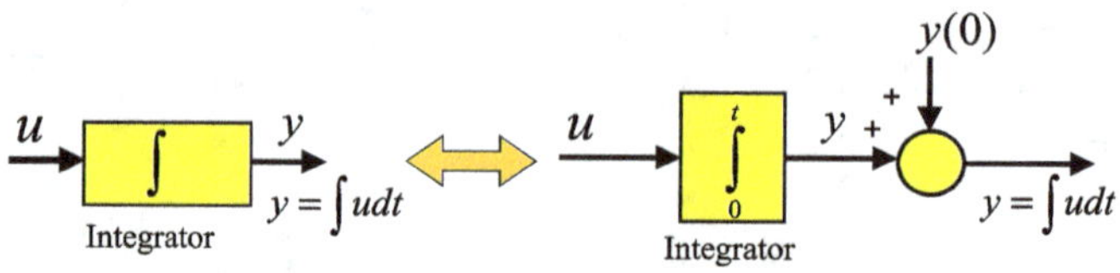

Figure 5.6 Detailed simulation diagram of an integral gain

The right side of (5.5) can be written in two parts as

$$y(t) = \int_{-\infty}^{t} u(\tau)d\tau = \int_{-\infty}^{0} u(\tau)d\tau + \int_{0}^{t} u(\tau)d\tau \tag{5.6}$$

$$y(t) = y(0) + \int_{0}^{t} u(\tau)d\tau \tag{5.7}$$

where the initial value $y(0)$ of the output y is defined as

$$y(0) = \int_{-\infty}^{0} u(\tau)d\tau \tag{5.8}$$

Equation (5.7) can be represented by a block diagram as in Figure 5.6. Note that the block diagram shown in Figure 5.6 can also be used to represent (5.4).

The system simulation is usually started at $t = 0$ s. If the system has a nonzero output at $t = 0$ s, this value is called the initial output value of the system. If the output variable is zero at $t = 0$ s then the initial value is set to zero as $y(0) = 0$. In this case, the block and simulation model representation of the integrator becomes the one given in Figure 5.5, in which Euler's method is used in the simulation code. In fact, the figure on the right in Figure 5.6 is a block diagram representation of Euler's method. The integral block already exists in the Simulink block library in both continues and discrete forms. Figure 5.6 depicts that an integrator block with an indefinite integral sign is equivalent to a definite integrator block and a summer.

A comparison and analysis of the responses of linear and integral gains in Figures 5.3 and 5.5 show that the output of the linear gain block depends upon the existence of the input signal while the integral block keeps giving a constant non-zero output even if its input is made zero after some time. If there is an input to the linear gain block then there is an output. If the input is zero then the output is zero. It can be concluded that a linear gain block gives an output if it has an input. The linear gain block does not keep any data from previous samplings. Therefore, the linear gain block is also called as *static* block *without memory*. It does not store information during the previous samplings to yield an output when the input is zero. On the other hand, the integral block stores data from the previous samplings and keeps giving a constant output when the input signal becomes zero for any reason. It is clear that the integrator block remembers, i.e., stores information. Therefore, the integral block is called as a *dynamic* block *with memory*.

In order to understand static and dynamic elements, let us consider a series connected RL circuit given in Figure 5.7. The voltage across the resistance R and inductance L can be written using Ohm's law as

$$u_{\mathrm{R}} = Ri \tag{5.9}$$

$$u_{\mathrm{L}} = L\frac{di}{dt} \tag{5.10}$$

The loop equation is then written using Kirchhoff's voltage law as

$$v(t) = u_{\mathrm{R}} + u_{\mathrm{L}} \tag{5.11}$$

$$v(t) = Ri(t) + L\frac{di(t)}{dt} \tag{5.12}$$

This is a first-order LTI differential equation and can be rewritten as follows by moving the highest-order term to the left side.

$$L\frac{di(t)}{dt} = v(t) - Ri(t) \tag{5.13}$$

The left side of (5.13) is equal to the arithmetical sum of two separate terms. Therefore, a summer is needed to add the terms on the right side so that the output becomes equal to the term on the left. This summation is shown in Figure 5.8. The parameter R is a constant and $i(t)$ must be obtained first to be multiplied by R.

The output of the summer is the derivation of current $i(t)$ multiplied by the constant L. Let us multiply the output of the summer by $1/L$ so that the derivation of current is separated as a single term. Then the derivation of current can be integrated to get the current itself, which is to be multiplied by R and added to the summer as a negative feedback signal as given in Figure 5.9.

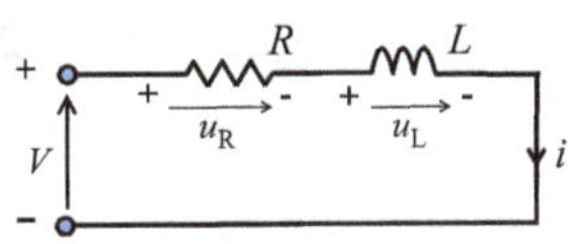

Figure 5.7 Series connected RL circuit

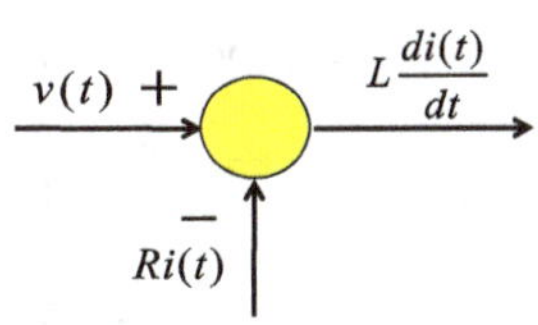

Figure 5.8 A summer adding two inputs to yield an output

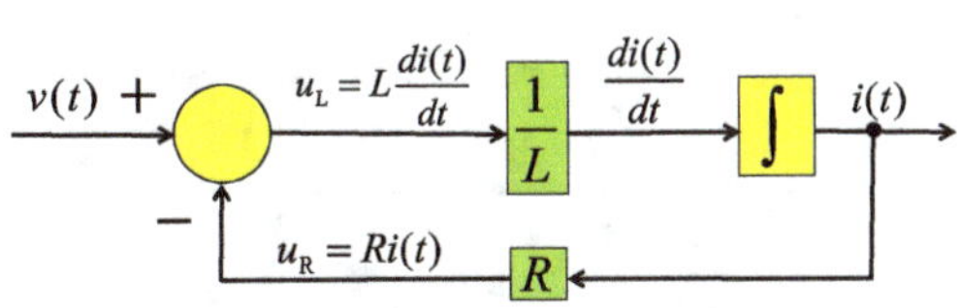

Figure 5.9 A detailed simulation diagram of a series RL circuit

Example 5.1

Simulate the series-connected RL circuit shown in Figure 5.9 in MATLAB/Simulink and observe the responses of the signals at the input and output of the static and dynamic elements. The numerical values are given as $v(t) = 50$ V, $R = 5$ Ohm, and $L = 2$ H.

Answer 5.1

Detailed simulation diagram shown in Figure 5.9 represents the mathematical model of the series-connected RL circuit shown in Figure 5.7. The mathematical model equation of this system is obtained as the first-order LTI differential equation given by (5.13). Both Equation (5.13) and the simulation diagram in Figure 5.9 are two different ways of representing the series RL circuit. The simulation diagram given in Figure 5.9 can be constructed in Simulink easily with the same look as shown in Figure E5.1(a).

The current flowing through the inductor in Figure 5.7 is equal to the signal at the output of the integrator in Figures 5.9 and E5.1(a). The signal at the input side of the integrator is the derivation of current. The input and output signals of the integrator are given in Figure E5.1(b). As depicted in this figure, the integrator output, which is the current, reaches a steady-state value of 10 A and remains constant while the integrator input signal is zero. The integrator is a *dynamic element* and keeps remembering the last output before the input becomes zero. The magnitude of $di(t)/dt$ is scaled down to half in Figure E5.1(b) in order to make the graphs more visible. The gain blocks $1/L$ and R in Figures 5.9 and E5.1(a) are *static* blocks and their inputs are multiplied by constant gains to yield the outputs. It is clearly understood that zero inputs to these static blocks result in zero outputs.

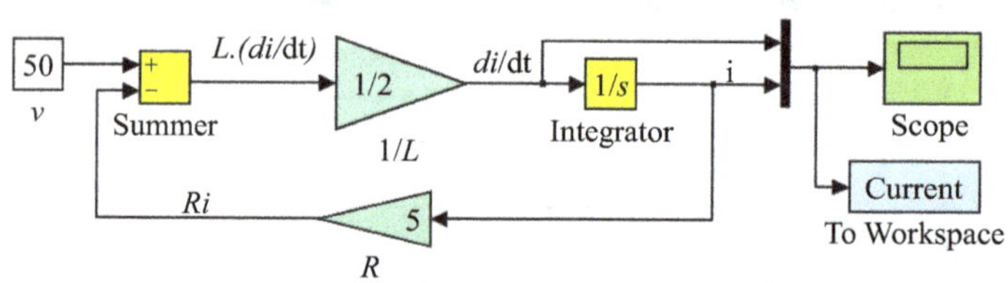

Figure E5.1(a) Simulink model of Series RL circuit in Example 5.1

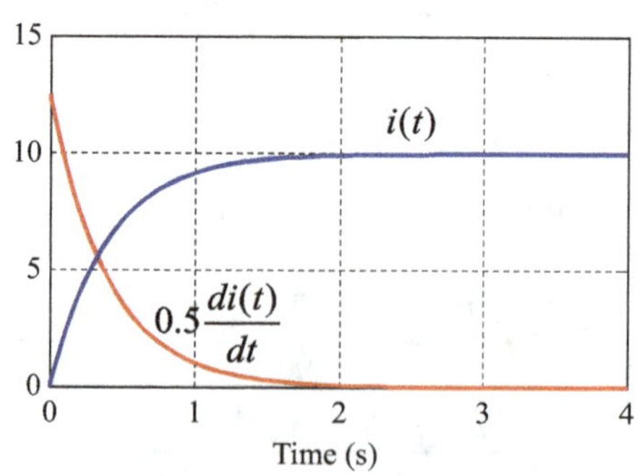

Figure E5.1(b) The input and output signals of the integrator for Example 5.1

This example shows that the integrator remembers the last output from the nonzero input case and keeps giving it as long as the simulation continues. Due to its memory, the integrator is considered as a *dynamic* element and very useful in automatic control systems because of this property. Since a controller generates the control input signal from the error, the integral controller keeps generating a constant control input signal when the error becomes zero. Recall that the final control input signal from the fuzzy logic controller is obtained by adding the change in control signal to the control signal value from the previous sampling as given by (4.26).

5.3 Modeling electrical systems

The modeling process is summarized in the previous section and a series connected RL electrical circuit is modeled as an example. It has shown that the model of a continuous time system is described by a set of differential and algebraic equations or in general by a set of integro-differential equations and algebraic equations. Converting these sets of equations into detailed simulation diagrams enables us to have an overall look at the system structure. As it is shown in Figure 5.9, the simulation diagram is a pictorial representation of the mathematical model of the system. Therefore, the simulation diagram approach is used extensively in this book. The use of block diagrams in Simulink for the design and analysis of control systems can be found in [5].

The mathematical model of the system is derived using the physical laws governing the input–output map. These physical laws are counted as the law of conservation of mass, the law of conservation of energy, the law of conservation of momentum, and theoretical laws such as Newton's law, Kirchhoff's law, Ohm's law, Ampere's law, Faraday's law, and Lenz's law. The physical laws used in the modeling of electrical systems are mainly Kirchhoff's law, Ohm's law, Ampere's law, Faraday's law, and Lenz's law.

The modeling of electrical systems starts with the circuit diagram, which is a representation of system wiring diagram or an equivalent circuit of a more complicated system showing the input–output relation. Then the physical laws are applied to drive the mathematical model equations in various forms such as

- Integro-differential equations,
- Simulation diagrams,
- Block diagrams,
- Signal flow graphs,
- Transfer functions, and
- State-space equations.

Resistance (R), inductance (L), and capacitance (C) are three basic elements of electrical systems. The voltage–current equations of these elements can be written as follows using Ohm's law.

$$v_R(t) = Ri(t) \tag{5.14}$$

$$v_L(t) = L\frac{di(t)}{dt} \tag{5.15}$$

$$v_C(t) = R\frac{1}{C}\int i(t)dt \tag{5.16}$$

These three basic equations are used to model the passive electrical networks and systems. The individual equations of each component are combined depending on the connection diagram. Kirchhoff's voltage law is applied to write the mesh equations if there are a series of connected elements. And Kirchhoff's current law is applied to write the node equations if there are parallel connections.

The representation of electrical system modeling in different forms will be shown through examples.

Example 5.2

An RC electrical circuit is shown in Figure E5.2(a). Assume that $v_i(t)$ is the input and $v_O(t) = v_C(t)$ is the output in this circuit, which represent a low pass filter.

Obtain the following mathematical models to represent input–output relation of this circuit.

(a) Integro-differential equations
(b) Simulation diagram
(c) Transfer function
(d) Block diagram

Use the numerical values $v(t)$ = 50 V, R = 5 Ohm, and C = 1000 μF and simulate this system in MATLAB and Simulink.

Answer 5.2

(a) Integro-differential equations

The circuit diagram is used to derive the integro-differential equations. Let us write the mesh equations using Kirchhoff's voltage law as

$$v_i(t) = v_R + v_C \tag{E5.2a}$$

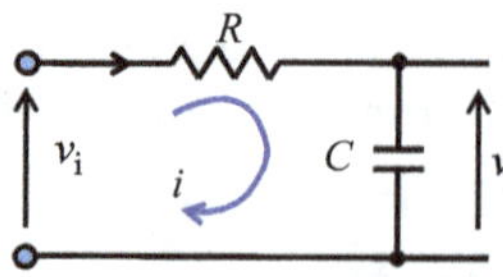

Figure E5.2(a) Circuit diagram of a low pass filter

Substituting (5.14) and (5.16) yields the first-order LTI integro-differential equation as

$$v_i(t) = Ri(t) + \frac{1}{C}\int i(t)dt \tag{E5.2b}$$

(b) Simulation diagram,

The integro-differential equation given in (E5.2b) can be rewritten as follows by moving the highest-order derivative term to the left side.

$$Ri(t) = v_i(t) - \frac{1}{C}\int i(t)dt \tag{E5.2c}$$

The left side of (E5.2c) is equal to the arithmetical sum of two separate terms. Therefore, a summer is needed to add the terms on the right side so that the output becomes equal to the term *Ri(t)* on the left. One of the terms to be added is the input voltage $v_i(t)$ and it is there, ready to be added. The second term on the right is a product of constant parameter $1/C$ and the integration of the current $i(t)$. Therefore, the current $i(t)$ is separated from the term $Ri(t)$ at the output of the summer, and then integrated. The simulation diagram representing (E5.2c) is given in Figure E5.2(b).

The output variable of the electrical circuit given in Figure E5.2(a) is the voltage across the capacitor and defined as

$$v_O(t) = \frac{1}{C}\int i(t)dt \tag{E5.2d}$$

The output voltage $v_O(t)$ is shown at the right end of simulation diagram in Figure E5.2(b).

(c) Transfer function

The Laplace transformation of (E5.2b) yields (E5.2e) for zero initial values.

$$V_i(s) = RI(s) + \frac{1}{Cs}I(s) \tag{E5.2e}$$

$$I(s) = \left(\frac{Cs}{RCs + 1}\right)V_i(s) \tag{E5.2f}$$

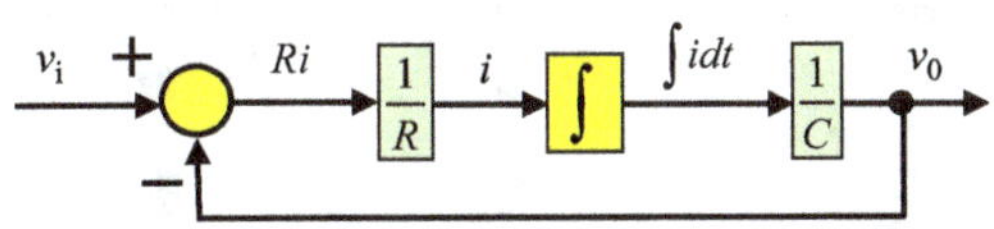

Figure E5.2(b) Simulation diagram of the low pass filter given in Figure E5.2(a)

The Laplace transformation of (E5.2d) yields (E5.2g).

$$V_O(s) = \frac{1}{Cs} I(s) \tag{E5.2g}$$

Rewriting (E5.2g) yields

$$I(s) = CsV_O(s) \tag{E5.2h}$$

Equating (E5.2f) and (E5.2h) yields the transfer function as

$$\frac{V_O(s)}{V_i(s)} = \frac{1}{RCs + 1} \tag{E5.2i}$$

(d) Block diagram
Transfer function given in (E5.2i) can be represented by a block diagram as in Figure E5.2(c).

Simulation of this system in Simulink can be done by creating the simulation diagram in Figure E5.2(b) in Simulink as in Figure E5.2(d).

Simulation of this system in MATLAB can be done using the transfer function with MATLAB functions *lsim* and *step* functions in MATLAB. Substituting numerical values into (E5.2i) yields (E5.2j) and simulation results as in Figure E5.2(e).

$$\frac{V_O(s)}{V_i(s)} = \frac{1}{2 \times 1000^{-6}s + 1} \tag{E5.2j}$$

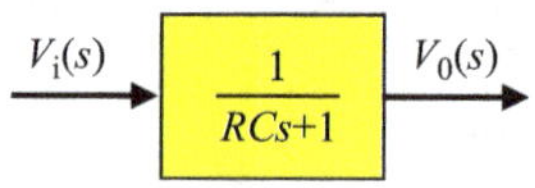

Figure E5.2(c) Transfer function block diagram of the low pass filter

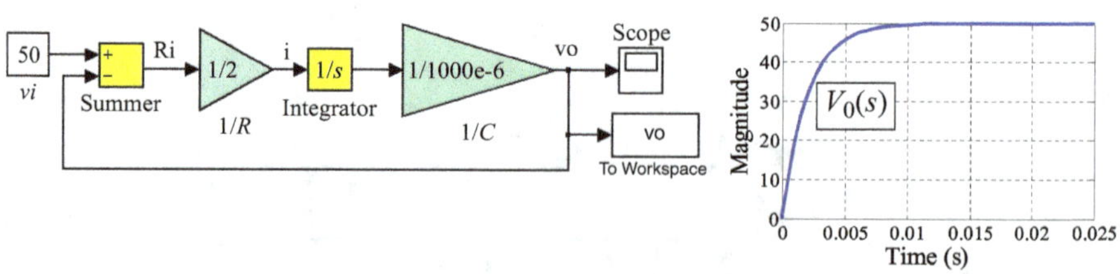

Figure E5.2(d) Simulation in Simulink

```
% Figure E05.02
clear, clc,clf
num=[1]; den=[2*1000e-6  1]; k=1;
for t1=0:0.001:0.012;
    U(k)=50;k=k+1;
end
t=0:0.001:0.012; V0=tf(num,den);
[y,t]=lsim(V0,U,t); plot(t,y);
xlabel('Time (sec)');
ylabel('Magnitude');grid; pause
step(num,den);grid
```

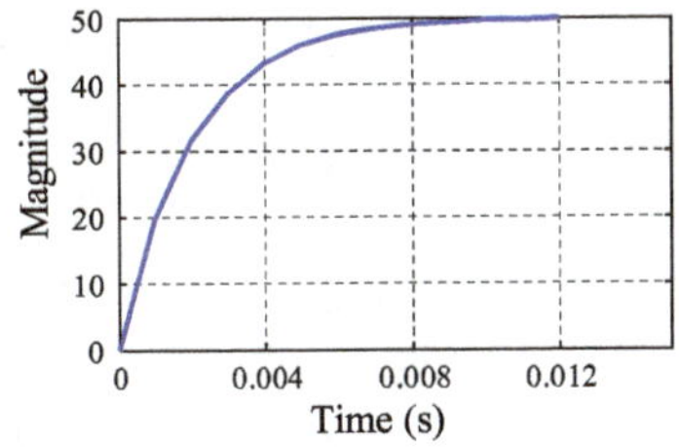

Figure E5.2(e) Simulation in MATLAB using lsim and step functions

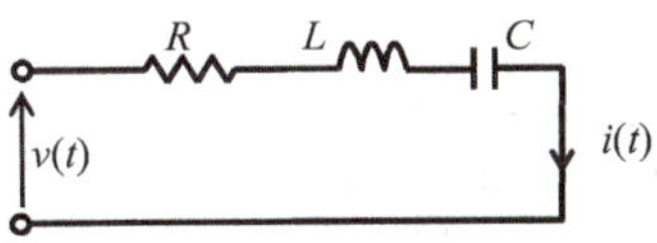

Figure E5.3(a) Series connected RLC circuit

Example 5.3

A series-connected RLC circuit is shown in Figure E5.3(a). Assume that applied voltage $v(t)$ is input and current $i(t)$ flowing through the circuit is the output.

Obtain the following mathematical models to represent input–output relation of this circuit.

(a) Integro-differential equations
(b) Simulation diagram
(c) State-space equations
(d) Transfer function
(e) Block diagram

Answer 5.3

(a) Integro-differential equations

The circuit diagram is used to derive the integro-differential equations. Let us write the mesh equations using Kirchhoff's voltage law as

$$v(t) = v_R(t) + v_L(t) + v_C(t) \tag{E5.3a}$$

Resistance (R), inductance (L), and capacitance (C) are three basic elements of this electrical circuit and their voltage equations are given in (5.14), (5.15), and (5.16), respectively. Substituting these equations into (E5.3a) we get

$$v(t) = Ri(t) + L\frac{di(t)}{dt} + \frac{1}{C}\int i(t)dt \qquad \text{(E5.3b)}$$

This is a second-order LTI integro-differential equation.

(b) Simulation diagram,

The first step to sketch simulation diagram is writing the integro-differential equation in proper order. This is done by writing the highest order derivative term in (E5.3b) to the left side and mowing all others to the wright as in (E5.3c).

$$L\frac{di(t)}{dt} = v(t) - Ri(t) - \frac{1}{C}\int i(t)dt \qquad \text{(E5.3c)}$$

As the second step, identify the input and output variables, which are $v(t)$ and $i(t)$, respectively, in this example. Then place the input to the left and the output to the right ends.

In the third step, place the summer right after the input variable and connect the input to the summer, usually as a positive input.

Check the other terms to be added to the summer and connect them as input signals using unidirectional arrows. These additional terms to be added may not be available clearly. Therefore, they should be separated from internal variables and then added to the summer, later.

Identify the output term from the summer as the fifth step. The output from the summer is the highest order derivative term placed to the left side of (E5.3c). Then separate the derivative part from the constant coefficient if there is any.

Starting from the highest derivative term, place integrators one after the other until the derivatives are over and the variable is obtained.

As the seventh and final step, connect the individual linear and integral gain blocks to satisfy (E5.3c). The final simulation diagram is obtained in Figure E5.3(b). Note that the derivative term di/dt is integrated until the current $i(t)$ is obtained. However, the current $i(t)$ is integrated again and multiplied by $1/C$ to obtain the voltage $v_C(t)$ across the capacitor.

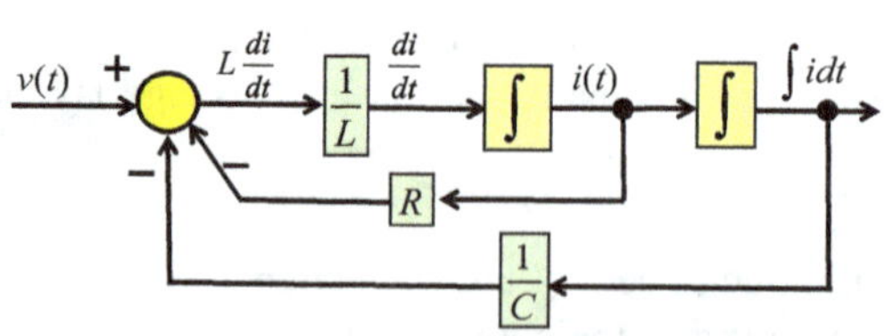

Figure E5.3(b) Simulation diagram of series connected RLC circuit-state-space equations

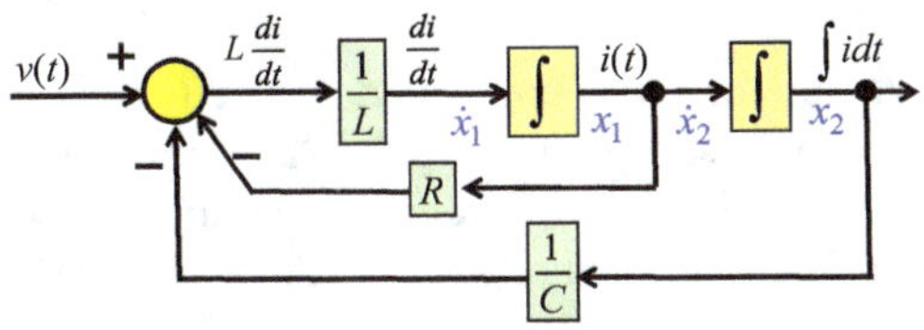

Figure E5.3(c) Simulation diagram used to obtain state-space equations

It is an easy task to obtain the state-space model equations from the simulation diagram. To do this, assign a variable at the output of each integrator. The input side of the integrator will be the derivation of that variable. Then, just write the state equation from the simulation diagram and put it in a matrix form.

The following equations can be written directly from Figure E5.3(c):

$$\dot{x}_1(t) = \frac{1}{L}v(t) - \frac{R}{L}x_1(t) - \frac{1}{LC}x_2(t) \tag{E5.3d}$$

$$\dot{x}_2(t) = x_1(t) \tag{E5.3e}$$

$$y(t) = x_1(t) \tag{E5.3f}$$

where $x_1(t) = i(t)$ and $x_2(t) =$ integration of $i(t)$. Then the state-space equations can be written in matrix format.

$$\begin{bmatrix} \dot{x}_1(t) \\ \dot{x}_2(t) \end{bmatrix} = \begin{bmatrix} -\frac{R}{L} & -\frac{1}{LC} \\ 1 & 0 \end{bmatrix} \begin{bmatrix} x_1(t) \\ x_2(t) \end{bmatrix} + \begin{bmatrix} \frac{1}{L} \\ 0 \end{bmatrix} u(t) \tag{E5.3g}$$

$$y(t) = \begin{bmatrix} 1 & 0 \end{bmatrix} \begin{bmatrix} x_1(t) \\ x_2(t) \end{bmatrix} \tag{E5.3h}$$

where $u(t) = v(t)$. The state-space model equations have a well-known form of abbreviation as given in (E5.3i), which is very useful to be used in MATLAB and Simulink. The abbreviated form of the state-space equations is also useful in additional manipulations

$$\begin{aligned} \dot{x}(t) &= Ax(t) + Bu(t) \\ y(t) &= Cx(t) + Du(t) \end{aligned} \tag{E5.3i}$$

where $\dot{x}(t)$ is called a state vector, $x(t)$ is a vector of state variables, $y(t)$ is the output variable and $u(t)$ is the input variable. The other input variable $w(t)$ a disturbance input. A is an $n \times n$ state matrix, B is the $n \times m$ input matrix, and D is the disturbance matrix.

Note that, if the places of $x_1(t)$ and $x_2(t)$ are switched in Figure E5.3(c), the state-space equations will be changed. In fact, a physical system may have many different simulation diagrams, but just one unique transfer function.

(c) Transfer function

The transfer function of series connected RLC circuit is obtained in a couple of ways. It can be obtained by applying Laplace transformation to (E5.3b) or (E5.3d) and (E5.3f) or (E5.3i). The transfer function is also obtained by using block diagram reduction methods to reduce Figure E5.3(b) into one single block.

Let us apply Laplace transformation to (E5.3b) with zero initial values.

$$V(s) = RI(s) + sLI(s) + \frac{1}{sC}I(s) \tag{E5.3j}$$

Then the transfer function is obtained as

$$\frac{I(s)}{V(s)} = \frac{s}{Ls^2 + Rs + \frac{1}{C}} \tag{E5.3k}$$

Let us apply Laplace transformation to (E5.3i) with zero initial values.

$$sX(s) = AX(s) + BU(s) \tag{E5.3l}$$

$$Y(s) = CX(s) + DU(s) \tag{E5.3m}$$

From (E5.3i),

$$X(s) = (sI - A)^{-1}BU(s) \tag{E5.3n}$$

Substituting (E5.3n) into (E5.3m) yields the transfer function in terms of state-space model parameters.

$$T(s) = \frac{Y(s)}{U(s)} = C(sI - A)^{-1}B + D \tag{E5.3o}$$

(d) Block diagram

Let us use the block diagram reduction approach to obtain the block diagram and transfer function of the given system. Replacing the integral sign with its Laplace transform 1/s, converts the simulation diagram in Figure E5.3(b) from the time domain to the *s* domain as shown in Figure E5.3(d). Note that the block diagram has been reshaped so that the output signal *i*(*t*) is located at the right end.

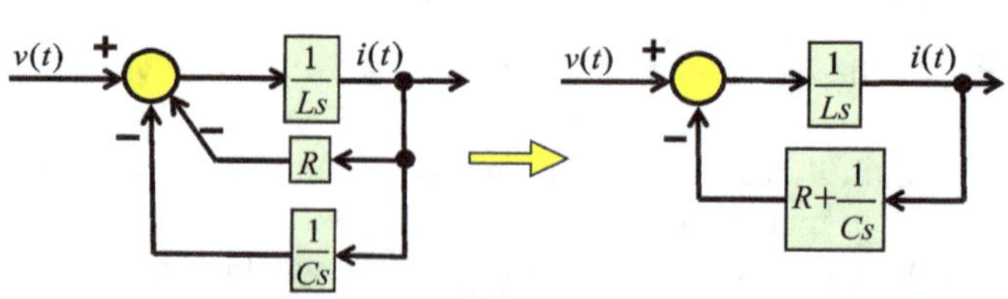

Figure E5.3(d) Application of block diagram reduction to the simulation diagram

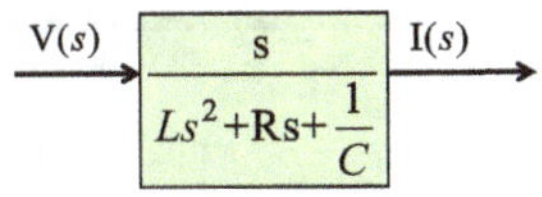

Figure E5.3(e) The transfer function of series RLC circuit as a single block diagram

After combining two parallel branches at the bottom of the block diagram on the left in Figure E5.3(d), the one with negative feedback on the right is obtained. With a last reduction, the final single block showing the relation between input and output is obtained as given in Figure E5.3(e). Note that this final block includes the transfer function of series connected RLC circuit as given in (E5.3k).

Example 5.4

Use the numerical values $v(t)$ = 50 V, R = 10 Ohm, L = 2 H, and C = 100 F and simulate the series connected RLC circuit modeled in Example 5.3.

(a) In MATLAB and
(b) In Simulink.
(c) Repeat (a) and (b) by adding an FLC to keep the current constant at a reference value.

Answer 5.4

a. and c. Simulation and FL control in MATLAB and Simulink

System simulation in MATLAB was given in Examples E4.7 and E4.8. The same MATLAB codes can be modified and used in this example as well. First, the system will be simulated without FLC, then the FLC will be added. The .m file used for the simulation in this example is given in Figure E5.4(a). This simulation is done without any controller and shows that the circuit current reaches its steady-state value of 5 A in approximately 1 s. A controller can be designed to control the circuit current at any reference value between 0 and 5 A. The FLC adapted from [6] and used in Example E4.8 and the FLC block adapted from [7] and given in Figure 4.25 are used to control the current of series RLC circuit in MATLAB and in Simulink, respectively. These simulation codes and block diagram are shown in Figures E5.4(c) and E5.4(d).

The function *runge.m* used in the first line of *while* … *end* loop, is a function or a subprogram developed by the author using Runge–Kutta algorithm. The details of the Runge–Kutta algorithm are given in Section 5.6 at the end of this chapter.

The settling time of the controlled current is longer than it is in the uncontrolled case. The response can be made faster with the controller. However, faster response results in overshoot and oscillations in the response.

```
%  TRANSIENTS OF Serşes RLC circuit
clear;clf;clc
% -----------------------Circuit data
V=50; R=10; L=2; C=100;
% ---------------   State Space model
A=[ -(R/L)   -1/(L*C)
      1       0  ];
B=[1/L 0;0   0 ];
C=[1   0];   D=[0 ] ;
U=[V   0];
% ---------- initials for simulation
 t0=0;         dt=0.001;     tend=2;
SA=size(A);   NL=SA(1);     NC=SA(2);
for n=1:NL;   x0(n)=0;end
% -------------- Data for controller
y0=0;    e10=0;    k=1;
%----------------- Simulation loop
while t0<tend-dt
   [x]=runge(A,B,U,x0,dt);
   x1(k)=x(1); x2(k)=x(2); t(k)=t0+dt; t0=t(k);
   for n=1:NL
    x0(n)=x(n); XX(k,n)=x(n);
   end
   k=k+1;
end;
%---------------------- Plots
plot(t,x1); xlabel('Time (s)');
ylabel('Current (A)');grid
```

Figure E5.4(a) MATLAB .m file used to simulate the RLC circuit

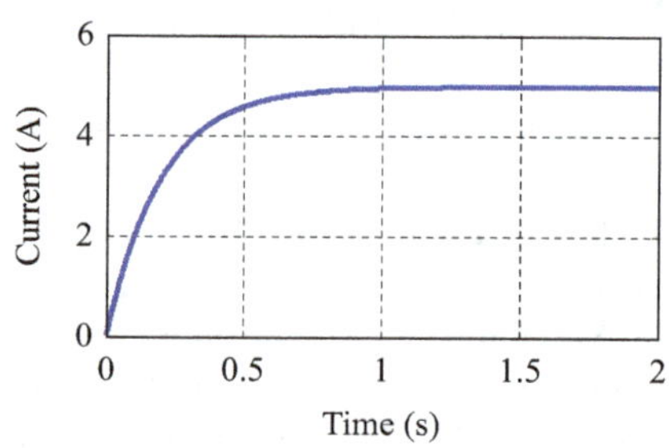

Figure E5.4(b) Time response of the circuit current

b. and c. Simulation and FL control in Simulink.

Simulation diagram of the given system in Simulink is shown in Figure E5.4(e), which is very similar to the simulation diagram shown in Figure E5.3(b). Simulation result in Simulink is given in Figure E5.4(f).

FLC controlled simulation diagram of the series-connected RLC circuit in Simulink is given in Figure E5.4(g). The FLC subsystem block [7,8] defined in Chapter 4 and shown in Figure 4.25 is used in Figure E5.4(g) as the controller and connected to the subsystem representing the simulation diagram of a series

```
% ---------------------Circuit data
V=50; R=10; L=2; C=100;
A=[ -(R/L)   -1/(L*C);   1      0  ]; B=[1/L 0;0  0 ];
C=[1  0];    D=[0 ] ;   Ua=[V  0];
dt=0.01; t0=0;  tend=4;
SA=size(A);  NL=SA(1);    NC=SA(2);
for n=1:NL; x0(n)=0; end
R0=2;  y0=0;    e10=0;    k=1;
% Input parameters for fuzzy logic controllers
    EMAX  = R0;          EMIN  = -EMAX;
    DEMAX = EMAX/50;     DEMIN = -DEMAX;
    DUMAX = 1;           DUMIN = -1;
%   Partition the universe of e into five fuzzy subsets
    TE=2*EMAX;  WE=2*pi/TE; AE=1;
    NBE=-TE/2;  NSE=-TE/4;  ZZE=0;      PSE=TE/4;      PBE=TE/2;
    x1E=NBE;    x2E=NSE;    x3E=ZZE;    x4E=PSE;       x5E=PBE;
%   Partition the universe of de into five fuzzy subsets
    TDE=2*DEMAX;  WDE=2*pi/TDE;  ADE=1;
    NBDE=-TDE/2;  NSDE=-TDE/4;   ZZDE=0;      PSDE=TDE/4;  PBDE=TDE/2;
    x1DE=NBDE;    x2DE=NSDE;     x3DE=ZZDE;  x4DE=PSDE;   x5DE=PBDE;
%   Partition the universe of su into five fuzzy subsets
    TDU=2*DUMAX; WDU=2*pi/TDU; ADU=1;
    NBDU=-TDU/2;  NSDU=-TDU/4;   ZZDU=0;     PSDU=TDU/4;  PBDU=TDU/2;
    x1DU=NBDU;    x2DU=NSDU;     x3DU=ZZDU;  x4DU=PSDU;   x5DU=PBDU;
%   Initial values for FLC variables
    ee=EMAX;          dee=0;    E=EMAX;     dE=0;
    e0=EMAX;          e(1)=0;   e(2)=0;     de=e(2)-e(1);
    DUSCALE=0.2;      U00=0;
% ***************       membership matrix
DU=[  NBDU   NBDU   NSDU   NSDU   ZZDU
      NBDU   NSDU   NSDU   ZZDU   PSDU
      NSDU   NSDU   ZZDU   PSDU   PSDU
      NSDU   ZZDU   PSDU   PSDU   PBDU
      ZZDU   PSDU   PSDU   PBDU   PBDU  ];
DULC=size(DU);  DUR=DULC(1);  DUC=DULC(2);
%**************  Iteration loop **************
while t0 < tend
    % ------------------------ Fuzzy Logic Controller
    E=limiter(EMIN,EMAX,ee);
    DE=limiter(DEMIN,DEMAX,dee);
    FSE=sinusoid5(E,WE,AE,NBE,NSE,ZZE,PSE,PBE);
    FSDE=sinusoid5(DE,WDE,ADE,NBDE,NSDE,ZZDE,PSDE,PBDE);
    FSDUTOP=0;    DDUTOP=0;
    for n=1:DUR
        for m=1:DUC
            FSDU(n,m)=min(FSE(n),FSDE(m)); DDU(n,m)=FSDU(n,m)*DU(n,m);
            DDUTOP=DDUTOP+DDU(n,m);        FSDUTOP=FSDUTOP+FSDU(n,m);
        End
    end
    % ---------------------------- defuzzification
      duu=DUSCALE*(DDUTOP/FSDUTOP);  du=limiter(DUMIN,DUMAX,duu);
      UU=U00+du;     U00=UU;         U=limiter(0,1,UU);
      U0 = U*Ua;
% ********solution of the equations *********
[x]=runge(A,B,U0,x0,dt);   I(k)=x(1);
r0(k)=R0;   e(k)=R0-I(k);   de(k)=e(k)-e0;
e0=e(k);    Vin(k)=U0(1);   duty(k)=U;
ee=e(k);    dee=de(k);   t(k)=t0+dt; t0=t(k);
    for n=1:NL;   x0(n)=x(n); XX(k,n)=x(n);   end
    k=k+1;
end;
plot(t,I,t,r0);xlabel('Time (sec)')
ylabel('Current (A)'); grid
```

Figure E5.4(c) MATLAB .m file used to control the circuit current by FLC

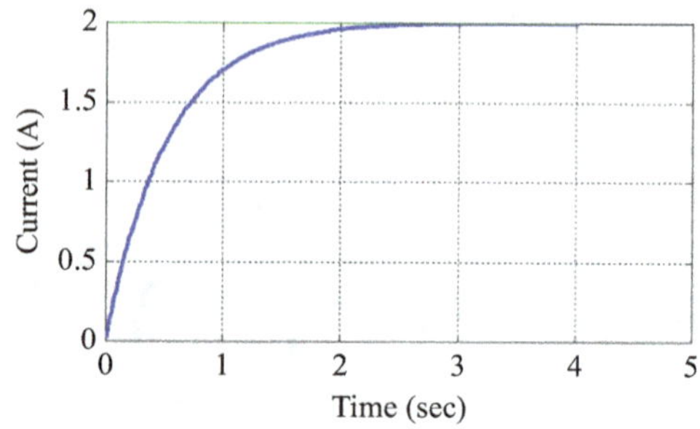

Figure E5.4(d) The circuit current is kept at 2 A by FLC in MATLAB

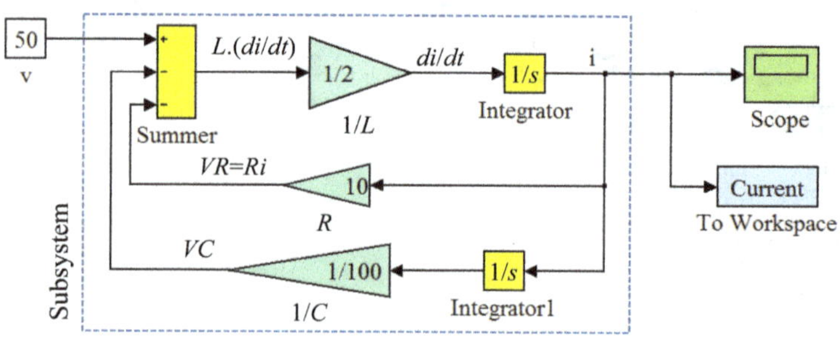

Figure E5.4(e) Simulation diagram of RLC circuit in Simulink

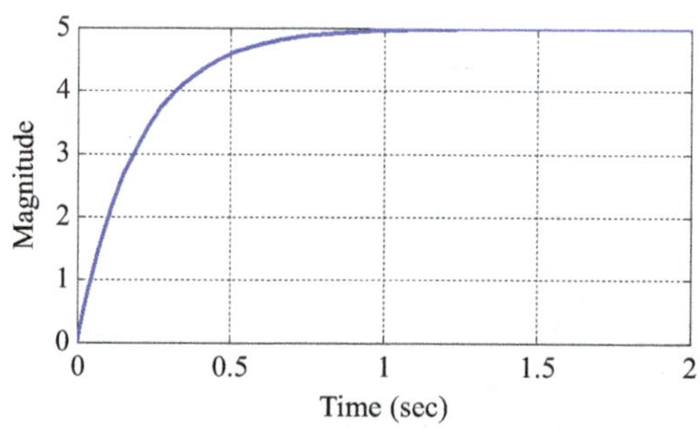

Figure E5.4(f) Time response of the circuit current in Simulink

connected RLC circuit. The inside view of the RLC subsystem is depicted by dashed rectangular in Figure E5.4(e). A DC chopper modeled by (4.27) is used as an actuator in Figure E5.4(g). Since the output variable to be controlled is the circuit current, it is used as a negative feedback signal.

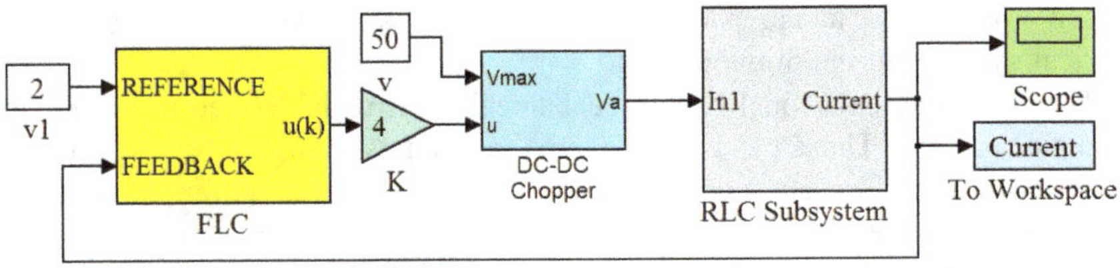

Figure E5.4(g) *Simulation diagram of FLC controlled circuit subsystem in Simulink*

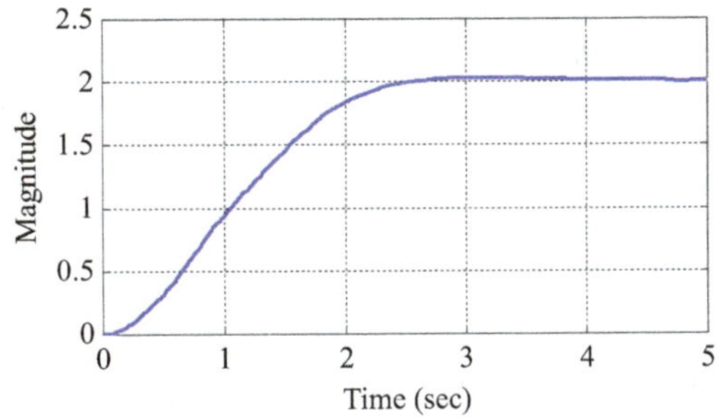

Figure E5.4(h) *The circuit current is kept at 2 A by FLC in Simulink*

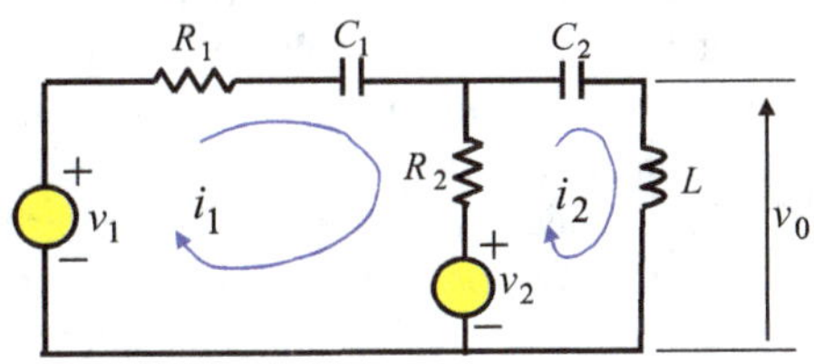

Figure PRA5.1 *Circuit diagram for Practice 5.1*

Practice 5.1

Develop a simulation diagram and state-space model for the electrical circuit given in Figure PRA5.1.

5.4 Modeling mechanical systems

The modeling process is summarized in Section 5.2. It has shown that the model of a continuous time system is described by a set of differential and algebraic equations or in general by a set of integro-differential equations and algebraic equations. Converting these sets of equations into detailed simulation diagrams enables us to have an overall look at the system structure. As it is shown for the modeling of electrical systems in the previous part.

The mathematical model of the mechanical systems is derived using the physical laws, as well. These physical laws are counted as the law of conservation of mass, the law of conservation of energy, the law of conservation of momentum, and theoretical laws such as Newton's law.

The modeling of mechanical systems starts with the schematic diagram, which is a representation of system's operational structure showing the input–output variables and interior parameters. Then the physical laws are applied to drive the mathematical model equations in the same forms as mentioned in Section 5.3 for electrical systems.

- Integro-differential equations,
- Simulation diagrams,
- Block diagrams,
- Signal flow graphs,
- Transfer functions, and
- State-space equations.

As mentioned in the previous section for electrical system modeling, simulation diagrams are extensively used in the modeling process of physical systems including mechanical systems as well. The simulation diagrams are an elegantly visual way of describing differential equations and seeing or visualizing a problem, which makes it more understandable. The simulation diagram model involving only integrators, summers, and coefficient multipliers is called *analog computer representation* since they are in a form suitable for analog computer simulation. Simulation diagrams show how components of a system are connected together physically without the complication of shafts, gears, dampers, and springs.

Mathematical modeling of mechanical systems for control is categorized into three parts as *linear motion models*, *rotational motion models*, and *transformation models*. Similar to the basic elements resistance (R), inductance (L), and capacitance (C), translational mechanical systems also have three basic elements mass (m), viscous friction (b), and spring (k). In rotational mechanical systems, the mass (m) is replaced by inertia (J), and the spring is replaced by stiffness (s). Mathematical modeling of mechanical systems will be discussed in two parts as linear motion and rotational motion models. More information about the modeling of mechanical systems can be found in [9–12].

5.4.1 Mechanical systems with linear motion

A rigid mechanical body makes a linear motion under an applied force. Therefore, it is assumed that the applied force is input and linear displacements are the output in the mathematical modeling of the basic mechanical elements.

If a force $f(t)$ (N) is applied to a mass m (kg), it moves linearly with a displacement $x(t)$ (m). The relation between the applied force and the displacement is written by invoking Newton's second law, which states that the vector sum of all external forces acting on a rigid body must be equal to the rate of change of momentum, $p(t)$.

$$\frac{dp(t)}{dt} = f(t) \tag{5.17}$$

The momentum depends on the amount of change in displacement of the mass m and is expressed as in (5.18)

$$p(t) = m\dot{x}(t) \tag{5.18}$$

Since the change in displacement is defined as velocity (m/s), the momentum in (5.18) can be written as

$$p(t) = mv(t) \tag{5.19}$$

The change in momentum is written by getting the derivation of both sides of (5.18) as in (5.20) or getting the derivation of both sides of (5.19) as in (5.21).

$$\dot{p}(t) = m\ddot{x}(t) \tag{5.20}$$

$$\dot{p}(t) = m\dot{v}(t) \tag{5.21}$$

where a dot over a variable indicates the first derivation and two dots indicate the second derivation.

Mathematical model equation of a linear motion mechanical system is written considering either force–displacement or force–velocity relations as given in (5.20) and (5.21), respectively. Equating (5.17) and (5.20) yields the force–position relation of mass, m as given in (5.22).

$$f(t) = m\ddot{x}(t) \tag{5.22}$$

where $f(t)$ is the sum of all the forces affecting the mass m. A rigid body, the mechanical mass, might be placed on a surface horizontally or vertically such that viscous friction resists the motion by generation a force as long as the mass moves with a nonzero velocity as shown in Figure 5.10. The viscous friction force always has a negative effect on the motion and is expressed as in (5.23).

$$f_b(t) = b\dot{x}(t) \tag{5.23}$$

As the mass might be placed on a surface, it might be tied to a fixed steady point or to another mass horizontally or vertically using a connector that shows a

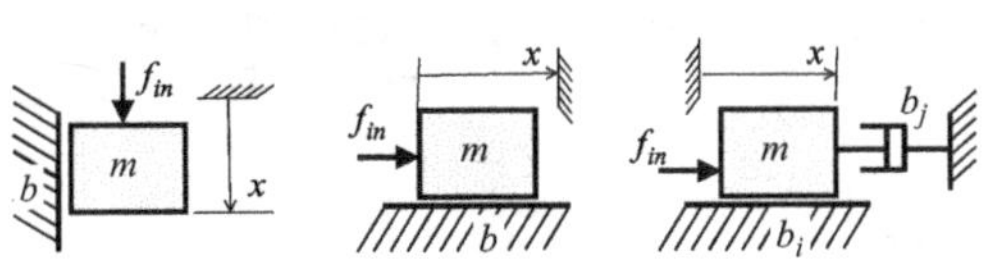

Figure 5.10 Viscous friction forces affecting the motion of a mass, m

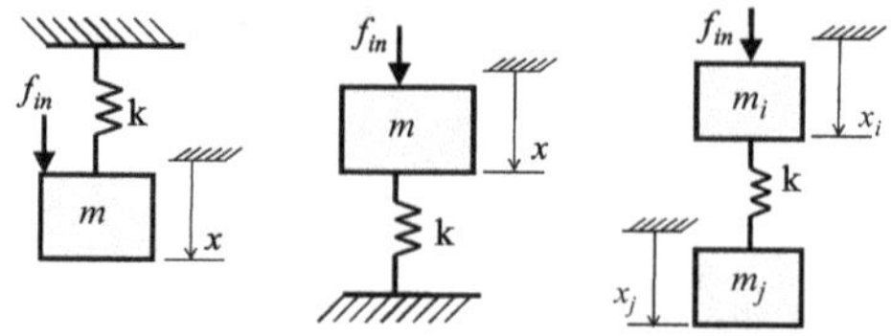

Figure 5.11 Spring forces affect the motion of a mass, m

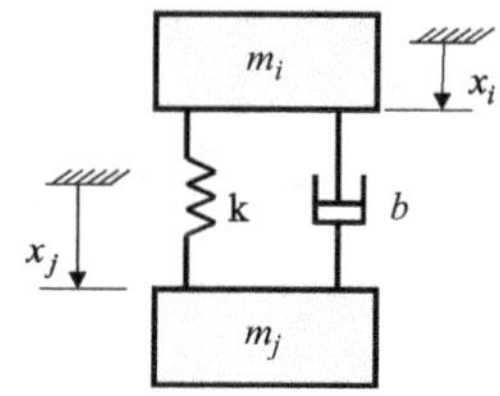

Figure 5.12 A two-mass system with spring and damper

spring-like behavior by resisting the motion of the mass as shown in Figure 5.11. Therefore, this force is called as spring force as defined in (5.24) and always shows a negative effect on the motion.

$$f_s(t) = kx(t) \tag{5.24}$$

The fixed steady points have a similar meaning as the ground in electrical systems. They are shown by hashed lines in Figures 5.10 and 5.11.

Spring and viscous friction forces may exist alone or at the same time depending upon the mechanical system structure and complexity. Figures 5.10 and 11 show some cases of how the viscous friction and spring might be represented in the schematic diagram of a mechanical system.

Mechanical systems and electrical systems both have three passive linear components: R, L, and C in electrical and the mass m, the spring k, and friction b in mechanical systems. Both systems are analogous. Two of the components namely the mass and the spring are energy storage elements (like inductor L and capacitor C) and the damper b is energy dissipater like the resistor.

Let us use the simple mechanical scheme given in Figure 5.12 to write the differential equation for a mass, spring, and damper system. The mechanical system scheme has two masses connected to each other by a spring and a damper, which is considered as viscous friction force.

Let x_i and x_j be the motion of a bodies m_i and m_j, respectively. Let k_i and b_i be the coefficients of the spring and damper attached to the mass m_i.

Then the opposing spring force affecting the motion of m_i is written as in (5.25) and affecting the motion m_j is written as in (5.26).

$$f_{si}(t) = k\left(x_i(t) - x_j(t)\right) \tag{5.25}$$

$$f_{sj}(t) = k\left(x_j(t) - x_i(t)\right) \tag{5.26}$$

Then the opposing damping force affecting the motions of m_i and m_j are written as in (5.27) and (5.28), respectively.

$$f_{bi}(t) = b(\dot{x}_i(t) - \dot{x}_j(t)) \tag{5.27}$$

$$f_{bj}(t) = b(\dot{x}_j(t) - \dot{x}_i(t)) \tag{5.28}$$

Many mechanical systems are analogous to multi-loop and multi-node electrical circuits. For example, assume that two loop currents $i_i(t)$ and $i_j(t)$ are passing through a resistance R in the opposite direction. Then consider writing the voltage across the resistance for loop i and for loop j. The resultant equations will be similar to those given in (5.25) and (5.26) with k replaced by R, x_i by i_i, and x_j by i_j. If the element is inductor L instead of resistor R, then the resultant equations become similar to those in (5.27) and (5.28).

Both electrical and mechanical systems are modeled by a set of simultaneous differential equations and a set of algebraic equations. The number of equations in electrical systems is equal to the number of independent loops or nodes. In mechanical systems, the number of equations of motion is equal to the number of *linearly independent* motions. Linearly independent motion implies that a point of motion in a system can still move if all other points of motion are held fixed. A motion is said to be dependent if it can be expressed as a linear combination of the other. Usually, the number of linearly independent motions is called the number of *degrees of freedom*.

Note the degrees of freedom do not preclude motions that are coupled. For example in multi-loop circuits, each loop current may depend upon other loop currents, but if one of the loops is open-circuited, the other loop currents will still exist if there is a voltage source in that loop. Similarly in a mechanical system with multi-degrees of freedom, one point of motion can be held while another point of motion moves under an applied force. Newton's law of motion is applied to each inertia element as long as the motion associated with the inertia element is independent. In other words, Newton's law of motion is not applied to an inertia element if its motion is linearly dependent on the motion of other inertia elements. The equation of motion may be obtained by using the principle of superposition. Consider each mass and obtain its equation of motion using Newton's law by assuming all other motions are fixed. If the mass m_i is considered, then the equation of its motion is obtained by assuming that the motion associated with the mass m_j is fixed. Similarly, for the mass m_j obtain the equation of its motion assuming that the motion associated with the mass m_i is fixed.

The following steps can be used to obtain a mathematical model of a mechanical system:

(I) Label input–output and internal variables.
(II) Consider each mass separately.
(III) Identify the cause–effect relationship. The applied force is usually the cause and motion is the resultant effect.
(IV) Choose a reference direction of motion.

(V) Recall that the spring and damping elements oppose the motion.
(VI) Write Newton's law with acceleration term to the left and all external forces to the right.
(VII) Sketch the simulation diagram.

Example 5.5

A single mass with a spring and viscous friction is given in Figure E5.5(a) and (b). The input force is applied vertically dawn in Figure E5.5(a) and vertically up in Figure E5.5(b).

Write the motion equations of this system for both figures assuming that

(a) The reference direction is vertically down
(b) The reference direction is vertically up

Answer 5.5

a.1 The reference direction is vertically down for Figure E5.5(a).

Let us follow the design steps given above.

- The input is the applied force $f_{in}(t)$ and the output is the position $x(t)$ in this system. There is another input force to this system due to gravity. The gravity force *mg* occurs naturally and is considered as a disturbance input to the system. The gravity force may be positive or negative depending on the reference direction. Since both gravity and reference directions are downward in this part, the gravity force becomes positive.
- There is only one mass. Therefore, this system has one degree of freedom.
- The applied force $f_{in}(t)$ causes the mass m to accelerate and move with a displacement x in meters. Direction of the applied force is downward, which is the reference direction. Therefore, the applied force is considered positive according to the reference direction.
- The spring and damping elements always oppose to the motion. So, the forces caused by the spring and damper will be negative.

$$f_s(t) = -kx(t) \tag{E5.5a}$$

$$f_b(t) = -b\dot{x}(t) \tag{E5.5b}$$

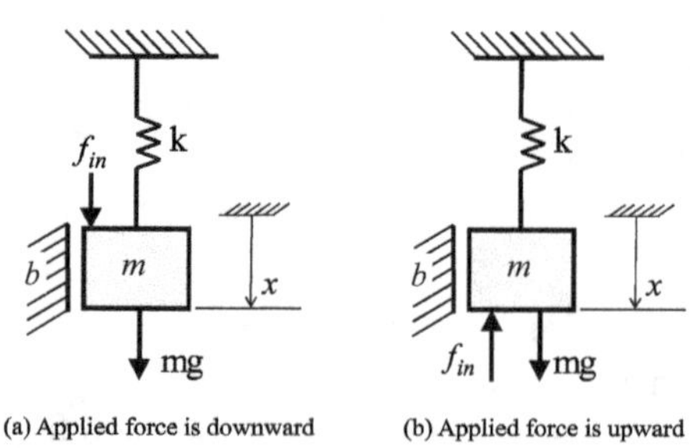

Figure E5.5 The single mass system for Example 5.5. (a) Applied force is downward (b) applied force is upward.

- The motion equation can be written as in (E5.5c) where Newton's law with acceleration term is written to the left and all external forces to the right

$$m\ddot{x}(t) = +f_{\text{in}}(t) + mg - kx(t) - b\dot{x}(t) \tag{E5.5c}$$

- The reference direction is vertically down for Figure E5.5(b).
  The reference direction is still the same. However, the input force is applied vertically up in the opposite direction of the reference. Therefore, the sign of the applied force is negative now resulting in a motion equation as in (E5.5d).

$$m\ddot{x}(t) = -f_{\text{in}}(t) + mg - kx(t) - b\dot{x}(t) \tag{E5.5d}$$

b.1 The reference direction is vertically up for Figure E5.5(a).
The reference direction is vertically up, the applied force is vertically down, the gravity force is naturally vertically down and the forces caused by the spring and damper are always negative. All forces in this condition are opposing to the reference. Therefore, the motion equation is written as in (E5.5e).

$$m\ddot{x}(t) = -f_{\text{in}}(t) - mg - kx(t) - b\dot{x}(t) \tag{E5.5e}$$

b.2 The reference direction is vertically up for Figure E5.5(b).
The reference direction is vertically up, the applied force is vertically up, gravity force is naturally vertically down and the forces caused by spring and damper are always negative. Therefore, the motion equation is written as in (E5.5f).

$$m\ddot{x}(t) = +f_{\text{in}}(t) - mg - kx(t) - b\dot{x}(t) \tag{E5.5f}$$

Example 5.6

A two-mass system is given in Figure E5.6. The input forces $f_1(t)$ and $f_2(t)$ are applied vertically dawn. The direction of the applied forces is also assumed to be the reference direction. Write the motion equations of this system.

Answer 5.6

The system in this example is a two-mass system with two degrees of freedom. Therefore, two motion equations can be written for this system by applying Newton's law to both systems separately.

The motion equation of the mass m_1:

$$m_1\ddot{x}_1 = f_1 + m_1g - k_1x_1 - b_1\dot{x}_1 - k_2(x_1 - x_2) \tag{E5.6a}$$

The motion equation of the mass m_2:

$$m_2\ddot{x}_2 = f_2 + m_2g - k_2(x_2 - x_1) \tag{E5.6b}$$

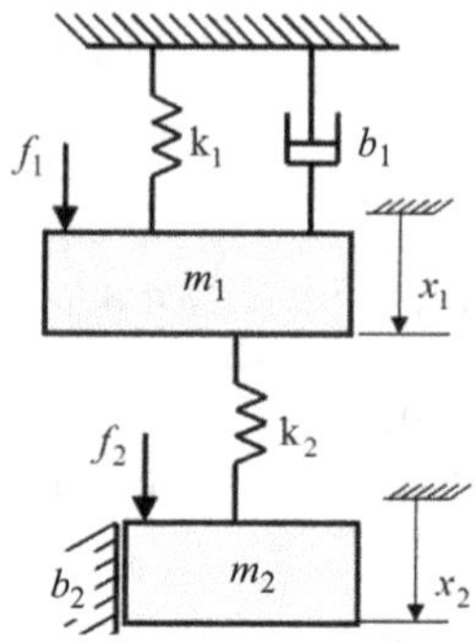

Figure E5.6 A two-mass system for Example 5.6

The spring k_2 between the masses m_1 and m_2 is a common element connecting them together both physically and mathematically.

Example 5.7

A linear motion system with one mass is given in Figure E5.7(a). The input force f_{in} is applied to move the system. The direction of the applied force is also assumed to be the reference direction.

(a) Write the motion equations of this system.
(b) Draw a simulation diagram for this system, and
(c) Write state-space equations of this system.

Answer 5.7

In this system, a rigid body with a mass m is moved linearly by applying a force f_{in}. While the body m is moving a viscous friction b shows a resistance to the motion. Therefore, the friction effect is considered as a resisting force with a negative effect on the motion. Since one of the ends of the rotational shaft is fixed, this end has a spring k, which also resists the motion. It is clear that both spring and frictional forces are negative being independent from the reference and applied force directions.

(a) Let us write the motion equations of this system.

$$m\ddot{x}(t) = +f_{\text{in}}(t) - kx(t) - b\dot{x}(t) \tag{E5.7a}$$

(b) Let us draw a simulation diagram for this system using the simulation diagram drawing approach given in Example 5.3. The highest order differential term is already written on the left in (E5.7a). Force f_{in} (kg) is the input to the system

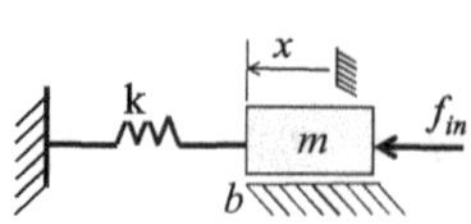

Figure E5.7(a) A rotational system with one inertia

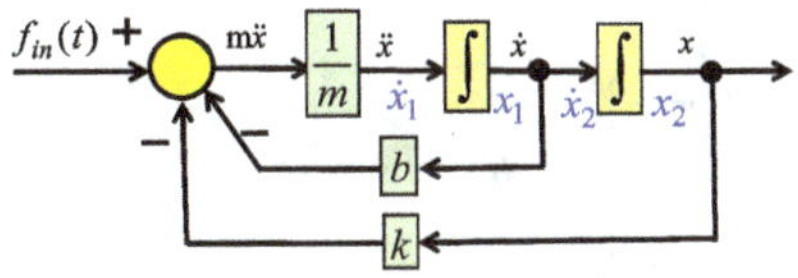

Figure E5.7(b) Simulation diagram of the system given by (E5.7a)

and linear displacement x (m) is the output. The first and the second derivations of the position x are the internal variables and are used to complete the simulation diagram as in Figure E5.7(b), so that (E5.7a) is represented completely.

(c) Let us write state-space equations of this system using the simulation diagram obtained in Figure E5.7(b).

The following equations can be written directly from Figure E5.7(b).

$$\dot{x}_1(t) = \frac{1}{m} f_{\text{in}}(t) - \frac{b}{m} x_1(t) - \frac{k}{m} x_2(t) \tag{E5.7b}$$

$$\dot{x}_2(t) = x_1(t) \tag{E5.7c}$$

$$y(t) = x_2(t) \tag{E5.7d}$$

where $x_1(t) = dx(t)/dt$, which is the velocity (m/s) of the system and $x_2(t)$ is the linear displacement $x(t)$ in meters. Then the state-space equations can be written in matrix format.

$$\begin{bmatrix} \dot{x}_1(t) \\ \dot{x}_2(t) \end{bmatrix} = \begin{bmatrix} -\frac{b}{m} & -\frac{k}{m} \\ 1 & 0 \end{bmatrix} \begin{bmatrix} x_1(t) \\ x_2(t) \end{bmatrix} + \begin{bmatrix} \frac{1}{m} \\ 0 \end{bmatrix} u(t) \tag{E5.7e}$$

$$y(t) = \begin{bmatrix} 0 & 1 \end{bmatrix} \begin{bmatrix} x_1(t) \\ x_2(t) \end{bmatrix} \tag{E5.7f}$$

where $u(t) = f_{\text{in}}(t)$. A comparison of the simulation diagram and state-space model equations of this example with those of Example 5.3 is suggested.

5.4.2 Mechanical systems with rotational motion

The systems with rotational motion are analogs with the linear motion systems. Such that, the mass m in linear motion is replaced by inertia J in rotational motion. The force f is replaced by torque T and the linear displacement x is replaced by angular displacement θ. The effect of the gravity force is included in the inertia and not considered as an additional disturbance input. Example 5.8 shows the modeling process of the rotational mechanic systems.

Example 5.8

A rotational system with one inertia is given in Figure E5.8(a). The input torque T_{in} is applied to rotate the system. The direction of the applied torque is also assumed to be the reference direction.

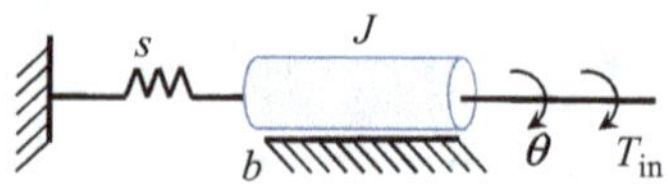

Figure E5.8(a) A rotational system with one inertia

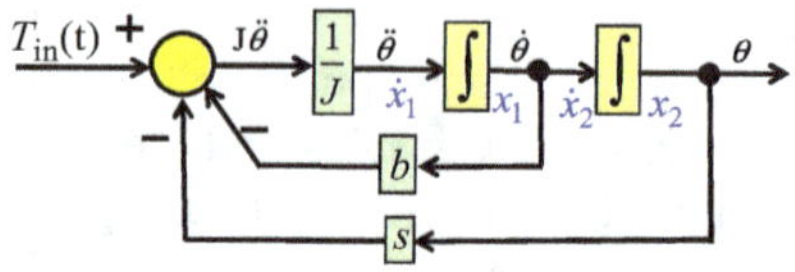

Figure E5.8(b) Simulation diagram of the rotational system given in (E5.8a)

(a) Write the motion equations of this system.
(b) Draw a simulation diagram for this system, and
(c) Write state-space equations of this system.

Answer 5.8

In this system a rigid body with an inertia J is rotated around an axis by applying a torque T_{in}. While body J is rotating viscous friction b shows a resistance to the rotational motion. Therefore, the friction effect is considered as a resisting torque with a negative effect on the motion. Since one of the ends of the rotational shaft is fixed, this end has a stiffness s, which also resists the rotational motion similar to the spring effect. It is clear that both stiffness and frictional torques are negatively independent of the reference and applied torque directions.

(a) Let us write the motion equations of this system.
Under these comments, the motion equation for the system described in this example is written as in (E5.8a).

$$J\ddot{\theta} = T_{in} - s\theta - b\dot{\theta} \tag{E5.8a}$$

(b) Let us draw a simulation diagram for this system using the simulation diagram drawing approach given in Example 5.3. The highest order differential term is already written on the left in (E5.8a). Torque T_{in} (Nm) is the input to the system and angular position θ (rad) is the output. The first and the second derivations of the position angle θ are the internal variables and are used to complete the simulation diagram as in Figure E5.8(b), so that (E5.8a) is represented correctly.
(c) Let us write state-space equations of this system using the simulation diagram in Figure E5.8(b).

The following equations can be written directly from Figure E5.8(b).

$$\dot{x}_1(t) = \frac{1}{J}T_{\text{in}}(t) - \frac{b}{J}x_1(t) - \frac{s}{J}x_2(t) \tag{E5.8b}$$

$$\dot{x}_2(t) = x_1(t) \tag{E5.8c}$$

$$y(t) = x_2(t) \tag{E5.8d}$$

where $x_1(t) = d\theta(t)/dt$, which is the rotational speed (rad/s) of the system and $x_2(t)$ is the rotational displacement $\theta(t)$ in radians. Then the state-space equations can be written in matrix format.

$$\begin{bmatrix} \dot{x}_1(t) \\ \dot{x}_2(t) \end{bmatrix} = \begin{bmatrix} -\frac{b}{J} & -\frac{s}{J} \\ 1 & 0 \end{bmatrix} \begin{bmatrix} x_1(t) \\ x_2(t) \end{bmatrix} + \begin{bmatrix} \frac{1}{J} \\ 0 \end{bmatrix} u(t) \tag{E5.8e}$$

$$y(t) = \begin{bmatrix} 0 & 1 \end{bmatrix} \begin{bmatrix} x_1(t) \\ x_2(t) \end{bmatrix} \tag{E5.8f}$$

where $u(t) = T_{\text{in}}(t)$. A comparison of the simulation diagram and state-space model equations of this example with those of Examples 5.3 and 5.7 is suggested.

Example 5.9

A rotational system with two inertias is given in Figure E5.9. The input torque T_{in} is applied to rotate the system. The direction of the applied torque is also assumed to be the reference direction. Write the motion equations of this system.

Answer 5.9

In this system, two rigid body with the inertias J_1 and J_2 are rotated around an axis by applying a torque T_{in}. While the bodies with the inertias J_1 and J_2 are rotating two viscous frictions b_1 and b_2 show resistance to the rotational motion. Besides, two stiffnesses affect the motion in the negative direction. The stiffness s_2 is a common element between two inertias and must be included in the equations of both motions. Under these comments, the motion equation can be written as in (E5.9a) and (E5.9b).

$$J_1\ddot{\theta}_1 = -b_1\dot{\theta}_1 - s_1\theta_1 - s_2(\theta_1 - \theta_2) \tag{E5.9a}$$

$$J_2\ddot{\theta}_2 = -b_2\dot{\theta}_2 - s_2(\theta_2 - \theta_1) \tag{E5.9b}$$

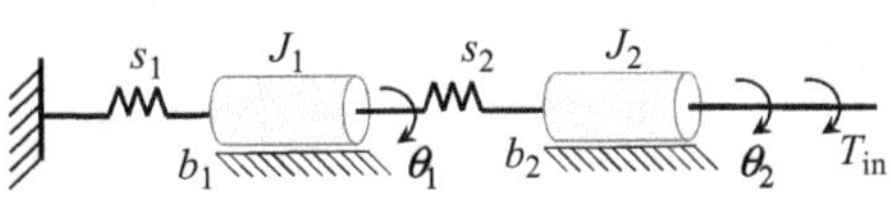

Figure E5.9 A rotational system with two inertias

Practice 5.2

Compare the simulation diagrams and state-space model equations of the series-connected RLC circuit in Example 5.3, the linear motion mechanical system in Example 5.7, and the rotational system given in Example 5.8. Comment on the similarities between the modeling of both systems, in terms of simulation diagrams and state-space model equations. Identify the elements that are analogous to each other. Is it possible to use equivalent electrical circuits to represent the mechanical systems' linear or rotational motion?

5.5 Modeling electromechanical systems

One of the most used and the most known electromechanical systems in electrical power engineering is the electrical machines. They are used in industry either as electrical generators or electrical motors. Even though an electrical machine itself is an electromechanical system, it might be used as a part of larger electromechanical systems. Electrical motors are used in various parts of the industry. DC machines are especially used widely in robotic systems, numerical control machines, anti-aircraft radar tracking control systems, antenna positioning systems, and electrical car drive systems. In this section of the book, the modeling of DC motors is discussed and their simulation models are derived.

A DC motor is based on fundamental laws of electromagnetism. If a conductor of length, l,carrying a current, i, is placed in a uniform magnetic field of flux density B, it will experience a force, f, given by Ampere's law as in (5.29) when the conductor is oriented at right angles to the magnetic field.

$$f = Bi\ell \tag{5.29}$$

In other words, if a 1-m conductor carrying one Ampere current is placed at right angles to a magnetic field with a flux density of one Tesla (Weber/m^2), a force of one Newton is generated affecting the conductor to move. In terms of units, this force can be written as 1 Newton = 1 Tesla × 1 A × 1 m. A schematic diagram of the described system is shown in Figure 5.13.

In Figure 5.13, a voltage v is applied to a conductor with the length. This voltage causes a current i to flow through the conductor. Since the length ℓ of the conductor is placed in a magnetic field B, a force f is generated. This form of operation is called an electrical motor.

If a moving conductor cuts a magnetic field of flux density, B, with a velocity v then an electromagnetic force (emf), e, is induced in the conductor given by Faraday's law of electromagnetic induction as in (5.30). This mode of operation is called an electrical generator.

$$e = B\ell v \tag{5.30}$$

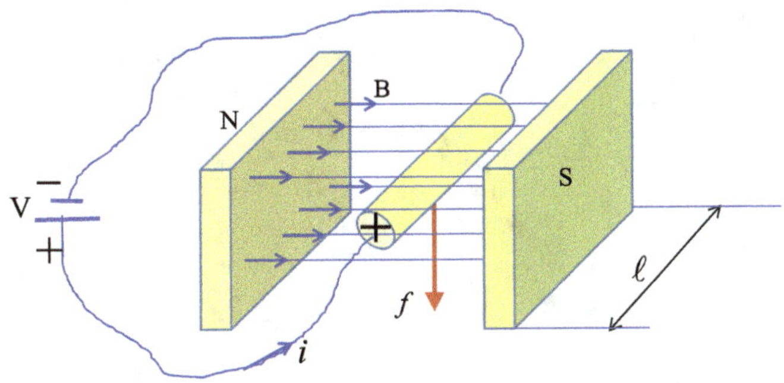

Figure 5.13 Force generation in a conductor

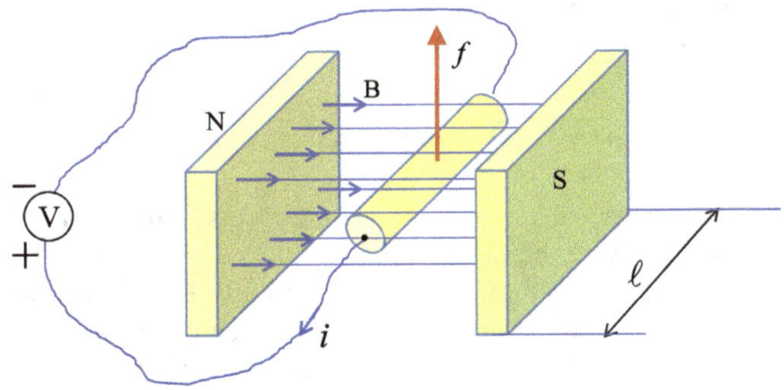

Figure 5.14 Voltage induction in a conductor

In the form of generator operation, a force f is applied to a conductor placed inside a magnetic field. With the effect of the applied force, the conductor moves in the magnetic field of flux density cutting the magnetic field lines. This cause and fact process is defined by Faraday's law and expressed as in (5.30). A schematic diagram of generator mode operation is depicted in Figure 5.14.

Now, let us place a single-turn coil in a uniformly distributed magnetic field and apply a voltage V so that a current i flows through the coil as depicted in Figure 5.15. Due to the flowing current, forces are generated on both sides of the coil as a result of Ampere's law. Since the coil is placed to rotate around an axis, the forces result in a torque T_e related to the distance d between two sides of the coil. The generated torque is expressed as in (5.31).

$$T_e = Bi\ell d \tag{5.31}$$

where ℓ is the length of the coil under the effect of magnetic field B, i is the current flowing in the coil and d is the distance between two sides of the coil. If the coil in Figure 5.15 has N turns instead of one, the torque equation becomes as in (5.32).

$$T_e = NBi\ell d \tag{5.32}$$

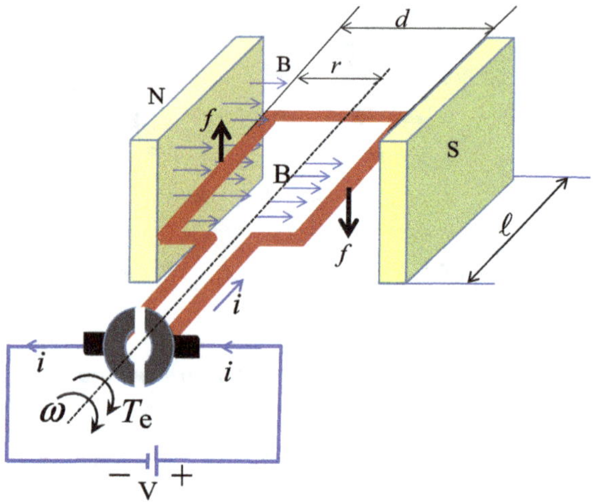

Figure 5.15 A simple electrical motor

As mentioned earlier, if a moving conductor cuts a magnetic field of flux density, B, with a velocity v then an emf, e, is induced in the conductor given by Faraday's law of electromagnetic induction as in (5.30). Let us expand this theory to a coil with two sides rotating around an axis. Since there are two sides of the coil, the induced voltage becomes twice that for a single conductor. Due to the rotation around an axis, the velocity is replaced by an angular speed in rad/s. Then (5.30) is replaced by (5.33) for a coil with N turn.

$$e = 2NB\ell\omega r \tag{5.33}$$

Substituting $r = d/2$ from Figure 5.15 for r, the emf is written as

$$e = NB\ell\omega d \tag{5.34}$$

The emf is an induced voltage in the coil of an electrical machine due to Faraday's law of induction. This induced voltage has a negative effect on the applied voltage V according to Lenz law and is called back emf.

In the case of an electrical machine, the coil is placed on a rotatable body called rotor, and for DC machines, it is called armature. The required magnetic field is either generated by permanent magnets or electromagnets. Therefore, the DC machines requiring electromagnets need additional coils to generate the magnetic field. Therefore, except for the permanent magnet (PM) DC machines, the other DC machines have two coils, one is the armature coil where the power or torque is generated, and the other one is the field coil used for excitation to generate the magnetic field. These two coils, armature coil and field coil, are either fed separately or from the same source. If they are fed separately, then this machine is called a separately excited DC machine. If both coils are fed from the same source, then the machine is named to represent the connection type of the coils according to each other. If they are connected in series, then the machine is called a series excited DC machine. If the field coil is connected parallel to the armature, then the machine is called shunt excited DC machine. The properties of series and shunt-connected field coils are different. In a series DC machine, the field coil is supposed to carry the same current flowing through the armature current. Since the armature current is usually a higher current required for excitation, the series-connected field coil should be able to carry the armature current. Therefore, it has a larger diameter and a smaller number of turns compared to the shunt or separately excited field coils, which have a smaller diameter and higher number of turns. Some DC machines have both series and shunt-connected coils. These machines are called compound-excited DC machines.

Equations (5.31)–(5.34) can be adapted for DC machines considering both armature and field coils along with electromechanical torque and speed.

A separately excited DC machine in motor operating conditions can be represented with a general equivalent electromechanical system given in Figure 5.16.

V_a is the applied DC voltage to the armature. Because of V_a, a current i_a starts flowing through the armature coil, which is represented by its resistance R_a and inductance L_a in Figure 5.18. The magnetic field is generated by applying the field voltage V_f to the field coil, which has a resistance R_f and inductance L_f. The voltage V_f causes the field current i_f to flow in the field coil to generate the required magnetic field B. Once the magnetic field is generated, the armature coil under the

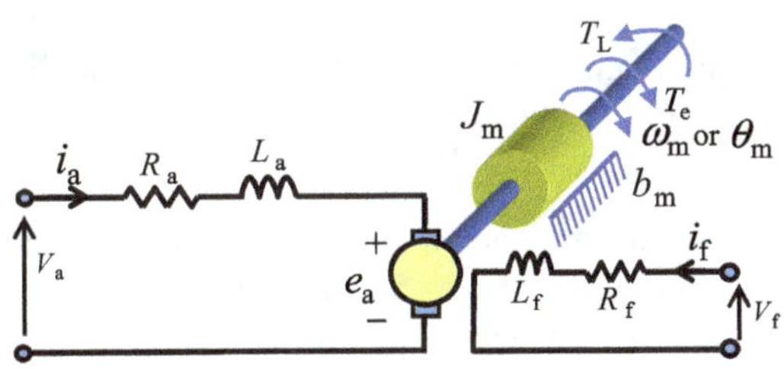

Figure 5.16 Electromechanical diagram of a separately excited DC motor

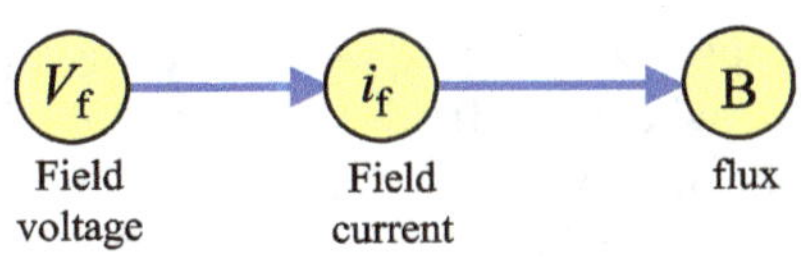

Figure 5.17 The cause-and-effect digraph of the field subsystem

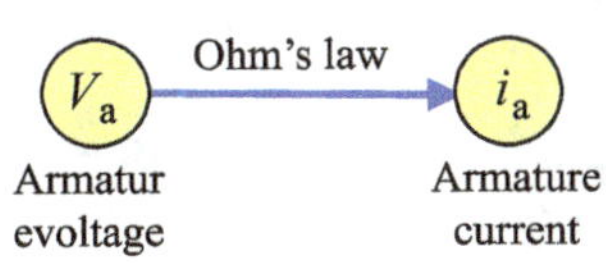

Figure 5.18 The cause-and-effect digraph of the armature subsystem

effect of the magnetic field generates a torque T_e due to current i_a flowing through, and the rotor where the armature coil is placed and represented by the inertia J_m, starts rotating with an angular velocity ω_m. When the armature coil starts rotating in the magnetic field, the back emf e_a is induced in the armature coil opposing the voltage V_a, which is the reason e_a is generated. The other parameters and variables in Figure 5.16 are the viscous friction b_m, rotor angle θ_m and T_L is the load torque.

In generator operating mode, the direction of armature current, i_a, and angular velocity, ω_m, change. Therefore, the electromechanical diagram given in Figure 5.16 can be used for both motor and generator operating conditions by taking the related directions into consideration.

The mathematical model equations of a separately excited DC motor are going to be written by developing a qualitative model based on the cause-and-effect relationships of the subsystem variables [1]. A qualitative model can be developed for DC motors based on Faraday's and Ampere's laws described above and given by (5.32) and (5.34). The qualitative model can be shown by a digraph obtained from cause-and-effect relationships. The state-space equations of a DC machine are obtained using armature circuit and torque equations, generally. This method may be a shorter approach going directly to the result. However, the use of cause-and-effect relationships will make a significant contribution to the better understanding of the subject instead of using the mathematical equations directly. Let us use the cause-and-effect relationships of the subsystems in separately excited DC motors to obtain a diagram representing the qualitative model.

Separately excited DC motor shown in Figure 5.16 can be represented by four subsystems as field subsystem, armature subsystem, mechanical subsystem, and electro-magnetic interaction subsystem.

Field subsystem

Assume that the motor is standstill and all initial conditions are zero. A field current i_f starts flowing in the field coil as soon as a field voltage V_f is applied to the circuit. The loop equation of the circuit is written as in (5.35) using Kirchhoff's voltage law.

$$\mathrm{V_f} = R_f i_f + L_f \frac{di_f}{dt} \tag{5.35}$$

The field current i_f causes a magnetic field generation due to the magnetizing characteristics of the core where the field coil is wounded. Due to the magnetic field, a flux density B occurs between the poles. This cause-effect process is depicted as a digraph in Figure 5.17.

The magnetic field density becomes constant if the field current is kept constant. In this case, the separately excited DC motor operates as if it is a PMDC motor.

Armature subsystem

Assume that the magnetic field density exists due to field voltage as shown in Figure 5.17 and the motor is stand still. If the armature voltage V_a is applied under these conditions, the armature current i_a starts flowing due to Ohm's law in armature circuit as represented by the digraph given in Figure 5.18.

When a voltage is applied to the armature of a DC motor at stand still, an armature current occurs due to resistance and inductance of the armature coil such that a loop equation is written as in (5.36) by applying Kirchhoff's voltage law.

$$\mathrm{V_a} = R_a i_a + L_a \frac{di_a}{dt} + e_a \tag{5.36}$$

where e_a is the back emf and it is equal to zero when the motor is stand still, i.e., when the speed is zero. Once the rotor starts rotating due to generated torque, e_a becomes nonzero and increases as the speed increases.

Mechanical subsystem

According to Ampere's law, an electromagnetic torque T_e is generated by the armature coil when there is an armature current as shown by digraph in Figure 5.18. The generated electromagnetic torque causes the rotor to rotate with an angular velocity of ω_m, which is in rad/s. A digraph representing this process is shown in Figure 5.19. Torque-speed equation of this mechanical subsystem is written as in (5.37) by applying Newton's second law.

$$T_e = b_m \omega_m + J_m \frac{d\omega_m}{dt} + T_L \tag{5.37}$$

Electromechanical interaction subsystem

If a DC voltage is applied to the field circuit, a field current occurs and a magnetic field is generated as described under the subtitle *field subsystem* above. When a magnetic field B is generated and the machine is stand still, the application of armature voltage causes the armature current i_a to flow in armature coil, which is under the effect of the magnetic field B. Since a current-carrying coil in a magnetic field generates torque according to Ampere's law as given in (5.32) and written for the motor as in (5.38).

$$T_e = NB\ell d i_a \tag{5.38}$$

A digraph representing the relationship between armature current i_a and generated torque T_e can be obtained as in Figure 5.20.

Due to the generated torque, the rotor starts rotating with an angular speed ω_m as shown in Figure 5.19. According to Faraday's law, if a coil is rotated in a magnetic field, a voltage is induced in this coil. Therefore, once the coil starts rotating due to the generated torque T_e in the magnetic field B, a back emf e_a is

Figure 5.19 The cause-and-effect digraph of the mechanical subsystem

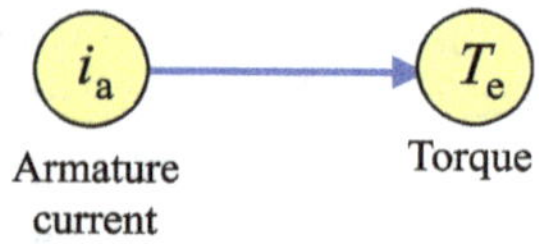

Figure 5.20 Digraph of the current–torque electromechanical interaction

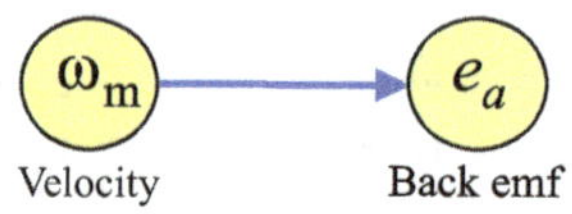

Figure 5.21 Digraph of the speed-back emf electromechanical interaction

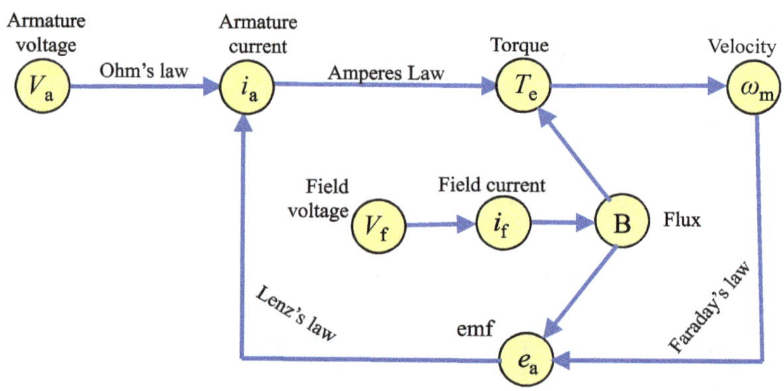

Figure 5.22 The cause-and-effect digraph of a separately excited DC motor

generated as given by (5.34), which is rewritten for the DC motor in (5.39).

$$e_a = NB\ell d\omega_m \tag{5.39}$$

A digraph representing the cause-and-effect relationship of angular velocity and the back emf is shown in Figure 5.21.

The induced voltage in the armature windings will be opposite direction to the voltage V_a leading to the production of first motion which is the reason for its formation in the first place. This event is called Lenz's law and it is actually a result of the principle of conservation of energy.

If the cause and effect relationships given in Figures 5.17–5.21 are combined, the cause-and-effect diagram representing the overall operation of a DC motor for both armature and field control cases is obtained as in Figure 5.22.

Since N is the number of turns of the armature coil, it is a constant quantity. Similarly, d is the rotor diameter and l is the rotor length, they depend upon the machine dimensions and are constants. Therefore, (5.38) and (5.39) can be written as in (5.40) and (5.41), respectively.

$$T_e = k_m k_f i_f i_a \tag{5.40}$$

$$e_a = k_m k_f i_f \omega_m \tag{5.41}$$

where

$$k_m = N\ell d \tag{5.42}$$

$$k_f = \frac{B}{i_f} \tag{5.43}$$

where k_f is the slope of the linear portion of the magnetizing curve between magnetic field B and excitation current i_f. Since both k_m and k_f are constants, they can be combined under a new constant k_e as in (5.44). Then (5.40) and (5.41) become as in (5.45) and (5.46)

$$k_e = k_m k_f \tag{5.44}$$

$$T_e = k_e i_f i_a \tag{5.45}$$

$$e_a = k_e i_f \omega_m \tag{5.46}$$

If the field current is kept constant then the magnetic field becomes constant. In this case, another constant parameter is defined as

$$k = k_e i_f \qquad (5.44)$$

$$T_e = k i_a \qquad (5.45)$$

$$e_a = k \omega_m \qquad (5.46)$$

If the magnetic field is kept constant, then the cause and effect diagram representing the field circuit can be removed from Figure 5.22 representing the digraph of a PMDC motor.

If the field current is not kept constant to have a more flexible control action, Figure 5.22 can be used for both armature and field-controlled separately excited machines. It should be noted that the use of parameters k_m, k_e, k_f, and k depends on the constancy of the field current. The cause-effect diagrams given in Figures 5.17 to 5.22 and equations from (5.35) to (5.46) can be used to model DC machines.

5.5.1 Modeling DC motors

Depending on application type, DC motors are controlled in terms of three variables, which are speed control, position control, and torque control. The control process of these variables is done through armature, field, or both armature and field. Therefore, the mathematical models of DC motors are obtained by considering the control objective and these three control methods. The models of DC motors given in this section are adapted from [13–16].

Armature-controlled DC motor

In this method, armature voltage is adjusted by keeping the magnetic flux density B constant, which means the field current is kept constant. Therefore, the equivalent electromechanical diagram of armature controlled DC motor becomes as in Figure 5.23 by removing the field circuit from Figure 5.16.

The electromechanical model of armature-controlled DC motor given in Figure 5.23 can be used for both armature-controlled separately excited DC motor and PMDC motor. The mathematical equations of the armature-controlled DC motor are obtained by considering the electrical system, electromechanical interactions, and mechanic system given in Figure 5.23. The use of cause-and-effect diagrams facilitates to obtain the mathematical equations. Given that excitation is constant, the overall cause and effect digraph in Figure 5.22 is converted to the one given in Figure 5.24. The mathematical model of the armature-controlled DC motor can be obtained by using this digraph.

The only difference between digraphs in Figures 5.22 and 5.24 is the non-existent of the field circuit and constant magnetic field. Therefore, armature circuit and mechanical equations given in (5.36) and (5.37), respectively, can be used

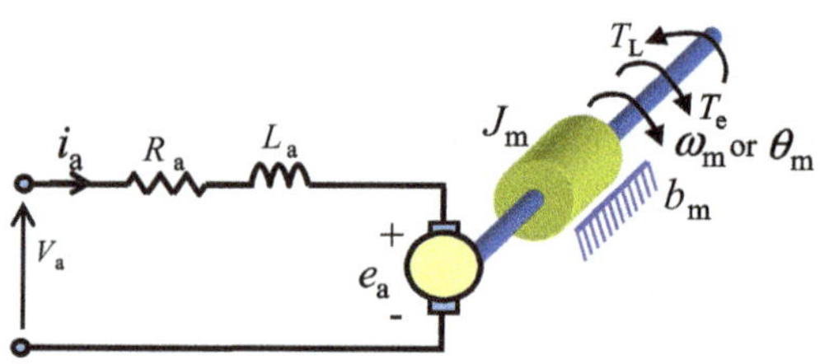

Figure 5.23 Electromechanical diagram of armature-controlled DC motor

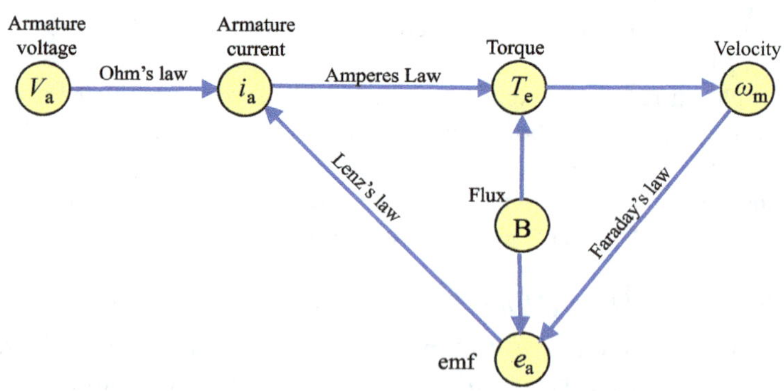

Figure 5.24 The overall cause and effect diagram of the armature-controlled DC motor

directly for the armature-controlled DC motor. Since the magnetic field is kept constant, Equations (5.45) and (5.46) should be used as electromechanical equations. Therefore, the mathematical model equations of an armature-controlled DC motor can be written as in (5.36), (5.37), (5.45), and (5.46), which are repeated by (5.47)–(5.50) in order to be used to obtain a simulation diagram for armature-controlled DC motor.

$$L_a \frac{di_a}{dt} = V_a - R_a i_a - e_a \tag{5.47}$$

$$J_m \frac{d\omega_m}{dt} = T_e - b_m \omega_m - T_L \tag{5.48}$$

$$T_e = k i_a \tag{5.49}$$

$$e_a = k\omega_m \tag{5.50}$$

The simulation diagram of armature-controlled DC motor can be obtained easily by using Equations (5.47)–(5.50). Since there are two first-order differential equations, there will be two integrators. The input to this system is the applied voltage V_a to the armature circuit and the rotor speed ω_m is the output since the speed control is the objective in this system. The load torque T_L is a disturbance input to the system. Since it is an armature-controlled DC motor, the speed control will be done through the armature voltage. The simulation diagram is shown in Figure 5.25. The position of the shaft can also be added to the simulation diagram by integrating the speed.

The state equations of the armature-controlled DC motor can be written directly from Figure 5.25.

$$\begin{bmatrix} \dot{x}_1 \\ \dot{x}_2 \\ \dot{x}_3 \end{bmatrix} = \begin{bmatrix} -\dfrac{R_a}{L_a} & -\dfrac{k}{L_a} & 0 \\ \dfrac{k}{J_m} & -\dfrac{b_m}{J_m} & 0 \\ 0 & 0 & 1 \end{bmatrix} \begin{bmatrix} x_1 \\ x_2 \\ x_3 \end{bmatrix} + \begin{bmatrix} \dfrac{1}{L_a} & 0 \\ -\dfrac{1}{J_m} & 0 \\ 0 & 0 \end{bmatrix} \begin{bmatrix} V_a \\ T_L \end{bmatrix} \tag{5.51}$$

where the variables x_1, x_2, and x_3 represent the armature current i_a, motor speed ω_m, and rotor position angle θ_m, respectively. The armature-controlled DC motor can be simulated either using the simulation diagram in Figure 5.25 or state-space Equation (5.51).

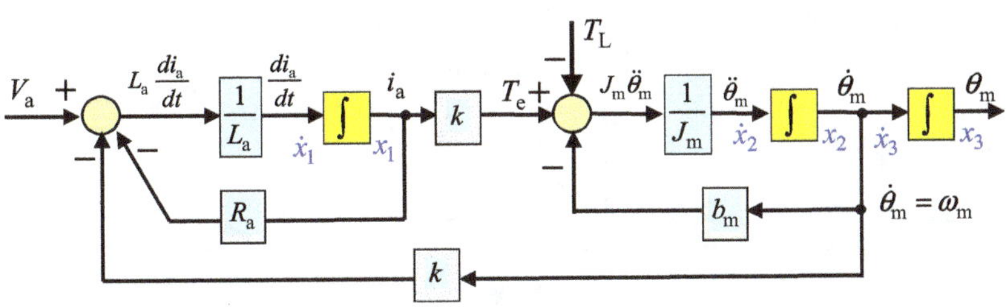

Figure 5.25 Simulation diagram of the armature-controlled DC motor

Computer simulation of armature-controlled DC motor has been given in Examples 4.7–4.10. The mathematical model of the PMDC motor is the same one obtained here. Therefore, similar examples are not repeated here.

Practice 5.3

Use the PMDC motor data given in Example 4.7 and simulate the simulation diagram given in Figure 5.25 and state-space model of (5.51) for the following cases.

(a) Develop an FLC for the speed control of this system and analyze the performance for step changes in reference signal.
(b) Repeat part (a) for ramp changes in the reference signal.
(c) Develop an FLC for the position control of this system and analyze the performance for step changes in reference signal.
(d) Repeat part (a) for ramp changes in the reference signal.
(e) Develop a simulation diagram for both armature and field-controlled separately excited DC motor.

Both armature and field controlled DC motor

In order to have more flexibility in controlling the speed, torque, and rotor position of separately excited DC motors, armature and field voltages are altered. Contrary to the just armature-controlled motor case, the field current is not constant anymore. Therefore, the magnetic field is not constant and it is a function of the field current i_f as shown in Figure 5.26 and given by (5.43) assuming that the relationship between excitation current and magnetic field is linear as shown by the dotted line with a slope of k_f.

The field control is generally used when the motor has to run above its rated speed. To understand the operation of field control, suppose that the DC motor is running at a constant speed. If the field current is reduced by reducing the voltage across the field coil, the flux density will be reduced. Reduction in flux will reduce the back emf instantaneously and will cause armature current to increase resulting in an increased motor speed. Consequently, the back emf will increase and a new equilibrium will be established at a higher speed. With field control, one can achieve as high a speed as

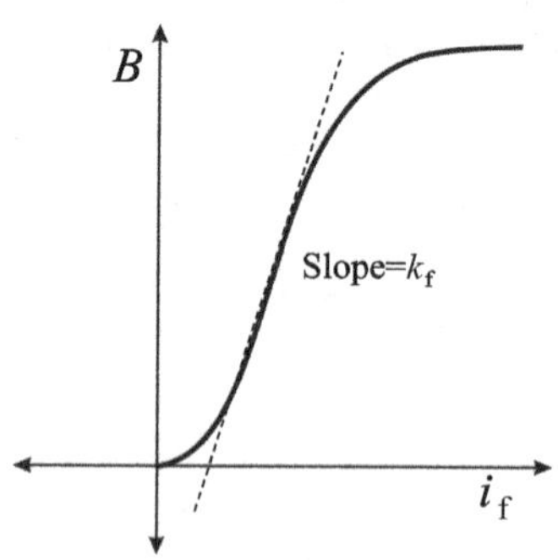

Figure 5.26 Magnetizing curve representing the B–i_f relationship

three times the rated speed. A DC motor with both field and armature control finds applications in traction systems. The inclusion of field control to the armature control helps to achieve speed control over a wide range of load torque variations.

Since the magnetic field is not constant, the electromechanical interaction equations given in (5.45) and (5.46) are used if both armature and field control are considered. Both armature and field circuits must be included in the modeling since the magnetic field is a function of the field current. The electromechanical circuit diagram of this operating case becomes the one given in Figure 5.16. Thus, the cause-and-effect diagram given in Figure 5.22 can be used to write mathematical model equations of the DC motor for the operating case under discussion. The field circuit, armature circuit, and mechanical torque equations can directly be written from (5.35), (5.36), and (5.37), respectively. Adding (5.45) and (5.46) as the electromechanical interaction equations, the mathematical model of an armature and field-controlled separately excited DC motor are obtained in the form of integro-differential equations as given by (5.52) to (5.56).

$$L_\mathrm{f}\frac{di_\mathrm{f}}{dt} = V_\mathrm{f} - R_\mathrm{f}i_\mathrm{f} \tag{5.52}$$

$$L_\mathrm{a}\frac{di_\mathrm{a}}{dt} = V_\mathrm{a} - R_\mathrm{a}i_\mathrm{a} - e_\mathrm{a} \tag{5.53}$$

$$J_\mathrm{m}\frac{d\omega_\mathrm{m}}{dt} = T_\mathrm{e} - b_\mathrm{m}\omega_\mathrm{m} - T_\mathrm{L} \tag{5.54}$$

$$T_\mathrm{e} = k_\mathrm{e}i_\mathrm{f}i_\mathrm{a} \tag{5.55}$$

$$e_\mathrm{a} = k_\mathrm{e}i_\mathrm{f}\omega_\mathrm{m} \tag{5.56}$$

where k_e is given in (5.44) as $k_\mathrm{e} = k_\mathrm{m}k_\mathrm{f}$ and k_m is given in (5.42) as $k_\mathrm{m} = N\ell d$. A qualitative model of armature and field-controlled DC motor is already obtained and shown in Figure 5.22 by a digraph, which is a macroscopic picture of the system. A generalized armature-field combined simulation diagram of a separately excited DC motor is obtained as shown in Figure 5.27 using (5.52)–(5.56).

A state-space model of the armature and field-controlled separately excited DC motor can be written directly from Figure 5.27. Let x_1, x_2, x_3, and x_4 be the state variables as shown in the output of each integrator in Figure 5.27. Since the inputs are the states of each variable, then the state-space equations can be written easily as given in (5.57).

$$\begin{bmatrix}\dot{x}_1\\ \dot{x}_2\\ \dot{x}_3\\ \dot{x}_4\end{bmatrix} = \begin{bmatrix} -\dfrac{R_\mathrm{a}}{L_\mathrm{a}} & -\dfrac{k_\mathrm{e}}{L_\mathrm{a}}x_3 & 0 & 0\\ 0 & -\dfrac{R_\mathrm{f}}{L_\mathrm{f}} & 0 & 0\\ \dfrac{k_\mathrm{e}}{J_\mathrm{m}}x_2 & 0 & -\dfrac{b_\mathrm{m}}{J_\mathrm{m}} & 0\\ 0 & 0 & 1 & 0\end{bmatrix}\begin{bmatrix}x_1\\ x_2\\ x_3\\ x_4\end{bmatrix} + \begin{bmatrix}\dfrac{1}{L_\mathrm{a}} & 0 & 0\\ 0 & \dfrac{1}{L_\mathrm{f}} & \\ 0 & 0 & -\dfrac{1}{J_\mathrm{m}}\\ 0 & 0 & 0\end{bmatrix}\begin{bmatrix}V_\mathrm{a}\\ V_\mathrm{f}\\ T_\mathrm{L}\\ 0\end{bmatrix} \tag{5.57}$$

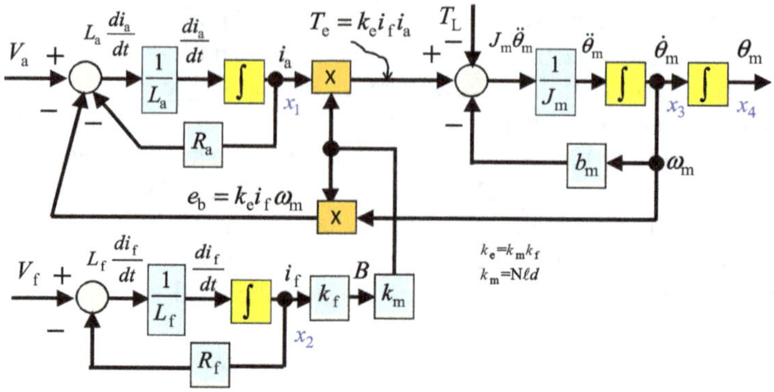

Figure 5.27 Simulation diagram of the armature and field-controlled DC motor

Figure 5.26 shows that there exists a nonlinear relationship between the flux density B and the current i_f in the field coil. Actually, the combined field and armature control system is nonlinear. This nonlinearity is depicted by multiplication blocks in Figure 5.27. The state matrix in (5.57) includes the variables x_2 and x_3 as multipliers. A DC motor with both field and armature control finds applications in traction systems. The inclusion of field control to the armature control helps to achieve speed control over a wide range of load torque variations.

Example 5.10

The unified armature and field-controlled DC motor can be modeled in Simulink as given in Figure E5.10(a) as a single block, whose inside view is shown in Figure E5.10(b). The inside view is categorized as armature circuit, field circuit, mechanical system, and electromechanical interaction system. Required data for the DC motor is given in the data subsystem as depicted in Figure E5.10(c).

Obtain the following:

(a) Develop a Simulink model for each one of the subsystems shown in Figure E5.10(b) as armature circuit, field circuit, mechanical subsystem, and electromechanical interaction system.
(b) Use the FLC model developed in Chapter 4 for the speed control of the system given in Figure E5.10(a).

Answer 5.10

(a) Simulink models of the subsystems given in Figure E5.10(b) as armature circuit, field circuit, mechanical subsystem, and electromechanical interaction system are given below.
(b) The use of the Simulink model of the FLC for the speed control of armature and field-controlled DC motor is shown in Figure E5.10(h).

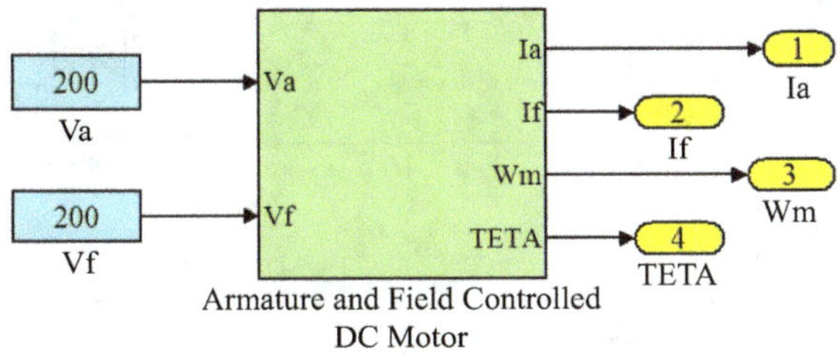

Figure E5.10(a) Simulink model of unified armature and field-controlled DC motor

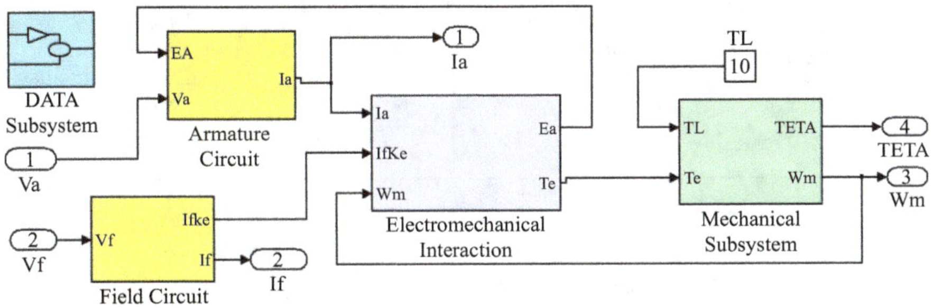

Figure E5.10(b) Subsystems of the Simulink model shown in Figure E5.10(a)

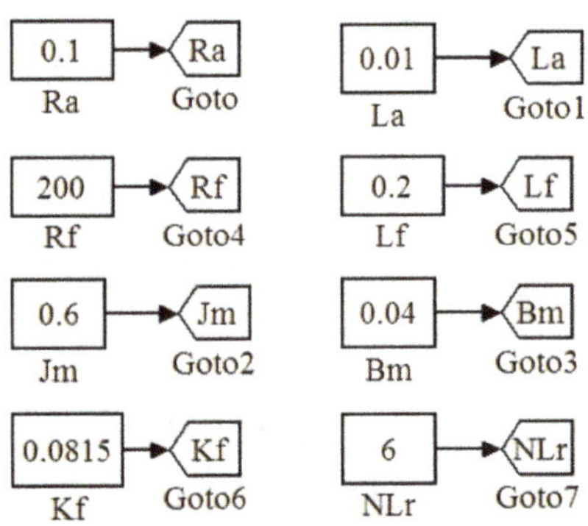

Figure E5.10(c) Inside view of the data subsystem shown in Figure E5.10(b)

The FLC block is used to control the speed of the motor. The control signal generated by the FLC is used in both armature and field control of the motor. Controlled speed response is shown in Figure E5.10(i). As shown in Figure E5.10(h), the same control signal is used in both armature and field control. Therefore, the same control signal results in the same voltage value in both circuits as an equilibrium voltage point. This approach restricts the

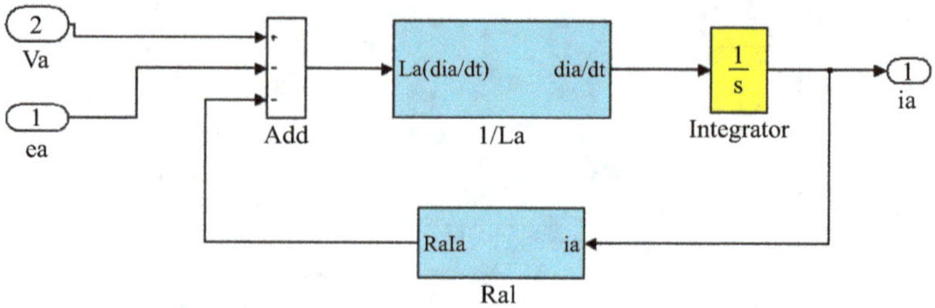

Figure E5.10(d) Simulink submodel of the armature circuit

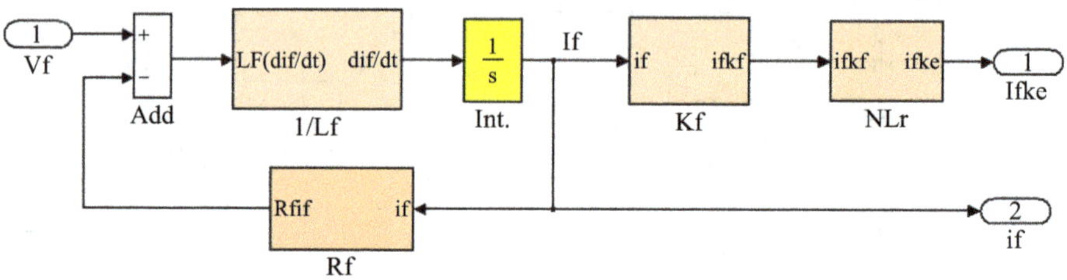

Figure E5.10(e) Simulink submodel of the field circuit

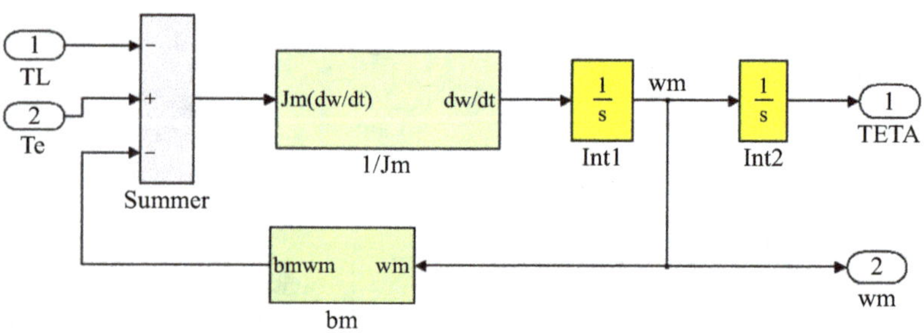

Figure E5.10(f) Simulink submodel of the mechanical subsystem

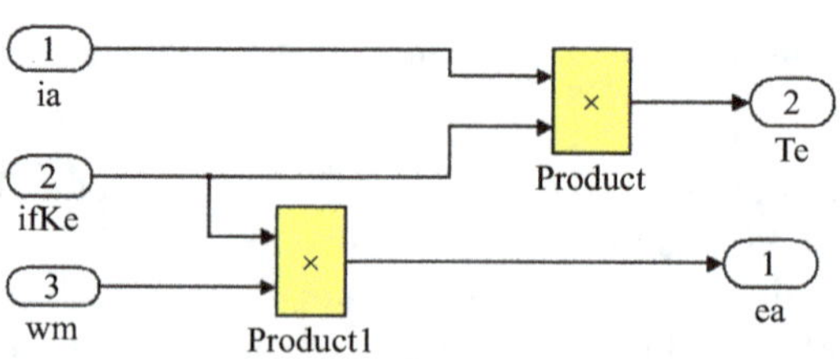

Figure E5.10(g) Simulink submodel of the mechanical subsystem

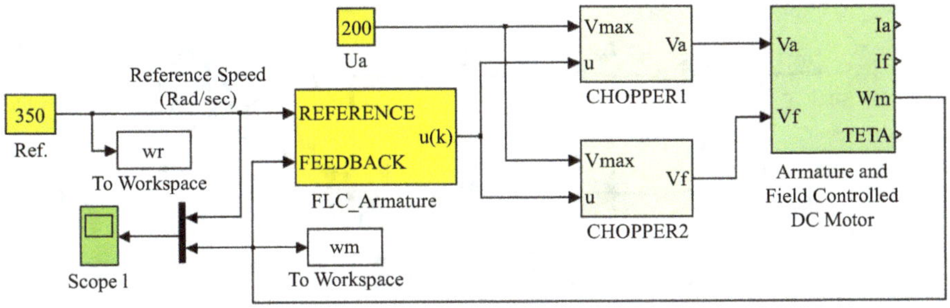

Figure E5.10(h) Simulink submodel of the FLC-controlled DC motor

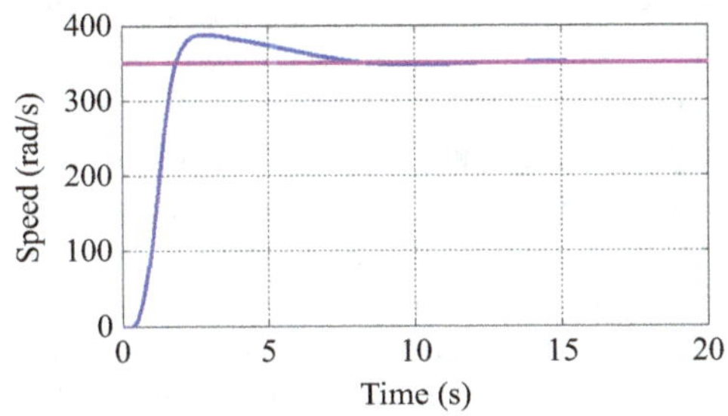

Figure E5.10(i) FLC controlled the speed response of the DC motor

flexibility of the controller. A better approach should be developed so that armature and field control are done separately. For example, in order to reach higher speeds above the rated value, the field voltage should be reduced while the armature voltage is increased.

Shunt-excited DC motor

Field coil in shunt-excited DC motor is connected in parallel to the armature circuit as shown in Figure 5.28. Two separate DC choppers or controlled rectifiers, which are represented by the controlled switches S_a and S_f, can be used to control the voltages to armature and field coils independently. One DC chopper or one controllable rectifier can be used to control both coil voltages at the same time, as well. Instead of one common controlled voltage, it is preferable to control both armature and field voltages independently.

The properties of the shunt-connected field coil are the same as that of separately excited machines. In other words, a separately excited DC motor can be used as shunt excited, too. The shunt resistance R_{fsh} has a larger value compared to armature resistance R_a such that a small amount of the total current i_m flows through the shunt excited coil while the majority of the total current flows in armature circuit. Therefore, the total motor current is written as in (5.58).

$$i_m = i_a + i_{fsh} \tag{5.58}$$

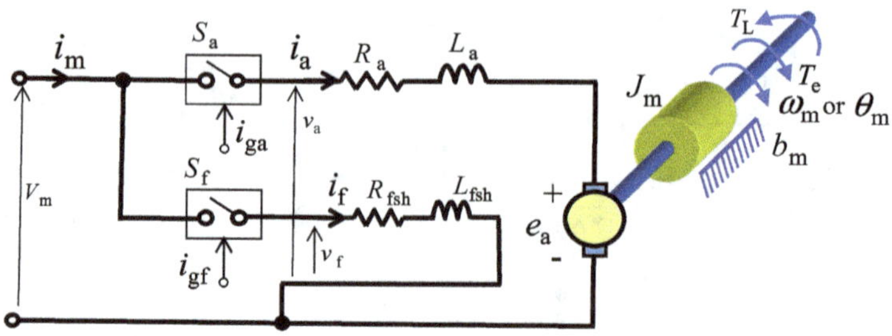

Figure 5.28 Electromechanical diagram of shunt excited DC motor

where $i_a > i_{fsh}$. If the controllable switches S_a and S_f in the armature and field circuits, respectively, are controlled separately, the operation of the shunt-excited DC motor becomes similar to that of a separately excited motor. Due to two separate controlled switches, armature, and field voltages can be controlled independently, resulting in different applied voltages for each circuit. Since lower field current results in higher speed while higher armature current generates higher torque and therefore higher speed, the control of the armature and field circuits separately, enable the motor to run above the rated speed. Therefore, proper control strategies should be developed for a more specific and flexible control action.

Series-excited DC motor

The field coil of a series-excited DC motor is different than separate or shunt-excited field coils. Contrary to the others, the series excited coil is bigger in diameter and has less number of turns with a smaller resistance. Since it is connected in series with the armature coil, the armature current flows through both coils. Therefore, the series-connected field coil should be able to carry the armature current. This implies that the armature current is also the excitation current. The changes in armature voltage affect both armature and field currents. The applied voltage to the series-excited DC motor is equal to the sum of armature and series field voltages as in (5.59).

$$v_m = v_a + v_f \tag{5.59}$$

The series-excited DC motor is controlled through the input voltage v_m by using a control device such as the controlled switch S_s in Figure 5.29.

The armature circuit equations of the series excited motor were rewritten as

$$(L_a + L_f)\frac{di_a}{dt} = V_a - (R_a + R_f)i_a - e_a \tag{5.60}$$

$$J_m\frac{d\omega_m}{dt} = T_e - b_m\omega_m - T_L \tag{5.61}$$

$$T_e = k_e i_f i_a \tag{5.62}$$

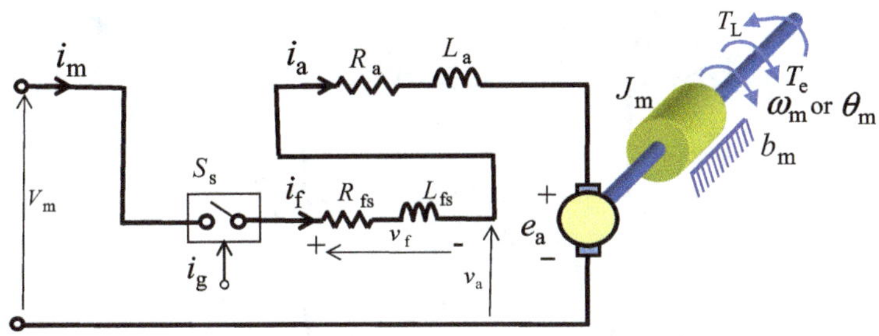

Figure 5.29 Electromechanical diagram of series-excited DC motor

$$e_a = k_e i_f \omega_m \tag{5.63}$$

$$i_f = i_a \tag{5.64}$$

where k_e is given in (5.44) as $k_e = k_m k_f$ and k_m is given in (5.42) as $k_m = N\ell d$. It should be noted that the field current is not constant and varies as the armature current changes due to changes in load torque and applied voltage. The mathematical model is nonlinear because of the terms $T_e = k_e i_a^2$ and $e_a = k_e i_f \omega_m$. Because of the nonlinearity, the speed, torque, or position control of series-excited DC motor require special designs.

Compound excited DC motor

The compound excited DC motor is a combination of the shunt and the series excited machines. The series and shunt excited coils result in a more flexible speed control by reducing the effect of the load torque changes. An electromechanical schematic diagram is given in Figure 5.30 with two control switches in armature and field circuits. The shunt excited coil is connected before the series coil in Figure 5.30. The series coil may be connected before the shunt coil depending on the use of purpose of the machine. The required magnetic field is generated both by the series field current, which is equal to the armature current and by the shunt field current. The polarity of the field coils must be taken into consideration during the connection of the field circuits so that the magnetic fields generated by the two coils do not conceal each other.

The mathematical model of the compound excited DC motor can be obtained using the cause-effect relationships of the submodels. The main difference between the compound excited machine and the others is the total effective magnetic field, which is the vectorial sum of the magnetic fields generated by series and shunt coils as in (5.65).

$$B = B_s + B_{sh} \tag{5.65}$$

where

$$B_s = k_{fs} i_{fs} \tag{5.66}$$

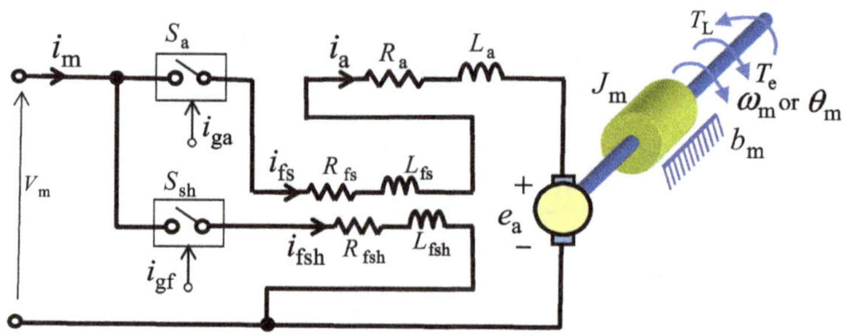

Figure 5.30 Electromechanical diagram of compound excited DC motor

$$B_{sh} = k_{fsh} i_{fsh} \tag{5.67}$$

The parameter k_{fs} is the slope of the magnetizing curve of $B_s = f(i_{fs})$ and k_{fsh} is the slope of the magnetizing curve of $B_{sh} = f(i_{fsh})$ similar to the magnetizing curve of $B = f(i_f)$ as shown in Figure 5.26. The full mathematical model of the compound-excited DC motor is obtained by combining the series and shunt-excited machine models.

A DC motor requires a power amplifier to drive it. High-speed, reliable, and inexpensive solid state semi-conductor devices such as diodes, thyristors, and recent devices such as gate turn-off thyristors (GTOs), bipolar junction transistors (BJTs), metal oxide field effect transistors (MOSFETs) and insulated gate bipolar transistors (IGBTs) have caused an important push up in the development of the DC motor drive amplifiers. These improvements in DC motor drive systems have increased the use of DC motors in industry and servo systems. The modeling of DC motors given in this chapter can be used to simulate them in MATLAB and Simulink by just integrating the models into the application.

5.5.2 Modeling AC motors

Mathematical modeling of DC motors is given in this chapter as examples of the modeling of electromechanical systems. Their modeling is straightforward and easy to follow the cause-and-effect relationships of the subsystems. However, the modeling of induction machines requires a reference frame to work out and unite the equations. The reference frame is used as a base for the equations to be transformed. Due to the complexity and large contest of the AC machines, their modeling is not going to be given in this book. However, in the chapters given later about the application of fuzzy logic controllers, the modeling and simulation studies of synchronous generators will be discussed.

5.6 Numerical solution of model equations

Modeling and simulation of electrical, mechanical, and electromechanical systems have been explained in previous sections of this chapter. The solutions of derived model equations in MATLAB and/or Simulink were given. MATLAB and Simulink are making the solutions easier by using subprogram codes as MATLAB functions for solutions. The users just need to put equations into proper forms such as transfer functions, state-space equations, or simulation diagrams. While MATLAB and Simulink make the solutions easier, the users generally do not care how those equations are solved in the background and they get away with the solution methods and their algorithms. This situation puts the engineers into difficulties if they need to solve model equations without using MATLAB and Simulink. Therefore, it would be better to include some numerical solution methods here, so that a solution code could be written in any programming language independent of MATLAB.

Two solution methods, which are Euler digital integration and the four-step Runge–Kutta algorithm are going to be explained here.

5.6.1 Euler digital integration

Euler digital integration given by (5.68) is a simple method to be applied in the solution of model equations of integral form as well as state-space form.

$$y_n = y_{n-1} + Tr_{n-1} \tag{5.68}$$

Euler digital integration is an iterative method based on sample-and-hold approach such that a new value is calculated at each iteration and held unchanged until the next sampling. Then the new calculated value from the next sampling takes over and is kept constant until the sampling comes after. Figure 5.31 shows the solution results from Euler integration and the actual curve. Where T is the sampling period or time increment at each iteration. As shown in Figure 5.31, the Euler solution remains unchanged during each sampling period of T until a new value is calculated in the next period.

There are other iterative methods such as trapezoid and Simpson rules to solve integro-differential equations. Trapezoid and Simpson rules are defined by (5.69) and (5.70), respectively. They give a better approximation compared to that of Euler's integration as could be understood from these definitions and from Figure 5.32. Although trapezoid and Simpson rules give more accurate solutions, they require more computer memory and iteration for the solution since they

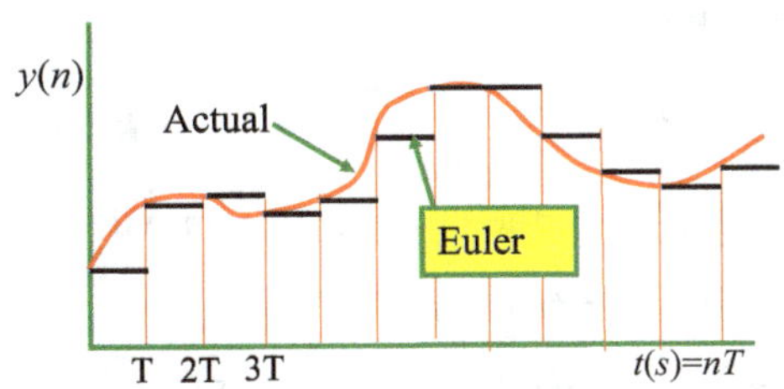

Figure 5.31 A comparison of the solution by Euler with the actual curve

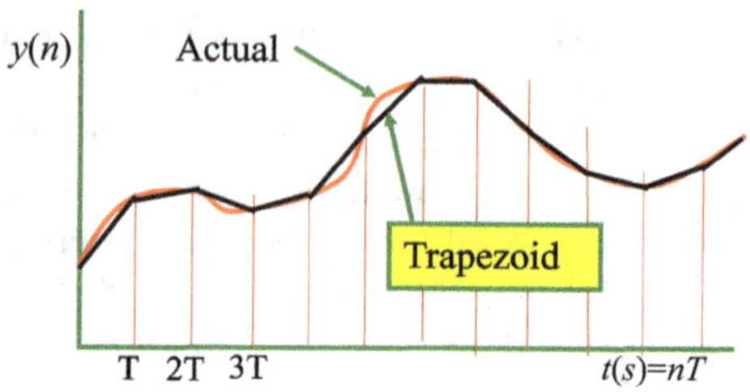

Figure 5.32 A comparison of the solution by trapezoid rule with the actual curve

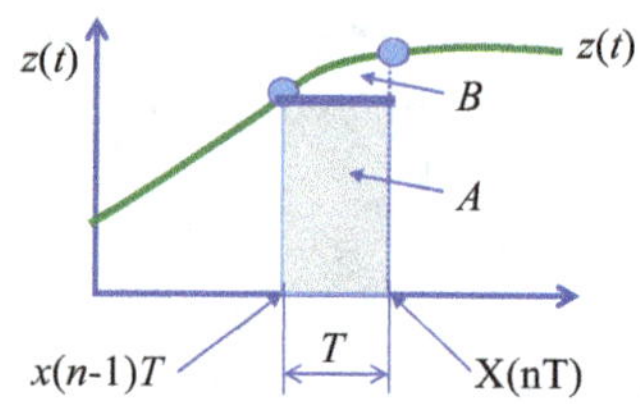

Figure 5.33 The rectangular area calculated by Euler's integration

require more data from previous samplings.

$$y_n = y_{n-1} + \frac{T}{2}(r_n + r_{n-1}) \tag{5.69}$$

$$y_n = y_{n-2} + \frac{T}{3}(r_n + 4r_{n-1} + r_{n-2}) \tag{5.70}$$

where y_n is the calculated output at the nth sampling, y_{n-1} and y_{n-2} are the calculated values from previous samplings, r_n is the current input at nth sampling, r_{n-1} and r_{n-2} are the input signals from previous samplings.

Equation (5.69) gives an output signal approximation for the trapezoid rule of integration as in Figure 5.32, which gives more accurate solution than Euler's integration.

Even if Euler's integration is not as accurate as the others, it is going to be explained and used in this section due to its simplicity and lower simulation time. Let us integrate the curve $z(t)$ in Figure 5.33 where the area under the curve can be obtained using (5.71) in the time domain.

$$x(t) = \int_0^t z(t)dt + x(0) \tag{5.71}$$

The equivalence of (5.71) in the discrete domain will be as in (5.72).

$$x(nT) = x[(n-1)T] + Tz[(n-1)T] \tag{5.72}$$

where $x(nT)$ is the sum of the calculated perpendicular areas under the curve until the nth sample. $x[(n–1)T]$ is the sum of the calculated perpendicular areas under the curve until the $(n-1)$th sample, and $Tz[(n-1)T]$ is the calculated area at the $(n-1)$th sample, which is shown by letter A in Figure 5.33. n is called iteration counter and T is the sampling period. Since T is a multiplier with all iteration instants, it can be removed from the equation to make it simpler.

$$x(n) = x(n-1) + Tz(n-1) \tag{5.73}$$

The second part of (5.73) is equal to the calculated area A during each sampling. The analytic solution at each iteration is equal to the sum of the areas as $A + B$. Therefore, area B is the calculation error of the Euler integration method. The calculation error can be made smaller if area B is reduced by selecting a smaller sampling time. However, the smaller sampling period results in more iteration and longer simulation time. Since the calculated area A is rectangular, the Euler method is also called rectangular integration.

Example 5.11

Use Euler integration to obtain a solution for (E5.11a). Assume that the initial condition $x(0) = 1$ and the sampling period is 0.1 s.

$$\dot{x}(t) + x(t) = 0 \tag{E5.11a}$$

P.S.: The analytical solution of this differential Equation is given as: $x(t) = e^{-t}$.

Answer 5.11

Let us reorganize (E5.11a) as in (E5.11b) and write it in discrete form as in (E5.11c)

$$z(t) = \dot{x}(t) = -x(t) \tag{E5.11b}$$

$$x(nT) = x[(n-1)T] + T\{-x[(n-1)T]\} \tag{E5.11c}$$

For $T = 0.1$ and $n = 1, \ldots, 10$, the solution is obtained as follows:

$$\begin{aligned}
&n = 1 \rightarrow x(0.1) = x(0) + Tx(0) = 1 - (0.1)(1) = 0.9\\
&n = 2 \rightarrow x(0.2) = x(0.1) + Tx(0.1) = 0.9 - (0.1)(0.9) = 0.81\\
&n = 3 \rightarrow x(0.3) = x(0.2) + Tx(0.2) = 0.81 - (0.1)(0.81) = 0.729\\
&\ldots\ldots\ldots\ldots\ldots\ldots\\
&n = 10 \rightarrow x(1.0) = x(0.9) + Tx(0.9) = 0.3487
\end{aligned} \tag{E5.11d}$$

Since the analytical solution is known, the accuracy can be checked for $t = 1$ s.

$$x(t) = x(1) = e^{-t} = e^{-1} = 0.3678 \tag{E5.11e}$$

The solution results are close to each other but not equal. This error could be reduced if the sampling period had been smaller such as 0.01 s. However, this time more iteration would have been needed to reach 1 s.

5.6.2 Euler integration for state-space equations

Euler integration can be used to solve state-space model equations, too. Let us consider integro-differential equation given in (5.74) and its discrete form in (5.75).

$$x_1(t) = \int_0^t \dot{x}(t)dt + x(0) \tag{5.74}$$

$$x(n) = x(n-1) + T\dot{x}(n-1) \tag{5.75}$$

where $x(0)$ is the initial value of the variable $x(t)$. The derivation $\dot{x}(t)$ in the integral part represents the state-space equations and is written as in (5.76) for continuous and (5.77) for discrete forms.

$$\dot{x}(t) = Ax(t) + Bu(t) \tag{5.76}$$

$$\dot{x}(n-1) = x(n-1) + Bu(n-1) \tag{5.77}$$

Since Euler integration is applied for digital simulation, (5.75) and (5.77) can be used together in order to solve state-space model equations. In order to obtain the solution numerically, the iteration starts by solving (5.77) and substituting the resultant solution into (5.75). The process is repeated in every iteration until the final result is obtained. This process is represented by the flowchart shown in Figure 5.34.

Example 5.12

Obtain a digital solution for the following state equations using Euler Integration. Assume all initial conditions as zero, approximation tolerance $\varepsilon = 0.001$ and use a sampling period of 0.01 s.

$$\begin{bmatrix} \dot{x}_1 \\ \dot{x}_2 \end{bmatrix} = \begin{bmatrix} -3 & 1 \\ -2 & 0 \end{bmatrix} \begin{bmatrix} x_1 \\ x_2 \end{bmatrix} + \begin{bmatrix} 0 \\ 1 \end{bmatrix} u \tag{E5.12a}$$

$$[y] = [1 \quad 0] \begin{bmatrix} x_1 \\ x_2 \end{bmatrix} + 0 \tag{E5.12b}$$

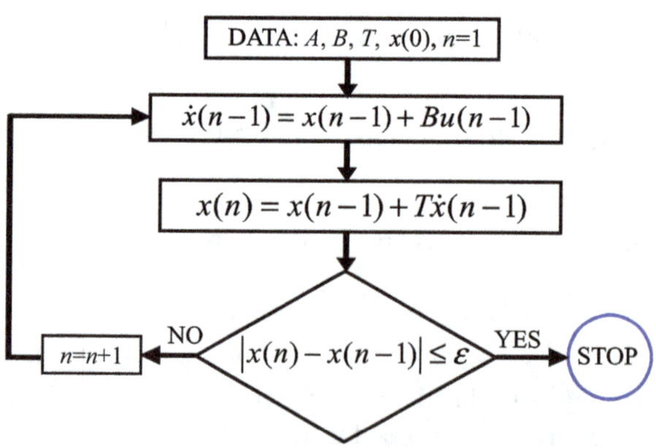

Figure 5.34 Flowchart of Euler's integration to solve state-space model equations

P.S.: Analytic solution of this second-order system is given below.

$$x_1(t) = \left(-e^{-t} + 2e^{-2t}\right)x_1(0) + \left(e^{-t} - e^{-2t}\right)x_2(0) + \frac{1}{2} - e^{-t} + \frac{1}{2}e^{-2t} \tag{E5.12c}$$

$$x_2(t) = \left(-2e^{-t} + 2e^{-2t}\right)x_1(0) + \left(2e^{-t} - e^{-2t}\right)x_2(0) + \frac{3}{2} - 2e^{-t} + \frac{1}{2}e^{-2t} \tag{E5.12d}$$

Answer 5.12

The MATLAB code given in Figure E5.12(a) can be used to solve the second-order state equations. Figure E5.12(a) is a software-coded version of the flowchart given in Figure 5.34. Any programming language can be used to realize the flowchart. The code shown in Figure E5.12(a) results in the plots given in Figure E5.12(b), where x_1 settles on 0.5 and x_2 settles on 1.5. These results comply with the analytic solution.

```
%**************************************************
% Example 5.12 - The Euler Integration
%**************************************************
clear
% DATA
  A=[ -3  1; -2  0];  B=[0; 1];  C=[ 1  0];  D=0;
  u=1;  T=0.01;
% Initial Values
  x10=0;  x20=0;  E=1000;  k=1;
% Iteration
while E > 0.0001
      dx1=A(1,1)*x10+A(1,2)*x20+B(1)*u;
      dx2=A(2,1)*x10+A(2,2)*x20+B(2)*u;
      x1(k)=x10+T*dx1;
      x2(k)=x20+T*dx2;
      E1=abs(x1(k)-x10);
      E2=abs(x2(k)-x20);
      E=max(E1,E2);
      x10=x1(k);    x20=x2(k);
      K(k)=k;       k=k+1;
end
plot(T*K,x1,T*K,x2);
xlabel('Time (s)');
ylabel('x1 and x2');grid
```

Figure E5.12(a) MATLAB code for Euler integration

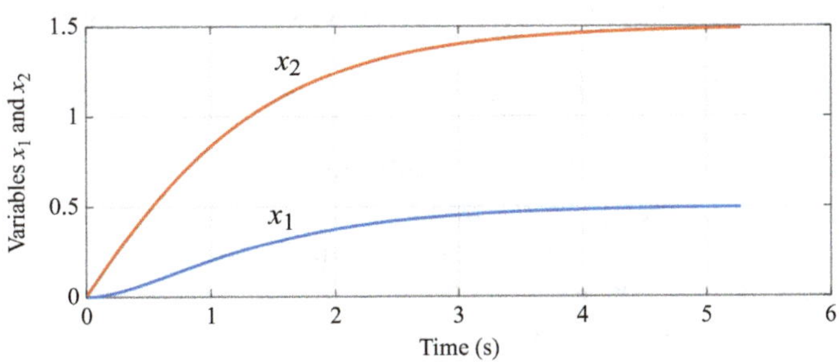

Figure E5.12(b) Solution results of Example 5.12 obtained by Euler integration

5.6.3 Runge–Kutta algorithm

There are several methods to be used to obtain approximate solutions of differential equations relating to both types of initial value and boundary value problems [18]. The Taylor series, Runge–Kutta, Adams–Bashforth, and Milne's methods are some of the well-known numerical methods used in the solution of differential equations [19]. Among those, the four-step Runge–Kutta method will be explained in this section.

Euler integration can be applied to solve ordinary differential equations more accurately by selecting a smaller iteration step size or sampling period. However, smaller step size requires more number of iterations to reach more accurate results. The smaller step size problem can be overcome by the Runge–Kutta method since it uses implicit and explicit iterative methods to reach the solution point. Especially four-step Runge–Kutta algorithm, which also includes Euler integration is very useful and widely used to solve higher-order ordinary differential equations including non-linear system equations, too.

The four-step Runge–Kutta algorithm uses four-line slope approximations to estimate the slope of the solution at time t_0. The line slopes are obtained by using (5.78)–(5.81).

$$a = f[x(t_0), t_0] \tag{5.78}$$

$$b = f\left[x(t_0) + a\frac{T}{2}, t_0 + \frac{T}{2}\right] \tag{5.79}$$

$$c = f\left[x(t_0) + b\frac{T}{2}, t_0 + \frac{T}{2}\right] \tag{5.80}$$

$$d = f[x(t_0) + cT, t_0 + T] \tag{5.81}$$

Then a weighted sum of the slopes *a*, *b*, *c*, and *d* gives the final estimate of $x(t_0 + T)$ as in (5.82).

$$x(t_0 + T) = x(t_0) + \frac{(a + 2b + 2c + d)T}{6} \tag{5.82}$$

$$x(t_0 + T) = x(t_0) + mT \tag{5.83}$$

where t_0 is the initial starting time and T is the step time or the sampling period. The other symbols are described below.

a: *a* is the slope at the beginning of the time step. This slope corresponds to Euler's method.

b: If slope *a* is used to halfway of the sampling time, slope *b* will be an estimate of the slope at the midpoint.

c: If the slope *b* is used halfway through the sampling time, then *c* becomes another estimate of the slope at the midpoint.

d: If the slope *c* is used all the way across the sampling time ($t_0 + T$), then *d* would be an estimate of the slope at the endpoint.

m: *m* is the weighted average slope.

5.6.4 Runge–Kutta algorithm for state-space equations

Let us adapt the four-step Runge–Kutta algorithm to the solution of state-space equations using the form (5.84).

$$\begin{bmatrix} \dot{x}_1 \\ \dot{x}_2 \end{bmatrix} = \begin{bmatrix} A_{11} & A_{12} \\ A_{21} & A_{22} \end{bmatrix} \begin{bmatrix} x_1 \\ x_2 \end{bmatrix} + \begin{bmatrix} B_{11} & B_{12} \\ B_{21} & B_{22} \end{bmatrix} \begin{bmatrix} u_1 \\ u_2 \end{bmatrix} \tag{5.85}$$

$$[y] = [C_{11} \quad C_{12}] \begin{bmatrix} x_1 \\ x_2 \end{bmatrix} + 0 \tag{5.86}$$

Since $\dot{x} = \frac{dx}{dt}$ and $dt = T$, then $\dot{x}_1$ and $\dot{x}_2$ can be written as in (5.87)

$$\begin{aligned} dx_1 &= dt[(A_{11}x_1 + A_{12}x_2) + (B_{11}u_1 + B_{12}u_2)] \\ dx_2 &= dt[(A_{21}x_1 + A_{22}x_2) + (B_{21}u_1 + B_{22}u_2)] \end{aligned} \tag{5.87}$$

where dx_1 and dx_2 are the changes of x_1 and x_2, respectively, over a sampling period. In order to obtain the solution for x_1 and x_2 the four-step Runge–Kutta algorithm is used as follows.

Table 5.1 has been developed for a second-order system. Since there were two variables we wrote two a slopes, two b slops, two c slopes, and two d slopes for the solution. If there were a 10th-order system, we would have to write 10 slope equations for each step of the Runge–Kutta algorithm. Therefore, a subprogram should be developed to get rid of repeating calculations and make things easier.

Table 5.1 Runge–Kutta algorithm for a second-order state-space model equations

Step	Formulation	Description
Step 1	$x_{11x} = x_{10}, x_{21} = x_{20},$ $a_1 = T[(A_{11}x_{11} + A_{12}x_{21}) + (B_{11}u_1 + B_{12}u_2)]$ $a_2 = T[(A_{21}x_{11} + A_{22}x_{21}) + (B_{21}u_1 + B_{22}u_2)]$	Initial values of x_1 and x_2 for step 1 Slope a_1 for the variable x_1 Slope a_2 for the variable x_2
Step 2	$x_{12} = x_{10} + \frac{a_1}{2}, x_{22} = x_{20} + \frac{a_2}{2},$ $b_1 = T[(A_{11}x_{12} + A_{12}x_{22}) + (B_{11}u_1 + B_{12}u_2)]$ $b_2 = T[(A_{21}x_{12} + A_{22}x_{22}) + (B_{21}u_1 + B_{22}u_2)]$	Initial values of x_1 and x_2 for the step 2 Slope b_1 for the variable x_1 Slope b_2 for the variable x_2
Step 3	$x_{13} = x_{10} + \frac{b_1}{2}, x_{23} = x_{20} + \frac{b_2}{2},$ $c_1 = T[(A_{11}x_{13} + A_{12}x_{23}) + (B_{11}u_1 + B_{12}u_2)]$ $c_2 = T[(A_{21}x_{13} + A_{22}x_{23}) + (B_{21}u_1 + B_{22}u_2)]$	Initial values of x_1 and x_2 for the step 2 Slope c_1 for the variable x_1 Slope c_2 for the variable x_2
Step 4	$x_{14} = x_{10} + c_1, x_{24} = x_{20} + c_2,$ $d_1 = T[(A_{11}x_{14} + A_{12}x_{24}) + (B_{11}u_1 + B_{12}u_2)]$ $d_2 = T[(A_{21}x_{14} + A_{22}x_{24}) + (B_{21}u_1 + B_{22}u_2)]$	Initial values of x_1 and x_2 for the step 2 Slope d_1 for the variable x_2 Slope d_2 for the variable x_2
Result	$x_1(n) = x_{10} + (a_1 + 2b_1 + 2c_1 + d_1)/6$ $x_2(n) = x_{20} + (a_2 + 2b_2 + 2c_2 + d_2)/6$	Value of x_1 from the iteration (n) Value of x_2 from the iteration (n)
Update	$x_{10} = x_1(n), x_{20} = x_2(n)$	Initial value updates of x_1 and x_2 for the next step $(n + 1)$

Equation (5.85) can be written in a closed form as in (5.88).

$$\dot{x}_{n\times 1} = A_{n\times n}x_{n\times 1} + B_{n\times m}u_{m\times 1} \tag{5.88}$$

where $\dot{x}_{n\times 1}$ is a state vector with the dimension $n \times 1$. A $n \times n$ is a $n \times n$ system matrix with a square structure, $B_{n \times m}$ is called the input matrix, whose dimension is $n \times m$. Finally $u_{m \times 1}$ is the input vector. After these definitions, a MATLAB function can be developed as in Figure 5.35.

```
function [x]=runge(A,B,U,X0,dt)
% -----------------------------------------------
% Runge-Kutta algorithm
% Copyrright - Ismail H. Altas, 2002
% A : (n,n) state matrix
% B : (n,n) input matrix
% U : (n,1) input vector
% X0 : initial values
% dt : time step (sampling period)
% -----------------------------------------------
SIZEA=size(A);  NL= SIZEA(1); NC= SIZEA(2);
% ------------  The first step of R-K -----------
for n=1:NL
    x0(n)=X0(n);
    AA1(n)=0;
end
for n=1:NL
    for m=1:NC
        AA(n,m)=A(n,m)*x0(m);
        BB(n)=B(n,n)*U(n);
     end
end
for q=1:NL
    AA1(q)=AA1(q)+ dt*(sum(AA(q,:))+sum(BB(q)));
end
% -----------------------------------------------
```

Figure 5.35(a) Input and the first step of the four-step Runge–Kutta algorithm

```
% ------------  The second step of R-K -----------
for n=1:NL
    x0(n)=X0(n)+AA1(n)/2;
    BB1(n)=0;
end
for n=1:NL
    for m=1:NC
        AA(n,m)=A(n,m)*x0(m);
        BB(n)=B(n,n)*U(n);
    end
end
for q=1:NL
    BB1(q)=BB1(q)+ dt*(sum(AA(q,:))+sum(BB(q)));
end
%-----------------------------------------------
```

Figure 5.35(b) The second step of the four-step Runge–Kutta algorithm

```
% -----  The third step of R-K ---------
for n=1:NL
     x0(n)=X0(n)+BB1(n)/2;
     CC1(n)=0;
end
for n=1:NL
     for m=1:NC
          AA(n,m)=A(n,m)*x0(m);
          BB(n)=B(n,n)*U(n);
      end
end
for q=1:NL
     CC1(q)=CC1(q)+ dt*(sum(AA(q,:))+sum(BB(q)));
end
%------------------------------------------------
```

Figure 5.35(c) The third step of the four-step Runge–Kutta algorithm

```
% ------  The fourth step of R-K ------------------
for n=1:NL
     x0(n)=X0(n)+CC1(n);
     DD1(n)=0;
end
for n=1:NL
     for m=1:NC
          AA(n,m)=A(n,m)*x0(m);
          BB(n)=B(n,n)*U(n);
      end
end
for q=1:NL
     DD1(q)=DD1(q)+ dt*(sum(AA(q,:))+sum(BB(q)));
end
%-----The final result from iteration n------
for n=1:NL
   x(n)=X0(n)+(AA1(n)+2*BB1(n)+2*CC1(n)+DD1(n))/6;
end
```

Figure 5.35(d) The fourth step and the result of the four-step Runge–Kutta algorithm

This algorithm has been written as a MATLAB function, which is called from other MATLAB codes as if it is a subprogram. Example 5.13 shows how the *runge.m* function is used.

Example 5.13

Use the same state-space equations given in Example 5.12 and obtain a digital solution for a unity step input by using the four-step Runge–Kutta method. Assume all initial conditions as zero, approximation tolerance $\varepsilon = 0.001$ and use a sampling period of 0.01 s.

$$\begin{bmatrix} \dot{x}_1 \\ \dot{x}_2 \end{bmatrix} = \begin{bmatrix} -3 & 1 \\ -2 & 0 \end{bmatrix} \begin{bmatrix} x_1 \\ x_2 \end{bmatrix} + \begin{bmatrix} 0 \\ 1 \end{bmatrix} u \tag{E5.13a}$$

$$[y] = [1 \quad 0] \begin{bmatrix} x_1 \\ x_2 \end{bmatrix} + 0 \tag{E5.13b}$$

The analytical solution of this system was given in Example 5.12.

Answer 5.13

The MATLAB .m file code and resultant plots are given in Figures E5.13(a) and E5.13(b). As can be seen from the resultant graphs are very similar to those obtained from Euler integration for the same problem.

```
% EXAMPLE 5.13
%----- input data -----------------------
A=[ -3   1
     -2   0];
B=[1 0
    0 1];
C=[1   0];    D=0;
u1=0;  u2=1;    U=[u1;u2];
dt=0.01;    t0=0;    k=1;
tend=6;     % Run simulation for 6s.
% -------- Initial conditions ------------
SIZEA=size(A);  NL=SIZEA(1); NC=SIZEA(2);
for n=1:NL
     x0(n)=0;
end
y0=0;
% --------- start iterative solution ----------------------
while t0<tend
        [x]=runge(A,B,U,x0,dt);      % Call function runge.m
        u(k)=U(2);    y(k)=x(1);     % Store input and output
        x1(k)=x(1);   x2(k)=x(2);    % Store variables x1 and x2
        t(k)=t0+dt;   t0=t(k);       % Update time
        for n=1:NL
            x0(n)=x(n);  XX(k,n)=x(n); % update initial values
        end
        k=k+1;                       % increase iteration counter
end;
% ------------- Plots   ---------------------------------------
plot(t,u,t,y,t,x1,t,x2 );
xlabel('Time (sec.)');  ylabel('u, y, x1, x2'); grid
```

Figure E5.13(a)   MATLAB .m-file code for Example 5.13

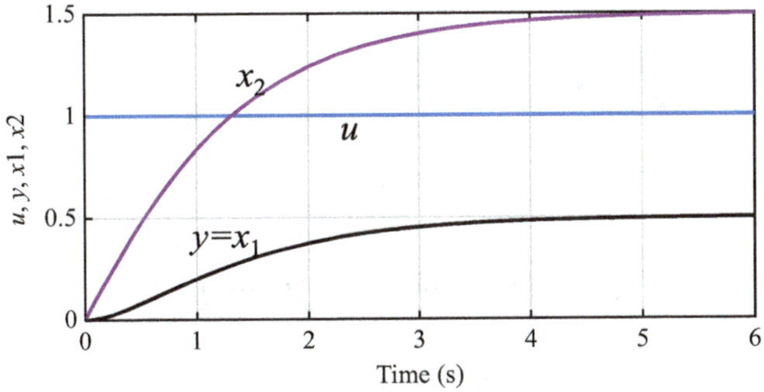

Figure E5.13(b) Variables x_1 and x_2

5.7 Problems

P.5.1 A LTI ordinary differential equation is given on the right. Assume all initial conditions as zero and $u_s(t) = 0$ for $t < 0$ and $u_s(t) = 1$ for $t \geq 0$.

$$\frac{d^2x(t)}{dt^2} + 15\frac{dx(t)}{dt} + 100x(t) = 100u_s(t)$$

(a) Obtain a simulation diagram (detailed block diagram) for his system.
(b) Obtain a state-space model for this system.

P.5.2 An operational amplifier circuit is given in Figure P5.2. The operational amplifier is assumed to be ideal and the values of capacitors and resistors are given as $C = 1$ μF, $R1 = 150$ kΩ, $R2 = 250$ kΩ, $R3 = 1$ kΩ, and $R4 = 100$ kΩ.
(a) Write integro-differential equations of this system.
(b) Draw a simulation diagram for this system.
(c) Obtain the transfer function $V0(s)/Vi(s)$ for this system.

P.5.3 A single input–single output linear mechanical system is given in Figure P5.3. The external force $f(t)$ is the input and the displacement $x(t)$ of the mass is the output. The initial position of the displacement corresponds to the equilibrium point when there is no external force.
(a) Apply Newton's second law and write the change of momentum equation of this system.

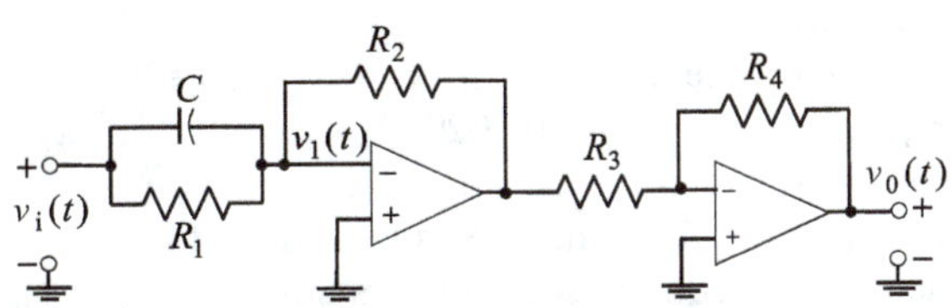

Figure P5.2 Figure for Problem P.5.2

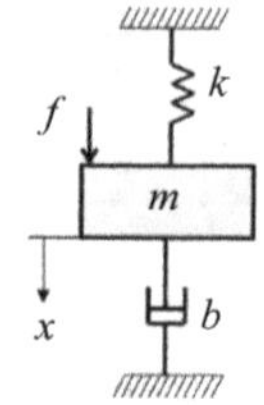

Figure P5.3 Figure for Problem P5.3

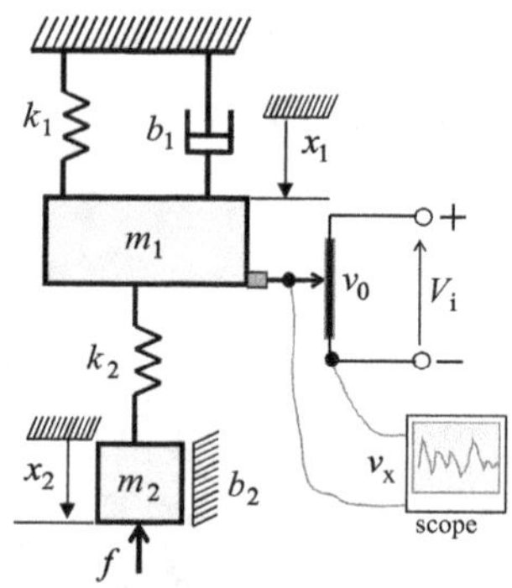

Figure P5.4. Figure for Problem P5.4

(b) Draw the simulation diagram of the system from the equation in part (a).
(c) Obtain state-space model equations of this system from part (b).
(d) Obtain Laplace transformation of the differential equation you obtained in part (a) and write it for $X(s)$. Assume that initial conditions are not zero.
(e) Obtain transfer function of this system from part (d). Assume that initial conditions are zero.

P.5.4 The system given Figure P5.4 is used to detect earthquakes by generating signals depending on the force F applied to mass m_2 as shown. A voltage V_i is applied to a potentiometer and the initial voltage to the scope is set to $v_x = v_0$. If any force F occurs due to earthquake both masses m_1 and m_2 move resulting in the displacements x_1 and x_2. k_1 and k_2 are spring constants, b_1 and b_2 are the friction constants. The change in voltage v_x is proportional to displacement x_1 as $v_x = v_0 + A_v x_1$.

System parameters are: m1 = 10 kg, $m_2 = 0.1$ kg, $g = 9.8$ m/s^2, $b_1 = 10$ Ns/m, $b_2 = 2$ Ns/m, $k_1 = 10$ N/m, $k_2 = 0.00205$ N/m, $Av = 0.2$, $V_0 = 5$ V, $V_i = 10$ V.
(a) Write integral-differential equations of this system.
(b) Draw a simulation diagram as F is input v_x is the output.
(c) Write state-space equations of this system.

(d) Design a fuzzy logic controller to decrease the effect of the force f while keeping v_x close to set value v_0.
(e) Use an FLC to track the step changes in v_x and obtain a relationship between the steady-state values F and V_x of the force and displacement.

P.5.5 A basic block diagram and electromechanical connection diagram of a door position control system driven by a permanent magnet DC motor (PMDCM) are adapted from [14] and shown in Figures P5.5(a) and P5.5(b). The difference between the measured door position angle θ_D and reference position θ_R is minimized by controller C to generate a control signal u, which is amplified by amplifier A to drive the door system. The purpose of the controller is to bring and keep the door at a reference position angle θ_R. A gear system depicted by G in Figure P5.5a is used to match the position angles of the motor and door shafts. All required data are given in [14] as $R_a = 1.4\ \Omega$, $L_a = 0.0805$ H, $k = 0.095$, $J_m = 0.0007432$ kgm^2, $B_m = 0.000431\ V_s$/rad, $V_a = 36$ V, and $N = 2.67$. Where k represents both torque and voltage constant because of the constant magnetic field. V_a is the rated motor voltage and N is the turn ratio of the gear subsystem. The mathematical model of PMDCM is the same one given in Figure 5.25 as the armature-controlled DC motor. Figure P5.5(b) shows the details of the gear system and the use of a potentiometer to detect the door position.

(a) Draw a control system block diagram for his door position system.
(b) Explain the purpose of using each block in detail.
(c) Show physical and informative parts of the block diagram.

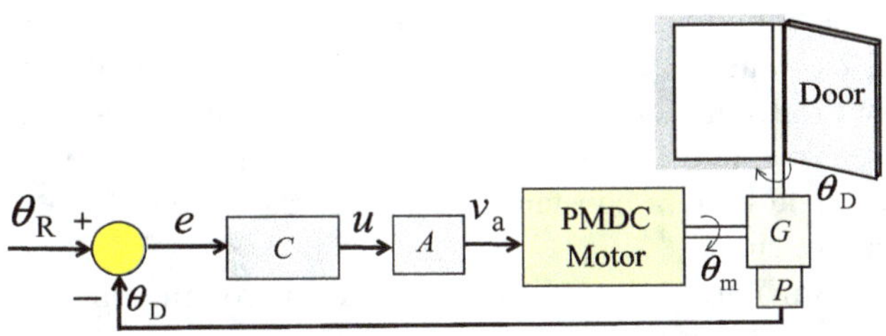

Figure P5.5(a) Figure for Problem P5.5

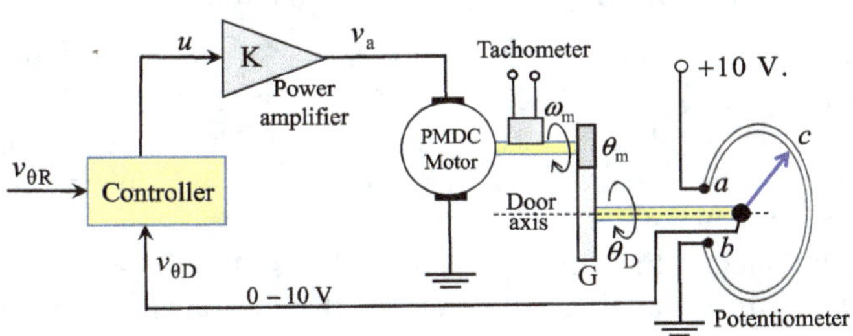

Figure P5.5(b) Details of gear system for Problem P5.5

(d) Use the FLC block developed in Chapter 4 and used in Example 5.10 to control the door position system. Assume that the maximum door position is π rad.

P.5.6 Develop an FLC-based control algorithm to control the speed of the DC motor given in Figure 5.10 such that the controller should be able to detect the changes in armature current, load torque, speed, and speed error. The approach developed by A.M. Sharaf as the one in [17] may be helpful for this problem.

P.5.7 Modify the FLC-controlled DC motor given in Example 5.10 such that the motor speed can be increased up to twice the rated speed, which is 400 rad/s. It should be remembered that the speed of a separately excited DC motor is increased with increasing armature current and decreasing field current. The effects of the armature and field currents should be taken into account while designing the controller.

After designing the FLC, perform the following analysis:

(a) While the field current is kept constant at its rated value, run the armature-controlled system simulation five times with 50 rad/s increments in reference speed starting at 100 rad/s, and record steady-state values of speed and armature current from each simulation run.
(b) While the armature current is kept constant at its rated value, run the field-controlled system simulation five times with 50 rad/s increments in reference speed starting at 100 rad/s, and record steady-state values of speed and field current from each simulation run.
(c) Plot speed versus armature current and speed versus field current. Then comment on the plots.

P.5.8 Develop a Simulink model for the shunt-excited DC motor given in Figure 5.28 by its electromechanical diagram. Assume that the parameters of the shunt excited motor are the same as those given for the separately excited motor in Example 5.10.

(a) Keep field current constant and control the speed of the motor through armature circuit.
(b) Keep armature voltage constant and control the speed of the motor through field circuit.
(c) Control the speed of the motor both through armature and field circuits.
(d) Repeat parts (a), (b), and (c) to control the rotor.

P.5.9 For the armature-controlled DC motor shown in Figure P5.9, the numerical values are given as $R_a = 20\ \Omega$, $L_a = 0.05$ H, $b_m = 0.1$ Nm s/rad, $J_m = 2$ Nm s^2/rad, $k_a = 1$ V s/rad, $k_t = 1$ Nm/A and $T_L = 0$ Nm. Assume that the input voltage v as input and motor speed as the output. Then simulate the system to obtain the following:

(a) Write integro-differential equations of this DC motor.
(b) Obtain the simulation diagram using the equations in part (a).

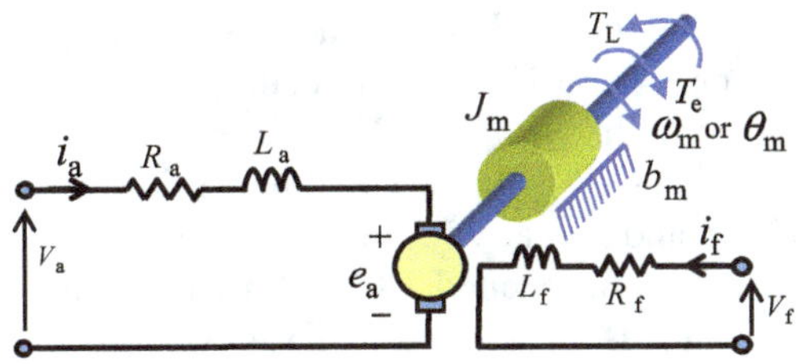

Figure P5.9 Figure for Problem P5.9

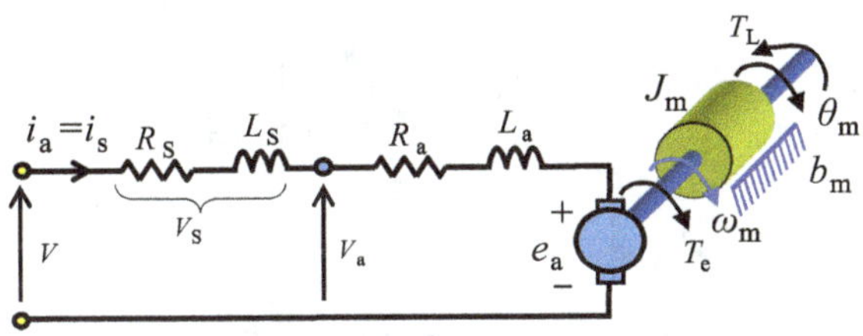

Figure P5.10 Figure for problem P5.10

(c) Obtain the state-space equations of this DC motor using the simulation diagram in part (b).

(d) Obtain the characteristic equation of this DC motor.

(e) Locate the poles of the DC motor in a complex plane.

(f) Obtain time constant, settling time, damping ratio, and percent overshoot of this system.

P.5.10 A series excited DC motor circuit diagram is shown in figure. Where the subscript s represents the parameters and variables of the series exited field circuit and the subscript a represents the parameters and variables of the armature circuit. The motor parameters are given as $R_a = 0.1\ \Omega$, $L_a = 0.01$ H, $R_f = 0.05\ \Omega$, $L_f = 0.02$ H, $b_m = 0.04$ Nm s/rad, $J_m = 0.6$ Nm s^2/rad, $k_f = 0.1$, $k_m =$ Nlr $= 6$ and $T_L = 0$ Nm.

The applied voltage V is the input and speed ω_m is the output. Determine the following:

(a) Cause-and-effect diagram
(b) Differential equations
(c) Simulation diagram
(d) State-space equations
(e) Transfer function

P.5.11 Develop a fuzzy logic-based speed and position control algorithm for the DC motor given in P5.9 and simulate the FL-controlled system in MATLAB and Simulink.

P.5.12 Develop a fuzzy logic-based speed and position control algorithm for the DC motor given in P5.10 and simulate the FL-controlled system in MATLAB and Simulink.

P.5.13 Use the series DC motor given in P5.10 in the door position control system described in P5.5 and repeat P5.5 for the series-excited DC motor.

P.5.14 Repeat P5.5 for a shunt-excited DC motor assuming that the motor parameters are the same as those given in Example 5.10.

P.5.15 Solve the following state-space equations using the four-step Runge–Kutta algorithm. Assume that input $u_1(t) = 100$, $u_2(t) = 0$, iteration time step $dt = T = 0.01$ s, and all initial conditions are zero. Simulate the system for 1 s starting $t = 0$ s.

$$\begin{bmatrix} \dot{x}_1 \\ \dot{x}_2 \end{bmatrix} = \begin{bmatrix} -4 & -3 \\ 1 & -8 \end{bmatrix} \begin{bmatrix} x_1 \\ x_2 \end{bmatrix} + \begin{bmatrix} 1 & 0 \\ 0 & 1 \end{bmatrix} \begin{bmatrix} u_1 \\ u_2 \end{bmatrix} \tag{P5.15a}$$

$$[y] = [1 \quad 0] \begin{bmatrix} x_1 \\ x_2 \end{bmatrix} \tag{P5.15b}$$

P.5.16 Series-connected RLC circuit given in Figure P5.16(a) can be represented by its simulation diagram shown in Figure 5.16(b) and state-space equations as in (P5.16a) and (P5.16b). The circuit parameters are given as $L = 2$ H, $R = 10$ Ohm; $C = 100$ F, and $v = 50$ V. Assume that all initial conditions are zero and use a sampling period of dt = 0.01 s to simulate this system as follows:

(a) Build Figure P5.16(b) in MATLAB/Simulink and simulate it to see and analyze the variables $x_1(t)$ and $x_2(t)$ where $x_2(t)$ is the current flowing in the circuit.

(b) Use *runge.m* function in MATLAB as in Example 5.13 and solve state-space equations to see and analyze the variables $x_1(t)$ and $x_2(t)$ where $x_2(t)$ is the current flowing in the circuit.

(c) Compare the results from parts (a) and (b).

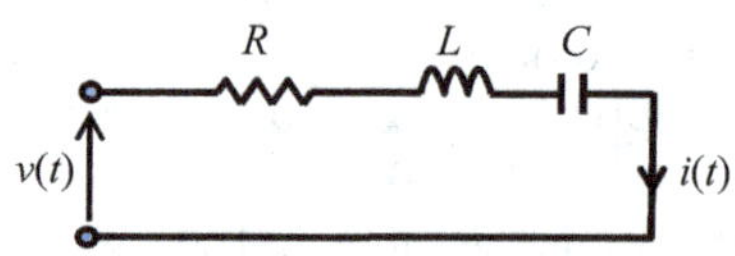

Figure P5.16(a) Circuit diagram of series connected RLC circuit

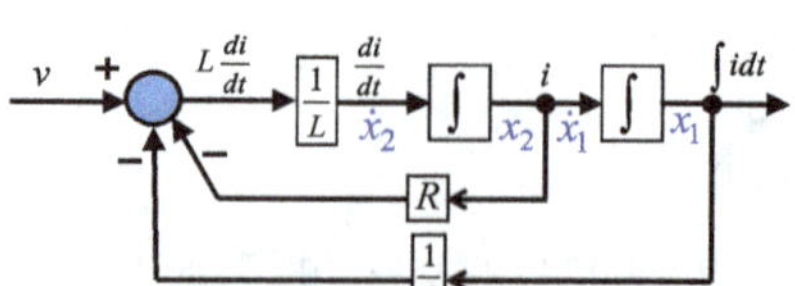

Figure P5.16(b) Simulation diagram of series connected RLC circuit

$$\begin{bmatrix} \dot{x}_1 \\ \dot{x}_2 \end{bmatrix} = \begin{bmatrix} 0 & 1 \\ -\dfrac{1}{LC} & -\dfrac{R}{L} \end{bmatrix} \begin{bmatrix} x_1 \\ x_2 \end{bmatrix} + \begin{bmatrix} 1 & 0 \\ 0 & \dfrac{1}{L} \end{bmatrix} \begin{bmatrix} 0 \\ v \end{bmatrix} \tag{P5.16a}$$

$$[y] = \begin{bmatrix} 0 & 1 \end{bmatrix} \begin{bmatrix} x_1 \\ x_2 \end{bmatrix} \tag{P5.16b}$$

References

[1] Doraiswami R., Stevenson M. and Diduch C., *Identification of Physical Systems: Applications to Condition Monitoring, Fault Diagnosis, Soft Sensor and Controller Design*. New York: John Wiley & Sons, 2014.

[2] Wellstead P.E. *Introduction to Physical System Modelling*. New York: Academic Press Ltd., 1979.

[3] Kailath T. *Linear Systems*. Englewood Cliffs, NJ: Prentice-Hall, 1980.

[4] Franklin G.F., Powell J.D. and Emani-Naemi A. *Feedback Control of Dynamic Systems*. Englewood Cliffs, NJ: Prentice-Hall, 2006.

[5] Bishop R.H. *Modern Control Systems Analysis and Design Using MATLAB and Simulink*. Reading, MA: Addison-Wesley, 1997.

[6] Altas I.H. and Sharaf A.M. 'A fuzzy logic power tracking controller for a photovoltaic energy conversion scheme'. *Electric Power Systems Research Journal*. 1992;25(3):227–238.

[7] Altas I.H, 'Bulanık Mantık Denetleyici: MATLAB/Simulink Ortamı için Bir Modelleme', *Otomasyon*, Bileşim Yayınları, İstanbul, Turkey, March 2007, Sayfalar: pp. 58–62.

[8] Altas I.H. and Sharaf A.M. 'A generalized direct approach for designing fuzzy logic controllers in MATLAB/Simulink GUI environment'. *International Journal of Information Technology and Intelligent Computing*. 2007; 1(4).

[9] Ogata K., *Modern Control Engineering*. Englewood Cliffs, NJ: Prentice-Hall, 1997.

[10] Kuo B.J. *Automatic Control Systems*. Englewood Cliffs, NJ: Prentice-Hall, 1995.

[11] Dorf R.C. and Bishop R.H. *Modern Control Systems*. Englewood Cliffs, NJ: Prentice-Hall, 2001.

[12] Nise N. S. *Control Systems Engineering*, 4th edn. New York: John Wiley & Sons, 2004.

[13] Altas I.H. 'A fuzzy logic controlled tracking system for moving targets'. 12th IEEE International Symposium on Intelligent Control; Istanbul, Turkey, July 1997. pp. 43–48.

[14] Altas I.H. 'A fuzzy logic based door position control system'. The Proceedings of the 3rd International Mechatronic Design and Modeling Workshop, 15–18 September 1997, Middle East Technical University, Ankara, Turkey, pp. 107–116.

[15] Altas I.H. 'A fuzzy logic controlled tracking system for moving targets'. 12th IEEE International Symposium on Intelligent Control, *ISIC'97*, 16–18 July 1997, Istanbul, Turkey, pp. 43–48.
[16] Altas, I.H. 'Endüvi Kontrollü DA Motorlarının Modellenmesinde Sebep-Sonuç Yaklaşımı'. *Otomasyon Dergisi*, Bileşim yayınları, Ocak 2008, pp. 186–190.
[17] Altaş I.H. and Sharaf A. M. 'Novel control strategies via load matching for maximum energy utilization in photovoltaic energy systems'. *2007 IEEE Canadian Conference on Electrical and Computer Engineering*, Vancouver, Canada.
[18] Tai-Ran Hsu T.-R. *Applied Engineering Analysis*. New York: John Wiley & Sons, 2018.
[19] Karris, S.T. *Numerical Analysis Using MATLAB and Excel*, 3rd edn. London: Orchard Publications, 2007.

Chapter 6
FLC in power systems

Application of fuzzy logic controller and decision maker to excitation control, load–frequency control, and power compensation are discussed in this chapter. Single- and multi-area control of power systems are also studied as examples in the chapter besides FL-based load flow analysis.

6.1 Introduction

Electrical power systems start with the generation of electrical power. Transmission, distribution, and utilization are the other parts of conventional electrical power networks. Each one of these four stages, generation, transmission, distribution, and utilization, has its own problems and control approaches in order to keep them in service. The synchronous generator driven by either a hydraulic or a steam turbine is the main unit in electrical power generation and the control of power systems means the control of a synchronous generator. However, the concept of power system control has been changed since the introduction of microgrids. In conventional power networks, the generation is usually away from the urban areas, and the distribution is either done by radial or ring networks. Due to the impedances of transformers, distribution, and transmission lines, the power network is considered as an infinite bus compared to the distribution networks and power dissipation units. Small load changes and fault currents in distribution feeders do not affect the rest of the power network due to the high impedances between the generator and distribution. Therefore, the large power networks are represented by infinite bus and power management is done in some specified locations in power transmission networks, not everywhere in distribution systems. As a result of the network structure, the equilibrium between generation and dissipation is obtained by controlling the generating units, which are the synchronous generators.

The power generation and dissipation in microgrids are side by side in various locations in the network resulting in a structure called distributed generation. The concept of microgrid and distributed generation has started with the increasing use of photovoltaic (PV) and wind energy systems, which are installed right in distribution areas. Therefore, any change in either generation or dissipation affects each other. When a large load is switched on or off the voltage of the bus where this switched load is connected, drops down or goes up being dangerous for the other loads connected to the same bus. The RMS voltage and the frequency of every load bus must be kept constant at their nominal values in order to feed and operate the other loads normally. Since the generation is distributed everywhere in the power distribution network, the control and management problems become more difficult and complicated in microgrids. The control of PV and wind systems are discussed in separate chapters later as a part of microgrids. Note that the control of microgrids with distributed generation requires measurement, transmission, and analysis of data from everywhere in distribution systems. In order to manage and control the generated and dissipated power, secure communication channels must be established between units all over the network. The classical control and management methods do not work for the microgrids. Thus, new methods are developed to deal with the control and management problems in distributed power systems leading to the concept of smart grids. The control of wind and PV energy systems are discussed in Chapters 10 and 11, respectively. FLCs are designed to control wind and PV power generation and utilization schemes as a part of microgrids.

The voltage magnitude and frequency of power systems are controlled by controlling the synchronous generators. It is well known that two types of controllers are employed to control synchronous generators. Excitation control is used for output voltage and governor control is used for load–frequency control. A schematic diagram of the synchronous generator control system is shown in Figure 6.1 including both excitation and load–frequency controls.

The excitation control of a synchronous generator is based on the magnitude control of the output voltages. As the voltage magnitude changes due to overloading or light loadings, the excitation control is employed to keep it at the desired reference value. Therefore, the output voltages are measured and fed back to the controller. Various methods of excitation control have been discussed in the literature [1–7]. In Figure 6.1, three-phase terminal voltages at the output terminals of the generator are converted to DC and used as the feedback signal. Then usually an

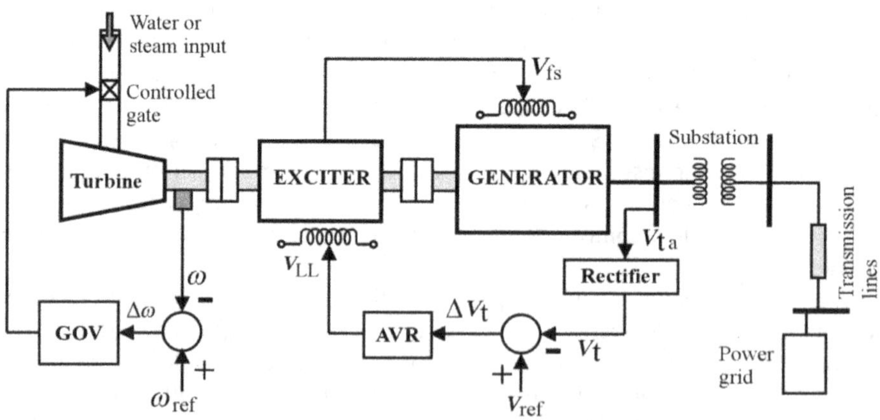

Figure 6.1 Excitation and load–frequency control diagram of a synchronous generator

automatic voltage regulator (AVR) is used as the control device in the excitation control process. The AVR control is combined with other methods such as power system stabilizers [2,8–10] or replaced by other controllers [11–13]. Some examples of modeling and controlling the excitation systems are given in the next section.

The governor (GOV) control is used to control the input of water or steam to the turbine as shown in Figure 6.1. The steam or water input to the turbine affects the speed and the mechanical power input of the generator. Therefore, the governor control is used as power–frequency or load–frequency control. The loading conditions of the synchronous generator affect the speed, which is directly related to the frequency of the voltages generated by the synchronous machine. Therefore, the speed, representing the frequency, is measured and used as the feedback signal in the governor to control the power–frequency equilibrium of the system. Load–frequency control is very important for power system stability and has been studied well in the literature [14–23]. The use of FLC in load–frequency control is discussed in the coming sections of this chapter.

The use of fuzzy logic controllers has been increased rapidly in power systems, such as in load–frequency control, bus bar voltage regulation, stability, load estimation, power flow analysis, parameter estimation, protection systems, and many other fields [11–14,24–33]. Fuzzy logic applications in power systems are given in [34–38] with a detailed survey.

A tremendous amount of papers and books on the control of power systems can be found in literature. Many of these papers and most of the books also cover the modeling of excitation systems, automatic voltage controllers (AVR), power system stabilizers (PSS), and load–frequency (L–F) controllers [39–43]. Although a mathematical model of the physical systems is not required in fuzzy logic controllers for real-time applications, it is necessary for digital simulation in the computer environment. Therefore, simulation models of excitation systems and L–F controllers are needed. Since the modeling of excitation systems and L–F controllers in power systems are well studied in the literature [1–7], detailed modeling of these systems is not going to be repeated here. Well-known excitation systems and L–F models from the literature are preferred to be used instead. The modeling of the fuzzy logic controller (FLC) used in this chapter is discussed and explained clearly in Chapter 7.

As it is well known, instantaneous load changes affect the bus voltage and its frequency causing their values to oscillate around the nominal operating points. When the magnitudes of these oscillations are large and natural damping takes a long time, the stability of the power system may be lost resulting in damages and power outages in the system. In order to overcome the effects of swinging frequency and voltage oscillations, some precautions must be taken such as using L–F controllers and PSS. Any load change in one of the L–F control areas affects the tie line power flow causing other L–F control areas to generate the required power to damp the power and frequency oscillations. The response time of the L–F controllers is very important to have the power system to gain control with increased stability margins. Therefore, the proposed L–F controller must reduce the response time as well as reduce the magnitude of the oscillations.

The voltage oscillation may cause increased and decreased bus voltages over the power system. If the durations of these higher or lower voltages are long enough, they may cause damages, misoperations, and even instabilities. Therefore, voltage controllers and power system stabilizers must be utilized to prevent these problems for a reliable power system. In this chapter, the application of fuzzy logic controllers to the excitation and load–frequency control systems is discussed.

6.2 Excitation control

The control of electrical power systems is done in two steps, which are load–frequency control and excitation control as depicted in Figure 6.1. In order to generate the required power, steam or water input to the turbine is controlled by load–frequency controller, which is called governor control. The emf, power factor, and magnitude of the generated current are adjusted by excitation control. For example, let us represent the synchronous generator by its per-phase Thevenin equivalent circuit as shown in Figure 6.2.

Assume that the generator is operating with a lagging power factor and generating both active and reactive powers. The terminal voltage $\underline{V}_t$ is constant because it is connected to an infinite bus. If the mechanical input power P_m is kept constant, the generated electrical power P_e becomes constant, too. Under these operating conditions, the phasor diagram given in Figure 6.3 is obtained from the per-phase equivalent circuit in Figure 6.2.

The active electrical power generated by this machine can be expressed by two different equations as in (6.1) and (6.2).

$$P_e = V_t I \cos\theta \tag{6.1}$$

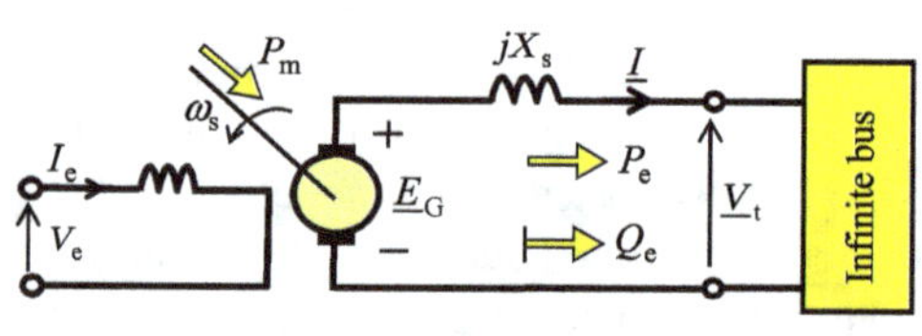

Figure 6.2 Per-phase Thevenin equivalent circuit of a synchronous generator

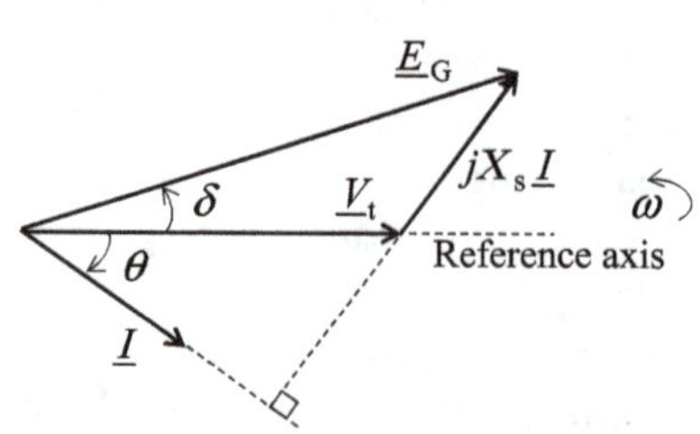

Figure 6.3 Phasor diagram of a synchronous generator with lagging power factor

$$P_e = \frac{E_G V_t}{X_s} \sin \delta \tag{6.2}$$

If the losses are ignored, the generated electrical power P_e becomes equal to mechanical input power P_m. Since mechanical input power is kept constant, the generated electrical power becomes constant as well. This assumption is usually considered correct and (6.3) is written.

$$P_e = P_m = \text{constant} \tag{6.3}$$

Since the terminal voltage $\underline{V}_t$ and synchronous reactance X_s are constant, it is obvious that (6.4) and (6.5) can be written from (6.1) and (6.2), respectively.

$$I \cos \theta = \text{constant} \tag{6.4}$$

$$E_G \sin \delta = \text{constant} \tag{6.5}$$

The meanings of the constants (6.4) and (6.5) can be understood better from Figure 6.4.

If the mechanical input power is kept constant and $\underline{E}_G$ is increased by adjusting the excitation voltage V_e, power factor angle θ, load angle δ, and current $\underline{I}$ are changed from values θ_1, δ_1, and $\underline{I}_1$ to new values θ_2, δ_2, and $\underline{I}_2$, respectively, as shown in Figure 6.5.

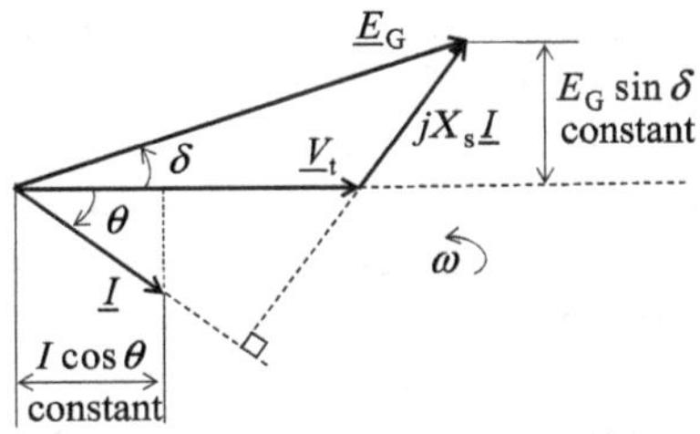

Figure 6.4 The phasor diagram with constant parameters and variables

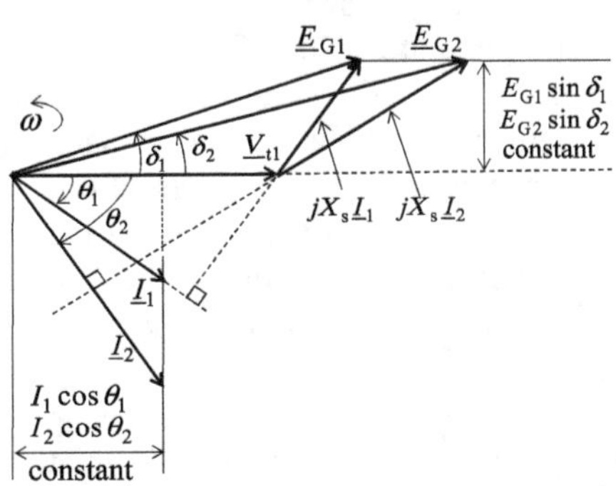

Figure 6.5 The excitation effect on the phasor diagram

Note that the constant parts represented by $E_G\sin\delta$ and $I\cos\theta$ are kept unchanged while the power factor angle, the load angle, and the current are changed. However, while the magnitude of the current is changed, $I\cos\theta$ is kept unchanged. This means that $\cos\theta$ gets smaller due to the increase in θ. Therefore, the power factor becomes smaller. Similarly, when E_G is increased, $\sin\delta$ gets increased as well, because $E_G\sin\delta$ is constant. Therefore, δ must be smaller if E_G is increased.

It can be concluded from the effects of the excitation depicted in Figure 6.5 if the excitation is increased, then the following should be done in order to keep the power system stable, while the terminal voltage and output power are constant.

(I) The load angle should be reduced,
(II) The current should be increased, and
(III) The power factor should be increased.

Similar conclusion can be made when the excitation is reduced. The operating mode described in Figure 6.5 is done for a stable operation. When the machine is operating at a stable point, the excitation is increased to operate the machine at a new stable point. This change from one stable operating point to another stable operating point causes some transients in the system variables such as frequency and voltage. Therefore, the effects of transients must be considered if there is a change in stable operation. Transient response of a step change in excitation may be an oscillating response, damped response, overdamped response, or critically damped response. Overshoot and settling time in these responses must be limited to maintain stability. In order to overcome these problems and maintain a stable operation excitation control is required.

An excitation system consists of a power source a voltage regulator and auxiliary control elements. In high power generation, the power to the excitation coils of the main generator is usually supplied from a DC or another synchronous generator coupled on the same shaft with the main generator as shown in Figure 6.1. In this case, three-phase windings of the exciter and excitation coils of the main generator are placed on the rotor of each machine so that the connection can be made without using slip rings and brushes. If another synchronous generator is used as the exciter, its three-phase voltages are converted to DC and applied to the excitation coils of the main generator. Therefore, the excitation of the main generator is controlled through the exciter generators. In small-size power systems, the excitation coils of the main generator are usually placed in the stator and the required excitation power is supplied from an external controllable DC source.

A voltage regulator is used to adjust the output of the exciter so that the generated voltage and reactive power are kept inside the described limits. Therefore, the voltage regulator is one of the important elements in excitation systems. As it is understood, the voltage regulator is nothing but a controller used the control the terminal voltage and reactive power generation of a synchronous generator through the exciter system. The response time of a voltage regulator is very important for the system stability. Due to the high impedance of the excitation coil, it is not possible to change the excitation current suddenly. Therefore, a delay is always present as a problem to be solved in excitation systems. Various types of excitation system models for stability studies in power systems are given by IEEE Std 421.5™-2005 [44]. Besides, IEEE committee reports [5,6] and many other studies in power system excitation modeling and control have been done in the literature. Since there are many details available in the literature, more theoretical background will not be repeated here. However, the modeling processes of mostly used excitation systems for control purposes are given next.

6.2.1 Excitation system modeling

Automatic control of a physical system is a continuous process and requires measurement, observation, and action continuously. A control signal for any changes affecting the operation of the exciter must be generated to keep tracking the set values, which are the terminal voltage and the reactive power of the generator.

Consider a typical exciter with a feedback structure as shown in Figure 6.6. In this system the main synchronous generator is excited by using a DC generator. The output voltage of the DC generator is directly applied to the field circuit of the synchronous generator. Therefore, the DC voltage applied to the field of the synchronous generator is adjusted in terms of the armature control of the DC generator. As shown in Figure 6.6, the armature voltage of the DC generator is determined by a voltage regulator using the difference between the reference voltage and the synchronous generator's terminal voltage, which is converted to DC to be used as a feedback signal. The excitation system in Figure 6.6 consists of six main elements as listed below.

- Synchronous generator, the main generator
- Rectifier and voltage transformers
- Difference amplifier
- Voltage regulator
- Power amplifier
- DC generator, the exciter

Since the terminal voltage of the synchronous generator in Figure 6.6 is to be controlled, it is used as the feedback signal. Three-phase terminal voltages at medium voltage levels are measured using a voltage transformer (VT) at low

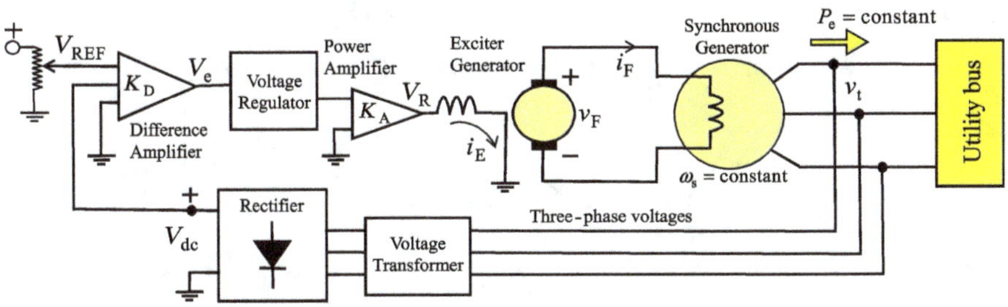

Figure 6.6 A schematic diagram of the excitation control system

voltage levels and converted to DC voltage, V_{dc}, which is the negative feedback voltage representing the terminal voltages. The connection diagram of VTs and the rectifier is given in Figure 6.7.

As shown in Figure 6.7, the DC voltage V_{dc} is proportional to the sum or average value of the three-phase RMS voltages. If the average RMS voltages are represented by V_t, then the DC voltage can be written as in (6.6).

$$V_{dc} = \frac{K_{dc}}{1 + \tau_{dc}s} V_t \tag{6.6}$$

where K_{dc} is the rectifier gain constant, τ_{dc} is a time constant representing the delay caused by VT and the rectifier. The time constant τ_{dc} in these systems usually has a small value such as around or less than 0.06 s.

The difference amplifier in Figure 6.6 compares the terminal voltage in terms of V_{dc} with the reference terminal voltage V_{REF} and yields an amplified voltage error V_e as in (6.7).

$$V_e = K_D(V_{REF} - V_{dc}) \tag{6.7}$$

The difference amplifier can be realized in different ways. If an electronic difference amplifier is used, the time constant becomes very small and can be ignored yielding (6.7) and the block diagram shown in Figure 6.8.

The voltage regulator and the power amplifier acting as an actuator in the system can be a rotational, a magnetic, or an electronic amplifier. Assume that any one of these amplifiers is represented by a proportional gain K_A and time constant

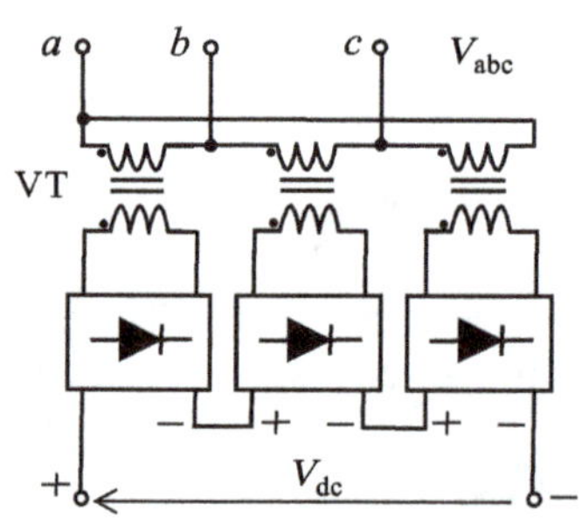

Figure 6.7 Connection diagram of VTs and rectifiers

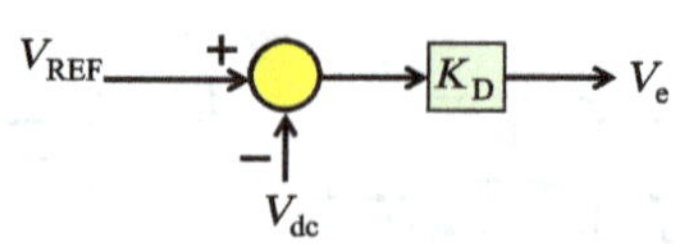

Figure 6.8 Block diagram representation of difference amplifier

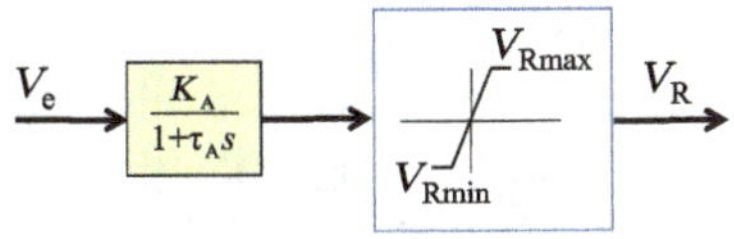

Figure 6.9 Block diagram representation of the regulator amplifier

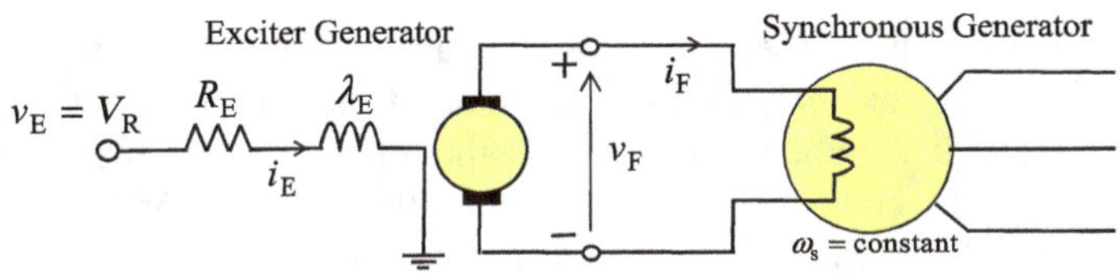

Figure 6.10 Separately excited DC generator used as the exciter

τ_A as in (6.8).

$$V_R = \frac{K_A}{1 + \tau_A s} V_e \tag{6.8}$$

The regulator amplifier has a transfer function similar to that of a low-pass filter. Knowing that each amplifier, whatever the type is, saturates and the modeling includes a saturation element to limit the output between predefined minimum V_{Rmin} and maximum V_{Rmax} values. The regulator and power amplifier with the saturation effect are represented by a block diagram as given in Figure 6.9.

The output voltage V_R of the regulator amplifier is the input voltage to the field of the separately excited DC generator, which is the exciter generator in the system. Since the exciter generator is coupled to the same shaft as the synchronous generator, the speed of both machines is constant. Therefore, the induced emf is a function of the field voltage V_R only. In order to show this relationship, let us write the voltage equation of the field coil of the synchronous generator as in (6.9) using Figure 6.10.

$$\dot{\lambda}_E + R_E i_E = v_E \tag{6.9}$$

where λ_E is the flux (wb/turn) encircling the field coil and R_E is the field coil resistance of the exciter generator. The excitation current of the exciter generator is represented by i_E and the excitation voltage by v_E.

Instead of encircling flux, let us rewrite (6.9) in terms of excitation field flux ϕ_E as in (6.10), which will be more useful.

$$N\dot{\phi}_E + R_E i_E = v_E \tag{6.10}$$

where the voltage v_E may be assumed to be constant. Thus (6.10) becomes a function of ϕ_E and i_E. The excitation current i_E in (6.10) depends on the operating point on magnetizing curve, which is not linearly related to the excitation voltage v_E.

Due to the saturation of the magnetizing curve, the flux ϕ_E is separated into two parts as leakage and air gap fluxes, ϕ_l and ϕ_a, respectively. Due to the saturation effect (6.10) is not a linear equation. Since the magnetizing curve is a plot of v_F versus i_E, the flux ϕ_E should be written in terms of v_F or i_E. If the main exciter runs at a constant speed, its voltage becomes linearly proportional to air gap flux ϕ_a as in (6.11).

$$v_F = k\phi_a \tag{6.11}$$

Now a relationship between air gap flux ϕ_a and ϕ_E should be established. As it was stated earlier, ϕ_E has two components. One of these components is called as leakage flux ϕ_l, which is about 10%–20% of the total flux and proportional to the air gap flux ϕ_a or the current i_E flowing through the field coil of DC exciter generator. Assume that the leakage flux ϕ_l and air gap flux ϕ_a are related to each other as in (6.12).

$$\phi_l = c\phi_a \tag{6.12}$$

where c is a constant parameter. Since the total flux can be written as in (6.13)

$$\phi_E = \phi_a + \phi_l \tag{6.13}$$

Then, (6.14) and (6.15) are obtained using (6.12).

$$\phi_E = (1 + c)\phi_a \tag{6.14}$$

$$\phi_E = \sigma\phi_a \tag{6.15}$$

where the σ is known as the dispersion constant and has values between 1.1 and 1.2. Substitution of (6.15) into (6.10) yields (6.16).

$$N(\sigma\dot{\phi}_a) + R_E i_E = v_E \tag{6.16}$$

The airgap flux can be written as in (6.17) using (6.11).

$$\phi_a = \frac{1}{k} v_F \tag{6.17}$$

Then its derivation becomes as in (6.18)

$$\dot{\phi}_a = \frac{1}{k} \dot{v}_F \tag{6.18}$$

Thus (6.16) can be rewritten as in (6.19).

$$N\left(\sigma\frac{1}{k}\dot{v}_F\right) + R_E i_E = v_E \tag{6.19}$$

If $(N\sigma/k)$ is defined as the exciter time constant in seconds as

$$\tau_E = \frac{N\sigma}{k} \tag{6.20}$$

Then (6.19) becomes as in (6.21).

$$\tau_E \dot{v}_F + R_E i_E = v_E \tag{6.21}$$

The last equation looks like a linear time-invariant ordinary differential equation. However, this equation is still not linear because the excitation voltage v_F of the synchronous generator is not linear with the exciter current i_E. The exciter voltage v_E is usually assumed to be constant in stability studies. In excitation control, the exciter voltage v_E is adjusted to change the exciter current i_E, and therefore the terminal voltage of the synchronous generator.

Equation (6.21) represents an exciter system with a separately excited DC machine is the exciter generator. Similar equation can be obtained for an exciter using the self-excited DC generator. More details can be found in [1,2,4].

The exciter model given in (6.21) is not linear due to the nonlinear relation between the exciter field current i_E and synchronous generator field voltage v_F. Therefore, the solution of (6.21) requires nonlinear solution methods. The relationship between i_E and v_F is usually represented by a saturated magnetizing curve, which includes a saturation function that is defined as in (6.22).

$$S_E = \frac{I_A - I_B}{I_B} \tag{6.22}$$

where I_A and I_B are shown in Figure 6.11.

From (6.22)

$$I_A = (1 + S_E) I_B \tag{6.23}$$

is obtained. Similarly, (6.24) is obtained from Figure 6.11.

$$E_A = (1 + S_E) E_B \tag{6.24}$$

where S_E is nonlinear and can be represented by any suitable nonlinear function [1]. If the slope of the air gap line is equal to $1/G$, the total saturated current becomes as

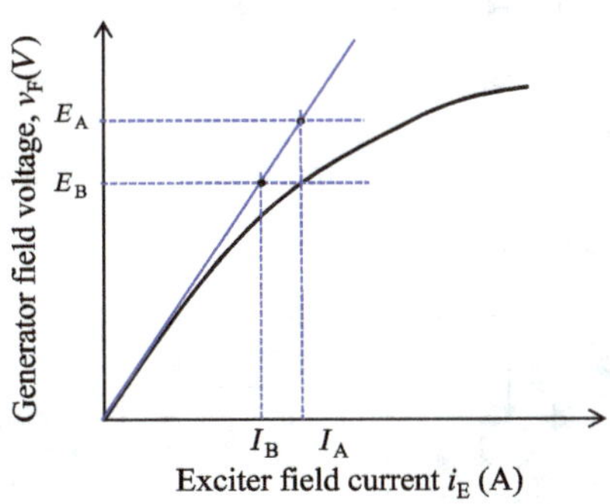

Figure 6.11 Saturation curve of the exciter

in (6.25).

$$i_E = Gv_F + Gv_F S_E \tag{6.25}$$

Substituting (6.25) into (6.21) yields (6.26).

$$\tau_E \dot{v}_F = v_E - R_E G v_F - R_E G v_F S_E \tag{6.26}$$

It should be noted that the output voltage v_F of the exciter is the field voltage of the synchronous generator and it causes the induced emf voltage E_{FD} in the stator windings of the synchronous generator as described by (6.27).

$$E_{FD} = \frac{L_{AD}}{\sqrt{3} r_F} v_F \tag{6.27}$$

where r_F is the total resistance in the field circuit of the generator. Finally, the exciter model is obtained as in Figure 6.12 using the last two Equations (6.26) and (6.27).

A normalization of the exciter system equations results in (6.28).

$$E_{FD} = \frac{1}{K_E + \tau_E s}(V_E - E_{FD} S_E) \tag{6.28}$$

where $K_E = R_E G - 1$. Thus, the block diagram of the exciter will be as in Figure 6.13.

The effect of the excitation voltage v_F of the synchronous generator on the terminal voltage is analyzed by Anderson and Fouad [1]. This analysis shows that the operational behavior of the generator is similar to that of a linear amplifier with a time constant τ_G having values between no load time constant τ'_{d0} and short circuit time constant τ'_d. Representing the generator amplifier gain by K_G and

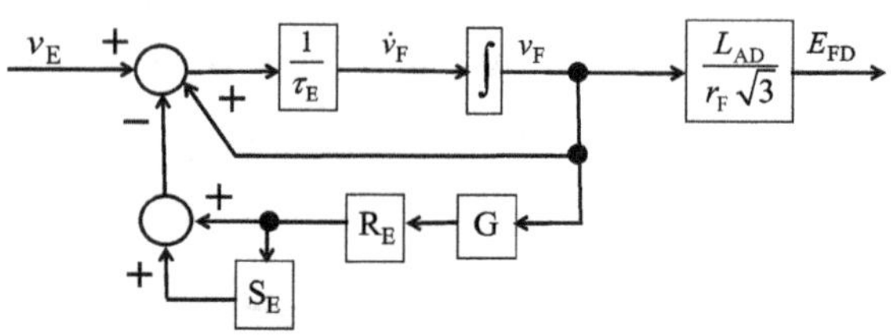

Figure 6.12 Simulation diagram of the exciter

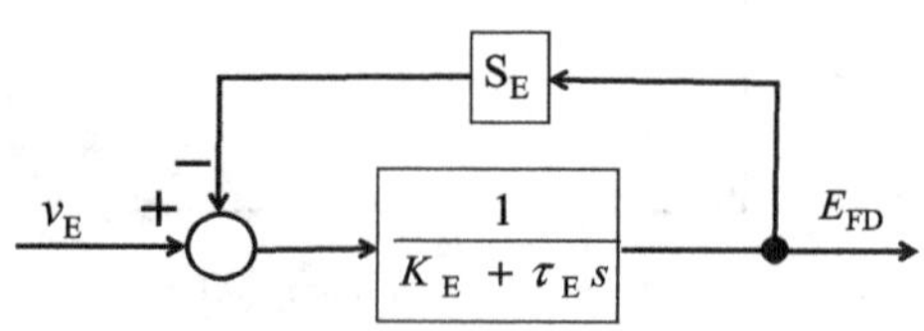

Figure 6.13 Normalized block diagram of the exciter

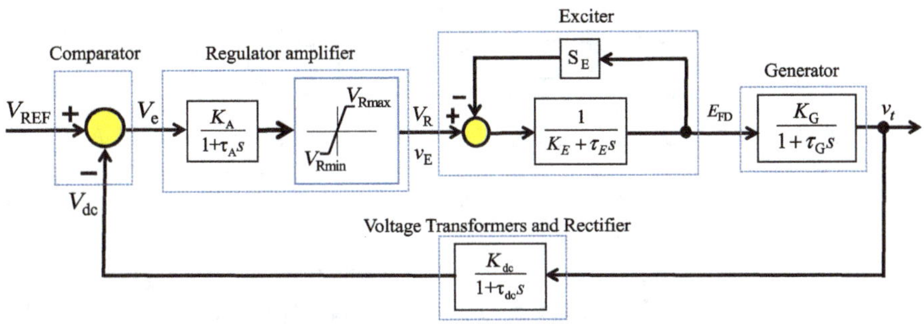

Figure 6.14 Block diagram of the overall exciter system model

including the saturation effect, the terminal voltage can be written as a function of the induced voltage E_{FD} as in (6.29).

$$V_t = \frac{K_G}{1 + \tau_G s} E_{FD} \tag{6.29}$$

The overall block diagram of the complete exciter system is obtained by combining the models of the subsystems included in the overall system. Therefore, the model of the VTs and rectifier given by (6.6), the model of the difference amplifier shown in Figure 6.8, the model of the regulator amplifier given in Figure 6.10, the exciter model in Figure 6.14, and the generator model given in (6.29) are combined in operational order to yield the complete model of the exciter system shown in Figure 6.14.

6.2.2 *State-space model of excitation systems*

In order to simulate the excitation control system, the block diagram given in Figure 6.14 can be simulated by constructing the same diagram in Simulink®. However, a state-space model will be more useful to simulate the system by applying numerical solution methods in any programming language such as C programming language. The state-space model of the excitation system can be obtained by various methods. One of these methods is the one converting the block diagram in Figure 6.14 into a simulation diagram and writing the state-space model equations directly from the simulation diagram. In order to get an overall simulation diagram, each transfer function block in Figure 6.14 is converted to a simulation diagram first and then combined to yield the general simulation diagram for the complete excitation system. For example, the transfer function of the regulator amplifier given by (6.8) can be rewritten as in (6.30) by ignoring the saturation block shown in Figure 6.9.

$$\tau_A s V_R = K_A V_e - V_R \tag{6.30}$$

Inverse Laplace transformation of (6.30) with zero initial conditions yields (6.31), which can be represented by the simulation diagram shown in Figure 6.15.

$$\tau_A \dot{v}_R = K_A v_e - v_R \tag{6.31}$$

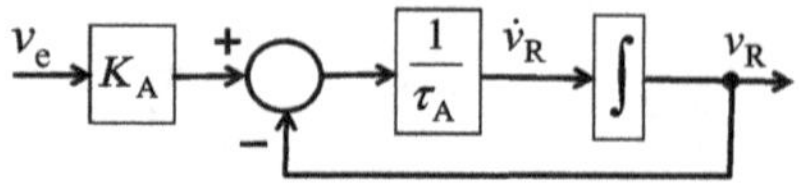

Figure 6.15 Block diagram of the overall exciter system model

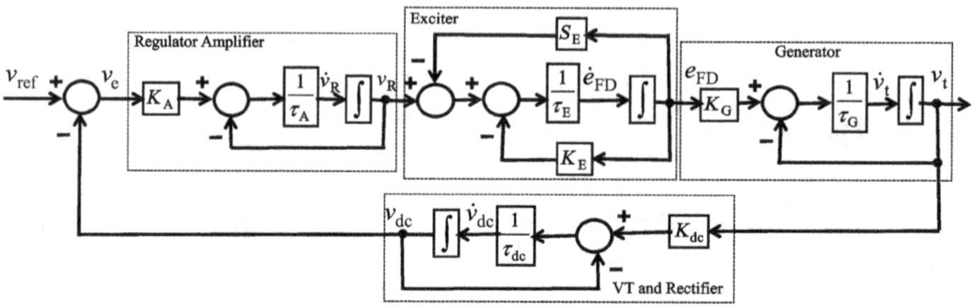

Figure 6.16 Simulation diagram of the overall exciter system with the regulator

If all of the transfer function blocks shown in Figure 6.14 are converted to their simulation diagrams and combined together keeping the input and output signals unchanged in order to obtain the complete simulation diagram given in Figure 6.16.

The state-space model equation of the excitation system with the regulator amplifier can be written as in (6.32) directly from the simulation diagram given in Figure 6.16.

$$\begin{bmatrix} \dot{v}_{dc} \\ \dot{v}_t \\ \dot{e}_{FD} \\ \dot{v}_R \end{bmatrix} = \begin{bmatrix} -\dfrac{1}{\tau_{dc}} & \dfrac{K_{dc}}{\tau_{dc}} & 0 & 0 \\ 0 & -\dfrac{1}{\tau_G} & \dfrac{K_G}{\tau_G} & 0 \\ 0 & 0 & -\dfrac{(S_E + K_E)}{\tau_E} & \dfrac{1}{\tau_E} \\ -\dfrac{K_A}{\tau_A} & 0 & 0 & -\dfrac{1}{\tau_A} \end{bmatrix} \begin{bmatrix} v_{dc} \\ v_t \\ e_{FD} \\ v_R \end{bmatrix} + \begin{bmatrix} 0 \\ 0 \\ 0 \\ \dfrac{K_A}{\tau_A} \end{bmatrix} v_{ref} \tag{6.32}$$

The regulator amplifier included in Figure 6.19 is a first-order amplifier and acts as a controller in the system. Therefore, the regulator amplifier is replaced by any other applicable type of controller such as PID or FLC if needed. Therefore, a simulation model without the regulator amplifier as given in Figure 6.17 may be required to be used. The blocks representing the regulator amplifier have been removed from Figure 6.16 to yield Figure 6.17, which includes rectifier and VTs, comparator, exciter, and the generator.

The state-space model of the exciter system without the regulator amplifier is given in (6.33), which can be written directly from Figure 6.17.

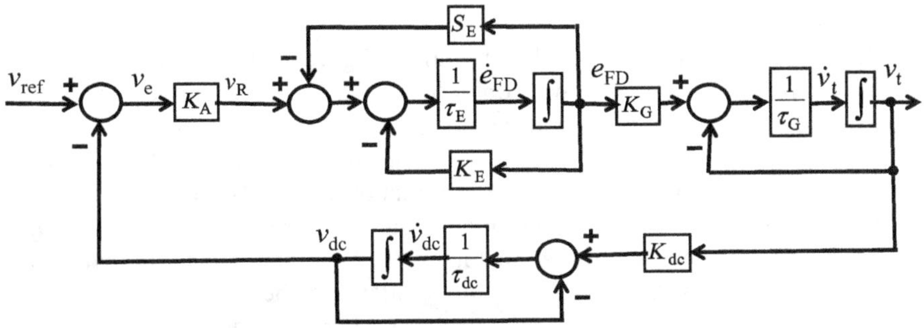

Figure 6.17 Simulation diagram of the overall exciter system without the regulator

$$\begin{bmatrix} \dot{v}_{dc} \\ \dot{v}_t \\ \dot{e}_{FD} \end{bmatrix} = \begin{bmatrix} -\dfrac{1}{\tau_{dc}} & \dfrac{K_{dc}}{\tau_{dc}} & 0 \\ 0 & -\dfrac{1}{\tau_G} & \dfrac{K_G}{\tau_G} \\ -\dfrac{K_A}{\tau_E} & 0 & -\dfrac{(S_E + K_E)}{\tau_E} \end{bmatrix} \begin{bmatrix} v_{dc} \\ v_t \\ e_{FD} \end{bmatrix} + \begin{bmatrix} 0 \\ 0 \\ \dfrac{K_A}{\tau_E} \end{bmatrix} v_{ref} \tag{6.33}$$

The simulation diagrams and state-space models of the excitation systems given in this section are just the conventional models of rotational exciters used in power systems. Static exciters are also used. The excitation system models discussed above are the models of the basic structures of these systems. They do not include any compensators, controllers, or power system stabilizers. Only the regulator amplifier is included in (6.32) and in Figure 6.16.

6.2.3 Fuzzy logic control of excitation systems

In order to damp oscillations faster and to make the power system more robust against the oscillations in system frequency and voltages, additional compensators or power system stabilizers are used. These compensators and stabilizers act like controllers and filters to damp the oscillation so that the system keeps operating at stable conditions. A power system stabilizer (PSS) gets the oscillating signal as an input and generates a damping signal, which is added to the summer where the error signal v_e is calculated. The output signal of the PSS is added as a feedback signal to damp the oscillations faster. A PSS can be designed to be used to damp voltage oscillations as well as frequency oscillations. Various PSS structures have been studied in literature including both conventional and intelligent methods. The use of fuzzy logic as controller, stabilizer, and compensator is discussed here with examples. The effect of adding an FLC to the exciter system is presented in Example 6.1.

Example 6.1

Develop a fuzzy logic controller for the system given in Figure 6.16 so that the voltage oscillations are reduced. The system parameters are given as $\tau_{dc} = 0.015$, $K_{dc} = 1.0$, $\tau_A = 0.05$, $K_A = 5$, $\tau_E = 0.5$, $K_E = -0.05$, SE = 0.4, $\tau_G = 1.0$, $K_G = 1.0$.

Answer 6.1

In order to develop an FLC for the system in Figure 6.16, the structure of the oscillations should be observed with only the regulator amplifier used as the controller. Then the fuzzy rules should be created to generate the actions so that the oscillations are reduced. Therefore, let us simulate the exciter system shown in Figure 6.17 or given by (6.33) without a regulator and any other controller. The response of unregulated and uncontrolled systems is given in Figure E6.1a. A comparison of the voltage error V_e in Figure E6.1a with the error signal $e(k)$ in Figure 4.16 shows that both responses are similar and the fuzzy rules developed using Figure 4.16 can be applied to the exciter systems as well.

As the next step, let us add an FLC to the exciter system as depicted in Figure E6.1b, such that the regulator is not included. The FLC block includes a MATLAB® .m file representing the FLC code.

The effect of adding an FLC to the exciter as a regulator is given in Figure E6.1(c). It is shown that the oscillations are reduced by the FLC. The new response does not oscillate and reaches a steady state value with minimum error.

The FLC used in this system is developed as a MATLAB .m file using a 25-rule table and sinusoidal membership functions. The fuzzy rule table used in this example is given in Table E6.1(a) and the fuzzy subsets used in input spaces of error V_e and error change dV_e and in the output space du are given in Figure E6.1(d). The fuzzy

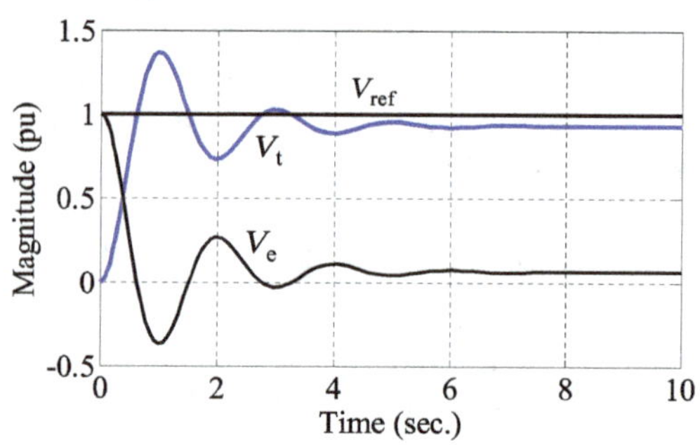

Figure E6.1(a) Responses of the exciter system without regulator and without controller

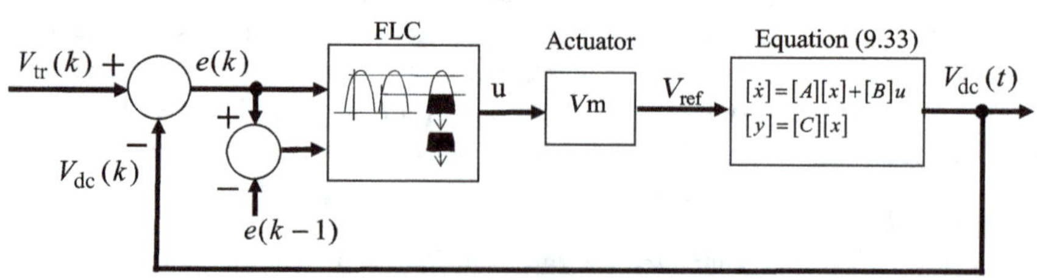

Figure E6.1(b) Block diagram of the exciter system with FLC

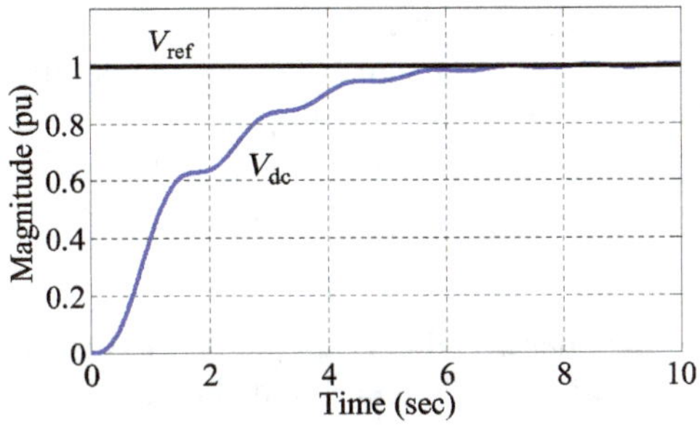

Figure E6.1(c) Response of the exciter system with FLC

Table E6.1(a) The fuzzy rule table used in Example 6.1

***e*(*k*)**	***de*(*k*)**				
	NB_{de}	**NS_{de}**	**ZE_{de}**	**PS_{de}**	**PB_{de}**
NB_e	NB_{du}	NB_{du}	NB_{du}	NB_{du}	ZE_{du}
NS_e	NB_{du}	NS_{du}	NS_{du}	ZE_{du}	PS_{du}
ZE_e	NS_{du}	NS_{du}	ZE_{du}	PS_{du}	PS_{du}
NS_e	NS_{du}	ZE_{du}	PS_{du}	PS_{du}	PB_{du}
PB_e	ZE_{du}	NS_{du}	NS_{du}	PB_{du}	PB_{du}

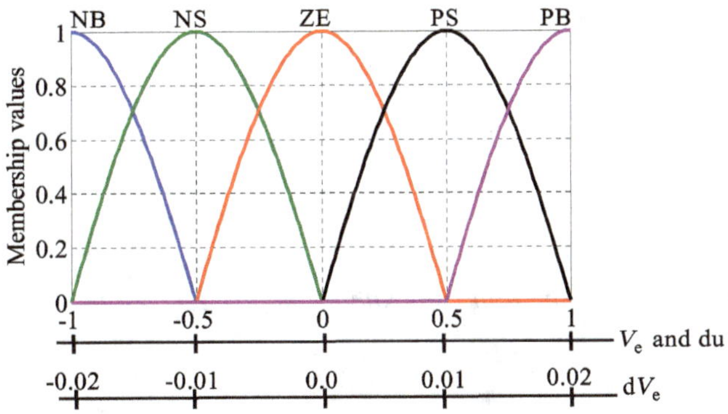

Figure E6.1(d) Fuzzy membership functions used in Example 6.1

subsets used in all input and output spaces are represented by *sinusoid* membership functions. The difference comes only from the boundaries of the universes. Since the terminal voltage of the synchronous generator is assumed to be 1.0 pu, the reference voltage input is also selected as 1.0. Therefore, the voltage error V_e is defined in the interval $\{-1, 1\}$. Depending on the sampling period, the interval for the change in error is much smaller than the interval of the universe of error V_e. In this example, the interval for the change in error is set to $\{-0.02, 0.02\}$. The boundaries of the change in control signal are up to the control signal u. Since the control signal is defined in the interval $\{0, 1\}$, than the change in control signal can be defined in $\{-1, 1\}$ same as the error signal. Mamdani fuzzy reasoning algorithm with the center of area defuzzification method is used as the fuzzy processor to control this system.

Example 6.2

Simulate the exciter system described in Example 6.1 in Simulink by using the FLC model developed for Simulink in Chapter 4 and shown in Figure 4.25.

Answer 6.2

Simulation diagram of the exciter system given in Figure 6.16 can be represented in Simulink as shown in Figure E6.2(a) with the FLC block added as a controller. Each block named as regulator amplifier, exciter generator, generator, and VT and rectifier shown in Figure E6.2(a) contains the related simulation models depicted in Figure 6.15 by dashed lines. The numerical values used for the parameters in Example 6.1 are also used in this example. A gain block used to scale the control signal for magnitude adjustment is added to the block diagram. The resultant plot of the exciter system in Simulink is given in Figure E6.2(b). The response of the output DC voltage V_{dc} in this example is similar to the one obtained in Example 6.1. Due to the numerical solution method and its algorithm used in Simulink being different in Simulink, the response from the Simulink model is smoother.

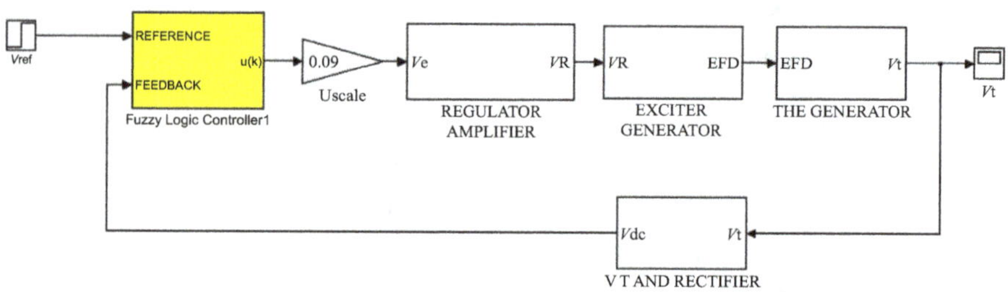

Figure E6.2(a) Simulink diagram of the exciter system with FLC

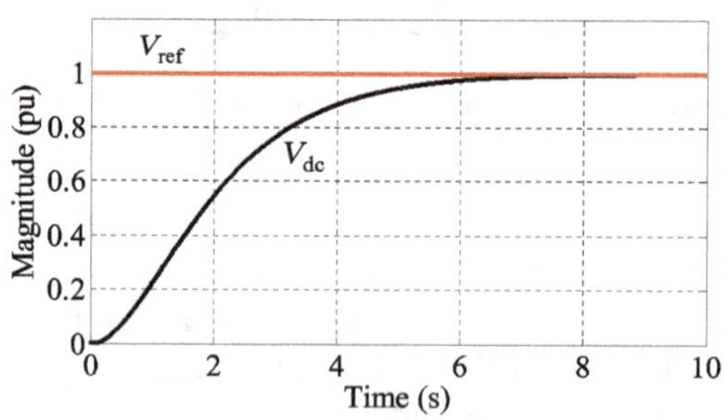

Figure E6.2(b) Response of the exciter system with FLC in Simulink.

Design Example 6.1

In order to increase the robustness of power excitation systems, usually a power system stabilizer (PSS) or a power system compensator (PSC) is added to the excitation systems. The outputs of both PSS and PSC are used as additional inputs to the summer where the reference and feedback signal from the system output are connected. Use Figures DE6.1(a) and DE6.1(b) as a benchmark for the design cases asked below.

(a) Simulate the exciter system shown in Figure 6.16 with data given in Example 6.1 without any additional controller or component.
(b) Simulate the exciter system shown in Figure 6.17 with data given in Example 6.1 without any additional controller or component.
(c) Develop a fuzzy PSS that uses the terminal voltage change as input in order to reduce the voltage oscillations and increase the stability of the system.
(d) Develop a fuzzy PSC that uses the terminal voltage as input in order to reduce the voltage oscillations and increase the stability of the system.

Compare the effects of fuzzy PSS and Fuzzy compensator in power excitation systems.

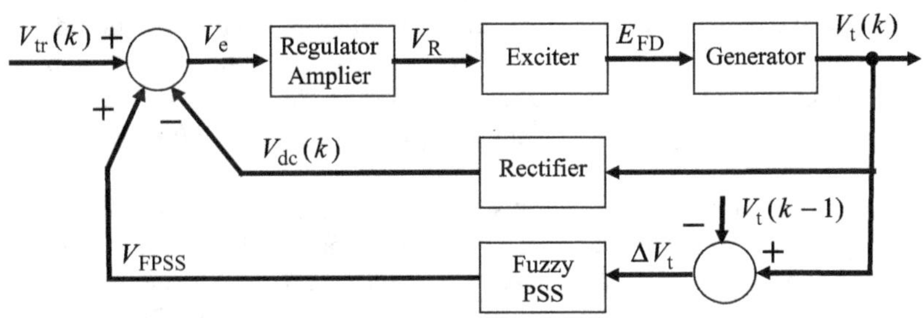

Figure DE6.1(a) A fuzzy PSS design diagram for Design Example 6.1

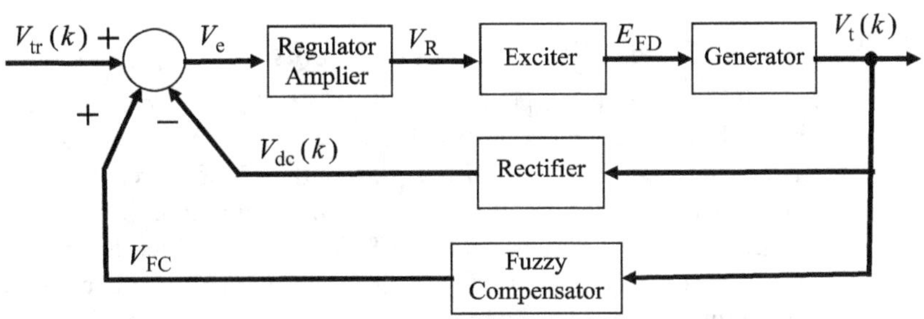

Figure DE6.1(b) A fuzzy compensator design diagram for Design Example 6.1

6.3 Load–frequency control

Load–frequency (L–F) control is an important task in electrical power system design and operation. Since the load demand varies without any prior schedule, the power generation is expected to overcome these variations without any voltage and frequency instabilities. Therefore, voltage and frequency controllers are required to maintain the generated power quality in order to supply power to the utility under constant voltage and frequency operating conditions. The voltage control is done by AVRs containing various types of compensators and controllers as discussed in the previous section. The frequency control is done by load–frequency controllers, which deal with the control of generator loadings depending on the frequency changes. Each single generation area includes a load–frequency controller to comply with the oscillations in frequency. Similarly, the load–frequency control has to be applied in each area of multi-area power systems [15–21].

The load–frequency controllers are designed to ensure the stable and reliable operation of power systems. Since the components of a power system are nonlinear, a linearized model around an operating point is used in the design process of L–F controllers. However, controllers based on linearized models are not capable of supporting parameter variations for stability. Therefore, alternative methods such as conventional controllers with improved dynamic models or adaptive and intelligent controllers are required. Some of the proposed methods in literature deal with system stability using fixed local plant models without considering the changes in some system parameters as some methods use decentralized control approaches based upon field testing and tuning in order to give robustness to the controllers. Generalized approaches [45] based on the concepts of discontinuous control, dual-mode control, and variable structure systems have also been proposed [46,47].

The variables of a single power system operating as a part of multi-area power affect the system stability of the other areas. Therefore, an L–F control process has to overcome the negative effects of system variables. The increasing interest and development in intelligent control techniques have become a solution to the control problems in power systems [11–15]. Due to their ability being able to model human decision-making processes and represent vague and uncertain data, fuzzy logic-based controllers have become an attractive choice in solving power system control problems [11–14].

6.3.1 Small-signal modeling of power systems

As it is well defined in the literature [1–10], changes in real power mainly affect the system frequency while the changes in reactive power have more effects on voltage. The frequency is more sensitive to real power than it is to reactive power. Therefore, the frequency oscillations due to the changes in active power demand are damped using load–frequency controllers (LFC), which are used to maintain a reasonable uniform frequency. The first step of control engineering consists of mathematical modeling. Although fuzzy logic controllers do not require mathematical models in real-time applications, they are needed for simulation purposes.

Different types of models such as the one called small signal model have been drawn for LFC systems [1–10].

The load on a power system is usually the sum of two separated parts: one is the independent load required by devices such as the ones in lighting and heating, and the other part is required by motor-type loads, which are sensitive to changes in frequency. The degree of sensitivity to frequency changes depends on the total combination of the speed–load characteristics of all driven devices. The speed–load characteristic of a composite load is approximated as in (6.34).

$$\Delta P_{\mathrm{e}} = \Delta P_{\mathrm{L}} + D\Delta\omega \tag{6.34}$$

where ΔP_{L} is the nonfrequency-sensitive load change, and $D\Delta\omega$ is the frequency-sensitive load change. D is a constant, expressed as the percentage change in load divided by percentage change in frequency. ΔP_{e} is the generated electrical power and is assumed to be equal to the total load power.

The Laplace transform of the generator model equation (the swing equation of a synchronous machine in this case), applied on a small perturbation, leads to (6.35).

$$\Delta\Omega(s) = [\Delta P_{\mathrm{m}}(s) - \Delta P_{\mathrm{e}}(s)]/(2Hs) \tag{6.35}$$

Substitution of (6.34) into (6.35) yields (6.36).

$$\Delta\Omega = \frac{\Delta P_{\mathrm{m}} - \Delta P_{\mathrm{L}}}{2Hs + D} \tag{6.36}$$

A linearized small signal model of LFC is shown in Figure 6.18 for a single area system, where H is the machine inertia in seconds and all τs are time constants.

The combination of (6.34) and (6.36) results in the block *Rotating mass and load* in Figure 6.18. The mechanical power generation, commonly known as the prime mover, may be either by hydraulic or steam turbine. The mathematical model of the turbine relates the changes in mechanical power output ΔP_{m} to changes in steam valve position ΔP_{V}. The most simple prime mover model can be approximated with a single time constant as in (6.37).

$$\Delta P_{\mathrm{m}} = \frac{1}{1 + \tau_{\mathrm{T}} s}\Delta P_{\mathrm{V}} \tag{6.37}$$

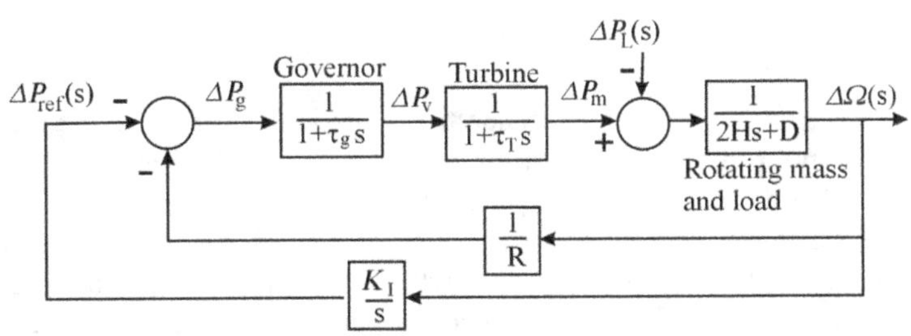

Figure 6.18 Block diagram of a single area load–frequency control

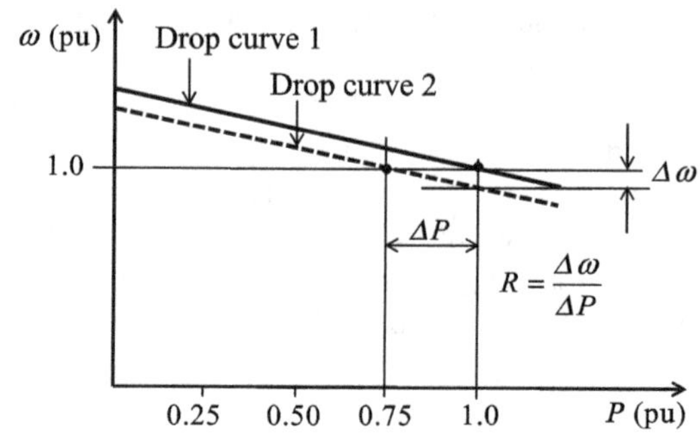

Figure 6.19 Steady-state frequency drop curve of the governor system

The governor model is derived from Figure 6.19, which is called the frequency drop curve. The frequency drops linearly with increasing demand power. A speed changer is employed to keep the frequency constant at different operation power values. For example, the speed changer is set to operate the system at nominal values $\omega = 1.0$ and $P = 1.0$ pu as depicted by *drop curve 1* (the solid line) in Figure 6.19. If demand power is reduced to 0.75 pu, then the speed changer adjusts the steam input to the turbine so that the generator operates at a new drop curve maintaining the frequency at $\omega = 1.0$. This new drop curve is named as *drop curve 2* (dashed line) in Figure 6.19.

As depicted in Figure 6.19, a change ΔP in power causes a change $\Delta\omega$ in frequency. The change in frequency due to the change in active power gives the slope of the drop curve and is represented by letter R, which is also called *governor speed regulation*. The slop R has a value around 5%–6% of a power change from no load to full load condition. The speed regulator compares a reference power with the frequency change due to the slope of the drop curve and yields a change in triggering power signal ΔP_{g} as given in (6.38) which is passed through an amplifier, such as a hydraulic amplifier, to generate a change in the power to the turbine valve opening as in (6.39). The amplifier is represented by a first-order block called governor in Figure 6.18.

$$\Delta P_{\mathrm{g}} = \Delta P_{\mathrm{ref}} - \frac{1}{R}\Delta\omega \tag{6.38}$$

$$\Delta P_{\mathrm{V}} = \frac{1}{1+\tau_{\mathrm{g}}s}\Delta P_{\mathrm{g}} \tag{6.39}$$

When the electrical load of the generator is suddenly increased, the mechanical power input cannot provide this electrical power instantly. At first, this extra energy will be taken from the kinetic energy stored in the rotating system. Then, this energy will moderate causing a decrease in the generator frequency. The change in speed of the shaft will affect the governor, which will adjust the turbine input valve, whereby a new steady-state will be reached. A mechanical construction, existing of the governor and a hydraulic amplifier, tends to the feedback loop with gain $1/R$, which can be seen in Figure 6.18 where another feedback loop is considered as an integral controller to reduce the frequency oscillations. By applying this integral controller, the steady-state frequency deviation, caused by the governor speed regulation, can be reduced to zero. This integral controller will be replaced by a fuzzy logic controller associated with a fuzzy decision unit in this section.

A group of generators is said to be coherent if they are coupled internally, swing in unison, and tend to have the same response characteristics. Such a group of generators is included in a control area that can be represented by a load–frequency control loop. As a simplification, a two-area system is considered here. The generating units are interconnected by a lossless tie line as shown in Figure 6.20.

The active power transmission from generating unit 1 to generating unit 2 is expressed as in (6.40).

$$P_{12} = \frac{E_1 E_2}{X_{12}} \sin \delta_{12} \tag{6.40}$$

where E_1 and E_2 are the effective values of back emfs of generating units 1 and 2, respectively. δ_{12} is the power angle and is equal to the difference between δ_1 and δ_2. X_{12} is the total reactance between the back emfs of both generating units. Since the effective voltages E_1, E_2, and total reactance X_{12} are assumed to be constant parameters, a plot of active power transmission versus the power angle is obtained as in Figure 6.21. If there is a small change in transmitted active power from initial power P_{120} to another power P_{12}, there will be a small change in power angle from the initial angle δ_{120} to the new value δ_{12} as shown in Figure 6.21.

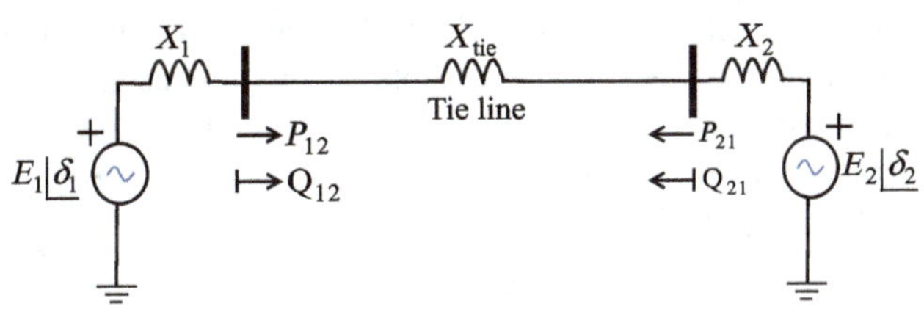

Figure 6.20 Power transmission between two areas over a tie line

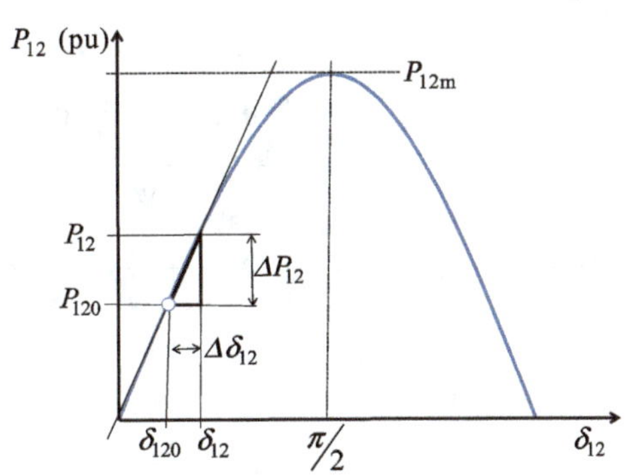

Figure 6.21 Active power transmission as a function of power angle

The change in the power flowing in the tie line is represented by ΔP_{12}, which is a function of the generated voltages in both generators and the phase angle difference, $\delta_{12} = \delta_1 - \delta_2$. A small deviation in the tie line flow can be linearized as in (6.41) and (6.42).

$$\Delta P_{12} = \left(\frac{dP_{12}}{d\delta_{12}}\right)_{\delta_{120}} \Delta\delta_{12} \tag{6.41}$$

$$\Delta P_{12} = P_{\mathrm{s}} \Delta\delta_{12} \tag{6.42}$$

The last equation can be rewritten as in (6.43) considering the changes in δ_1 and δ_2.

$$\Delta P_{12} = P_{\mathrm{s}}(\Delta\delta_1 - \Delta\delta_2) \tag{6.43}$$

where P_{s} is the slope of power angle curve at the initial operating angle δ_{120}. The tie line power flow begins whenever a load increment or decrement occurs in one of the areas.

An LFC system should be able to keep frequency approximately at the nominal value, maintain the tie line flow at around the scheduled value and each area should absorb its own load changes. In order to reach these goals, a controller should be added to each area along with the frequency deviation feedbacks over frequency bias factors, B_1 and B_2, as depicted in Figure 6.22. The power flow signal in the tie line is called as the area control error (ACE) to be used by the applied controller.

The small signal simulation model of two-area power system given in Figure 6.22 uses a classical integral controller. In order to compare the results from the FL controller with a decision maker, a classical integral controller is used to simulate the system first. The parameters R, B, D, $1/H$, P_{s}, and K_is are constant and defined earlier in the text. All τs are time constants. The variables x_1 to x_9 are used for: $x_1 = \Delta\omega_1$, $x_2 = \Delta\omega_2$, $x_3 = \Delta P_{\mathrm{m1}}$, $x_4 = \Delta P_{\mathrm{m2}}$, $x_5 = \Delta P_{\mathrm{V1}}$, $x_6 = \Delta P_{\mathrm{V2}}$, x_7 = output from controller in area 1, x_8 = output from controller in area 2, and $x_9 = \Delta P_{12}$. Steady-space matrices can easily be written from Figure 6.22.

$$A = \begin{bmatrix} -\frac{D_1}{2H_1} & 0 & \frac{1}{2H_1} & 0 & 0 & 0 & 0 & 0 & -\frac{1}{2H_1} \\ 0 & -\frac{D_2}{2H_2} & 0 & \frac{1}{2H_2} & 0 & 0 & 0 & 0 & \frac{1}{2H_2} \\ 0 & 0 & -\frac{1}{\tau_{\mathrm{T1}}} & 0 & \frac{1}{\tau_{\mathrm{T1}}} & 0 & 0 & 0 & 0 \\ 0 & 0 & 0 & -\frac{1}{\tau_{\mathrm{T2}}} & 0 & \frac{1}{\tau_{\mathrm{T2}}} & 0 & 0 & 0 \\ -\frac{1}{R_1\tau_1} & 0 & 0 & 0 & -\frac{1}{\tau_{\mathrm{g1}}} & 0 & \frac{1}{\tau_{\mathrm{g1}}} & 0 & 0 \\ 0 & -\frac{1}{R_2\tau_2} & 0 & 0 & 0 & -\frac{1}{\tau_{\mathrm{g2}}} & 0 & \frac{1}{\tau_{\mathrm{g2}}} & 0 \\ B_1K_{I1} & 0 & 0 & 0 & 0 & 0 & 0 & 0 & K_{\mathrm{I1}} \\ 0 & -B_2K_{\mathrm{I2}} & 0 & 0 & 0 & 0 & 0 & 0 & K_{\mathrm{I2}} \\ P_{\mathrm{s}} & -P_{\mathrm{s}} & 0 & 0 & 0 & 0 & 0 & 0 & 0 \end{bmatrix} \tag{6.44}$$

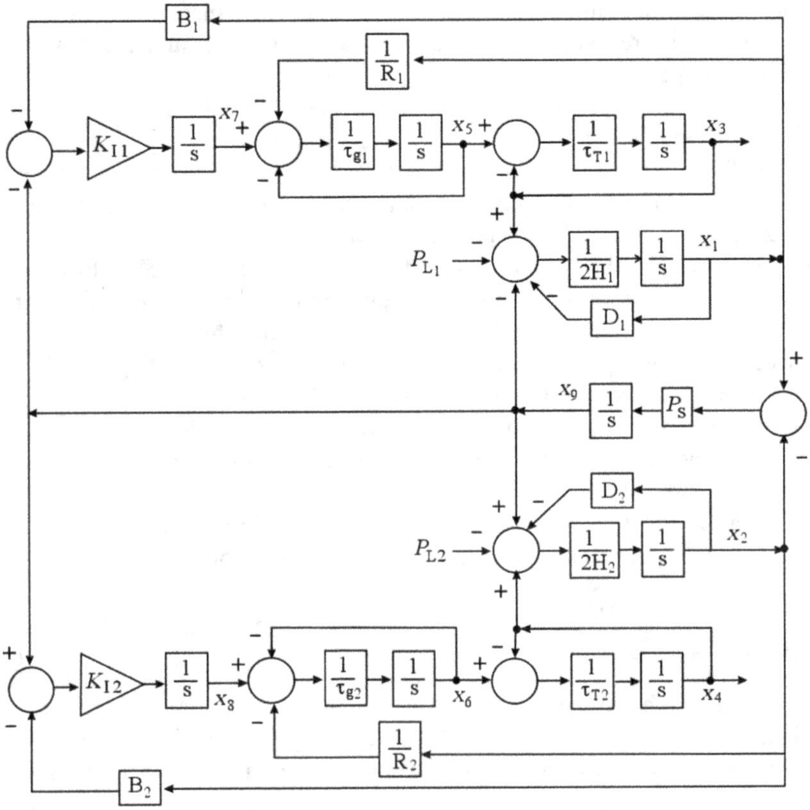

Figure 6.22 Simulation diagram of a two-area system

Since there are two areas, the steady-state equation can be written with two input matrices B_1 and B_2 as in (6.45).

$$B_1 = \begin{bmatrix} -1 \\ 0 \\ 0 \\ 0 \\ 0 \\ 0 \\ 0 \\ 0 \\ 0 \end{bmatrix}, \quad B_2 = \begin{bmatrix} 0 \\ 1 \\ 0 \\ 0 \\ 0 \\ 0 \\ 0 \\ 0 \\ 0 \end{bmatrix}, \quad \dot{x} = Ax + B_1u_1 + B_2u_2 \tag{6.45}$$

The input variable vector, x, is a column vector with elements from x_1 to x_9. The inputs u_1 and u_2 are the changes ΔP_{L1} and ΔP_{L2}, respectively, in load powers of each area.

The system contains two controllers, one for each plant. The input signals to the controllers are called area-controlled errors (ACE), and are obtained as, $e_1 = -B_1x_1 - x_9$ and $e_2 = -B_2x_2 + x_9$.

The symbol Δ is used to represent the changes in actual signals. Since this is a small signal model, all signals expressed by Δ are zero in the equilibrium point of operation. Therefore, the system can be triggered by changes in any of the variables. Under normal operating conditions, the source of this triggering signal is the change ΔP_{L} in load demand in power systems. It is the fact that the triggering signals here are not absolute values, but relative values according to a selected reference level. In order not to lead to any confusion between this definition and the more common one used in small signal modeling, in which the Δ refers to the derivative. A visual representation of this notation is shown in Figure 6.23 where ΔP_{L} is presented to explain the way the symbol Δ is used. The coordinate system with the origin 0 is the original one with the variables at stand-still. If a step change occurs in ΔP_{L} power demand at time t_0, then the coordinates move to the one with the origin marked as $0'$.

As seen in Figure 6.23, the system load changes from P_{L0} to P_{L1} at the instant t_0. Therefore, a triggering signal is produced if the condition given by (6.47) exists.

$$\text{if}\, t < t_0 \Rightarrow P = P_{\mathrm{L0}} \tag{6.46}$$

$$\text{if}\, t > t_0 \Rightarrow P = P_{\mathrm{L1}} \tag{6.47}$$

In this case, the scale of the axis stays equal with the data being moved to the coordinate system with the origin $0'$. The process explained and given by (6.46) and (6.47) shows how the changes in load power are represented as the triggering signal of the system.

6.3.2 Fuzzy logic controller design for LFC

Referring to the FLC developed and discussed in Chapter 4, it is known that the FLC performs the same actions as a human operator by adjusting the input variables, only looking at the system output. Since the frequency oscillations are to be damped in this problem, the frequency deviations, $\Delta\omega$, and changes in frequency deviations, $\Delta(\Delta\omega)$, are the input signals of the FLC. When there is no oscillation,

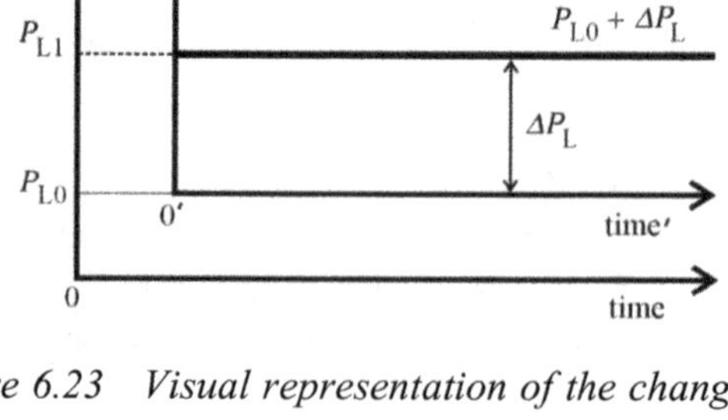

Figure 6.23 Visual representation of the change ΔP_L

the frequency remains constant and its deviation becomes zero. Control action is not needed under these operating cases. However, a control action is needed when there are frequency oscillations, which means that the frequency deviation and its change are not zero requiring control action. Any change in load powers P_{L1} and P_{L2} in Figure 6.22 causes the frequency changes $\Delta\omega=\Delta\omega_1=x_1$ for area 1 and $\Delta\omega=\Delta\omega_2=x_2$ for area 2. In order to damp the frequency oscillations, the generated power is either increased or decreased. Therefore, a power change is added to the previous value of the power to make the generation and dissipation equal to each other. Then the question is how the amount of power change is determined and generated. As shown in Figure 6.24, the amount of change $\Delta(\Delta P(k))$ in generated power is determined by a fuzzy logic inference system and added to the previous value $\Delta(\Delta P(k-1))$ to yield the current output $\Delta(\Delta P(k))$. It should be noted that this is nothing but the digital implementation of an integrator, using Euler integration. In other words, the FLC acts as an integrator. The FL rules in the FLC are developed to yield a similar but more effective output than an integrator gives. The difference between a fuzzy logic controller and an integral controller is the procedure used to calculate $\Delta(\Delta P(k))$. The fuzzy processor used in this procedure is given next.

Fuzzy processor for load–frequency control

As shown in Figure 6.24, there are two inputs to the fuzzy processor. The first one is the change in angular velocity, the other one is the change of it, $\Delta(\Delta\omega(k))$. As the angular velocity is supposed to be constant, the change of this velocity can be considered as a disturbance in the system and should be reduced to zero as soon as possible. These two inputs are fuzzified and converted to fuzzy membership values that are used in the rule base in order to execute the related rules so that output can be generated. The fuzzy rule base, or the fuzzy decision table, is the unit mapping of two crisp inputs, the just mentioned ones, to the fuzzy output space defined on the universe of $\Delta(\Delta P(k))$. To simplify the following text, the iteration counter, k, will be omitted from now on.

The time response of the disturbed change in angular velocity for an impulse input can be represented by the generalized impulse response error of a second-order system. The impulse response is considered here for the reason that when a uniform step in the angular velocity, ω, appears, the derivative of ω will be one for a short time and then zero later on. Since the intention is to design a fuzzy logic controller with a better performance (shorter settling time, less overshoot) than the

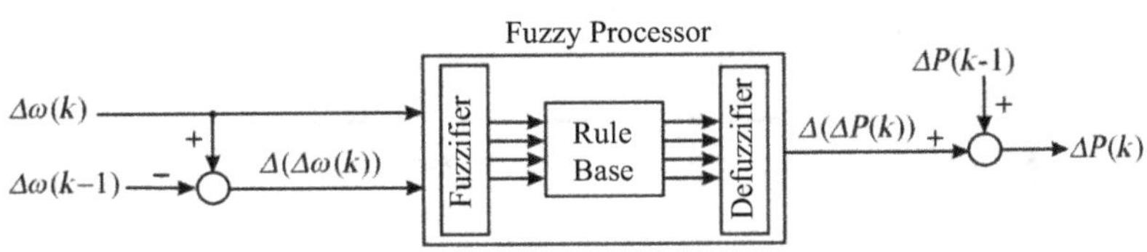

Figure 6.24 The FLC system used to determine the change in generated power

classical controller, the response signal of this system is taken as a reference to construct the rule table. The response signal, generated by the classical load–frequency controller for an impulse input change is given in Figure 6.25. The plot is obtained by using the matrices composed of (6.40) and (6.41) in the previous section for the classical integral controller.

The fuzzy rules represent the knowledge and abilities of a human operator who makes necessary adjustments to operate the system with minimum error and fast response. It is necessary to observe the behaviors of the error signal $\Delta\omega$ and its change $\Delta(\Delta\omega)$ on different operating regions in order to model the actions a human operator would take. The decision is based upon different operating cases, deciding whether the change, $\Delta(\Delta P)$, in the controller output should be increased or decreased according to the inputs of the fuzzifier. This controlled output is the required change in the input of the system.

As been told, the result shown in Figure 6.25 is useful to construct the rule table. The values from these graphics will be used to define the initial fuzzy set intervals, which will be modified to reach a better structure with improved performance.

Let us first try to develop an initial rule base (with only 3 fuzzy sets), which will be extended to a 5 fuzzy sets base. According to the signs of $\Delta\omega$ and $\Delta(\Delta\omega)$, the sign of $\Delta(\Delta P)$ is decided whether to be positive or negative. A summary of all possible situations, or so-called operation regions, is given in Table 6.1.

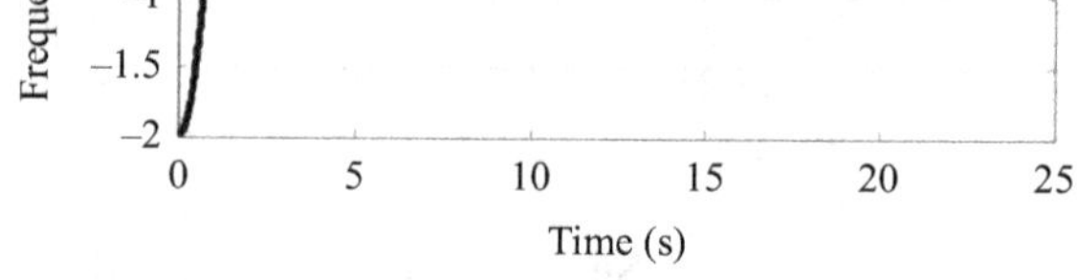

Figure 6.25 Impulse response of the system without any controller

Table 6.1 Output decision table

	Operating regions								
$\Delta\omega$	+	0	−	−	0	+	−	+	0
$\Delta(\Delta\omega)$	−	−	−	+	+	+	0	0	0
$\Delta(\Delta P)$	+	−	−	−	+	+	−	+	0

The sign of $\Delta(\Delta P)$ should be positive if ΔP has to be increased and it should be negative otherwise. This simple rule is applied as in Table 6.1 to determine the sign of $\Delta(\Delta P)$. A verbal expression of these rules is: "The error is positive and decreasing towards zero. Therefore, $\Delta(\Delta P)$ is set to positive to reduce the error." This example expresses the first column of Table 6.1. Similar reasoning can be applied to the other columns. A "programming language translation" of this table gives the expression.

IF$\Delta\omega$ is zero THEN $\Delta(\Delta P)$ takes the sign of $\Delta(\Delta\omega)$, ELSE $\Delta(\Delta P)$ takes the sign of $\Delta\omega$.

Table 6.1 shows that each one of $\Delta\omega$, $\Delta(\Delta\omega)$, and $\Delta(\Delta P)$ has three different options for the signs to be assigned. They are either positive, negative, or zero. With this knowledge, an initial rule decision table with nine rules can be formed like in Table 6.2, where N means negative, Z zero, and P positive. The main part in the middle represents the rules as well as the signs of $\Delta(\Delta P)$.

From Table 6.2, some logical reasoning should be considered. A closer look at this table shows that in some cases, there is a transition from negative to positive, without passing through zero. Therefore, an adjustment in the initial rule table leads to another table without this inconvenience. The influence of $\Delta(\Delta\omega)$ must stay, and a symmetric solution has to be advised, so the modified rule table is the one that can be found in Table 6.3.

A next step in designing the FLC is the extension of the nine rules table shown in Table 6.3 to a 25 rules table, as in Table 6.4. In this last table, two extra sets are added which gives us 5 fuzzy sets for each one of the input spaces. A nine-rule fuzzy decision table may be sufficient for some applications. However, many applications require more rules than nine rules.

Table 6.2 Initial rule table for Δ(ΔP)

Δω	**Δ(Δω)**		
	N	**Z**	**P**
N	N	N	N
Z	N	Z	P
P	P	P	P

Table 6.3 A nine-rule symmetrical table for Δ(ΔP)

Δω	**Δ(Δω)**		
	N	**Z**	**P**
N	N	N	Z
Z	N	Z	P
P	Z	P	P

In Table 6.4, the negative and positive parts of the corresponding spaces are portioned into additional subsections as negative large (NL), negative small (NS), positive small (PS), and positive large (PL). The zero part in each universe is represented by just a single fuzzy set zero. Now with the input sets extended to five regions, a similar extension can be made for the output values of $\Delta(\Delta P)$. The result of this logical and symmetric extension is also shown in Table 6.4. This rule table is the final one that is going to be used in the FLC for the problem discussed here.

Table 6.4 is the final fuzzy rule table constructed to be used. Both input spaces in this table are partitioned into five subsections yielding a 25-rule table for the output decision space, which is also portioned into five fuzzy subsections. As suggested, the initial limits of the fuzzy sets will be derived from the impulse response of the system without any controller given in Figure 6.25. The set values can be derived immediately from Figure 6.25, but a plot of $\Delta\omega$ versus $\Delta(\Delta\omega)$ is preferred to identify the upper and lower limits of $\Delta\omega$ and $\Delta(\Delta\omega)$, clearly. A visualization of the definition of these fuzzy sets can be seen in Figure 6.26 where

Table 6.4 A nine-rule symmetrical table for Δ(ΔP)

Δω	Δ(Δω)				
	NL	**NS**	**ZE**	**PS**	**PL**
NL	NL	NL	NS	NS	ZE
NS	NL	NS	NS	ZE	PS
ZE	NS	NS	ZE	PS	PS
PS	NS	ZE	PS	PS	PL
PL	ZE	PS	PS	PL	PL

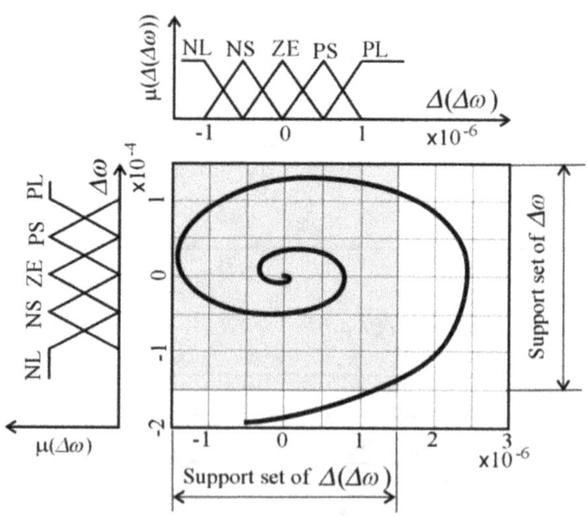

Figure 6.26 Partitioning input spaces into five fuzzy subsections

triangular fuzzy sets are used to partition the support sets of the input universes into fuzzy sub categories.

As given in Figure 6.26, the upper and lower limits of the support sets are different for $\Delta\omega$ and $\Delta(\Delta\omega)$. The interval of the latter is much smaller yielding different scaling for a proper operation of the controller. The fuzzy sets are initially defined in the intervals $\{-1.5 \times 10^{-4}, -1.5 \times 10^{-4}\}$ for $\Delta\omega$, and $\{-1.5 \times 10^{-6}, -1.5 \times 10^{-6}\}$ for $\Delta(\Delta\omega)$.

The combination of Table 6.4 and Figure 6.26 gives a well-defined summary of the fuzzy sets and the rules that can be applied to problems similar to the one being discussed here. The next step in the design process is fuzzy reasoning as discussed below.

Fuzzy reasoning

The crisp universes of $\Delta\omega$, $\Delta(\Delta\omega)$, and $\Delta(\Delta P)$ have been partitioned into five subsections as NL, NS, ZE, PS, and PL as explained earlier. These five fuzzy subsections in all three universes are represented by triangular fuzzy membership functions as described by (2.7) in Chapter 2. The peaks of triangular fuzzy membership functions are placed evenly in all three universes. Referring to (2.7) and Figure 2.4, the parameters of triangular fuzzy subsets used in simulations of FLC controlled load–frequency problem are summarized in Table 6.5. The use of Table 6.5 to define the fuzzy subsets in MATLAB or in Simulink is depicted in Figure 6.27 where a generalized reference universe X is used to show the concept.

Table 6.5 A nine-rule symmetrical table for $\Delta(\Delta P)$

	NL	NS	ZE	PS	PL
$\Delta\omega$	-1.5×10^{-4}	-0.75×10^{-4}	0.0	0.75×10^{-4}	1.5×10^{-4}
$\Delta(\Delta\omega)$	-1.5×10^{-6}	-0.75×10^{-6}	0.0	0.75×10^{-6}	1.5×10^{-6}
$\Delta(\Delta P)$	-1.0	-0.5	0.0	0.5	1.0

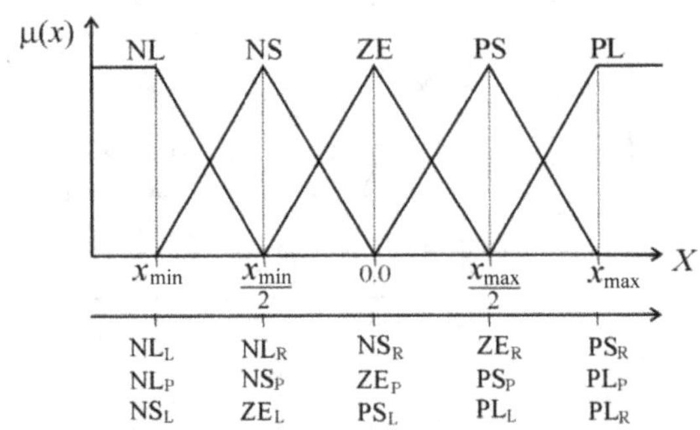

Figure 6.27 Partitioning input spaces into five fuzzy subsections

The subscripts L, P, and R represent left, peak, and right, respectively. The abbreviations such as NS_L, NS_P, and NS_R used in Figure 6.27 represent the parameters of triangular fuzzy subset negative small (NS). Therefore, (2.7) can be written as in (6.48) for the fuzzy subset NS.

$$\mu_{NS}(x) = \max\left(\min\left(\frac{x - NS_L}{NS_P - NS_L}, \frac{NS_R - x}{NS_R - NS_P}\right), 0\right) \tag{6.48}$$

Besides the definition of triangular fuzzy membership function, it is required to calculate the membership degree of $\Delta(\Delta P)$. The following example is given to clarify the method used to obtain the membership values in output space.

Example 6.3

The frequency change $\Delta\omega$ and its change $\Delta(\Delta\omega)$ are measured as $\Delta\omega = 0.45 \times 10^{-4}$ and $\Delta(\Delta\omega) = 0.15 \times 10^{-4}$ for a sampling instant. Obtain the required change $\Delta(\Delta P)$ in active power generation to damp the frequency oscillations by using a fuzzy decision process with the rule Table 6.4. Assume that the input and output universes have the lower and upper limits as given in Figure 6.26.

Answer 6.3

The fuzzy subsets used to partition the input and output spaces into five sub categories are shown in Figure E6.3a. The same fuzzy subsets with different parameters are used in three universes. The numerical values for each universe are given in Figure E6.3(a).

When $\Delta\omega = -0.9 \times 10^{-4}$ it intercepts with the fuzzy sets NL and NS in the input universe of $\Delta\omega$, and when $\Delta(\Delta\omega) = 0.45 \times 10^{-6}$ it intercepts with the fuzzy sets ZE and PS in the input universe of $\Delta(\Delta\omega)$ as shown in Figure E6.3(a). The horizontal lines are drawn through the intercepting points of $\Delta\omega$ on NL and NS in Figure E6.3(a) gives the membership values of $\Delta\omega$ on fuzzy sets NL and NS, respectively, while the horizontal line passing through the intercepting points of $\Delta(\Delta\omega)$ on ZE and PS gives the membership values of $\Delta(\Delta\omega)$ on these fuzzy sets,

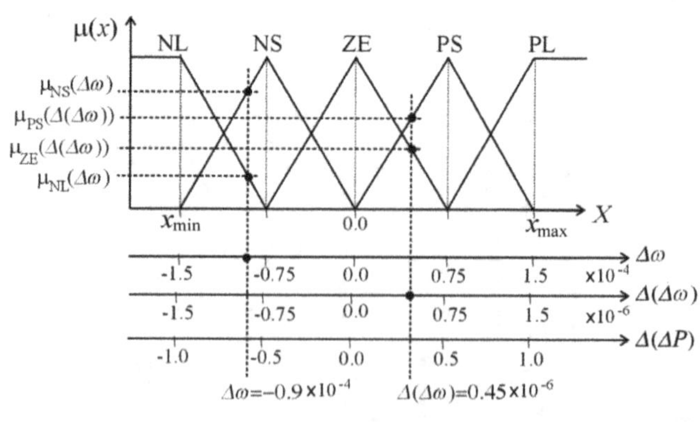

Figure E6.3(a) Input data fuzzification for Example 6.3

respectively. A MATLAB function implementing (6.48), such as the one presented in Figure 2.5 in Chapter 2, can be used to obtain the membership values of $\Delta\omega = -0.9 \times 10^{-4}$ and $\Delta(\Delta\omega) = 0.45 \times 10^{-6}$ in related fuzzy subsets as follows:

$$\mu_{\text{NL}}(\Delta\omega) = \text{triangle01m}(-1.5, -1.5, 0.75, 0.9) = 0.2 \tag{E6.3a}$$

$$\mu_{\text{NS}}(\Delta\omega) = \text{triangle01m}(-1.5, 0.75, 0.0, 0.9) = 0.8 \tag{E6.3b}$$

$$\mu_{\text{ZE}}(\Delta\omega) = \mu_{\text{PS}}(\Delta\omega) = \mu_{\text{PL}}(\Delta\omega) = 0.0 \tag{E6.3c}$$

$$\mu_{\text{NL}}(\Delta(\Delta\omega)) = \mu_{\text{NS}}(\Delta(\Delta\omega)) = 0.0 \tag{E6.3d}$$

$$\mu_{\text{ZE}}(\Delta(\Delta\omega)) = \text{triangle01m}(-0.75, 0.0, 0.75, 0.45) = 0.4 \tag{E6.3e}$$

$$\mu_{\text{PS}}(\Delta(\Delta\omega)) = \text{triangle01m}(0.0, 0.75, 1.5, 0.45) = 0.6 \tag{E6.3f}$$

$$\mu_{\text{PL}}(\Delta(\Delta\omega)) = 0.0 \tag{E6.3g}$$

where the multipliers $\times 10^{-4}$ and $\times 10^{-6}$ are omitted in (E6.3a)–(E6.3g) since the use of these multipliers does not affect the membership values.

The membership values of $\Delta\omega$ and $\Delta(\Delta\omega)$ on fuzzy subsets NL, NS, ZE, and PS are evaluated by Table 6.4 using the min operator as in (E6.3h) to yield the fuzzy membership values or the weighting factors at the output space for each rule as depicted in Table E6.3(a).

$$\mu_{\text{NS}}(\Delta(\Delta P)) = \min[\mu_{\text{NL}}(\Delta\omega), \mu_{\text{PS}}(\Delta(\Delta\omega))] \tag{E6.3h}$$

$$\mu_{\text{NS}}(\Delta(\Delta P)) = \min[0.2, 0.6] = 0.2 \tag{E6.3i}$$

There are four active rules for the given values of $\Delta\omega$ and $\Delta(\Delta\omega)$ as depicted in Table E6.3(a). All of the other rules are zero and they have no effect in the output. Four active rules are expressed as follows.

R1: If $\Delta\omega$ is NL and $\Delta(\Delta\omega)$ is ZE, **then** $\Delta(\Delta P)$ is NS.
R2: If $\Delta\omega$ is NL and $\Delta(\Delta\omega)$ is PS, **then** $\Delta(\Delta P)$ is NS.
R3: If $\Delta\omega$ is NS and $\Delta(\Delta\omega)$ is ZE, **then** $\Delta(\Delta P)$ is NS.
R4: If $\Delta\omega$ is NS and $\Delta(\Delta\omega)$ is PS, **then** $\Delta(\Delta P)$ is ZE.

The numerical values are obtained by employing a fuzzy decision process and are shown in Table E6.3(a). The fuzzification results of input data $\Delta\omega = -0.9 \times 10^{-4}$ and $\Delta(\Delta\omega) = 0.45 \times 10^{-6}$ from (E6.3a) to (E6.3g) are carried on to the

Table E6.3(a) Weighting factors of each rule at the output space Δ(ΔP)

Δω = 0.45 × 10⁻⁶	**Δ(Δω) = −0.9 × 10⁻⁴**				
	0.0/NL$_P$	**0.0/NS$_P$**	**0.4/ZE$_P$**	**0.6/PS$_P$**	**0.0/PL$_P$**
0.2/NL$_P$	0.0/NL$_P$	0.0/NL$_P$	0.2/NS$_P$	0.2/NS$_P$	0.0/ZE$_P$
0.8/NS$_P$	0.0/NL$_P$	0.0/NS$_P$	0.4/NS$_P$	0.6/ZE$_P$	0.0/PS$_P$
0.0/ZE$_P$	0.0/NS$_P$	0.0/NS$_P$	0.0/ZE$_P$	0.0/PS$_P$	0.0/PS$_P$
0.0/PS$_P$	0.0/NS$_P$	0.0/ZE$_P$	0.0/PS$_P$	0.0/PS$_P$	0.0/PL$_P$
0.0/PL$_P$	0.0/ZE$_P$	0.0/PS$_P$	0.0/PS$_P$	0.0/PL$_P$	0.0/PL$_P$

input universes of Table E6.3(a). Then the weighting factor for each rule is obtained using (E6.3h) and written into the corresponding active rule cell of the output universe Δ(Δ*P*). The final crisp decision is obtained by implementing the center of area method for defuzzification as in (E6.3j).

$$\Delta(\Delta P)_{\mathrm{O}} = \frac{\sum_{i=1}^{n} \mu_{\mathrm{FS}}(\Delta(\Delta P)) \cdot (\mathrm{FS_P})_i}{\sum_{i=1}^{n} \mu_{\mathrm{FS}}(\Delta(\Delta P))} \tag{E6.3j}$$

where the subscription FS corresponds to the fuzzy partition subsets defined in output space. The abbreviation $\mathrm{FS_P}$ is the crisp Δ(Δ*P*) corresponding to the peak of the triangular membership function. Defuzzification for this problem is done as in (E6.3k).

$$\Delta(\Delta P)_{\mathrm{O}} = \frac{\mu_{\mathrm{NS1}}(\Delta(\Delta P))\mathrm{NS_P} + \mu_{\mathrm{NS2}}(\Delta(\Delta P))\mathrm{NS_P} + \mu_{\mathrm{NS3}}(\Delta(\Delta P))\mathrm{NS_P} + \mu_{\mathrm{ZE1}}(\Delta(\Delta P))\mathrm{ZE_P}}{\mu_{\mathrm{NS1}}(\Delta(\Delta P)) + \mu_{\mathrm{NS2}}(\Delta(\Delta P)) + \mu_{\mathrm{NS3}}(\Delta(\Delta P)) + \mu_{\mathrm{NS}}(\Delta(\Delta P))} \tag{E6.3k}$$

$$\Delta(\Delta P)_{\mathrm{O}} = \frac{(0.2)(-0.5) + (0.2)(-0.5) + +(0.4)(-0.5) + (0.6)(0.0)}{(0.2) + (0.2) + (0.4) + (0.6)}$$

$$\Delta(\Delta P)_{\mathrm{O}} = \frac{(0.2)(-0.5)+(0.2)(-0.5)++(0.4)(-0.5)+(0.6)(0.0)}{(0.2)+(0.2)+(0.4)+(0.6)}$$

$$\Delta(\Delta P)_{\mathrm{O}} = \frac{-0.4}{1.4} = -0.285$$

Example 6.4

A small signal state-space model of a two-area power system with integral control is given in (6.40) and (6.41) and represented by the simulation diagram shown in Figure 6.22. Reconsider the same two-area power system and obtain a new simulation diagram and state-space model to be used with fuzzy logic controllers instead of integral control.

Answer 6.4

The state-space matrices of a two-area power system for the fuzzy logic controller are obtained by replacing the integral controllers with the fuzzy ones as in Figure E6.4(a). Since the variables x_7 and x_8 in Figure 6.22 are generated by integral controllers, they will be eliminated after the replacement of integral controllers by FLC in each control area. Therefore, the order of the system is reduced from 9 to 7 as in (E6.4a) and (E6.4b).

$$A = \begin{bmatrix} -\frac{D_1}{2H_1} & 0 & \frac{1}{2H_1} & 0 & 0 & 0 & -\frac{1}{2H_1} \\ 0 & -\frac{D_2}{2H_2} & 0 & \frac{1}{2H_2} & 0 & 0 & \frac{1}{2H_2} \\ 0 & 0 & -\frac{1}{\tau_{\mathrm{T1}}} & 0 & \frac{1}{\tau_{\mathrm{T1}}} & 0 & 0 \\ 0 & 0 & 0 & -\frac{1}{\tau_{\mathrm{T2}}} & 0 & \frac{1}{\tau_{\mathrm{g1}}} & 0 \\ -\frac{1}{R_1\tau_1} & 0 & 0 & 0 & -\frac{1}{\tau_{\mathrm{g1}}} & 0 & 0 \\ 0 & -\frac{1}{R_2\tau_2} & 0 & 0 & 0 & -\frac{1}{\tau_{\mathrm{g2}}} & 0 \\ P_s & -P_s & 0 & 0 & 0 & 0 & 0 \end{bmatrix} \tag{E6.4a}$$

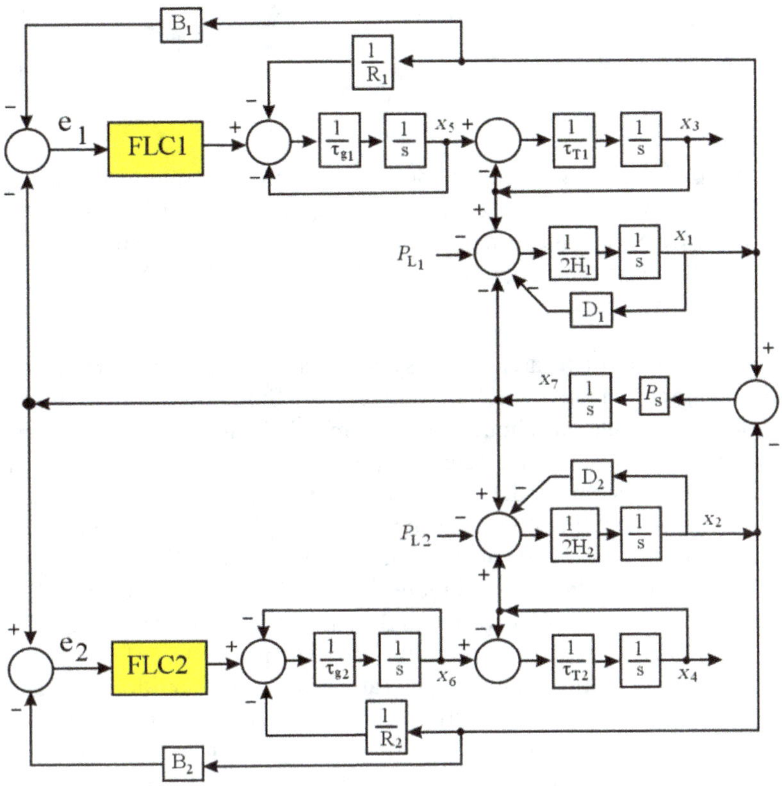

Figure E6.4(a) Two-area power system block diagram wit fuzzy logic controller

The input matrix B and the input variable matrix U in state-space model equations are split into submatrices as in (E6.4b) and (E6.4c) to be used in the simulation with fuzzy logic controller. The submatrices B_1, B_2, U_1, and U_2 express the influence of the fuzzy logic controller as B_3, B_4, U_3, and U_4 express the load difference input.

$$B_1 = \begin{bmatrix} 0 \\ 0 \\ 0 \\ 0 \\ \dfrac{1}{\tau_{g1}} \\ 0 \\ 0 \end{bmatrix}, \quad B_2 = \begin{bmatrix} 0 \\ 0 \\ 0 \\ 0 \\ 0 \\ \dfrac{1}{\tau_{g2}} \\ 0 \end{bmatrix}, \quad B_3 = \begin{bmatrix} -\dfrac{1}{2H_1} \\ 0 \\ 0 \\ 0 \\ 0 \\ 0 \\ 0 \end{bmatrix}, \quad B_4 = \begin{bmatrix} 0 \\ -\dfrac{1}{2H_2} \\ 0 \\ 0 \\ 0 \\ 0 \\ 0 \end{bmatrix} \tag{E6.4b}$$

The steady-state equations become as in (E6.4c)

$$\dot{x} = Ax + B_1u_1 + B_2u_2 + B_3u_3 + B_4u_4 \tag{E6.4c}$$

Example 6.5

Develop a small signal Simulink model for the FLC-controlled two-area power system described in Example 6.4 and perform a simulation using the parameters given in Table E6.5(a). Assume that the slope of the power angle curve at the initial operating point is $P_s = 2$.

Answer 6.5

The Simulink model of two-area power systems can be obtained by transferring the simulation diagram in Figure E6.4(a) directly to Simulink with the parameters given. Instead of a detailed simulation diagram of Figure E6.4(a), the block diagram given in Figure 6.18 for a single-area power system can be expanded to a two-area AGC power system model. A second copy of the block diagram given in Figure 6.18 can be combined with the first one using the tie line power angle slope P_s as the common element to obtain a unified two-area AGC power system as shown by the Simulink diagram in Figure E6.5(a). A similar two-area AGC power system model is given in [4] with integral controllers, which are replaced by the FLCs here. In order to show the operating performance of the FLC developed in Chapter 4 of this book, the data given in Table E6.5(a) is taken from [4] and used here for comparison purposes. The PID controllers with the parameters $K_P = 0$, $K_I = 0.3$, and $K_D = 0$ in [4] are replaced by the FLCs as shown in Figure E6.5(a). The simulation results are given in Figure E6.5(b) where the FLC responses for both areas have less oscillations and shorter settling time.

Actually, it should be noted that the comparison made here is not given to show which controller is better. The comparison is given to show that the proposed FLC model for the MATLAB/Simulink environment works and gives acceptable results. The resultant responses from FLC- and PID-controlled systems are given in Figures E6.5(b) and E6.5(c).

The FLCs used in this example and shown in Figure E6.5(a) by blocks FLC1 and FLC2 are identical and the detailed modelings are discussed in Chapter 4.

Table E6.5(a) Parameters for Example 6.5

Parameter	Area 1	Area 2
D	0.6	0.9
H	5	4
R	0.05	0.0625
τ_g	0.2	0.3
τ_T	0.5	0.6
B	20.6	16.9
ΔP_L	0.2	0

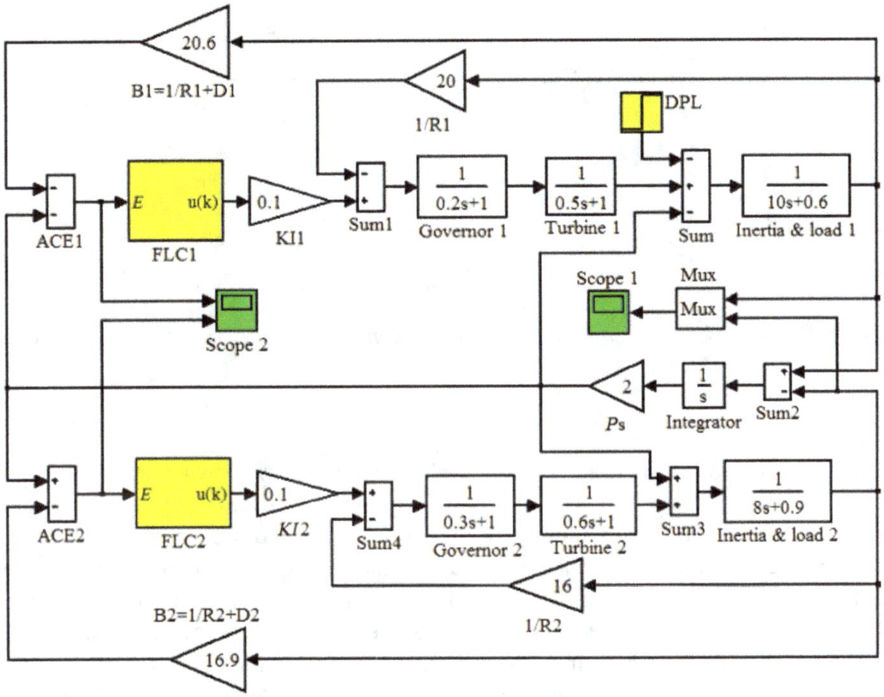

Figure E6.5(a) Simulink model of two-area AGC power system with FLC

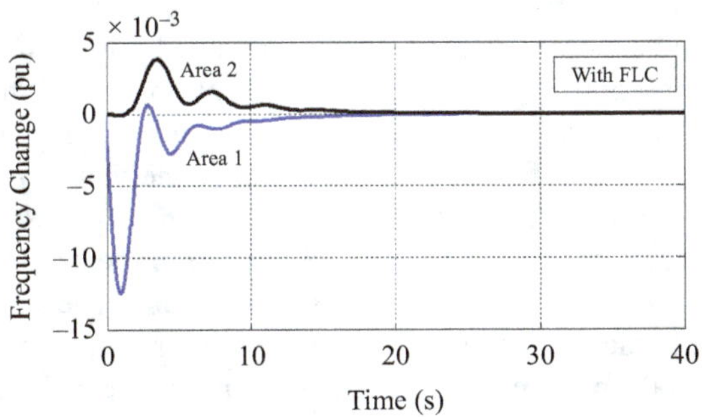

Figure E6.5(b) Load–frequency controller responses in a two-area power system using FLC

Design Example 6.2

In two-area AGC power systems, the frequency of both areas is affected by the demand power change that occurred in one of the areas. However, the frequency of the area where the demand power is changed gets affected faster with a larger change then the frequency of the other area. For example, if a change occurs in the

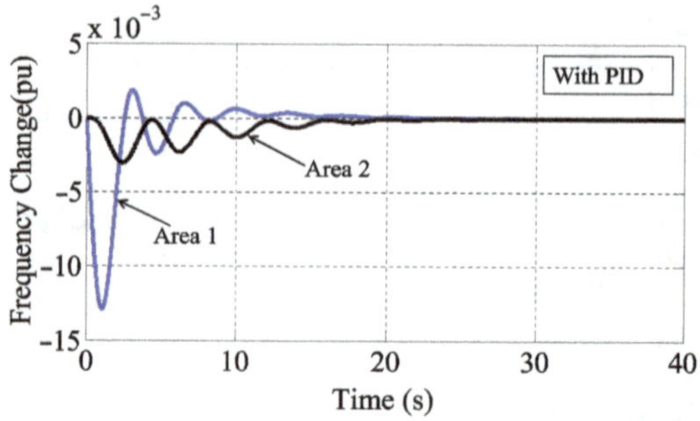

Figure E6.5(c) Load–frequency controller responses in a two-area power system PID

demand power of area 1, its angular velocity will slow down earlier than the angular velocity of area 2. Therefore, the prime movers in area 1 will act faster and try to generate more mechanical power to supply the required change while the prime movers in area 2 start later to supply the change in demand. Due to this retardation, too much extra mechanical power will be delivered to both plants that cause the large oscillations in the frequencies. The AGC systems shown in Figure 6.22 with integral control and in Figure E6.4(a) with FLC use the frequency change signals at the same time without considering the delay between the instant the frequency deviations start.

In order to increase the two-area AGC system with FLC, design a fuzzy decision unit (FDU) such that this unit should compare the angular velocity of areas 1 and 2, and then send extra control signals to the fuzzy logic controllers of both areas [14]. If the velocity drop is bigger due to demand power increment in area 1, an extra and larger control signal should be sent to area 1 to generate more mechanical power than it is in area 2 where a smaller extra signal should be sent. These additional signals are factors used to multiply the values in the universes of the change in the control signal, which will result in a quicker change of extra mechanical power for area 1 and only a small change in mechanical power for area 2 if the disturbance of area 1 is bigger at that moment. These two extra control signals are also influenced by the value of the angular velocities of both areas in such a way that if both frequency changes ($\Delta\omega$s) are little, both factors will be very small too.

Design a fuzzy decision unit for the system simulated in Example 6.5 to improve the performance of the FLC-based AGC using the information provided above.

6.4 FLC in power compensation

The transmitted or distributed power in AC systems is a complex function with real and imaginary parts. The power converted to usable energy by electrical machines

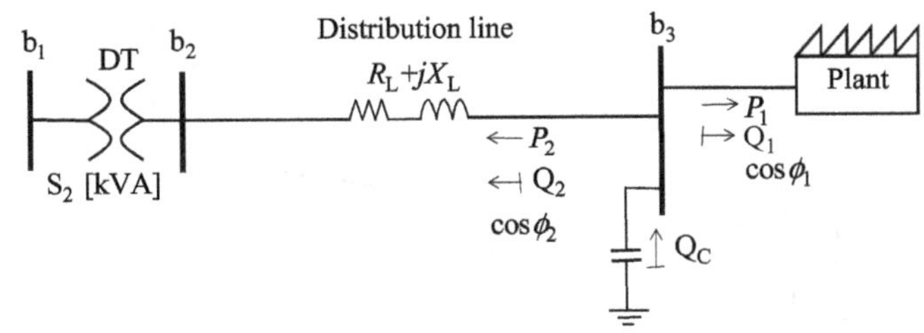

Figure 6.28 Reactive power generation by capacitors on the load site

and other electrical devices is called *active power*. An electrical heater, for example, dissipates only active power. Electrical motors and devices with a coil require another type of power called *reactive power*. Reactive power has a magnetizing effect and is required to generate magnetic field for the electrical machines, which do not have a magnetizing coil or permanent magnet. Especially asynchronous machines, transformers, and coils require reactive power to operate. The complex AC power is a combination of these active and reactive powers, which are the real and imaginary components as given in (6.49).

$$\underline{S} = P + jQ \tag{6.49}$$

where $\underline{S}$ is the complex power in VA, P is the active power in W and Q is the reactive power in VAr. Active power is generated by electrical generators in power stations and transmitted to the user sites by overhead lines and distribution networks. Reactive power is also generated and transmitted to the user in the same way. However, the reactive power can be generated by capacitors, too. Since it is able to be generated by capacitors on the user site, then there is no need to transmit all of the reactive power required by loads from power stations to the loads. Except for a smaller part needed by transformers in transmission and distribution systems, all of the reactive power dissipated by loads can be generated by capacitors right on the load side of the power network as shown in Figure 6.28.

The reactive power compensation in power systems is done for different purposes. It is done for power factor improvement, harmonic elimination, and bus voltage control. Power factor improvement and bus voltage control applications of reactive power compensation are discussed in this section.

6.4.1 Power factor improvement

The use of capacitors to generate the reactive power requirement of the load and improve distribution line power factor closer to unity is called power factor improvement. With a small, such as 5%, reactive power loading, the power transmission and distribution lines are prevented from overloading and higher power losses. Assume that the required active and reactive powers in a manufacturing plant are P_1 and Q_1, respectively. The corresponding power factor for P_1 and Q_1 is $\cos\phi_1$ as depicted by the power triangle in Figure 6.29. However, the utility can

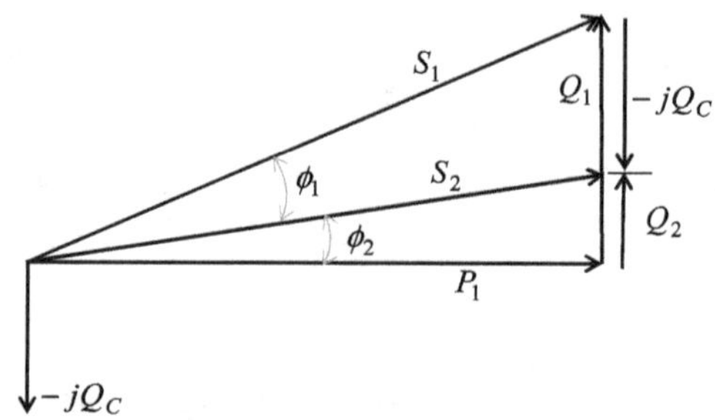

Figure 6.29 Power triangle before and after the compensation

supply a reactive power up to the amount of Q_2 under the power factor cosϕ_2 only. In order to reduce the transmitted reactive power down to Q_2, a reactive power Q_C must be generated right on the power input bus of the factory as given in Figure 6.28.

Using the power triangles before and after the compensation in Figure 6.29, the following can be written.

$$\tan\phi_1 = \frac{Q_1}{P_1} \tag{6.50}$$

$$\tan\phi_2 = \frac{Q_1 - Q_C}{P_1} \tag{6.51}$$

Required compensation power to increase the power factor up to cosϕ_2 can be obtained from (6.50) and (6.51) as in (6.52).

$$Q_C = P_1(\tan\phi_1 - \tan\phi_2) \tag{6.52}$$

Required reactive power to be generated by capacitors for the power factor cosϕ_2 can be obtained by using (6.52). This is a practical equation to be used to obtain capacitor power. As the active and reactive power dissipation change on the load side, the power factor cosϕ_1 of the load bus also changes requiring a change in capacitor power Q_C so that the desired power factor cosϕ_2 on the feeder side remains unchanged. In order to add or remove capacitor groups from the compensator unit to keep the feeder side power factor at cosϕ_2 when there are load changes, a power factor controller should be designed. A single-line diagram of a power factor control system is given in Figure 6.30. Measured three-phase active power, line current, and phase-to-phase voltages are used to calculate three-phase apparent power, reactive power, and power factor on the feeder side. Calculated power factor cosϕ_2 is compared with the reference power factor cosϕ_r to generate a difference, which is called error. Magnitude and the sign of error are then used by a fuzzy decision maker (FDM) to decide how much reactive power should be added or removed from the capacitor groups.

The magnitude of feeder side apparent power, reactive power, and power factor can be calculated from the measured current, voltage, and active power as in

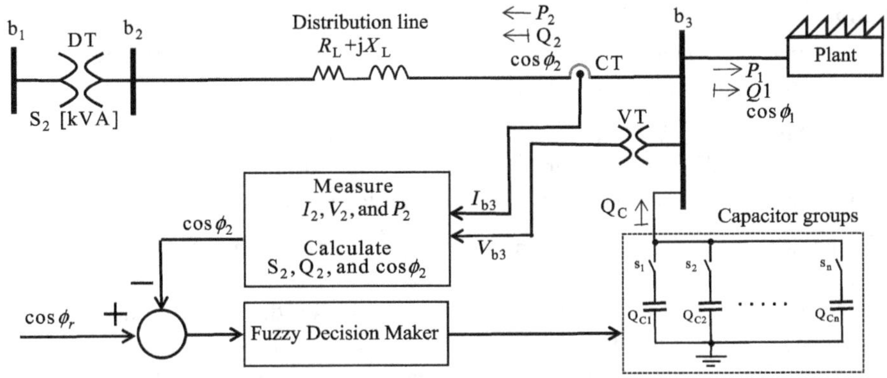

Figure 6.30 A power factor control scheme

(6.53), (6.54), and (6.55).

$$S_2 = \sqrt{3} I_2 V_2 \tag{6.53}$$

$$Q_2 = \sqrt{S_2^2 - P_2^2} \tag{6.54}$$

$$\cos\phi_2 = \frac{P_2}{S_2} \tag{6.55}$$

Design Example 6.3

Assume that the distribution line shown in Figure 6.30 has a length of 10 km. The line impedance is $\underline{Z}_L = (0.196 + j0.320)$ Ohm/km. The line-to-line voltage at the receiving end is 6 kV, load power at the receiving end is $S_1 = 30$ MVA at a lagging power factor of 0.75. A 5% voltage drop is allowed in this distribution feeder.

(a) Solve this problem by hand calculation and find active and reactive power components flowing through line and related power factor.
(b) Obtain the voltage at the sending end and active power loss in the line.
(c) Calculate the required capacitor power to increase the power factor up to 0.95 and repeat parts (a) and (b) to observe the change in power loss in the line.
(d) Simulate this system in MATLAB/Simulink without the compensator. Plot Voltage, current, and power waveforms at both ends of the transmission line.
(e) Connect the required capacitor bank you have calculated in part (c) to the load bus (bus 3) and simulate the system. Observe the changes in bus voltages, power flows, line power factor, and line losses.
(f) Add a 10 MVA load with a lagging power factor of 0.80 and repeat parts (b)–(e).
(g) Remove the load you have connected in part (f) and another power of 5 MW. Then repeat parts (b)–(e).

Design Example 6.4

Assume that the manufacturing plant has asynchronous motors characterized by their powers and power factors as follows.

Category 1: 20 kW at a lagging power factor of 0.80.
Category 2: 50 kW at a lagging power factor of 0.70.
Category 3: 100 kW at a lagging power factor of 0.75.

The capacitors are available in 50 kVAr groups enabling 50 kVAr switching steps per phase.

Design a fuzzy logic decision maker (FDM) for the problem described in Design Example 6.3 such that the lagging power factor of the power carried by the distribution line is kept between 0.92 and 0.98 by switching the capacitor banks on or off under the operating cases given below. Assume increasing and decreasing capacitors in steps of 50 kVAr.

(a) 5 of 20 kW motors, 5 of 50 kW motors, and 2 of 100 kW motors are turned off.
(b) 5 of 20 kW motors, 15 of 50 kW motors, and 1 of 100 kW motors are turned on.

6.4.2 Bus voltage control

The load bus voltage decreases due to heavy loading or increases due to light loadings. In both cases, the voltage at the load bus must remain at its rated value so that the loads operating at this bus keep running without getting harmed because of over-voltage or without losing performance because of under-voltage conditions.

The power stations where the electrical power is generated are generally far away from the load sites in conventional power grids. Therefore, the utility acts as an infinite source to the users at the end of distribution lines such that the effects of sudden load changes along or at the end of the feeders are negligible. The voltage profile along a radial feeder shows a reduction in magnitude due to line losses. Switching on or off a couple of houses or houses along a street does not have a considerable large effect on the voltage profile of the whole neighborhood. High impedance between electrical generators and distribution feeders is high enough to compensate for these changes at user sites. Although the changes in loads are not large enough to affect the stability of power systems, the bus voltages along a radial feeder may be affected by these changes. Higher load power causes more voltage drop along the power line. In order to keep the bus voltages at nominal values capacitor groups are connected to reduce the power losses so that the voltage drops are also reduced.

The capacitors connected to the load bus for power factor improvement cause a voltage rise in the bus they are connected to. Since the capacitor current is leading the bus voltage it causes a voltage rise due to line reactance as in (6.56).

$$V_{\text{rise}} = I_{\text{C}}X_{\text{L}} \tag{6.56}$$

where I_{C} is the rated total current drawn by the capacitor groups connected to bus 3 in Figure 6.30 and X_{L} is the reactance of the distribution line. The capacitive current is generated by the capacitor groups as a function of the total capacitor power and the bus voltage as in (6.57).

$$V_{\text{rise}} = I_{\text{C}} X_{\text{L}} \tag{6.57}$$

Since the rated total capacitor current is obtained as in (6.58)

$$I_{\text{C}} = \frac{Q_{\text{C}}}{\sqrt{3} V_{\text{LL}}} \tag{6.58}$$

The voltage rise can be expressed as a function of three-phase reactive power Q_{C} and line-to-line voltage V_{LL} as in (6.59).

$$V_{\text{rise}} = \frac{X_{\text{L}}}{\sqrt{3}} \left(\frac{Q_{\text{C}}}{V_{\text{LL}}} \right) \tag{6.59}$$

Note that the line reactance is constant. V_{LL} is the phase-to-phase voltage before the voltage rise and Q_{C} is three-phase reactive power causing the voltage rise. The voltage rise is usually expressed in percentages as in (6.60) and (61) for line-to-line and line-to-neutral voltages, respectively.

$$V_{\text{LL-rise}}\% = \frac{X_{\text{L}}}{\sqrt{3}} \left(\frac{Q_{\text{C}}}{V_{\text{LL}}^2} \right) 100 \tag{6.60}$$

$$V_{\text{LN -rise}}\% = \frac{X_{\text{L}}}{\sqrt{3}} \left(\frac{Q_{\text{C}1\phi}}{V_{\text{LN}}^2} \right) 100 \tag{6.61}$$

where $Q_{\text{C}1\phi} = Q_{\text{C}}/3$ and $V_{\text{LL}} = \sqrt{3} V_{\text{LN}}$.

In order to keep the bus voltage at its rated value, the required change in the bus voltage is calculated and used in a controller or in an FDM to decide how much increment or decrement should be made in reactive power. An algorithm shown in Figure 6.31 can be applied for bus voltage control.

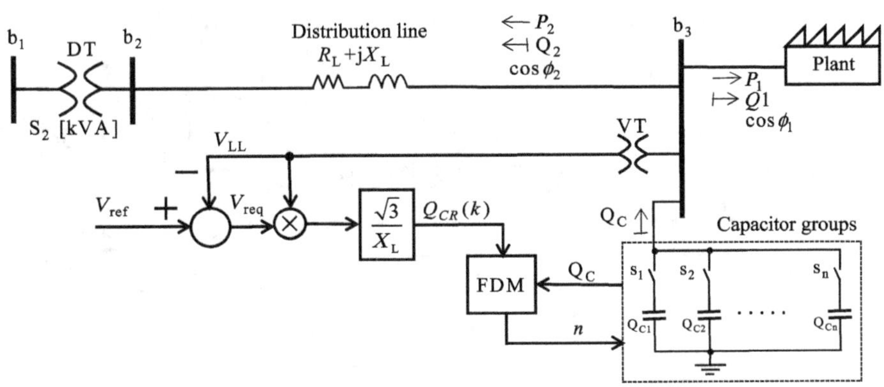

Figure 6.31 A power bus voltage control scheme

The required voltage rise can be calculated as the difference between the reference bus voltage and actual bus voltage as in (6.62).

$$V_{\text{req}} = V_{\text{ref}} - V_{\text{LL}} \tag{6.62}$$

The sign of resultant voltage rise may be positive or negative depending upon the loading condition of the power line. Positive sign means a voltage increment is necessary while negative sign means a decrement is required. In order to increase or decrease the bus voltage, more capacitors should be connected to or disconnected from the capacitor groups. The required change in capacitor groups can be calculated from (6.59) as in (6.63).

$$Q_{\text{CR}} = \frac{\sqrt{3}}{X_{\text{L}}} V_{\text{LL}} V_{\text{req}} \tag{6.63}$$

where Q_{CR} is the required total reactive power to be generated by the capacitor groups to raise the bus voltage by the amount of V_{req}. Therefore, Q_{CR} is used with the current Q_{C} to find out how many capacitor groups to be connected or disconnected to have a desired bus voltage.

Example 6.6

A three-phase short distribution line has a length of 10 km. The line impedance is $\underline{Z}_{\text{L}}$ = (0.149 + j0.289) Ohm/km. The line-to-line voltage at the receiving end is expected to be 6 kV and plant powers are $\underline{S}_{\text{P1}}$ = (0.8 + j0.6) MVA and $\underline{S}_{\text{P2}}$ = (0.6 + j0.4) MVA. A 5% voltage drop is allowed in this transmission system. The sending end voltage is set to 6.3 kV.

(a) Find load bus voltage, line power flows, and line power factor when there is no capacitor connected to the load bus b_3. Check the voltage drop in line whether it is less than 5%.
(b) Find the required capacitor power to be connected to bus b_3 to increase the line power factor to 0.98 lagging.
(c) Calculate the voltage rise at bus b_3 after the capacitors calculated in part (b) are connected. Check the voltage drop in line whether it is less than 5%.

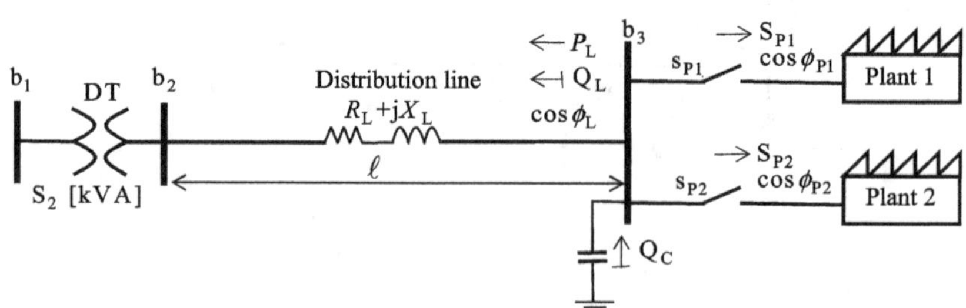

Figure E6.6 The power system to be studied in Example 6.6

Answer 6.6

The voltage at bus b2 can be written as in (E6.6.1)

$$\underline{V}_{b2} = \underline{V}_{b3} + (R_L + jX_L)\underline{I}_L \tag{E6.6.1}$$

where the voltages are phase to neutral and $\underline{I}_L$ is the line current and can be calculated using three-phase apparent power and line-to-line voltage as in (E6.6.2)

$$\underline{I}_L = \frac{S_{PT}}{\sqrt{3}\ \ V_{b3LL}} \underline{|-\phi_T} \tag{E6.6.2}$$

The line current can be calculated using per phase apparent power and line-to-neutral voltage as in (E6.6.3)

$$\underline{I}_L = \frac{S_{PT1\phi}}{V_{b3}} \underline{|-\phi_T} \tag{E6.6.3}$$

Since the voltage at bus b3 is not known yet, the line current is expressed as a function of the voltage at bus b3. Let the phase A to neutral voltage at bus b3 be the reference to reduce the complexity of the problem. Then (E6.6.1) is written as in (E6.6.4)

$$V_{b2}\underline{|\delta} = V_{b3}\underline{|0^{\circ}} + \left(Z_L\underline{|\theta_Z^{\circ}}\right)\left(\frac{S_{PT1\phi}}{V_{b3}}\underline{|-\phi_T^{\circ}}\right) \tag{E6.6.4}$$

$$V_{b3}V_{b2}\underline{|\delta} = V_{b3}^2\underline{|0^{\circ}} + \left(Z_L\underline{|\theta_Z^{\circ}}\right)\ \left(S_{PT1\phi}\underline{|-\phi_T^{\circ}}\right) \tag{E6.6.5}$$

Rewriting (E6.6.5) yields (E6.6.6)

$$V_{b3}^2\underline{|0^{\circ}} - V_{b3}\frac{6.3}{\sqrt{3}}\underline{|\delta} + (3.2515\underline{|62.7257})\ \ (0.5735\underline{|35.5377^{\circ}}) = 0 \tag{E6.6.6}$$

where two unknowns, which are the line-to-neutral voltage V_{b3} at bus b3 and phase angle δ of the line-to-neutral voltage V_{b2} at bus b2 are obtained from the solution of (E6.6.6) as $V_{b3} = 3.3486$ kV and $\delta = 7.82^{\circ}$, respectively.

(a) Line-to-line voltage at load bus b3 is obtained as $V_{b3LL} = 3.3486\sqrt{3} = 5.8$ kV

Line power flows and line power factor when there is no capacitor connected to the load bus b_3 are calculated as follows.

$\underline{S}_{P1} = 0.8 + j0.6$ MVA, $\underline{S}_{P1} = 1\underline{|36.87^{\circ}}$ MVA, $\cos\phi_{P1} = 0.8$ lagging
$\underline{S}_{P2} = 0.6 + j0.4$ MVA, $\underline{S}_{P2} = 0.7211\underline{|33.69^{\circ}}$ MVA, $\cos\phi_{P2} = 0.8321$ lagging

Total apparent power and power factor in the power line will be

$\underline{S}_{PT} = 1.4 + j1.0$ MVA, $\cos\phi_{PT} = 0.8137$ lagging and $\phi_{PT} = 35.5377^{\circ}$

Voltage drop is calculated as the difference between sending and receiving end voltages.

$$\underline{\Delta V} = \underline{V}_{b2} - \underline{V}_{b3} \tag{E6.6.7}$$

Substitution of numerical values yields

$$\underline{\Delta V} = \frac{6.3}{\sqrt{3}}\underline{|7.82^{\circ}} - \frac{5.8}{\sqrt{3}}\underline{|0^{\circ}} = 0.5569\underline{|62.7257^{\circ}kV}$$

Percentage voltage drop is obtained as 7.9449% using (E6.6.8).

$$\Delta V\% = \frac{V_{b2} - V_{b3}}{V_{b2}} \times 100 \tag{E6.6.8}$$

It is obvious that 7.9449% is a high percentage since the allowable voltage drop is given as 5%.

(b) Required capacitor power to be connected to bus b_3 to increase the line power factor to 0.98 lagging is obtained directly by using (6.52).

Power factor angle before the compensation is

$$\phi_{PT} = \cos^{-1}\phi_{PT} = \cos^{-1}(0.8137) = 35.5377^{o}$$

Required power factor angle after the compensation is

$$\phi_{PT2} = \cos^{-1}\phi_{PT2} = \cos^{-1}(0.98) = 11.478^{o}$$

Substituting power factor angles before and after the compensation into (6.48) yields the capacitive power to be connected to bus b3.

$$Q_C = 1.4(\tan(35.5377) - \tan(11.478)) = 0.7157\ MVAr$$

(c) The voltage rise at bus b_3 after the capacitors with a total power of 715.7 kVAr are connected is calculated using (6.60).

$$V_{LL-rise}\% = \frac{2.89}{\sqrt{3}}\left(\frac{0.7157}{5.8^2}\right)100 = 3.5995$$

The new line-to-neutral voltage at bus b3 becomes as follows

$$V_{3LNnew} = \frac{5.8}{\sqrt{3}} + 0.0355 \times \frac{5.8}{\sqrt{3}} = 3.4675\ \text{kV}$$

Then the new line-to-line voltage becomes

$$V_{3LLnew} = \sqrt{3}\ \ (3.4675) = 6.0059 \cong 6\ \text{kV}$$

The new voltage drop in the power line is equal to

$$\Delta V_{new}\% = \frac{6.3 - 6.0}{6.3} \times 100 = 4.6683$$

After the compensation, the voltage drop has been reduced to 4.6683%, which is smaller than the acceptable limit given as 5%.

Design Example 6.5

Repeat Design Example 6.3 and calculate the voltage rise for the following cases.

(a) After connecting the capacitors to increase the power factor up to 0.95 lagging.
(b) After adding a 10 MVA load with a lagging power factor of 0.80 when the capacitors calculated in (a) are connected.
(c) After removing the load you have connected in part (b) and another power of 5 MW when the capacitors calculated in (a) are connected.
(d) Simulate this system in MATLAB/Simulink for the operating cases described above.

Design Example 6.6

Design a fuzzy logic decision maker (FDM) for the problem described in Design Example 6.4 such that the load bus voltage is kept at 6 kV when the lagging power factor of the power carried by the distribution line is kept between 0.92 and 0.98 by switching the capacitor banks on or off under the operating cases given below. Assume increasing and decreasing capacitors in steps of 50 kVAr.

(a) 5 of 20 kW motors, 5 of 50 kW motors, and 2 of 100 kW motors are turned off.
(b) 5 of 20 kW motors, 15 of 50 kW motors, and 1 of 100 kW motors are turned on.

6.5 FLC in load flow analysis

The main duty of a power system is to supply the required load power at rated voltage and frequency whenever it is needed. The power system must be secure and reliable, which requires system information at all times. Power flow (PF) analysis is an operational overlooked map of the interconnected power system providing information on bus voltages and active–reactive power flows in transmission lines. This information on both bus voltages and the power flows in transmission lines is used for planning, operating, and management purposes as well as for recovering the system after the occurrence of unexpected faults and abnormal operating conditions. Therefore, it is an important task to have an accurate model of the power system to perform the PF analysis [48].

Interconnected power systems are modeled in various ways for PF solutions. Usually, node or branch equations are used. Since the PF equations are nonlinear, iterative methods such as Gauss–Seidel (G–S), Newton–Raphson (N–R), and Fast Decoupled (FD) PF solution methods are preferred. Besides these classical PF solution methods, many alternative works have appeared in the literature. A Continuation Power Flow method was described in [49] for steady-state voltage stability analysis. In this study, the authors developed a new approach based on classical N–R Power Flow solution by injecting sub-parameters into the system so that the convergence of the N–R solution is guaranteed and the PF solution is obtained continuously for changing load conditions. Compensation-based power flow method is given in [50] for weakly meshed distribution and transmission networks. The authors show that this method works well, especially in radial and weak distribution networks. A three-phase power flow method for real-time distribution system analysis is proposed in [51]. This method also uses a compensation-based PF approach for weak distribution networks rather than power transmission systems. A power flow solution method is described in [52] for the systems with uncertain inputs. The method uses interval arithmetic to include the uncertain data in the solution. A fuzzy set approach is proposed in [53] by considering the inputs of the power system as fuzzy variables. In this study, the bus powers, bus voltages, and bus voltage angles are modeled as fuzzy variables and evaluated in an N–R power flow algorithm. In reference [30], fuzzy logic is used to adjust the variables of the power system during the power flow solution. The line impedances, bus voltage angles, and transformer tap positions are selected as the parameters to be adjusted within the limits of the power flowing in the lines and the bus voltage magnitudes. This study also deals with the acceleration of the iteration convergence. In order to reduce the processing time of the PF solution in classical PF methods, a fuzzy logic-based approach is used to update a correction factor in [29] where the PF solution is based upon the modification of fast decoupled PF solution method. A method consisting of some of the details that are ignored for simplicity in classical PF modelings of power systems is given in [54]. The fuzzy subsets are used to include the reactive power boundaries and load uncertainties for more accurate PF modeling. Another fuzzy logic-based PF solution algorithm is described in [55] by including the fuzziness of loads and generated power in the modeling, which results in the PF solutions as fuzzy outputs.

Fuzzy logic is going to be used to develop a fuzzy-based acceleration factor in this section of the book. The Fuzzy Logic Acceleration Factor (FLAF) has been developed to speed up the PF solution by modifying classical G–S, N–R, and FD power flow solution algorithms. With the FLAF algorithm, large initial steps are processed as the step sizes are reduced as the solution point comes closer. Besides, the ability of the fuzzy decision-making process remains the direction of the iteration toward the convergence point. The proposed FLAF algorithms are compared with the original PF algorithms for performance analysis.

6.5.1 Power system modeling for power flow

It is not intended in this section to explain how the PF equations are obtained. This information can be found in any book related to power system analysis. Therefore, only the final equations are discussed and the application of the FLAF algorithm is explained.

A 30-bus IEEE test system is shown in Figure 32, which is adapted from [4] and used in the simulations to test and compare the results with the other ones. All line and bus data are given in [4]. The final forms of the PF equations for G–S, N–R, and FD solution algorithms are summarized below and the use of the FLAF algorithm is described afterward.

6.5.2 FL accelerated Gauss–Seidel algorithm

Gauss–Seidel power flow algorithm is based on the calculation of the bus voltages iteratively. Starting with the initial guesses, the iteration continues until the

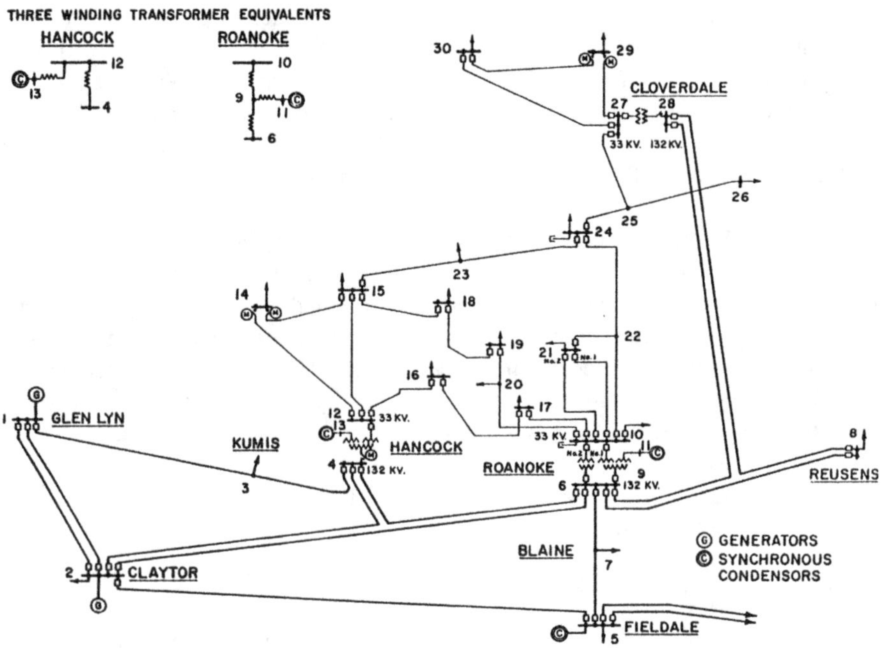

Figure 6.32 IEEE 30-bus test system [4]

difference between two consecutive values is within a preset precision value.

$$\left|\Delta P_i^{(k)}\right| \leq \varepsilon \tag{6.64}$$

$$\left|\Delta Q_i^{(k)}\right| \leq \varepsilon \tag{6.65}$$

The bus voltages are updated at each iteration as

$$V_i^{(k+1)} = V_i^{(k)} + \alpha \Delta V_i^{(k)} \tag{6.66}$$

where V_i^k is the voltage of the ith bus for the kth iteration, V_i^{k+1} is the newly updated voltage of the same bus, and ΔV_i^k is the change in the ith bus voltage from iteration $(k-1)$ to k. α is called the acceleration factor and is used to speed up the solution. The best value of α in classical PF solution algorithms is obtained by using optimization techniques and kept constant during the iterative PF solution. Large values of α may cause the system to reach the convergence point faster. However, there is a high risk of skipping the solution point with larger α. Smaller α may result in a closer catch-up to the convergence point, but the required iteration number will tremendously be increased this time. Therefore, the use of a flexible acceleration factor may be a solution to speed up the system when the solution is far away from the convergence and slow it down when the solution approaches the convergence point.

Using the changes $\Delta P_i^{(k)}$ and $\Delta Q_i^{(k)}$ in active and reactive power, respectively, as the inputs to the fuzzy logic decision maker (FLDM) for the ith bus, a flexible acceleration factor is generated. This new acceleration factor generated by FLDM is represented by α_b to replace α in (6.66). Since α_b is obtained by a fuzzy decision-making process it is called a Fuzzy Logic Acceleration Factor (FLAF). The process of FLDM to generate the FLAF, which is also represented by α_b, is shown in Figure 6.33.

The FLDM consists of three stages namely *fuzzifier*, *fuzzy rules*, and *defuzzifier*. The power changes are fuzzified in the first stage and converted to membership values, which are evaluated by fuzzy rules in the second stage. Since the output from the fuzzy rule process is fuzzy, the third stage is used a defuzzifier in order to convert the fuzzy decision to a crisp value. The fuzzification stage uses triangular fuzzy membership functions to fuzzify crisp inputs as shown in Figure 6.34 where TS, S, M, B, and TB stand for Too Small, Small, Medium, Big, and Too Big, respectively.

The absolute total power change $\Delta(PQ)^{(k)}$ is the sum of the absolute active and reactive power changes and obtained as

$$\Delta(PQ)_i^k = \left|\Delta P_i{}^{(k)}\right| + \left|\Delta Q_i{}^{(k)}\right| \tag{6.67}$$

Due to the triangular fuzzy membership functions used in the fuzzification stage shown in Figure 6.34, there will be maximum of two active fuzzy rules for each iteration. For example, there are two active fuzzy rules because of two interceptions of the crisp input $\Delta(PQ)^{(k)}$ with the fuzzy subsets S and TS in Figure 6.34

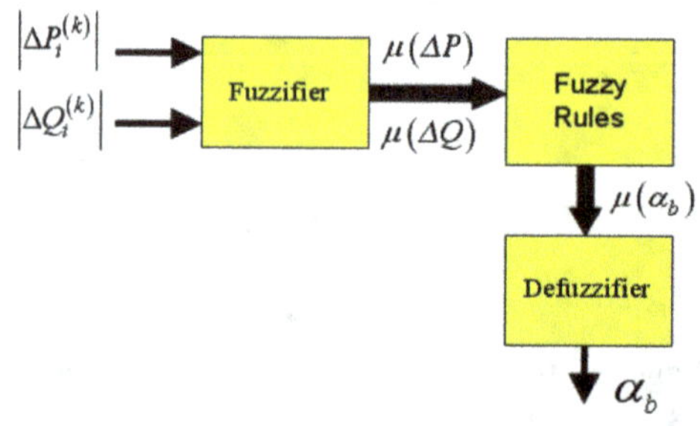

Figure 6.33 Fuzzy decision maker used as FLAF

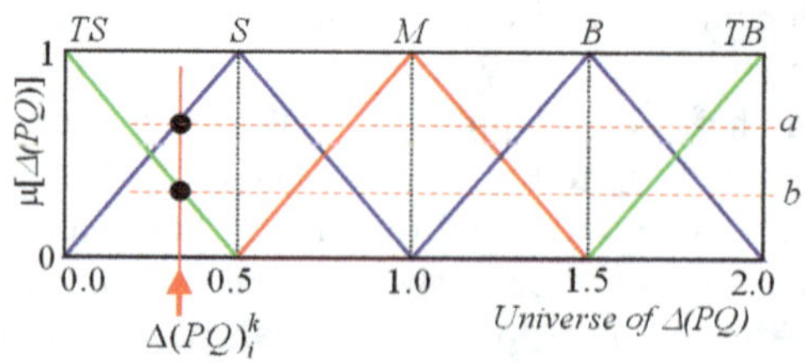

Figure 6.34 The fuzzy subsets and the fuzzification of the power change

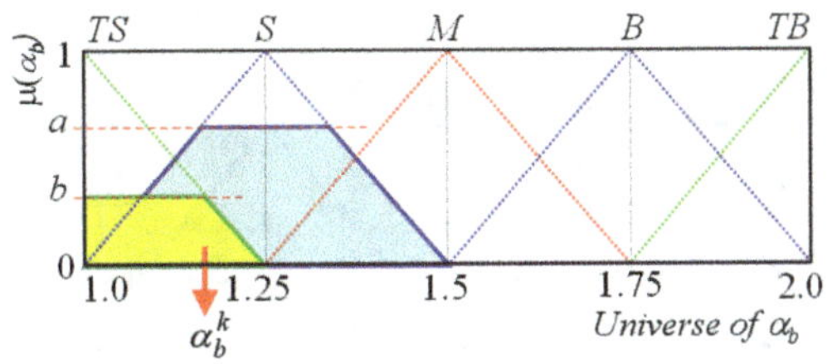

Figure 6.35 The fuzzy output surface of the FLDM

resulting in the fuzzy membership values a and b as

$$a = \mu_{\mathrm{S}}[\Delta(PQ)_i^k] \tag{6.68}$$

$$b = \mu_{\mathrm{TS}}[\Delta(PQ)_i^k] \tag{6.69}$$

The following five rules are defined in the rule base of the FLDM.
Rule 1: If the input is TS, then the output is TS.
Rule 2: If the input is S, then the output is S.
Rule 3: If the input is M, then the output is M.
Rule 4: If the input is B, then the output is B.
Rule 5: If the input is TB, then the output is TB.

The rule base implies that the output takes the same action as the input. The only difference is the amount of change which is determined by the upper and lower limits of the crisp universes. These limits are given in Figure 6.34. for the power change and in Figure 6.35 for the acceleration factor.

The final crisp output α_b^k of the FLDM is obtained by applying the center of area method [56]. It can be seen that the FLAF can have any value in the interval [1,2] depending on the value of the absolute total power change at bus *i*. The value of the acceleration factor FLAF may differ from bus to bus for all iterations.

6.5.3 FL accelerated Newton–Raphson algorithm

As well known, the classical N–R PF algorithm has already many advantages over the classical G–S PF algorithm. However, the N–R algorithm can be improved more by applying a flexible FLAF to the solution. As in the G–S algorithm, the convergence of the N–R algorithm also uses the absolute power changes as given by (6.64) and (6.65). Similar to (6.66), the acceleration factor is included in the updated equations of power angle and bus voltages as

$$\delta_i^{(k+1)} = \delta_i^{(k)} + \alpha_b \Delta\delta_i^{(k)} \tag{6.70}$$

$$\left|V_i^{(k+1)}\right| = \left|V_i^{(k)}\right| + \alpha_b \Delta\left|V_i^{(k)}\right| \tag{6.71}$$

The same five fuzzy rules given for G–S algorithm can be used for N–R algorithm as well. The fuzzy subset for input and the output spaces are also defined

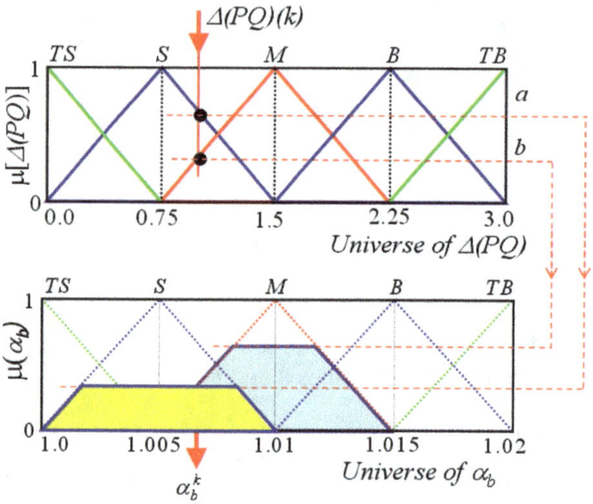

Figure 6.36 Graphical representation of the fuzzy reasoning for the FLAF in N–R PF algorithm

by the triangular membership function as given in Figure 6.36 where the FLDM works in the same way as given in Figure 6.33.

6.5.4 FL accelerated fast decoupled algorithm

A similar approach is also applied here to get a faster solution. However, the fuzzy logic acceleration factor is injected at two points of the FD algorithm because of the inclusion of active and reactive power changes in the solution as decoupled variables of two separate equations. Therefore, two FLAFs (α_{bP} and α_{bQ}) are generated in this section. One from active power change and the other one from the reactive power change. These acceleration factors are then used to speed up the solutions to find the change in power angle $\Delta\delta$ and the change in bus voltage magnitude$\Delta|V|$.

$$\Delta\delta = -\alpha_{bP}[\mathbf{B}']^{-1}\frac{\Delta P}{|V|} \tag{6.72}$$

$$\Delta|V| = -\alpha_{bQ}[\mathbf{B}'']^{-1}\frac{\Delta Q}{|V|} \tag{6.73}$$

where $\mathbf{B}'$ and $\mathbf{B}''$ are the imaginary parts of the bus admittance matrix $\mathbf{Y}_{\text{bus}}$.

Each one of the FLAFs α_{bP} and α_{bQ} are generated using the FLDM process depicted in Figure 6.33. In order to obtain α_{bP}, the absolute value of the active power change $|\Delta P_i{}^{(k)}|$ at bus i is used as the crisp input variable and to obtain α_{bQ}, the absolute value of the reactive power change $|\Delta Q_i{}^{(k)}|$ at bus i is used as the crisp input variable. Therefore, the fuzzy reasoning described in Figure 6.36 is done twice for α_{bP} and α_{bQ} by assigning the boundaries of $|\Delta P|$ from 0.0 to 1.2 and the boundaries of $|\Delta Q|$ from 0.0 to 3.5. The boundaries of the resultant acceleration factors α_{bP} and α_{bQ} are chosen from 0.6 to 1.2 and from 0.45 to 1.25.

6.5.5 Evaluations of the effect of FLAF

The IEEE 30-bus load flow test system is simulated using Gauss–Seidel, Newton–Raphson, and Fast Decoupled algorithms without and with the fuzzy logic-based acceleration factor. Figures 6.37 and 6.38 show the convergence error change with respect to the iteration count for the cases without and with the fuzzy logic acceleration factor (FLAF), respectively. The convergence error changes for N–R

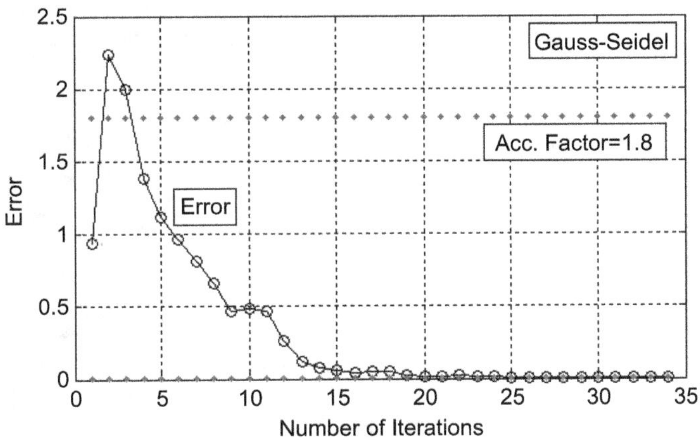

Figure 6.37 Convergence error change during the iteration of the classical Gauss–Seidel algorithm with α = 1.8 constant

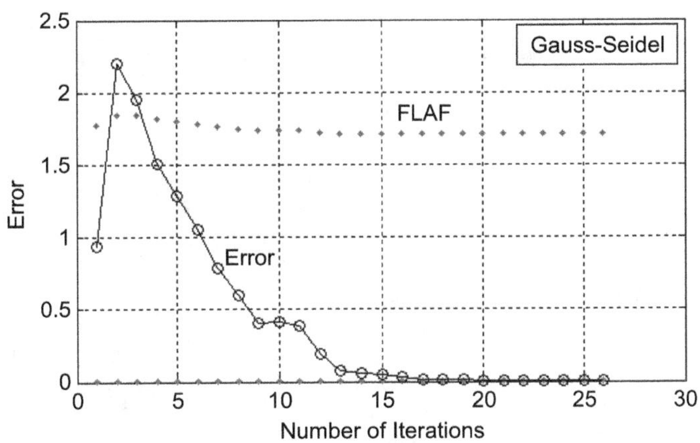

Figure 6.38 Convergence error and the FLAF changes during the iteration of the Gauss–Seidel algorithm

algorithm are shown in Figures 6.39 and 6.40 without and with the FLAF. Similar results for FD algorithm are given in Figures 6.41 and 6.42. Figure 6.43 shows a comparison of the results from G–S algorithm for the cases with and without the use of FLAF. Results obtained when the FLAF is used give slightly less iteration and better convergence than the results obtained with the constant acceleration factor. There is no significant improvement in the result of N–R and FD algorithms. Since both N–R and FD converge in a couple of iterations for the IEEE 30-bus test system, the effect of using FLAF is not visible. In fact, the FLAF shows its advantages if the number of busses is increased to over 100.

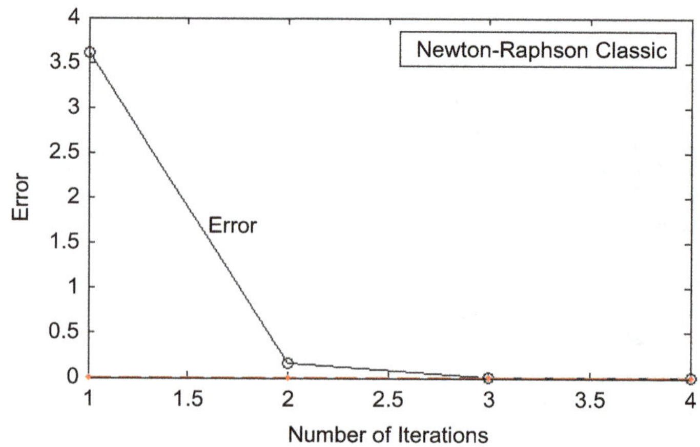

Figure 6.39 Convergence error change during the iteration of the classical Newton–Raphson algorithm

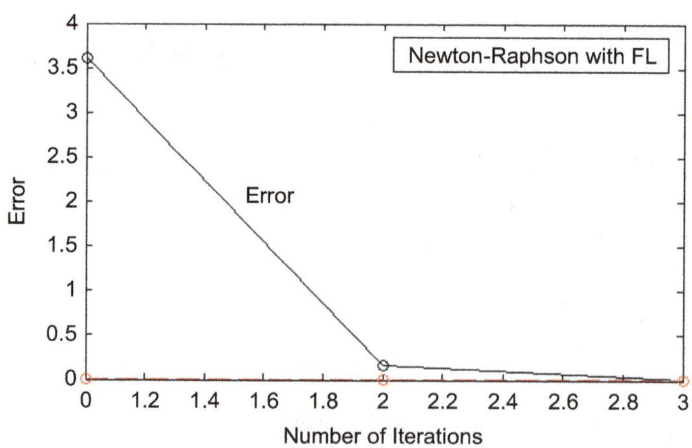

Figure 6.40 Convergence error change during the iteration of the Newton–Raphson algorithm with the injected FLAF

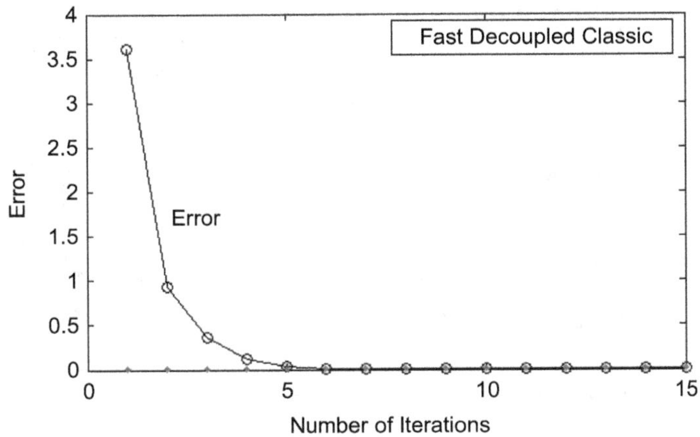

Figure 6.41 Convergence error change during the iteration of the classical Fast Decoupled algorithm

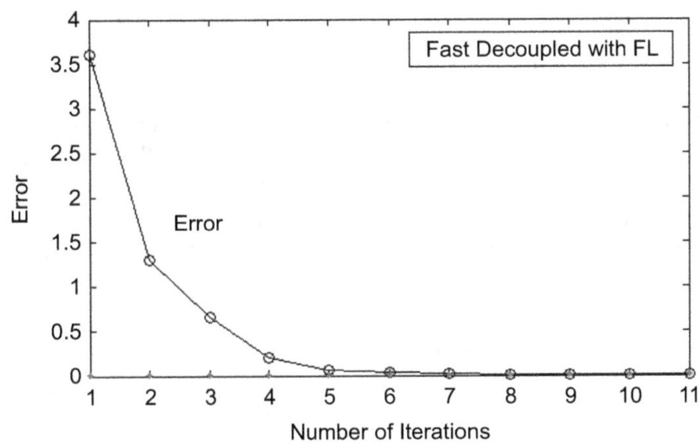

Figure 6.42 Convergence error change during the iteration of the Fast Decoupled algorithm with the injected FLAF

In order to increase the speed of power flow analysis algorithms, a dynamic acceleration factor is generated using the FL decision-making process and applied to Gauss–Seidel, Newton–Raphson, and Fast Decoupled load flow solution methods. The effect of the FLAF is more clear in Gauss–Seidel algorithm, because this algorithm requires more iterations to solve the IEEE 30-bus test system compared to the other two methods discussed here. Even if the effect of the FLAF is not visible for the problems requiring less than 10 iterations, it is more effective for the problems that require more iteration. Besides the iteration reduction, the FLAF can be used to ensure the convergence whatever the starting values are.

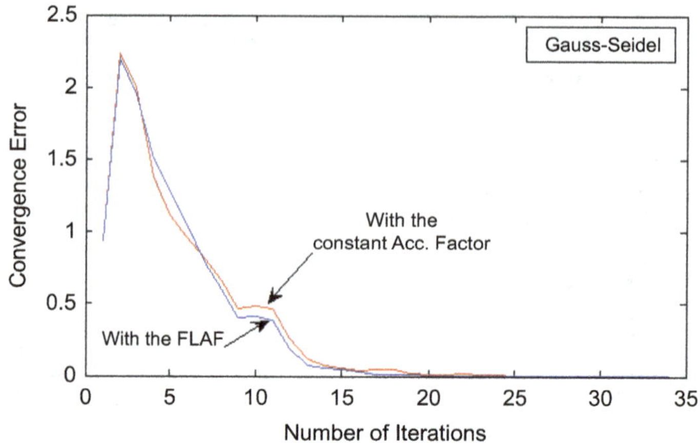

Figure 6.43 Comparison of the G–S algorithm with and without the fuzzy logic acceleration factor

6.6 Problems

P.6.1 Describe the two main control systems for synchronous generators used to generate power in power systems. Write the names of the control approaches, and describe them by sketching-related figure representations. Describe power–power angle curves of round rotor synchronous machines, and explain why small power angle values such as 30° are preferred in practice.

P.6.2 Design an FLC to replace the regulator amplifier of the exciter system given in Figure 6.16 to regulate the output voltage of the system for the conditions given below. Use the data given in Example 6.1.

(a) There are +5% step changes in reference voltage.
(b) There are −5% step changes in reference voltage.
(c) Repeat parts (a) and (b) for 10% step changes in reference voltage.
(d) Repeat the first three operating conditions for impulse changes in reference voltage instead of step changes.

P.6.3 A linear approach is used to model the saturation curve of the exciter in Section 6.2.1. Develop a model to represent the non-linearity of the saturation curve given in Figure 6.11 so that the excitation system modeling will be more accurate.

P.6.4 Instead of a separately excited DC generator is used as the exciter in Figure 6.10. Use a controlled rectifier as an exciter instead of a DC generator and remodel the excitation system of the synchronous generator. Assume that the field coil of the synchronous generator is in the stator and three-phase windings are on the rotor. After the modeling is complete, use an FLC to control this exciter system similar to the one described in Example 6.1.

P.6.5 An isolated power system exciter system is shown in Figure P6.5(a) where the amplifier gain is set to 1. Neither a regulator amplifier nor any other compensator is included. All system data are given in Figure P6.5a.

(a) Obtain a transfer function relating power system terminal voltage V_t to reference voltage V_{ref}.
(b) Check the stability of this exciter system using the Routh–Hurwitz method and determine a proper numerical value for amplifier gain, K, such that the system remains stable.
(c) Apply a unit step reference input and observe the time response of the output voltage v_t.
(d) Design a fuzzy logic controller for the given exciter to dampen the voltage oscillations with zero steady-state error.

P.6.6 A single-area AGC power system is shown in Figure P6.6(a). All system data are given in the figure.

(a) Obtain a transfer function relating the reference power change ΔP_{ref} to the frequency change $\Delta\omega$.
(b) Apply a unit step change in power demand and observe the time response of the frequency change $\Delta\omega$. Assume that the reference power change is zero.
(c) Design a fuzzy logic controller for the given exciter to dampen the frequency oscillations and keep the frequency unchanged.

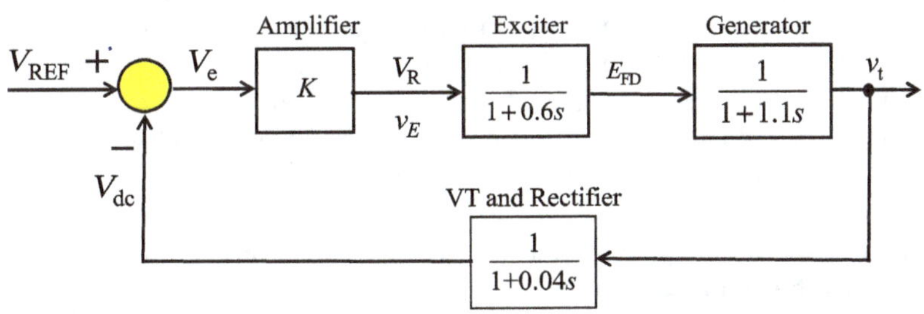

Figure P6.5(a) Power system exciter model without any regulator or stabilizer

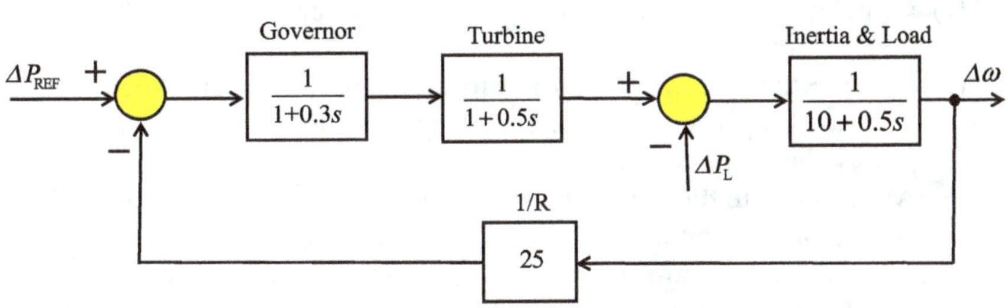

Figure P6.6(a) A single area AGC power system to be used in P.6.6

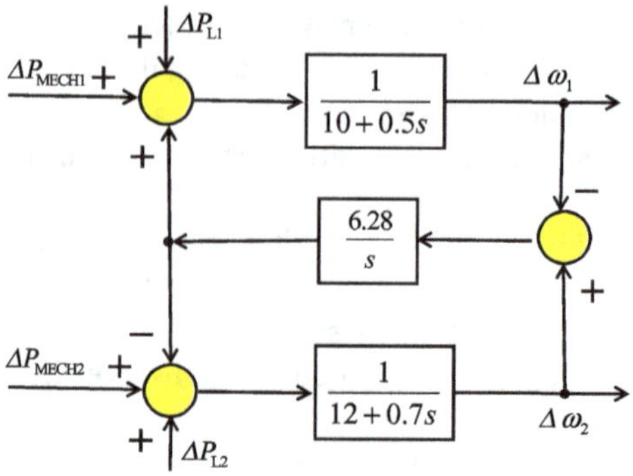

Figure P6.7(a) A two-area AGC power system to be used in P.6.7

P.6.7 A two-area AGC power system is shown in Figure P6.7(a). All system data are given in the figure.

(a) Assume that all mechanical power input changes are zero when a step change of 0.2 pu occurs in demand power ΔP_{L1} and analyze the frequency change responses by developing a Simulink model of the system.

(b) Develop a fuzzy logic controller or use the one developed in Chapter 7 in this two-area AGC system to dampen the frequency oscillations and control the mechanical power input changes of the units in both areas.

P.6.8 Design a fuzzy decision maker for the two-area AGC power system given in Problem P.6.7 so that the units in both areas are activated to overcome the frequency oscillations by adjusting the mechanical power inputs on time when a change occurs in demand powers of any one of the areas.

P.6.9 A large manufacturing plant is operated at a voltage of 10 kV and a demand power of 6 MW at a power factor of 0.75 lagging. The impedance of the power line is $0.4 + j1.3$ Ohm/phase.

(a) Calculate the apparent and reactive power driven by the load when the load bus voltage is 10 kV.

(b) Calculate the voltage at the sending end of the power line. Assume that this voltage at sending end is kept constant for the rest of the operating cases.

(c) Calculate the required reactive power to improve the power factor to 0.98 lagging.

(d) Assume that the capacitors are to be connected in grounded wye and there are only 100 kVAr increments in capacitor size. Write an order list to buy the required capacitors.

(e) Also, calculate the power line current and line I^2R losses before and after adding the capacitors.

P.6.10 Assume the same power line and the manufacturing plant described in Problem P.6.9. If the reactive power generated by the capacitors to improve the power factor to 0.98 lagging is 4073.2 kVAr.

(a) Obtain the voltage rise at the load bus.
(b) Obtain the new voltage at the load bus.
(c) Obtain the voltage rise as a percentage of the nominal bus voltage.
(d) If a voltage rise of 2% is allowed at the load bus, what changes in capacitor power should be made.
(e) What will be the value of line power factor after the changes made to the capacitors in part (d).

P.6.11 Simulate the system described in Example 6.6 using MATLAB/Simulink and observe the operational behavior of the system in parts (a), (b), and (c) of Example 6.6.

P.6.12 Repeat Problem P.6.11 if Plant 2 in Example 6.6 is turned off when the system is operating under the conditions described in part (b) of Example 6.6.

References

[1] Anderson P.M. and Fouad A.A. *Power System Control and Stability*. Piscataway, NJ: IEEE Press, 1993.
[2] Sauer P.W. and Pai M.A. *Power Systems Dynamics and Stability*. Englewood Cliffs, NJ: Prentice-Hall, 1998.
[3] Crenshaw M.L., Cutler J.F., Wright G.F. and Reid W.J. 'Excitation system models for power system stability studies'. *IEEE Committee Report, IEEE Transaction on Power Apparatus and Systems*. 1981;PAS-100(2):494–509.
[4] Saadat H. *Power System Analysis*. New York: McGraw-Hill, pp. 556–566, 1999.
[5] IEEE Committee Report. 'Excitation system models for power system stability studies'. *IEEE Transaction on Power Apparatus and Systems*. 1981; PAS-100(2):494–509.
[6] IEEE Committee Report. 'Computer representation of excitation systems'. *IEEE Transaction on Power Apparatus and Systems*. 1968;PAS-87(6):1460–1464.
[7] Demello F.P. and Concordia C. 'Concepts of synchronous machine stability as affected by excitation control'. *IEEE Transaction on Power Apparatus and Systems*. 1969;PAS-88(4):316–329.
[8] Larsen E.V. and Swann D.A. 'Applying power system stabilizers. Part I: general concepts'. *IEEE Transaction on Power Apparatus and System*. 1981; PAS-100(6):3016–3024.
[9] Ekinci S. and Demirören A. 'Modeling, simulation, and optimal design of power system stabilizers using ABC algorithm'. *Turkish Journal of Electrical Engineering & Computer Sciences*. 2016;24(3): 1532–1546.
[10] Kamwa I., Grondin R. and Trudel G. 'IEEE PSS2B versus PSS4B: the limits of performance of modern power system stabilizers'. *IEEE Transactions on Power Systems*. 2005;2(2):903–915.
[11] Corcau J.I. and Stoenescu E. 'Fuzzy logic controller as a power system stabilizer'. *International Journal of Circuits, Systems and Signal Processing*. 2007;1(3):266–273.

[12] Chaturvedi D.K. and Malik O.P. 'Neurofuzzy power system stabilizer'. *IEEE Transactions on Energy Conversion*. 2008;23(3): 887–894.

[13] Hoang P. and Tomsovic K. 'Design and analysis of an adaptive fuzzy power system stabilizer'. *IEEE Transactions on Energy Conversion*. 1996;11 (2):455–461.

[14] Altas I.H. and Neyens J. 'A fuzzy logic decision maker and controller for reducing load frequency oscillations in multi-area power systems'. *IEEE Power Engineering Society General Meeting*, Montreal, Canada, June 2006.

[15] Dimond H.M. and Lunge G.S. 'Continuous load–frequency control for interconnected power systems'. *Transactions of the American Institute of Electrical Engineers*. 1948;67(2):1483–1490.

[16] Bengiamin N.N. and Chan W.C. 'Multilevel load–frequency control of interconnected power systems'. *Proceedings of the Institution of Electrical Engineers*. 1978;125(6): 521–526.

[17] Rahmani M. and Sadati N. 'Hierarchical optimal robust load–frequency control for power systems'. *IET Generation, Transmission & Distribution*. 2012;6(4):303–312.

[18] Liu F., Song Y.H., Ma J., Mei S. and Lu Q. 'Optimal load–frequency control in restructured power systems'. *IEE Proceedings-Generation, Transmission and Distribution*. 2003;150(1):87–95.

[19] Christie R.D. and Bose A. 'Load frequency control issues in power system operations after deregulation'. *IEEE Transactions on Power Systems*. 1996; 11(3):1191–1200.

[20] Pan C.T. and Liaw C.M. 'An adaptive controller for power system load–frequency control'. *IEEE Power Engineering Review*. 1989;9(2):45–46.

[21] Ersdal A.M., Imsland L. and Uhlen K. 'Model predictive load–frequency control'. *IEEE Transactions on Power Systems*. 2016;31(1):777–785.

[22] Eremia M. and Shahidehpour M. (eds.) 'Active power and frequency control'. *Handbook of Electrical Power System Dynamics: Modeling, Stability, and Control*. Wiley-IEEE Press, 2013, pp. 291–339. DOI: 10.1002/9781118516072.ch3

[23] Feliachi A. 'Optimal decentralized load frequency control'. *IEEE Power Engineering Review*. 1987;PER-7(5): 44–45.

[24] Varma K.A., Mohanta D.K. and Reddy M.J.B. 'Applications of type-2 fuzzy logic in power systems: a literature survey'. *12th International Conference on Environment and Electrical Engineering*; 2013, pp. 282–286.

[25] Baysal Y.A. and Altas I.H. 'A fuzzy reasoning approach for optimal location and sizing of shunt capacitors in radial power systems'. IEEE Energy Conversion Congress and Exposition (ECCE); Montreal, Canada, September 2015, pp. 5838–5842.

[26] Song Y.H. and Johns A.T. 'Applications of fuzzy logic in power systems. I. General introduction to fuzzy logic'. *Power Engineering Journal*. 1997;11 (5):219–222.

[27] Song Y.H. and Johns A.T. 'Application of fuzzy logic in power systems. II. Comparison and integration with expert systems, neural networks and genetic algorithms'. *Power Engineering Journal*. 1998;12(4):185–190.

[28] Song Y.H. and Johns A.T. 'Applications of fuzzy logic in power systems. III. Example applications'. *Power Engineering Journal*. 1999;13(2): 97–103.

[29] Vlachogiannis J.G. 'Fuzzy logic application in load flow studies'. *IEE Proceedings—Generation, Transmission and Distribution*. 2001;148(1):34–40.

[30] Lo K.L., Lin Y.J. and Siew W.H. 'Fuzzy-logic method for adjustment of variable parameters in load–flow calculation'. *IEE Proceedings—Generation, Transmission and Distribution*. 1999;146(3): 276–282.

[31] Gupta A., Jain D.K. and Dahiya S. 'Management of power exchange between hybrid microgrids using intelligent control'. IEEE 6th International Conference on Power Systems (ICPS); 2016, pp. 1–6.

[32] Gajalakshmi P. and Rajesh S. 'Fuzzy modeling of power flow solution'. *INTELEC 07—29th International Telecommunications Energy Conference*; 2007, pp. 923–927.

[33] Dixit S., Srivastava L. and Agnihotri G. 'Power flow analysis using fuzzy logic'. *IEEE Power India Conference*; 2006, pp. 7.

[34] El-Hawary M.E. *Electric Power Applications of Fuzzy Systems*. Piscataway, NJ: IEEE Press, 1998, pp. 112–146.

[35] Liu C.C., Pierce D.A. and Song H. 'Intelligent system applications to power systems'. in *IEEE Computer Applications in Power*. 1997;10(4):21–24. doi: 10.1109/67.625369.

[36] Ibrahim W.R.A. and Morcos M.M. 'Artificial intelligence and advanced mathematical tools for power quality applications: a survey'. *IEEE Transactions on Power Delivery*. 2002;17(2):668–673.

[37] Pandey S.K., Mohanty S.R. and Kishor N. 'A literature survey on load–frequency control for conventional and distribution generation power systems'. *Renewable and Sustainable Energy Reviews*. 2013;25:318–334.

[38] Suganthi L., Iniyan S. and Samuel A.A. 'Applications of fuzzy logic in renewable energy systems—a review'. *Renewable and Sustainable Energy Reviews*. 2015;48:585–607.

[39] Hiyama T., Miyazaki K. and Satoh H. 'A fuzzy logic excitation system for stability enhancement of power systems with multi-mode oscillations'. *IEEE Transactions on Energy Conversion*. 1996;11(2):449–454.

[40] Guan X., Luh P.B. and Prasannan B. 'Power system scheduling with fuzzy reserve requirements'. *IEEE Transactions on Power Systems*. 1996;11 (2):864–869.

[41] Liu W.H.E. and Guan X. 'Fuzzy constraint enforcement and control action curtailment in an optimal power flow'. *IEEE Transactions on Power Systems*. 1996;11(2):639–645.

[42] Mori H. and Kobayashi H. 'Optimal fuzzy inference for short-term load forecasting'. *IEEE Transactions on Power Systems*. 1996;11(1):390–396.

[43] LaMeres B. and Nehrir M.H. 'Fuzzy logic based voltage controller for a synchronous generator'. *IEEE Computer Applications in Power*. 1999: 46–49.

[44] IEEE Recommended Practice for Excitation System Models for Power System Stability Studies. IEEE Std 421.5-2005, IEEE Power Engineering Society, 2006.

[45] Malik O.P., Kumar A. and Hope G.S. 'A load frequency control algorithm based on a generalized approach', *IEEE Transactions on Power Systems*. 1988;3(2):375–382.

[46] Chan W.C. and Hsu Y.Y. 'An optimal variable structure stabilizer for power system stabilization'. *IEEE Power Engineering Review*. 1983;PER-3(6):49.

[47] Korelic J. and Jezernik K. 'Predictive variable-structure system control of unity power factor rectifiers'. *IET Power Electronics*. 2013;6(8):1608–1617.

[48] Bergen A.R. and Vittal V. *Power Systems Analysis*, 2nd ed. Englewood Cliffs, NJ: Prentice-Hall, 2000.

[49] Ajjarapu V. and Christy C. 'The continuation power flow: a tool for steady state voltage stability analysis'. *IEEE Transactions on Power Systems*. 1992; 7(1):416–423.

[50] Shirmohammadi D., Hong H.W., Semlyen A. and Luo G.X. 'Compensation-based power flow method for weakly meshed distribution and transmission networks'. *IEEE Transactions on Power Systems*. 1988;3(2):753–762.

[51] Carol S.C. and Shirmohammadi D. 'Three-phase power flow method for real-time distribution system analysis'. *IEEE Transactions on Power Systems*. 1995;10(2):671–679.

[52] Wang Z., Alvarado F.L. 'Interval arithmetic in power flow analysis'. *IEEE Transactions on Power Systems*. 1992;7(3):1341–1349.

[53] Satpathy P.K., Das D. and Gupta P.B.D. 'Power flow analysis using fuzzy set approach'. *The Institution of Engineers (India) Publications*. 2004;85(1):55–59.

[54] Bijwe P.R., Hanmandlu M. and Pande V.N., 'Fuzzy power flow solutions with reactive limits and multiple uncertainties'. *Electric Power Systems Research*. 2005;76:145–152.

[55] Miranda V., Manuel M.A.C.C. and Saraiva J.T. 'Fuzzy load flow—new algorithms incorporating uncertain generation and load representation', *Proceedings of PSCC'90—10th Power System Computation Conference*, August 1990, pp. 621–627.

[56] Altas I.H. 'A fuzzy logic controlled tracking system for moving targets', *12th IEEE International Symposium on Intelligent Control, ISIC'97*, July 16–18, 1997, Istanbul, Turkey, pp. 43–48.

Chapter 7

FLC in wind energy systems

After a general introduction about wind energy systems, the application of fuzzy logic control and decision-making processes to wind energy conversion systems and their interfacing to utility are given in this chapter. Problems and control issues in wind energy conversion systems are studied and the utilization of fuzzy logic in solving these problems are shown.

7.1 Introduction

Wind is one of the renewable clean energy sources. It can be possibly operated without producing carbon dioxide, sulfur dioxide, particulates, or any other type of air pollution. Wind energy offers strong ecological concerns and advantages over conventional energy solutions, especially in remote and isolated areas. Due to its positive impacts on energy demand, the installations of wind energy systems (WES) have been rapidly growing for the last decades.

A wind energy conversion (WEC) system converts the kinetic energy available in the wind to mechanical energy that can be used as a prime mover to drive a rotational electrical generator [1]. Generally, a WEC system consists of a wind turbine, gearbox, generator, interfacing converter, required control systems, and of course a tower. The WEC systems can be installed as either standalone schemes for supplying power to locally isolated users in remote areas or installed as a part of the electric grid system.

As the share of WEC systems in electrical power generation keeps increasing, new challenging problems come to the present. Interfacing the WEC systems to the utility grid is not an easy task as it is in conventional hydraulic or steam-based electrical power generation systems. Changes in wind speed affect the amount of generated power as well as the magnitude and frequency of the generated voltage. Besides, the switchable loads in off-grid applications affect the voltage and the frequency at the load bus. An analysis of the operational characteristics shows that the WEC systems are very complicated and require high-level controllers to deal with the problems. Fuzzy logic controllers are considered as one of the solution approaches in these systems. Fuzzy logic decision-makers are also used as energy management tools in WEC interfacing systems. The pitch angle of the blades, excitation of asynchronous generators, voltage and frequency of the interfacing units, and maximum power extraction are the main tasks that are required to be controlled.

In addition to the control of the main operational process, the power quality and security aspects could also be considered. Therefore, effective active filters such as flexible AC transmission system (FACTS) devices are needed to be designed and utilized. Energy management algorithms for energy sustainability and economic system operation are applied. Fuzzy decision-based management systems can be used to develop effective and dynamic power management on both the generation and user sides. The harmonics and power factor correction problems can be solved easily and in the cheapest ways by employing the conventional passive power filter (PPF) with the fixed AC power capacitor banks as described in Chapter 6. However, the use of PPFs may pose hidden series and parallel resonance risks in the AC grid system with harmonic sources. The power resonance condition causes the harmonic voltage and current at a given bus to be increased and results in damage to other equipment connected to the same bus.

Nowadays, FACTS devices are used in various sections of power systems for power factor improvements and harmonic mitigation including those of renewable energy systems, too. FACTS or other active power filters are realized either as voltage source or current source devices. WEC systems require fast and reliable power conditioners for reactive power requirements, harmonic eliminations, and bus voltage stability. The power conditioners consisting of generation control, interfacing control, dissipation control, and power quality control must be fast enough to comply with any changes in wind speed and demand power.

Studies in the literature show that researchers are dealing with all problem aspects in various ways. A literature search enlightens us to see which problems are to be solved, which methods have been applied to solutions, and which new methods can be applied for better solutions. Although there are so many researches in pitch angle control [1–7], generator excitation control [8–12], AC–DC–AC link control [13–15], inverter control [16–18], interfacing device control [19–21], power quality and harmonic filtering control [22,23], and power management [24–26]. Besides, the studies in electrical engineering point of view, the blade size and structure, turbine and generator sizing, and blade materials are important topics under study [27–29].

In this chapter, after the basic theory of electrical power generation by WEC systems, some examples based on the author's research and thesis done by his graduate students are given.

7.2 Wind turbine

A WEC system consists of mainly a wind turbine, an electrical generator, and interfacing devices to load or to the utility grid. Besides the main units, turbine blades, gear trains, excitation capacitors, converters, and controllers are the other important elements of WEC systems. A generalized schematic diagram of a WEC system is given in Figure 7.1. The wind speed effect, the wind power, to the wind turbine is controlled by pitch angle control of the turbine blades. Turbine speed is increased or decreased by a gear train to have the electrical generator run at a speed that is suitable for generating electrical power.

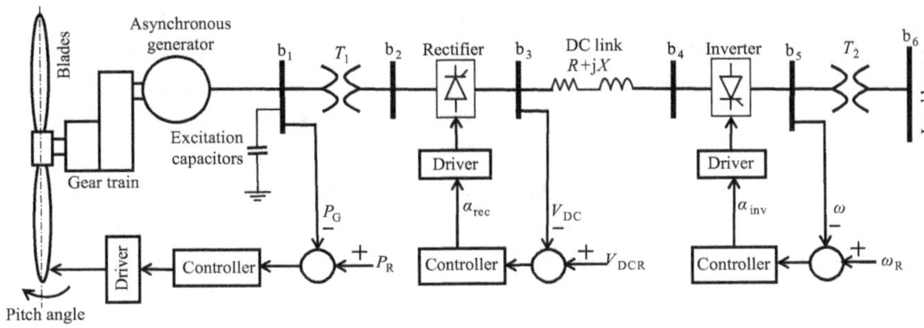

Figure 7.1 General schematic diagram of a WEC system

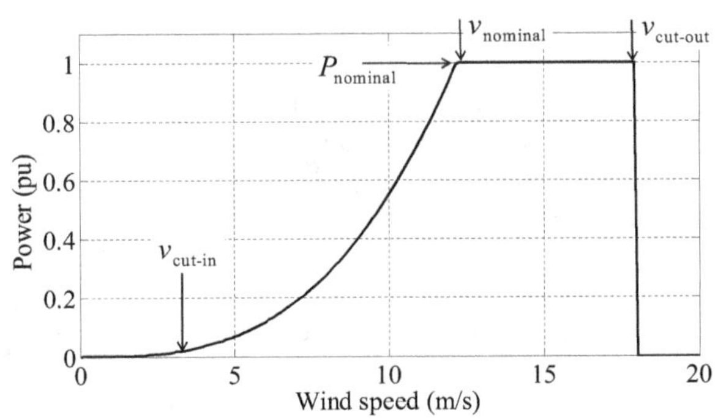

Figure 7.2 Wind power as a function of wind speed

In an electrical power system, the frequency and voltage magnitudes are two important system variables that are kept constant on the user side. The frequency and voltage magnitude are two critical design parameters of commercial devices. Every commercial device or appliance is designed to be operated at their rated voltages with the rated frequencies. Therefore, either the speed of electrical generator has to be kept constant or additional interfacing converters have to be employed and controlled to keep the voltage and the frequency constant. Since the wind speed is variable due to its nature, it is not an easy task to keep it constant by adjusting pitch angle or gear train. Therefore, power electronic converters must be used to have a more stable, flexible, and easier-to-control voltage and frequency profile.

A WEC system mainly has wind speed, voltage, and frequency controllers. The wind speed is adjusted by controlling the pitch angle of the blades. This method is usually applied to match the wind turbine speed with the electrical power demand. The mechanical power from a wind turbine is determined by the wind speed characteristics. The turbine generates more mechanical power at higher speed and less power at lower speeds. A typical wind speed characteristic showing the generated power is given in Figure 7.2.

Wind turbines convert wind power into mechanical power in terms of wind speed. Due to its inertia a wind turbine is able to generate a mechanical output power if the wind speed is greater than a value called cut-in speed as shown in Figure 7.2. The mechanical structure and strength of the turbine are limited to stand up to a speed called cut-out speed. The turbine can generate power at wind speeds between cut-in and cut-out speeds. Output power from the turbine is zero if the wind speed is below cut-in or above cut-out values. The speed above the cut-out value becomes dangerous for the mechanical structure and rotating shaft. Therefore, the rotation of the wind turbine blades and the shaft are mechanically locked to prevent damage. The mechanical power output of the turbine is a cube of the wind speed and must be limited at its rated value at higher speeds by stall control, active stall control, or pitch angle control [30]. During stall control, the blade position is fixed to reduce the speed of the blades. In active stall control, the blade angle is adjusted and a stall is obtained along the blades. By pitch control, the blades are turned around the blade axis so that the wind effect on the blades becomes minimal.

Wind power depends on the amount of air (volume), speed of air (velocity), and mass of air (density) flowing through an area. The kinetic energy of a mass m moving with a speed of v is expressed as in (7.1).

$$E_{\mathrm{K}} = \frac{1}{2} m v^2 \tag{7.1}$$

Since the energy per unit of time is defined as the power, the wind power can be written as in (7.2).

$$P_{\mathrm{W}} = \frac{1}{2} \frac{dm}{dt} v^2 \tag{7.2}$$

where $\dot{m} = dm/dt$ is called mass flow rate and calculated using fluid flow equations as in (7.3).

$$\frac{dm}{dt} = \rho A v \tag{7.3}$$

where v is the wind speed (m/s), ρ is the air mass density (kg/m^3) and A is the swept area (m^2) of the rotor blades and defined by the radius of the turbine rotor. Substituting (7.3) into (7.2) yields the wind power equation as a function of air density, rotor swept area, and cube of velocity as given in (7.4).

$$P_{\mathrm{W}} = \frac{1}{2} \rho A v^3 \tag{7.4}$$

Equation (7.4) gives theoretically available wind power to the turbine. This is the ideal wind power available. The wind turbine can transfer only some part of this available wind power to usable mechanical power at the output. Therefore, the output power of a wind turbine is expressed as in (7.5).

$$P_{\mathrm{T}} = \frac{1}{2} \rho A v^3 C_{\mathrm{p}} \tag{7.5}$$

where $C_p = P_T/P_W$ is called the turbine power coefficient and represents the power conversion efficiency of the wind turbine. A possible maximum value of C_p = 16/27 is called the Betz limit and gives an efficiency of 59%, which is between 40% and 45% in practice [31]. The power coefficient C_p is defined as a function of blade tip speed ratio, λ, and blade pitch angle, β. It is obtained experimentally for the turbine used. Various functions and approaches based on curve fitting techniques to polynomial equations [32–38] have appeared in the literature. The function given in (7.6), which is widely used in literature for simulation purposes, is suggested in [39] to represent $C_p(\lambda,\beta)$.

$$C_p(\lambda,\beta) = C_1\left(\frac{C_2}{\lambda} - C_3\beta - C_4\right)e^{\left(-\frac{C_5}{\lambda}+C_6\right)} \tag{7.6}$$

where λ is the blade tip speed ratio (TSR), which is defined as in (7.7).

$$\lambda = \frac{nR\omega_r}{v} \tag{7.7}$$

where n is the gear ratio, R is the radius of the turbine rotor, and ω_r is the angular velocity of the rotor. The wind speed can be kept constant at turbine-rated speed as depicted in Figure 7.2 or at other desired speeds by active stall control or pitch angle control. At various constant wind speeds, the mechanical power or torque to the electrical generator varies with the angular velocity of the rotor. Equations (7.5)–(7.7) can be used to obtain turbine output power versus rotor angular speed curves as shown in Figure 7.3, which are also called power curves of wind turbines. The power–speed curves in Figure 7.3 are obtained using a MATLAB® file as in Figure 7.4 with a MATLAB function given in Figure 7.5.

The turbine output curves are used as a reference benchmark in order to simulate WEC systems. As shown in Figure 7.3, the maximum power from the wind turbine is increasing as the wind speed increases. Therefore, a maximum power point tracking controller is required to get the maximum power available when the wind speed changes. The output power from the wind turbine is the input

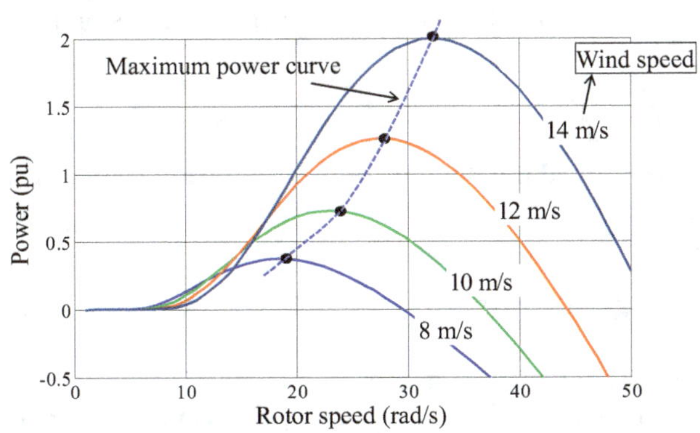

Figure 7.3 Turbine output power curves at various wind speeds

```
% WEC Modelling
clear; clc
wrmax=50;  R=3;    RO=1.225;  NG=1;     PBASE=10000;  BETA =0;
C1=0.73;   C2=151; C3=0.4;    C4=13.65; C5=18.4;      C6=0.0055;
vw=8;
[P1,W1]=windpower(wrmax,R,RO,NG,PBASE,BETA,C1,C2,C3,C4,C5,C6,vw);
vw=10;
[P2,W2]=windpower(wrmax,R,RO,NG,PBASE,BETA,C1,C2,C3,C4,C5,C6,vw);
vw=12;
[P3,W3]=windpower(wrmax,R,RO,NG,PBASE,BETA,C1,C2,C3,C4,C5,C6,vw);
vw=14;
[P4,W4]=windpower(wrmax,R,RO,NG,PBASE,BETA,C1,C2,C3,C4,C5,C6,vw);
plot(W1,P1,W2,P2,W3,P3,W4,P4); grid
xlabel('Rotor Speed (rad/s)')
ylabel('Power (pu)')
```

Figure 7.4 MATLAB file used to plot wind turbine power curves

```
function
[P,W]=windpower(wrmax,R,RO,NG,PBASE,BETA,C1,C2,C3,C4,C5,C6,vw);
k=1;
for wr=0:1:wrmax
    A=pi*(R^2);
    LAMDA(k)=NG*R*wr/vw;
    CA=C2/LAMDA(k);
    CB=C5/LAMDA(k);
    CP(k)=C1*(CA-C3*BETA-C4)*exp(-CB+C6);
    PM(k)=0.5*CP(k)*A*RO*(vw^3);
    Pmpu(k)=PM(k)/PBASE;
    WM(k)=wr;
    k=k+1;
end
P=Pmpu; W=WM;
```

Figure 7.5 MATLAB function used to obtain wind power curves

mechanical power to the electrical generator. The driving torque of the electrical generator is directly obtained from the mechanical power using (7.8).

$$T_m = \frac{P_T}{\omega_r} \tag{7.8}$$

where T_m is the mechanical input torque of the electrical generator used in the WEC system.

7.3 Electrical generator

The wind power is converted to electrical power by electrical generators. The mechanical torque given in (7.8) is generated by the wind turbine due to wind speed and is used as the input torque to the generator. Various types of electrical generators are used in WEC systems. Asynchronous or induction, synchronous, permanent magnet synchronous, and direct current (DC) generators are the most used ones. Synchronous and DC generators require additional power sources for their excitation units. Besides, their speed–torque characteristics show that they are not suitable to generate the same electrical power at different speeds.

Generating the same power is an important task in power systems as long as the power demand does not change. Therefore, electrical power generation should not be affected by the wind speed changes easily. If the wind speed is kept constant by pitch angle or stall control, then synchronous generators can be used to feed variable loads at a constant frequency. However, wind speed is a variable quantity and changes randomly. For small changes, its effect on the wind turbine may be kept constant by stall and/or pitch angle control. However, during larger stepwise changes in wind speed, the constant speed moves to a new level as depicted in Figure 7.3. In this case, the same power should be generated at the new speed level in order to keep feeding the loads.

Induction generators, which are also called asynchronous generators, are generally preferred to be used in WEC systems because they can generate the same power at different speeds. Therefore, in order to have a constant power output from the electrical generator at different speed levels, induction generators are the most suitable generator types. As shown by speed–power characteristics in Figure 7.6, an induction machine can generate the same electrical power under different speed levels.

Although the same electrical power is to be generated at different speed levels, the frequency of the generated voltage will be changed as the speed changes. Thus, a DC link is placed between the induction generator and AC load bus for voltage and frequency controls as shown in Figure 7.1. Since the electrical variables and their control are the major topic in this book, an asynchronous generator with a DC link interfacing equipment to the load bus or utility grid is considered as the electrical generator in the WEC system here.

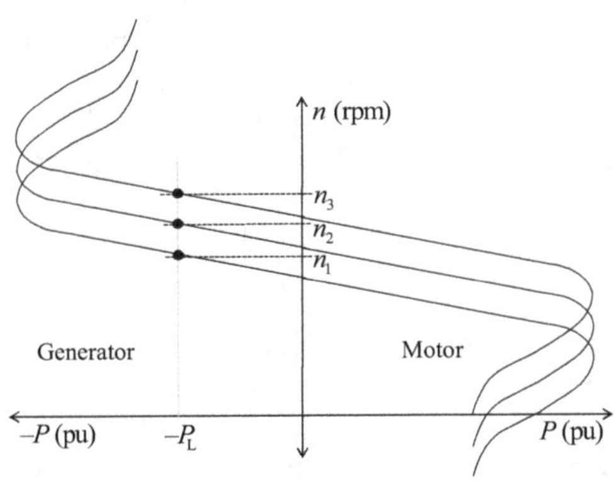

Figure 7.6 Speed–power characteristics of an asynchronous machine

7.3.1 Dynamic modeling of induction generator

In order to analyze the dynamic performance and transients as well as the steady-state response of the wind-driven three-phase induction machine, a mathematical model representation must be obtained. Numerous methods have been developed to analyze the transient performance of induction machines [40–44]. A three-phase induction machine can be represented by its voltage equations corresponding to the stator and rotor windings in terms of machine variables under the following assumptions [42,43]:

(i) Both rotor and stator have three-phase balanced windings.
(ii) Three-phase windings are sinusoidally distributed in space and displaced 120° apart.
(iii) Flux magnetomotive force (mmf) and flux waves are distributed sinusoidally.
(iv) Effects of teeth and slots are neglected.
(v) Air gap is uniform.
(vi) Saturation, hysteresis, and eddy current effects are ignored.
(vii) Rotor windings are short-circuited so that the rotor terminal voltages are zero.

Under these assumptions, the resistances and the self-inductances of three-phase windings in stator and rotor can be considered independent of time and rotor displacement angle. Hence the following equalities are written.

$$R_s = R_a = R_b = R_c \tag{7.9}$$

$$L_s = L_a = L_b = L_c \tag{7.10}$$

$$R_r = R_d = R_e = R_f \tag{7.11}$$

$$L_r = L_d = L_e = L_f \tag{7.12}$$

where the stator resistance R_s is equal to the individual resistances R_a, R_b, and R_c of phases a, b, and c, respectively. The stator inductance L_s is equal to the phase inductances L_a, L_b, and L_c of the phases a, b, and c. Similarly, the rotor resistance R_r and inductance L_r are equal to phase resistances and inductances R_d, R_e, R_f and L_d, L_e, L_f of the phases d, e, and f, respectively. The mutual inductances between the stator windings and between the rotor windings themselves are also regarded as equal and independent of time and rotor position angle due to the assumptions of the balanced three-phase distribution of the windings a, b, c and d, e, f.

$$M_s = M_{ab} = M_{ac} = M_{ba} = M_{ca} = M_{cb} \tag{7.13}$$

$$M_r = M_{de} = M_{df} = M_{ed} = M_{fd} = M_{fe} \tag{7.14}$$

where M indicates mutual inductance and subscripts ab, ac, …, de, df, …, etc. indicate the phases between which the mutual inductance occurs. The mutual inductances between the stator and rotor windings are functions of rotor displacement angle θ, and therefore, time t, since the rotor position is time dependent.

The generalized equations of a three-phase induction machine are usually written for a machine with a wound stator and wound rotor because the short-circuited bars of a squirrel cage rotor can be assumed equivalent to short-circuited rotor windings. Therefore, the generalized equations of a wound rotor induction machine are valid for induction machines with squirrel cage rotors.

Under the assumptions made, a three-phase induction machine can be represented as in Figure 7.7 by the three-phase stator and rotor windings, which yield the voltage equations given in (7.15)–(7.20).

For the stator:

$$v_a = R_s i_a + p\lambda_a \quad (7.15)$$

$$v_b = R_s i_b + p\lambda_b \quad (7.16)$$

$$v_c = R_s i_c + p\lambda_c \quad (7.17)$$

For the rotor:

$$v_d = R_r i_d + p\lambda_d \quad (7.18)$$

$$v_e = R_r i_e + p\lambda_e \quad (7.19)$$

$$v_f = R_r i_f + p\lambda_f \quad (7.20)$$

where $p = d/dt$ and λ_a, λ_b, λ_c, $\ldots$, λ_f, are the derivative operator and flux linkages, respectively. If these equations are written in terms of self- and mutual inductances instead of flux linkages, the voltage equations of a three-phase induction machine in a three-phase system are obtained as in (7.21). The voltage and current vectors, and the impedance matrix of Equation (7.21) may be represented in a simpler form with submatrices as shown in (7.22).

$$\begin{bmatrix} v_a \\ v_b \\ v_c \\ v_d \\ v_e \\ v_f \end{bmatrix} = \begin{bmatrix} R_a + pL_a & pM_{ab} & pM_{ac} & pM_{ad} & pM_{ae} & pM_{af} \\ pM_{ba} & R_a + pL_a & pM_{bc} & pM_{bd} & pM_{be} & pM_{bf} \\ pM_{ca} & pM_{cb} & R_a + pL_a & pM_{cd} & pM_{ce} & pM_{cf} \\ pM_{da} & pM_{db} & pM_{dc} & R_r + pL_r & pM_{de} & pM_{df} \\ pM_{ea} & pM_{eb} & pM_{ec} & pM_{ed} & R_r + pL_r & pM_{ef} \\ pM_{fa} & pM_{fb} & pM_{fc} & pM_{fd} & pM_{fe} & R_r + pL_r \end{bmatrix} \begin{bmatrix} i_a \\ i_b \\ i_c \\ i_d \\ i_e \\ i_f \end{bmatrix} \quad (7.21)$$

$$\begin{bmatrix} v_s \\ v_r \end{bmatrix} = \begin{bmatrix} Z_{ss} & Z_{sr} \\ Z_{rs} & Z_{rr} \end{bmatrix} \begin{bmatrix} i_s \\ i_r \end{bmatrix} \quad (7.22)$$

where v_s and v_r are the subarrays containing three-phase stator and rotor voltages v_a, v_b, v_c, and v_d, v_e, v_f, respectively. Similarly, the subarrays i_s and i_r contain three-phase stator and rotor currents i_a, i_b, i_c, and i_d, i_e, i_f, respectively. The stator submatrix Z_{ss} and rotor submatrix Z_{rr} are 3×3 matrices which are independent of rotor position angle and time. However, the matrices Z_{sr} and Z_{rs} include the mutual inductances between stator and rotor windings which are not stationary with respect to each other. Therefore, 3×3 submatrices Z_{sr} and Z_{rs} consist of time-dependent mutual inductances that vary as the rotor position angle changes as a function of time. These variable mutual inductances of the three-phase machine are defined as follows:

$$M_{ad} = M_{da} = M_{sr} \cos \theta \tag{7.23}$$

$$M_{ae} = M_{ea} = M_{sr} \cos \left(\theta + \frac{2\pi}{3} \right) \tag{7.24}$$

$$M_{af} = M_{fa} = M_{sr} \cos \left(\theta - \frac{2\pi}{3} \right) \tag{7.25}$$

$$M_{bd} = M_{db} = M_{sr} \cos \left(\theta - \frac{2\pi}{3} \right) \tag{7.26}$$

$$M_{be} = M_{eb} = M_{sr} \cos \theta \tag{7.27}$$

$$M_{bf} = M_{fb} = M_{sr} \cos \left(\theta + \frac{2\pi}{3} \right) \tag{7.28}$$

$$M_{cd} = M_{dc} = M_{sr} \cos \left(\theta + \frac{2\pi}{3} \right) \tag{7.29}$$

$$M_{ce} = M_{ec} = M_{sr} \cos \left(\theta - \frac{2\pi}{3} \right) \tag{7.30}$$

$$M_{cf} = M_{fc} = M_{sr} \cos \theta \tag{7.31}$$

where $M_{sr} = M_{rs} = M$ is the peak value of the mutual inductance between any pair of stator and rotor windings. The mutual inductance equations given above show that the impedance matrices $Z_{sr} = Z_{rs}$ are equal and vary as the rotor rotates.

Since the stator and rotor windings are not stationary, the solution of Equation (7.21) or (7.22) becomes difficult. Hence, this three-phase equation system is transferred to a reference system such that all stator and rotor variables are represented in a two-phase stationary system, called the d–q axis system. The transformation process from the rotating three-phase system to the stationary two-phase system can be found in [42,43]. After the transformation to a d–q axis model, the circuit representation of the three-phase induction motor windings is changed from Figures 7.7 and 7.8 resulting in d–q axis model equations as given in (7.32).

$$\begin{bmatrix} v_{ds} \\ v_{qs} \\ v_{dr} \\ v_{qr} \end{bmatrix} = \begin{bmatrix} R_s + L_s p & 0 & Mp & 0 \\ 0 & R_s + L_s p & 0 & Mp \\ Mp & Mp\theta & R_r + L_r p & L_r p\theta \\ -Mp\theta & Mp & -L_r p\theta & R_r + L_r p \end{bmatrix} \begin{bmatrix} i_{ds} \\ i_{qs} \\ i_{dr} \\ i_{qr} \end{bmatrix} \tag{7.32}$$

where $p = d/dt$ is the derivative operator and θ is the rotor angle which is the angle between phase a of the stator and phase d of the rotor. Then, $p\theta$ in equation (7.32) is the first derivative, $d\theta/dt$, of the rotor angle.

The d–q axis model equations of the three-phase induction machine are solved for the currents i_{ds}, i_{qs}, i_{dr}, and i_{qr}. The rotor speed $\omega_m = p\theta = d\theta/dt$ is another unknown in equation (7.32) if the machine is operated as a motor. However, the

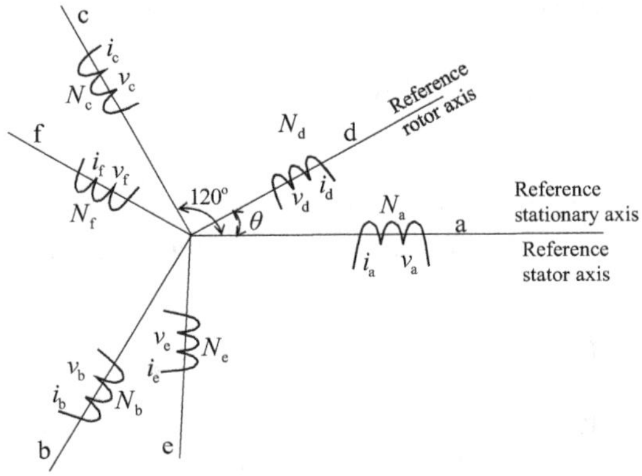

Figure 7.7 Circuit representation of induction machine in rotating three-phase frame

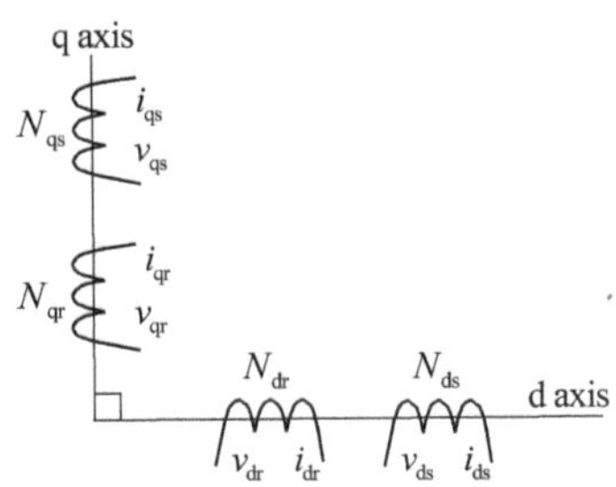

Figure 7.8 Stationary d–q axis representation of induction machine

machine is operated as a generator in this analysis. Therefore, the rotor speed is known because it is determined by the wind speed. Besides the voltage equations, an electromagnetic interaction equation is required to tie up the electrical and mechanical quantities in the same frame. The electromagnetic interaction equation is obtained from the torque–speed equation of the dynamic system. For a rotational system, (7.33) is written, ignoring rotation and friction losses.

$$T_e = T_m - J_m \frac{d\omega_m}{dt} = T_m - J_m \frac{d^2\theta}{dt^2} \tag{7.33}$$

where T_e is developed generator torque (Nm), T_m is mechanical input torque (Nm) from the turbine, J_m is rotational inertia (kg m^2), θ is rotor position angle (radius), and ω_m is the rotor speed (radius/seconds). In this system, the rotor speed ω_m and input torque T_m are known inputs from the wind turbine. It should be noted that the developed torque T_e is approximately equal to the mechanical input torque T_m during steady-state operation. The developed torque T_e can be written as a function of d–q system currents as in (7.34).

$$T_e = \frac{P}{2}\left[i_{\mathrm{dq}}\right]^T [G]\left[i_{\mathrm{dq}}\right] \tag{7.34}$$

where P is the number of the poles of the machine, $[i_{\mathrm{dq}}]$ is the d–q system current array as in Equation (7.32), and $[i_{\mathrm{dq}}]^T$ is its transpose. $[G]$ is a matrix containing only the terms with $p\theta$ in the generalized d–q system equation (7.32) as given in (7.35).

$$[G] = \begin{bmatrix} 0 & 0 & 0 & 0 \\ 0 & 0 & 0 & 0 \\ 0 & M & 0 & L_{\mathrm{r}} \\ -M & 0 & -L_{\mathrm{r}} & 0 \end{bmatrix} \tag{7.35}$$

After replacing the current arrays and matrix $[G]$ in (7.34), the induced torque is obtained as a function of d–q currents i_{ds}, i_{qs}, i_{dr}, and i_{qr}, rotor inductance L_{r}, and the peak mutual inductance M between the stator and the rotor windings as in (7.36).

$$T_e = \frac{P}{2}M\left[i_{\mathrm{ds}}i_{\mathrm{dr}} - i_{\mathrm{qr}}i_{\mathrm{ds}}\right] \tag{7.36}$$

Thus, the substitution of (7.36) into (7.33) yields (7.37).

$$\frac{P}{2}M\left[i_{\mathrm{ds}}i_{\mathrm{dr}} - i_{\mathrm{qr}}i_{\mathrm{ds}}\right] = T_m - J_m\frac{d\omega_m}{dt} \tag{7.37}$$

This is an ordinary differential equation with only one independent variable, time. After some manipulations and replacing $p\theta$ by the speed ω_m, Equations (7.32) and (7.37) are modified to the simultaneous first-order ordinary differential equation forms as given in (7.38)–(7.42).

$$\frac{di_{\mathrm{ds}}}{dt} = \frac{1}{L_{\mathrm{s}}L_{\mathrm{r}} - M^2}\left(-R_{\mathrm{s}}L_{\mathrm{r}}i_{\mathrm{ds}} + \omega_m M i_{\mathrm{qs}} + MR_{\mathrm{r}}i_{\mathrm{dr}} + \omega_m L_{\mathrm{r}}i_{\mathrm{qr}} + L_{\mathrm{r}}v_{\mathrm{ds}} - Mv_{\mathrm{dr}}\right) \tag{7.38}$$

$$\frac{di_{\mathrm{qs}}}{dt} = \frac{1}{L_{\mathrm{s}}L_{\mathrm{r}} - M^2}\left(R_{\mathrm{s}}Mi_{\mathrm{ds}} + \omega_m M i_{\mathrm{qs}} - R_{\mathrm{r}}L_{\mathrm{r}}i_{\mathrm{dr}} + \omega_m L_{\mathrm{r}}i_{\mathrm{qr}} - Mv_{\mathrm{ds}} + L_{\mathrm{r}}v_{\mathrm{dr}}\right) \tag{7.39}$$

$$\frac{di_{\mathrm{dr}}}{dt} = \frac{1}{L_{\mathrm{s}}L_{\mathrm{r}} - M^2}\left(R_{\mathrm{s}}Mi_{\mathrm{ds}} + \omega_m M i_{\mathrm{qs}} - R_{\mathrm{r}}L_{\mathrm{s}}i_{\mathrm{dr}} + \omega_m L_{\mathrm{r}}i_{\mathrm{qr}} - Mv_{\mathrm{ds}} + L_{\mathrm{s}}v_{\mathrm{dr}}\right) \tag{7.40}$$

$$\frac{di_{\mathrm{qr}}}{dt} = \frac{1}{L_{\mathrm{s}}L_{\mathrm{r}} - M^2}\left(-\omega_m M i_{\mathrm{ds}} + R_{\mathrm{s}}Mi_{\mathrm{qs}} - \omega_m L_{\mathrm{r}}i_{\mathrm{dr}} - L_{\mathrm{r}}L_{\mathrm{s}}i_{\mathrm{qr}} - Mv_{\mathrm{qs}} + L_{\mathrm{s}}v_{\mathrm{qr}}\right) \tag{7.41}$$

$$\frac{d\omega_m}{dt} = \frac{T_m}{J_m} - \frac{PM}{2J_m}\left[i_{\mathrm{ds}}i_{\mathrm{dr}} - i_{\mathrm{qr}}i_{\mathrm{ds}}\right] \tag{7.42}$$

In order to solve the d–q model equations of the three-phase induction machine given above, the voltages v_{ds}, v_{qs}, v_{dr}, and v_{qr} must be known besides the machine parameters. Since a squirrel cage induction machine is used, the rotor voltages v_{dr} and v_{qr} become zero. Only the stator voltages v_{ds} and v_{qs} are required to be known.

Since excitation capacitors are used in self-excited induction generators the stator voltages are equal to the excitation voltages of the capacitors. The d–q axis stator voltages are obtained from the three-phase voltages v_a, v_b, and v_c by applying phase conversion methods as described in [40–44].

7.3.2 Self-excited induction generator

All electrical machines require a magnetic field for energy conversion. If a current-carrying conductor is placed in a magnetic field, a force is generated as a result of Ampere's law. This is the basic principle for motor operation of rotating electrical machines. If a conductor moves in a magnetic field with a speed, then a voltage is induced in this conductor according to Faraday's law of electromagnetic induction. This is the basic principle for generator operation of electrical generators with linear or rotational motion.

DC and synchronous machines are built with excitation windings to be used to generate the required magnetic field. However, induction machines do not have an excitation winding. They generate a magnetizing current by using the reactive power component they draw from the AC source when they are operated as motor. Therefore, induction motors dissipate both active and reactive power, which are supplied by the AC source such as the AC utility. If the same induction machine is operated as a generator, its rated reactive power must be supplied so that the generator can produce and deliver its rated active power to the load bus.

Usually, there are two special cases for the induction generators in terms of the excitation process: utility-connected case and utility-isolated case. In utility-connected WEC systems, the induction machine is operated as a motor below the synchronous speed first, then the speed is increased above the synchronous speed to continue with the generator operator. In this process, the induction machine gets both active and reactive power from the utility while operating as a motor. When the generator operation is started, the machine continues to receive reactive power from the utility and begins to give active power to the utility as shown in Figure 7.9(a). However, transferring reactive power to the WEC system by the utility is not

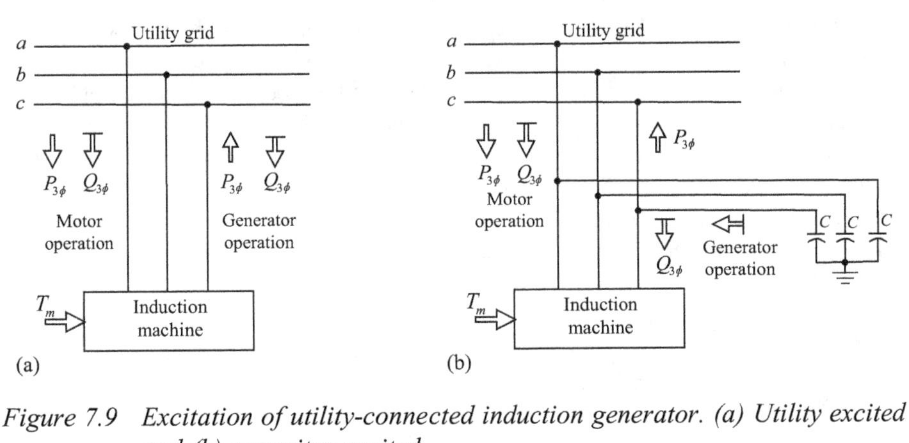

Figure 7.9 Excitation of utility-connected induction generator. (a) Utility excited and (b) capacitor excited.

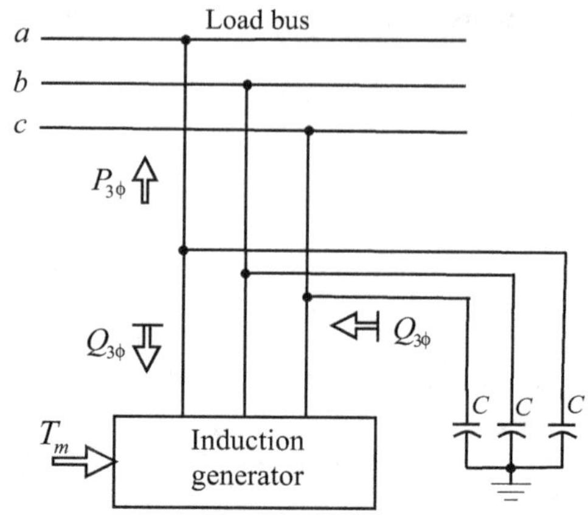

Figure 7.10 Excitation of utility-isolated induction generator

preferable due to line losses and distribution system performance. Therefore, the required reactive power is generated by capacitor groups installed next to the WEC system as depicted in Figure 7.9(b).

In utility-isolated WEC systems using induction generators, the reactive power must be supplied from capacitors since there is no utility or other sources around as given in Figure 7.10. In this case, capacitors are used for both excitation and for the reactive power demand of the loads. It should be noted that Figures 7.9 and 7.10 are used to show only the active and reactive power flow direction and sources. The interfacing devices, such as rectifiers and inverters, are used to establish a DC link between the generator and utility or the load bus for voltage and frequency control purposes.

As the above explanations show, an induction generator should be considered with its excitation unit. Since the excitation unit requires a capacitor group to be connected across the output terminals of the induction generator, this capacitor should be considered in dynamic modeling. Since the bars in the rotor of squirrel cage induction machines are short-circuited and the ends of three-phase rotor winding of the wound rotor induction machines are assumed to be short-circuited for the dynamic modeling process, then the circuit representation of induction machine in stationary d–q frame becomes as in Figure 7.11, which yields the per phase equivalent circuit representation of stationary d–q axis model of induction generator as in Figure 7.12.

The excitation capacitor per phase is placed across the per phase equivalent stationary d–q axis circuit. They are initially charged by the remnant voltages of the three-phase windings due to the magnetizing effects of the magnetic core. The voltages across the capacitors in Figure 7.12 can be written as in (7.43) and (7.44).

$$v_{\mathrm{qc}} = \frac{1}{C}\int i_{\mathrm{qc}}dt + v_{\mathrm{qc0}} \tag{7.43}$$

$$v_{\mathrm{dc}} = \frac{1}{C}\int i_{\mathrm{dc}}dt + v_{\mathrm{dc0}} \tag{7.44}$$

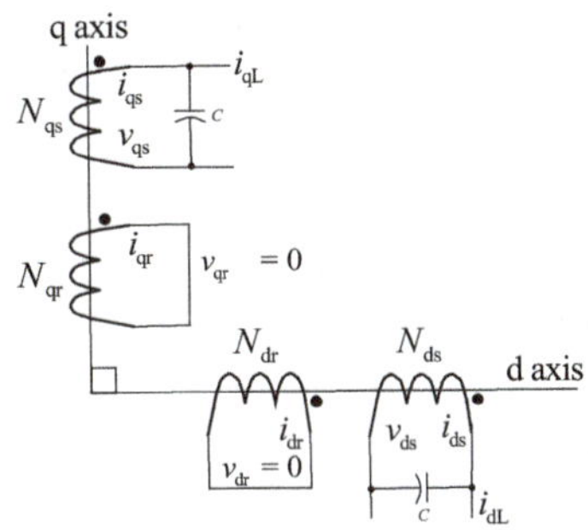

Figure 7.11 Stationary d–q axis representation of capacitor excited induction generator

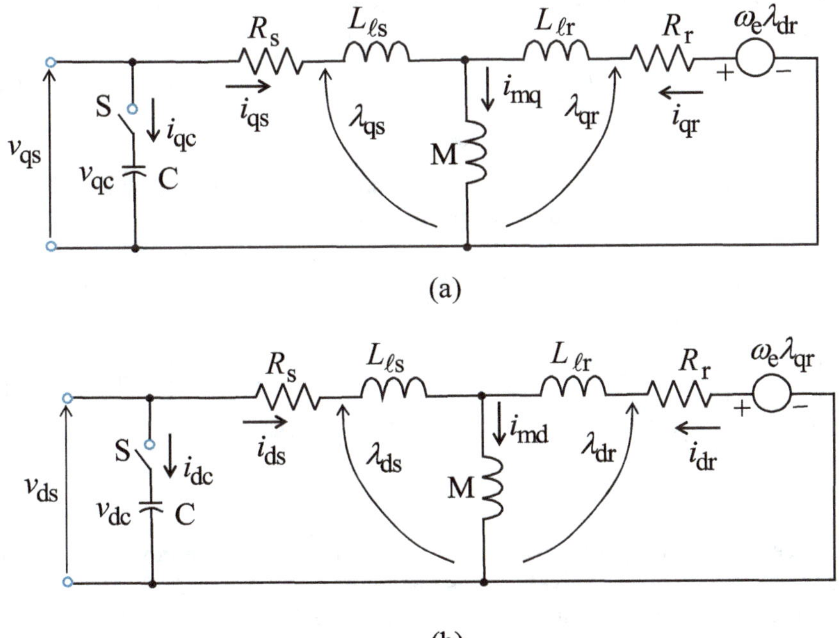

Figure 7.12 Stationary d–q circuit representation of self-excited induction generator. (a) q-Circuit and (b) d-circuit.

where during no load operating condition $i_{dc} = i_{ds}$ and $i_{qc} = i_{qs}$. The voltages v_{qc0} and v_{dc0} are the initial capacitor voltages along q and d exes. It has been assumed the capacitors are charged to these values initially. The rotor flux linkages λ_{qr} and

λ_{dr} in Figure 7.12 are obtained as in (7.45) and (7.46), respectively.

$$\lambda_{qr} = Mi_{qs} + L_r i_{qr} + \lambda_{qr0} \tag{7.45}$$

$$\lambda_{dr} = Mi_{ds} + L_r i_{dr} + \lambda_{dr0} \tag{7.46}$$

where λ_{qr0} and λ_{dr0} are the residual flux linkages and are equal to the initial values of the flux linkages λ_{qr} and λ_{dr}. The stator and rotor inductances are expressed as in (7.47) and (7.48), respectively.

$$L_s = L_{ls} + M \tag{7.47}$$

$$L_r = L_{lr} + M \tag{7.48}$$

where L_{ls} and L_{lr} are the leakage self-inductances of stator and rotor windings. The terms $\omega_e\lambda_{dr}$ and $\omega_e\lambda_{qr}$ in Figure 7.12 are the electrical rotor speed voltages along d- and q-axis, respectively. These are the generated voltages of the induction generator due to the angular electrical speed ω_e, which is obtained as a product of the number of pole pairs and mechanical speed ω_r of the rotating shaft. These voltages are expressed as in in (7.49) and (7.50).

$$\omega_e\lambda_{qr} = \omega_e\left(Mi_{qs} + L_r i_{qr}\right) + \omega_e\lambda_{qr0} \tag{7.49}$$

$$\omega_e\lambda_{dr} = \omega_e\left(Mi_{ds} + L_r i_{dr}\right) + \omega_e\lambda_{dr0} \tag{7.50}$$

The last terms $\omega_e\lambda_{qr0}$ and $\omega_e\lambda_{dr0}$ in (7.49) and (7.50) are the residual or initially induced voltages along q- and d-axis of the rotor, respectively.

Stationary d–q axis model equations of self-excited induction generator can be obtained as in (7.51) using the equivalent circuits in Figure 7.12 and Equations (7.45–7.50).

$$\begin{bmatrix} v_{qs} \\ v_{ds} \\ 0 \\ 0 \end{bmatrix} = \begin{bmatrix} R_s + pL_s & 0 & pM & 0 \\ 0 & R_s + pL_s & 0 & pM \\ pM & -\omega_e M & R_r + pL_r & -\omega_e L_r \\ \omega_e M & pM & \omega_e L_r & R_r + pL_r \end{bmatrix} \begin{bmatrix} i_{qs} \\ i_{ds} \\ i_{qr} \\ i_{dr} \end{bmatrix} + \begin{bmatrix} 0 \\ 0 \\ -\omega_e\lambda_{dr0} \\ \omega_e\lambda_{qr0} \end{bmatrix} \tag{7.51}$$

Since the terminal voltages v_{qs} and v_{ds} in Figure 7.12 are equal to capacitor voltages v_{qc} and v_{dc}, respectively, the substitution of (7.43) and (7.44) into (7.51) yields the final form of stationary d–q axis model equations of a self-excited induction generator as given in (7.52).

$$\begin{bmatrix} 0 \\ 0 \\ 0 \\ 0 \end{bmatrix} = \begin{bmatrix} R_s + pL_s + \dfrac{1}{pC} & 0 & pM & 0 \\ 0 & R_s + pL_s + \dfrac{1}{pC} & 0 & pM \\ pM & -\omega_e M & R_r + pL_r & -\omega_e L_r \\ \omega_e M & pM & \omega_e L_r & R_r + pL_r \end{bmatrix} \begin{bmatrix} i_{qs} \\ i_{ds} \\ i_{qr} \\ i_{dr} \end{bmatrix} + \begin{bmatrix} v_{qc0} \\ v_{dc0} \\ -\omega_e\lambda_{dr0} \\ \omega_e\lambda_{qr0} \end{bmatrix} \tag{7.52}$$

The solution of (7.52) yields d–q axis currents of the rotor and stator of the induction machine using the initial capacitor and residual voltages. Since the terminal voltages v_{qs} and v_{ds} are equal to the capacitor voltages v_{qc} and v_{dc}, respectively, (7.43) and (7.44) can be used to calculate terminal voltages, as well.

The d–q modeling of induction generator is given for no load condition in this section. Under loading conditions, the stator current becomes equal to the sum of load and capacitor currents.

7.4 FLC examples in WEC systems

A general schematic diagram of a WEC system is given in Figure 7.1 where control blocks are also included. Stall and pitch angle are controlled on the wind turbine side for demand power matching. The excitation capacitor switching is controlled to adjust the reactive power, and therefore the terminal voltage of the induction generator. Depending on the interfacing scheme to loads or to the utility additional controllers are also used in WEC systems. Rectifier and inverter control in AC/DC/AC interfacing units are also required for voltage magnitude and frequency control. Besides the main operating components, there may be additional design criteria that require controllers. For example, harmonic elimination, power conditioning, load power matching, and maximum power tracking are some of these additional designs that require controllers. Some examples of the control applications in WEC systems are given in this section.

Example 7.1

A utility-isolated single-phase micro-WEC system has been adapted from [45] and is shown in Figure E7.1(a). The WEC system is directly connected to a single-phase load bus by a transmission line. In order to deliver improved quality power to the load bus to feed linear, nonlinear, and motor-type loads, a dynamic voltage regulator based on the works by Sharaf [46–49] is to be designed and controlled using a fuzzy-tuned PI controller. Develop the required models and simulate the system using the wind turbine model available in MATLAB/Simulink®.

Answer 7.1

The wind turbine used in this example has 10 m/s wind speed and 10 kW power output to derive a 10 kW, three-phase, 380 V, 50 Hz squirrel cage induction generator (SCIG). The dynamic voltage regulator (DVR) is a switched power filter (SPF) type and is placed between bus b1 and bus b2 as shown by the single line diagram in Figure E7.1(a). A detailed wiring diagram of DVR/SPF is given in Figure E7.1(b).

Three-phase stator windings of the SCIG used in the WEC system of this example are operated as a single-phase SCIG. Therefore, two capacitors, C_s, were placed in series with the two of the wye-connected stator windings as depicted in Figure E7.1(b). The capacitor group C_e across the output terminals of the SCIG is used for the excitation.

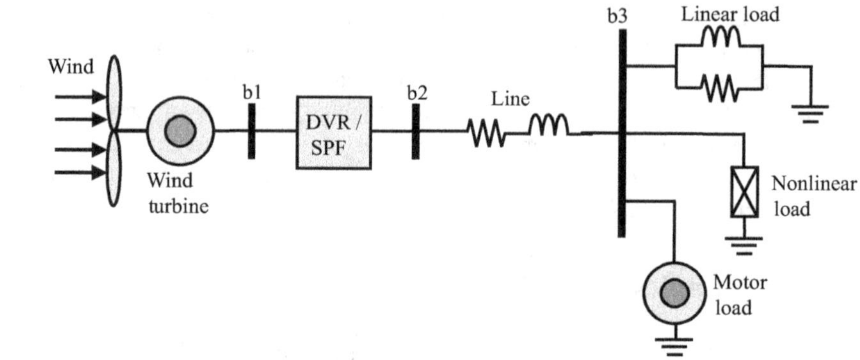

Figure E7.1(a) Single line diagram of the system for Example 7.1

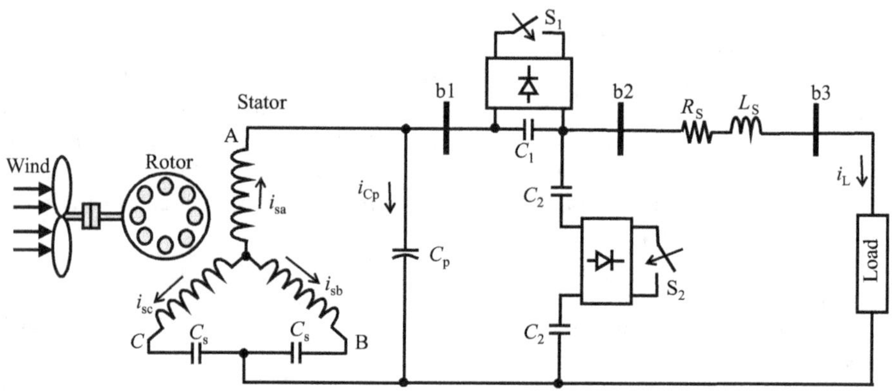

Figure E7.1(b) Detailed circuit diagram of the WEC system

An open circuit diagram of the DVR/SPF is placed between buses b_1 and b_2. DVR/SPF diagram has two filtering capacitors (C_1, C_2) and two switches (S_1, S_2). A transmission line with the parameters $R_L = 1\ \Omega$ and $L_L = 1$ mH is connected between busses b_2 and load bus b_3. Since the modelings of wind turbines and induction generators are given in previous sections, only the control of the DVR/SPF is discussed in this section. However, the capacitors connected in series with the wye-connected stator windings and depicted by C_s in Figure E7.1(b) must be included in the voltage equations of the induction generator. Besides, the output voltage of the generator is just a single-phase voltage and the excitation capacitors are also single phase.

DVR/SPF is controlled in terms of the switching instants of the switches S_1 and S_2, where S_1 is the negation of S_2 as $S_1 = \overline{S}_2$. The voltage magnitude of the load bus or bus b_2 is controlled by controlling the switching sequences of S_1 and S_2. Therefore, voltage and current readings at bus b_2 are used as input signals to the control process of the DVR/SPF as shown in Figure E7.1(c).

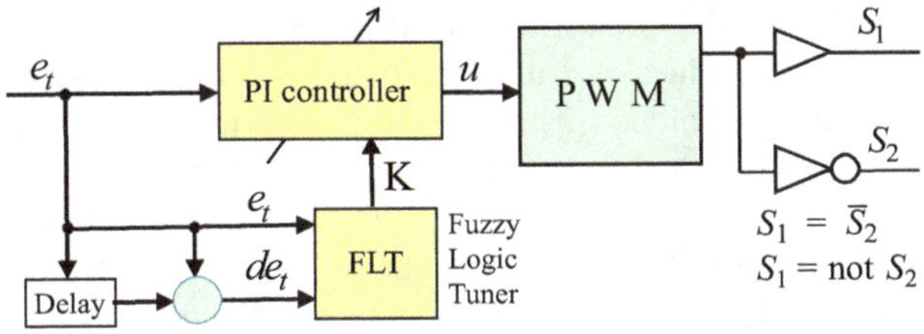

Figure E7.1(c) The control subsystem generates switching signals

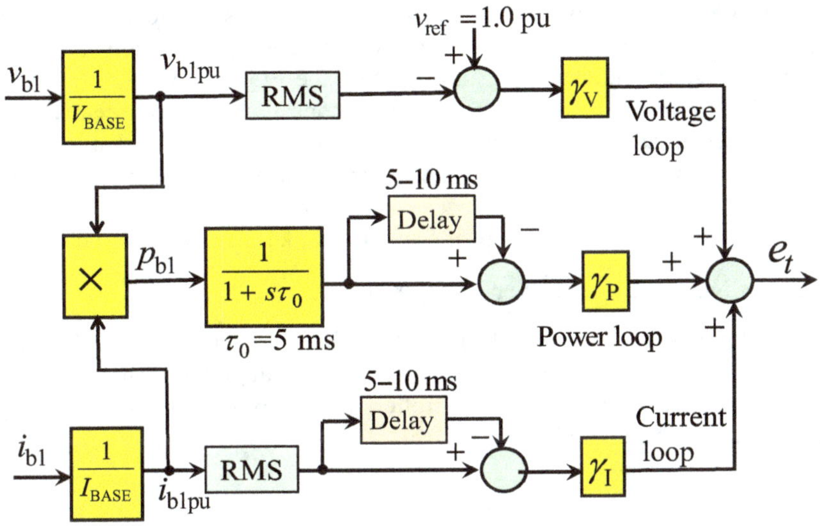

Figure E7.1(d) Block diagram of tri-loop dynamic error detector

The output signal of the PI controller is employed to adjust the reference pulse width modulation (PWM) voltage, to be compared with a fixed carrier signal to produce two complementary pulses, which are used as the external control signals for the ideal insulated gate bipolar transistor (IGBT) switches [45] as depicted in Figure E7.1(c).

The control subsystem includes a tri-loop dynamic error detector, which is based on Sharaf's ideas presented in [46–49]. The dynamic error detector uses instantaneous current and voltage signals as inputs and generates their error signals as bus voltage error e_v, power change error $e_{\Delta P}$ and current change error $e_{\Delta i}$, which are summed up to yield a total error e_t as given in Figure E7.1(d). The voltage error is calculated as the difference between the reference pu bus voltage and the pu value of the measured pu RMS bus voltage. The voltage error loop is used for voltage stabilization. The instantaneous pu value of the power is calculated from the instantaneous pu values of the bus voltage and current and then passed through a low pass filter. The power change error is then obtained as the difference between two consecutive values of the filtered power. The current change error is the difference between two consecutive RMS pu values of the bus current and is used to detect any sudden electrical load excursions or wind velocity variations. All the values of scaling and time delay of the tri-loop error collector were selected by an offline guided trial and error method to ensure a fast response and minimized total error, which is the sum of all three basic loops.

The total error e_t is used by a PI controller to drive a pulse width modulator in order to generate the switching signals S_1 and S_2. As shown in Figure E7.1(c), the total error e_t is also used by a fuzzy logic tuner (FLT) in order to tune the controller gains K_p and K_i.

The FLC described in Chapter 4 and given in Figures 4.24 and 4.25 can also be modified and used in this system as a parameter tuner for the PI controller. A block diagram showing the modification done to adapt the FLC in Figure 4.24 to the tuning process in this example is given in Figure E7.1(e). The FLT is used to adjust a common gain factor K so that the effects of the PI controller gains K_P and K_I are tuned.

The error e_t and its change de_t are used by the FLT to generate the required change $dK(k)$ for the common factor K of the PI controller. Two consecutive values of the total error are used to generate the error change de_t so that a decision table based on the error and the change in error can be generated to store information about the effects of the changes in controller parameters. The decision table includes implications in terms of fuzzy rules like the one given below.

If e is NE and de is ZE then dK is dKN

The output is the required change dK of the common factor K. The online fuzzy tuning process of the controller gains for the changing operating conditions gives a sort of adaptivity to the PI controller. The fuzzy logic gain tuner is represented by the block FLT in Figure E7.1(c).

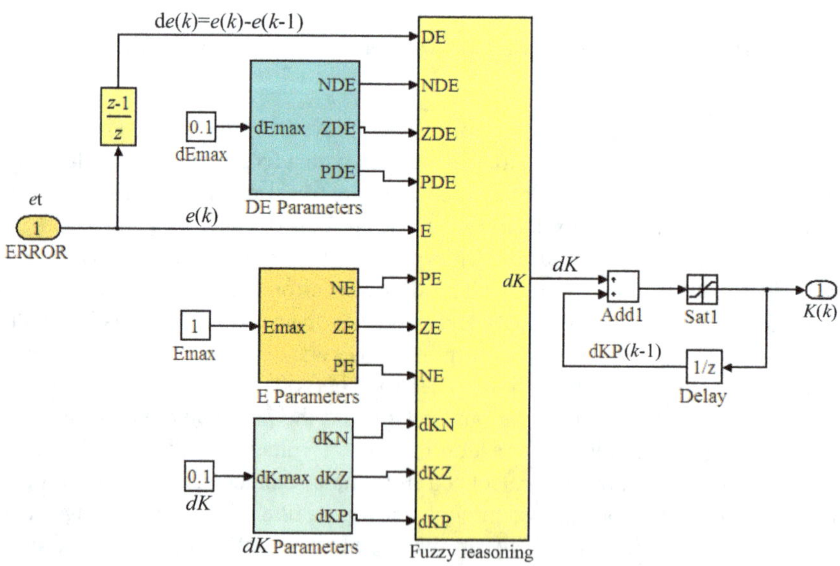

Figure E7.1(e) Simulink block structure of the FL tuner

The crisp universes of the error and change in error are partitioned into three fuzzy subsets each using triangular fuzzy membership functions. Since the amplitude of the change in error is affected by the sampling period, it is much smaller than the error. Therefore, the maximum value of the change in error is selected as 10 times smaller than the error. The fuzzy subsets used to represent the subcategories of the universes *e*, *de*, and *dK* are defined as shown in Figures E7.1(f)–E7.1(h).

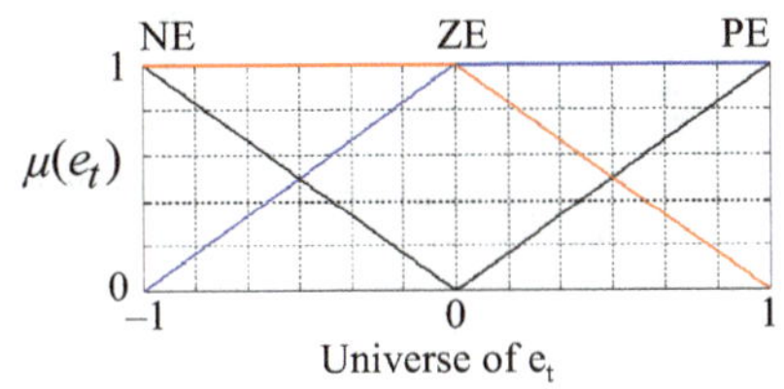

Figure E7.1(f) Fuzzy subsets in the universe of error e_t

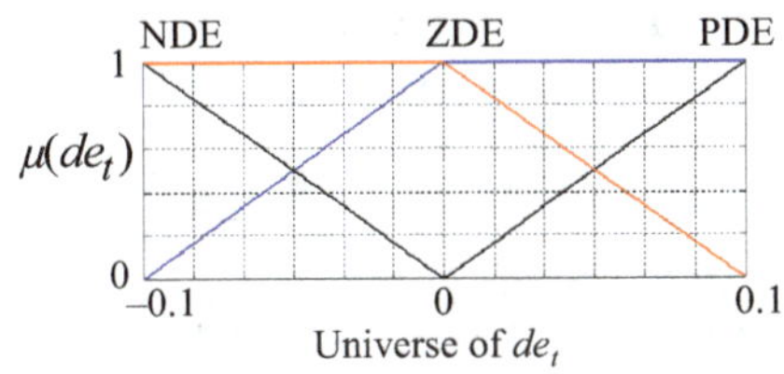

Figure E7.1(g) Fuzzy subsets in the universe of error de_t

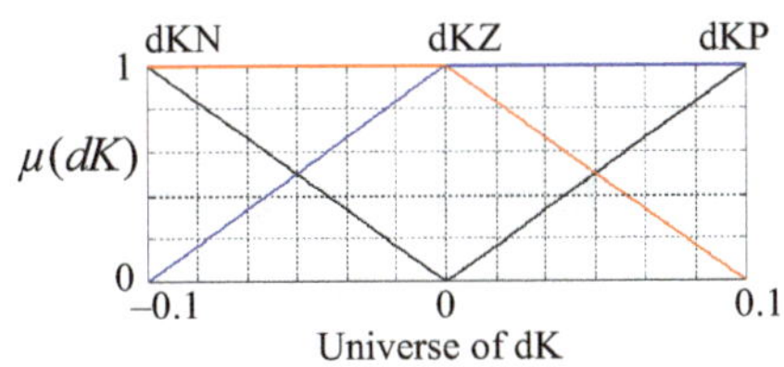

Figure E7.1(h) Fuzzy subsets in the universe of error dK

A nine-rule table is used in the fuzzy reasoning block of the FLT given in Figure E7.1(e). The output is the change *dK*, which is added to the value of K from the previous sampling in order to obtain the new value of the tuning factor. If the change *dK* is zero, then the previous value of *K* remains unchanged. This action is nothing but the digital representation of Euler's integration. Therefore, the FLT generates a change in the output signal to tune the gains K_P and K_I of the PI controller.

The variations of effective voltage and current at bus b_3 are shown in Figures E7.1(i) and E7.1(j). Bus b_3 is the load bus and effective voltage is required to be 220 V. It is shown that this voltage is reached at its steady-state value of 220 V in 9 s. Current value is about 40 A. Active and reactive power variations at bus b_3 are shown in Figures E7.1(k) and E7.1(l), respectively. Steady-state active power is 7.5 kW and reactive power is 3.4 kVAr.

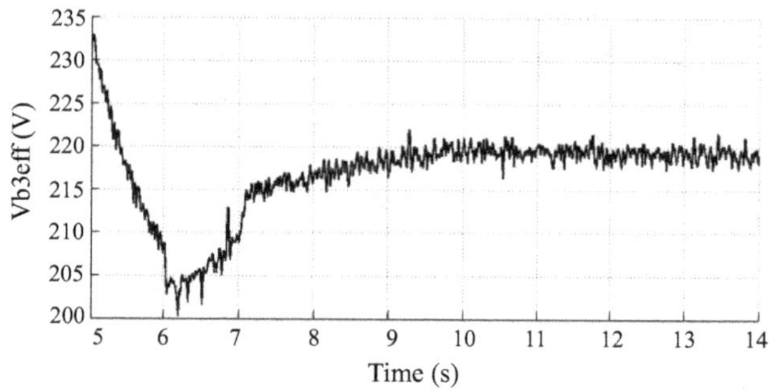

Figure E7.1(i) Effective value of voltage at bus b_3

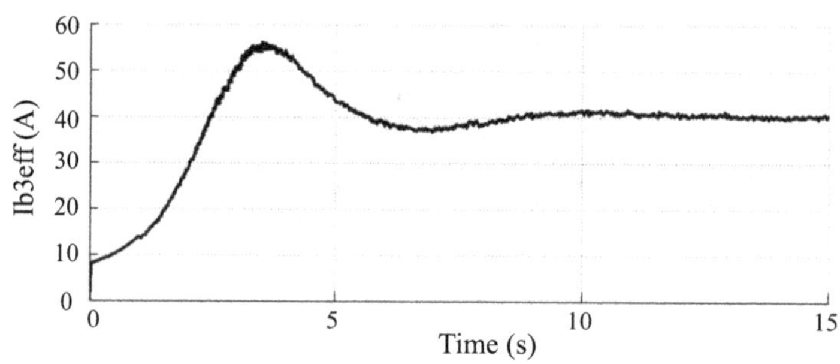

Figure E7.1(j) Effective value of current at bus b_3

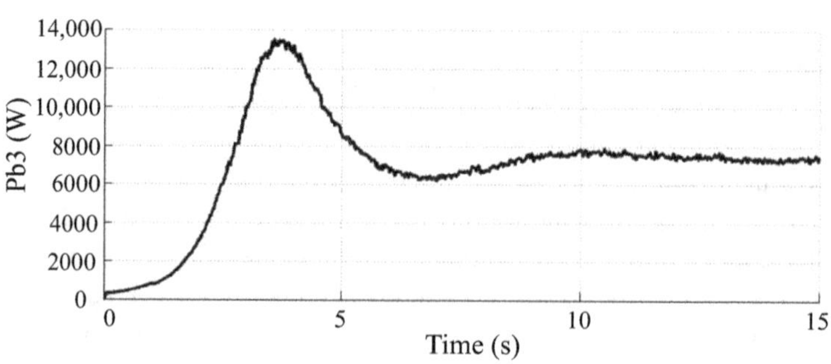

Figure E7.1(k) Active power at bus b_3

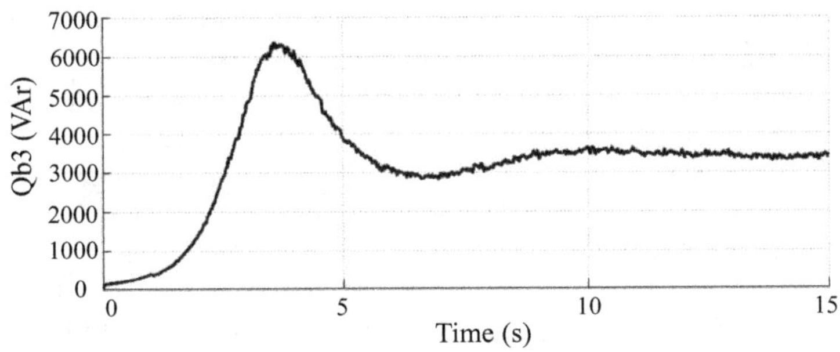

Figure E7.1(l) Reactive power at bus b_3

In this example, single-phase loads are fed from a wind turbine by operating a three-phase induction generator in single-phase mode. A dynamic switched power filter device with fuzzy logic tuned PI controller is used for dynamic load voltage stabilization and clean energy.

Example 7.2

A wind-utility grid interface scheme is adapted from [46] and given in Figure E7.2(a). A voltage source converter (VSC)-based hybrid power filter compensator (HPFC) scheme is used for effective reactive power compensation and harmonic reduction in distribution grid networks with the dispersed wind energy interface. A multi-loop dynamic error-driven PI controller equipped with an auxiliary DC side voltage tracking loop is used to ensure robust and fast control of the HPFC connected to bus 4.

The wind turbine has a nominal speed of 15 m/s, nominal output power of 3.6 MW, and a rotational speed 1.2 pu of its synchronous speed. The induction generator used in Figure E7.2(a) has a base nominal power of 3.6 MVA and a nominal voltage of 4160 V with a nominal frequency of 60 Hz. Stator and rotor resistances are assumed to be equal and given as 0.019 pu. Similarly, stator and rotor inductances are also equal and both are given as 0.06 pu. A 441 μF capacitor bank is used for the self-excitation of the induction generator. Three transformers are installed in the WEC-utility interface system operating at 60 Hz. Nominal power and voltages of these transformers are given in Figure E7.2(a). The transformers marked as T1, T2, and T3 are connected as Y/Y, Y/Δ, and Y/Y, respectively. The length of each section of the distribution feeder is 3 km with a resistance of 0.25 Ω/km and an inductance of 0.93 mH/km. Three linear type loads are connected to buses b_2, b_4, and b_5 with a power factor of 0.8 (lagging). The active and reactive powers of the linear loads are 1.2 MW and 0.9 MVAr, respectively. A 1.6 MW and 1.2 MVAr nonlinear load is connected at bus 7. Besides the linear and nonlinear loads, a 0.6 MVA, 4160 V, 60 Hz three-phase induction motor is also connected to bus 6 as a motor type load with the same stator and rotor pu resistances and inductances given for the induction generator.

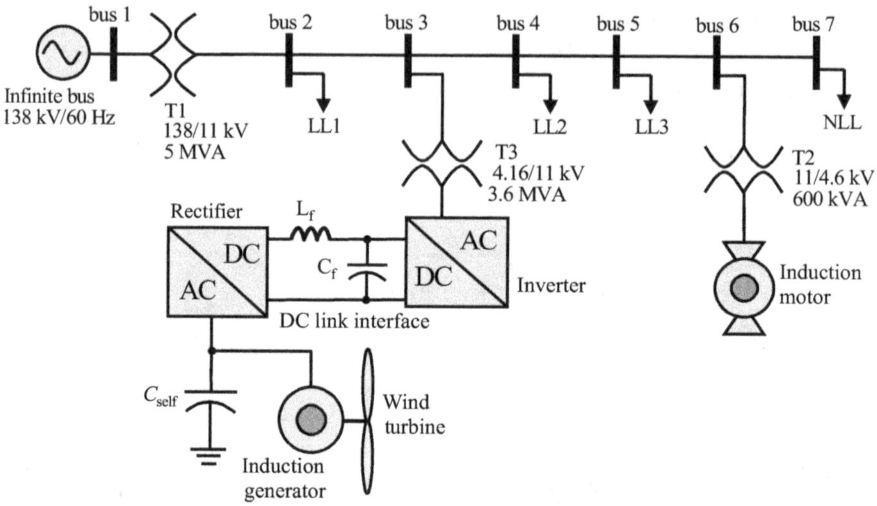

Figure E7.2(a) Utility interfaced WEC system for Example 7.2

Except for the WEC system interfaced to 11 kV distribution network at bus 3 and 138 kV main feed-in substation at bus 1, there is no other auxiliary generation in the system. A step-down transformer is also used at the main feed-in substation and at bus 6 where a large 4160 V/600 kVA induction motor is fed via an 11/4.16 kV step-down transformer.

(a) Develop a Simulink model for the 11 kV distribution system described above with only 138 kV main feed-in substation at bus 1 connected as the power source.
(b) Adopt and connect the Simulink model of the WEC system available in MATLAB/Simulink to the utility at bus 3 as the second power source in addition to 138 kV main feed-in substation at bus 1.
(c) Develop a Simulink model for dynamic power filter compensator (DPFC) given in Figure E7.2(b) to be connected at bus 5.
(d) Modify and adapt the FLC developed in chapter 4 and given by Figure 4.25 to be used to control the DPFC connected at bus 5.
(e) Show the effectiveness of the FLC-controlled DPFC scheme under electric load disturbances.

Answer 7.2

(a) A Simulink model for the 11 kV distribution system described above can be developed as the one given in Figure E7.2(c) with only 138 kV main feed-in substation at bus 1 is connected as the power source. All system parameters used in Simulink modeling shown in Figure E7.2(c) are given in the description part of this example.

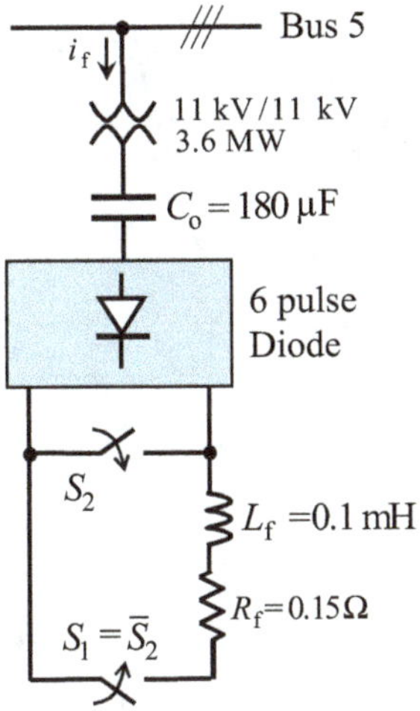

Figure E7.2(b) A schematic of the dynamic power filter compensator

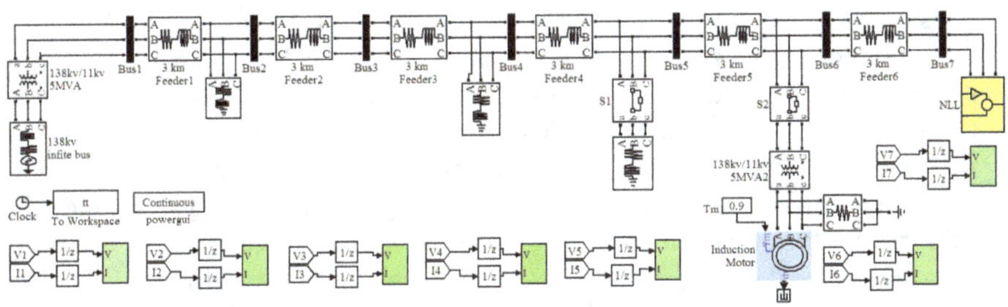

Figure E7.2(c) A schematic of the dynamic power filter compensator

(b) The integration of the WEC system Simulink model to the utility at bus 3 as a second power source is given in Figure E7.2(d). It should be noted that this configuration is not the only choice. Many different types of utility integration can be established.

The wind turbine equations given in (7.1)–(7.7) in Section 7.2 can be used for Example 7.2 with the coefficients $C1 = 0.5176$, $C2 = 116$, $C3 = 0.4$, $C4 = 5$, $C5 = 21$, and $C6 = 0.0068$. The pitch angle is assumed to be constant at the value of 0. The base wind velocity is selected to be 12 m/s and the base shaft rotational speed is set to be 1.2 times the generator's synchronous speed. The turbine power characteristics are similar to those given in Figure 7.3. The induction generator is a self-excited low-cost squirrel cage type and is fully employed for a wind energy scheme. Typical two-axis reference frame (d–q) model can be used to represent the three-phase induction generator.

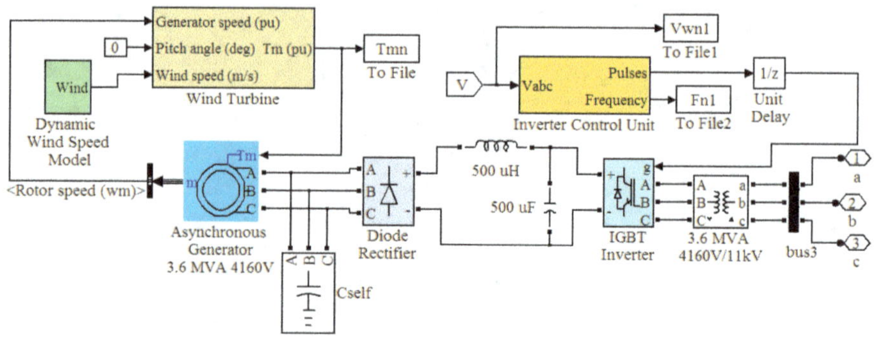

Figure E7.2(d) Integration of the WEC system Simulink model to the utility grid

Assuming that the rigid grid is an infinitely rich source of active and reactive power, the connection of unlimited excitation capacitors can stabilize the electrical grid resulting in almost constant voltage and frequency operation at the point where the WEC system is interfaced. However, the WEC systems are usually interfaced with distribution systems, which are considered weak due to low short circuit levels [46]. Therefore, WEC systems connected to AC utility grids must meet the host utility integration requirements and adhere to strict safety, isolation, and protection regulations [47]. As a result, necessary interface interconnection apparatuses, power conditioning devices, and grid protection gears should be employed and fully utilized to guarantee the reliability of the grid system and high power quality. A DC link with a rectifier and an inverter is used as the interfacing unit between the WEC system and the utility grid. This arrangement helps to reduce the required reactive power burden on the self-excitation capacitor bank under low-speed conditions since the power factor at the uncontrolled diode rectifier input is almost unity [46]. The inverter in the DC link of the WEC system is controlled by a d–q–based PI controller as explained in [46].

The wind speed model consists of four basic key components, namely the mean wind speed, a wind speed ramp, a wind gust, and the turbulence component. The eventual wind speed to be applied to the wind turbine is the summation of all four key components, which are explained in [46]. The effects of the WEC system added as a second power source to the 11 kV distribution network are depicted in Figure E7.2(d).

(c) A Simulink model for DPFC is given in Figure E7.2(e) to be connected at bus 5.

In order to improve the power quality of the distribution network given in Figure E7.2(a) the DPFC is connected to bus 5, which is located before motor and nonlinear type loads. Figure E7.2(e) depicts a schematic diagram of the DPFC, which consists of a six-pulse diode bridge, two switches, a series R–L branch, and a capacitor (C_o) connected in series with a three-phase coupling transformer. The coupling transformer can help avoid any high frequency near

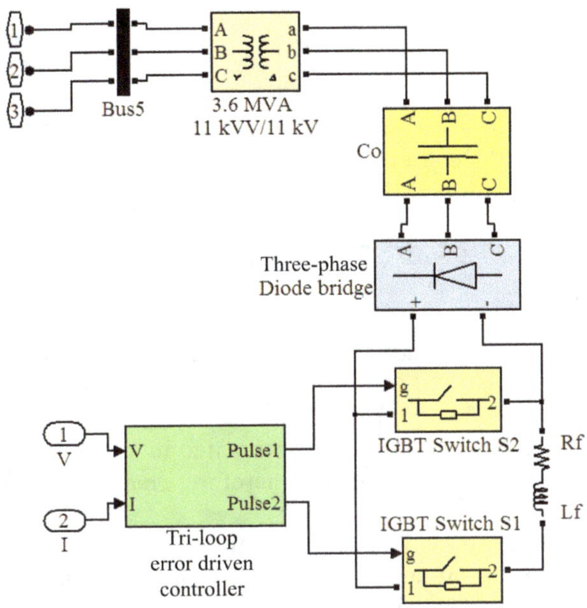

Figure E7.2(e) Simulink model of the DPFC and integration to the utility grid

resonance conditions. Switch S1 is the negation of switch 2. They are triggered simultaneously to generate the required filtering depending upon the voltage and current at bus 5.

The IGBT switches have a forward voltage of 0.3 V and a conducting resistance of 0.001 Ω. A snubber resistance of 0.1 MΩ is also connected to the IGBT circuitry. The resistance R_f and inductance L_f between the switches are selected as 0.15 Ω and 0.1 mH, respectively. The value of the series connected coupling capacitor (Co) is selected as 180 μF. A 3.6 MVA, 60 HZ 11 kV/11 kV, *Y*/*Y*-connected three-phase transformer is used for coupling the DPFC to the utility bus 5. The switching instants can be determined by controlling the width of the pulses. A tri-loop error-driven control structure presented in [46,48] is used in this example together with the FLC described in Section 4.9 of Chapter 4.

(d) Now, let us modify and adapt the FLC developed in Chapter 4 and given in Figure 4.25 to be employed to control the DPFC connected at bus 5. The FLC is used in a tri-loop dynamic error-driven control scheme based on Sharaf's ideas in [46] to control the DPFC. The total error signal is a collection of the RMS voltage error, change in RMS current, and RMS values of instantaneous current ripples. Therefore, the controller will act whenever there is a change or a ripple in the current besides the voltage error. A schematic diagram of the tri-loop dynamic error-driven controller with the FLC and discrete PWM generator to generate the switching signals S_1 and S_2 is given in Figure E7.2(f).

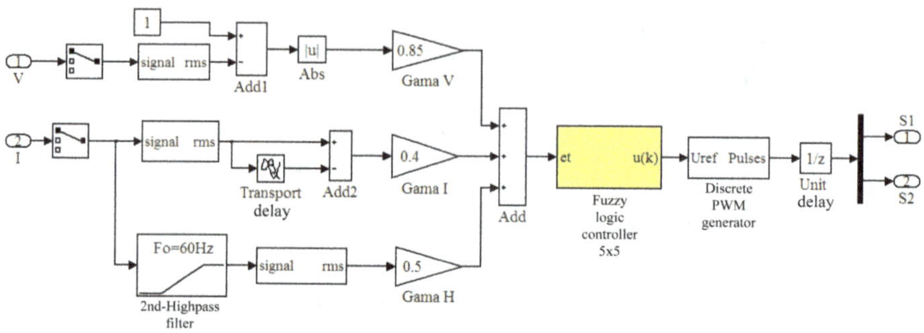

Figure E7.2(f) Simulink model of the DPFC and integration to the utility grid

The switching signals S_1 and S_2 are generated in a way that one of them is turned off when the other one is on. Therefore, one switch is the negation of the other one.

An FLC with 25 rules is used to control the DPFC by determining the switching instants of the switches S_1 and S_2. The FLC rules used in Example 7.2 are the same rules given in Table 4.7, which is a generalized rule table that can be applied to different problems. The FLC rules are developed using the instantaneous error and its change over one sampling. Since the rule table (Table 4.7) is also based upon the same approach of using the error and its change, the same table is applicable to the problem in Example 7.2. A generalized Simulink block diagram of the FLC with the input parameters is given in Figure E7.2(g). Triangular membership functions are used to represent the fuzzy subsets in input spaces error (e) and its change (de) besides the output space for the change du. The maximum boundary values of the input and output universes are given in Figure E7.2(g). The peaks of the fuzzy subsets in each universe are evenly distributed using equal-sided and right-angle triangles. The output du from the fuzzy reasoning block is added to the previous value of the control signal u to obtain the new value of the control signal. This procedure gives an integral action to the FLC.

(e) The effect of the FLC-controlled DPFC scheme under electric load disturbances is shown by the simulation results in Figures E7.2(h) and E7.2(i) for voltage and power factor profiles at bus 5.

The DPFC is connected at bus 5 in a radial distribution grid network with the wind energy interfaced at bus 3 was digitally simulated and validated in the MATLAB/Simulink software environment. Full digital simulation and validation testing were carried out without and with the FLC-controlled DPFC located at bus 5 for a full test duration of 1.0 s in order to show the dynamic voltage stabilization, harmonic reduction, and reactive power flow.

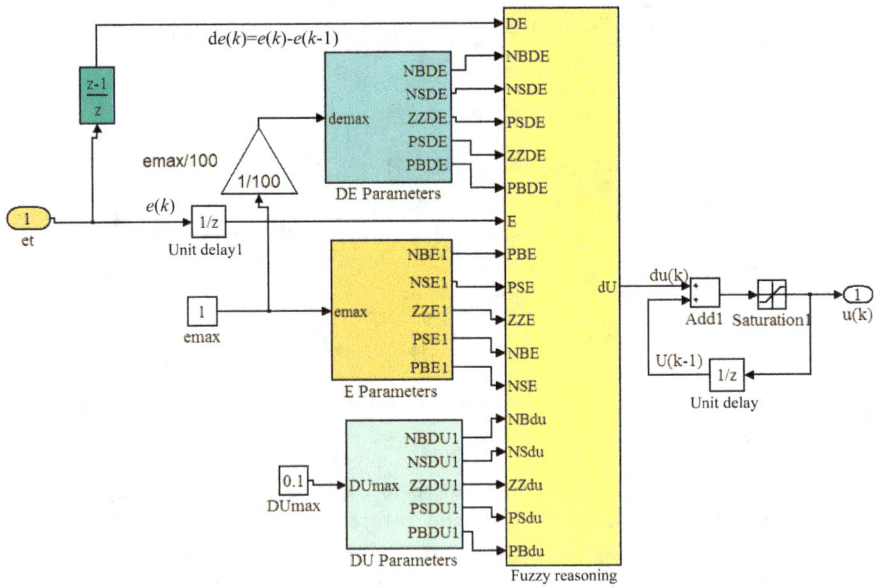

Figure E7.2(g) *The FLC block diagram with input parameters*

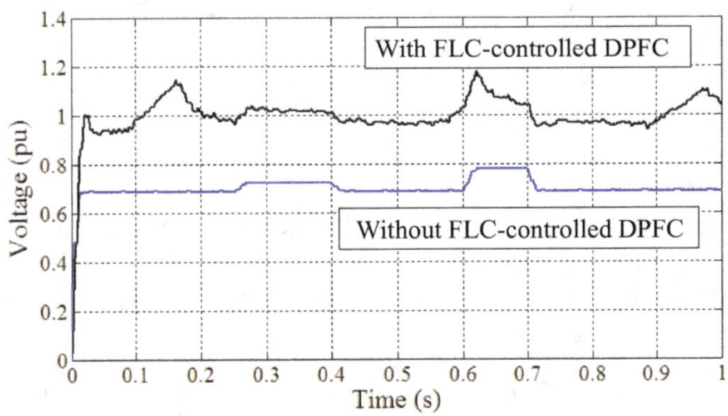

Figure E7.2(h) *Voltage transients at bus 5 without and with the FLC-controlled DPFC*

The dynamic performance of the FLC-controlled DPFC device was tested under two load-switching disturbance sequences. As the first operating condition an 11 kV, 1.2 MW, and 0.9 MVAr linear load at bus 5 is switched on at 0.25 s and switched off at 0.4 s of the digital simulation. A 600 kW, 4.16 kV, 60 Hz three-phase induction motor and a 500 Ohms resistive load are switched on and off at the same instants of 0.7 s and 0.8 s, respectively.

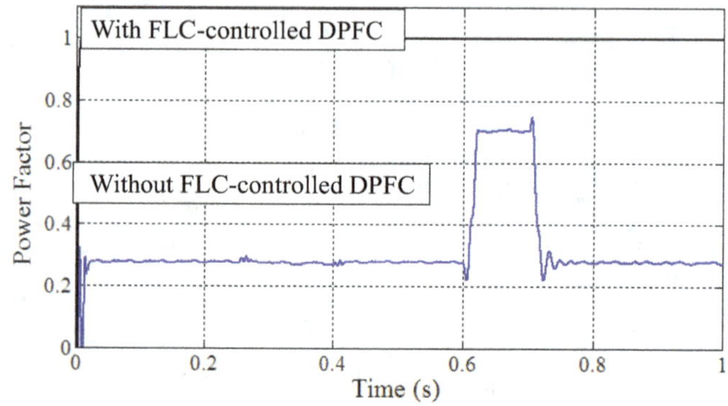

Figure E7.2(i) Power factor at bus 5 without and with the FLC-controlled DPFC

The results without the FLC-controlled DPFC show that the voltage at bus 5 is about 0.7 pu instead of the reference value of 1 pu. The power factor is below 0.3 lagging instead of being 1.0. The linear load causes a small increment in voltage at bus 5 when the additional loads are connected. The additional active load power increases the active power to reactive power ratio of the system resulting in more increment in the power factor. Both voltage and power factor profiles are controlled and kept around the reference values when the FLC-controlled DPFC is employed. Especially the power factor control is done very effectively such that it does not get affected by the changes in loads. Although there are some oscillations in voltage during the switching instants, it remains at 1.0 pu in a steady state.

Design Example 7.1

The power generated by a WEC system will be connected to the distribution system of a town by 5 km power cable with the parameters $0.2 + j0.1$ Ohm/km. A 20 MW power at 66 kV will be transmitted from WES to the distribution center. The distribution voltage is 6.3 kV.

(a) Draw a one-line diagram to represent the power generation and distribution system described above.
(b) Obtain the sizing of the WES considering that the three-phase asynchronous wind energy generating units available are 5 MW (0.8 lagging). 2 MW (0.85 lagging) and 1 MW (0.8 lagging). The rated voltages of all units are equal and 15 kV at 50 Hz.
(c) Obtain the required capacitor group sizing for the wind asynchronous machines to be operated as a generators at their rated values. Assume that the available capacitors groups have the ratings as 250 kVAr, 500 kVAr, and 1000 kVAr at 50 Hz with 8.66 kV operating voltage. Obtain all possible capacitor sizing for the generator selections you obtained in (a).

(d) Assume that 10 wind turbine and generator units will be used and only 250 kVAr capacitors are available. Draw a three-phase detailed wiring circuit diagram of capacitor groups for just one machine. Decide the connection type of the capacitor groups. Should they be connected in delta or wye?
(e) Draw a single-line diagram of the wind energy and transmission system including required capacitors, transformers, and transmission lines. Represent the distribution system by a bus bar.
(f) Develop a Simulink model for this system and analyze the voltage, current, and power profiles.
(g) Design a power compensator and a suitable controller to overcome the load-switching effects as well as wind speed excursions.
(h) Design a power management algorithm for the system in this problem including both generation and demand side requirements such that the energy utilization is optimum.

Design Example 7.2

A water pumping system for agricultural purposes is planned to be powered by a WES. There are three 2 kW wind generators (SCIG) in WES. Each generator requires 600 VAr for excitation. The capacitor groups to be used are available as 75, 100, 125, and 150 VAr units.

Three water pumping motors are three-phase asynchronous motors with 1 kW electrical input power at 0.8 lagging power factor. The applied voltage to the motors must be 50 Hz, 380 V phase to phase. An additional 1 kW of power is required by a shelter built on the agricultural site. All of the appliances in the shelter operate with 220 V phase to neutral voltage.

(a) Draw a single-line diagram of this system showing the WES, the bus where the WES is connected, the load bus and loads connected to the generator bus. Show all required devices and components including the required devices and connections to keep a constant frequency, and constant magnitude AC voltage at load terminals.
(b) Determine the capacitor sizing for the excitation of each induction generator and draw a three-phase solidly grounded Wye connection diagram of the capacitors.
(c) Determine the capacitor sizing for the compensation of each pumping motor.
(d) Determine the capacitor sizing for the compensation of the appliances in the shelter.
(e) Develop a Simulink model for this system and analyze the voltage, current, and power profiles.
(f) Design a power compensator and a suitable controller to overcome the load-switching effects as well as wind speed excursions.
(g) Design a power management algorithm for the system in this problem including both generation and demand side requirements such that the energy utilization is optimum.

7.5 Problems

P7.1 A three-phase synchronous generator generates voltages with a frequency of 50 Hz when operated at 750 rpm. A per phase Thevenin equivalent circuit of this generator is given in Figure P7.1 where $X_s = j0.7$ pu, $\underline{V}_t = 1.0\angle 0°$ pu, $\underline{I} = 1.0\angle{-30°}$ pu and pu excitation current is defined as $I_F = E_O + 0.00138\ e^{4.28E_O}$. The excitation circuit is fed from a DC voltage source to generate the required magnetic field and mechanical power input is supplied from a wind turbine.

Assume that the wind speed is constant generating a constant rotor speed of 750 rpm.

(a) Find the back electromagnetic force (emf) E_0 in pu.
(b) Find the required excitation current IF in pu.
(c) If the gearbox between the generator and wind turbine has a turn ratio of 1/4 from the generator to wind turbine, what is the wind speed?
(d) Develop a Simulink model for the system described in this problem including wind turbine speed and generator excitation control for voltage regulation and power–frequency control.

P7.2 Replace the synchronous generator in P7.1 with the DC machine described in Example 5.10 and perform the following.

(a) Develop a Simulink model for the wind turbine–DC generator combination.
(b) Modify the FLC described in Chapter 4 and use it to control the terminal voltage of the DC generator via the field circuit at constant wind speed.
(c) Repeat part (b) by giving a step change to the wind speed while the system is being operated at a steady state.

P7.3 Replace the synchronous generator in P.7.1 with a three-phase, 20 HP, 220/380 V, 50 Hz, 6 pole squirrel cage induction motor with an efficiency of 80%. The induction machine is to be operated as a self-excited generator. Assume that the wind speed is constant at 15 m/s.

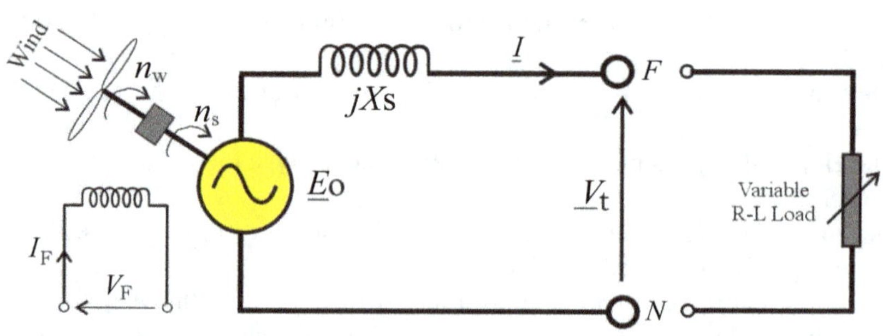

Figure P7.1 Per phase equivalent circuit of a synchronous generator

(a) Develop a Simulink model for the wind turbine-induction generator combination.
(b) Modify the FLC described in Chapter 4 and control the switching sequences of the self-excitation capacitors to maintain a terminal RMS voltage.
(c) Repeat part (b) by giving a step change to the wind speed while the system is being operated at a steady state. Observe the voltage frequencies when the wind speed is changed.
(d) Establish a DC link using controllable AC/DC and DC/AC converters for constant voltage and frequency operation.
(e) Use an FLC to control AC/DC and DC/AC converters in part (d).

P7.4 A three-phase, 20 HP, 220/380 V, 50 Hz, 6 pole squirrel cage induction machine is used in a wind power generation system. The induction machine has a 50 A nominal current and a power factor of 80% under nominal power operating conditions.
(a) What would be the synchronous speed of this induction machine when it is operated as a generator?
(b) What would be the value of Δ-connected excitation capacitors for each phase when the induction generator is operated at its rated power to feed a resistive load?
(c) What would be the value of *Y*-connected excitation capacitors for each phase while the induction generator is feeding a load with 80% lagging at its rated values?

P7.5 A WEC system is to be installed in a farmhouse where there is no utility grid. The farm has a 2 kW R–L linear AC load with 0.80 lagging power factor, a three-phase 3 kW, 50 Hz, 220/380 V AC motor operating at 0.70 lagging power factor, lights with a total power of 1.5 kW and heating with a total power of 5 KW. Diversity is ignored.
(a) Obtain the total active power required by the farmhouse
(b) Obtain the required total reactive power required by the farmhouse
(c) Obtain the sizing of the WES considering that the wind energy generating units available are 2 KW (0.8 lagging). 500 W (0.85 lagging) and 1 KW (0.9 lagging) units are also available.
(d) Obtain the required Wye-connected optimal capacitor group sizing for the wind asynchronous machine to be operated as a generator at its rated values. Assume that the available capacitors groups have the ratings as 0.5 kVAr, 1 kVAr, and 1.5 kVAr at 50 Hz with 220 V and 380 V.
(e) Obtain the required Wye-connected capacitor sizing to supply reactive power to the loads. Make your selection so that the reactive power supplied to the load can be controlled with the highest range as the loads are turned on/off randomly.
(f) Obtain the total Wye connected capacitor sizing required for generators and loads together.

P7.6 A farmhouse gets its electricity from a WEC system 2 km away. There are two wind generators in the WEC system. Each generator requires 900 VAr for excitation and capacitor groups used in each phase are built up with 100 VAr units.

The farmhouse requires 6 kW active power and 3 kVAr reactive power. 1 kVAr capacitor groups are used to supply this reactive power to the loads.

(a) Draw a single-line diagram of this system showing the WEC system, the bus where the WEC system is connected, a 2 km line causing 3% voltage drop under full loading, a load bus, and loads connected to the load bus.
(b) Add the single-line drawings of capacitor groups described above to your drawing in part (a).
(c) Draw three-phase solidly grounded Wye connected diagrams of these capacitor groups you draw in part (b).
(d) Develop a Simulink model for the system described above.

P7.7 Single-line diagram of a 10 kV power distribution system is shown in Figure P7.7. The distribution system is powered by a 30 kV infinite bus. A 500 kW linear type load with a 0.8 lagging power factor is connected to bus b_3 and two other linear loads are connected to bus b_4. One of these loads is operating at 1500 kW with 0.90 lagging power factor, and the other one is operating at 1000 kW with 0.85 lagging power factor. The lengths and parameters of the power lines between the buses b_2–b_3 and busses b_3–b_4 are given in Figure P7.7.

(a) Obtain an impedance diagram for this system.
(b) Develop a Simulink model for this system.
(c) Connect a 3 MW, 0.80 lagging, 10 kV, 50 Hz three-phase WEC system using an induction generator to bus b2 and simulate the system.
(d) Design a dynamic power filter (DPF) to bus b2 and simulate the system. Tune the DPF by designing an FLC so that the bus voltage remains at 1.0 pu nominal value while the power factor is increased above 0.95.

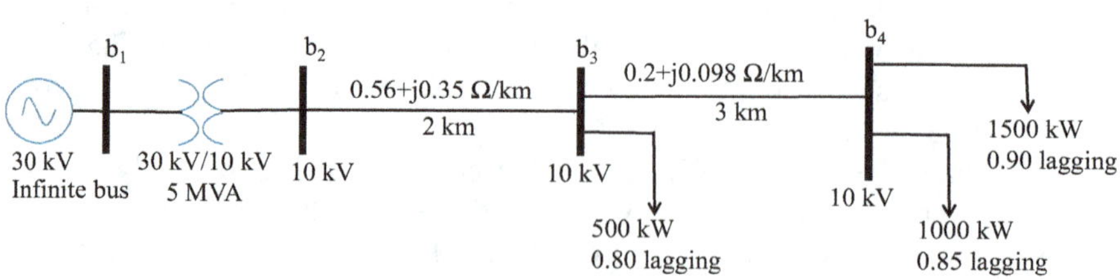

Figure P7.7 Sample distribution system for Problem 7.7

References

[1] Lin W.M., Hong C.M., Lee M.R., Huang C.H., Huang C.C. and Wu B.L. 'Fuzzy sliding mode-based control for PMSG maximum wind energy capture with compensated pitch angle'. *2010 International Symposium on Computer Communication Control and Automation (3CA)*; May 2010, pp. 397–400.

[2] Senjyu T., Sakamoto R., Urasaki N., Funabashi T., Fujita H. and Sekine H. 'Output power leveling of wind turbine generator for all operating regions by pitch angle control'. *IEEE Transactions on Energy Conversion*. 2006; 21(2), 467–475.

[3] Luo Q., Yang Q., Han C. and Cheng P. 'Pitch angle controller of variable-speed wind turbine based on L1 adaptive control theory'. *2014 International Conference on Mechatronics and Control (ICMC)*; July 2014, pp. 955–960.

[4] Buffoni M., Galletti B., Ferreau J., Fagiano L. and Mercangoez V. 'Active pitch control of tethered wings for airborne wind energy'. *2014 IEEE 53rd Annual Conference on Decision and Control (CDC)*; December 2014, pp. 4893–4898.

[5] Imran R.M., Hussain D.A. and Soltani M. 'DAC with LQR control design for pitch regulated variable speed wind turbine'. *2014 IEEE 36th International Telecommunications Energy Conference (INTELEC)*; 2014, pp. 1–6.

[6] Garasi P., Watanabe M. and Mitani Y. 'Power smoothing of wind turbine generator using fuzzy-PI pitch angle controller'. *2014 Australasian Universities Power Engineering Conference (AUPEC)*; 2014, pp. 1–5.

[7] Kumar H., Gupta A., Pachauri R.K. and Chauhan Y.K. 'PI/FL based blade pitch angle control for wind turbine used in wind energy conversion system'. *2015 International Conference on Recent Developments in Control, Automation and Power Engineering (RDCAPE)*; 2015, pp. 15–20.

[8] Chi S., Liang-Deng H. and Sheng A. 'Research on a static excitation control method for the high-power induction generator with a H-bridge cascaded converter topology'. *17th International Conference on Electrical Machines and Systems (ICEMS)*; 2014, pp. 107–113.

[9] Wang L. and Chen S.S. 'Design of a DC excitation control for an autonomous wound-rotor induction generator'. *IEEE Transactions on Industry Applications*. 2014;50(4):2807–2817.

[10] Liyanage D. and Rajakaruna S. 'Transient analysis of a single-phase variable speed cage induction generator with active excitation control'. *2013 Australasian Universities Power Engineering Conference (AUPEC)*; 2013, pp. 1–6.

[11] Hazra S. and Sensarma P.S. 'Self-excitation and control of an induction generator in a stand-alone wind energy conversion system'. *IET Renewable Power Generation*. 2010;4(4):383–393.

[12] Wekhande S. and Agarwal V. 'Wind driven self-excited induction generator with simple de-coupled excitation control'. *Conference Record of the 1999 IEEE Industry Applications Conference. Thirty-Fourth IAS Annual Meeting*; 1999, pp. 2077–2083.

[13] Cong W., Jun W., Guanglin S. and Liang Y. 'Research on control strategy for high power AC–DC–AC cascaded converter'. *2014 IEEE Conference and Expo Transportation Electrification Asia-Pacific (ITEC Asia-Pacific)*; 2014, pp. 1–5.

[14] Song P. and Lin J. 'A hybrid AC–DC–AC matrix converter with a boost circuit'. *9th International Conference on Electronic Measurement and Instruments*; 2009, pp. 2-416–2-421.

[15] Maol H., Zhang J., Lee F.C. and Boroyevich D. 'Zero-voltage-transition DC-link techniques for three-phase AC–DC–AC PWM converters'. *Proceedings of APEC 97—Applied Power Electronics Conference and Exposition*; 1997, pp. 692–698.

[16] Jabeen H.N., Nagarjuna G., Deepa S.N. and Anoop K. 'A new control strategy for grid interfacing inverter in wind energy applications'. *International Conference on Energy Efficient Technologies for Sustainability*; 2013, pp. 1238–1242.

[17] Prutianu F. and Popescu V. 'Control of single phase inverters for wind energy conversion using PWM techniques'. *9th International Symposium on Electronics and Telecommunications*; 2010, pp. 95–98.

[18] Zhu N. and Liang H. 'Implementation and control of dual grid connected inverter for wind energy conversion system'. *32nd Annual Conference on IEEE Industrial Electronics (IECON)*; 2006, pp. 2861–2866.

[19] Khan M.S. and Iravani M. R. 'Hybrid control of a grid-interactive wind energy conversion system'. *IEEE Transactions on Energy Conversion*. 2008; 23(3):895–902.

[20] Soderholm L.H. 'Interfacing small wind systems to rural power distribution systems'. *IEEE Transactions on Industry Applications*. 1984;IA-20(2):439–442.

[21] Wang L., Zhou L., Cheung H., Hamlyn A., Mander T., Lee I., *et al.* 'Adaptive protection and control strategy for interfacing wind generators to distribution grids'. *International Conference on Electrical Machines and Systems*; 2008, pp. 2478–2483.

[22] Bhattacharjee C. and Roy B.K. 'Advanced fuzzy power extraction control of wind energy conversion system for power quality improvement in a grid tied hybrid generation system'. *IET Generation, Transmission and Distribution*. 2016;10(5):1179–1189.

[23] Tourou P., Schael M. and Sourkounis C. 'Power quality impact on grid-connected wind energy conversion system'. *MedPower 2014 Conference*; 2014, pp. 1–5.

[24] Meng W. and Wang X. 'Distributed energy management in smart grid with wind power and temporally coupled constraints'. *IEEE Transactions on Industrial Electronics*. 2017.

[25] Guo L., Liu W., Li X., Liu Y., Jiao B., Wang W., *et al.* 'Energy management system for stand-alone wind-powered-desalination microgrid'. *IEEE Transactions on Smart Grid*. 2016;7(2):1079–1087.

[26] Hui J., Bakhshai A. and Jain P. 'Power management supervisory control algorithm for standalone wind energy systems'. *IEEE 36th International Telecommunications Energy Conference (INTELEC)*; 2014, pp. 1–6.

[27] Omoware W.D., Maheri A. and Azimov U. 'Aerodynamic analysis of flapping-pitching flat plates'. *3rd International Symposium on Environmental Friendly Energies and Applications (EFEA)*; 2014, pp. 1–5.

[28] Li N. and Balas M.J. 'Adaptive flow control of wind turbine blade using microtabs with unsteady aerodynamic loads'. *IEEE Green Technologies Conference (GreenTech)*; 2013, pp. 134–139.

[29] Merugula L.A., Khanna V. and Bakshi B.R. 'Comparative life cycle assessment: reinforcing wind turbine blades with carbon nanofibers'. *Proceedings of the 2010 IEEE International Symposium on Sustainable Systems and Technology*; 2010, pp. 1–6.

[30] Chen Z. and Blaabjerg F. 'Wind turbines—a cost effective power source'. *Przeglad Elektrotechniczny*. 2004;80(5):464–469.

[31] Twidell J. 'Wind turbines: technology fundamentals'. *Renewable Energy World*. 2003;6(3):102–111.

[32] Raj M.S.M., Alexander M. and Lydia M., 'Modeling of wind turbine power curve'. *IEEE PES Innovative Smart Grid Technologies*; December 2011, pp. 144–148.

[33] Liu X. 'An improved interpolation method for wind power curves'. *IEEE Transactions on Sustainable Energy*. 2012;3(3):528–534.

[34] Lydia M., Selvakumar A.I., Kumar S.S. and Kumar G.E.P. 'Advanced algorithms for wind turbine power curve modeling'. *IEEE Transactions on Sustainable Energy*. 2013;4(3): 827–835.

[35] De Kooning J.D., Gevaert L., Van de Vyver J., Vandoorn T.L. and Vandevelde L. 'Online estimation of the power coefficient versus tip-speed ratio curve of wind turbines'. *39th Annual Conference of the IEEE Industrial Electronics Society (IECON)*; 2013, pp. 1792–1797.

[36] Shokrzadeh S., Jozani M.J. and Bibeau E. 'Wind turbine power curve modeling using advanced parametric and nonparametric methods'. *IEEE Transactions on Sustainable Energy*. 2014;5(4):1262–1269.

[37] Goudarzi A., Davidson I.E., Ahmadi A. and Venayagamoorthy G.K. 'Intelligent analysis of wind turbine power curve models'. *IEEE Symposium on Computational Intelligence Applications in Smart Grid (CIASG)*; 2014, pp. 1–7.

[38] Panahi D., Deilami S. and Masoum M.A.S. 'Evaluation of parametric and non-parametric methods for power curve modelling of wind turbines'. *9th International Conference on Electrical and Electronics Engineering (ELECO)*; 2015, pp. 996–1000.

[39] Heier S. *Grid Integration of Wind Energy Conversion Systems*. New York: John Wiley & Sons; 1998.

[40] Park R.H. 'Two-reaction theory of synchronous machines generalized method of analysis—Part I'. *Transactions of the American Institute of Electrical Engineers*. 1929;48(3):716–727.

[41] Krause P.C. 'Simulation of symmetrical induction machinery'. *IEEE Transactions on Power Apparatus and Systems*. 1965;PAS-84(11):1038–1053.
[42] Krause P.C. *Analysis of Electric Machinery*. New York: McGraw-Hill; 1986.
[43] Hancock N.N. *Matrix Analysis of Electrical Machinery*. Oxford: Pergamon Press; 1974.
[44] Bose B.K. *Power Electronics and AC Drives*. Englewood Cliffs, NJ: Prentice-Hall; 1986.
[45] Yanmaz K., Mengi O.O. and Altas I.H. 'Design and implementation of a smart STATCOM in small scale wind turbines'. *1st International Black Sea Congress on Environmental Sciences (IBCESS-2016)*; Giresun, Turkey, September 2016.
[46] Sharaf A.M., Wang W. and Altas I.H. 'A novel modulated power filter compensator for distribution networks with distributed wind energy'. *International Journal of Emerging Electric Power Systems*. 2007;8(3):1–20.
[47] Sharaf A.M., Wang W. and Altas I.H. 'A novel hybrid active filter compensator for stabilization of wind-utility grid interface scheme'. *International Transactions on Electrical Energy Systems*. 2010;20(3):306–326.
[48] Sharaf A.M., Elbakush E. and Altas I.H. 'A dynamic error driven PID controller for PMDC motor drives fed by non-constant voltage sources'. *IEEE Canadian Conference on Electrical and Computer Engineering*; Vancouver, Canada, 2007.
[49] Sharaf A.M., Elbakush E. and Altas I.H. 'A wind energy conversion scheme utilizing FACTS'. *IEEE Canadian Conference on Electrical and Computer Engineering*; Vancouver, Canada, 2007.

Chapter 8
FLC in PV solar energy systems

Application of fuzzy logic control and decision-making processes in photovoltaic (PV) solar systems are given in this chapter. Modeling and characteristic properties of PV solar arrays are viewed in terms of maximum power point tracking (MPPT), sun tracking, voltage control, battery charging, interfacing issues, and management of the generated power. As new to this chapter, shaded effects of PV systems will be added and simulation examples will be given related to MPPT control under shading effects.

8.1 Introduction

The use of photovoltaic (PV) cells has emerged as an important solution to the growing energy crisis and the need for conservation and load management during the last decades. Due to their high initial costs, solar PV cells were not a very attractive alternative to electricity users until the early 2000s. However, the installation costs have dropped tremendously during the last decade such that the price of electricity from PV systems has been competitive with that of electricity from conventional sources.

The use of PV cells started in the 1950s with the introduction of silicon cells in 1941 [1]. The capability of silicon cells to convert light directly to electricity stimulated new research on solar PV cells, so the PV solar cell has become a power source to be used in spacecraft since the late 1950s [1]. Unfortunately, the use of PV solar cells in small-scale and special commercial applications gained importance only after the energy crisis started in the mid-1970s due to increasing oil and natural gas prices [2]. Today, PV solar cells are being used in a wide range of application areas from small power appliances to utility grid-connected large-scale solar power stations [3–5].

Solar energy from the sun is provided in the form of solar irradiation and may be converted to useful energy. The conversion process is done either by natural means or by technical processes. With natural means, solar energy is expressed in the forms of winds, renewable organic fossil fuels, and ocean temperature differences. The technological conversion process, on the other hand, involves manufactured devices and systems to convert solar energy into thermal or electrical energy.

The main part of a PV energy system is the solar PV cell. The solar PV cell is merely a P–N junction diode. Most of the commercially available solar cells today are made of silicon which is the second most plentiful element on the earth [2]. Besides the silicon cells, copper–cadmium sulfide (Cu_2S/CdS) and gallium arsenide (GaAs) cells are two other types of solar cells used in early PV systems [2]. The PV cell manufacturing technologies have been developed tremendously during the last two decades affecting the efficiencies as well. Nowadays, PV cells are manufactured in the form of crystalline, multi-crystalline, thin film, amorphous, microcrystalline, organic, and multijunction device technologies. The efficiency of crystalline Si cells has been increased up to 25.6%. Thin film and multi-crystalline Si cells have an efficiency range from 10.5% to 21.2%. Gallium arsenide-based III- and V-type cells have an efficiency of up to 28.8%. Amorphous and micro-crystalline Si cells have an efficiency of around 11%. Organic thin-film cells have been manufactured with an efficiency of about 11%. The highest efficiency of applicable PV cells belongs to InGaP, GaAs, and InAs cells with an efficiency of 37.9%. A detailed list of solar PV cell types and efficiencies is given in [6].

The performance of a PV power system depends on the operating conditions as well as the solar cell and panel design quality. The output voltage, current, and power of a PV panel vary as functions of solar irradiation level, temperature, and load current. Therefore, the effects of solar irradiation, temperature, and load current must be considered in the design of PV panels so that any change in the temperature and solar irradiation levels should not adversely affect the PV output power to the load/utility which is either a power company utility grid or any stand-alone electrical type load. To overcome the undesired effects of the variable temperature and solar irradiation on the output power of PV systems, different control strategies have been applied. These strategies are mainly divided into two groups: (1) controlling the sun input to the PV panel and (2) controlling the power output from the PV panel. The combination of these two groups also may be considered. Both groups may include electric or thermal energy storage systems or auxiliary power sources that supply electricity during the night or on cloudy days. Sun input to the PV systems is kept as high as possible either by rearranging the solar panel configurations of PV arrays with respect to the changes in weather conditions or by designing and controlling the position of sun-tracking solar collectors. The sun tracking method is generally used with concentrator arrays and is not within the scope of this book. A literature list showing some projects and research related to the sun-tracking systems is given in reference [7].

Studies related to the maximum output power tracking of PV panels deal mostly with the isolation requirements and interfacing problems to the electrical utility grid or with the problems of direct coupling to loads. Since the electricity obtained from a PV system is DC and from a utility grid is AC, an inverter is needed. A simple DC/AC conversion, however, is not sufficient to ensure effective interfacing because the converters generate harmonics and cause additional voltage distortions. Therefore, such a PV array–utility grid interface system should be designed so that the integrity of the complete system is maintained and improved. The maximum power utilization of a stand-alone PV array under varying conditions in temperature and solar irradiation levels is also considered as one of the research areas in PV systems.

The output electric power level and energy produced from a stand-alone PV array feeding a single or a group of resistive and motor-type loads can be adjusted by connecting the solar panels in different configurations of series and series/parallel groups. When the input solar power to the PV array changes due to varying ambient temperature and solar irradiation levels, the output power also tends to change. But the load being fed usually requires a maintained constant power level. In order to keep the PV array output electric power constant or limit power changes, the number of solar panels in series or parallel strings is continuously modified. More panels are either added to or removed from the PV series/parallel matrix array. If all of the panels in an array are in use and the output power is not sufficient, then some of the electrical loads are removed by employing a power management algorithm to operate the priority loads more efficiently. For both cases, either changing array configuration or removing loads, the PV array should be operated at its highest utilization efficiency so that all of the solar energy that is converted to electricity is used. If a PV array is operated under its maximum output power level, some of the solar power which is already converted to useful electrical energy is wasted. Therefore, a PV array should be operated at such a voltage and current level that its maximum output power is always available. If the load is connected directly to the PV array and there is no interface controller, the array operates at any arbitrary voltage and current levels that match the power requirement of the changing load. In this case, the array may supply less power than its maximum power and operate with low utilization efficiency wasting the available energy. In addition, in direct coupling schemes, if the load power requirement is more or less than the array output power, the array may not generate any power at all. Besides, any change in the temperature and irradiation levels will affect the load power directly and will make the maintenance of stable operation very difficult or even impossible. If an electrical motor is used as load, it may run at lower or higher speeds depending on the PV array operating zone, which may be either over-voltage or over-current operating zone. The motor may not run at all if the array output power is very low, or the motor may run at very high speeds causing electrical and mechanical damage if the array output power is too high. Therefore, a PV energy conversion scheme should not only satisfy the maximum power output condition of the PV array but should also be designed to ensure protection from the danger of over-voltage and over-current operation. These problems can be solved by using a matching-controllable interface device for the coupling of the PV array to the electrical-load system. By controlling the interface semiconductor converter device, the output power of the PV array can be controlled easily so that the load is operated at its desired voltage, current, and power levels. In a fixed array topology configuration, the load power lies in the PV array output power range which varies depending on the solar irradiation level and temperature changes. If the PV array power does not match the load power demand, i.e., if the array current and voltage are too high or too low to be controlled at the existing level of temperature and solar irradiation, a feedback controller can be employed for topology reconfiguration of the solar cells to get an output power that can be controlled to match that required by the load.

The main problems in PV energy systems are usually counted as sustainability, changes in solar irradiation and ambient temperature, shading effects, positioning, and sun-tracking, maximum power operating point tracking, cooling, cleaning, and integrating to loads and utility grids. The modeling, maximum power point tracking (MPPT), and power management of PV-powered systems are given here by adapting from the previous works of the author [8–13]. The management of PV-powered energy systems is important as is the modeling and MPP tracking. For the sustainability of the PV energy systems, both load and generating side management gain importance. Battery backup units are also used for sustainability. However, the effects of the changes in solar irradiation level and temperature have to be tracked for the MPPT and load matching. The MPPT always gets affected from whether changes as well as from the load changes. Therefore, a load-matching control algorithm is always required to extract the maximum power from the PV system. Battery backups may be a solution for the MPPT tracking in stand-alone applications. The utility on the other hand can be used as a power pool to be filled up for the maximum power operating and extracting all the available power from the PV system.

Proper interface controllers for maximum power tracking and utility interfacing of PV energy systems are discussed in this chapter. Therefore, after discussing the modeling issues of the PV cells and arrays, some MPPT algorithms will be explained and examples will be given using the MATLAB® and Simulink® modeling of the PV cell and array. Some control and management approaches will also be discussed.

8.2 PV cell models

PV solar cells are devices that convert solar energy directly to useful electrical energy by creating equal numbers of positive and negative charges from the absorbed solar photon energy. The negative and positive charges then are separated to develop photovoltage and photocurrent. To separate the negative (electrons) and positive (holes) charges, semiconductor materials such as silicon, copper–cadmium sulfide, and gallium arsenide are the most suitable ones and are the ones that are used widely to manufacture PV solar cells.

The operation of a PV solar cell is similar to that of a positive–negative (P–N) junction diode where the electrons flow from the N layer to the P layer of the PV cell via an external connection circuit and rejoin the holes. Since this book deals with the utilization of PV cells and arrays rather than their structural properties, the internal structures and physical properties of the PV solar cells are not going to be given here. However, the equivalent circuit and mathematical modeling of the PV cells and arrays are discussed and given in MATLAB and Simulink environments for the system simulation. A detailed modeling of the PV cells including the effects of the weather conditions is given next.

A solar cell converts solar energy to electric output power only if there is a load connected across the P–N layers. Without any load or any connection between P–N layers via an outside path, the electrons cannot flow, and therefore no photocurrent is generated. Output power generation is not achieved if the load resistance is too small or the P–N layers are shorted via a wire with negligible resistance. In this case, where P–N layers are shorted, the potential difference or the photovoltage between P and N layers becomes zero while the photocurrent has its maximum value. Contrary to the conventional dc power sources, the current–voltage relationship of PV solar cells shows a nonlinear behavior. Therefore, solar cell power sources cannot be represented by classical DC current or voltage sources. Typical current–voltage (*I–V*) and power–voltage (*P–V*) characteristics of a PV solar cell are shown in Figure 8.1 where the values at the maximum power point are the base values of the per-unit system. If the solar irradiation level and temperature change, the maximum power point values will also change resulting in new base values for the per-unit system.

In order to simulate a PV solar cell array, an equivalent circuit model is needed. Different circuit models of PV cells have been studied over the years [14,15]. A review of PV cell models used in steady-state and transient analysis is given in [15]. In this reference, the authors reviewed seven different static PV cell models based on the equivalent circuit and current equation given in Figure 8.2, and equation (8.1), respectively.

$$I_C = I_{ph} - I_0\left[\exp\left(\frac{e}{kT_C}(V_C + R_S I_C)\right) - 1\right] - \frac{V_C + R_S I_C}{R_{SH}} \tag{8.1}$$

where I_C: cell output current (A)

I_{ph}: photocurrent, function of irradiation level, and junction temperature (A)
I_0: reverse saturation current of the diode D (A)

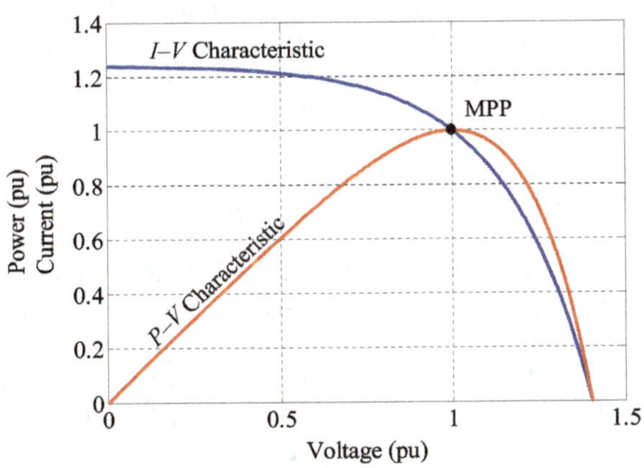

Figure 8.1 Typical I–V and P–V characteristics of a photovoltaic cell

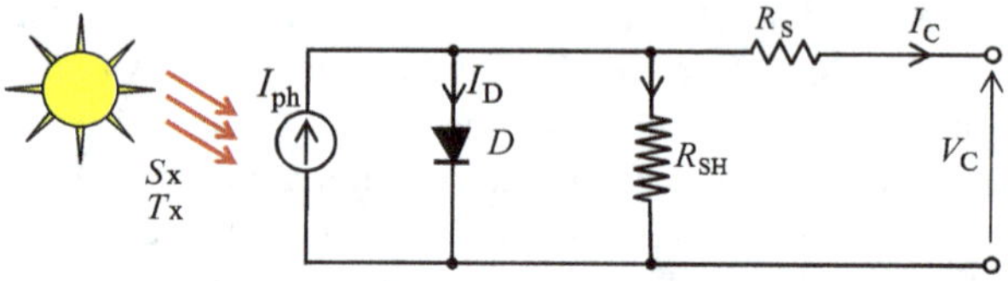

Figure 8.2 Generalized static equivalent circuit of a photovoltaic cell

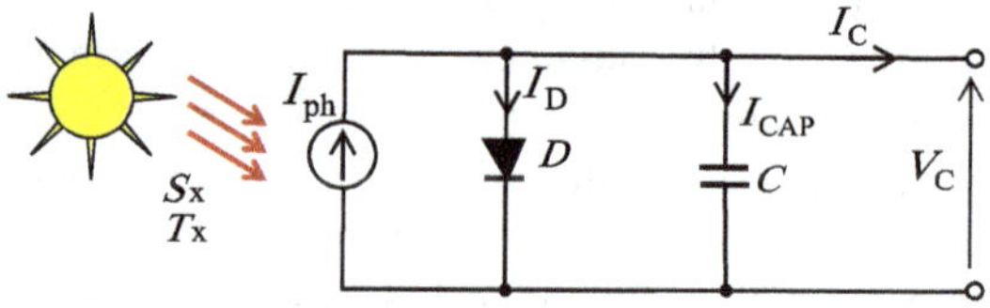

Figure 8.3 Dynamic equivalent circuit of a photovoltaic cell

V_C: cell output voltage (V)
R_S: series resistance of the cell (Ohm)
R_{SH}: parallel resistance of the cell (Ohm)
e: electron charge (C)
k: Boltzmann constant (J/K)
T_C: Reference cell operating temperature (°C)

Both k and T_C should have the same temperature unit, either Kelvin or Celsius.

A polynomial equation representation of PV cell I–V characteristics is also given in [14,15]. However, the comparison of a seventh-order polynomial equation with the equations derived from (8.1) shows that the polynomial equation results in the highest operating and maximum power point errors [15]. A static and dynamic model comparison of the PV cells is also given in [15].

A PV cell dynamic model with a capacitor connected parallel to shunt resistance or across the array output terminals [15] as shown in Figure 8.3 is compared with the static models. The dynamic model equivalent circuit given in Figure 8.3 has a cell output current I_C defined as in (8.2).

$$I_C = I_{ph} - I_0\left[\exp\left(\frac{e}{kT_C}(V_C)\right) - 1\right] - C\frac{dv_C}{dt} \tag{8.2}$$

It is concluded that the time responses of the PV cell current and voltage from both static and dynamic models are basically the same and there is no need to incorporate the capacitance into the PV cell model. Therefore, instead of including a derivative term in the PV cell modeling, a static equation with a similar response is usually preferred. Much research has appeared in the literature using the static models of the PV cell in place of the dynamic model [15–19]. Since the shunt resistance R_{SH} in Figure 8.2 is much greater than series resistance R_S, the last term in (8.1) becomes very small compared to the value of I_C. Therefore, the last term in

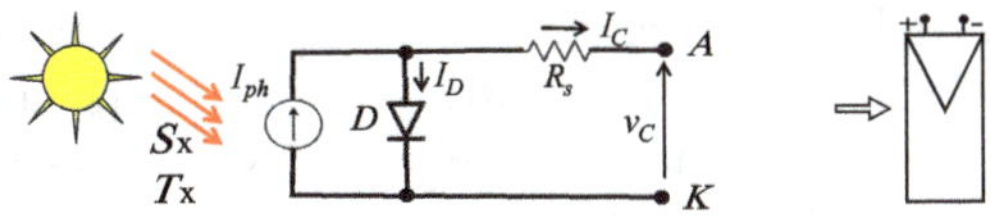

Figure 8.4 Simplified equivalent circuit and symbol of a photovoltaic cell

(8.1) is usually ignored for simplicity resulting in (8.3) and the equivalent circuit shown in Figure 8.4 which is one of the static models discussed in [15] and widely used in literature. Figure 8.4 also includes the PV cell symbol, which is also used in the literature.

$$I_C = I_{ph} - I_0\left[\exp\left(\frac{e}{kT_C}(V_C + R_S I_C)\right) - 1\right] \tag{8.3}$$

The current I_D in Figures 8.2–8.4 is the internal current and flows across the cell's semiconductor junction as a function of the cell's absolute temperature, terminal voltage, and external load current. The equation for I_D is given as in (8.4).

$$I_D = I_0 \exp\left(\frac{e}{kT_C}(V_C + R_S I_C)\right) - I_0 \tag{8.4}$$

Using the nonlinear equation form $V_C = f(I_C)$ instead of the form $I_C = f(V_C)$ as in (8.1) and (8.3) makes it easier for computer PV source emulation because the cell current I_C is determined from the load current and is used to calculate the corresponding single cell output voltage V_C as in (8.5).

$$V_C = \frac{AkT_C}{e}\ln\left(\frac{I_{ph} + I_0 - I_C}{I_0}\right) - R_S I_C \tag{8.5}$$

The curve fitting factor A is used to adjust the I–V characteristics obtained from (8.5) to the tested I–V characteristics of the PV cell. Equation (8.5) gives the voltage of a single solar cell which is then multiplied by the number of the cells connected in series to calculate the full panel voltage. It should be mentioned that problems would occur if the short circuit currents of the PV cells in series are not identical or very close. If the differences between these short circuit currents are great, the PV cell with the smaller short circuit current will break down and act as a diode load resulting in a reverse voltage across its terminals [2]. To protect the cell with a smaller short circuit current from the reverse voltage, a bypass diode is connected across that cell as an alternative path for the current [2]. In this book, the breakdown problems are not considered since the cells connected in series and in parallel are assumed to be identical and there are no shading effects.

Before being used in (8.5), the cell current I_C in a single panel system is obtained by dividing the full panel current, which is effectively equal to the electric load current, by the number of the cells connected in parallel. It should be noted that (8.5) is valid only for a certain cell junction operating temperature T_C at a certain solar irradiation level S_C. If the temperature and solar irradiation levels change, the voltage and current outputs of the PV cell will vary following this change. Therefore, the effects of the changes in cell temperature and solar irradiation levels should also be included in the final PV cell modeling. A method to include these effects in the PV cell modeling is given by Buresch [2]. According to this method, a reference solar cell model is obtained for a known temperature at a known solar irradiation level, and then this model is modified to include the changing solar irradiation level and temperature effects. The procedure for obtaining the reference PV cell model is given in the next section.

8.2.1 Reference I–V characteristics of a PV panel

The PV cells are usually available in the form of PV panels. Therefore, the measurements of *I–V* characteristics of a single PV cell may be possible in a laboratory environment only. Since PV panels are commercially available and easily accessible for testing, it is better to model the measurable PV panels rather than the PV cells. PV panels include series and parallel connected identical cells. Therefore, the *I–V* characteristics of a PV cell are nothing but a scaled version of *I–V* characteristics of a panel.

The reference *I–V* characteristics are usually obtained by testing the panel under a certain solar irradiation level S_R and cell operating temperature T_R. Then the reference *I–V* characteristics of a single panel are used to represent the *I–V* characteristics of the PV array consisting of parallel connected panel strings which in turn are made up of panels identical to the reference panel connected in series. Since the array connection configuration is usually changed depending on the weather and load conditions, it is preferred to use a single panel characteristic as a reference rather than the array characteristics. The reference *I–V* curve used in this section was also obtained by testing one of the PV panels, which was available in the Electrical Engineering Power Laboratory at The University of New Brunswick, Canada when the author was working on his PhD thesis during the early 1990s [8]. In order to obtain an approximate mathematical model representation of the reference *I–V* characteristics of the solar panel simulation model, the following steps are followed.

1. Set up the connection diagram shown in Figure 8.5. Vary the load resistance R_L, connected across the output terminals of the solar PV panel, from highest (open circuit) to lowest (short circuit) value and record voltage V_R and current I_R for different load levels.
2. Identify the short circuit current I_{SC} and the open circuit voltage V_{OC} from the data obtained in step 1 as $I_{SC} = I_R$ when $V_R = 0.0$ and $V_{OC} = V_R$ when $I_R = 0.0$.
3. The photocurrent, I_{ph}, of the solar cell, was assumed to be equal to the short circuit current, I_{SC}, since the reverse saturation current of a diode is negligible compared to the short circuit current.
4. Measure and record the reference solar irradiation level S_R and panel operating temperature T_R. Assume that these values remain constant during the test.
5. Using a curve fitting process, the values of panel series resistance, R_S, diode reverse saturation current, I_0, and curve fitting factor, A, in (8.5) are

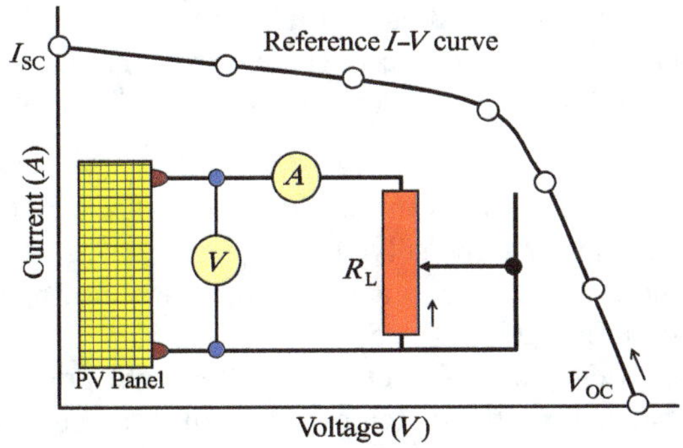

Figure 8.5 Reference I–V curve measurement of a PV panel

determined. Then they are used in the mathematical model representation of the reference *I–V* characteristics.

The values of R_S, I_0, and A in (8.5) can also be determined using three measured data points around the knee of the *I–V* characteristics. Substitution of the voltage and current values from these three data points results in three nonlinear equations to be solved for three unknowns.

Example 8.1

Current and voltage readings from a PV panel testing are listed in Table E8.1(a). The panel power is also calculated and listed in the same table. Solar irradiation level and

Table E8.1(a) Current, voltage, and power readings from PV panel testing

Voltage (V)	Current (A)	Power (W)
0.00	5.40	0.00
6.80	5.21	35.428
17.28	4.75	82.08
19.30	4.50	86.85
20.60	4.25	87.55
22.12	4.00	88.48
23.40	3.75	87.75
24.72	3.50	86.52
25.46	3.25	82.745
26.14	3.00	78.42
26.88	2.75	73.92
27.50	2.50	68.75
28.00	2.25	63.00
28.60	2.00	57.20
29.12	1.75	50.96
29.50	1.50	44.25
29.76	1.25	37.20
30.10	1.00	30.10
30.16	0.89	26.84
31.20	0.00	0.00

operating temperature are measured as 100 mW/cm^2 and 20 °C, respectively, and assumed to be unchanged during the rest. The electron charge and Boltzmann constant are given as $e = 1.106 \times 10^{-19}$ Coulomb and $k = 273 \times 0.138 \times 10^{-23}$ J/°C, respectively.

(a) Select three data points around the knee of the I–V characteristic and obtain three nonlinear equations to be solved.
(b) Apply a trial-and-error approach and obtain the unknown parameters R_S, I_0, and A using measured data and (8.5).
(c) Plot measured and estimated I–V curves on the same graph and analyze how they match by calculating the matching errors of three nonlinear equations.

Answer 8.1

(a) Selected three data points are listed in Table E8.1(b). Since the PV panels are preferred to be operated around their maximum power points, current and voltage data corresponding to the maximum power will be the most usable values. Therefore, data given in Table E8.1(b) are selected at and around the maximum power point of the I–V curve.

Substitution of voltage and current data in Table E8.1(b) into (8.5) yields three nonlinear equations as given in (E8.1a), (E8.1b), and (E8.1c).

$$17.28 = A(0.0681)\ln\left(\frac{I_0 - 11.88}{I_0}\right) - (4.75)R_S \tag{E8.1a}$$

$$22.12 = A(0.0681)\ln\left(\frac{I_0 - 16.72}{I_0}\right) - (4.0)R_S \tag{E8.1b}$$

$$25.46 = A(0.0681)\ln\left(\frac{I_0 - 20.06}{I_0}\right) - (3.25)R_S \tag{E8.1c}$$

(b) A trial-and-error approach results in the unknown parameters as $R_S = 0.01$ Ohms, $I_0 = 0.051$ A, and $A = 99$.
(c) A plot of measured and estimated I–V curves is depicted in Figure E8.1.

An analysis of how the measured and estimated I–V curves match can be done by calculating the matching errors of three nonlinear equations given in (E8.1a), (E8.1b), and (E8.1c). The resultant errors are given in Table E8.1(c).

Table E8.1(b) Selected data points from PV panel testing

Voltage (V)	Current (A)	Power (W)
17.28	4.75	82.08
22.12	4.00	88.48
25.46	3.25	82.745

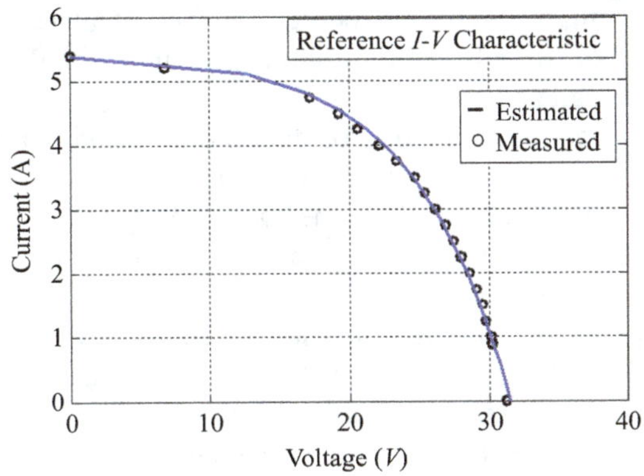

Figure E8.1 Measured and estimated I–V curves

Table E8.1(c) Error analysis of the estimated parameters

Voltage (V)	Current (A)	Errors (V)	Error %
17.28	4.75	−0.3478	1.97
22.12	4.00	−0.4219	1.87
25.46	3.25	0.1004	0.39

An average error is calculated as −0.2231 V, which is 1.046%. This is a reasonably small error. However, it can be made much smaller by using theoretical curve fitting or optimization methods.

8.2.2 The effects of changes in solar irradiation and temperature

PV panels generate electricity from the sunlight, which carries the solar irradiation to PV panels. The solar irradiation arrives at the panels as photons and is converted to electrical power in terms of photocurrents. Therefore, the solar irradiation level is an important factor in the generated power. However, it depends upon the availability of sunlight that varies from morning to evening, from day to day, from month to month, and from season to season. The sunlight also carries heat that causes the PV panel to get heated. The load current also causes heat in the panels by increasing the temperature in P–N junction of the PV cells. The changes in both solar irradiation level and operating temperature affect the performance of PV panels in terms of generated voltage, current, and therefore power.

The generated photocurrent of a PV cell or a PV panel is affected by the changes in solar irradiation and operating temperature resulting in changes in voltage and power, as well. Therefore, the mathematical model of a PV cell or a panel must include the changing effects of solar irradiation and temperature levels. The model must reflect any changes in solar irradiation and temperature so that the generated photocurrent, voltage, and power can be calculated correctly. Model accuracy is the first key to analyze, control, manage, and track the maximum power generated by the PV panels.

Let (8.5) be the benchmark model for the known operating temperature T_R and known solar irradiation level S_R. Since all the cells in a panel are assumed to be identical and they carry the same current with the same output voltage, the junction temperatures of all the cells are assumed to be equal too. The ambient temperature is also the same for all the cells because they are in the same location. Therefore, the effects of the changes in solar irradiation level and operating temperature in a PV cell will be similar to those in a PV panel. When the ambient temperature and irradiation levels change, the operating temperature T_P of the panel in the panel also changes resulting in a new output voltage and a new photocurrent value. The PV panel operating temperature varies as a function of solar irradiation level and ambient temperature. The variable ambient temperature T_X affects the panel output voltage and panel photocurrent. These effects are represented in the model by the temperature coefficients [2] C_{TV} and C_{TI} for panel output voltage and panel photocurrent, respectively, as in (8.6) and (8.7).

$$C_{TV} = 1 + \beta_T(T_R - T_X) \tag{8.6}$$

$$C_{TI} = 1 + \frac{\gamma_T}{S_R}(T_X - T_R) \tag{8.7}$$

where the coefficients β_T and γ_T indicate the slopes of the changes in PV panel voltage and current, respectively, due to variable temperature. The values of β_T and γ_T vary from panel to panel and are obtained experimentally. Normally, the value of β_T is between 0.004 and 0.006 [2], and the value of γ_T is between 0.02 and 0.1. These values are assumed to be 0.004 and 0.06 for β_T and γ_T, respectively, in the examples of this chapter. T_R in (8.6) and (8.7) is the known reference ambient temperature during the panel testing and it is assumed to have a value of 20 °C. Equations (8.6) and (8.7) are used to obtain the modified model of the panel for another ambient temperature T_X. Even if the ambient temperature does not change significantly during the daytime, the solar irradiation level changes depending on the amount of sunlight and passing clouds causing a change in the panel photocurrent and operating temperature which, in turn, affects the panel output voltage. If the solar irradiation level changes from S_{X1} to S_{X2}, the panel operating temperature will also change from T_{X1} to T_{X2} resulting in a change in the photocurrent from I_{ph1} to I_{ph2}. Thus, the effect of the changing solar irradiation level on the panel photocurrent and output voltage is expressed via two correction factors C_{SV} and C_{SI}, which are obtained using the method in [2] and given in (8.8) and (8.9).

$$C_{SV} = 1 + \beta_T \alpha_S(S_X - S_R) \tag{8.8}$$

$$C_{SI} = 1 + \frac{1}{S_R}(S_X - S_R) \tag{8.9}$$

where S_R with a value of 100 mW/cm^2 is the benchmark reference solar irradiation level during the panel testing to obtain the modified panel model. S_X is the new level of solar irradiation. The coefficient α_S indicates the temperature change due to changing solar irradiation level and is expressed as in (8.10)

$$\alpha_S = \frac{\Delta T_R}{S_X - S_R} \tag{8.10}$$

The value of α_S is different for nonidentical cells or panels and obtained experimentally. It is assumed to be $\alpha_S = 0.2$ (°C cm^2/mW) in the examples of this chapter. Using correction factors C_{TV}, C_{TI}, C_{SV}, and C_{SI}, the new values of the panel output voltage V_{PX} and photocurrent I_{PHX} are obtained for the new temperature T_X and solar irradiation S_X as in (8.11) and (8.12).

$$V_{PX} = C_{TV} C_{SV} V_{PR} \tag{8.11}$$

$$I_{PHX} = C_{TI} C_{SI} I_{PHR} \tag{8.12}$$

where C_{TV} and C_{TI} represent the effects of the ambient temperature variations as given in (8.6) and (8.7), and C_{SV} and C_{SI} represent the effects of the solar irradiation changes as given in (8.8) and (8.9). V_{PR} and I_{PHR} are the benchmark reference panel output voltage and reference panel photocurrent, respectively.

The reverse saturation current I_0 of the diode in the equivalent circuit model is much smaller than the photocurrent I_{PH} during the short circuit condition and can be neglected. Thus it can be assumed that I_{PH} is approximately equal to the short circuit current I_{SC} [15]. Therefore, for each new value I_{PHX} of the photocurrent, there will be a new value I_{SCX} of the short circuit current on the *I*–*V* characteristics.

Example 8.2

A single PV panel is represented by the reference model equation (8.5) with the reference model parameters given in Example 8.1 as $A = 99$, $R_S = 0.01$ Ohms, $I_0 = 0.051$ A, reference solar irradiation $S_R = 100$ mW/cm^2 and reference temperature $T_R = 20$ °C. Open circuit voltage and short circuit current are given as $V_{OCR} = 31.20$ V and $I_{SCR} = 5.4$ A, respectively. Temperature coefficients are $\beta_T = 0.004$ and $\gamma_T = 0.06$. Solar irradiation coefficient $\alpha_S = 0.2$ °C cm^2/mW.

(a) Write a MATLAB function to simulate the PV panel model including the effects of changes in solar irradiation and temperature levels.
(b) Obtain the reference *I*–*V* characteristic using reference data and MATLAB function developed in part (a).
(c) Simulate and observe how the *I*–*V* and *P*–*V* characteristics of this PV panel change if the solar irradiation level is increased from 100 mW/cm^2 to 150 mW/cm^2 and then to 200 mW/cm^2 while the temperature level is kept constant.
(d) Simulate and observe how the *I*–*V* and *P*–*V* characteristics of this PV panel change if temperature T_C increases from 0 °C to 20 °C and then up to 40 °C while the solar irradiation level is kept constant.

Answer 8.2

(a) A MATLAB function given in Figure E8.2(a) can be used to simulate the PV panel model including the effects of changes in solar irradiation and temperature levels.

The function called *pvpanel.m* returns the panel voltage V by using three inputs, which are solar irradiation S_X, operating temperature T_X, and panel current I.

(b) Reference *I–V* characteristics can be obtained using the reference MATLAB function *pvpanel.m* developed in part (a). A MATLAB code used to plot reference *I–V* characteristics is given in Figure E8.2(b). The function zlimiter.m shown in Figure E8.2(b) is a zero limiter and is used to prevent negative values of V_P.

(c) *I–V* and *P–V* characteristics of the PV panel in use change as shown in Figure E8.2(c) when the solar irradiation level is increased from 100 mW/cm^2 to 150 mW/cm^2 and then to 200 mW/cm^2 while the temperature level is kept constant. The same MATLAB code given in Figure E8.2(b) can be used for this part to see the effects of changing solar irradiation on *I–V* and *P–V*

```
function V=pvpanel(SX,TX,I)
q=1.106e-19;      k=273*0.138e-23;
A=99;             Rs=0.01;
Io=0.051;         IPH=5.4;
TR=20;            SR=100;
BETA=0.004;       GAMMA =0.06;
ALFA=0.2;
CTV=1+BETA*(TR-TX);
CTI=1+(GAMMA*(SX-SR)/SR);
CSV=1+BETA*ALFA*(SX-SR);
CSI=1+((SX-SR)/SR);
IPHX=CTI*CSI*IPH;
VPR=(A*k*TR/q)*log((IPHX-I+Io)/Io)-I*Rs;
V=CTV*CSV*VPR;
```

Figure E8.2(a) MATLAB function pvpanel.m to be used as reference function

```
ISCR=5.4;   IPHR=ISCR;
SR=100;     TR=20;
dS=0;       dT=0;
% ---------------------
n=1;  IP=0;  VP=VOCR;
Sx=SR+dS;        Tx=TR+dT;
while VP > 0
  VP1=pvpanel(Sx,Tx,IP);
  VP=zlimiter(VP1);
  Ip1(n)=IP;
  Vp1(n)=VP;
  Pp1(n)=IP*VP;
  IP=IP+0.01;
  n=n+1;
end
```

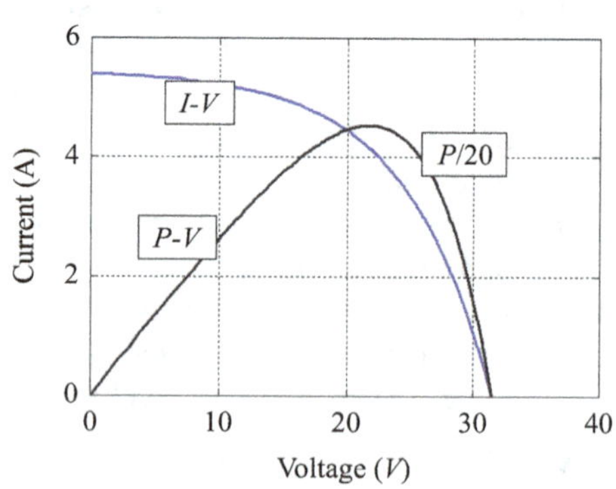

Figure E8.2(b) MATLAB function pvpanel.m to be used as reference function

characteristics by assigning 0, 50, and 100 to dS as the changes in solar irradiation while keeping dT = 0 for three cases.

The effects of solar irradiation level are very important on the generated panel power. As the solar irradiation level changes from morning to evening, so does the generated power. If the load demand power is constant all the day, the generated panel power does not match the load. It may be equal sometimes but smaller or higher other times. If the generated maximum power marked as PM1, PM2, and PM3 in Figure E8.2(c) is not tracked and used as the solar irradiation changes, some of the generated power will be lost. Therefore, the maximum power generated by the panel should be tracked and used or stored to be used when needed.

(d) *I–V* and *P–V* characteristics of the PV panel used in this example change as shown in Figure E8.2(d) if the temperature increases from 0 °C to 20 °C and then up to 40 °C while the solar irradiation level is kept constant.

The effects of temperature changes on the generated panel power are smaller than those of solar irradiation. However, the effects of the changes in operating temperature affect the panel output voltage besides the generated power. High temperatures have negative effects on both panel power and voltage. This

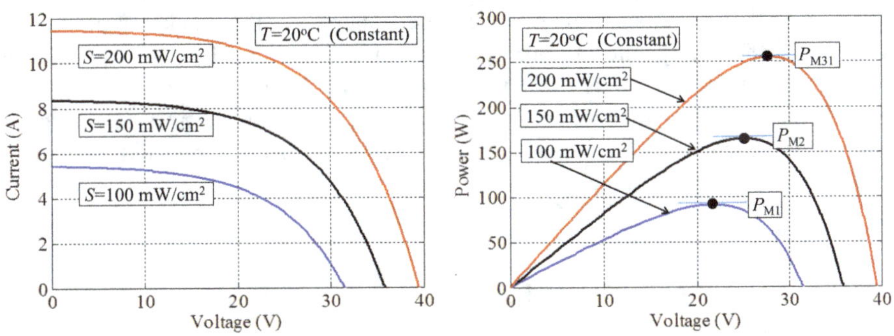

Figure E8.2(c) Solar irradiation effects on I–V and P–V characteristics

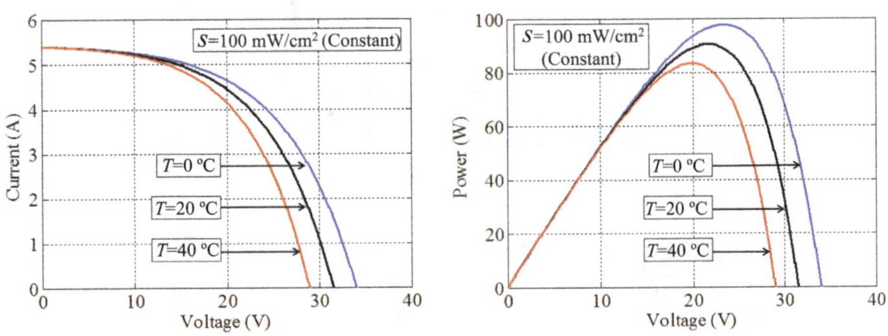

Figure E8.2(d) Temperature effects on I–V and P–V characteristics

analysis shows that cooler locations with higher solar irradiation levels are better for installing the PV power stations.

Design Example 8.1

Convert the MATLAB function *pvpanel.m* in Example 8.2 to a Simulink model and simulate the system for the following:

(a) Repeat parts (b), (c), and (d) in Example 8.2 using the Simulink model of the PV panel.
(b) Use PV panel Simulink model as the power source for DC motor given in Example 8.10 and repeat Example 8.10.

8.2.3 PV panel modeling in Simulink

A general Simulink GUI block of the PV panel model is given in Figure 8.6. The block has four inputs and one output. Solar irradiation and temperature are two external inputs to the model. They are represented by a slider-type input element in Simulink in order to be able to change the values during the simulation by dragging the slider gain adjustments of these blocks named as *variable temperature* and *variable solar irradiation*. The other two inputs N_S and N_P are the number of panels in series and the number of series strings in parallel, respectively. $N_S = 1$ and $N_P = 1$ mean that there is just one single panel. Therefore, the block can be used as PV array model too, by entering proper values for N_S and N_P. The output port of the model given in Figure 8.6 is the PV panel or array voltage to be applied to the loads. A measurement bus bar is used to measure and plot the time responses of array output current and voltage as the output port to be connected to load is represented by the array voltage V_A.

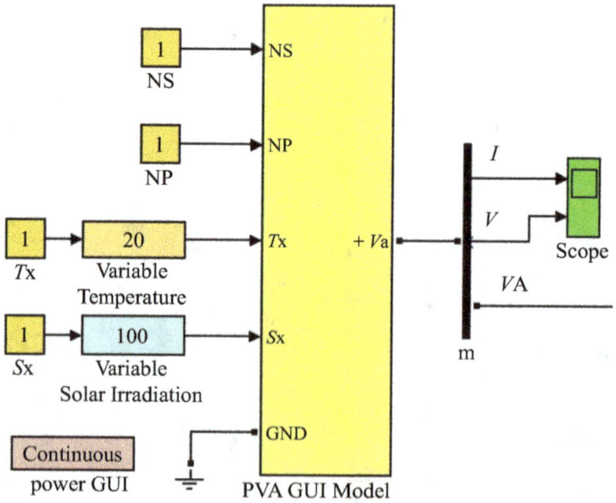

Figure 8.6 Simulink block of a single PV panel

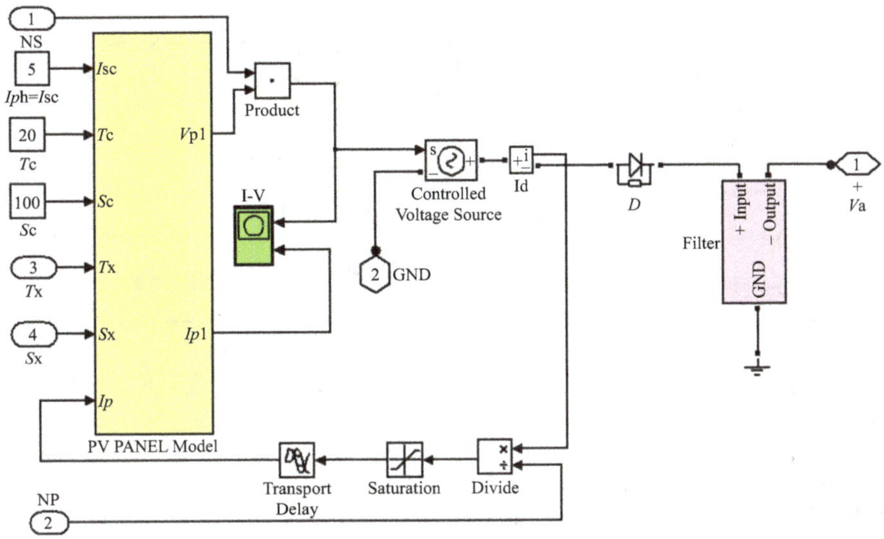

Figure 8.7 PV panel subsystem of Simulink GUI block of PV panel

The block called *PVA GUI Model* is the last stage of the Simulink modeling of the PV array. This block contains submodels such as a PV Panel Model block, a Controlled Voltage Source, a Diode, and a Filter. These subsystems of the PVA GUI Model are shown in Figure 8.7. The subblock called *PV Panel Model* contains the digital simulation model of a single PV panel described by Equations (8.5)–(8.12) including the changing effects of solar irradiation and temperature. The controlled voltage source is used to convert the numerical value of the PV array voltage to an analog-type circuitry source voltage. A diode (D) is connected in series with the output port to prevent the reverse current flow. A filter is connected to the output port to maintain a stable voltage. The filter contains a series of R–L and parallel C elements. The PV array consists of just one panel, however, the number of panels connected in series can be increased to have a desired voltage output while the number of parallel connected series strings is increased to match the required load power. A reference short circuit panel current, which is assumed to be equal to photocurrent is entered as 5 A for the PV panel under study. The reference temperature and solar irradiation levels are entered as 20 °C and 100 mW/cm^2 assuming that the reference *I–V* characteristic is obtained under these operating conditions. The effects of the changes in temperature and solar irradiation levels are included by the input ports T_X and S_X, respectively. Since the digital model of the PV panel includes the panel current as another input, the load current is measured and divided by the number of parallel connected series strings to yield the panel current I_P to the PV Panel Model block.

The inside view of *PV Panel Model* is depicted in Figure 8.8. The *PV Panel Model* includes two blocks. One of them is the initial reference model of PV panel

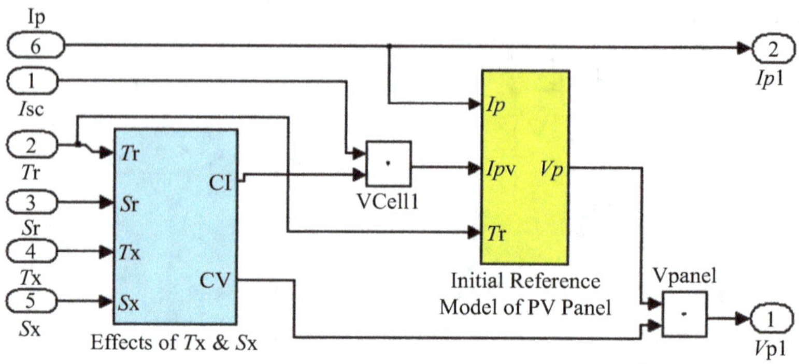

Figure 8.8 PV panel subsystem of Simulink GUI block of PV panel

given in (8.5), and the other one called *Effect of Tx and Sx* represents the submodels of the changing effects of temperature and solar irradiation levels as defined by (8.6)–(8.12).

Example 8.3

A PMDC motor and a resistance-type load are to be fed from a PV system. The PMDC motor has a nominal operating voltage of 36 V and all the other parameters are given in Problem P8.5 in Chapter 8. The resistive load gives 150 W under 40 V.

Use the Simulink model of the PV panel given in Example 8.2 and simulate this PV-powered PMDC motor and resistive load system. It is known that the motor current is 1.5 A under a full speed of 356 rad/s and full voltage of 36 V. Assume that the motor is operated at 150 rad/s under a load torque of 0.55 Nm using a fuzzy logic controller.

(a) Plot time responses of current and voltage.
(b) Plot *I–V* and *P–V* curves of this system under loading conditions.
(c) Plot time responses of the motor speed.
(d) Increase solar irradiation level by 50% at half time of the total simulation and observe the first three responses. How they are affected by this change in solar irradiation?

Answer 8.3

The reference PV panel given in Example 8.2 has an open-circuit voltage of around 30 V and a short-circuit current of about 5 A. Therefore, at least two panels should be connected in series. The resistive load drives 3.75 A under 40 V and the motor current will be up to 1.5 A. Since the short circuit current of the PV panel is 5 A under the reference conditions, one series string will be sufficient to supply the required power. Therefore, $N_S = 2$ and $N_P = 1$ will be sufficient to be used as the PV array for the load described.

The Simulink model of the PV-powered PMDC motor and resistive load scheme is given in Figure E8.3(a). The output voltage V_A of the PV array is applied to a parallel connected resistive load and FLC-controlled PMDC motor.

The PVA GUI Model is the same model discussed above in this section. The FLC-controlled PMDC motor block, given in Figure E8.3(b), includes the simulation models of the FLC and PMDC motor, both were discussed in Chapters 4 and 5, respectively. DC Chopper and Discrete PWM Pulse Generator are available in the Simulink SimPower toolbox.

The time responses of PV array current and voltage, *I–V* and *P–V* characteristics, motor speed, and current and voltage simulation results are all given in Figures E8.3(c)–E8.3(h).

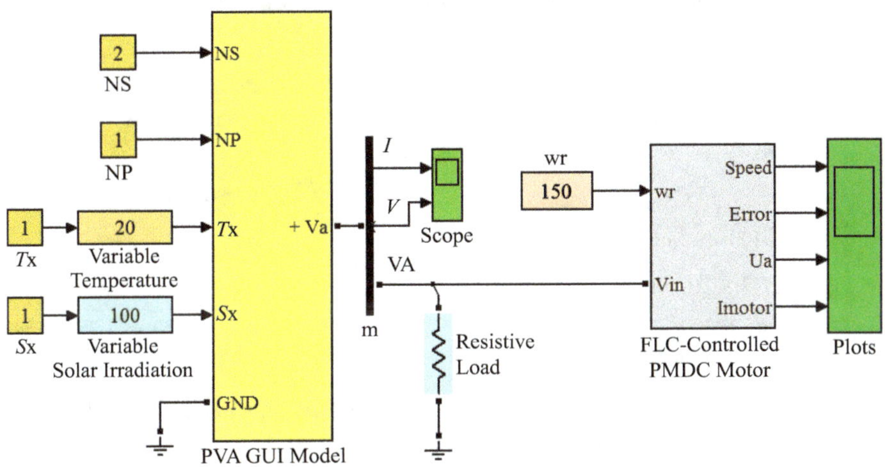

Figure E8.3(a) PV-powered DC load system

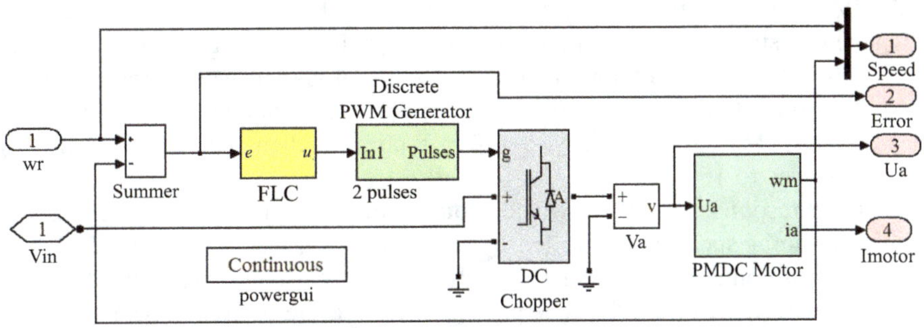

Figure E8.3(b) PV-powered DC load system

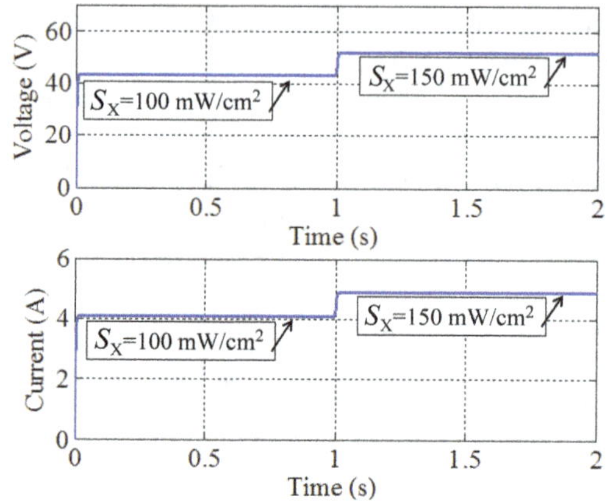

Figure E8.3(c) Time responses of PV array voltage and current

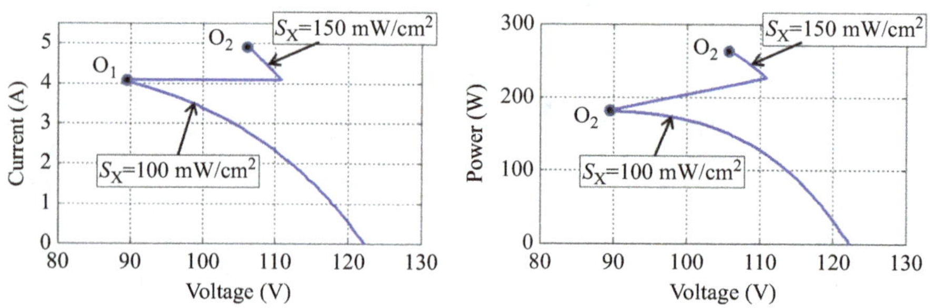

Figure E8.3(d) PV array I–V and PV characteristics under operation

P.5.1 Time responses of PV array voltage and current are shown in Figure E8.3(c) including the effect of a step change in load irradiation level from 100 mW/cm^2 to 150 mW/cm^2. An increment in solar irradiation level causes increments in both voltage and current, and of course in power, too.

P.5.2 PV array *I–V* and *P–V* curves are shown in Figure E8.3(d). As stated in the question part, the simulation starts with a solar irradiation level of 100 mW/cm^2 and after some time (at $t = 1$ s) it is increased to 150 mW/cm^2. As depicted in Figure E8.3(d), PV array operates at operating point O_1 on the *I–V* and *P–V* curves when $S_X = 100$ mW/cm^2 and then operates at O2 after S_x increases to 150 mW/cm^2.

P.5.3 Time responses of PV-powered motor speed are shown in Figure E8.3(e). The motor speed is controlled by the FLC developed in Chapter 5. Motor speed is controlled with a zero steady-state error when the solar irradiation is 100 mW/cm^2. However, the speed goes above the reference when the irradiation level is increased to 150 mW/cm^2 because the generated power becomes much greater than load power under this condition and the PV panel

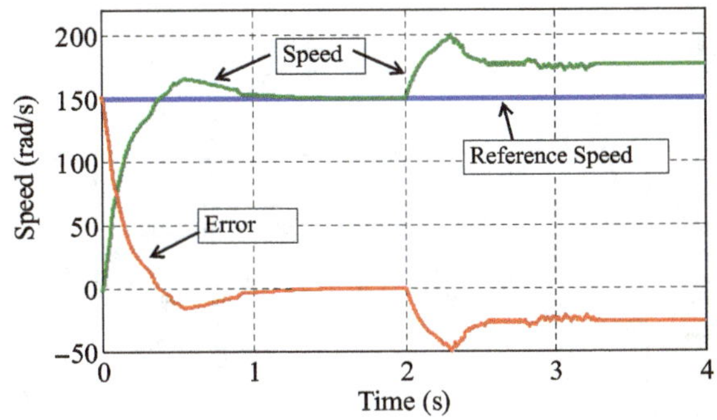

Figure E8.3(e) PV-powered PMDC motor speed responses

operates at over-voltage zone where the voltage is greater than the maximum power–voltage. The PV array output voltage becomes greater than it is supposed to because of insufficient load current. If the additional power is managed to be stored or used by the other loads and the load bus voltage is controlled and kept constant, then the speed control of the motor would be easier to track the reference.

P.5.4 The effect of increasing solar irradiation level by 50% at the half time of the total simulation is included in parts (a), (b), and (c) above. The *I–V* and *P–V* characteristics change their locations and operating points move up to new values. This example does not include any MPPT controller and MPP operation depends on generated PV power to load power matching.

Example 8.4

The PV array Simulink model described in Section 8.2.3 is going to be used as a power station to feed three-phase linear RLC type loads connected to a 230 V, 60 Hz load bus.

(a) Use additional blocks available in the Simulink SimPower block library and operate three-phase loads after converting DC output voltage of the PV array to the required three-phase voltages at 60 Hz. Use a three-phase isolation transformer with a turns ratio equal to 1 and observe the voltage waveforms before and after the isolation transformer.
(b) Find the total harmonic distortion (THD) of the AC voltages.

Answer 8.4

(a) Since the output voltage from a PV array is DC, an inverter must be used to convert the DC voltages to AC at 60 Hz. Therefore, a three-phase inverter is connected to the output of the PV array as shown in Figure E8.4(a). The inverter's control signals are generated by a fixed six-pulse PWM generator to yield 60 HZ voltages. After converting the DC voltages to AC, a three-phase isolation transformer is placed on the path between the inverter and the three-phase linear

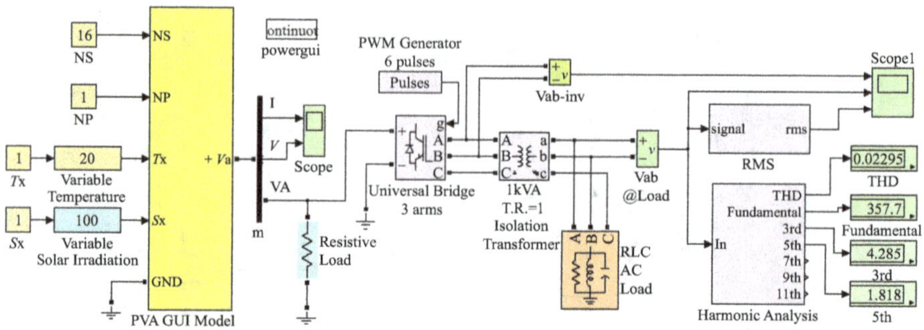

Figure E8.4(a) PV array as a three-phase AC generator

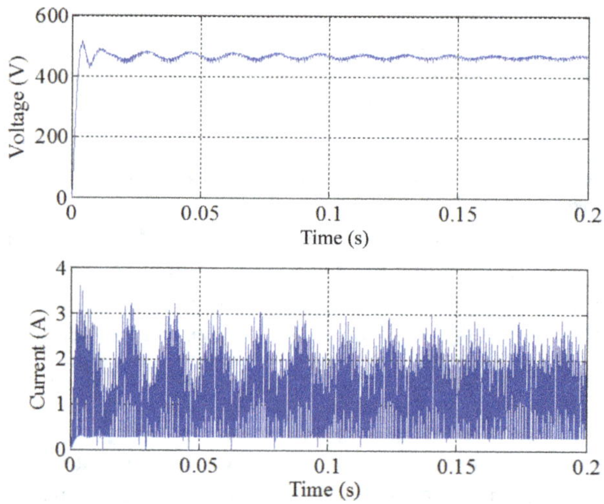

Figure E8.4(b) PV array voltage and current at DC side

RLC load. Line-to-line voltage voltages before and after the isolation transformer are measured for comparison. A three-phase RLC load is connected at the output terminals of the isolation transformer. Active, inductive reactive, and capacitive reactive power ratings are 500 W, 300 VAr, and 200 VAr, respectively.

PV array output voltage and current are shown in Figure E8.4(b). the output voltage oscillates around 470 V. The array current fluctuates between zero and 2.5 A with a DC average. However, the switching effects of the inverter are reflected in the array current. A smooth current waveform can be obtained by designing proper filters and controlling the operating power.

Three-phase line-to-line voltages before and after the isolation transformer are given in Figure E8.4(c). The AC voltage waveforms are in the form of positive and negative pulse groups as a result of the pulse width modulated gating signals of the inverter. Due to the smoothing filter effect of the isolation

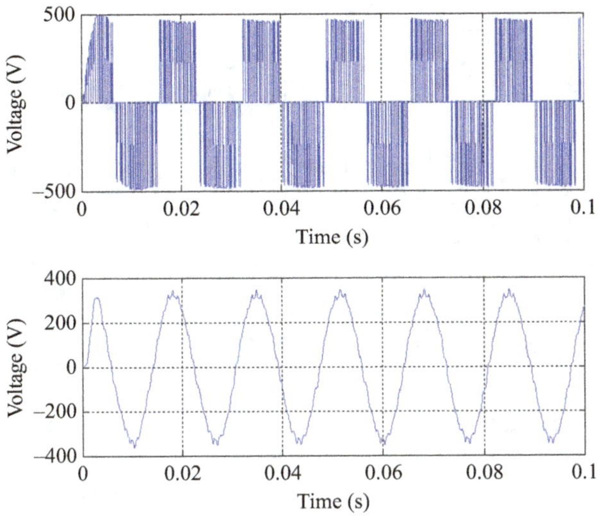

Figure E8.4(c) Line-to-line AC voltages before and after the transformer

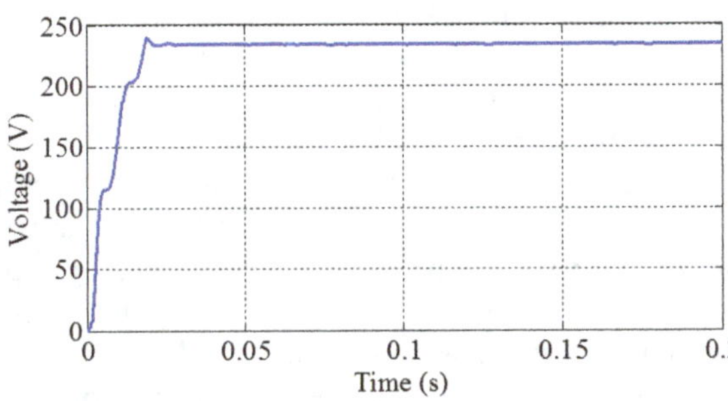

Figure E8.4(d) RMS value of the line-to-line AC

transformer coil, the alternating pulses are converted to a sinewave. However, the converted voltage waveforms are not pure sinewaves. They include harmonic components with a total harmonic distortion (THD) of 3.9.

(b) The total harmonic distortion (THD) of the AC voltages is obtained as THD = 0.03908, which is about 3.9%. Besides the THD, the fundamental, third, and fifth harmonic components are measured as 331.8 V, 0.9876 V, and 1.095 V, respectively.

The RMS value of the line-to-line voltages is given in Figure E8.4(d). The RMS voltage oscillates around 234–235 V.

8.2.4 A PV array emulator

Besides the computer simulation models, emulation models of PV arrays are also developed to implement the characteristic behaviors of the arrays by feeding the

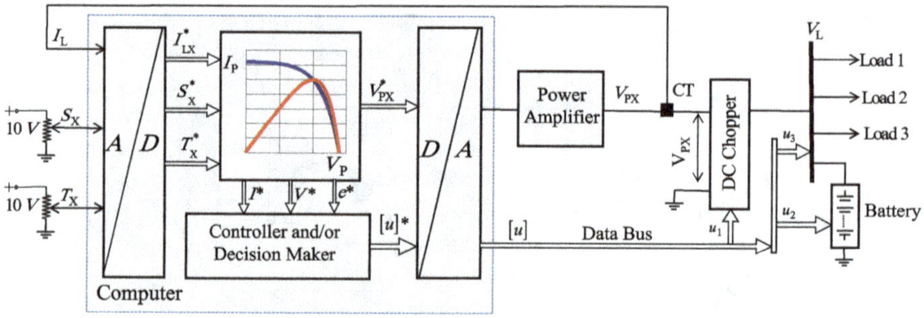

Figure 8.9 An emulator model for PV arrays

real loads. A PV emulator model was developed by the author in the early 1990s [8]. In this PV emulator, the simulation model of the PV array discussed in the previous section is stored in a computer and the variables affecting the operating characteristics are entered as real-time measurements. Potentiometers are used to enter and change the effects of solar irradiation and temperature levels while the load current is measured and fed back to the PV array model stored in the computer so that a PV array output voltage is generated using (8.5)–(8.12). A schematic diagram of the emulator model is given in Figure 8.9.

The value of the solar irradiation level used in the PV array computer emulator model is determined using a variable analog voltage input, 0–10 V to the computer via analog-to-digital converters. This reference voltage is then multiplied by 25 to yield a reference solar irradiation level of 0–250 mW/cm^2. The value of the array operating temperature is also entered using a variable analog input from 0 V to 10 V via a potentiometer. The analog voltage is multiplied by 5 digitally in the computer to obtain a temperature change from 0 °C to 50 °C, which may be used to obtain temperatures below zero, as well. The solar irradiation level and the temperature inputs are represented by reference analog voltages because an emulation model of the PV array is the subject instead of a real one. If an actual PV array is used, then the solar irradiation level and the temperature must be measured instantly using solar irradiation instruments such as the radiometers *pyranometer* and *pyrheliometer*, and temperature measurement instruments such as *thermometer* or *thermocouple*. The corresponding PV array voltage is then calculated using equations from (8.5) to (8.12).

The load current is measured as a real-time variable using a current transformer and fed back to the PV array simulator as the third input variable besides the solar irradiation level and operating temperature. Three analog input variables are converted to analog signals by analog to digital (A/D) converters and used in the digital model of the PV array stored in the computer as programmed codes. The digital values of the signals are represented by a superscript * in Figure 8.9. The power amplifier in Figure 8.9 is also a part of the emulator. The voltage and current gain settings of the amplifier correspond to the number of series-connected panels in a string and parallel-connected series strings, respectively. Therefore, V_{PX} is the output voltage of the PV panel emulator for the input variables S_X, T_X, and I_{LX}, where I_{LX} is the load current drawn from the PV array.

A digital controller and/or a decision maker may be included in the PV emulation model to control or manage the power, voltage, or current of the PV array. Assume that the PV array is to be operated at its maximum power point. In order to do this operation, one needs to measure or estimate the maximum power and use it as the reference to be tracked. Assume that e^* in Figure 8.9 is the error obtained as the difference between the reference maximum power point (RMPP) and the current operating power of the panel. This error is also called maximum power operating error (MPPE) and is minimized by the controllers so that an MPP operation is satisfied. Besides the MPPE, array current and voltage are also used in controllers and decision-makers for load and battery backup management.

As explained in the previous section, the solar PV array voltage is obtained as a function of the array load current along with the solar irradiation level and operating temperature. Since the array current is directly proportional or equal to the load current, control of the load current means control of the array current, and therefore the array voltage, and the array power. In order to control the load voltage, a DC chopper has been used as a variable voltage interface transformer in the scheme. The required load voltage that gives the load current corresponding to the maximum power operating point of the PV array, is obtained by controlling the chopper duty cycle. The PV emulator scheme may also contain some additional external components such as filters to eliminate current ripples and harmonics.

8.3 Maximum power point (MPP) search in PV arrays

PV array systems are designed to operate at their maximum output power levels for any temperature and solar irradiation level at all the time to reach high efficiency and energy utilization. The performance of a PV array system depends on the operating conditions as well as the solar cell and array design quality. The output voltage, current, and power of a PV array vary as functions of solar irradiation level, temperature, and load current. Therefore, the effects of these three quantities must be considered in the design of PV arrays so that any changing temperature and solar irradiation levels should not adversely affect the PV array output power to the load/utility, which is either a power company utility grid or any stand-alone electrical type load. To overcome the undesired effects of the variable temperature and solar irradiation on the output power of PV systems, two different control strategies have usually been applied:

(i) Controlling the sun input to the PV array, and
(ii) Controlling the power output from the PV array.

The combinations of these two groups also may be considered. Both groups may include electrical or thermal energy storage systems or auxiliary power sources that supply electricity during the night or on cloudy days. Sun input to the PV systems is kept as high as possible either by rearranging the solar cell configurations of PV arrays with respect to the changes in weather conditions or by designing and controlling the position of sun-tracking solar collectors. The sun tracking method is generally used with concentrator arrays and is not within the scope of this book. Studies related to the maximum output power tracking of PV arrays deal mostly with the isolation requirements and interfacing problems of direct coupling to loads. The output electric power level and energy produced from a stand-alone PV array feeding a single or a group of resistive and motor-type loads can be adjusted by connecting the solar cells in different configurations of series/ parallel groups. When the input solar power to the PV array changes due to Varying ambient temperature and solar irradiation levels, the output power from the PV array also changes. However, the load being fed usually requires a maintained constant power level. Therefore, if the PV array output power exceeds the power requirement of the load connected, either additional loads should be put into service or the power over the requirement should be stored to be used when it is needed. If the PV array output power is below the power requirement of the load, then either the required additional power is supplied from the storage or some of the loads are put off the service to match the power of the PV array. For all the cases, the operation of the PV array at its maximum output power is necessary in order to operate it with maximum efficiency.

The MPPT of a PV array is achieved by maximizing its output power to load. To obtain a maximized output power, controllers are used to minimize the error between the operating power and the reference maximum power, which varies with the changing temperature and solar irradiation level. Therefore, the value of the reference maximum power must be determined for the changing temperature and solar irradiation level before it is compared with the operating power.

Various methods to find the maximum power point of PV arrays have been introduced and classified as follows:

(i) Using look-up tables representing the previously determined maximum power point data for different temperature and solar irradiation levels [20–22].
(ii) Simulating the PV array characteristic equations to search for the new maximum power point if any change occurs in temperature or solar irradiation level [23].
(iii) Using an online search technique for the maximum power point regardless of the changes in the temperature and solar irradiation level [16,24].

The first two methods require a PV array model and the measurement of temperature and solar irradiation levels. The third method, however, requires neither the measurement of temperature and solar irradiation nor a PV array model. The latter one has been studied by the author in his earlier research [8–13] and will be described in the next sections.

The maximum power point operation of a PV array is achieved by maximizing its output power to load. The controllers are used to minimize the error between the operating power and the reference maximum power which varies with the changing temperature and solar irradiation level. Therefore, the value of the reference maximum power must be determined for the changing temperature and solar irradiation level before it is compared with the operating power.

8.3.1 MPP by lookup tables

Using the PV panel model given in the previous sections, the *I–V* and *P–V* characteristics of the PV array are obtained for different temperature and solar irradiation levels. The maximum power point values from these characteristics are then found to yield the look-up table for the maximum power points at different temperatures and solar irradiation levels. The look-up tables containing the MPP values of PV array power, P_M, current, I_M, and voltage, V_M, as functions of temperature and solar irradiation are represented by the third-order polynomial equations whose coefficients can be found by the least square curve fitting of the known P_M, I_M, and, V_M values. Polynomial equations given in (8.13), (8.14), and (8.15) for the maximum power points, maximum power point currents, and maximum power point voltages, are adopted from [8] with the parameters given in Table 8.1.

$$\begin{aligned}P_M &= \left(X_{P1} + X_{P2}S_X + X_{P3}S_X^2 + X_{P4}S_X^3\right)\\&\quad+\left(X_{P5}T_X + X_{P6}T_XS_X + X_{P7}T_XS_X^2 + X_{P8}T_XS_X^3\right)\\&\quad+\left(X_{P9}T_X^2 + X_{P10}T_X^2S_X + X_{P11}T_X^2S_X^2 + X_{P12}T_X^2S_X^3\right)\\&\quad+\left(X_{P13}T_X^3 + X_{P14}T_X^3S_X + X_{P15}T_X^3S_X^2 + X_{P16}T_X^3S_X^3\right)\end{aligned} \tag{8.13}$$

$$\begin{aligned}I_M &= \left(X_{C1} + X_{C2}S_X + X_{C3}S_X^2 + X_{C4}S_X^3\right)\\&\quad+\left(X_{C5}T_X + X_{C6}T_XS_X + X_{C7}T_XS_X^2 + X_{C8}T_XS_X^3\right)\\&\quad+\left(X_{C9}T_X^2 + X_{C10}T_X^2S_X + X_{C11}T_X^2S_X^2 + X_{C12}T_X^2S_X^3\right)\\&\quad+\left(X_{C13}T_X^3 + X_{C14}T_X^3S_X + X_{C15}T_X^3S_X^2 + X_{C16}T_X^3S_X^3\right)\end{aligned} \tag{8.14}$$

Table 8.1 Polynomial coefficients for (8.13), (8.14), and (8.15)

X_{Pi}, $i = 1, \ldots, 16$	X_{Ci}, $i = 1, \ldots, 16$
−0.0129586	−0.00075382
0.00525487	0.00107311
4.70429×10^{-5}	3.06706×10^{-7}
1.27305×10^{-7}	5.10227×10^{-10}
3.67328×10^{-5}	7.01662×10^{-6}
-1.42439×10^{-5}	-9.43174×10^{-8}
-1.3688×10^{-7}	2.39102×10^{-8}
3.74931×10^{-10}	-1.43316×10^{-10}
4.84305×10^{-8}	-1.70453×10^{-8}
-1.96064×10^{-8}	7.3108×10^{-9}
-1.21402×10^{-10}	-3.45127×10^{-10}
2.69142×10^{-13}	2.68109×10^{-12}
-7.76782×10^{-11}	-1.63084×10^{-9}
6.37178×10^{-12}	8.97043×10^{-11}
-1.779×10^{-13}	7.98484×10^{-13}
9.32998×10^{-16}	-1.66707×10^{-14}

After the maximum power points for a range of solar irradiation and temperature are obtained from (8.13) and the MPP currents for the same solar irradiation and temperature levels from (8.14), they are used to obtain the MPP voltages using (8.15).

$$V_M = \frac{P_M}{I_M} \tag{8.15}$$

8.3.2 MPP search algorithm based on measurements of S_X and T_X

MPP is also determined using search algorithms. The flow chart of a search algorithm is given in Figure 8.10, which requires the measurements of solar irradiation S_X and temperature T_X. Two consecutive values of temperature and two consecutive values of solar irradiation level are compared in this search algorithm. If the change in temperature or solar irradiation level is greater than a reasonable change, ε, then a new reference MPP is calculated by simulating the PV array *I–V* characteristics, defined by (8.5), for the new temperature and solar irradiation level.

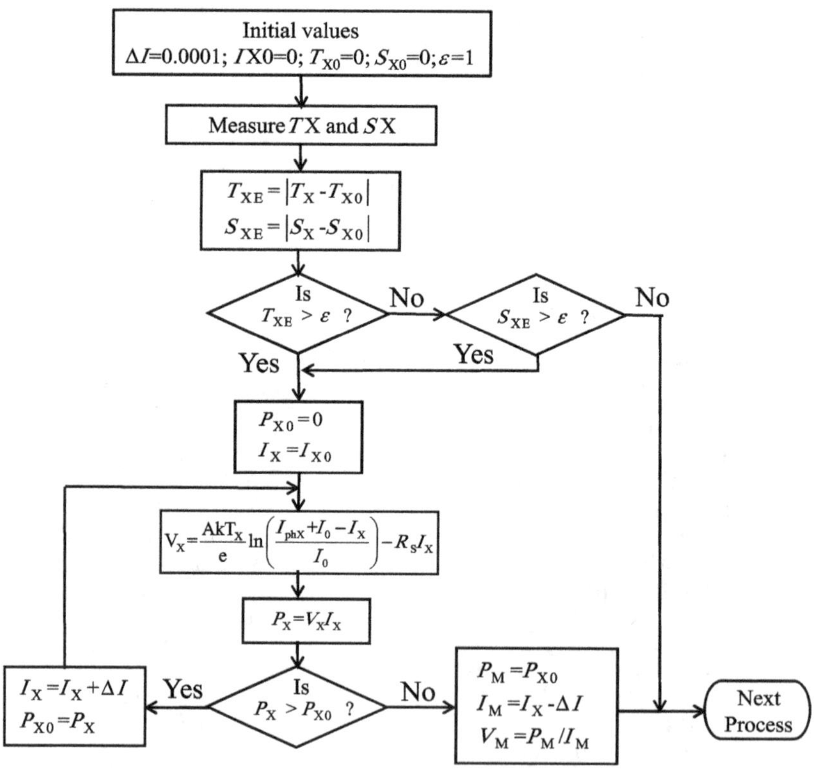

Figure 8.10 Flowchart of MPP search algorithm based on the measurements of S_X and T_X

Starting with a small initial value, the array current is increased step by step until the maximum power is obtained. At every step, the array power is calculated and compared with the previous value until the new value becomes smaller than the previous one. This means that the previous value of the power is the new reference maximum power, P_M, for the new temperature and solar irradiation level, and the corresponding array current, I_M, and array voltage, V_M, are the reference current and voltage representing the new maximum power point of the PV array *I–V* characteristics. After the new MPP quantities P_M, I_M, and V_M are obtained, the initial values of temperature and solar irradiation level, T_{X0} and S_{X0}, respectively, are replaced by the current values T_X and S_X to be used as initial values for the next temperature and solar irradiation change.

Apparently, the temperature and the solar irradiation level must be measured continuously to determine the change. Besides, in this method, the program executes only the iteration loop of the search algorithm during the determination of the MPP and introduces a time delay in calculating the controller output and transferring data between analog and digital systems. If the change in temperature or solar irradiation level occurs frequently, the delay caused by the MPP search method will affect the analog-to-digital and digital-to-analog data translation process resulting in instability in the system. However, the method is still applicable if the time constants of temperature and solar irradiation levels are much greater than the computation time of the computer used. This method also introduces an MPP error depending on array current increments.

8.3.3 MPP search algorithm based on voltage and current measurements

Figure 8.11 shows the flowchart of another MPP search method that does not require the measurement of temperature T_X and solar irradiation level S_X. This algorithm uses only the measurement of array voltage and current and calculates the operating power. The operating power is increased by connecting more loads or battery backup units to the system until the increment in operating power is stopped and started to decrease. This search algorithm is given in Figure 8.11 and described below.

The algorithm given in Figure 8.11 searches the MPP of the PV array *I–V* characteristics with an error, e_{PT}. Initially, the reference values of MPP power, P_M, and current, I_M, are set to zero, and voltage, V_M, is equated to the open circuit voltage, V_{OC}, of the PV array. At each sampling instant, the difference, e_{P1}, between the reference initial value of P_M, and current operating power, P_A, is calculated and compared with the assumed MPP error, e_{PT}. If the difference, e_{P1}, is smaller than the acceptable error, ε_P, the currently assigned initial values P_M, I_M, and V_M are maintained unchanged. If the power error, e_{P1} is greater than ε_P, then the algorithm searches for a new MPP. If the operating power P_A is greater than the current MPP power value, P_M, then the value of P_A is assigned as new P_M. If the operating power, P_A, is smaller than the current MPP power, P_M, then the operating current, I_A, and voltage, V_A, are compared with the MPP current, I_M, and voltage, V_M, respectively. If I_A is greater than I_M or if V_A is smaller than V_M, the operating power point (OPP) values P_A, I_A, and V_A are assigned as the new reference MPP values. Otherwise, the algorithm continues with

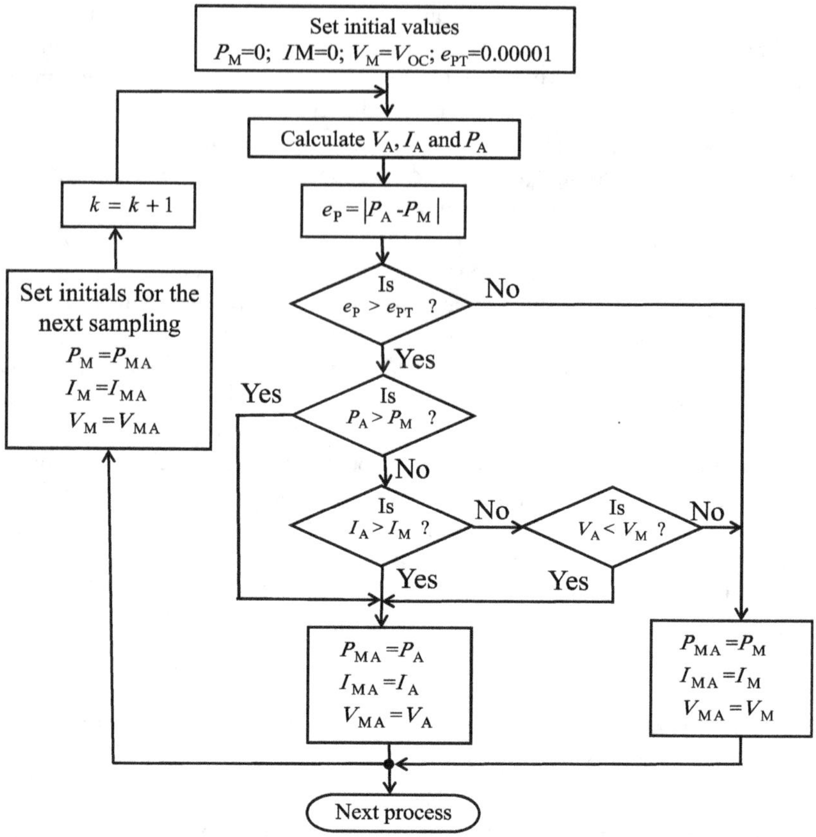

Figure 8.11 Load-based MPP search algorithm without the measurement of S_X and T_X

the current MPP values P_M, I_M, and V_M. This search algorithm is based on the operating power of the load. The PV array output voltage, V_A, is adjusted by the algorithm to increase or decrease the operating current of the load to the level of PV array MPP current, I_M, which yields the MPP power P_M. This algorithm will not be able to determine the PV array MPP if the load power or current is much smaller than the PV array MPP power and current. In this case, additional loads should be connected to increase the PV array current so that the PV array can be operated at the MPP. If the reference MPP is changed due to changing temperature and solar irradiation level, the algorithm adjusts the array voltage and finds the new MPP.

8.3.4 MPP search algorithm based on online repetitive method

Another MPP search algorithm for different solar irradiation and ambient temperature levels using the information only from the current and voltage measurements is addressed in this section. The algorithm is adapted from [11] and depicted by the flow chart shown in Figure 8.12.

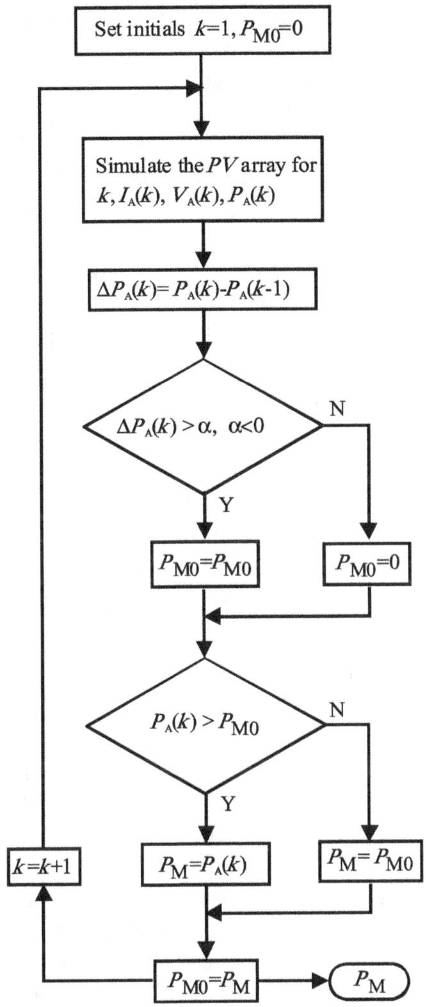

Figure 8.12 The online repetitive MPP search algorithm

When the simulation starts, a small initial value is assigned as the maximum power ($P_{M0} = 0$). Then the PV array is simulated to obtain the values of the array current, voltage, and power for the current kth sampling instant. The difference between two consecutive values of the power is calculated to be compared with a *sensitivity constant* ($\alpha < 0$). The sensitivity constant is one of the key parameters of the algorithm because it is used to detect the negative change in operating power due to less solar irradiation level or load switch off and therefore initialize the maximum power determined before searching a new MPP. Normally, the algorithm operates on a straight-up basis such that it searches for the peak point of the P–V curve starting from a small value and replacing it with a higher one each time.

Since there are no local minima or maxima, it is sufficient to look only for the peak value of the available power. After the initialization, the controller part of the MPPT acts to switch on more loads or backup batteries for charging until the new MPP is reached. If some loads are switched off without any change in solar irradiation level, the power will have a negative change causing the initialization of the MPP. In this case, if there are no other loads to be switched on, additional backup batteries will be switched on by the MPPT controller until the MPP is obtained again.

The positive change in operating power will cause the algorithm to update the MPP to the new higher value. Therefore, it is not required to detect positive changes in power. The MPP search algorithm given in Figure 8.12 is not responsible for the PVA to be operated at MPP. Once the MPP is determined by the algorithm described in this section, it is used as the reference power to be tracked. Therefore, the MPPT has a second part that acts as a controller for tracing the MPP. The initialization of the MPP search makes the algorithm to search the MPP from the beginning starting from a small power. During each initialization, the search restarted from the beginning by repeating the same search process. Therefore, this MPP search algorithm is called the *repetitive MPP search algorithm* by the author.

8.4 Maximum power point tracking of PV arrays

The MPPT consists of two problems to be dealt with. One of these problems is to determine the maximum power point (MPP) for the current solar irradiation level and temperature and then track it as the solar irradiation level and/or temperature change. For different solar irradiations, the PVA generates different peak power points to be determined and used as reference power values to be tracked. This problem and its solutions have been discussed in the previous section.

It should be noted that the MPPs are not fixed at a certain location. They can be anywhere on the *P–V* space depending on the light levels. Therefore, the first problem of the MPPT is to determine the locations of MPPs. Once the MPP for the current solar irradiation level is determined, then the second part of the MPPT problem starts. The second part of the problem is more likely a control problem and the goal is to operate the PVA at the MPP by managing loads and backup batteries. MPPT control approaches and algorithms developed and published earlier by the author will be discussed next.

8.4.1 Constant maximum power angle approach

The current–voltage space (*I–V* space) of a PV array is separated into three subspaces as maximum power zone, over-voltage zone, and over-current zone. The *maximum power zone* is represented by a maximum power line passing through the origin and the maximum power point. The zone where the PV array operating voltage V_A is greater than the maximum power point voltage V_M ($V_A > V_M$), is called *over-voltage zone*. Similarly, the zone where the PV array operating current

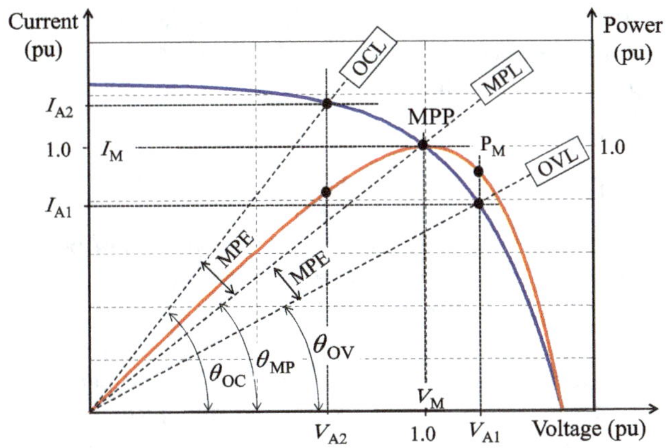

Figure 8.13 MPP operation lines and maximum power error definitions

I_A is greater than the maximum power current I_M ($I_A > I_M$) is called the *over-current zone.* The maximum power line separates over-voltage and over-current zones from each other as depicted in Figure 8.13. Three power lines are used to represent three operating conditions. The maximum power line (MPL) represents the MPP operation while over-voltage line (OVL) and over-current line (OCL) represent over-voltage and over-current operating cases.

The PV array is operated at its MPP by minimizing the maximum power operating error (MPE), θ_E, which is the difference between MPL and OVL or the difference between MPL and OCL. As shown in Figure 8.13, MPL is the line passing through the maximum power point (MPP) and the origin of the PV array *I–V* and *P–V* characteristics. Similarly, OVL is the line connecting any operating point in *over-voltage zone* with the origin and OCL is the line connecting any operating point in *over-current zone* with the origin. Assuming the voltage axis of the *I–V* characteristics as the reference axis, the maximum power angle θ_{MP}, over-voltage angle θ_{OV}, and over-current angle θ_{OC} are defined as the angles measured from the voltage axis to the maximum power line, to over-voltage line and to over-current line, respectively.

If the voltage and current at the MPP are denoted by V_M and I_M, and if the voltage and current at the operating point on the over-voltage zone are denoted by V_{A1} and I_{A1}, then the maximum power angle, over-voltage angle, and over-current angle are defined as in (8.16), (8.17), and (8.18), respectively.

$$\theta_{MP} = \tan^{-1}\left(\frac{I_M}{V_M}\right) \tag{8.16}$$

$$\theta_{OV} = \tan^{-1}\left(\frac{I_{A1}}{V_{A1}}\right) \tag{8.17}$$

$$\theta_{OC} = \tan^{-1}\left(\frac{I_{A2}}{V_{A2}}\right) \tag{8.18}$$

The maximum power error (MPE), θ_E is then found as in (8.19) if the operating point is on the over-voltage zone or as in (8.20) if the operating point is on the over-current zone of *I–V* characteristics.

$$\theta_E = \theta_{MP} - \theta_{OV} \tag{8.19}$$

$$\theta_E = \theta_{MP} - \theta_{OC} \tag{8.20}$$

According to Equations (8.19) and (8.20), the maximum power error, θ_E, is positive if the operating point is in the over-voltage zone and it is negative if the operating point is in the over-current zone.

If the PV array maximum power point quantities P_M, V_M, and I_M are selected as the base values for the power, voltage, and current, respectively, the power, voltage, and current are expressed in terms of a per-unit system, then the maximum power angle θ_{MP} given in (8.16) becomes a constant angle as in (8.21) since $I_M = 1$ pu and $V_M = 1$ pu.

$$\theta_{MP} = \tan^{-1}\left(\frac{1}{1}\right) = 45^\circ = \frac{\pi}{4}\,\text{rad} \tag{8.21}$$

where $I_{BASE} = I_M$ and $V_{BASE} = V_M$. Since the power, voltage, and current at MPP are considered as the base values, whenever the MPP changes due to the changes in solar irradiation and temperature levels the base values are updated such that PM, IM, and VM are always 1.0 pu resulting in a constant maximum power angle $\pi/4$ rad as given in (8.21). The maximum power angle is always kept at $\pi/4$ rad by updating the base values every time the MPP changes and is used as the reference in the control process. Therefore, this MPPT approach is called as *the constant power angle approach*. Any operating power angle rather than $\pi/4$ rad, results in a maximum power error. Since over-voltage operating angle, θ_{OV}, becomes 0 rad for $I_A = 0$, and $\pi/2$ rad for $V_A = 0$, the maximum power operating error varies between $-\pi/4$ rad and $+\pi/4$ rad for a maximum power angle $\theta_{MP} = \pi/4$ rad. If the value of the load current (array current), does not match the MPP current I_M, then the array operates at a different point rather than MPP. For an operating current $I_{A2} > I_M$, the array operates in *the over-current* zone resulting in lower output voltage V_{C2} with an operating power angle $\theta_{OC} > \theta_{MP}$. For an operating current $I_{A1} < I_M$, the operating power angle becomes less than the maximum power angle as $\theta_{A1} < \theta_M$, and the array operates in *the over-voltage* zone; which means that the array current is too small for MPP operation. The MPE, θ_E, varies between 0 and $\pi/4$ rad in the over-voltage zone, and between $-\pi/4$ rad and 0 in the over-current zone. The MPE, θ_E, is then minimized by the controllers so that the operating power point is moved toward and over MPP. Neither the over-current nor the over-voltage operating zone is preferred as only the MPP operation is needed to get almost the maximum available power output from the PV array. Therefore, the use of the error signal θ_E, as an input to the controller is suggested. θ_E is defined as the difference between $\pi/4$ rad and the operating power angle, which is either θ_{OV} or θ_{OC} as given in (8.19) and (8.20).

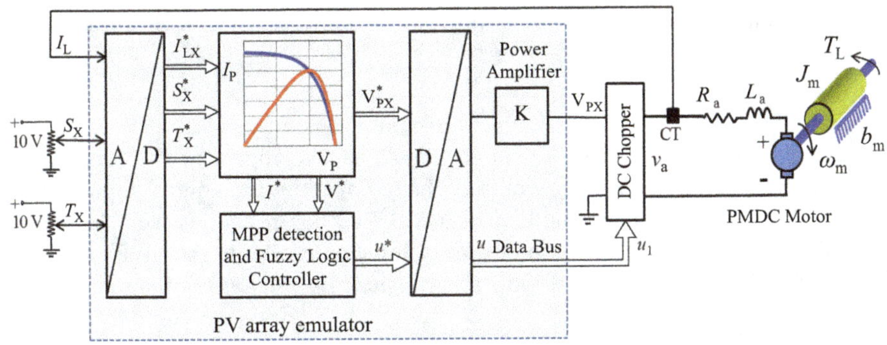

Figure 8.14 PV emulator feeding a PMDC motor

The constant maximum power angle approach has been employed in a fuzzy logic-based MPP tracking (MPPT) scheme shown in Figure 8.14. The PV array emulator is used to supply power to a chopper-controlled PMDC motor. The chopper is used to adjust the voltage va to the motor so that the motor current is adjusted to operate the PV array at its maximum available power under a certain solar irradiation level.

The PV emulator model is used under real-time conditions to drive a permanent magnet direct current (PMDC) motor. The digital model of the PV array uses real-time load current, solar irradiation level, and temperature inputs and generates a real-time output voltage V_{PX}. The MPPT is done by a straightforward process as described below.

(i) Get digital values of operating power point voltage V^* and current I^*.
(ii) Detect maximum power point using voltage V^* and current I^* in one of the methods given in Section 8.3.
(iii) Determine voltage V_M, current I_M, and power P_M at MPP.
(iv) Use V_M, I_M, and P_M as the base values, and convert Ampere–Volt (I–V) and Watt–Volt (P–V) curves to pu I–V and pu P–V curves.
(v) Set the reference maximum power angle θ_{MP} to $\pi/4$ rad.
(vi) Calculate the operating power angle $\theta_X(k)$ for the kth sampling using (8.17) or (8.18) as in (8.22).

$$\theta_X(k) = \tan^{-1}\left(\frac{I_A^*}{V_A^*}\right) \tag{8.22}$$

It should be noted that the operating power angle $\theta_X(k)$ in (8.22) will be equal to the over-voltage angle $\theta_{OV}(k)$ as in (8.17) if $\theta_X(k) < \pi/4$ rad and over-current angle $\theta_{OC}(k)$ as in (8.18) if $\theta_X(k) > \pi/4$ rad.

(vii) Calculate the maximum power angle error for the kth sampling using (8.23).

$$\theta_E(k) = \frac{\pi}{4} - \theta_X(k) \tag{8.23}$$

(viii) Calculate the change in maximum power angle error for the *k*th sampling using (8.24).

$$d\theta_{\mathrm{E}}(k) = \theta_{\mathrm{E}}(k) - \theta_{\mathrm{E}}(k-1) \tag{8.24}$$

(ix) Use the error $\theta_{\mathrm{E}}(k)$ and its change $d\theta_{\mathrm{E}}(k)$ as two inputs to the fuzzy logic controller and generate the required change $du(k)$ in the control input signal. The MPPT can be controlled by classical PID controllers as well. However, only the design and application of fuzzy logic controller to the MPPT problem will be discussed and used with *the constant power angle approach* in this section.

Once the MPP is determined, the process continues with MPP tracking, which is done by reducing the maximum power angle $\theta_{\mathrm{E}}(k)$. Let us use a Fuzzy Logic Controller for the MPPT. In this case, the tracking error $\theta_{\mathrm{E}}(k)$ and its change $d\theta_{\mathrm{E}}(k)$ will be two input signals to the FLC to ensure minimum tracking angle error. The output space of the FLC is the change $du(k)$ in control signal, which is used to control DC chopper. The input and output variables of the FLC are depicted in Figure 8.15. The output variable $du(k)$ of the FLC is the required change in control signal. This change is added to the previous control signal to obtain the new value. If the tracking error and its change are zero, then the change $du(k)$ will be zero showing no effect on the control signal. The final signal $U(k)$ is used to generate PWM pulses in order to adjust the load voltage by controlling a DC chopper.

The FLC used in MPPT can be designed by assigning the lower and upper boundaries of the input and output universes. Since the reference maximum power angle is $\pi/4$ rad, then the maximum power angle error becomes $\pm\pi/4 = \pm 0.785$ rad. Therefore, the universe of maximum power error, θ_{E}, can be defined between -0.75 and 0.75 rad. Depending on the sampling period, the boundaries of the universe of change in maximum power angle error, $d\theta_{\mathrm{E}}$, becomes much smaller than that of maximum power angle error. The universes of discourse for input and output spaces are given in Figure 8.16.

Both input and output spaces are partitioned into seven subcategories as Negative Big (NB), Negative Medium (NM), Negative Small (NS), Zero (ZE), Positive Small (PS), Positive Medium (PM), and Positive Big (PB). The same type of trapezoidal membership functions are used as the fuzzy subsets in all universes.

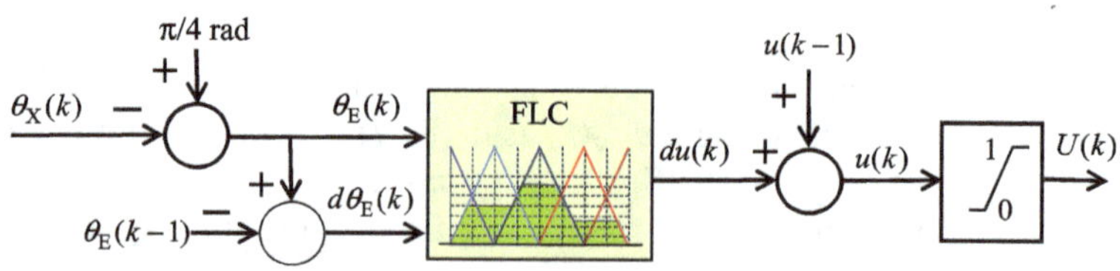

Figure 8.15 Input and output signal routing of the FLC used for MPPT in PV system

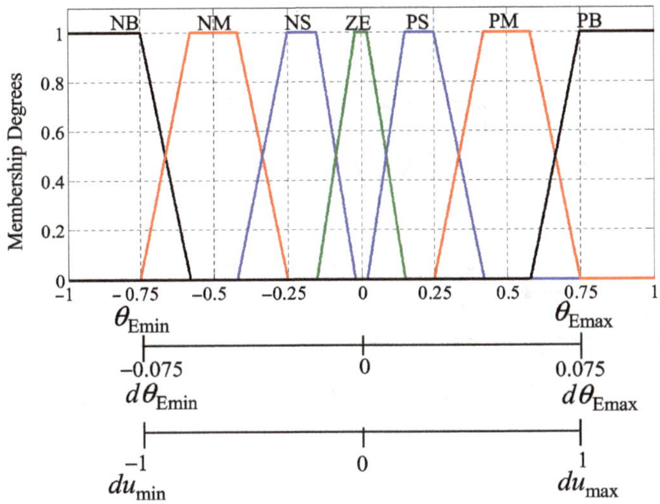

Figure 8.16 The universe of discourse for the I/O spaces of FLC

Table 8.2 A 49-rule table for the FL-controlled MPPT

	$d\theta e(k)$						
$\theta e(k)$	NB_{de}	NM_{de}	NS_{de}	ZE_{de}	PS_{de}	PM_{de}	PB_{de}
NB_e	NB_{du1}	NB_{du2}	NM_{u3}	NM_{du4}	NS_{du5}	NS_{du6}	ZE_{du7}
NM_e	NB_{du8}	NM_{du9}	NM_{du10}	NS_{du11}	NS_{du12}	ZE_{du13}	PS_{du14}
NS_e	NM_{du15}	NM_{du16}	NS_{du17}	NS_{du18}	ZE_{du19}	PS_{du20}	PS_{du21}
ZE_e	NM_{du22}	NS_{du23}	NS_{du24}	ZE_{du25}	PS_{du26}	PS_{du27}	PM_{du28}
PS_e	NS_{du29}	NS_{du30}	ZE_{du31}	PS_{du32}	PS_{du33}	PM_{du34}	PM_{du35}
PM_e	NS_{du36}	ZE_{du37}	PS_{du38}	PS_{du39}	PM_{du40}	PM_{du41}	PB_{du42}
PB_e	ZE_{du43}	PS_{du44}	PS_{du45}	PM_{du46}	PM_{du47}	PB_{du48}	PB_{du49}

The fuzzy rules yield the change $du(k)$ in control signal $u(k)$ using the inputs, which are the maximum power angle error $\theta_E(k)$ and its change $d\theta_E(k)$, in a Mamdani fuzzy reasoning algorithm. The FLC rules are heuristically developed and given in Table 8.2.

Since the stages of fuzzification, fuzzy rule processing, reasoning, and defuzzification of the FLC are explained in Chapter 4, they are not going to be repeated here. The procedure of the FLC is the same as Mamdani fuzzy reasoning algorithm with the center of area defuzzification method. This FL-based MPPT is similar to the one studied and tested in [8] by the author.

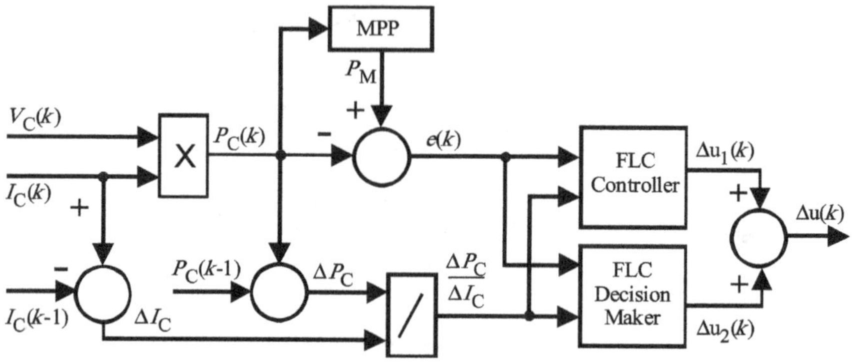

Figure 8.17 Dual FL operational block diagram of the MPPT control scheme

8.4.2 Online load-matching approach

An online load-matching approach for the MPPT in PV arrays is described in this section. As mentioned earlier, after detecting the MPP, it is supposed to be tracked by operating power. A two-stage FLC is designed to keep the operating power of the PV array at its maximum power available. A general block diagram of the MPPT consisting of MPP detector and tracking controller is given in Figure 8.17. Two-stage FLC uses the maximum power operating error and the ratio of the power change to the current change in order to keep the operating point at the maximum. The maximum power operating error is the difference between the maximum and operating power values. Two-stage FLC generates the control signal, u, which is used in the switching process of load and backup batteries. The dual fuzzy logic control scheme comprises one fuzzy logic controller (FLC) and one fuzzy decision maker (FDM). The total control action is the output sum of these two fuzzy processing units. Therefore, it is called as a *two-stage FLC*.

The MPPT described here is based on load and battery backup switching. If the operating power is less than the maximum power determined by the MPP search process, then the loads and backup batteries are switched on or off. The first action may be thought to be taken to increase the array current by switching on additional loads. However, it is not that clear and easy to give a decision when the switching should be on or off since the lower operating power condition occurs both due to high and low currents, which means both heavy and light loads.

As shown in Figure 8.18(a) and (b), when the system starts to operate from a stand-still, initially the array current was zero and the array voltage is the open circuit voltage. Therefore, the operating power is also zero resulting in a large maximum power error as shown by operating region 1 in Figure 8.18(b). The initial error is equal to the maximum power and gets smaller along the operating region 2. Once the operating power becomes equal to the maximum power, the error becomes zero as depicted by region 3 in Figure 8.18(a) and (b). If the load current keeps increasing the operating power becomes smaller and moves toward region 4 due to heavy load. If some of the loads are not switched off, then the operating

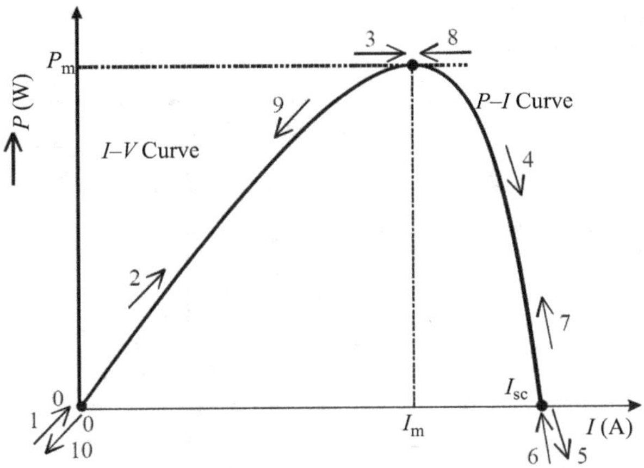

(a) MPP search on the P–I characteristic.

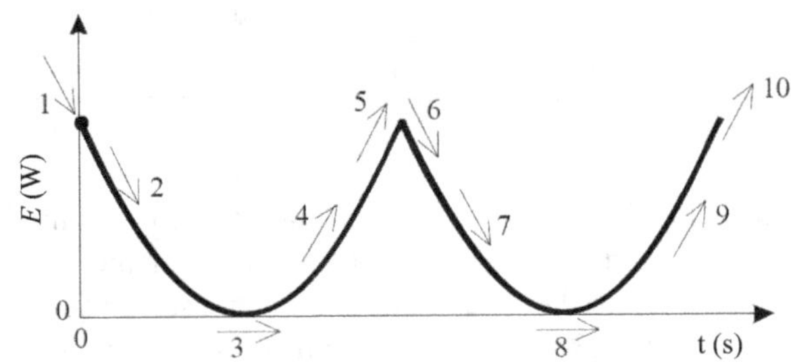

(b) Time response of the MPP operating error.

Figure 8.18 Analysis of operating regions of the PV array for MPPT. (a) MPP search on the P–I characteristic. (b) Time response of the MPP operating error.

power drops down to zero resulting a short-circuited PV array operation as shown by region 5 in Figure 8.18(a) and (b). When some of the loads are switched off, then power will be back starting from region 6 and moving along region 7 toward region 8. Operating regions 8 and 3 are the same MPP and the PV array should be operated at these regions with the correct switching of loads and/or backup batteries. Otherwise, the operating power point moves to the right or left side of the MPP current I_m. If the current is smaller than I_m, then the load is lighter. If the current is greater than I_m then the load is heavy.

As explained above, the maximum power operating can be achieved by either increasing or decreasing the load. However, this process does not guarantee the MPP operation because an optimum load value must be connected across the PV array with the proper switching. In order to get a proper load matching, a fuzzy logic-based MPPT controller is developed and utilized here. The switching commands are generated by the two-stage FLC using a set of operating fuzzy rules, which are obtained using the operating regions shown in Figure 8.18(a) and (b).

Table 8.3 Required maximum power point tracking actions for different locations of the operating power point on the P–V curve

	Operating Regions									
	1	2	3	4	5	6	7	8	9	10
P	0	+	+	+	0	0	+	+	+	0
I	0	+	+	+	I_{sc}	I_{sc}	+	+	+	0
V	V_{oc}	+	+	+	0	0	+	+	+	V_{oc}
ΔP	+	+	0	-	-	+	+	0	-	-
ΔI	+	+	+	+	+	-	-	-	-	-
ΔV	-	-	-	-	-	+	+	+	+	+
ΔP /ΔV	-	-	0	+	+	+	+	0	-	-
ΔP/ΔI	+	+	0	-	-	-	-	0	+	+
e	+	+	0	+	+	+	+	0	+	+
Δu_1	+	+	0	-	-	-	-	0	+	+

In order to determine the control action to be taken, the operating behavior of the PV array should be analyzed. The variation of the array voltage, current, and power can be used to decide whether the load ought to be increased or decreased. The signs of the output signals generated by FLC are determined by the signs of voltage, current, and power changes as described below and summarized in Table 8.3.

The following steps for each operating region describe how the control decisions are made.

Region 1: The starting point. Power and current start increasing as the voltage decreases. Therefore, ΔP and ΔI are positive, ΔV is negative, and error is positive. Thus, the load should be increased by assigning a positive change Δu_1.

Region 2: ΔP and ΔI are positive, ΔV is negative, and error is positive. Therefore, Δu_1 remains positive to increase the load.

Region 3: The peak point. ΔP and error become zero for a short time. ΔI and ΔV are still positive and negative, respectively. Therefore, the load should remain as it is by making Δu_1 zero.

Region 4: ΔP becomes negative. ΔI and ΔV are still positive and negative, respectively, while the error becomes positive. The load requires more power than the PVA can give. Therefore, it should be decreased by assigning a negative Δu_1.

Region 5: The power and voltage are zero resulting in a positive error. The load is too much so the PVA is on the stage of short circuit. ΔP and ΔV are negatives while ΔI is positive. Therefore, the load must be reduced to end the short circuit operation and get to the optimum operating region. Hence Δu_1 must be negative.

Region 6: ΔP, ΔV, and the error are positive while ΔI is negative. The load has started to be decreased by assigning a negative Δu_1.

Region 7: ΔP, ΔV, and the error are still positive while ΔI is negative. The load is decreasing and should be decreased more by keeping Δu_1 as negative.

Region 8: The peak point. ΔP and the error becomes zero again for a short time while ΔI is negative and ΔV is positive. Therefore, the load should remain as it is by making Δu_1 zero.

Region 9: ΔP and ΔI are negative while ΔV and error are positive. Therefore, the load should be increased with positive Δu_1.

Region 10: The power and current is zero. ΔP and ΔI are negative while ΔV and error are positive. The PVA is open-circuited and there is no load connected. Therefore, the load must be connected and increased with positive Δu_1.

As shown in Table 8.3, the sign of Δu_1, which determines the action to be taken for MPPT, is similar to the sign of $\Delta P/\Delta I$. Therefore, the other rows can be eliminated in Table 8.3 yielding only the last three shaded rows. According to the last three shaded rows in Table 8.3, the control signal Δu_1 has the same signs as the ratio $\Delta P/\Delta I$. Therefore, only the signs of $\Delta P/\Delta I$ might be sufficient to maintain MPP operation. However, Table 8.3 does not give information about what happens if ΔI or both ΔP and ΔI are zero. The change in current may be zero while power is changing. The power may remain constant below maximum power or the current may remain constant below or above the optimal value resulting in a maximum power operating error. Thus, the information about the error signal is also needed to generate the required control signal. Therefore, the error signal and the ratio of the power change to the current change are used as two major crisp input signals to both the FLC and FDM parts of the two-stage FLC to generate the output signals for switching. An interpretation of the three shaded rows in Table 8.3 yields the FLC rule table given in Table 8.4 where the fuzzy subsets are defined as negative big (NB), negative small (NS), zero (Z), positive small (PS), and positive big (PB).

The signs of the FDM output signals are also determined in a similar way to cover the gaps left by the FLC due to undefined operating conditions when ΔI or both ΔP and ΔI are zero. The error signal is always positive or zero since the operating power cannot be greater than the maximum power. However, it becomes negative for a short time during the initialization of the MPP search algorithm when any change occurs in the power generated by the PV array is below the *sensitivity constant* α, which is defined in the section called "MPP search." Thus the row representing the negative values of the error is added to the rule table given in

Table 8.4 The rules generated for the FLC

		$\Delta P/\Delta I$				
		NB	**NS**	**Z**	**PS**	**P**
e	PB	Z	PS	PS	PB	PB
	PS	NS	Z	PS	PS	PB
	Z	NS	NS	Z	PS	PS
	NS	NB	NS	NS	Z	PS
	NB	NB	NB	NS	NS	Z

Table 8.5 The rules generated for the FDM

		ΔP/ΔI						
		NB	NS	NUC	Z	PUC	PS	P
e	PB	NB	NS	NB	Z	PB	PS	PB
	PS	NB	NS	NS	Z	PS	PS	PB
	Z	NB	NS	Z	Z	Z	PS	PB
	NS	Z	Z	Z	Z	Z	Z	Z
	NB	Z	Z	Z	Z	Z	Z	Z

Table 8.4 for the FLC and Table 8.5 for the FDM. The operating cases that are not included in the FLC are called negative undefined case (NUC) and positive undefined case (PUC), and are defined as in (8.25) and (8.26), respectively.

$$\text{NUC}: \lim_{\Delta P \to 0^-} \frac{\Delta P}{\Delta I}, \quad \text{for} \quad \Delta I = 0 \text{ or } \lim_{\Delta I \to 0^-} \frac{\Delta P}{\Delta I}, \quad \text{for} \quad \Delta P = 0 \tag{8.25}$$

$$\text{PUC}: \lim_{\Delta P \to 0^+} \frac{\Delta P}{\Delta I}, \quad \text{for} \quad \Delta I = 0 \text{ or } \lim_{\Delta I \to 0^+} \frac{\Delta P}{\Delta I}, \quad \text{for} \quad \Delta P = 0 \tag{8.26}$$

As can be seen in Table 8.5, the structure of the rule table for the FDM is different than the rule table obtained for the FLC as in Table 8.4. Both negative and positive undefined values of the ratio $\Delta P/\Delta I$ are included in Table 8.5 on both sides of the column representing zero. The actions to be taken in the output space of Δu_2 for NUC and PUC are similar to those of NS and PS, respectively. The fuzzy decision is made with respect to the values of e and $\Delta P/\Delta I$ using the verbal structure:

If input then output (8.27)

The Simulink model of the fuzzy decision maker (FDM) is similar to that of the FLC. The only difference is the addition of NUC and PUC columns as shown in Table 8.5. Therefore, the Simulink model of the FDM is not going to be explained since it will be a repeat of the FLC part.

Since commercially available loads are generally manufactured to be operated with a voltage, not greater than the one written on their nameplates or suggested by the manufacturers, a load bus should be created to match the load voltage. A DC–DC chopper can be used to control the PV array output voltage and create a DC load bus with a 40 V constant voltage. In order to have a 40 V average output voltage from the DC chopper, the input PV array voltage must be over 40 V. This is a limitation for the PV array. However, it can be solved using boost-up converters. In order to have a desirable operation, either the PV array should generate a voltage over 40 V or the backup batteries should be used to supply the voltage necessary. Since the backup batteries are used to supply constant voltage and required power for emergency or to the most important loads, it is not necessary to simulate or implement the whole scheme with the MPPT. Therefore, the simulation can be applied when there is sufficient sunlight to generate power that needs to be tracked.

The loads connected to the DC bus may have their own controllers for special operating cases such as the PMDC motor speed control.

Example 8.5

Two types of load are connected to the load bus of the scheme as shown in Figure E8.5(a). One of them is an R–L load and the other one is a PMDC motor load. The R–L load is connected to the load bus directly without any controller. However, the PMDC motor is connected through a speed controller. The motor is modeled to drive a fan-type load with a torque–speed characteristics defined as $T_L = k_1 + k_2\omega_m + k_3\omega_m^2$, where ω_m is the motor speed in rad/s and k_1, k_2, and k_3 are constants. Besides these two loads, a switchable R–L load and backup batteries are also connected. The latter ones are used for maximum power matching to the MPP by the MPPT controller.

(a) Develop a Simulink model of the system given in Figure E8.5(a).
(b) Develop a Simulink model of the MPPT described in Section 8.4.2.
(c) Design an integral controller to control the DC load bus voltage to keep it at 40 V.
(d) Use an FLC to control the speed of the PMDC motor such that the speed remains constant under changing conditions of the solar irradiation.
(e) Simulate models in parts (a), (b), and (c) and observe the performance of the models under the changing effects of solar irradiation level.

Answer 8.5

(a) The MPPT is tested by simulating the system given in Figure E8.5(b), which is a Simulink representation of Figure E8.5(a). The MPPT block is placed right after the PV array. An adjustable load is connected after the MPPT block so that the MPP can be determined by *the online load-matching approach*. The MPPT block, shown in Figure E8.5(c) in detail, generates a control signal for the adjustable load so that the load is controlled to match the MPP under the

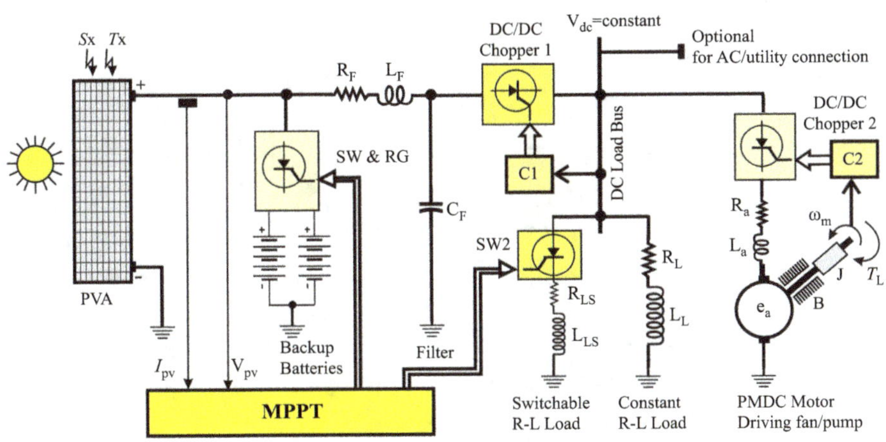

Figure E8.5(a) An MPPT application in stand-alone PV array system

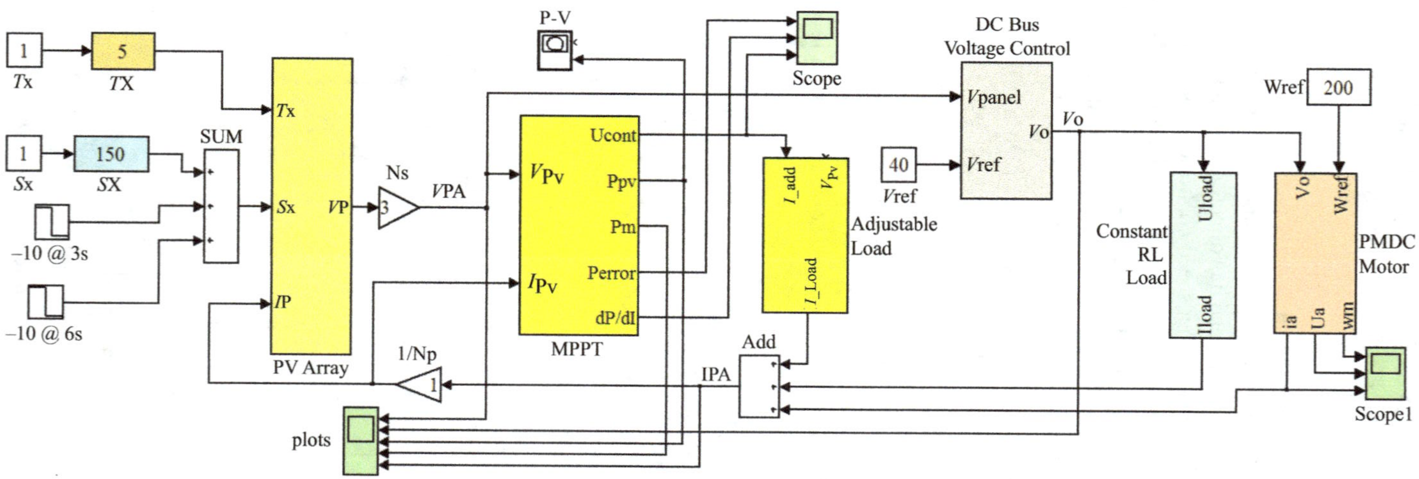

Figure E8.5(b) Simulink model of the PV array system given in Figure E8.5(a)

current operating conditions. Then a constant 40 V DC load bus is established to supply power to a constant R–L type load and a PMDC motor. The DC load bus voltage is controlled and kept constant at 40 V by using an integral controller with the integral gain set to 0.1. The speed of the PMDC motor is also controlled using the FLC developed in Chapter 7 and modified here to meet the speed control specifications of the motor. Since the rated voltage of the PMDC motor is 36 V, the load bus voltage is kept constant at 40 V. The average value of the output voltage of the chopper 1 is compared with 40 V reference voltages and the difference is used by the controller C1 depicted in Figure E8.5(a) to generate the required chopper duty cycle so that the DC bus voltage is kept at 40 V. The duty cycle of the chopper 2 is determined by controller C2 using the error between the reference and the running speeds of the motor.

The total current, I_{PA}, drawn by all the loads and backup batteries being charged is the current drawn from the PV array. This total current is divided by the number of PV panels in parallel to find the current of a single panel so that it can be used to simulate a PV panel model to generate a voltage, which is then multiplied by the number of the panels in series to obtain the PV array output voltage, V_{PA}.

(b) Simulink model of the MPPT is given in Figure E8.5(c). PV array voltage and current are the only two inputs to the MPPT block. The power of the PV array is calculated as the product of the array voltage and current, then used by the MPP detector to find the MPP for the current operating conditions. The power at MPP is then used as the reference power to be tracked by the FLC and FLDM as described in Section 8.4.2. The maximum power point operating error and the rate of dP/dI are the inputs to the FLC and FLDM to generate a control signal Ucont to ensure the MPPT. The performance of the MPPT is shown by the simulation results in part (e).

(c) An integral controller is applied to keep the DC bus voltage constant at 40 V. As the input voltage of the DC–DC chopper in Figure E8.5(a) changes due to solar irradiation, temperature, and loading conditions, the integral controller with a gain of 0.1 is employed to keep the DC output voltage at the reference

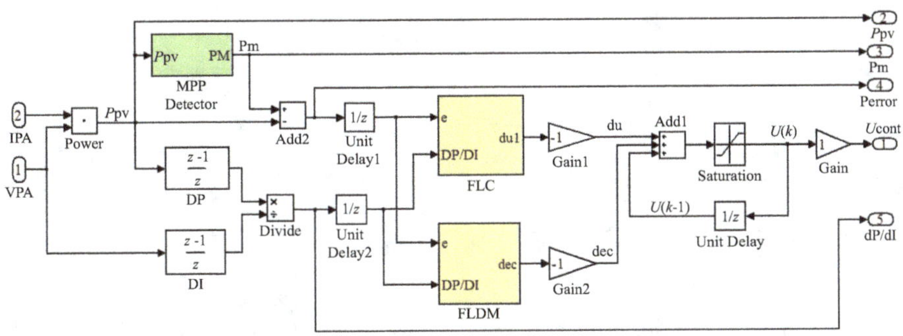

Figure E8.5(c) Inside view of the MPPT block

value of 40 V. Therefore, a 40 V constant DC load bus is generated for the loads to be connected. The performance of the system is given by the simulation results in part (c) below.

(d) A 36 V PMDC motor, whose parameters are given in Example 7.1, is connected to the DC load bus and is supposed to be operated at a constant speed of 200 rad/s. The motor speed should not be affected by the changes in solar irradiation level and loads. Therefore, the FLC block developed in Chapter 7 is used in this example for the speed control of the motor. Since the simulation models of the PMDC motor have been given in Chapter 8, it is not going to be repeated here. The simulation results, however, are given and discussed next in part (e).

(e) The simulation results of the models described in parts (a), (b), and (c) are given in Figures E8.5(d), E8.5(e), and E8.5(f). The effects of the changes in solar irradiation level are observed by giving some stepwise increments and decrements. In the first part of the simulation, the solar irradiation level is started from 150 mW/cm^2 and reduced to 130 mW/cm^2 with 10 mW/cm^2 steps at the time instants 3 s and 6 s. The values of both temperature and solar irradiation levels are arbitrary inputs to the PVA without any measurement and are normally unknown.

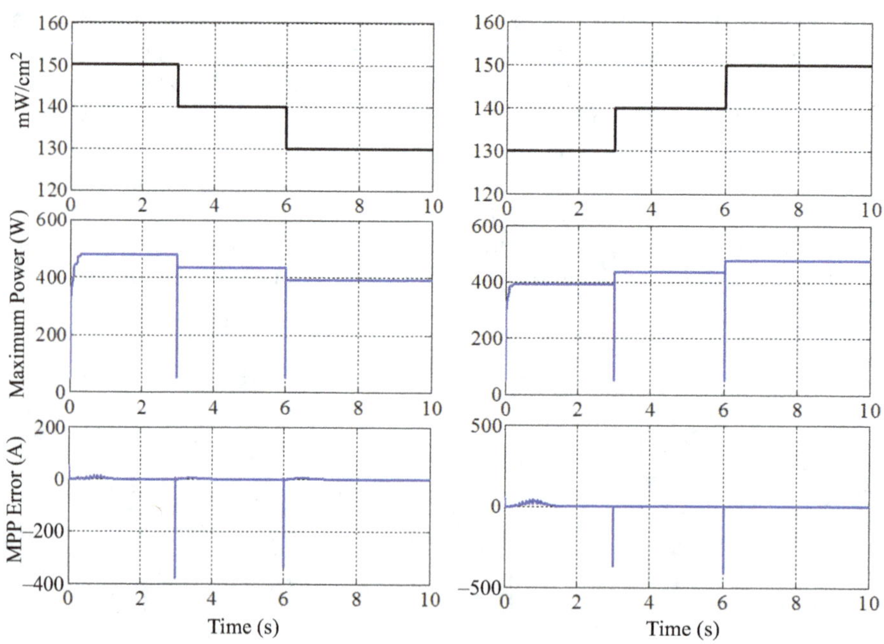

Figure E8.5(d) The changing effects of the solar irradiation level in PV array power

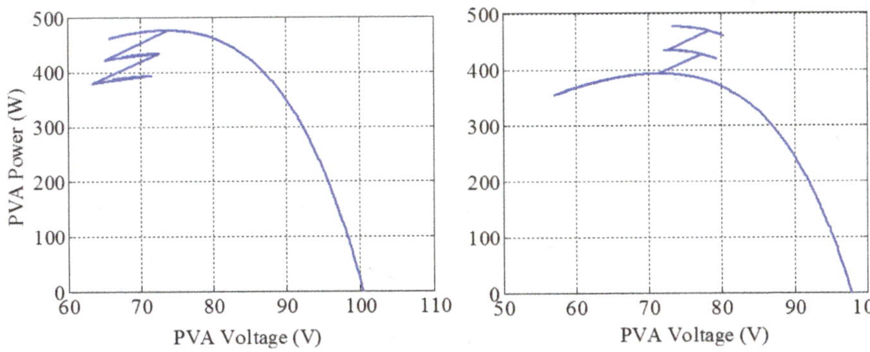

Figure E8.5(e) The path of the operating power point on the PV curves

The solar irradiation level is initially set to 150 mW/cm^2. When the simulation is started, the MPP search algorithm detects the maximum power available. While maximum power detection is underway, the FLC and FDM generate the required control signals so that the PV array operating power tracks the MPP with a minimum error as shown on the left side in Figure E8.5d. At $t = 3$ s, the irradiation level drops to 140 mW/cm^2. Since the negative change is below the sensitivity constant, α, the MPP search algorithm is initialized to search for the new MPP. Once the new MPP is determined, then the two-stage MPPT operates the PV array at this new MPP with the minimum error possible. The effects of the increasing solar irradiation level are given on the right side in Figure E8.5(d). The new MPP is detected as soon as a change occurs in the irradiation level and the MPPT operates to track and operate the system at maximum power available. The zero MPP error shows the performance of the MPPT presented in Section 8.4.2.

The path of the operating power point on the *P–V* curve is shown in Figure E8.5(e). As the solar irradiation level is decreased, the operating power point moves on the new *P–V* curve below while it moves on the new curve above for the case of increasing solar irradiation level. When the operating power point moves on the new *P–V* curve due to changing solar irradiation levels, the MPPT acts and brings the operating power point to the maximum power point.

The DC load bus voltage, motor voltage, and motor speed responses are given in Figure E8.5(f) for decreasing and increasing effects of solar irradiation.

The decreasing irradiation effects are given on the left and increasing irradiation effects are given on the right side in Figure E8.5(f). The DC load bus voltage is controlled and kept constant at 40 V by eliminating the effects of the changes in solar irradiation. In order to operate the motor at a constant speed of 200 rad/s, the motor voltage is adjusted by an FLC-based speed controller. The speed response of the PMDC motor given in Figure E8.5f, shows the performance of the FLC-based speed control. The speed remains constant at 200 rad/s without getting affected by the changes in solar irradiation. The motor voltage also remains unchanged to yield the same speed response under the changing conditions.

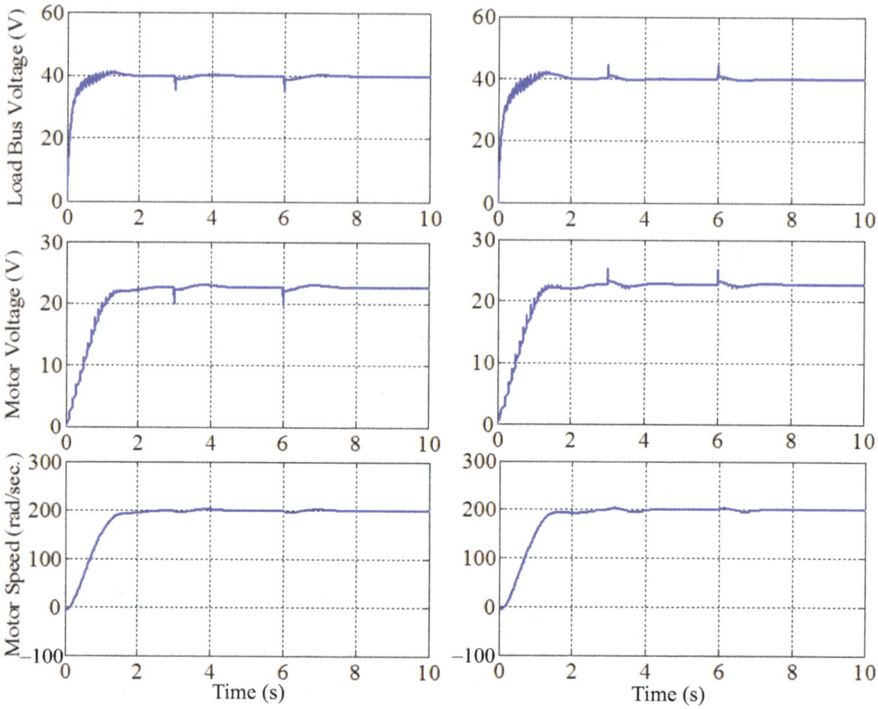

Figure E8.5(f) The responses of DC load bus voltage and motor speed

Design Example 8.2

Modify the PV array models described in Sections 8.2.2 and 8.2.3 to include the shading effects on the PV array characteristics as well as MPPT control algorithms.

Design Example 8.3

A water pumping system for agricultural purposes is planned to be powered by PV panels. Three separately excited DC motors are used for the pumping. The motors are identical and have the nameplate values of 220 V, 4 poles, 2 HP, and 80% efficiency. The excitation voltages of the motors are 220 V DC. The PV array is to be designed so that three motors can be powered at the same time when there is sufficient sunlight as well as individually when there is lower sunlight.

Besides the water pumping motors, a small shelter next to the water pumping system used by the farmers for living and resting is also powered by the same PV arrays. Since there is no sunlight at night, a backup battery unit is used to supply the required power during nights. There is a 500 W refrigerator, 200 W lights, a 500 W water heater, and a 300 W TV in the shelter. All equipment in the shelter operate with 12 V DC.

(a) Obtain the total active power to be supplied by the PV array.
(b) Obtain total reactive power required by the loads.

(c) Draw a single-line diagram to supply electricity to the pumping motors and the shelter from the PV array. Consider all main devices and equipment in this single-line diagram.
(d) Obtain the sizing of the PV array System considering that PV panels are available with 30 V open circuit voltage, 6 A short circuit current, and 120 Wp each under the best solar irradiation level.
(e) Draw the detailed wiring diagram of the PV-powered pumping and shelter system.
(f) Develop a Simulink model for the system described above and operate the PV array at its MPP by designing proper controllers.

Design Example 8.4

The electricity of a fully equipped single house is going to be supplied from PV panels mounted on the roof. The house is also connected to the utility grid and gets electricity from the utility when the sun power is not sufficient. The installed power capacity of the house is 8 kW with a demand power of 5 kW. The appliances require 220 V, 50 Hz single-phase AC voltages.

(a) Design a PV system to supply power to this house such that all the required power is supplied from the PV system ensuring sustainability without considering the utility.
(b) Develop a Simulink model for the whole system and simulate it for performance testing.
(c) Operate the PV array at the MPP by using storage batteries and storing the additional power generated more than the required amount.

8.5 Shading effects in PV arrays

PV arrays consisting of identical or unidentical PV modules are studied in this section. Although the shading effect starts from PV cells, which are connected in series and parallel to make up a module, the shading effect on the PV modules is considered here instead of that of PV cells. For commercial applications, usually, PV modules are available rather than single PV cells. Therefore, this section has focused on shading effects on the modules rather than PV cells. Besides, it is not easy to access the connection ports for connecting any bypass and/or protection diodes in order to eliminate the shading effects because the PV modules are manufactured with industrial processes.

Even though the same type of PV modules are always preferred to be used in PV arrays, characteristics of the PV panels or modules connected in series/parallel in an array may change due to aging, dusting, shading, and/or replacing with unidentical panels [25]. Connecting unidentical PV panels in series or in parallel gives the same effect as shading does [26]. Therefore, the shading effect on PV arrays can be understood by the analysis of an array that consists of unidentical panels. However, let us understand the *I–V* and *P–V* characteristics of a PV array with identical PV panels first.

8.5.1 PV arrays with identical PV panels

The same type of PV panels are preferred to be connected in series and parallel to make up the PV arrays. This is because the currents in series connected panels and the voltages in parallel connected panels must be equal to each as shown in Figure 8.19(a) and (b). The operating points of the panels in the PV array having the same type of PV panels will be the same as given in Figure 8.20 resulting in a smoother operation and easier control of the system. If the panels with different *I–V* characteristics are used in series and parallel combinations their operating points will be different affecting the efficiency and operating properties of all PV installations.

In Figure 8.19, the figure on the left shows a series connected two identical PV panels, and the figure on the right shows the same panels connected in parallel. In a series connection, the same current flows through the panels. Since the panels are identical, their *I–V* characteristics become the same and they overlap as given in

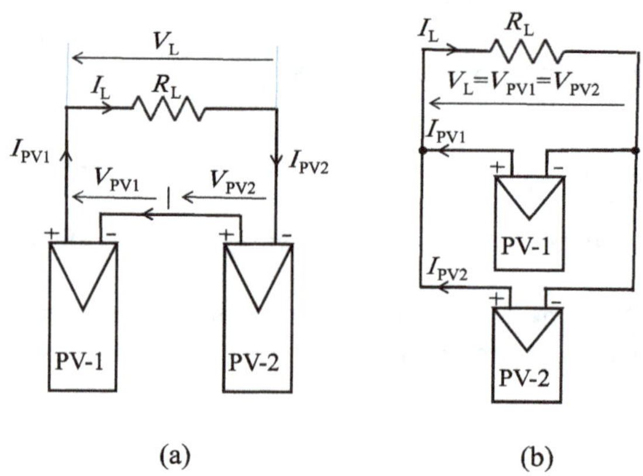

Figure 8.19 (a) Series and (b) parallel configurations of two identical PV panels

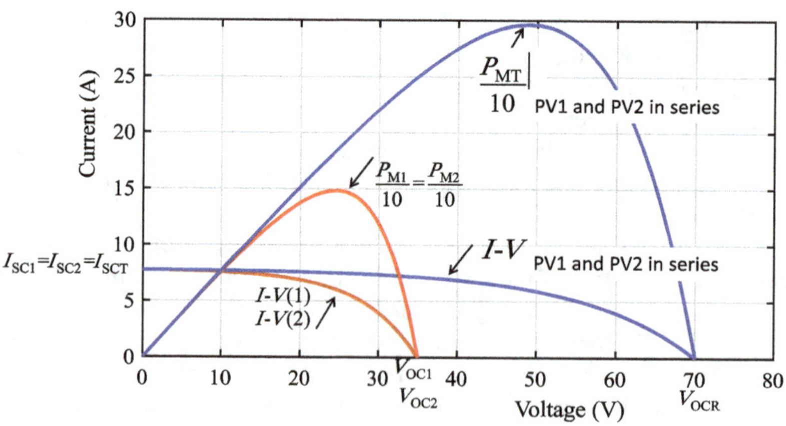

Figure 8.20 I–V characteristics of two identical PV panels connected in series

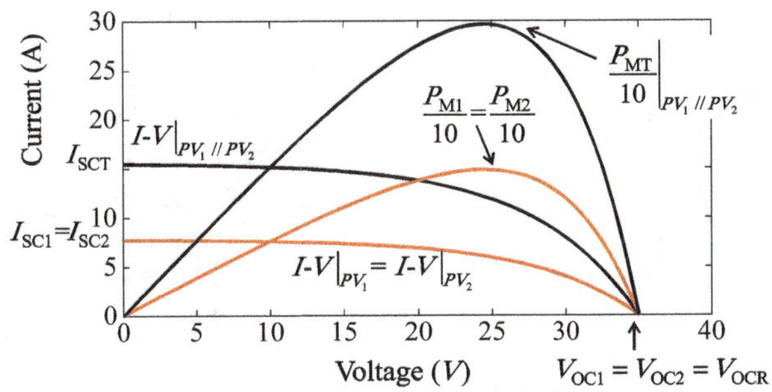

Figure 8.21 I–V characteristics of two identical PV panels connected in parallel

Figure 8.20. The voltage V_L across the resistance R_L is the sum of the voltages of two panels and the total characteristics will be the expansion of *I–V* characteristic of one array toward the right for voltage and upward for the power as shown in Figure 8.20.

Parallel connected two identical PV panels are given in the figure on the right in Figure 8.19, In parallel connection, the same voltage appears across both of the panels and resistive load *R*. However, the current flowing in the resistance R_L would be equal to the total current of the panels connected in parallel. Since the panels are identical, their *I–V* characteristics become the same and they overlap as given in Figure 8.21. Since the current I_L drawn by the resistance R_L is the sum of the currents of two panels, the total characteristics will be the expansion of *I–V* characteristic of one array toward up for current and power as shown in Figure 8.21. It should be noted that the voltage axis remains unchanged.

Even though the same type of PV panels are used in a PV array, the characteristics of the PV panels connected in series/parallel in an array may change due to aging, dusting, shading, and/or replacing with unidentical panels. Connecting unidentical PV panels in series or in parallel gives the same effect as shading does. Therefore, the shading effect on PV arrays can be understood by the analysis of an array that consists of unidentical panels [27,28].

8.5.2 PV arrays with unidentical PV panels

As mentioned earlier, the characteristics of the PV panels or modules connected in series/parallel in an array may change due to aging, dusting, shading, and/or replacement with unidentical panels [25–30]. Connecting unidentical PV panels in series or in parallel gives the same effect as shading does [31–33]. Now, let us see what happens if unidentical PV modules are connected in series or in parallel. Consider series and parallel connected PV modules in Figure 8.19 to be unidentical such that the open circuit voltage and short circuit current of module PV-2 are 10% and 20% more than that of module PV-1, respectively. If the modules PV-1 and PV-2 are connected in series, their individual and combined *I–V* characteristics would be as the ones given in Figure 8.22. Since two PV modules are connected in series as depicted in Figure 8.19, the module currents have to be equal to each other while the total load voltage would be the sum of the module voltages as given by the combined *I–V* characteristics in Figure 8.22.

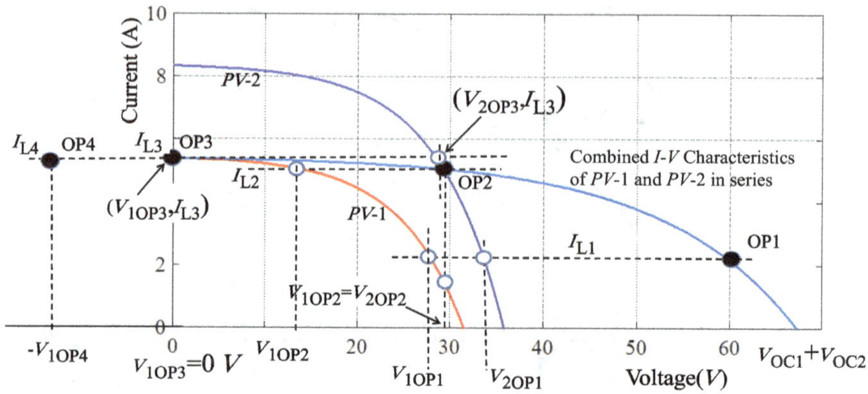

Figure 8.22 I–V characteristics of two unidentical PV panels connected in series

Let us analyze the *I–V* curves for the given four operating points. At operating point 1 (OP1), The module currents are equal to the load current, and the total load voltage is equal to the sum of the module voltages as in (8.28) and (8.29), respectively.

$$I_{L1} = I_{PV1} = I_{PV2} \tag{8.28}$$

$$V_{L1} = V_{1OP1} + V_{2OP1} \tag{8.29}$$

where VL1 is the voltage across the load resistance R_L, V_{1OP1}, and V_{2OP1} are the voltages of models PV-1 and PV-2, respectively, at OP1. At operating point 2 (OP2), the load current and module currents will be equal to I_{L2} while the module voltages are equal ($V_{1OP2} = V_{2OP2}$), which yield load current and total load voltage for OP2 as in (8.30) and (8.31), respectively.

$$I_{L2} = I_{PV1} = I_{PV2} \tag{8.30}$$

$$V_{L2} = V_{1OP2} + V_{2OP2} \tag{8.31}$$

At operating point 3 (OP3), the operating voltages and currents correspond to the points indicated by (V_{2OP3}, I_{L3}) and by (V_{1OP3}, I_{L3}). At this point, the operating voltage of module 1 is zero ($V_{1OP3} = 0$ V) and the load voltage V_{L3} is equal to the voltage of module 2, only.

$$V_{L3} = 0 + V_{2OP2} \tag{8.32}$$

It should be noted that $V_{1OP3} = 0$ V corresponds to the short circuit current of module PV-1 and the load current flowing through series-connected modules cannot be greater than the short circuit current. Therefore, the maximum load current would be equal to the smallest short circuit current of the series-connected modules. Now, let us increase the load current more by decreasing the load resistance value. Since the load current will remain equal to the short circuit current of module PV-1, the polarity of the PV-1 operating voltage reverses and moves toward operating point 4 (OP4) located at the negative side of the voltage axis. The total load voltage under this operating condition becomes as in (8.33).

$$V_{L2} = -V_{1OP2} + V_{2OP2} \tag{8.33}$$

It is clear that the load is supplied by PV-2 only if the operating voltage of PV-1 is zero. The load voltage will be reduced if the voltage of PV-1 is negative and the load voltage will be zero if the voltage of PV-1 is negative and equal to the positive voltage of PV-2. This means that PV-1 acts as another load connected in series with the resistive load and consumes some or maybe all energy generated by the module PV-2.

Module PV-1 has a smaller short circuit current because its *I–V* characteristics differ from that of module PV-2 or it gets lesser solar irradiation than module PV-2. This analysis shows that the PV modules under shading effects may have smaller short circuit currents than those of the modules without shading. Since the modules under shading act as if they are loads and consume the power generated by the other modules that are not under shading, some precautions must be taken to prevent power losses and increase efficiency. One solution to this problem is using bypass and blocking diodes. Thus, the PV modules under shading will be bypassed as if they do not exist. Besides, the blocking diodes will prevent the reverse current flow through the healthy PV modules [34–36]. A connection diagram of bypass and blocking diodes is shown in Figure 8.23. Blocking and bypass diodes are

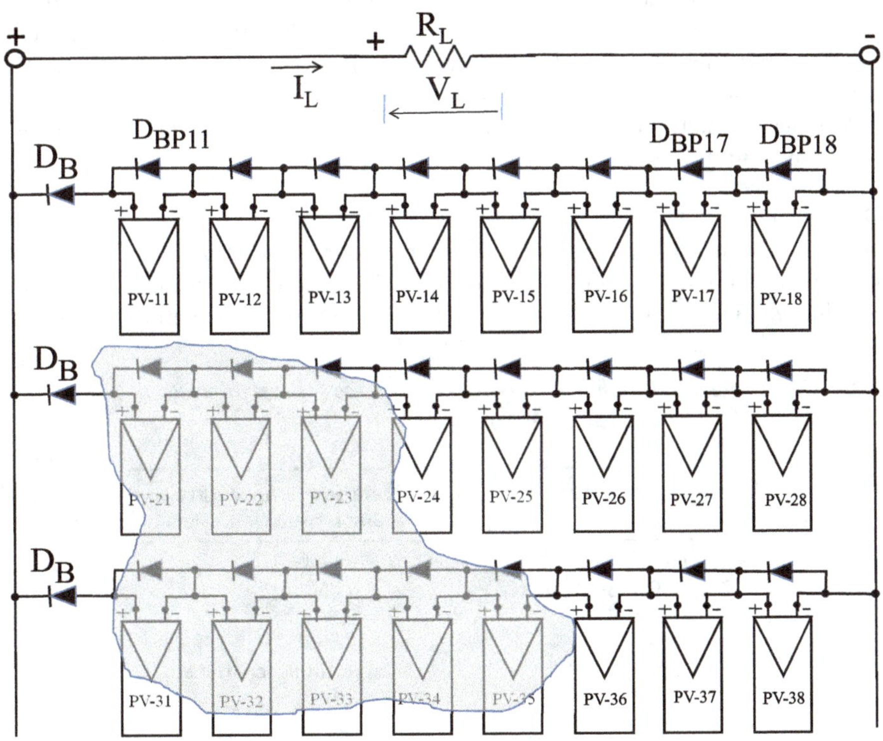

Figure 8.23 A 24-module PV array with fully and partially shaded sections

represented by DB and DBP_{xy}, respectively. The letters x and y in subscripts have been used to depict the locations of the PV modules in the array. For example, x indicates the row number and y is the column number. Similarly, the PV numbers on the PV symbols in Figure 8.23 are also showing the locations of the modules in matrix format.

If a PV module is under shading effect and acts as a load, the bypass diode becomes active and lets the load current continue to flow bypassing the shaded PV module. In this case, the unshaded modules in a row operate normally. However, the shaded modules cause a reduction in the power of related rows.

Example 8.6

Two identical PV modules are connected in series as given in Figure 8.19(a). The panels have the same characteristic data as given in Table E8.6. Use the approximate model of the PV array represented by Figure 8.4 and Equation (8.5) and establish the series-connected PV array.

A MATLAB simulation software code of the approximated model of a PV array is given in Example 8.2. This simulation model includes the effects of the changes in temperature and solar irradiation levels. Modify the MATLAB .m file code given in Example 8.2 to analyze the series connected two modules for the following operating conditions.

(a) Two identical modules are connected in series and there is no shading effect.
(b) Two identical modules are connected in series. Module 1 is under 150 mW/cm^2 solar irradiation level while Module 2 can receive 100 mW/cm^2 since it is under the shade.
(c) Module 2, which is under the effect of the shading is bypassed and only module 1 is feeding the load.

Answer 8.6

Since two identical PV modules are connected in series total open circuit voltage would be the sum of the open circuit voltages of individual modules. Similarly, the operating voltage across the load would be the sum of the operating voltages of

Table E8.6 Characteristic data for the PV array used in Example 8.6

Characteristic data	Data for temperature and solar irradiation effects
$q = 1.106e^{-19}$ C	$\beta = 0.004$
$k = 273 \times 0.138e^{-23}$ J/°C	$\gamma = 0.06$
$R_s = 0.01$ Ohm	$\alpha = 0.2$
$I_0 = 0.051$ A	
IPH = 5.4 A	Curve fitting coefficient
$T_R = 20$ °C	$A = 99$
$S_R = 100$ mW/cm^2	

the individual modules. The module currents would be equal to each other due to the series connection. Both voltage and current equations for series-connected PV modules are defined from (8.28) to (8.33). The MATLAB .m file code given in Figure E8.6(a) can be used for the solution of this example.

(a) Two identical modules are connected in series and there is no shading effect. Under this condition, both PV modules are fully operative and *I–V* and *P–V* characteristics are obtained as in Figure E8.6(b).

```
% 2 identical panels ara in series
% This file can be modified and used for
% Figure 8.22 and Example E8.6
% -----------------------------------
clear
ISC1=5.4; IPH1=ISC1; VOC1=31.2;
SR=150; TR=20;
dS1=0; dT1=0;
dS2=-50; dT1=0;
% --------------------
Sx1=SR+dS1;   Tx1=TR+dT1;
Sx2=SR+dS2;   Tx2=TR+dT1;
VOC1=pvpanel(Sx1,Tx1,0);
VOC2=pvpanel(Sx2,Tx2,0);
VP1=VOC1;
VP2=VOC2;
VLOC=VOC1+VOC2;
% --------------------
[I1,V1,P1]=panelx(Sx1,Tx1,VOC1);
[I2,V2,P2]=panelx(Sx2,Tx2,VOC2);
% --------------------
IL=I1*(max(I2)/max(I1));
VL=V1*(VLOC/VOC1);
V1MAX=max(V1);
V2MAX=max(V2);
VLMAX=V1MAX+V2MAX;
VOMAX=2*VOC1;
VL2=VL*(VLMAX/VOMAX);
PLO=2*V1.*I1;
PL1=(2*V1).*IL;
PL2=IL.*VL2;
subplot(211)
plot(V1,I1,V2,I2,VL,I1,VL2,IL); grid
subplot(212)
plot(V1,P1,VL2,PL2,VL,PLO);grid
```

```
function V=pvpanel(SX,TX,I)
q=1.106e-19;
k=273*0.138e-23;
A=99;        Rs=0.01;
Io=0.051;    IPH=5.4;
TR=20;       SR=100;
BETA=0.004;  GAMMA =0.06;
ALFA=0.2;
CTV=1+BETA*(TR-TX);
CTI=1+(GAMMA*(SX-SR)/SR);
CSV=1+BETA*ALFA*(SX-SR);
CSI=1+((SX-SR)/SR);
IPHX=CTI*CSI*IPH;
VPR=(A*k*TR/q)*log((IPHX-I+Io)/Io)-I*Rs;
V=CTV*CSV*VPR;
```

```
function [I,V,P]=panelx(Sx,Tx,VP)
n=1; IP=0;
while VP > 0
  Vp=pvpanel(Sx,Tx,IP);
  VP=zlimiter(Vp);
  I(n)=IP;
  V(n)=VP;
  P(n)=I(n)*V(n);
  IP=IP+0.01;
  IP1=zlimiter(IP); IP=IP1;
  n=n+1;
end
```

Figure E8.6(a) MATLAB .m file for Example 8.6

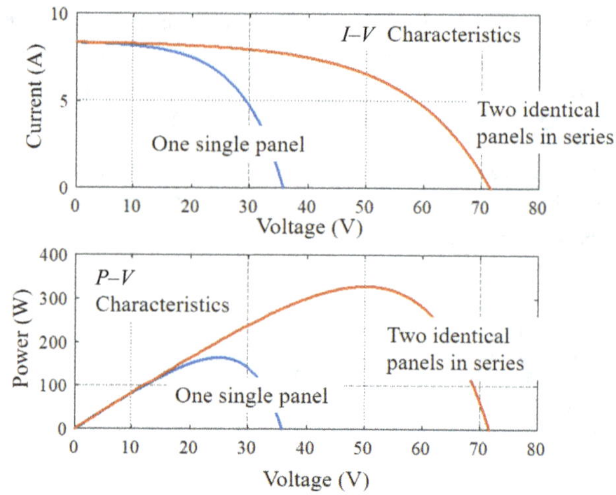

Figure E8.6(b) Two identical panels are in series. There are no shading effects

Since there is no shading, the load voltage would be $V_L = 2V_{PV1}$ and $I_L = I_{PV1} = I_{PV2}$. The power can be obtained as $P_L = (V_L)(I_L)$.

In the case of two identical PV modules connected in series, the total power and load voltage are doubled. This operation mode corresponds to operating a PV array without shading effects as depicted in Figure E8.6(b).

(b) Two identical modules are connected in series. Module 1 is under 150 mW/cm^2 solar irradiation level while Module 2 can receive 100 mW/cm^2 since it is under the shade.

If Module 2 is under the shade effect and receives lesser solar irradiation than that of Module 1, its short circuit current would be smaller and the magnitude of Module 2 current would shrink toward the *x*-axis. Since the open circuit voltage of Module 2 under the shade becomes smaller than that of during unshaded operation, the total open circuit voltage of the PV array with one PV module under the shading effect would be smaller than that of the array during unshaded operation. Since the short circuit current of Module 2 under shading is smaller, the total current of the series connected modules could not be greater than the short circuit current of Module 2 as shown in Figure E8.6(c). Curve (*I–V*)1 represents the current–voltage characteristics of Module 1, which has no shading effects. Curve (*I–V*)2 is the *I–V* characteristics of Module 2, which is partially shaded. Curves (*I–V*)3 and (*I–V*)4 belong to the PV array made of two series connected PV modules. Curve (*I–V*)3 is the *I–V* characteristics of the PV array with unshaded two modules connected in series and curve (*I–V*)4 is the *I–V* characteristics of the same PV array when one of the modules is partially shaded. The *P–V* curves of the same PV array are given in the second part of Figure E8.6(c). The curve (*P–V*)3 is the *P–V* curve of the PV array when there is no shading effect on both of the modules. The *P–V* curve of the PV

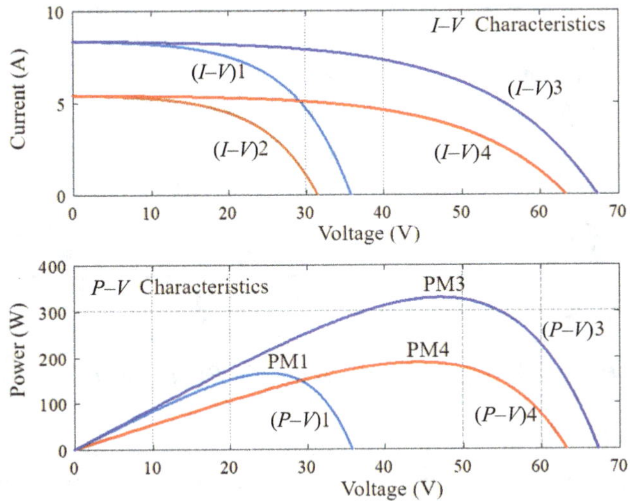

Figure E8.6(c) *Two identical panels are in series. Module 2 is under shading effects.*

array with partially shaded Module 2 is represented by the curve (*P*–*V*)4. Due to the shading effect, the total load voltage and the current have decreased resulting in a decrement the total power. Therefore, the peak power PM3 without shading condition has dropped down to PM4 when one of the modules is partially shaded.

(c) Module 2, which is under the effect of the shading is bypassed and only module 1 is feeding the load.

If Module 2 is bypassed using a bypass diode, then it will act as if it is a wire with zero resistance. Therefore, PV module 1 will be the only source connected across the load resulting in the *I*–*V* characteristics (*I*–*V*)1 and *P*–*V* characteristics (*P*–*V*)1 in Figure E8.6(c).

As it can be understood from Example 8.6, the maximum operating point of a PV array changes if there is full or partial shading over the modules of the PV array. The shading may exist on some modules in a row with series-connected modules or on multiple rows connected in parallel. Depending on connection configuration and shading, more than one maximum power point may occur [37]. Therefore, the challenge of operating a PV array under shading effects becomes harder than the array without shading or uniformly shaded arrays [38]. Since there are more than one maximum power point, the search algorithms may stuck on a local maximum and lose the actual maximum.

8.5.3 MPPT in PV arrays under shading

The use of bypass diodes causes a PV characteristic with multiple local maxima, as shown in Figure 8.24. For a PV array under partial shading to produce maximum

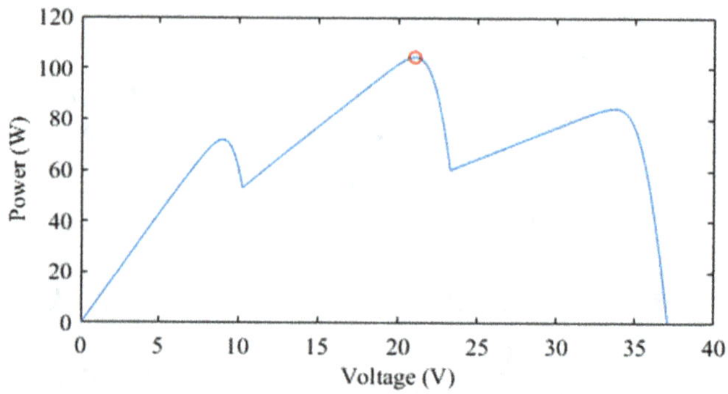

Figure 8.24 Power–voltage characteristics of a PV array under partial shading

power, it must be operated at current–voltage operating point corresponding to the highest global maximum power point by using an MPPT. Some of the maximum power point trackers such as the ones called Reset and Find (RSF) algorithm [10,11], the traditional Perturb and Observe (P&O) algorithm, Incremental Conductance (IC) algorithm, and switching for reconfiguration (SFR) have been introduced by the literature [39–44]. Under homogeneous solar irradiation, RSF and the P&O algorithm are simpler and faster than the IC algorithm. However, all these three algorithms have the disadvantage of getting stuck at the local maximum power point in case of partial shading. In the switching for reconfiguration of the modules, the efficiency can be improved, but the costly hardware requirement has a negative impact on this method. Research studies [45–53] show that the optimization algorithms used for MPPT have advantages and disadvantages in terms of a number of parameters, convergence time, and tracking capability.

8.5.4 MPPT optimization methods for PV arrays under shading

Optimization can be defined as the problem of finding suitable values for the system variables that maximize or minimize the objective function defined under constraints. Optimization methods are classified into two categories as deterministic and stochastic approaches. The deterministic approach includes linear and nonlinear programming methods, while the stochastic approach includes heuristic and meta-heuristic methods [54]. Two population-based meta-heuristic methods namely Symbiotic Organism Search (SOS) and Whale Optimization Algorithms (WOA) are introduced in this section using an example, which is adapted from [57].

Symbiotic Organism Search (SOS) uses a meta-heuristic algorithm such as Artificial Bee Colony and Particle Swarm Optimization to mimic natural phenomena such as the foraging behavior of honeybee swarms and animal swarm behavior. This algorithm searches for the most suitable organism by simulating symbiotic interactions within a paired organism relationship. The SOS starts with an initial population, referred to as an ecosystem, where a group of organisms is randomly generated in the search space. Each organism in the ecosystem is associated with a fitness value toward the desired goal. To generate new solutions for the next iteration, SOS is governed by mimicking the biological interactions of mutualism, commensalism, and parasitism. Every organism interacts with each other in all phases until termination criteria are satisfied. For example, in the mutualism phase, X_i is an organism that pairs with the ith member of the ecosystem, and organism X_j is randomly selected to interact with X_i. Both organisms generate new solutions based on the symbiosis given in (8.34) and (8.35) in order to increase their survival advantage.

$$X_{inew} = X_i + \text{rand}(0, 1)(X_{\text{best}} - \text{Mutual}_{\text{v}}\text{BF}_1) \tag{8.34}$$

$$X_{jnew} = X_j + \text{rand}(0, 1)(X_{\text{best}} - \text{Mutual}_{\text{v}}\text{BF}_2) \tag{8.35}$$

where BF_1 and BF_2 are utility factors representing whether each organism benefits partially or fully from the interaction and are randomly selected as either 1 or 2. The mutual vector representing the relationship characteristic between X_i and X_j is given in (8.36).

$$\text{Mutual}_{\text{v}} = (X_i + X_j)/2 \tag{8.36}$$

X_{best} in (8.34) and (8.35) represents the highest degree of adaptation and reflects the mutual effort to reach a goal that increases the survival advantage of the organism [55].

Whale Optimization Algorithms (WOA) is based on humpback whales, which have a unique hunting method known as bubble net. In the first maneuver, humpback whales dive to a depth of about 12 m and then start swimming toward the surface by forming spiral-shaped bubbles around their moment path. This maneuver is called as upward spiral. The other maneuver, known as multiple loops consists of coral loop, lobtail, and capture loop. The spiral bubble network feeding maneuver has been mathematically modeled for the implementation of the whale optimization algorithm. For example, in the journey encirclement phase, WOA assumes that the current solution is close to optimum because the position of the optimal design is unknown. Once the best agent is identified, the other agents update their positions relative to the best agent using (8.37) and (8.38).

$$\vec{\text{D}} = \vec{\text{C}}.\vec{\text{X}}^*(k) - \vec{\text{X}}(k) \tag{8.37}$$

$$\vec{X}(k+1) = \vec{X}^*(k) - \vec{A} \cdot \vec{D} \tag{8.38}$$

where k is the iteration counter, $\vec{\text{X}}^*$ is the position vector of the most optimal solution obtained, $\vec{\text{D}}$ and $\vec{\text{X}}$ are the position vector. The vectors $\vec{\text{A}}$ and $\vec{\text{C}}$ are computed as in (8.37) and (8.38), respectively.

$$\vec{\text{A}} = 2\vec{\text{a}} \cdot \vec{\text{r}} - \vec{\text{a}} \tag{8.39}$$

$$\vec{\text{C}} = 2 \cdot \vec{\text{r}} \tag{8.40}$$

where $\vec{a}$ is linearly decreased from 2 to 0 across iterations and $\vec{r}$ is a random vector in the interval [0, 1] [56].

An application of the SOS-WOA hybrid optimization method for the MPPT problem under partial shading of PV arrays has been studied by the author and his PhD student Büşra Özgenç in [57,58] and will not be repeated here.

8.6 Problems

P.5.1 A PV array system is used to supply power to single phase 220 V, 50 Hz AC loads. The PV panels used in the array are identical and each one has 30 V open circuit voltage with 150 Wp power maximum when there is maximum solar irradiation (sunlight) available.

(a) Draw a basic circuit diagram model for this PV array system and show all required components used in the system such that the energy sustainability and PV array–AC load system interfacing are also considered.

(b) Explain the purpose of each component you used in your circuit.

P.5.2 A single PV panel is represented by the reference model equation (8.5) with the reference model parameters given as $A = 100$, $R_S = 0.02$ Ohms, $I_0 = 0.001$ A, reference solar irradiation $S_R = 100$ mW/cm^2, and reference temperature $T_R = 20$ °C. Open circuit voltage and short circuit current are given as $V_{OCR} = 40$ V and $I_{SCR} = 6$ A, respectively. Temperature coefficients are $\beta_T = 0.005$ and $\gamma_T = 0.06$. Solar irradiation coefficient $\alpha_S = 0.22$ °C cm^2/mW.

(a) How the I–V characteristics of this PV panel change if the solar irradiation level is increased from 100 mW/cm^2 to 130 mW/cm^2 and then to 160 mW/cm^2? Assume temperature T_A = constant at reference temperature 20 °C.

(b) How the I–V characteristics of this PV panel change if temperature T_A is decreased from 20 °C to 10 °C and then down to 0 °C. Assume that the solar irradiation level is constant at a reference value of 100 mW/cm^2.

(c) Assume that you have PV panels with a 150 Wp power. What would be the PV array sizing if you want to connect it directly to a 1 kW load that operates with 220 V DC.

P.5.3 The equivalent circuit given in Figure 8.2 is used to represent a PV panel. In the circuit, I_{ph} is the generated photocurrent, I_D is the reverse leakage current of the diode, R_{SH} is the shunt resistance representing the I^2R losses due to shunt leakage currents, R_S is the series resistance representing the I^2R losses due to load current. I_C is the panel output current, which is equal to the load current, and V_C is the panel's output terminal voltage.

When this equivalent circuit is developed, only the I^2R-type heat losses are considered. There are two main equations for this equivalent circuit. The current equation, $I_C = I_{ph} - I_0 - I_{sh}^2 R_{sh}$ and the voltage equation, $V_C = V_{ph} - R_S I_C$, where V_{ph} is the generated photovoltage. The PV panels are connected in series/parallel combinations to build a PV array.

Consider and add the following additional facts to the equivalent circuit diagram that may be effective on the PV array operation performance.

(a) If some parts of the PV panels are shaded, what changes should be added to the equivalent circuit?
(b) If some parts of the PV array are shaded, what changes should be added to the equivalent circuit?
(c) If there are some discontinuities between the series- and shunt-connected PV panels, how do you represent these effects on the circuit diagram?

P.5.4 Draw a single-line diagram representing the connection diagram of the power transfer from wind and solar PV systems to the utility grid.

(a) Show all of the required components and devices in your diagram.
(b) Give a short description of each component and device you used in the scheme.

P.5.5 Five identical PMDC motors are being fed from a PV system. The power flowing from the PV system to the load bus where the motors are connected is controlled by a DC chopper. The duty cycle C of the DC chopper may have values between 0 and 1. The chopper can be modeled as a power transformer with the output voltage proportional to the input voltage as $V_O = CV_I$.

The rated armature voltage and currents of the identical PMDC motors are given as $V_a = 120$ V and $I_a = 2$ A, respectively. The armature resistances are 0.1 Ω and inductances are 0.001 H. PV panel data is given as $T_A = 30$ °C, $S_A = 100$ mW/cm^2, $R_S = 0.001$, $I_D = 0.0002$ A, $I_{SC} = 0.25$ A and $I_{PH} = I_{SC}$. The electron charge $e = 1.6 \times 10^{-19}$ C and Boltzmann constant $k = 1.38 \times 10^{-23}$ J/K. The I–V curve fitting factor $A = 80$. Assume constant temperature and irradiation during the operation

(a) Draw an operational block diagram for the system described above.
(b) Assume $C = 1$ and find the number of the PV panels to be connected in series and the number of branches to be connected in parallel when all of the motors are in use.
(c) Develop an FLC to control and keep the DC load bus voltage constant when the motors are turned on and off randomly.
(d) Develop an MPPT to track the maximum power output of the PV array and store the extra power in the batteries when some of the motors are off or the generated power is more than the total power of the motors.
(e) Develop a Simulink model for the system described above and simulate all the conditions asked.

P.5.6 A PV array system is used to supply power to a 110 V DC motor and single phase 220 V, 50 Hz AC loads. The PV panels used in the array are identical and each one has 30 V open circuit voltage with 150 Wp power maximum when there is maximum solar irradiation (sunlight) available.

(a) Draw a basic circuit diagram model for this PV array system and show all required components used in the system such that the energy sustainability and PV array to DC and AC load system interfacing are also considered.

(b) Explain the purpose of each component you used in your circuit.
(c) Develop a Simulink model and simulate this system.

P.5.7 A farmhouse gets its electricity from a PV array 200 m away. The farmhouse requires 6 kW active power and 3 kVAr reactive power. 3 1 kVAr capacitor groups are used to supply this reactive power to the loads.
(a) Draw a single-line diagram of this system showing the PV array, the bus where the PV system is connected, 200 m line, load bus, and loads connected to the generator bus.
(b) Add the single-line drawings of capacitor groups required for this system to your drawing in part (a).
(c) Draw three phases of solidly grounded Wye-connected diagrams of these capacitor groups you draw in part (b).

P.5.8 A PV system will be connected to a 380 V 50 Hz utility grid. Draw a one-line diagram showing all required connections and control blocks. Describe how this system is being operated.

P.5.9 In a 200 house countryside, all houses are separate from each other and have PV panels installed on their roofs. The power utility company buys all of the generated PV power from the houses and sells back energy to them. Answer the following for this system.
(a) What is the name of these sorts of power distribution systems?
(b) What are the main problems in these systems?
(c) What are the main processes in order to manage these systems?
(d) What sort of power quality problems occur in these sorts of systems?
(e) What can be done to improve the power quality?

P.5.10 A PV system is made of five parallel branches with six series connected PV panels in each branch. All PV panels are identical with open circuit voltages, short circuit currents, and maximum power is given as 40 V, 8A, and 300 W/p, respectively.
(a) Determine the total power capacity of this system in terms of voltage, current, and power.
(b) Reconnect whole panels to feed a 220 V, 50 Hz house with the same power range you have obtained in part (a).
(c) Redesign this PV system so that you are supposed to use a 48 V charge regulator to supply power to the house in part (b).
(d) Draw a connection diagram for part (c).

References

[1] Green M.A. *Solar Cells—Operating Principles, Technology, and System Applications*. Englewood Cliffs, NJ: Prentice-Hall, 1982.
[2] Buresch M. *Photovoltaic Energy Systems Design and Installation*. New York: McGraw-Hill, 1983.

[3] Carrasco J.M., Franquelo L.G., Bialasiewicz J.T., *et al.* 'Power-electronic systems for the grid integration of renewable energy sources: a survey', *IEEE Transactions on Industrial Electronics*. 2006;53:1002–1016.

[4] Grau T., Huo M. and Neuhoff K. 'Survey of photovoltaic industry and policy in Germany and China', *Energy Policy*. 2012;51:20–37.

[5] Eltawil M.A. and Zhao Z. 'Grid-connected photovoltaic power systems: technical and potential problems—a review', *Renewable and Sustainable Energy Reviews*. 2010;14:112–129.

[6] Green M.A., Emery K., Hishikawa Y., Warta W. and Dunlop E.D. 'Solar cell efficiency tables (Version 45)', *Progress in Photovoltaics: Research And Applications*. 2015;23:1–9

[7] Ugehes R.O.H, 'Optimum control of sun tracking solar collectors', *Proceedings of the First Workshop on the Control of Solar Energy Systems for Heating and Cooling*, 1978, pp. 69–74.

[8] Altas I.H. 'Control strategies for maximum power tracking and energy utilization of a stand-alone photovoltaic energy system', PhD Thesis, University of New Brunswick, Faculty of Engineering, Department of Electrical Engineering, Fredericton, New Brunswick, Canada, 1993.

[9] Altas I.H. and Sharaf A.M. 'A fuzzy logic power tracking controller for a photovoltaic energy conversion scheme', *Electric Power Systems Research Journal*. 1992;25:227–238.

[10] Altas I.H. and Sharaf A.M. 'A novel on-line MPP search algorithm for PV arrays', *IEEE Transactions on Energy Conversion*. 1996;11:748–754.

[11] Altas I.H. and Sharaf A.M. 'Novel maximum power fuzzy logic controller for photovoltaic solar energy systems', *Renewable Energy*. 2008;33:388–399.

[12] Altas I.H. and Sharaf A.M. 'Novel control strategies via load matching for maximum energy utilization in photovoltaic energy systems', *2007 IEEE Canadian Conference on Electrical and Computer Engineering*, Vancouver, Canada.

[13] Altas I.H. and Sharaf A.M. 'A photovoltaic array simulation model for MATLAB–Simulink GUI environment', *2007 International Conference on Clean Electrical Power*, Capri, Italy, 2007, pp. 341–345, https://doi.org/10.1109/ICCEP.2007.384234.

[14] Rauschenbach H.S., *Solar Cell Array Design Handbook—The Principles and Technology of Photovoltaic Energy Conversion*. Washington, DC: Van Nostrand Reinhold Company, 1980.

[15] Yao Y., Ramshaw R.S. and Bustamante E. 'A review of photovoltaic models', *Canadian Conference on Electrical and Computer Engineering*, 25–27 September 1991, Quebec City, Quebec, Canada, pp. 56.3.1–56.3.4.

[16] Wasynczuk O. 'Dynamic behavior of a class of photovoltaic power systems', *IEEE Power Engineering Review*. 1983;PER-3:36–37, https://doi.org/10.1109/MPER.1983.5519293.

[17] Altas I.H. and Sharaf A.M. 'A solar powered permanent magnet DC motor drive scheme', *Proceedings of the 17th Annual Conference of the Solar Energy Society of Canada*, 21–26 June 1991, Toronto, Ontario, pp. 65–70.

[18] Roger J.A. 'Theory of the direct coupling between D.C. motors and photovoltaic solar arrays', *Solar Energy*. 1979;23:193–198.

[19] Bustamante E., Ramshaw R.S., and Yao Y. 'An alternative model of a photovoltaic array', *Canadian Conference on Electrical and Computer Engineering*, 25–27 September 1991, Quebec City, Quebec, Canada, pp. 56.2.1–56.2.4.

[20] Malathy S. and Ramaprabha R. 'Maximum power point tracking based on look up table approach', *Advanced Materials Research*. 2013;768:124–130.

[21] Ilyas A., Khan M.R. and Ayyub M. 'Lookup table based modeling and simulation of solar photovoltaic system', *Annual IEEE India Conference (INDICON-2015)*, 17–20 December 2015.

[22] Charfi S. and Chaabene M. 'A comparative study of MPPT techniques for PV systems', *5th International Renewable Energy Congress (IREC)*, 2014, pp. 1–6.

[23] Bose B.K., Szczesny P.M. and Steigerwald R.L. 'Microcomputer control of a residential photovoltaic power conditioning system', *IEEE Transactions on Industry Applications*. 1985;IA-21:1182–1191.

[24] Harashima F., Inaba H., Kondo S. and Takashima N. 'Microprocessor-controlled SIT inverter for solar energy system', *IEEE Transactions on Industrial Electronics*. 1987;IE-34:50–55.

[25] Raiker G.A., Reddy S.B., Ramamurthy P.C., Umanand L., Abines S.G. and Vasisht S.G. 'Solar PV interface to grid-tie inverter with current referenced boost converter', *2018 IEEE 13th International Conference on Industrial and Information Systems (ICIIS)*, Rupnagar, India, 2018, pp. 343–348, https://doi.org/10.1109/ICIINFS.2018.8721313.

[26] Bidram A., Davoudi A. and Balog R.S. 'Control and circuit techniques to mitigate partial shading effects in photovoltaic arrays', *IEEE Journal of Photovoltaics*. 2012;2:532–546, https://doi.org/10.1109/JPHOTOV.2012.2202879.

[27] Bharadwaj P. and John V. 'Subcell modelling of partially shaded solar photovoltaic panels', 2017 IEEE Energy Conversion Congress and Exposition (ECCE), Cincinnati, OH, USA, 2017, pp. 4406–4413, https://doi.org/10.1109/ECCE.2017.8096758.

[28] Ramaprabha R. and Mathur B.L. 'Modelling and simulation of Solar PV array under partial shaded conditions', *2008 IEEE International Conference on Sustainable Energy Technologies*, Singapore, 2008, pp. 7–11, https://doi.org/10.1109/ICSET.2008.4746963.

[29] Díaz-Dorado E., Cidrás J. and Carrillo C. 'Discretized model for partially shaded PV arrays composed of PV panels with overlapping bypass diodes', *Solar Energy*. 2017;157:103–115.

[30] Torres J.P.N., Nashih S.K., Fernandes C.A.F. and Leite J.C. 'The effect of shading on photovoltaic solar panels', *Energy Systems*. 2018;9:195–208, https://doi.org/10.1007/s12667-016-0225-5.

[31] Laamami S., Benhamed M. and Sbita L. 'Analysis of shading effects on a photovoltaic array', *2017 International Conference on Green Energy*

Conversion Systems (GECS), Hammamet, Tunisia, 2017, pp. 1–5, https://doi.org/10.1109/GECS.2017.8066212.

[32] Sebbagh T., Kelaiaia R. and Zaatri A. 'An experimental validation of the effect of partial shade on the *I–V* characteristic of PV panel', *The International Journal of Advanced Manufacturing Technology*. 2018;96:4165–4172, https://doi.org/10.1007/s00170-018-1858-4.

[33] Patel H. and Agarwal V. 'MATLAB-based modeling to study the effects of partial shading on PV array characteristics', *IEEE Transactions on Energy Conversion*. 2008;23:302–310, https://doi.org/10.1109/TEC.2007.914308.

[34] Ahmada R., Murtaza A.F., Sher H.A., Shami U.T. and Olalekan S. 'An analytical approach to study partial shading effects on PV array supported by literature', *Renewable and Sustainable Energy Reviews*. 2017;74:721–732.

[35] Balasubramanian I.R., Ganesan S.I. and Chilakapati N. 'Impact of partial shading on the output power of PV systems under partial shading conditions', *IET Power Electronics*. 2014;7:657–666, https://doi.org/10.1049/iet-pel.2013.0143.

[36] Krishna G.S. and Moger T. 'Reconfiguration strategies for reducing partial shading effects in photovoltaic arrays: state of the art', *Solar Energy*. 2019;182:429–452.

[37] Santos P dos., Vicente E.M. and Ribeiro E.R. 'Reconfiguration methodology of shaded photovoltaic panels to maximize the produced energy', *XI Brazilian Power Electronics Conference*, Natal, Brazil, 2011, pp. 700–706, https://doi.org/10.1109/COBEP.2011.6085266.

[38] Sokolov M. and Shmilovitz D. 'Load line emulation based maximum power point tracking', *2008 IEEE Power Electronics Specialists Conference*, Rhodes, Greece, 2008, pp. 4098–4101, https://doi.org/10.1109/PESC.2008.4592595.

[39] Keskin Y., Başoğlu M.E., Tekdemir I.G. and Çakir B. 'Fotovoltaik Sistemlerde Değiştir-Gözle ve Artan İletkenlik Algoritmalarının Karşılaştırılması', *Elektrik-Elektronik-Bilgisayar ve Biyomedikal Mühendisliği Sempozyumu (ELECO)*, Kasım 2014, Bursa.

[40] Jalil F., Saxena R., Ansari M.S. and Ali N. 'Reconfiguration of photovoltaic arrays under partial shading conditions', *Second International Innovative Applications of Computational Intelligence on Power, Energy and Controls with their Impact on Humanity (CIPECH)*, November 2016, India, pp. 193–200.

[41] Babu T.S., Ram J.P. and Dragičević T., 'Particle swarm optimization based solar PV array reconfiguration of the maximum power extraction under partial shading conditions', *IEEE Transactions on Sustainable Energy*. 2018;9:74–85.

[42] Bosco M.J. and Mabel M.C. 'A novel cross diagonal view configuration of a PV system under partial shading condition', *International Solar Energy Society*. 2017;158:760–773.

[43] Belhaouas N., Ait Cheikh M.S., Agathoklis P., *et al.* 'PV array power output maximization under partial shading using new shifted PV array arrangements', *Applied Energy*. 2017;187:326–337.

[44] Fathy A. 'Recent meta-heuristic grasshopper optimization algorithm for optimal reconfiguration of partially shaded PV array', *International Solar Energy Society*. 2018;171:638–651.

[45] Ishaque K., Salam Z., Taheri H. and Shamsudin A. 'Maximum power point tracking for PV system under partial shading condition via particle swarm optimization', *IEEE Applied Power Electronics Colloquium (IAPEC)*, April 2011, Malaysia, pp. 5–9.

[46] Ram, J.P. and Rajasekar N. 'A novel flower pollination based global maximum power point method for solar maximum power point tracking', *IEEE Transactions on Power Electronics*. 2017;32:8486–8499.

[47] Jubaer A. and Zainal S. 'A maximum power point tracking for PV system using cuckoo search with partial shading capability', *Applied Energy*. 2014; 119:118–130.

[48] Guo L., Meng Z., Sun Y. and Wang L. 'A modified cat swarm optimization based maximum power point tracking method for photovoltaic system under partially shaded condition', *International Journal of Energy*. 2018;144:501–514.

[49] Oshaba A.S., Ali E.S. and Abd Elazim S.M. 'MPPT control design of PV system supplied SRM using BAT search algorithm', *Sustainable Energy, Grids and Networks*. 2015;2:51–60.

[50] Rezk H., Fathy A. and Abdelaziz A.Y. 'A comparison of different global MPPT techniques based on meta-heuristic algorithms for photovoltaic system subjected to partial shading conditions', *Renewable and Sustainable Energy Reviews*. 2017;74:377–386.

[51] Jiang L.L. and Maskell D.L. 'A uniform implementation scheme for evolutionary optimization algorithms and the experimental implementation of an ACO based MPPT for PV system under partial shading', *IEEE Symposium on Computational Intelligence Applications in Smart Grid (CIASG)*, December 2014, USA, pp. 1–8.

[52] Manickam C., Raman G.P., Raman G.R., Ganesan S.I. and Chilakapati N. 'Fireworks enriched P&O algorithm for GMPPT and detection of partial shading in PV systems', *IEEE Transactions on Power Electronics*. 2017; 32:4432–4443.

[53] Mohanty S., Subudhi B. and Ray P. 'A grey wolf assisted perturb & observe MPPT algorithm for a PV system', *IEEE Transactions on Energy Conversion*. 2017;32:340–347.

[54] Mashwani W., Haider R. and Brahim Belhaouari S. 'A multiswarm intelligence algorithm for expensive bound constrained optimization problems'. *Complexity*. 2021;2021:1–18. https://doi.org/10.1155/2021/5521951.

[55] Cheng M.Y. and Prayogo D. 'Symbiotic organisms search: a new metaheuristic optimization algorithm', *Computers & Structures*. 2014;139:98–112.

[56] Mirjalili S. and Lewis A. 'The whale optimization algorithm', *Advances in Engineering Software*. 2016;95:51–67.

[57] Özgenç B. and Altaş, İ.H. 'Comparison of MPPT algorithms in PV systems'. *2019 Innovations in Intelligent Systems and Applications Conference*

(ASYU), Izmir, Turkey, 2019, pp. 1–5, https://doi.org/10.1109/ASYU48272.2019.8946396.

[58] Özgenç B. and Altaş, İ.H. 'Implementation of a new maximum power point tracking algorithm for partially shaded PV systems', *7th International Istanbul Smart Grids and Cities Congress and Fair (ICSG)*, İstanbul, Turkey, 2019, pp. 119–123.

Chapter 9
Microgrids and energy management

This chapter has been re-written and re-organized. Subsections about distributed generation, distributed networks, and microgrids have been added to this chapter. Day-ahead forecasting of power to be generated from sun and wind will be added and combined with energy management. Both generation and demand side management and control issues are considered including cybersecurity problems and solution approaches. The fuzzy decision-making and controllers developed in previous chapters will be used in energy management systems in this chapter.

9.1 Introduction

Electrical energy systems can be categorized into five sections as planning, generation, transmission, distribution, and operation. Energy management is considered as one of the parts of operation besides protection, maintenance, and marketing. Energy management deals with keeping generation and consumption in balance by considering system stability, sustainability, efficiency, and cost optimization. The conventional power systems have been divided into several control areas and controlled to maintain stability under overloading, light loading, and faulted operating conditions as mentioned in Chapter 6. The area control is based on automatic generation control of power systems for load matching and stable operation. Power and energy management in conventional power systems have been studied and discussed widely in the literature [1–5] and will not be repeated here. Energy management in distributed power systems is discussed, instead.

The main difference between conventional and distributed power systems is the location of generating units. The power generating units in distributed power systems are located right inside the user distribution networks while they are located far away from the user sites in conventional systems requiring transmission lines to carry the power to where it is needed. A general one-line diagram of conventional power systems is given in Figure 9.1.

The electrical power is generated in large power stations away from distribution networks. During transmission of the generated power from power stations to the distribution networks at user sites, the power transmission lines are interconnected to each other creating a power grid, which acts as an infinite power

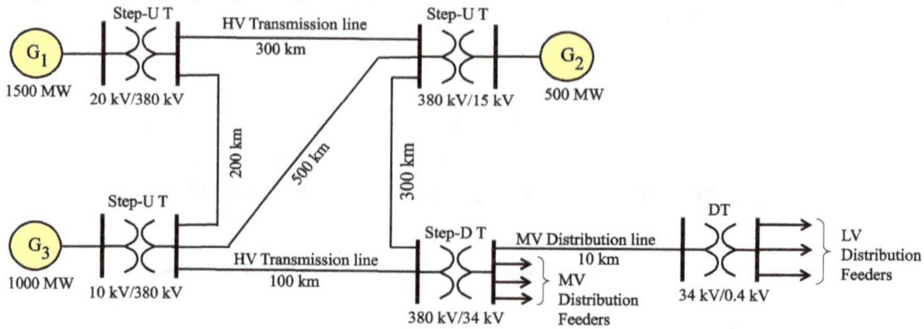

Figure 9.1 One-line diagram of a conventional power transmission and distribution system

source for the distribution networks. Therefore, small or large changes in the distribution network do not affect the infinite power grid except for sudden power outages of big distribution systems. Besides, the total impedances of generators, step-up transformers, transmission lines, step-down transformers, sub-transmission lines, distribution transformers, and distribution lines are large enough to compensate for the sudden changes with small magnitudes between generating units and consumption. Contrary to the conventional ones, small power generation units are placed inside the distribution networks resulting in a different generating system that is called *distributed generation* (DG). The distribution network containing the distributed generation is called *distributed power network*. An active distributed network is basically made up of a collection of DG systems and various loads and is called *microgrid*. A microgrid is operated at a distribution voltage level to supply power for some specific areas or small communities.

The high impedance of long power transmission systems is useful for damping oscillations and small fluctuations in voltage and frequency. However, high impedances between electrical sources and loads cause high power losses, which are not wanted. On the other hand, since there are no high transmission or distribution impedances in microgrids the transmission losses become much smaller compared to the conventional power networks. Besides, the environmental issues could be reduced since the electrical power is generated right where it is consumed. Since there are no high voltage transformers, towers, and lines, the investment costs of microgrids become lower than those of conventional ones [6,7]. The power losses are reduced because of the close distance of distributed power generators to each other and to the consumption units, but transient responses become much faster and affect the voltage and frequency oscillations [8–10], which cause stability problems [11]. Since the distribution lines have small impedances compared to the long transmission lines, the power sources in microgrids cannot be considered as infinite sources anymore because the loads get affected by the changes in distributed generation units and vice versa. Therefore, additional precautions must be taken in distributed generation systems to prevent voltage and frequency oscillations, eliminate harmonics, and increase power quality. Thus, various control and management methods have been developed to deal with these problems [12–14]. Generation and load-side power management are mainly done for energy efficiency, sustainability, power loss minimization, voltage stability, power quality, and operational problems. Control issues of voltage, frequency, power generation, and active power conditioners are also considered as a part of the power management.

Decision-makers are used with the controllers to manage the system operation in terms of power and energy management. The decision-makers act as expert operators to decide what to do in case of events that require actions based on technical and operational knowledge. Therefore, fuzzy decision-makers are used to store the expert's knowledge and decision-making abilities as fuzzy rules, which are processed by fuzzy reasoning algorithms.

A microgrid consists of subsections as electrical power generation units, distribution system apparatus, consumption devices, data transmission, filtering for power quality, management system, communication, controller, and cybersecurity [15,16]. Similar to conventional power systems, the generation and consumption must be kept in a balanced condition for the stability of microgrids. Both generation and consumption have to be managed and controlled. A good energy or power management system starts with correct and usable data before the actions are taken. Therefore, forecasting of the energy to be generated from renewable sources at least 1 day ahead is very important in order to plan and determine the supply and consumption amounts and locations for energy efficiency and sustainability. It has been clear that cybersecurity issues gain importance since any misguided or altered data cause the system to collapse and become inoperable. Hence, the topics related to day-ahead energy forecasting, cybersecurity, and energy management have been included in this chapter to discuss the microgrids with these future aspects.

9.2 Microgrids

As mentioned earlier, a conventional power network consists of generators, step-up transformers, transmission, step-down transformers, subtransmission, distribution transformers, distribution, and consumption as shown in Figure 9.2. Since the beginning of the 2000s, the use of renewable energy sources, especially wind and solar PV, has increased rapidly, resulting in wind and PV installations near consumption units as given on the right side of Figure 9.3. With this change in the conventional power network, wind, and PV become distributed sources in addition to emergency power supply units and other local supplies.

In Figures 9.2 and 9.3, SuT and SdT have meant for step-up and step-down transformers, respectively. DB has been used for the distribution bus-bar and DT is for the distribution transformer. In order to overcome operational problems in distributed networks, an effective power and energy management system, which requires measurement, communication, data processing, and control, must be

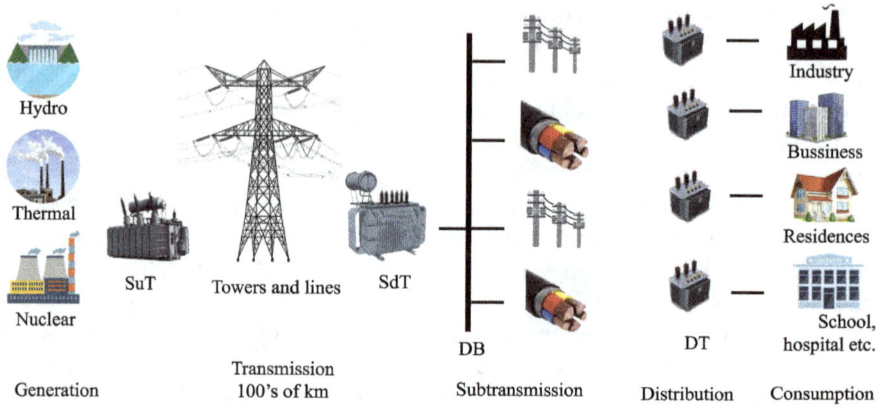

Figure 9.2 A conventional power system from generation to consumption

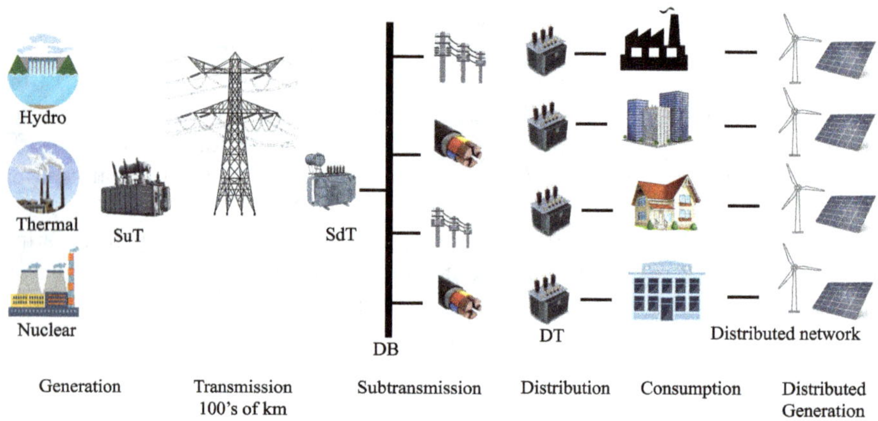

Figure 9.3 Addition of wind and PV sources to conventional power systems

applied. Inclusion of this power and management system in distribution networks gives a new ability to the network and this network is called as a microgrid.

An active distributed network is basically made up of a collection of DG systems and various loads and is called microgrid. A microgrid is operated at a distribution voltage level to supply power for some specific areas or small communities such as the ones listed below:

- Housing sites,
- University campuses,
- Business areas or trade centers,
- Industrial sites,

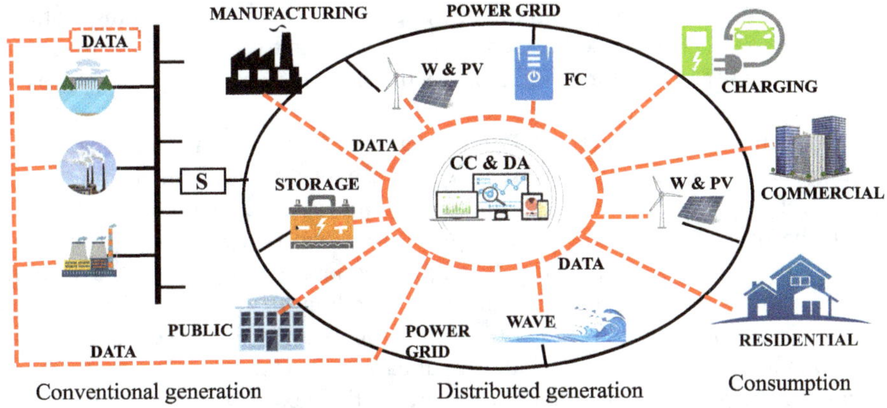

Figure 9.4 A utility-connected microgrid

- Municipal regions,
- Local suburbans.

A schematic diagram of the distributed power network is shown in Figure 9.4, which consists of only renewable energy sources such as solar PV, wind, wave, and fuel cells besides energy storage devices and car charging units. A distributed network also includes other electricity-generating units, which are given below [17,18]:

- Solar photovoltaic/solar thermal
- Wind energy (individual turbines small turbines)
- Micro hydro
- Tidal/wave
- Geothermal
    - Fuel cells
    - Energy storage systems
    - Gensets (variety of fuels)
    - Combined heat and power units
    - Microturbines

This book intends to focus on renewables. Thus, although gensets, combined heat and power units, and microturbines are considered as distributed resources, they are not shown as distributed generating units in Figure 9.4 since they use non-renewable resources. Micro-hydro power generation systems are one of the important renewable sources and are also used as distributed generators. Even if they are not shown in Figure 9.4, it should be noted that the micro-hydropower generating units are always a part of distributed networks and microgrids. The distributed sources used in the microgrid shown in Figure 9.4 are solar PV, wind energy, wave energy, fuel cells, energy storage units, and car charging stations. The car charging stations are generally considered as consumption units because they get the charging power from the power grid. However, car charging stations with storage devices can also be used as power sources to support the power grid during the peak consumption hours when the power from the other sources is insufficient.

The dotted lines in Figure 9.4 are being used to represent the transmission of data, which is collected continuously from every component in microgrid. The collected data is transferred to the data acquisition (DA) and control center (CC) to be processed by energy management system (EMS). After processing the data, the EMS generates some decisions to operate and manage the system. These decisions are sent back to the related components, where the data is collected in the beginning, in order to operate the microgrid more efficiently and securely. The EMS may also use additional data such as meteorological ones for forecasting the power to be able to be generated the next day from the renewable resources.

Microgrids are operated in two modes as utility connected and islanding modes. In utility-connected mode, the microgrid is connected to the utility over an interfacing device such as block *S* in Figure 9.4. If the switch *S* is on, the microgrid operates in utility-connected mode and receives power from the utility. If the switch *S* is off, the microgrid is operated in islanding mode with no electrical connection to the utility. In this case, the consumption in microgrid zone is supplied only from the distributed sources located in microgrid area.

The distributed power sources located in a microgrid district have different operating characteristics. Some of them have DC voltage output, and some of them have AC output. However, the common power grid has to be either DC or AC. This means that there might be two types of microgrids in terms of operating voltage as AC microgrids and DC microgrids. The power sources with different characteristics need a point of common coupling (PCC) to be connected to each other. This PCC would be either a common DC bus or an AC bus. Hence, power electronic interface devices are required to connect the distributed power sources to PCC. Now, let us zoom into the connection points to see the interfacing of the power sources closely.

9.3 Interfacing distributed sources to AC microgrids

Considering various types of renewable energy-based power sources, the common power grid could be AC or DC. Therefore, it would be better to look at the interfacing issue separately for AC microgrids and DC microgrids.

Renewable energy-based power sources such as solar PV and fuel cells have DC output. Wind and wave energy systems may have DC or AC outputs depending upon the type of generators used. Storage devices and car charging stations with storage units have DC outputs, too. These storage devices might be charged from renewable energy sources or from utility grids having fossil fuel-based conventional power sources. In this book, renewable energy systems are the main subject. Thus, only the interfacing of the storage devices charged from renewable energy sources is considered here.

9.3.1 Solar PV systems

Solar PV systems have DC output voltages. However, the magnitude of this DC voltage may not be constant due to changing meteorological conditions. Therefore, a voltage regulator or a DC–DC converter has to be used to have a constant DC voltage as given in Figure 9.5. If a backup battery unit is used then a battery charger with regulator function might be used instead. Besides, an inverter must be used to complete the connection of the PV system to common AC grid [19,20]. The PV system may also be directly connected to the input side of the inverter if the input DC voltage range is large enough. However, direct connection of the PV array output to the inverter is usually not preferred because of the possibility that the voltage magnitudes might be out of the input voltage range of the inverter. Another interfacing scheme of the PV arrays to the common power bus is shown in figure 9.6. The differences between Figures 9.5 and 9.6 are the replacement of the charge regulator by a DC chopper, the addition of the charge controller, and the inclusion of the AC–AC interfacing unit inside three-phase inverter. The AC–AC interfacing unit is a device that is used to check the synchronization conditions and connect two AC systems to each other. This device is generally included in the inverter and synchronizes the inverter output to the common power grid.

The variable output voltage V_{PV} of the PV panel is regulated by the charge regulator to get the voltage V_{DC} as a straight line as depicted in Figure 9.5. The same process can be done by using a DC–DC converter, which is operated as a DC chopper and shown in Figure 9.6. Digital simulation models of DC chopper and three-phase inverter are given later in this section. The DC chopper is controlled to obtain a constant DC voltage as the input voltage to three-phase inverter and to the charge controller. Therefore, the chopper output voltage V_{DC} is used as the feedback voltage to the DC chopper controller. The reference voltage used in the DC chopper controller is the desired input voltage of three-phase inverter. The instantaneous three-phase voltages of the common AC grid are the reference signals of the inverter controller. The controller detects the

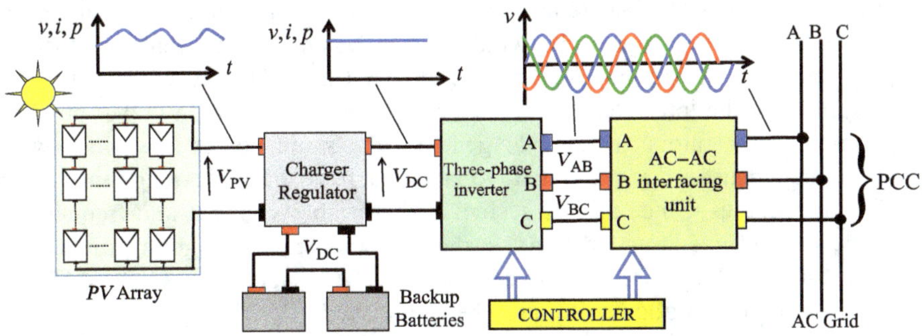

Figure 9.5 Interfacing of a PV array to the common AC grid using a charger regulator

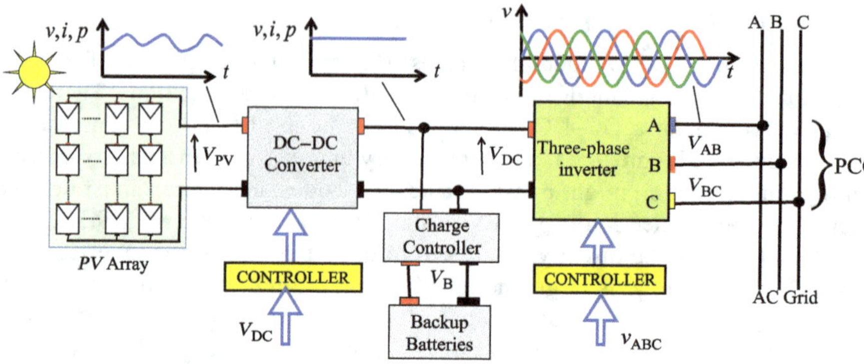

Figure 9.6 Interfacing of a PV array to the common AC grid using DC–DC converter

frequency, phase order, voltage magnitudes, and phase differences of the AC power grid and use them to adjust the inverter output voltages for synchronization and stable operation.

9.3.2 Fuel cells and storage devices

Fuel cells and storage devices are DC sources and their interfacing process to the AC power grid is the same as that of solar PV systems [20]. After the fuel cell or storage devices either a voltage regulator or a controllable DC–DC converter is used to obtain a constant DC voltage, which is the input DC voltage of the inverter. In some cases, if the DC source voltage is compatible with the input voltage of the inverter, then a DC voltage regulator or a DC–DC converter may not be needed [20]. Since the microgrid integration of solar PV system also represents the integration of all other DC sources, separate explanations will not be given.

9.3.3 Wind energy systems

Wind energy systems (WES) are usually considered with asynchronous generators, which generate AC voltages. The wind energy systems have been discussed in Chapter 7 and the same things will not be repeated here. However, the WES is a good example of the interfacing of AC sources to the AC common power bus-bar. Asynchronous generators are able to generate the same power at various wind speeds, which causes the generated voltages to have variable magnitudes and variable frequencies. Since an AC system with variable voltages and frequencies cannot be synchronized to another AC system with constant voltage magnitude and frequency, an AC–DC–AC link between these two AC systems that have different properties, would be a good solution to be established [19–21].

The asynchronous generator is converted to a DC source by using a full-bridge diode or a controllable rectifier. Controllable rectifier is usually preferred to control the magnitude of DC voltage to be applied to the inverter as given in Figure 9.7.

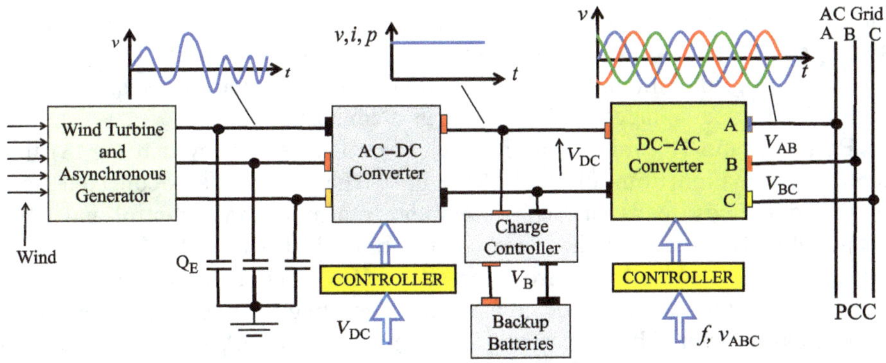

Figure 9.7 Interfacing of a WES to the common AC grid using AC–DC–AC link

The magnitude of the controlled DC voltage is set to a value that gives the desired RMS voltage magnitude of the three-phase AC voltages at the output of the inverter. After adjusting the voltage magnitude by using the rectifier, then the inverter is controlled to give three-phase voltages at the frequency to comply with the common AC grid. Excitation capacitors (Q_E) are being used to supply the required reactive power to the asynchronous generator for excitation purposes.

9.4 Interfacing with DC microgrids

Power distribution in DC microgrids is done by DC common DC power network. A part of a DC microgrid is given in Figure 9.8 where all distributed sources are

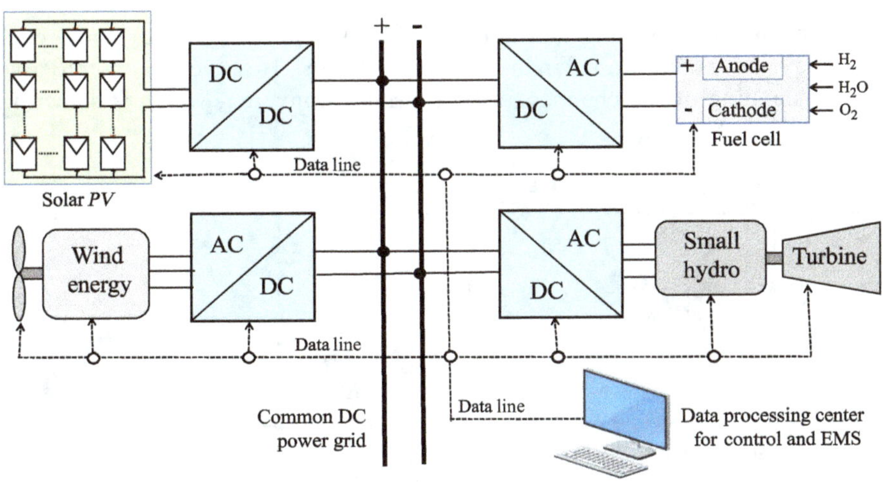

Figure 9.8 A DC microgrid with distributed generation units

connected to a common DC power grid. DC–DC power electronic converters are used to control DC source voltages so that they can be connected to the same common DC grid. Similarly, AC–DC power electronic converters are used to control AC source voltages for DC power grid connection.

Both AC and DC microgrids have energy management system (EMS) that requires information from all parts of the grid. Therefore, data is collected from distributed sources, loads, and converters and transmitted to a control and EMS center, which may have either local or centralized structure. Besides, meteorological data is also collected from the sites where solar PV, wind energy, and small hydro systems are located and used for forecasting the power to be generated the next day or later. This forecasting is an important part of the energy management system, which will be discussed later in this chapter.

As depicted in Figures 9.5–9.8, both AC and DC microgrids require power electronic converters for interfacing distributed sources to the microgrids. Therefore, it would be good to look at some of the power electronics converters used in the interfacing of distributed sources to the grids.

9.5 Modeling DC–DC converters (choppers)

DC–DC converters, which are also called as DC Choppers have been used widely in renewable energy systems in order to obtain a controlled DC voltage at the output from a constant or variable DC input voltage. DC–DC chopper converters are basically categorized into three operational modes [22–25] such as:

- DC–DC buck converters (step-down converters)
- DC–DC boost converters (step-up converters)
- DC–DC buck-boost converters (step-down and step-up converters)

Mathematical modeling and control of buck, boost, and buck-boost converters have been well studied in the literature [20–25]. However, in order to give an understanding of the choppers, a type B buck converter, which is shown in

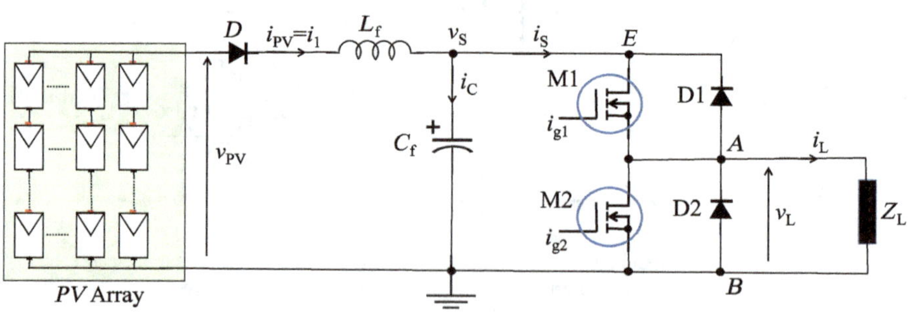

Figure 9.9 A type B DC–DC buck converter circuit

Figure 9.9, is going to be discussed here. Even though the DC–DC converter given in Figure 9.9 is a type B chopper, it is operated in type A mode as a step-down converter. As is well discussed in [26], a type B chopper operates in two quadrants where the load current may be positive or negative, but the chopper output voltage can only be positive.

Since a type B chopper cannot be operated with discontinuous load current, the MOSFETS M1 and M2 are turned on and off simultaneously so that the continuity of the load current is satisfied. There may be a small interval between the end of one gating signal and the beginning of the other; however, it has been concluded in [26] that this interval may not be considered for power circuit analysis of the chopper. The gating signals, the input/output currents, and the output voltage of a type B chopper supplied from a constant DC source with an input filter are given in Figure 9.10 to explain the operation of a type B chopper. The waveform of the PV array current, i_{PV}, is also given to show how the PV array current waveform will differ from that of a constant DC power source.

As shown in Figure 9.10, the load current i_L has a continuous waveform with four parts each one is conducted by different MOSFETs and diodes. In duration T_{ON}, the MOSFET M1 is turned on, and the MOSFET M2 is turned off so that the voltage across the input terminals of the chopper is applied directly to the load

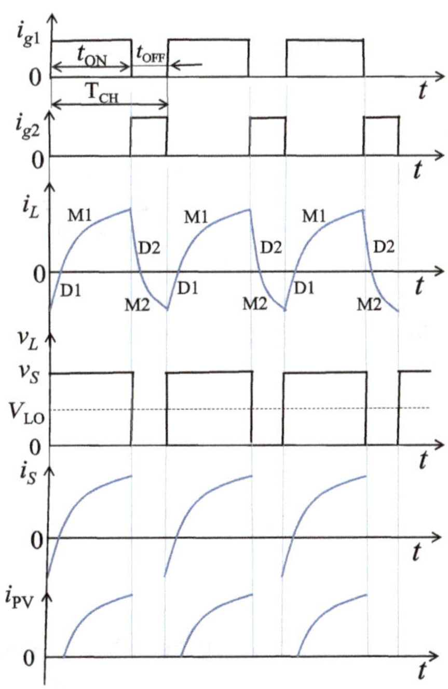

Figure 9.10 Basic waveforms of a type B DC–DC buck converter

terminals marked as A and B. The conducting period T_{ON} of M1 and D2 is determined by applying a controller to get the desired DC output voltage v_{L}. For an ideal DC–DC buck converter the output voltage would have a square wave shape as depicted by v_{L} in Figure 9.10. Due to the filter capacitor, the load current continues to flow in the opposite direction during the blocking mode operation of the chopper. Thus, the current i_{S} has negative parts, but i_{PV} does not because of diode D. The average value, V_{AV}, of the chopper output voltage can be calculated by the integral of the instant value of v_{L} over the chopping period T_{CH} as given in (9.1).

$$V_{\mathrm{AV}} = \frac{1}{T_{\mathrm{CH}}} \int_0^{t_{\mathrm{ON}}} v_{\mathrm{L}}(t)dt \tag{9.1}$$

$$V_{\mathrm{AV}} = \frac{t_{\mathrm{ON}}}{T_{\mathrm{CH}}} V_{\mathrm{S}} \tag{9.2}$$

where V_{S} is the input DC voltage of the chopper circuit. If the input L–C filter is not used, V_{S} would be equal to the PV array voltage v_{PV}. Since the highest value of T_{ON} cannot be greater than T_{CH}, the ratio $t_{\mathrm{ON}}/T_{\mathrm{CH}}$ would be between 0 and 1. The ratio $t_{\mathrm{ON}}/T_{\mathrm{CH}}$ is called chopper duty cycle and is represented by the letter D.

$$0 \leq D = \frac{t_{\mathrm{ON}}}{T_{\mathrm{CH}}} \leq 1.0 \tag{9.3}$$

The chopping period T_{CH} is determined by the carrier wave frequency and kept constant as long as the carrier frequency remains unchanged. Therefore, the duty cycle D is defined as a function of the conducting period t_{ON}, which is obtained by comparing the carrier wave with the control signal u as shown in Figure 9.11. If the control signal u is limited between 0 and 1, then the duty cycle D becomes linearly related to the control signal u. Hence, the conducting period t_{ON} is determined directly by the duty cycle D or the control signal u.

$$t_{\mathrm{ON}} = DT_{\mathrm{CH}} \tag{9.4}$$

$$t_{\mathrm{ON}} = uT_{\mathrm{CH}} \tag{9.5}$$

Substituting (9.5) into (9.29) yields the average output DC voltage as

$$V_{\mathrm{AV}} = uV_{\mathrm{S}} \tag{9.6}$$

An application and detailed model analysis of the chopper given in Figure 9.9 have been discussed in [27].

In Figure 9.9, the potential of point A is higher than point B, and the potential of E is higher than A during the conducting period of M1. Thus, the load current i_{L}

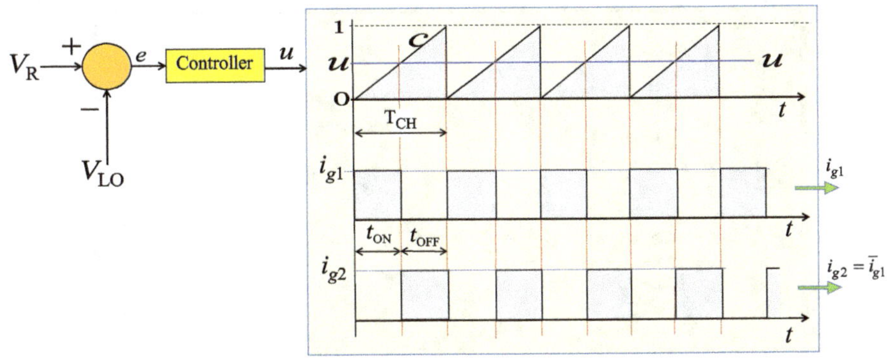

Figure 9.11 Generating the gating signals by a controller

flows through the load circuit and MOSFET M1 and becomes equal to the chopper input current, i_S. When M1 is turned off and M2 is turned on, the voltage across the load terminals, *A* and *B*, becomes zero. However, due to the energy stored in inductance, capacitance, and other elements with inertia, the current i_L does not become zero instantly and continues to flow through the diode D1. Since the MOSFET M2 is already in the conducting mode, the load current, i_L, continues to flow through M2 as soon as it becomes negative. Diode D2 starts transferring the negative load current i_L to the input side of the chopper when M1 is on and M2 is off. This operating sequence is repeated for each chopping cycle, T_{CH}.

As seen in Figure 9.9, the source current, i_S, becomes negative during the conducting period of D2. This negative source current may be reasonable for a normal DC power source. However, the current of a PV array always flows in one direction where the output voltage is always positive. Therefore, the waveform of the PV array current, i_{PV}, differs from the waveform of the current i_s which is flowing from the filter to the load through the chopper.

Example 9.1

Develop a MATLAB® .m file code to generate the gating signals i_{g1} and i_{g2} from a sawtooth carrier signal and a step type control signal $u(t)$ as shown in Figure 9.1. Assume that the related data is given as follows:

Control signal: $u(t) = 0.6$
Chopping frequency: $F_{CH} = 100$ Hz
Chopping period: $T_{CH} = 1/F_{CH}$
Number of chopping periods to be displayed: $nT_{CH} = 10$
Sampling frequency: $f_s = nT_{CH} \times F_{CH}$

Answer 9.1

An alternative MATLAB .m file code for Example 9.1 is given in Figure E9.1(a), which yields the plotting in Figure E9.1(b). In real applications, the control signal $u(t)$ is generated by a controller and it may have transients before reaching its

```
% EXAMPLE 9.1
clear; clc; clf;
FCH=100;          % Chopping frequency
TCH=1/FCH;        % Chopping period
nTCH=10;          % Number of TCH to be displayed
fs = nTCH*FCH;     % Sampling frequency
Tend = nTCH*TCH;  % Ending time (sec.)
n=1; t=0;
while t <= Tend
      x= sawtooth(2*pi*50*t);
      y(n)=(1+x)/1.90;
      u(n)=0.6;
      if u(n) >= y(n)
         YY(n)=1;  XX(n)=0;
      else
         YY(n)=0;  XX(n)=1;
      end
      tt(n)=t; t=t+(1/fs); n=n+1;
end
subplot(311); plot(tt,y,tt,u); grid
subplot(312); plot(tt,1.1,tt,YY); grid
subplot(313); plot(tt,1.1,tt,XX); grid
```

Figure E9.1(a) A MATLAB .m file code to generate gating signals i_{g1} and i_{g2}

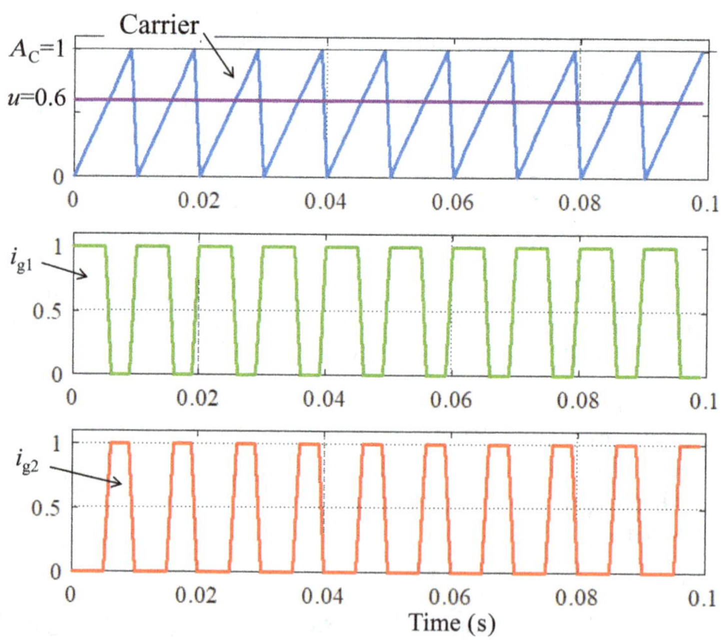

Figure E9.1(b) Gate signals i_{g1} and i_{g2} generated by using .m file given in Figure E9.1(a)

steady-state value. The control signal might have values between 0 and 1 and adjusts the width of the pulses until it reaches the final steady-state value.

Design Example 9.1

Develop a detailed simulation model for the chopper circuit given in Figure 9.9 by utilizing the following steps.

(a) Replace the PV array with a 50 V DC power source and redraw the circuit.
(b) Use a 50 Ohms resistance as the load.
(c) Write circuit equations while MOSFET M1 is operating in conducting mode and MOSFET M2 is operating in blocking mode.
(d) Write circuit equations while MOSFET M1 is operating in blocking mode and MOSFET M2 is operating in conducting mode.
(e) Combine the equations you wrote in parts (c) and (d) for the solution of the whole system equations.
(f) Design a fuzzy logic controller in order to control the load voltage v_L in Figure 9.9 and keep it at a desired constant value of 30 V.
(g) Write a solution algorithm in MATLAB as an .m file and simulate this system.
(h) Plot resultant time responses of load voltage, control error, load current, and gate signals.

9.6 Modeling DC–AC converters (inverters)

A simulation model of an inverter converting the PV array voltage to three-phase AC voltages is studied here. Considering its basic operating principles, the inverter may be treated as a device that converts input DC voltage V_{DC} into AC voltages v_{a0}, v_{b0}, and v_{c0} as represented in Figure 9.12. Assuming that the inverter is controlled only by the frequency control signal which is determined by the controllers using the common power grid frequency requirements. It should be noted that in the case of induction motor control, a constant *V*/*F* ratio scheme is required to ensure satisfactory operation and avoid saturation.

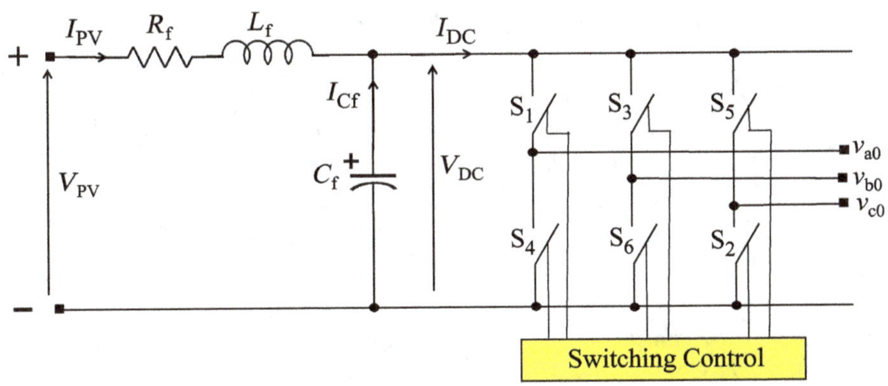

Figure 9.12 A simple representation of the inverter including the input filter

The simulation model of the PWM inverter was developed based on reference frequency control using a simulated triangular carrier wave at a constant frequency of 1000 Hz. The triangular wave, as shown in Figure 9.13, is represented in Equation (9.7), which is a truncated Fourier series.

$$v_C(t) = \frac{A_C}{2} \sum_{n=1(\text{odd})}^{9} \frac{4A_C}{n^2\pi^2} \cos(\omega_C n t) \tag{9.7}$$

where $\omega_C = 2\pi F_C$, $v_C(t)$: carrier wave voltage (V), A_C: peak value of the carrier triangular wave (V), and F_C: carrier frequency (1000 Hz).

Other analog and digital techniques for carrier wave simulation can also be utilized [26]. The resultant value of the carrier wave from (9.7) is compared with the reference three-phase sinusoidal voltage waveforms which are operated at the controlled reference frequency, $F_R = \omega_R/2\pi$, and are expressed as

$$v_{aR}(t) = A_R \sin(\omega_R t) \tag{9.8}$$

$$v_{bR}(t) = A_R \sin\left(\omega_R t + \frac{2\pi}{3}\right) \tag{9.9}$$

$$v_{cR}(t) = A_R \sin\left(\omega_R t + \frac{4\pi}{3}\right) \tag{9.10}$$

where A_R and $\omega_R = 2\pi F_R$ are, respectively, the peak value and the frequency of the reference three-phase voltages v_{aR}, v_{bR}, and v_{cR} shown in Figure 9.13. The reference frequency F_R of the reference phase signals is taken as 50 Hz. A modulation index is defined as $m = A_C/A_R = 1.1$, where A_C is the peak value of the carrier wave, and A_R is the peak value of the reference three-phase voltage signals as depicted in Figure 9.13. The carrier wave, v_C and phase a reference, v_{aR} are compared to yield the pulsating phase a voltage, v_{a0}, as shown in Figure 9.14, where v_{a0} is described in

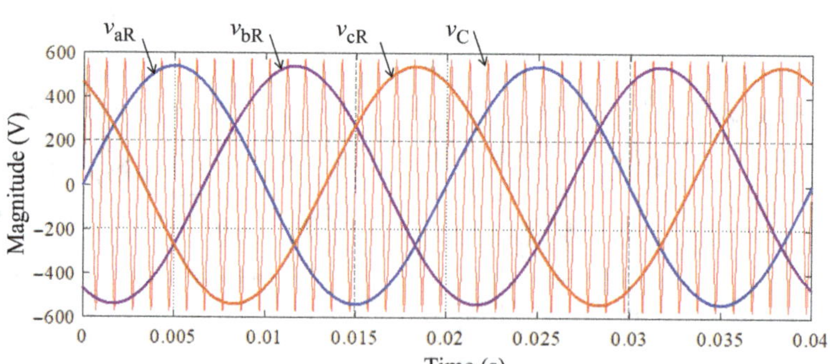

Figure 9.13 Carrier wave v_C and reference signals v_{AR}, v_{BR}, and v_{CR} for three phases A, B, and C

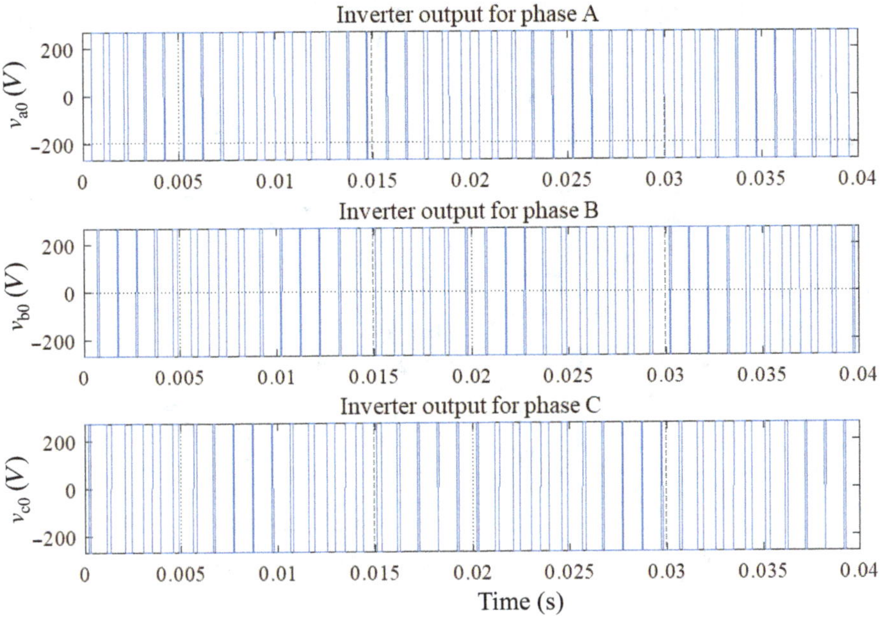

Figure 9.14 Inverter output voltages as pulses generated from carrier and reference waves

terms of the dc input voltage, V_{dc}, of the inverter as follows:

$$v_{a0} = 0.5\ V_{dc}, \quad \text{for } v_{aR} > v_C, \quad \text{where } v_{aR} > 0 \tag{9.11}$$

$$v_{a0} = 0.5\ V_{dc}, \quad \text{for } v_{aR} < -v_C, \quad \text{where } v_{aR} > 0 \tag{9.12}$$

Similarly, the comparison of the carrier wave, v_C, with the phase references v_{bR} and v_{cR}, for the phases b and c, respectively, results in the pulsating phase voltages v_{b0} and v_{c0} for these phases. The width of the pulses also represents the conduction periods of the power electronic switches shown in Figure 9.12.

The pulsating phase voltages are used to obtain the line-to-line voltages v_{ab}, v_{bc}, and v_{ca} as in (9.13), (9.14), and (9.15). The resultant line-to-line voltages are shown in Figure 9.15. Each one of these line-to-line voltage waveforms has a sinewave at reference signal frequency F_R besides other components at the multiples of the reference frequency of 50 Hz.

$$v_{ab}(\mathrm{t}) = v_{a0}(t) - v_{b0}(t) \tag{9.13}$$

$$v_{bc}(\mathrm{t}) = v_{b0}(t) - v_{c0}(t) \tag{9.14}$$

$$v_{ca}(\mathrm{t}) = v_{c0}(t) - v_{a0}(t) \tag{9.15}$$

The MATLAB m file used to obtain Figures 9.13–9.15 is given in Figure 9.16.

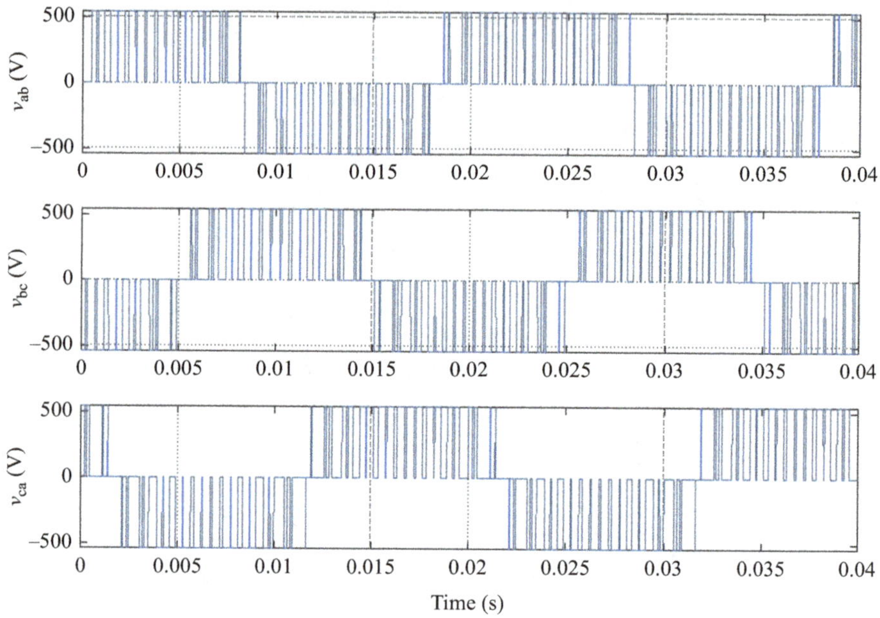

Figure 9.15 Line-to-line output voltages of the three-phase PWM inverter

Example 9.2

Develop a simulation flowchart in order to generate pulsating output signals v_{a0}, v_{b0}, and v_{c0} of a PWM inverter from a triangular carrier signal and reference sinusoid phase voltages.

Answer 9.2

The flowchart given in Figure E9.2 can be used to simulate the PWM inverter.

The flow chart for the inverter is shown in Figure E9.2, where V_{DC} is the DC voltage at the input terminals of the inverter, shown as V_{DC} in Figure 9.12. V_C is the carrier voltage having a triangular waveform at carrier frequency F_C = 1000 Hz, and V_R is one of the reference sinusoid voltages v_{aR}, v_{bR}, or v_{cR}, at the reference frequency F_R which is taken as 50 Hz. V_0 represents the pulsating phase output voltage corresponding to one of the phase voltages v_{a0}, v_{b0}, or v_{c0}.

Design Example 9.2

Develop a MATLAB .m file code to simulate a PWM inverter by generating the pulsating output signals v_{a0}, v_{b0}, and v_{c0} from triangular carrier signal, which is defined by (9.7) and reference sinusoid phase voltages defined by (9.8), (9.9), and (9.10). Assume that the related data is given as follows:

DC voltage input (V_{DC}) = 540 V;
The peak value of the reference signals (Ar) = 540;

```
%-----  Figure 9.13, 9.14 and 9.15.
% Generating Inverter output voltages
clear;   clc;   clf
disp('   Hit enter to continue   ')
% ------------------------------------------
Vd=540; Ar=540;   Fr=50;   Fc=1000;   CYR=2;
% ------------------------------------------
N=9;           Vtr=0;
Aar=Ar;        Abr=Ar;        Acr=Ar;
Ac=1.1*Ar;     dt=0.00001;    k=1;
Tc=1/Fc;       Tr=1/Fr;       Tend=Tr*CYR;
for t=0:dt:Tend-dt
   for i=1:2:N
       Vtr=Vtr+((sin(i*pi/2))/i^2)*sin(2*pi*i*t/Tc);
   end
   Vc(k)=(8*Ac/pi^2)*Vtr;
   Var(k)=Aar*sin(2*pi*t/Tr);
   Vbr(k)=Abr*sin((2*pi*t+2*pi/3)/Tr);
   Vcr(k)=Abr*sin((2*pi*t-2*pi/3)/Tr);
% ------------------------------------
   Vao(k) = pulses(Var(k),Vc(k),Vd);
   Vbo(k) = pulses(Vbr(k),Vc(k),Vd);
   Vco(k) = pulses(Vcr(k),Vc(k),Vd);
% ------------------------------------
   Vab(k)=Vao(k)-Vbo(k);
   Vbc(k)=Vbo(k)-Vco(k);
   Vca(k)=Vco(k)-Vao(k);
   tt(k)=t;    k=k+1;    Vtr=0;
end
plot(0,Ac,tt,Vc,tt,Var,tt,Vbr,tt,Vcr)
xlabel('TIME (sec)'); ylabel('magnitude'); grid; pause;clc;clf
subplot(311); plot(tt,Vao)
title('Inverter output for phase A')
xlabel('TIME (sec)'); ylabel('magnitude'); grid
subplot(312); plot(tt,Vbo)
title('Inverter output for phase B')
xlabel('TIME (sec)'); ylabel('magnitude'); grid
subplot(313); plot(tt,Vco)
title('Inverter output for phase C')
xlabel('TIME (sec)'); ylabel('magnitude'); grid; pause; clf
subplot(311); plot(tt,Vab)
title('Inverter output for phase AB')
xlabel('TIME (sec)'); ylabel('magnitude'), grid
subplot(312); plot(tt,Vbc)
title('Inverter output for phase AC')
xlabel('TIME (sec)'); ylabel('magnitude'); grid
subplot(313); plot(tt,Vca)
title('Inverter output for phase BC')
xlabel('TIME (sec)'); ylabel('magnitude'); grid
```

```
function Vx0=pulses(Vxr,Vc,Vd)
   if Vxr > 0.0
      if Vxr >= Vc
         Vx0 = 0.5*Vd;
      else
          Vx0 = -0.5*Vd;
      end
   end
   if Vxr <= 0.0
      if Vxr <= Vc
         Vx0 = -0.5*Vd;
      else
         Vx0 = 0.5*Vd;
      end
   end
```

Figure 9.16 The MATLAB .m file used to simulate a three-phase PWM inverter

The peak value of the carrier signals (Ac) = 1.1Ar;
Reference frequency (Fr) = 50 Hz;
Carrier wave frequency (Fc) = 1000 Hz;
Number of reference signal period to be displayed (C_{YR}) = 2;
Number of iterations, n, in (9.7) = 9;

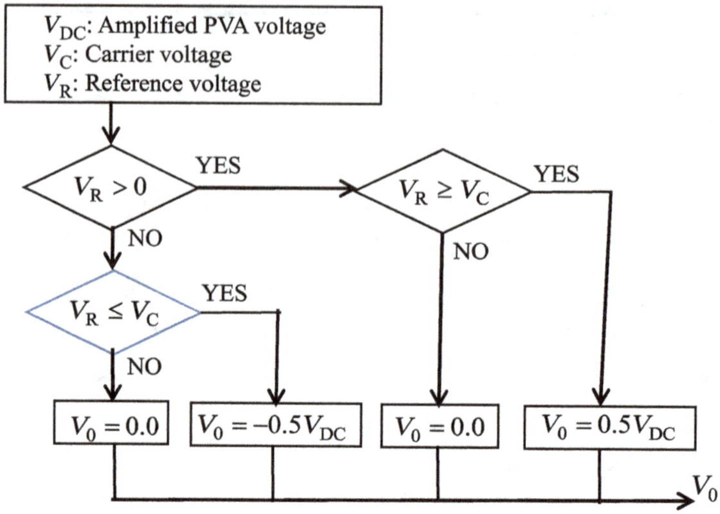

Figure E9.2 Basic flow chart of the inverter model

Sampling period $dt = 0.0001$;
Simulation time $(T_{end}) = C_{YR}(1/F_R)$.

9.7 Distributed generation and control

Electrical power generation from wind and PV power systems keeps increasing with the new installations in many countries. The renewable energy units are installed as large power generating stations, which are usually called wind farm and/or PV farm. The PV systems are also installed on the roofs of residential sites. A schematic one-line diagram of generalized distributed power system is given in Figure 9.17. The generated power from the distributed generation units is collected in a power bus to be distributed to the loads. The residential loads with PV units on their roofs may be powered directly from these units. However, it is usually preferred to sell the generated power to the utility and buy the required amount back from the utility in order to use the generated power more efficiently.

As shown in Figure 9.17, the power generating units and the loads are all connected to the same distribution network such that any sudden changes in loads or in generating units affect the power network. Therefore, load dissipation and power generation are monitored for system management and control purposes. Data is measured from every load and generation side and transmitted to a center where the management and control decisions are made. This control approach is called *centralized control* in distributed generation systems. The centralized control has been used widely in distributed generation [12–15]. However, the centralized control requires secure and fast data transmission from everywhere in the network.

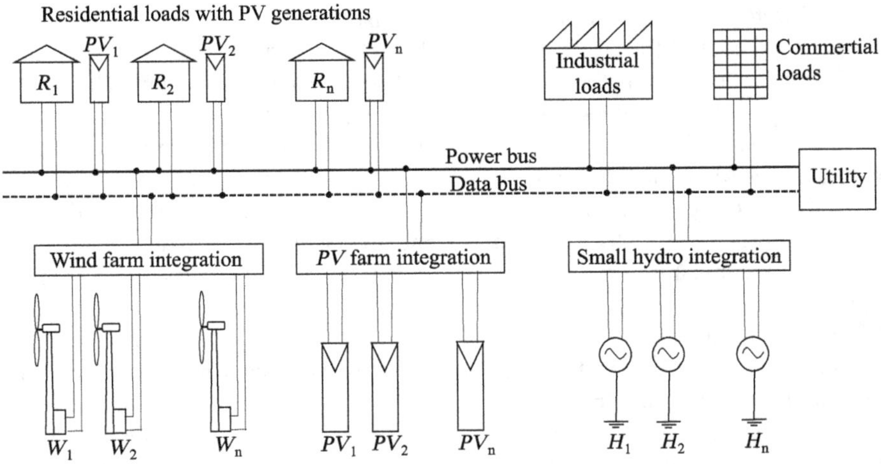

Figure 9.17 A distributed generation with centralized control

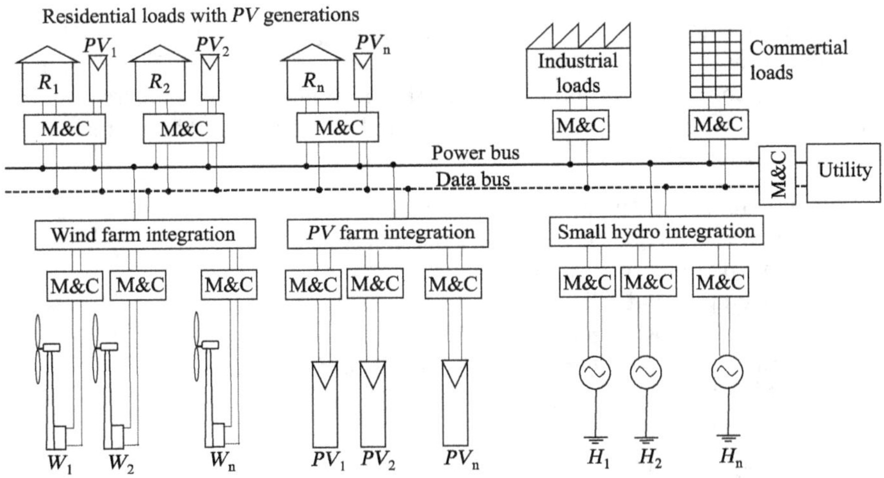

Figure 9.18 A distributed generation with distributed control

The speed of transmitted data is very important in real-time control as data security is very important for both control and management.

The development of smart device technologies has opened a new gate to a control approach that is applicable in remote sites of distributed networks. This method is called *distributed control* and can be used to control load and generation sites right where they are. Since data is used by the controller at the measurement site, it will be fast enough for real-time control as well as for fast decision-making and management. Since data communication is minimized, the sensitive data will be more secure with the distributed control. Figure 9.18 shows a schematic diagram of *distributed control* in power systems. In a distributed control scheme, the measurement and control (M&C) are done locally at every loading and generating unit.

Every data is not transferred to the central storage and control unit. The control and management are processed separately for each load and generation depending on local demand. Smart plugs, switches, and decision-makers are placed on the loads in terms of appliances and devices.

Distributed control has a more flexible structure compared to that of the centralized control. The distributed control and management approach enables the power utility company to reach the customers by offering alternative energy prices to run their loads at shifted times to prevent overloading and power outages because of insufficient generation.

Both schemes given in Figures 9.17 and 9.18 have data measurement, communication, control, and management. Both schemes include decision-making processes and smart devices to operate the distribution network. The distributed power systems with data communication, control, and management algorithms are called *smart grids*. Figures 9.17 and 9.18 show only the power and data measurement and transmission paths for simplicity to focus only on the measurement and control.

The distributed power networks and smart grids are more complicated including measurement devices, protection devices, switchgear connections, transformers, control elements, management algorithms, and computers. An analysis of these two control schemes of distributed generation yields the major problems as follows:

- Integration of the generating units to the distribution network
- The control of the generating units and loads
- Power quality and voltage regulation
- Stable operation and reliability
- Optimum locations of generating units
- The management of generation, distribution, and load sites
- Protection of distribution systems
- Energy backup and storage systems
- Data communication
- Cyber security in data transmission

Many other issues can be added to these listed ones. For example, the integration of distributed generation to utility and loads includes many other sub-issues such as power electronic converters, power filters, and their controls. Each one of these issues requires a solution so that the system can be operated smoothly.

9.8 Energy management in a renewable integration system

Solar PV, wind, hydrogen fuel cells, and small hydropower systems are used separately as well as together with hybrid systems. Even if renewable energy is environmentally friendly with unlimited sources, its implementation schemes require interface units that affect both installation costs and the quality of power transferred to load or utility grid. Power electronic converters are used for interfacing PV, wind turbines, fuel cells, and microturbines with utility. Switching and control techniques of the interfacing devices are important for power quality as well as for efficient and maximum power utilization. In fact, the use of power electronics is growing rapidly not only in renewable energy systems but also in power systems and promises to change the landscape of future power systems in terms of generation, operation, and control.

It is widely accepted that distributed generation is a very important energy option in the near future. Most of the distributed energy resources require a power electronic converter to interface with the load and utility. The most important application areas of power electronics can be considered as distributed generation, flexible AC transmission systems (FACTS), and power quality. FACTS are important due to two main reasons: (1) deregulation of utility requires precise control of power flow by FACTS and (2) growing energy demand coupled with difficulty in constructing more transmission lines requires FACTS devices to enable increased power flow in existing lines. FACTS are one of the important topics that should be considered as a part of a distributed generation system that consists of the hybrid wind/PV/fuel cell and small hydropower generating units.

9.8.1 Centralized control of distributed renewable energy systems

The control of distributed generation is done mainly by two approaches, which are called centralized control and distributed control. In the centralized control approach, all the measurements are transferred to a central computer unit, and the control or management decisions are made centrally by processing the algorithms in this computer. The centralized control of distributed renewable energy systems is discussed using a distributed renewable energy integration scheme shown in Figure 9.19, which includes hybrid wind, photovoltaic (PV), small hydro, and fuel cell (FC) renewable energy schemes. The integration consists of a FACTS scheme topology, controller structure, and interface converters for:

(i) Maximum energy utilization
(ii) Low-cost implementation
(iii) Conversion interface scheme robustness, effective control to minimize the impact of the renewable energy source on the host electrical grid, or satisfactory performance for stand-alone operation
(iv) Effective coordination and protection schemes for distributed energy systems.

The control schemes may include classical PID, fuzzy, neural fuzzy, and other intelligent approaches with adaptivity. Derivative free heuristic optimization methods such as genetic algorithms, particle swarm optimization, ant colony, etc. may be

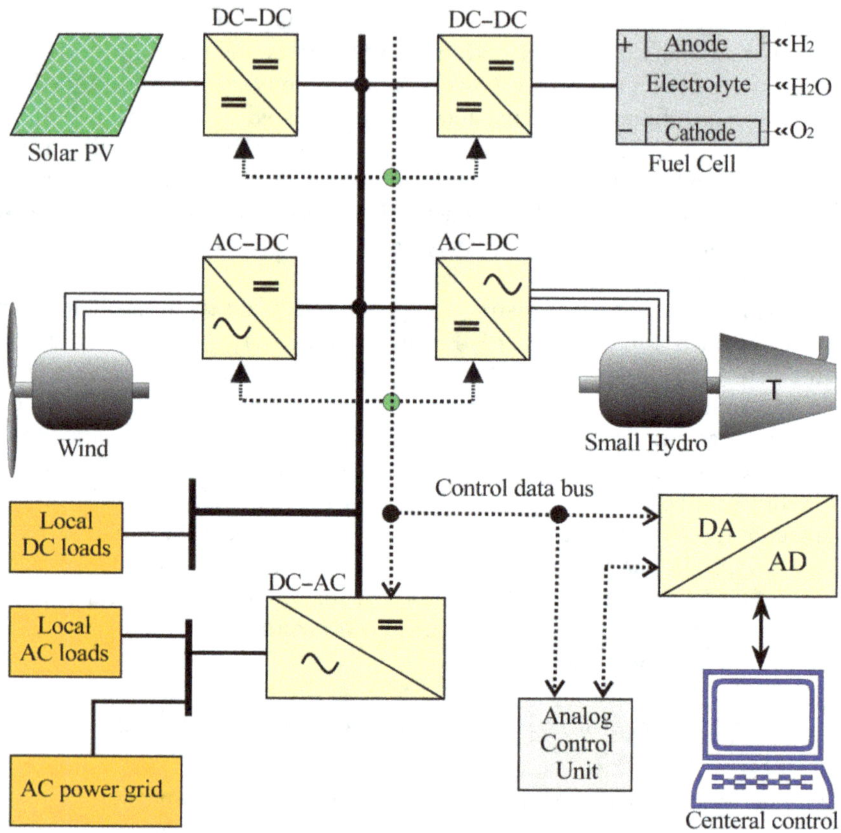

Figure 9.19 Distributed renewable energy system

included for optimum parameter tuning of the controllers and decision-makers for maintaining power quality, system stability, and operational improvements.

The stand-alone wind scheme is fitted with a novel modulated voltage/frequency (*V*/*F*) stabilization network, while the PV unit is fully controlled for combining maximum available PV solar energy utilization. Matching criteria are utilized to ensure a fully integrated hybrid wind, PV, FC, and small hydro renewable, low cost, and low maintenance functional system design.

The sample distributed renewable energy system is controlled by the centralized control approach. Therefore, all data from generating units and loads are collected and stored in a computer to be used for control and management purposes. The dotted lines in Figure 9.19 represent the data flow from measurement sites to central control unit and vice versa. In this system, it has been assumed that all four generating units, which are PV, wind, FC, and small hydro are located at the same site so that their output powers can be collected on a common DC bus, which acts as DC power pool. The FC in this system is used as an energy storage unit beside the backup batteries.

The energy management of the system given in Figure 9.19 is done by defining the primary duties of the generating units and the loading conditions. The availability of renewable sources is also another key parameter. The energy management system can be separated into three parts:

- Source management
- Load management
- Source and load management

Source management deals with the scheduling and using the electricity from renewable sources without any scheduling on the load side. Whenever a load is turned on or off, the generation is managed and controlled to supply the required power. Since the generated power from renewable sources depends on the availability of the generation, a decision has to be made under the current circumstances. A strategy to design the generation side management system can be developed as follows by assuming that getting energy from the utility grid is the last choice.

(i) The FC should be kept charged if the generation from the other sources is sufficient.
(ii) The power from the PV system should be used at its maximum capacity as long as there is enough sunlight.
(iii) The wind power should be used to feed the loads if the PV system fails to supply.
(iv) The FC should be used if the power generation from PV and wind fails to supply.
(v) Small hydro should be used if PV, wind, and FC fail.
(vi) The utility grid should be used if all the others are not sufficient.

After deciding the priority use of the sources, a fuzzy rule-based system can be developed to process the source-side management system. Possible operating cases for the source management in the sample distributed system are summarized by the fuzzy rules given below.

If load is VL and PV is SU then PV is source and others to storage
Else if load is VL and W is SU then W is source and others to storage
Else if load is VL and H is SU then H is source and others to storage
Else if load is VL and FC is SU then FC is source and others to storage
Else if load is VL and (PV and W) are SU then (PV and W) are source and others to storage
Else if load is VL and (PV and H) are SU then (PV and H) are source and others to storage
Else if load is VL and (PV and FC) are SU then (PV and FC) are source and others to storage
Else if load is VL and (W and H) are SU then (W and H) are source and others to storage
Else if load is VL and (W and FC) are SU then (W and FC) are source and others to storage

Else if load is VL and (H and FC) are SU then (H and FC) are source and others to storage

Else if load is VL and (PV and W and H) are SU then (PV and W and H) are source and nothing to storage.

Else if load is VL and (PV and W and FC) are SU then (PV and W and FC) are source and nothing to storage.

Else if load is VL and (PV and H and FC) are SU then (PV and H and FC) are source and others to storage.

Else if load is VL and (W and H and FC) are SU then (W and H and FC) are source and nothing to storage.

Else if load is VL and (PV and W and H and FC) are SU then (PV and W and H and FC) are source and nothing to storage

Else if load is VL than UG is source and nothing to storage.

The similar rules are written for the cases when load is LO, ME, HI, and VH.

The abbreviations used in the fuzzy rules to define the load power requirement are described as very low(VL), low (LO), medium (ME), high (HI), and very high (VH). The sufficiency of the sources is represented by the fuzzy subset sufficient (SU). The source names are abbreviated as PV, W, H, FC, and UG for photovoltaic, wind, hydro, fuel cell, and utility grid, respectively. The verbal fuzzy variables are represented by triangular fuzzy subsets as depicted in Figures 9.20 and 9.21.

The decision-maker in this sample case is using a Sugeno-type fuzzy reasoning algorithm. There are two input spaces, which are load demand power and renewable source power. The input space representing demand power is shown in Figure 9.20. The universe of renewable source power from PV, wind, hydro, and FC are shown in Figure 9.21 as the second input space.

The input space representing the renewable sources has four subspaces PV power, wind power, hydropower, and FC power. The crisp axis represents the generated power by the renewable sources are in per unit (pu). The base values of these universes are the maximum generated power value possible. For example, the base value of the PV power is the highest power that can be generated by the PV

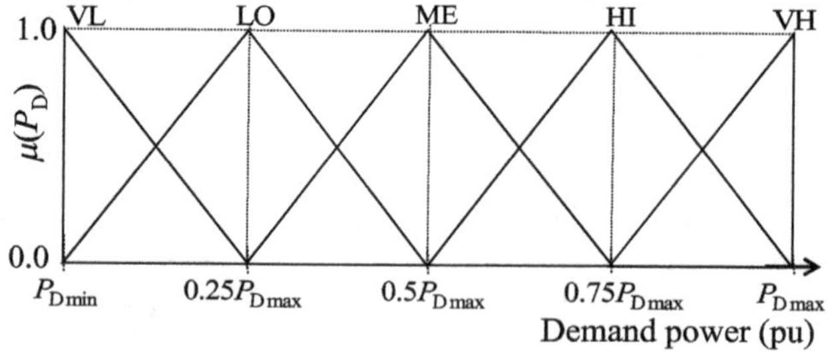

Figure 9.20 One of the input spaces represents the load demand power

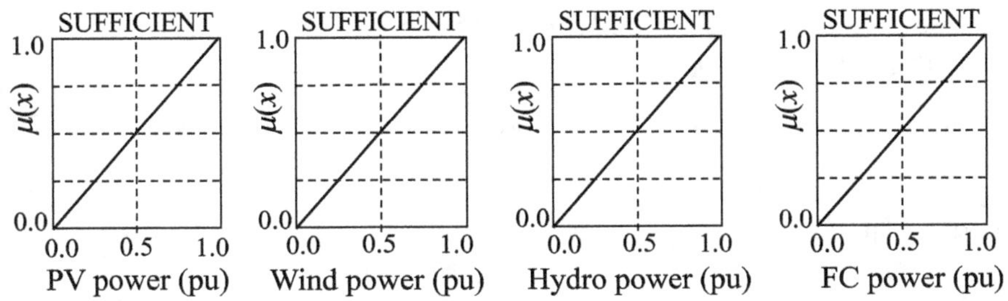

Figure 9.21 The input space represents the renewable source power

unit. Similarly, the base power for the wind energy conversion (WEC) system is the maximum power that the WEC system can generate. The similar base values are assigned for hydro and FC powers.

Depending on demand power size and availability of the renewable source power the decision on the output space is made. For example the decision in the output space for the rule

If load is VL and PV is SU then PV is source and others to storage

will be written as in (9.16) and (9.17).

$$P_G = w_{pv}P_{PV} \tag{9.16}$$

$$P_S = w_w P_W + w_h P_H \tag{9.17}$$

where P_G is the required generated power, P_{PV} is the PV power available at the moment, and w_{pv} is a weighting factor equal to the membership degree of P_{PV} in the fuzzy subset *sufficient*. P_S is the power to be stored, which is equal to the weighted sum of the powers P_W and P_H from wind and hydro, respectively. The weighting factors w_w and w_h are the membership values of the wind and hydro powers, respectively, in the fuzzy subset *sufficient*. The decision in the output space for the rule

Else if load is VL and (PV and H) are SU then (PV and H) are source and others to storage

can be written as in (9.18) and (9.19).

$$P_G = \left(P_{D\max} - w_{pv}P_{PV}\right) + w_h P_H \tag{9.18}$$

$$P_S = (1 - w_h P_H) + w_w P_W \tag{9.19}$$

where P_{Dmax} is the maximum value of the demand power, which is supplied from PV first with the amount determined by the weighting factor w_{pv} and the rest is completed by the hydro with the amount defined by w_h. The output of the fuzzy rule given below can be calculated by using (9.20) and (9.21).

Else if load is VL and (PV and H and FC) are SU then (PV and H and FC) are source and others to storage

$$P_G = \left(P_{D\max} - w_{pv}P_{PV}\right) + \left[\left(P_{D\max} - w_{pv}P_{PV}\right) - w_h P_H\right] + w_{fc}P_{FC} \quad (9.20)$$

$$P_S = (1 - w_h P_H) + (1 - w_{fc}P_{FC}) \quad (9.21)$$

Since all of the rules can be implemented in the same way, there is no need to develop and write all output equations for the rest of the rules. The reader can easily develop the output space equations for the remaining rules by following the pattern used in (9.16)–(9.21).

Example 9.3

A hybrid renewable power management system is given in Figure E9.3(a), which has been adapted from the PhD thesis [28] done by Onur Ozdal Mengi under the supervision of the author.

The system consists of a PV unit, a wind power generating unit, and a utility grid as hybrid electrical sources. These three generating units are connected to a DC power pool over the required converters. A backup battery group is also connected to PV system in order to store the extra generated solar power when all generated power from the PV is not delivered to the load. AC load types are considered in the system and they are connected to AC power bus, which is fed from

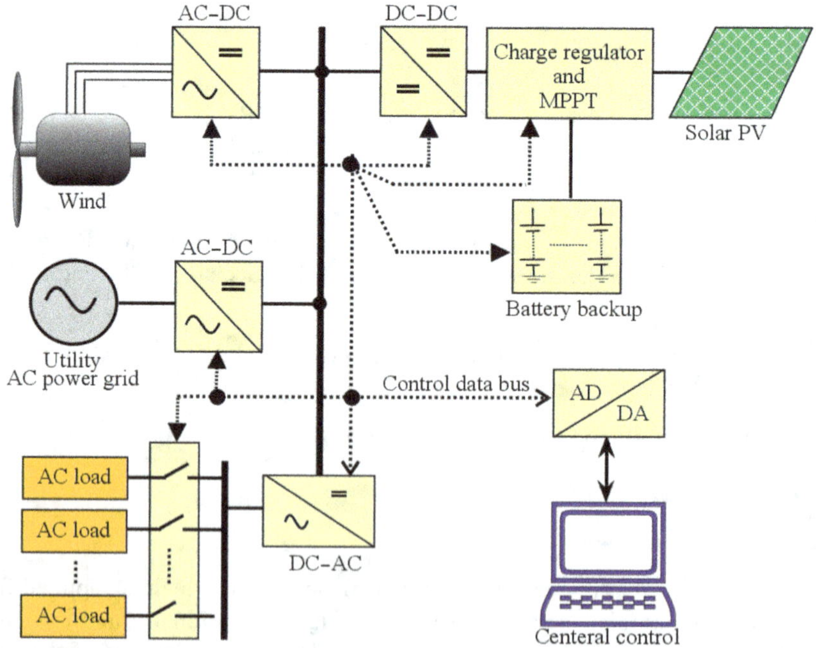

Figure E9.3(a) Renewable power generation system and energy management

the common DC power pool. Data collected from the source side and load side is transferred to a computer to be evaluated for the decision-making process of the power management system.

Design a Power Management Algorithm (PMA) to operate both Photovoltaic Energy System (PES) and Wind Energy System (WES) at the maximum power they can generate under various environmental conditions while maintaining the power supply demand of the load side at the required amount. Therefore, the PMA should include maximum power point tracking of PES and WES, energy storage, utility connection, and load switching. Besides, the power quality issues such as harmonic elimination, voltage sags, voltage increments, frequency deviations, and voltage magnitude should be included in the management system, as well. It is obvious that there are too many inputs and parameters in the management decision algorithms. Use a fuzzy logic-based decision-making algorithm for this multi-input, multi-objective power management system.

Answer 9.3

In the solution of this example, let us zoom into the connection diagrams of each generating unites one by one starting with the PV setup. Theoretical, characteristic, and structural details of wind and PV systems are not going to be repeated here since they are discussed in Chapters 7 and 8. Therefore, only the connection details will be given and power management design will be discussed. Since the system used in this example is realized in the laboratory environment and tested in real time, the system description and details are based on the experimental setup used in testing of the power management system.

The power flow path from PV power generation unit to the load is shown in Figure E9.3(b). Four PV modules connected 2 in series and 2 series branches in parallel are used here to get a PV array with 42 Volts and 5 Amperes maximum under good weather conditions.

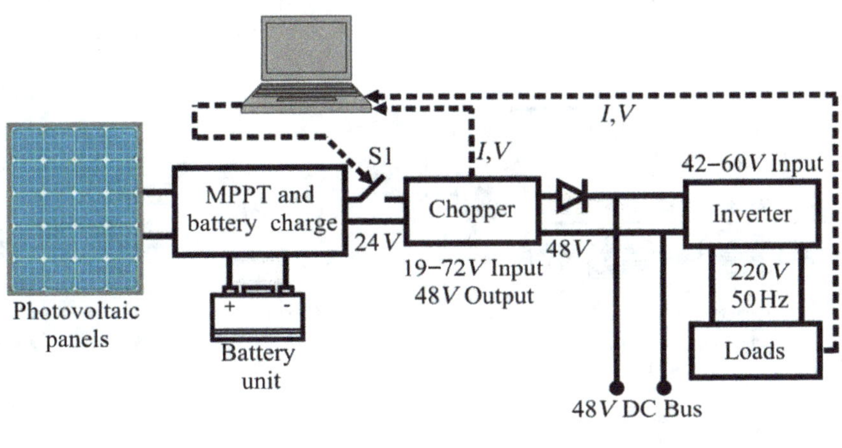

Figure E9.3(b) The PV power generation part of the system

PV array and battery groups are connected to each other with a device including a battery charging regulator and maximum power point tracker. When the sunlight is not sufficient, the batteries step in and supply the necessary power to the loads. The MPPT is used to transfer the maximum generated power from the PV array and charge the batteries if any power left after feeding the load. The batteries are also charged from the DC power bus when the sunlight is insufficient. A battery charge regulator with 9/24/36/48 V and 30 A is used as a charging interface device. Total peak power generated by the PV array under good weather conditions is about 320 Wp. Since the value of the generated voltage from the PV array changes depending upon sunlight, a DC chopper (19–72 V DC input voltage) is used to keep the DC voltage from the PV panels at 48 V, which is the magnitude of the DC bus voltage, which is inverted to 220 V, 50 Hz AC voltage by a boost up inverter. In Figure E9.3(b), the current and voltage data from the loads are measured and transferred to the computer besides the input voltage and output current of the chopper to be used in decision-making process.

The WES emulator is established by coupling two squirrel cage asynchronous machines together as seen in Figure E9.3(c). The machine on left is a 5 kW induction motor used as the prime mower representing the wind turbine. A 3.5 kW induction machine, the second one on the left, is the generator part of the wind emulator model. The power of the prime mower is selected higher than the one used as the generator in order to operate the generator at overpower conditions so that the analysis of the operating cases can be expanded.

The machine operating as prime mover and representing the wind turbine is controlled with a V/f speed controller. Depending on the wind speed, the output voltage of the generator changes between 320 V and 400 V. Three-phase voltage magnitudes obtained from the generator are reduced to a lower level using a step-down transformer with a ratio of 380 V/36 V. Therefore, transformer output voltages change between 30 V and 38 V. The output voltage magnitude of three-phase full-bridge diode rectifier is calculated as $U_O = 1.654U_{R\max}$, where U_O is rectifier output voltage in Volts and $U_{R\max}$ is the maximum value of single-phase voltage in

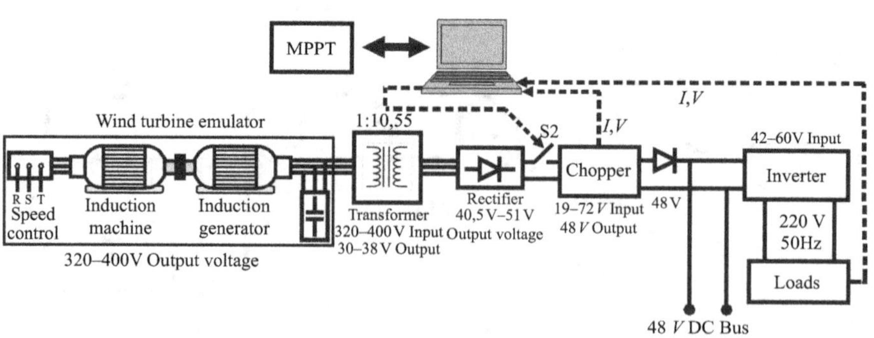

Figure E9.3(c) The wind power generates part of the system

Volts. Full-bridge rectifier output voltage changes from 40.5 V to 51 V. A DC chopper is used to keep this variable voltage at 48 V constant DC to be connected to common DC bus, which has a 48 V constant DC voltage.

The machine used as the generator is a three-phase, four-pole, 3.5 kW, 50 Hz 1415 rpm induction machine with a power factor of $\cos\varphi = 0.80$. In order to operate this induction machine as generator at its rated values, the required reactive power is calculated as 2.625 kVAr. This reactive power can be supplied by connecting 19.27 μF on each phase of a delta-connected configuration. However, 22 μF and 33 μF values are used in the testing to get a wide speed range.

Three-phase output voltages of the induction generator are rectified and kept constant at 48 V by controlling a DC chopper to be connected to common DC bus. Then an inverter is used to convert this 48 V DC voltage to 230 V, 50 Hz AC to operate the loads. During this operation, the load current, load voltage, chopper input voltage, and chopper output current are measured and transferred to the computer.

In addition to PES and WES systems, the utility is also considered as a power source in Figure E9.3(a). Therefore, two separate cases become available to be analyzed. The first case includes the consideration of PES and WES systems only to analyze the off-grid connection by supplying power to individual loads. The second case deals with utility-connected renewable energy schemes when the power from renewable sources is not sufficient. Details of the utility interface to 48 V DC bus and loads are shown in Figure E9.3(d).

As shown in Figure E9.3(d), the utility voltage is converted to 48 V DC constant value and connected to 48 V common DC bus as it is done for wind and PV systems. Data collected from various parts of the system is transferred to a central computer and used for control and decision-making processes. The interfacing between the real system and the computer is established by NI USB 6259 data acquisition card.

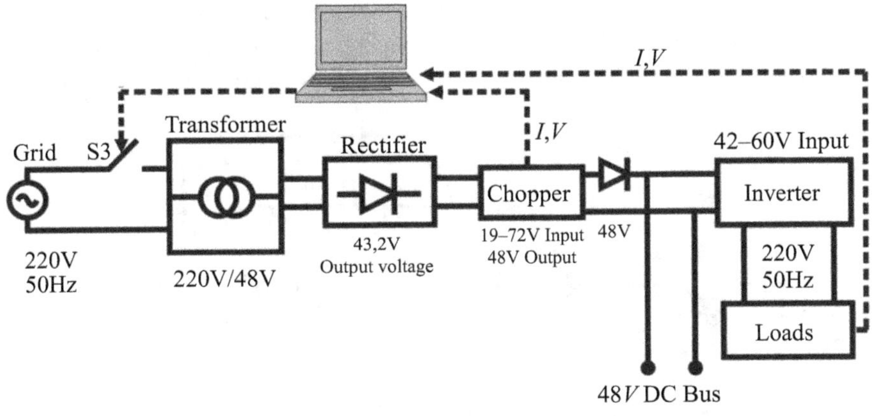

Figure E9.3(d) Utility grid interfacing

The electrical power generated by renewable sources such as WES and PES is affected by environmental conditions resulting in problems in load part. When there is no sun or the weather is cloudy, the power amount to be generated by PES changes. Accordingly, wind does not blow at the same speed all the time and shows a discontinuous behavior. In other words, the power and energy generated by renewable sources are randomly variable. Therefore, methods are developed to predict loading and generating conditions [29–31] at least 1 day earlier so that some actions can be taken on time. Conventionally the power generation is planned to supply the required load power. The users do not want to turn their loads off because the sources are not sufficient due to weather conditions. Therefore, a power management system is required for both the generation and load sides.

A power management algorithm (PMA) and decision-making process are described in this example to prevent problems like voltage sags and discontinuities that occur due to either weather changes or sudden load changes. The intelligent decision-making algorithm manages the energy storage and usage switching patterns so that energy sustainability is guaranteed.

General block diagram of the power management scheme is given in Figure E9.3(e). The total generated power from the wind is used as the primary supply power, which is used to supply base load power. As long as the wind power is sufficient, the secondary power system is kept off the load, and additional generated power is stored. When the wind power is less than the required load power then the secondary power system, which is solar and/or utility grid, is activated. The service order of the secondary power system goes as solar PV, storage, and utility grid. Since the utility grid requires additional payment, it is put at the end of the list and the priority is given to the wind and PV arrays. The load power is calculated as $P_{\text{Load}} = P_{\text{W}} + 300$ W, where P_{W} is the generated wind power and 300 W is the bias power supplied from backup units.

In order to solve sustainability and power quality problems, the power transfer from renewable sources to load must be managed in a proper way. Therefore, a PMA has been developed to prevent power discontinuity, overvoltage, and undervoltage operations. The power management system is automated in an efficient way by switching on or off the sources and backup units. For example, if the wind power is sufficient enough to feed the load, then there is no need for the auxiliary

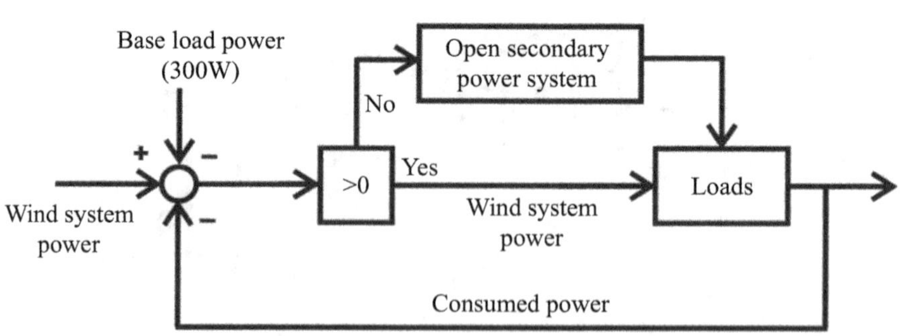

Figure E9.3(e) Basic block diagram of the power management system

sources of PV, backup batteries, and utility. If the wind power decreases, the gap is filled by PV first, then batteries, and then the utility. The over-generated power is stored and used only when needed.

The overall energy system in this example is established experimentally using the setup diagram shown in Figure E9.3(f). In this system, the electrical power is generated by wind generators and PV solar panels. The utility is reserved as an auxiliary source to be used when needed. The power from the PV system is used to supply power to the load when the wind power is not sufficient and to charge the batteries when there are sufficient wind and sun power. Data collected from various parts of the overall system is transferred to the computer to be analyzed and used.

The main objective of employing a power management algorithm in power systems where renewable energy is the priority supply is to have the power ready to be used and feed the load continuously. For this reason, the peak power value from both wind and PV solar panels must be calculated. The MPPT device used for PV solar panels tracks the maximum power of the PV system and makes it available to be used. However, the peak power value of the wind energy must be defined. Since the changing environmental conditions affect the amount of generated power, a strategy is applied to avoid this effect by leaving a base power in the system. The base power is the power that must be supplied all the time for the loads with

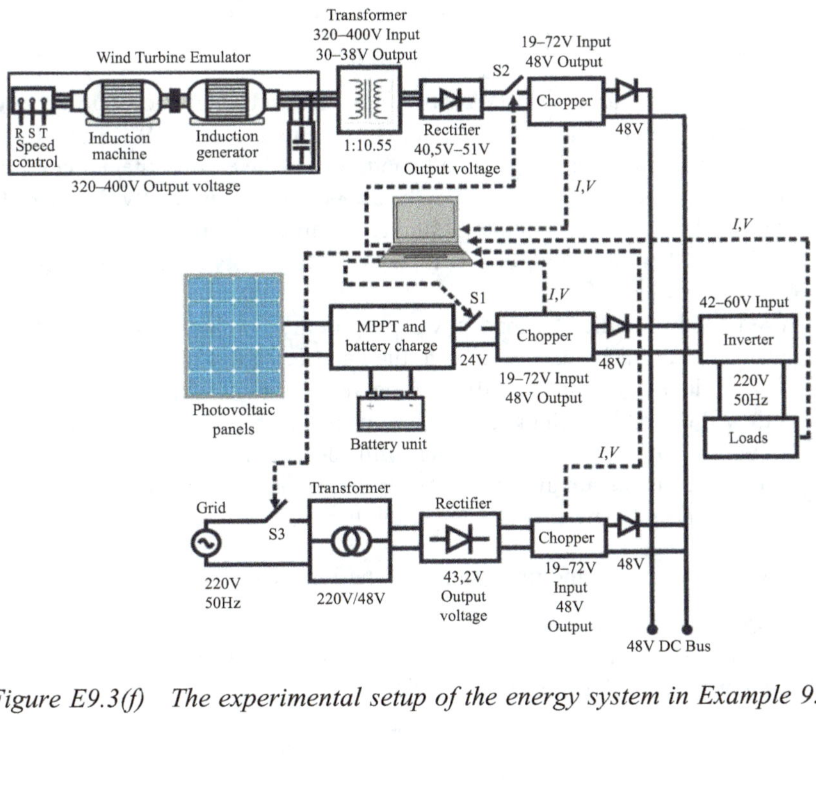

Figure E9.3(f) The experimental setup of the energy system in Example 9.3

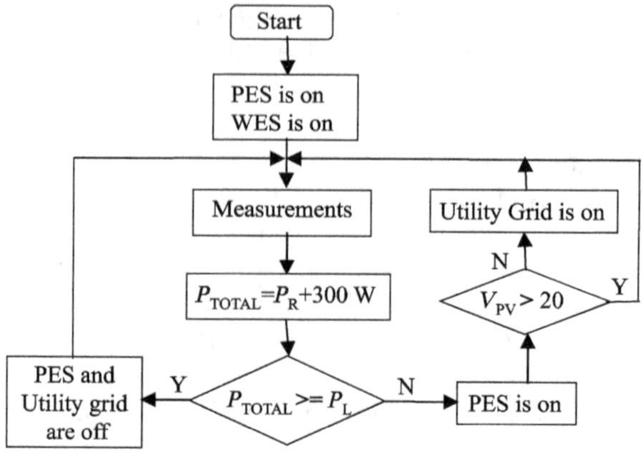

Figure E9.3(g) Simplified flowchart of power management algorithm

non-stop operating behaviors. The base power in the proposed system is defined as 300 W, which can be easily changed inside the software if desired. The system is designed to feed maximum of 1 kW, which is more than three times the base power. If the environmental conditions are sufficient, which means there are enough sun and wind to operate the wind and PV systems at their highest capacity, which is assumed to be 2 kW for the test system used.

The main operational principles of the system are summarized by a flow chart given in Figure E9.3(g) and listed as follows:

(i) Initially, the system is started with both solar and wind energy in service.
(ii) After the transients are over and measurements are done, WES or PES will be kept working according to load condition. If the environmental conditions are not suitable for PES or WES to be operated individually, both of them will be operated. When the total power generated from PES and WES is not sufficient, then the utility grid will take over to supply the rest of the required demand power.
(iii) If WES can handle the load power requirement alone, it will operate. PV solar panels will be used to charge the batteries only.
(iv) If WES does not generate sufficient power, PV solar panels will engage and both will operate together. If both are insufficient, then the grid will engage.
(v) If there is no wind, PV solar panels will feed the loads.
(vi) When there is no sun and the batteries are empty or the battery cannot meet the required power demand, the grid will begin to operate.

These steps bring up the importance of the following:

1. The operating time of each unit.
2. Turn off time of each unit.
3. The amount of load at present conditions.

Since only the wind energy will operate constantly, these lists of rules are processed by taking the measurements from wind energy system into consideration. 300 W power will be supplied permanently as the base power in the system. The wind turbine emulator is operated at various speeds in order to represent the generated wind power properly. The emulator has been set up so that the generated power will not be sufficient if the speed drops below 35 Hz. During this condition, the PV system will automatically be switched on to feed the load depending on the availability of sunlight.

The Chopper input voltage changes as the induction motor driver frequencies change. For instance, when the load power is 500 W, the generated voltage becomes 47.3 V at 44 Hz and 45.4 V at 43 Hz. It will decrease down to 31.5 V at 37 Hz. The relationship between voltage and frequency is used to estimate the operating frequency. Measurements of chopper output current and input voltage values in wind generating systems are used to detect induction motor driver frequency, dissipated power at that instant, and maximum demand power at that driver frequency. These findings are also used to determine switch on or switch-off times of PV panels. It should not be forgotten that 300 W power will always be in the system as the base power to be supplied.

Similarly, when the PV panels are on, the generated voltage from the PV panels to the chopper decreases as load power increases. Using the measured values from this operating case, it is possible to find out how much power should be generated by the panels to feed the load. Input voltage of the chopper in PV system will decrease as the batteries discharge. The batteries are assumed to be insufficient to feed the load when their voltage drops below 21 V. If the sunlight is not enough to generate the required power to the load, the batteries also will not be usable with an output voltage of less than 21 V. The wind turbine may start getting sufficient wind to operate while PV panels are feeding the loads. In this case, some of the load power is supplied from the WES while PV panels keep feeding the remaining loads and charging the batteries. It is important to get measurements of these operating cases continuously in order to respond immediately to sudden load changes. If a sudden load change causes the demand power to increase above 300 W, the additional power is transferred from the WES. If the WES is not able to supply all the required power, then PES is the next option to feed the load. This action is taken to prevent system failure during sudden changes in load demand power or environmental conditions. The grid connection is established when neither the sun nor the wind is sufficient enough to supply 300 W base power demand. Actually, the utility is always an option whenever the WES, PES, and backup batteries do not have enough power to feed the loads. Flow chart of the power management algorithm described above is given in Figure E9.3(g).

Since the wind conditions are not certain and not predictable easily, the power generated by the wind energy system becomes uncertain including the maximum generated power as well. Therefore, a Fuzzy Reasoning algorithm is developed to determine the maximum power generated by WES. A block diagram of the Fuzzy Reasoning applied to determine the maximum power of the WES is depicted in Figure E9.3(h). The Fuzzy decision-maker (FDM) applied here is using the Mamdani Fuzzy Reasoning algorithm discussed and modeled in Chapter 4. The Simulink® modeling of fuzzy decision-maker used in this example is similar to that of the fuzzy logic controller model given in subsection 4.9 of Chapter 4.

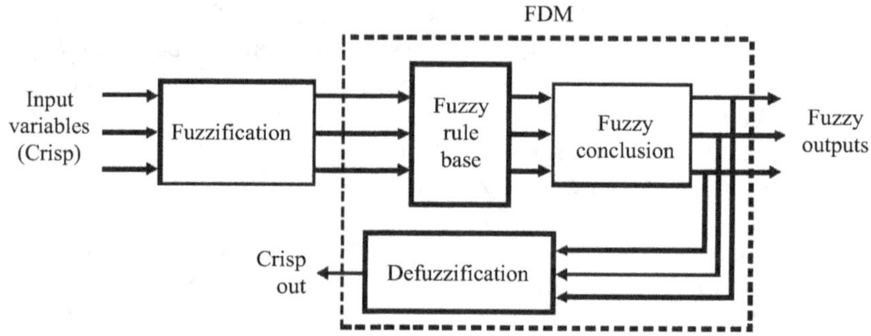

Figure E9.3(h) Fuzzy decision-maker

An FDM usually gets fuzzy inputs and evaluates them in the rule base system, which is set up earlier representing the input–output relations of the uncertain system in terms of fuzzy membership functions and fuzzy rules. A fuzzy reasoning is the evaluation of the rules to yield fuzzy conclusions from fuzzy inputs–fuzzy rules interactions as explained in Chapter 4.

The input values to the FDM are the current and voltage measured from the WES in this example. The generated rules are used to relate the input voltage and current with the power of the WES depending upon the wind speed conditions. The fuzzy reasoning algorithm in the FDM is used to obtain maximum power generation from WES for uncertain and unpredictable speed conditions. A 49-rule FDM is modeled in MATLAB/Simulink environment similar to the modeling of the fuzzy logic controller (FLC) discussed in Chapter 4 and modeled in Simulink as given in subsection 4.9. The maximum wind power is determined by FDM using the chopper's input voltage and chopper output current. The wind turbine, a three-phase transformer, a three-phase bridge rectifier, and a DC/DC converter are all assumed as a whole system. All of the calculations are based on the values of chopper input voltage and chopper output current in the WEC part of the whole system. Since the chopper output voltages are kept constant at 48 V DC, the main variable at the output terminals of the choppers or at the common DC bus is the current at the output terminals of the choppers. The output current of the chopper used to control the voltage of WES is the variable that reflects the changes on WES power. Therefore, the active power generated by the WES is obtained as $P_{DC/DC} = V_{DCbus} I_{DC/DCo}$, where V_{DCbus} is the DC bus voltage and equal to the chopper output voltage. This voltage is kept constant at $V_{DCbus} = 48$ V. $I_{DC/DCo}$ is the copper output current.

All of the MPPT calculations are based on V_{DCbus} and $I_{DC/DCo}$ measurements. As the wind speed changes, the wind power generation changes resulting in changes in current and voltage of the wind system. Therefore, input voltage $V_{DC/DCI}$ and output current $I_{DC/DCo}$ of the chopper in wind system also change. These two measurements are used to determine the peak power and voltage frequency, which gives the speed of the wind generator. The WES is tested under different loading at various speeds and power profile is stored as look-up tables to be used later for control purposes. The look-up table is obtained in terms of fuzzy rules as the fuzzy relation matrix between loading and speed conditions. Actually, the relation matrix represents the expert knowledge and gives the decision about what to do depending on the variables in input spaces.

The online data collected and transferred to computer is used to determine the amount of load power demand supplied from the PV/Wind sources. Meantime the data representing the WES quantities are used by FDM to determine the maximum power generated by the WES for the instant the measurements are made. The maximum power values of both WES and PV panels are used in the power management algorithm. As stated earlier the FDM used in the management of wind energy systems has two input and one output space. Chopper input voltage and output current are two input spaces and active power is the output space. The input and output spaces are partitioned into seven fuzzy subsections represented by triangular fuzzy membership functions as shown in Figure E9.3(i).

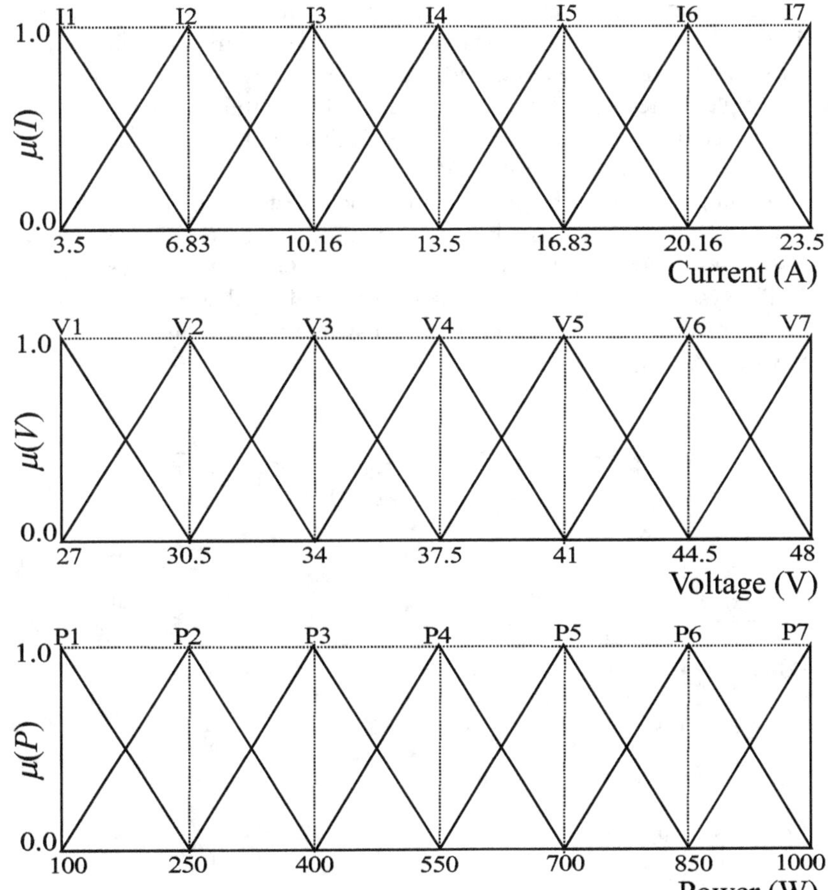

Figure E9.3(i) Fuzzy subset of input and output spaces used in power management of WES

The input universes of current and voltage are defined in the intervals {3.5, 23.5} A and {27, 55} V, respectively, and partitioned into seven subclasses as I1 to I7 and V1 to V7. Similarly, the output space is defined in active power interval {100, 1000} W and separated into seven fuzzy subsections from P1 to P7. The power generated by the WES varies as a function of the wind speed. Therefore, the maximum power obtained from the WES changes depending on the wind speed levels and must be determined for different speed levels as the wind speed changes.

The fuzzy rules used in the FDM are shown in Tables E9.3a and E9.3b. These rules are generated experimentally by testing the FDM over 100 times. The wind speed is set by changing the driver frequency of the induction machine used as the prime mover to represent the wind turbine. The rules listed in Table E9.3(c), are obtained from testing the system at various frequencies and loading conditions. The final forms of the fuzzy rules used in the FDM are modified and given in Tables E9.3(d) and E9.3(e). Instead of using curve fitting approaches, Tables E9.3 (d) and E9.3(e) are used as look-up tables stored in yjr central computer. This data is used for maximum power point tracking (MPPT) in wind energy systems. The MPTT is implemented as a Simulink model in the central computer and used to generate the required control signals. The Simulink model of the power management algorithm is given in Figure E9.3(j).

The Simulink model of the power management system shown in Figure E9.3(j) consists of the subsystems described as R_I_ok: wind system current, R_V_ok: wind system voltage, R_P_ok: wind system power, G_V_ok: PV system voltage, G_I_ok: PV system current, Y_P_ok: loads power, Pwind_Max: calculated maximum wind power, Grid_Switch: grid system switch (on/off) and PV_Switch: PV system switch (on/off).

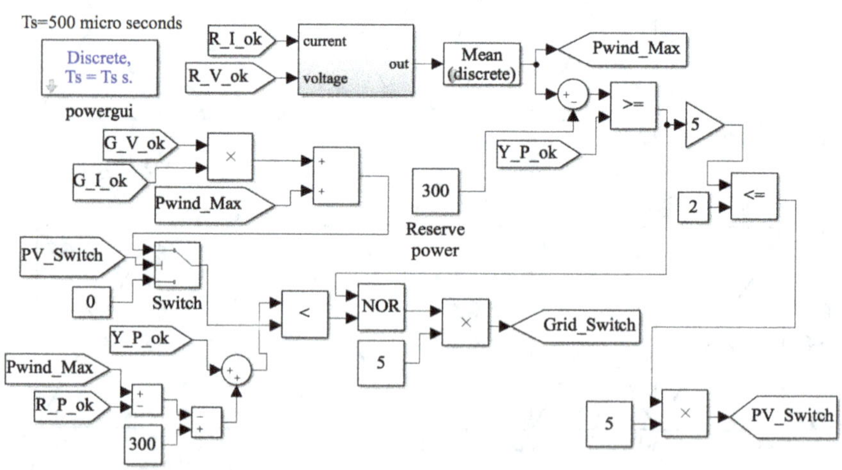

Figure E9.3(j) Simulink model of the power management system used in Example 9.3

Table E9.3(a) FDM rules for 0–700 W power

Current (A)	Voltage (V)	Frequency (Hz)	Peak Power (W)	7 Rules 18 Output FDM	7 Rules 7 Output FDM
I1→3.5 (100 W)	V1→27	32.1	150	P1	I1→3.5
I1→3.5	V2→30.5	33.5	250	P2	I1→3.5
I1→3.5	V3→34	34.8	350	P3	I1→3.5
I1→3.5	V4→37.5	36.6	500	P4	I1→3.5
I1→3.5	V5→41	38.5	650	P5	I1→3.5
I1→3.5	V6→44.5	40.3	900	P6	I1→3.5
I1→3.5	V7→48	42.2	1000	P7	I1→3.5
I2→6.83 (250 W)	V1→27	33.6	250	P2	I2→6.83
I2→6.83	V2→30.5	34.6	350	P3	I2→6.83
I2→6.83	V3→34	36	450	P3	I2→6.83
I2→6.83	V4→37.5	37.6	550	P4	I2→6.83
I2→6.83	V5→41	39.4	750	P5	I2→6.83
I2→6.83	V6→44.5	41.2	925	P7	I2→6.83
I2→6.83	V7→48	43	1000	P7	I2→6.83
I3→10.16	V1→27	35.4	400	P3	I3→10.16
I3→10.16	V2→30.5	35.9	450	P3	I3→10.16
I3→10.16	V3→34	37.3	550	P4	I3→10.16
I3→10.16	V4→37.5	38.8	650	P5	I3→10.16
I3→10.16	V5→41	40.2	850	P6	I3→10.16
I3→10.16	V6→44.5	42	1000	P7	I3→10.16
I3→10.16	V7→48	43.8	1000	P7	I3→10.16
I1→3.5	V1→27	32.1	150	P1	I1→3.5
I1→3.5	V2→30.5	33.5	250	P2	I1→3.5
I1→3.5	V3→34	34.8	350	P3	I1→3.5
I1→3.5	V4→37.5	36.6	500	P4	I1→3.5
I1→3.5	V5→41	38.5	650	P5	I1→3.5
I1→3.5	V6→44.5	40.3	900	P6	I1→3.5
I1→3.5	V7→48	42.2	1000	P7	I1→3.5

Table E9.3(b) FDM rules for 700–1000 W power

Current(A)	Voltage(V)	Frequency(Hz)	Peak Power (W)	7 Rules 18 Output FDM	7 Rules 7 Output FDM
I5→16.83 (700 W)	V1→27	–	–	–	P4
I5→16.83	V2→30.5	39.2	700	P12	P5
I5→16.83	V3→34	39.8	850	P15	P6
I5→16.83	V4→37.5	41	900	P16	P6
I5→16.83	V5→41	–	1000	P18	P7
I5→16.83	V6→44.5	–	1000	P18	P7
I5→16.83	V7→48	–	1000	P18	P7
I6→20.16 (850 W)	V1→27	–	–	–	P5
I6→20.16	V2→30.5	40.2	850	P15	P6
I6→20.16	V3→34	41.3	950	P17	P7
I6→20.16	V4→37.5	–	1000	P18	P7
I6→20.16	V5→41	–	1000	P18	P7

(Continues)

Table E9.3(b) (Continued)

Current(A)	Voltage(V)	Frequency(Hz)	Peak Power (W)	7 Rules 18 Output FDM	7 Rules 7 Output FDM
I6→20.16	V6→44.5	–	1000	P18	P7
I6→20.16	V7→48	–	1000	P18	P7
I7→23.5 (1000 W)	V1→27	–	–	–	P6
I7→23.5	V2→30.5	–	–	–	P6
I7→23.5	V3→34	–	1000	P18	P7
I7→23.5	V4→37.5	–	1000	P18	P7
I7→23.5	V5→41	–	1000	P18	P7
I7→23.5	V6→44.5	–	1000	P18	P7
I7→23.5	V7→48	–	1000	P18	P7

Table E9.3(c) FDM rules

Voltage	Current						
	I1	I2	I3	I4	I5	I6	I7
V1	P1	P2	P3	P3	P4	P5	P6
V2	P2	P3	P3	P4	P5	P6	P6
V3	P3	P3	P4	P5	P6	P7	P7
V4	P4	P4	P5	P6	P6	P7	P7
V5	P5	P5	P6	P7	P7	P7	P7
V6	P6	P7	P7	P7	P7	P7	P7
V7	P7	P7	P7	P7	P7	P7	P7

Table E9.3(d) Collected data from wind turbine 38–44 Hz

Driver frequency (Hz)		444	443	442	441	440	339	338
P_m (W)	I_C (A)	V_C (V)	V_C (V)	V_C (V)	V_C (V)	V_C (V)	V_C (V)	V_C (V)
1000	222.6	440.3	337.5	334.6				
900	220.4	441.5	338.8	336.2	332.4			
800	118	443.3	440.8	338.5	335.2	332		
700	115.8	444.9	442.7	440.4	337.8	335	330	
600	113.6	446.3	444.2	442	339.5	337	334.8	332.5
500	111.3	447.3	445.4	443.4	441.2	338.8	336.7	334.7
400	99.2	448.4	446.5	444.7	442.5	440.3	338.4	336.4
300	77	449.7	447.8	445.8	443.8	441.7	440	338
200	44.8	550.8	448.9	447	445	443	441.1	339.3
100	22.5	551.8	550	447.9	446	444	442.2	440.4
0	00.4	553.2	551.4	449.4	447.5	446	443.7	441.6

Table E9.3(e) Collected data from wind turbines 31–37 Hz

Driver frequency (Hz)		**337**	**336**	**335**	**334**	**333**	**332**	**331**
P_m **(W)**	I_C **(A)**	V_C **(V)**	V_C **(V)**	V_C **(V)**	V_C **(V)**	V_C **(V)**	V_C **(V)**	V_C **(V)**
1000	222.6							
900	220.4							
800	118							
700	115.8							
600	113.6							
500	111.3	331.5						
400	99.2	333.7	331.2					
300	77	335.5	333.7	330.8				
200	44.8	337	335.2	333	330.6	227.2		
100	22.5	338.3	336.5	334.4	332.1	229.7	227	
0	00.4	440	337.7	336.3	333.7	331.6	229.3	00

Now let us analyze some operating cases.

Case 1

As the first operating case, the system is analyzed when both wind and PV panels are on without any power management is applied. For this case, the power measurements from WES, PES, and load terminals are given in Figure E9.3(k).

It can be seen that, when a load power of 800 W is on, WES cannot feed the loads from $t = 40$–70 s and from $t = 86$–89 s, because no power control is applied. The load power drops down to zero for some instants, especially from the 40th second to the 70th second because the load voltage drops to zero during these instants due to insufficient wind speed. Actually, the PV system generates enough

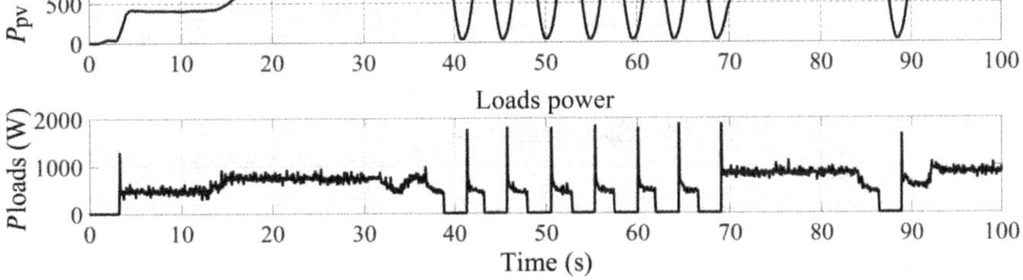

Figure E9.3(k) Power responses of WES, PES, and loads for Case 1

power during these low wind speed periods to feed the load. However, this PV-generated power is not transferred to the load during low and insufficient speed periods because there is neither a power control nor a power management system in this case to transfer the required power from the other sources including the utility grid. If a decision-maker or a power management system had been used, these load power discontinuities might have been eliminated.

Case 2

A fuzzy decision-maker (FDM) and fuzzy-based power management algorithm (PMA) are used this time in order to transfer the required load power from the other available sources if the wind system fails to do so. The power responses from wind, PV, utility grid, and load terminals are given in Figure E9.3(l) to show the behaviors of the WES, PES, and utility grid when there are arbitrary changes in load power for operating case 2. The WES system feeds the load as long as the wind power is sufficient. The FDM is used to extract the maximum available power from the wind system. When the load power is increased above the wind power generation capacity around the 20th second, the PMA acts and starts transferring the additional power from the PV system. At the instant of 45th seconds, the WES and PES become insufficient to feed the load due to the weather conditions. At this instant, the PMA decided to transfer the required power from the utility grid. The wind power comes back at 68th seconds and starts feeding the load. Since the load power demand is not higher than the generated wind power, all requirements are supplied only from the WES. When the time is around 88th seconds, the load power demand increases while there is a decrement in wind power. The PMA decides to get the required power from the PV system again for this operation case.

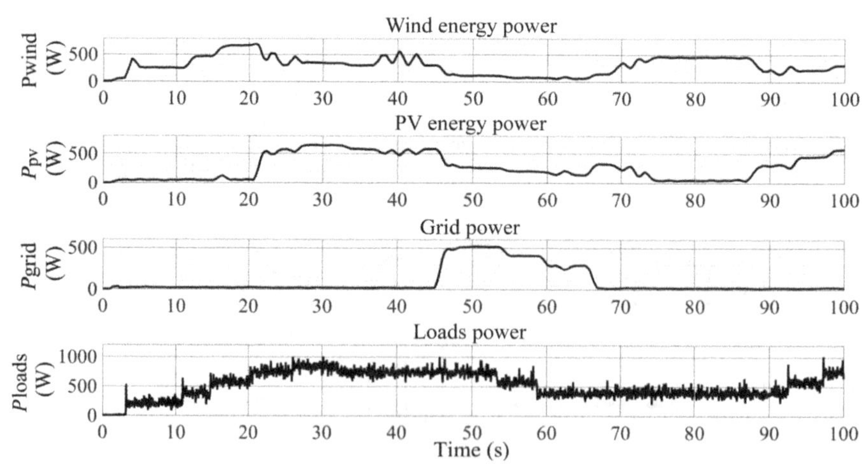

Figure E9.3(l) Power responses of WES, PES, grid, and loads with FDM for Case 2

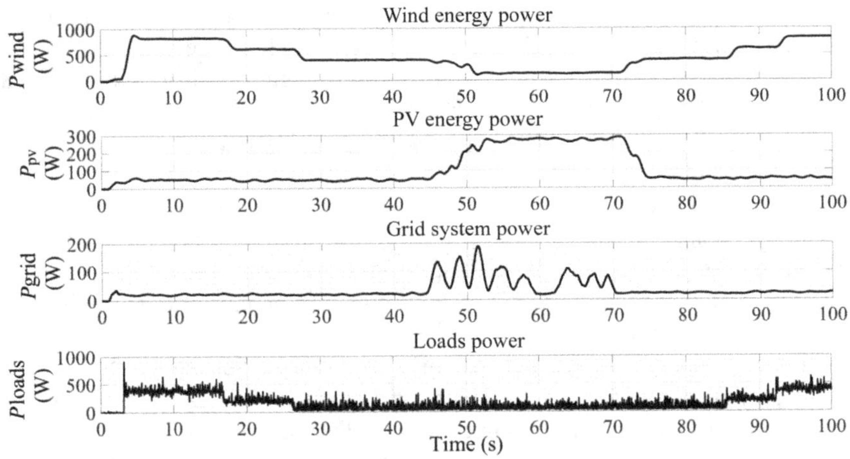

Figure E9.3(m) *Power responses of WES, PES, grid, and loads with FDM for Case 3*

Case 3

Both FDM and PMA are active in this case too. The power responses from sources and load terminals are plotted in Figure E9.3(m) for loading conditions different than those of Case 2. Both resistive and motorized loads are considered in this case. The active power responses are measured at the WES, PES, utility grid, and load terminals showing the performance of the power management algorithm.

The load power is high at startup in Figure E9.3(m). So is the wind power. Therefore, the load is fed from the WES. From the 20th second to the 50th second, the wind power decreases and drops below the 300 W bias power. Since the load also decreases following a similar decrement pattern as the load, the powers from PV and utility are not used. However, once the wind power drops below the 300 W bias power, which must be supplied continuously for permanent loads, this 300 W is supplied from PV and grid so that the bias power is maintained. After the 70th second, the wind power starts increasing and feeding the load. As long as the load demand power is supplied by the WES, no power is transferred from PV and grid to the load. When the generated power from the PES is not needed by the load, then it is used to charge the batteries, which is not discussed here.

9.8.2 Distributed control of renewable energy systems

The centralized control of distributed generation has been applied widely for power management in distributed power systems. However, as the distributed power system installations speed up and start getting larger by covering wide areas, the centralized control becomes harder to apply. As the distributed generation and distributed power systems turn to smart grids, measurements at every generation and every load become ineligible. Some of these measurements are used for analyzing, planning, pricing, and maintenance purposes, some others are used for power management and control purposes. Transferring all the data from far away points to a center and sending the decisions back to where the measurements are made and actions to be taken, cause delay and misoperation of the system besides cyber data security and danger of data loss. In order to reduce operational difficulties and take faster decisions and actions on time, distributed control algorithms are developed and applied [32–34].

Distributed power systems require not only generating side management but also load side management, as well. The generation side management is applied to supply the required load power from various sources or units in the same source site while load-side management deals with energy efficiency, energy saving, loss reduction, and optimum use of available power. Therefore, load-side management is an important task in distributed networks and smart grids. Since the loads are distributed in all parts of the power distribution network, distributed control gains importance. Since the generating units are also distributed along with the loads in distributed power networks, distributed control becomes a powerful alternative approach for generation and load-side management systems.

Let us consider the distributed generating network shown in Figure 9.22. The network is connected to the utility grid and consists of residential loads. The houses in the residential areas have PV panels installed on their roofs and they get their electricity from the distributed generating network. The power distribution company buys the PV-generated electricity from the houses and sells it back to the houses in the frame of an agreed management and a pricing plan. The priority source in this network is the PV system. The utility grid is used only when the PV power is not sufficient.

Each house in the network has a different load demand curve (LDC) and they may dissipate the higher power at different instants. The load demand curve of the

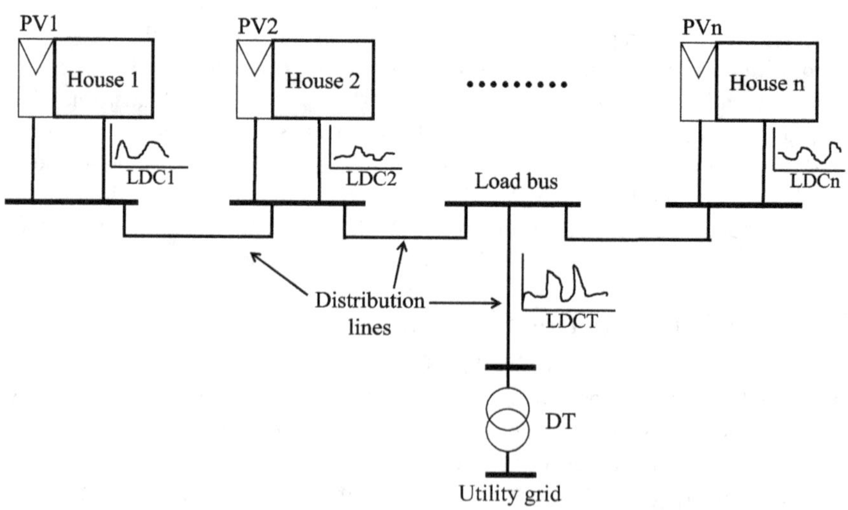

Figure 9.22 A sample distributed generating network

transformer (LDCT) depends on the availability of the PV power. The LDCT may be zero or very low if the PV power is sufficient to supply all the load demand power. Therefore, the LDCT also represents the availability of PV power generation.

In order to use the generated PV power more efficiently with minimum distribution losses, both generating side load side management algorithms must be applied. One of the methods for minimizing the losses is moving the loads at peak times to the lowest dissipation periods. This approach is called *load shifting*, which is done by classifying the loads according to their importance. Some household appliances may not be so important and they may be postponed to be used later. For example, ironing, vacuuming, dishwashing, cloth washing, and driers may be postponed and operated during low power demand hours of the network. Some loads must be operated at certain times and cannot be postponed or delayed. Some loads may be operated all the time without any other options. Therefore, the load management goes as far as to the plug where the loads are connected to get power. Actually, one can say that this load control and management approach leads us to intelligent loads, in which all management algorithms are stored in a processor mounted on the device itself. Since each smart device includes its own energy management algorithms by generating user options for energy savings, distributed control becomes an important alternative in energy management systems.

Example 9.4

This example is adapted from PhD thesis [35] of Recep Cakmak, who was a PhD student of the author. In a distributed power system, the network is overloaded during some hours of the day while the loading is very low during other times. Therefore, the power company wants to design a decision-maker so that the operating times of some loads can be shifted from heavy loading hours to light loading hours.

Design a fuzzy decision-maker (FDM), such that the shifting or delay time should be generated when a load demand is entered. For example, if the predicted loading in a time ahead is lower than the current load power, then some loadings should be postponed to be operated during these light loading hours.

Consider the following in your design:

1. The inputs to the fuzzy decision-maker should be current demand power and the difference between current load and estimated future load demand.
2. Partition the output space into their fuzzy subsets as low, medium, and high.

Answer 9.4

The FDM uses demand power and the difference between demand power and predicted future power as two inputs. Demand power may be categorized as low (LW), average (AV), and high (HI). Since the difference may be positive, negative, and zero depending on demand and predicted power values, the universe of power difference can be portioned into five fuzzy subsets as negative high (NH), negative low (NL), zero (ZE), positive low (PL) and positive high (PH). The output space representing the delay time is divided into three fuzzy subsets as short (SH), medium (ME), and long (LN). A block diagram of the FDM is given in Figure E9.4(a). The FDM block includes the fuzzy rule table given in Table E9.4(a), which includes 15 fuzzy rules for decision-making.

The fuzzy decision-maker given in Figure E9.4(a) uses the Mamdani fuzzy reasoning algorithm with two crisp inputs and one crisp output. The current power demand (P_D) is represented by variable x and the difference (ΔP) between current load power and predicted future depend power is represented by variable y in Figure E9.4(a). The output is the delay time t_D is represented by variable z.

If the current demand power is low (LW) and the difference power ΔP is negative high (NH) then this load demand can be delayed for a long time (LN) because negative ΔP means that the future power demand will be smaller than the current power dissipation so that the operating time of this power demand can be shifted ahead with a long delay. If the current power demand is high (HI) and difference power ΔP is positive high (PH) then this load demand should not be delayed because the predicted future power demand will already be high. If this high power demand is postponed to a later time ahead the load demand will be

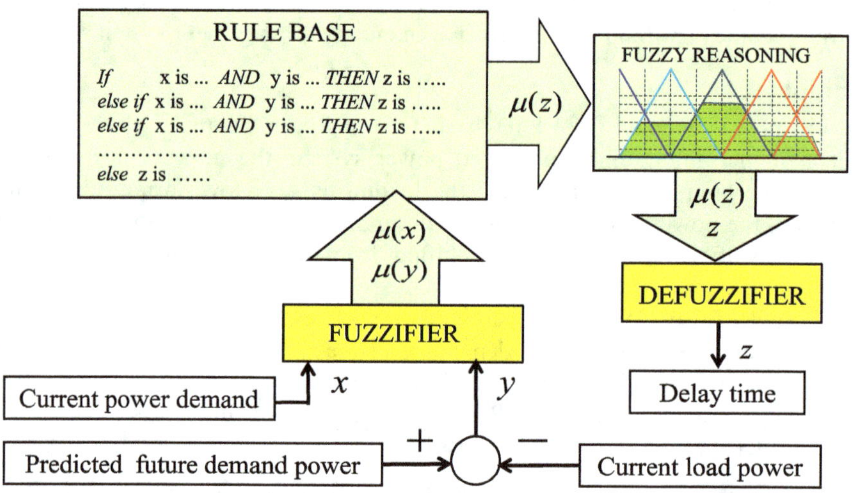

Figure E9.4(a) The fuzzy decision-maker algorithm for Example 9.4

Table E9.4(a) The fuzzy rules used in FDM in Example E9.4

	ΔP				
P_D	**NH**	**NL**	**ZE**	**PL**	**PH**
LW	LN	ME	SH	ZE	ZE
AV	LN	ME	SH	ZE	ZE
HI	LN	SH	ZE	ZE	ZE

much higher such that the generated power might be insufficient as the network faces overloading operating conditions, which cause overheating and higher losses. The other fuzzy rules are generated with a similar point of view to represent the experience of a power network operator.

Let us assume that the current power demand is 2300 W and the difference power ΔP is negative 400 W and try to find the decision about how long the load demand can be delayed.

The fuzzy decision process for this example is depicted in Figure E9.4(b).

A Mamdani fuzzy reasoning algorithm is applied for this example. The final crisp conclusion is obtained using the center of area defuzzification method as follows:

$$t_D = \frac{(0.3)(20) + (0.7)(40) + (0.2)(0) + (0.2)(20)}{0.3 + 0.7 + 0.2 + 0.2} = 27.14$$

The MATLAB .m file code of the fuzzy decision process depicted in Figures E9.4(a), E9.4(b), and E9.4(c) is given in Figure E9.4(d).

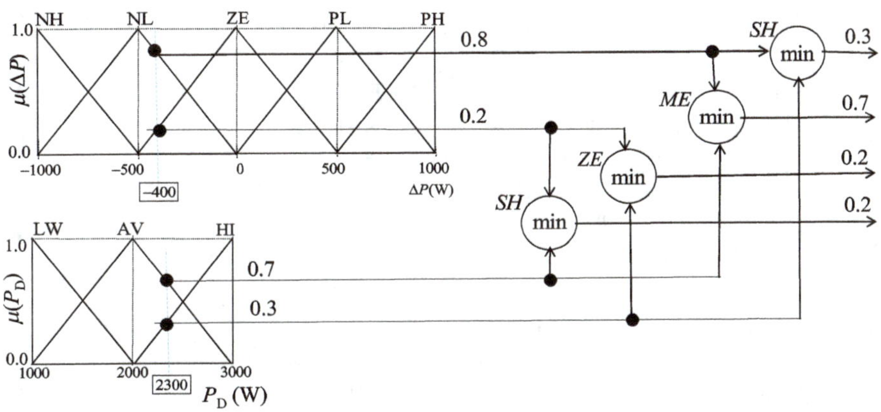

Figure E9.4(b) The fuzzy decision-maker process for Example 9.4

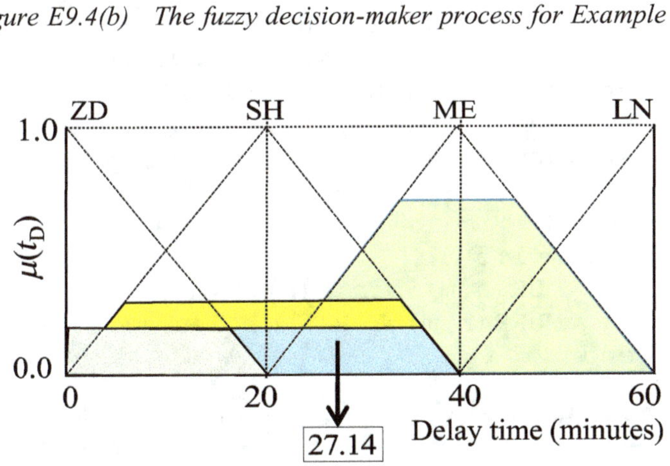

Figure E9.4(c) Defuzzification of the fuzzy decision for Example 9.4

```
% Example 9.4  - Mamdani Fuzzy reasoning
clear; clc;
% The input variables
PD=2300;   DP=-400;
%triangular fuzzy subset parameters
% For input space DP
NHL=-1000;  NHT=NHL;   NHR=-500;
NLL=NHL;    NLT=NHR;   NLR=0;
ZEL=NLT;    ZET=NLR;   ZER=500;
PLL=ZET;    PLT=ZER;   PLR=1000;
PHL=ZER;    PHT=PLR;   PHR=PHT;
% For input space Pdemand
LWL=1000;   LWT=LWL;   LWR=2000;
AVL=LWT;    AVT=LWR;   AVR=3000;
HIL=LWR;    HIT=AVR;   HIR=HIT;
% For output space time delay
ZDL=0;      ZDT=0;     ZDR=20;
SHL=ZDT;    SHT=ZDR;   SHR=40;
MEL=SHT;    MET=SHR;   MER=60;
LNL=MET;    LNT=MER;   LNR=LNT;
% --------------------------------
R=[LNT   MET   SHT   ZDT    ZDT
   LNT   MET   SHT   ZDT    ZDT
   LNT   SHT   ZDT   ZDT    ZDT ];
% --------------------------------
X(1)=triangle01m(LWL,LWT,LWR,PD);
X(2)=triangle01m(AVL,AVT,AVR,PD);
X(3)=triangle01m(HIL,HIT,HIR,PD);
% --------------------------------
Y(1)=triangle01m(NHL,NHT,NHR,DP);
Y(2)=triangle01m(NLL,NLT,NLR,DP);
Y(3)=triangle01m(ZEL,ZET,ZER,DP);
Y(4)=triangle01m(PLL,PLT,PLR,DP);
Y(5)=triangle01m(PHL,PHT,PHR,DP);
% --------------------------------
for n=1:3
    for m=1:5
        MUR(n,m)=min(X(n),Y(m));
        MURR(n,m)=MUR(n,m)*R(n,m);
    end
end
z=sum(sum(MURR))/sum(sum(MUR))
```

Figure E9.4(d) MATLAB .m file of the fuzzy decision in Example 9.4

Design Example 9.3

A residential site gets its electricity from PV panels installed on the roofs of the houses. Statistics show that the peak power dissipation in this site is between 19:00 and 22:00 hours in the evening. On the other hand, solar PV panels operate more efficiently and generate electrical power from 07:00 to 18:00 during the daytime.

In order to use the electricity from the PV system instead of the utility grid, which uses fossil fuel-based sources, the power company decided to apply some pricing tariffs so that the customers will prefer to use electricity during the daytime when the power from PV systems is available instead of using from the utility during evenings. The tariff applied by the power company is described as follows:

(i) The electricity pricing is low during high power generation from the PV systems.
(ii) The electricity pricing is low when the power dissipation in the distribution network is low.
(iii) The electricity pricing is high when the power demand and dissipation is high.

Develop a fuzzy decision-maker and a power management algorithm for this system using predicted load demand and predicted solar power capacity in a 2-day projection so that the customers get the electricity at lower price.

Design Example 9.4

Based on the renewable energy integration scheme given in Figure 9.18, develop a simulation model in MATLAB/Simulink environment for the following:

1. Investigation of novel interface converter topology and V/F stabilization network to ensure secure and reliable electrical energy source.
2. Investigate the effects of load changes on the voltage of the AC load bus and develop a FACTS-based dynamic power filter to reduce the effects of load changes.
3. Investigate the effects of the power sources given in Figure 9.4 during weather changes and develop solutions for energy sustainability.

Design Example 9.5

Investigate the following as separate projects:

1. Design and analyze a pollution-free distributed generation system including solar, wind, fuel cell, and small hydro units. Develop specifications/recommendations of a number of commercial renewable energy models for possible market commercialization and remote site installations.
2. Investigate and demonstrate ways of using the electricity and heat produced by fuel cells and develop a fuel cell simulation model.
3. Design distributed energy schemes to offer solutions to many pressing energy and electric power problems, including blackouts and brownouts, energy security concerns, power quality issues, tighter emission standards, transmission bottlenecks, and the desire for greater control over energy costs.

4. Develop a CAD and Engineering software package based on intelligent approaches for sizing, matching, and computer validation of different tracking and operating control strategies.
5. Design, prototype, test and validate four novel renewable energy schemes utilizing wind/PV/fuel cell/small hydro for stand-alone and utility grid-connected applications
6. Design a prototype implementation of a novel efficient nonlinear control system to ensure maximum available Wind/PV energy tracking and utilization.

Design Example 9.6

Develop a load flow solution approach for the distributed power systems. Consider the robustness and power flow stability issues. Assume an interconnected distributed network and the operating voltage of the buses are all the same. However, the PV generation and power dissipation by the houses are not constant, and they vary for changing weather conditions and load demands.

9.9 Day-ahead forecasting in energy management systems

The forecasting is making inferences about the future using current and past information. The determination of endogenous and exogenous variables is the most important part of the forecasting tool. A forecaster solves the linear or nonlinear relationship between features and targets.

Forecasting theory is used in many research areas in literature. But here it is focused on renewable energy technology and energy demand areas. The power generation forecast is an expression of the expectation of electricity production from one or more power generation plants in the near future. The value called production is the power that can be produced and the energy that is produced for each time interval. The energy demand or load forecast is an estimate of the amount of electrical energy that will be consumed in the near future.

The power forecasting of renewable energy sources (RES) is critical for energy generation, transmission, distribution, and the electric markets. The power forecasting is one of the core steps for scheduling of the generation, transmission, and distribution of electrical energy as well as the maintenance of power systems, management of the demand response, and financial planning [36–41]. Although the power forecasting is important to manage the power system planning and operations, inaccurate forecasting might cause insufficient or over-power generation, which yields voltage and frequency fluctuations, power quality problems, and the inability to manage the system economically and efficiently [37–39]. Besides, inaccurate forecasting might also cause power system instability. The forecasted power is also used for the trading of electric power. Incorrect energy forecasting might result in economic losses and even bankruptcy of energy companies. For this reason, companies try to keep their estimation error at low values to avoid these penalties. In addition, incorrect estimation of the power to be generated by renewable resources affects the trading profit of the companies [42].

The reliable operation of power systems relies on the planning of energy generation and consumption. Traditional energy sources can be easily controlled and managed with experience. However, with the widespread integration of renewable energy sources, which have variable and intermittent generation characteristics, electricity grids are becoming more complex, and power systems are becoming more difficult to manage. The correct planning required for the reliable operation of power systems fed by renewable resources depends on the correct estimation of the energy to be obtained from these resources and the load to be consumed by the consumer units.

The time span used to forecast is called the forecasting horizon, which can be divided into four categories as very short term, short term, medium term, and long term [43]. Microgrids are controlled and managed by very short-term forecasts with a time horizon of a few seconds to a few minutes. The short-term horizons might go up to 72 h ahead and they are used to schedule and control power system operations, economic dispatch, unit commitment, and electric market operation. Medium-term forecasting is from a few days to 1 week ahead and long-term forecasting is from a few months to a year or more. Medium and long-term horizons are typically used for network maintenance and planning. The forecast horizon is also categorized to describe the forecast horizon subsections including intra-hour, intra-day, and day-ahead [44]. Intra-hour involves the forecast horizon from a few seconds to an hour and it is used for regulation and safety of the system operation. The intra-day forecast horizon is being done for 1–6 h and it finds applications in electricity trading and electrical load control. The day-ahead forecasts cover a duration of 6–48 h. These forecasts are used for unit commitment, scheduling, and dispatching of electricity. The forecast horizon plays an important role in the accuracy of the forecast. Prior to the design of a forecasting model, the appropriate time horizon for the forecast should be selected. It has been shown that increasing the forecast horizon reduces the accuracy of the model [44].

9.9.1 Data using for renewable energy source power and demand forecasting

The PV and wind powers are related to solar radiation data and wind speed data, respectively. PV module power can be calculated using the solar radiation, module temperature, module efficiency, and module area as in (9.22), where η is the module efficiency, A is the module surface area in m^2, S_{PVM} is the solar radiation on module surface and T_{PVM} is the module operating temperature.

$$P_{\text{PVM}} = \eta A S_{\text{PVM}}[1 - 0.005 \times (T_{\text{PVM}} - 25)] \tag{9.22}$$

There are a number of factors that affect the energy produced by the wind. The most important factor is the wind speed. The rotational kinetic energy formula is used to calculate the energy generation by a wind turbine as explained in Chapter 7 and given in (9.23).

$$P_{\text{wind}} = 0.5 C\rho A v^3 \tag{9.23}$$

where P_{wind} is the wind power, ρ (kg/m^3) is the air density, A (m^2) is the swept area by the wings of the wind turbine, v (m/s) is the wind speed in, and C is Betz limit (0.59) [45].

The power produced from renewable energy sources is affected by meteorological parameters such as temperature, humidity, cloud cover, and precipitation. In the literature, forecasting of renewable energy generation is done using univariate or multivariate features. The multivariate approach also uses some variables (temperature, cloud cover, etc.) that are correlated with the main predictor variable [46–50], while the univariate approach uses only historical data [51–55].

On the other hand, the forecasting can be carried out using either a direct or an indirect model. Direct methods use solar or wind power data directly. Other variables associated with solar or wind power may also be included in this method. In the indirect method, the forecasting is done with forecasted data or variables. For example, PV power predictions use predicted solar irradiance data and wind power predictions use predicted wind speed data.

Data such as historical load, seasonal information, whether the day of the week is a working day or a holiday, and temperature information for the forecast day are used to forecast the electricity load [56–61].

9.9.2 *Methods for renewable energy source power and demand forecasting*

A number of different algorithms and models have been used in the literature for the forecasting of renewable energy and electric load demand. In the last few years, machine learning algorithms [36,38,42,56,57,62,63] and in particular deep learning algorithms [46,51,64–68] have come to the foreward. Although these methods are used alone, hybrid structures are also coming to the attention [38,47,69]. In addition, there are studies where the data are used as raw data, and there are also studies where the data are separated into many frequencies using data decomposition methods [47,70–72].

Machine learning (ML) refers to technologies that have the ability to learn automatically from experience or datasets and is a subset of artificial intelligence (AI). ML algorithms fall into three categories: supervised learning, unsupervised learning, and reinforcement learning. Forecasting problems are a type of regression problem. They can be solved using supervised learning algorithms. Artificial neural network (ANN), k-nearest neighbors (k-NN), and support vector machine (SVM) are some of the ML techniques. On the other hand, there is another technique called deep learning (DL), which is a subset of machine learning and uses improved versions of ANNs to solve problems. The main difference between a DL and a typical neural network is the number of hidden layers and the roles they play. Convolutional neural networks (CNN), gated recurrent unit (GRU), recurrent neural network (RNN), and long short-term memory (LSTM) architecture are some of the recently developed DL techniques.

9.9.3 *Artificial neural network (ANN)*

Artificial neural networks are mathematically simple models of the brain and the nervous system of humans. They work in a similar way to computer networks, but unlike programs written on computers, they have features such as classification, generalization, optimization, and learning, as well as performing a specific task. By evaluating the feature and target data, they can successfully learn their relations with each other.

The brain is the center of the biological nervous system. It is the structure that receives, interprets, and makes decisions. The neural networks in the brain are made up of billions of nerve cells called neurons. Neurons consist of the cell body, synapses, dendrites, and axons as shown in Figure 9.23. The nucleus is located in the cell. Dendrites transmit signals from other neurons to which they are connected to the nucleus. The nucleus collects the signals from the dendrite and transmits them to the axon. These collected signals are processed by the axon and sent to the connections to transmit the new signals produced to other neurons.

Like biological neurons, artificial neurons consist of parts that receive input information, collect and process these signals, and produce outputs. The artificial neuron, which is the most basic cell of artificial neural networks, consists of inputs (dendrites), weights (synapses), summation function, activation function, and output (axon) parts. Neurons receiving one or more inputs produce a single output, and this output value can be an input value for other neurons or a network output result. Figure 9.24 shows the mathematical model of an artificial neuron model. Input values are information that is transmitted to the cell from the outside world or from other neurons.

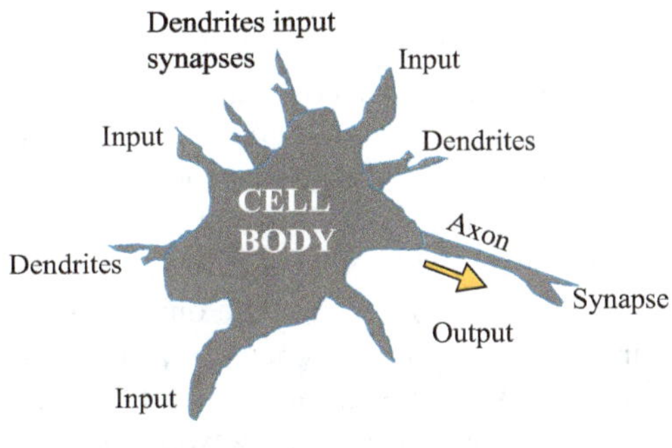

Figure 9.23 The structure of a biological neuron

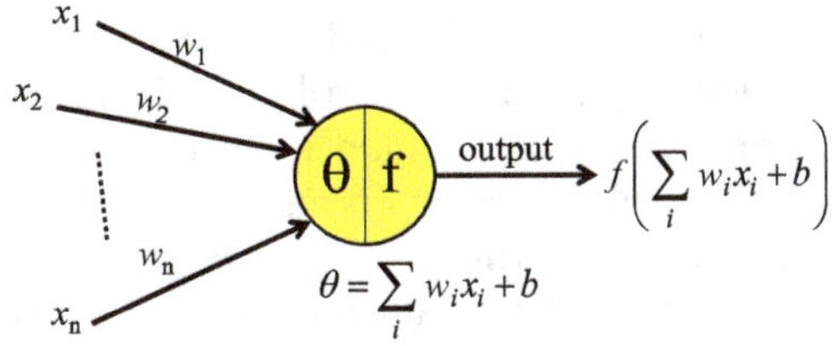

Figure 9.24 Mathematical model of a neuron cell

Each neuron can take one or more input values. For example, $x_1, x_2, \ldots$ and x_n values shown in Figure 9.24 represent the inputs to the neuron. The connections between all the neurons in the neural network are made by means of weights, which are coefficients that store the information of the neural network. In Figure 9.24, the weights are represented by $w_1, w_2, \ldots$ and w_n. These weights can be positive and/or negative real numbers. The iterative sum value refers to the scalar product of the weight and the input vectors. The *net* value of the cell is the summation of the iterative sum and bias, b, values as given in (9.24) and shown in Figure 9.24.

$$net = \sum_i w_i x_i + b \tag{9.24}$$

By processing the *net* value of the neuron, the activation function calculates an output value for the neuron. The output value of a neuron cell can be the input value of other cells or terminates as the output of the current cell. The neuron's output is limited to certain values by the activation function, which can be linear or nonlinear. In particular, nonlinear activation functions have allowed artificial neural networks to tackle complex problems. The activation function that is most commonly used in multi-layer perceptron models is the sigmoid function. Using the sigmoid activation function, the output value o of a neuron can be calculated as in (9.25).

$$o = f(net) = \frac{1}{1 + e^{-net}} \tag{9.25}$$

9.9.4 Multi-layer artificial neural network

Artificial neural networks are formed by combining artificial neurons. Depending on the number of layers, it is analyzed as single-layer and multi-layer ANN. The basic structure of a multi-layer artificial neural network consists of an input layer, one hidden layer, and an output layer. The general structure of a multi-layer artificial neural network with a single hidden layer is shown in Figure 9.25. The input layer is the layer where the input values from the outside world feed the artificial neural network. The input layer has as many neuron cells as the number of inputs, x, from the outside world. These inputs are transmitted from the input layer to the first hidden layer by being multiplied with the w weighting factors. For example, the vector X represents the values of the variables in the input layer in Figure 9.25. As depicted by x_n, there are n input variables.

The hidden layer transmits the information received from the input layer to the next layer as the products of its cell outputs and the related weighting factors of that layer. The number of hidden layers varies according to the problem. In Figure 9.25, t represents the number of neurons in the hidden layer. The output layer processes the information coming from the hidden layers and produces the output value of the network. The outputs produced are sent to the outside world. However, the output value produced is sent back to the artificial neural network if the network structure has feedback. In Figure 9.25, y_m represents the output signals and m represents the number of neurons in the output layer.

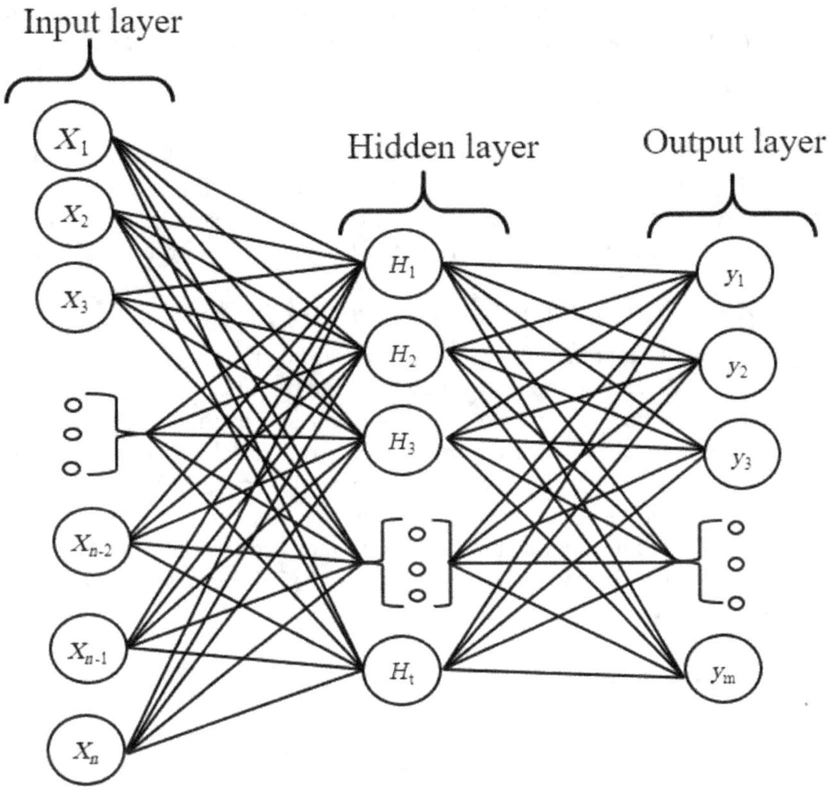

Figure 9.25 Single hidden layer artificial neuron network

9.9.5 Learning process in a neural network

ANN learning is the process of adjusting the values of the weights in the network in order to perform a particular function. The weight values where the information is stored are randomly assigned at the beginning of the process, the input samples are sent and taught to the network by changing the weighting factor values. Once the weights have been set correctly, test sample outputs are produced by the ANN. The accuracy of the outputs obtained gives information about the learning ability of the network. To determine the relationship between input and output data of a problem with artificial neural networks, patterns should be taught to the network using a learning algorithm. If the weights in the network are at the most appropriate values, it is possible for the network to learn this relationship.

The weighting factors are trained by updating them with the addition of a change Δw that includes a learning process as given in (9.26) and (9.27).

$$w_i(new) = w_i(old) + \Delta w_i \tag{9.26}$$

$$\Delta w_i = \alpha \delta x_i \tag{9.27}$$

where α is the learning constant and δ is the error of the related neuron. Equations (9.26) and (9.27) can be combined and written for a multi-layer ANN as in (9.28).

$$w^i_{jk}(new) = w^i_{jk}(old) + \alpha E^{i+1}_k x_{jk} \tag{9.28}$$

where i is a layer number, j is the jth neuron in layer i and k is the kth neuron in layer $(i + 1)$. Therefore, w'^{i}_{jk} represents the weighting factor that is connecting neuron j of layer i to the neuron k of the layer $(i + 1)$. The learning algorithms such as perceptron, delta, or back propagation might be used to train the weightings.

9.9.6 Training an artificial neural network for forecasting

Initially, the weights are randomly assigned for training the network. The randomly assigned weights determine the starting point of the training. Deciding how long to continue training is one of the most important issues in artificial neural networks. There are two options for completing the training process. The first of these is to set a certain amount of fault tolerance (mean absolute error, MAE) and continue training until it reaches a lower error value than that error value. The other option is to choose a fixed training number. Here, the trainer accepts the error that will be obtained as a result of the specified training number [73]. The training flowchart of the network is shown in Figure 9.26.

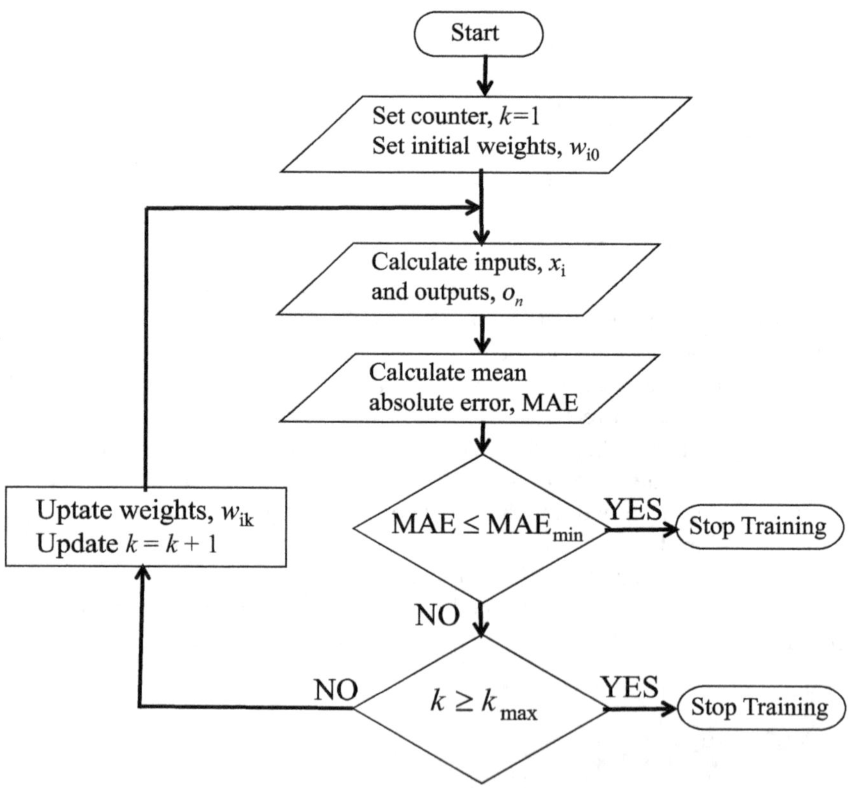

Figure 9.26 The training flowchart of the neural network

Example 9.5

Design a three-hidden layer neural network for forecasting day-ahead (24 h) solar irradiation in the province of Trabzon (latitude: 40.991150, longitude: 39.784964) for the total number of days on the horizontal plane [74]. Consider relative humidity and temperature data, which is adapted from [75]. Use data from the year 2015 to train ANN and use randomly selected sunny, cloudy, and rainy day data from 2016 to test the ANN.

Answer 9.5

This example and its solution are adapted from the conference paper [76], which belongs to the author and his PhD student Sibel Çevik Bektaş.

The designed model has two stages: (1) selecting the most similar day with *similar day selection algorithm* and (2) forecasting. The flowchart of the designed model is given in Figure E9.5(a). A similar day selection algorithm is used to determine the inputs of the neural network. The current temperature, humidity, and a 24-h solar irradiation data of the day, which is similar to the day whose solar irradiation is to be forecasted are used as the inputs to the network. The next 24-h solar irradiation data is the output of the network as the forecasted data. Using the temperature and humidity data for each day, the Euclidean norm (Euclidean distance) is calculated to find the day that is most similar to the day whose solar irradiation is to be forecasted. For each day, the similar day selection algorithm calculates the Euclidean norm with respect to all days. The day that gives the minimum distance is determined as the similar day of the relevant day. The Euclidean norm relation is used as in (9.26) to calculate

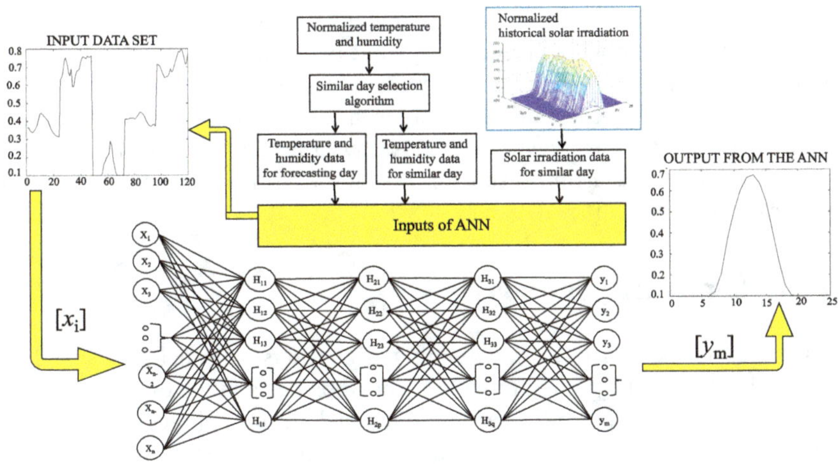

Figure E9.5(a) The flowchart of designed model for Example 9.5

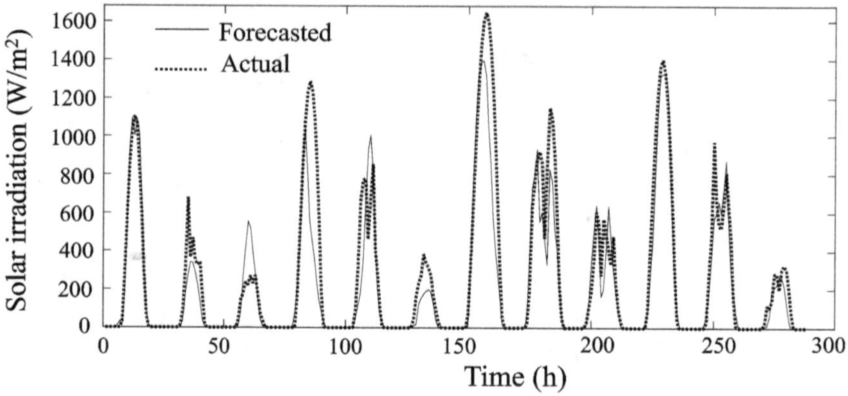

Figure E9.5(b) The forecasting and real value for test samples

the minimum distance.

$$d_i = \left[\sum_{j=1}^{T} (Y_j - X_{ij})^2\right]^{0.5} \quad i = 1, 2, \ldots, n \tag{9.26}$$

where Y_1, Y_2, ..., Y_T is the vector expressing hourly temperature, and humidity values of the days; $X_1, X_2, \ldots, X_T$ is the vector expressing hourly temperature and humidity values of another day. The distance vector calculated for all days and the day that gives the minimum value in $D = [d_1, d_2, \ldots, d_n]$ represents the most similar day among all the days where n is the day number.

As can be seen in Figure E9.5(b), hourly temperature and humidity data for the forecasting day, and hourly temperature, humidity, and irradiation data for the similar day are used as inputs of the ANN. The outputs are the hourly solar irradiation data. A total of 120 inputs and 24 outputs are used in this model. In addition, the ANN has three hidden layers. Sigmoid-type activation functions are used in the *backpropagation* learning algorithm The results are obtained after 20,000 iteration. Figure E9.5(b) shows that the algorithm gives better forecasting results, especially in sunny days. However, in cloudy days, the results are not as accurate as in sunny days. The forecasting result for 12 samples in mean absolute percentage error (MAPE) criteria is 10.94%.

9.10 Cybersecurity in microgrids

Cybersecurity is the practice of defending digital systems, networks, and data against a wide range of dynamic threats. It encompasses a range of measures, from basic best practices to advanced technologies and strong policies. As the digital landscape evolves, cyber threats become increasingly complex, necessitating constant adaptation and innovation by cybersecurity professionals. This synopsis provides an overview of the challenges associated with protecting digital infrastructures. Critical infrastructure is essential for the well-being of modern societies and its smooth operation is crucial for public welfare and economic stability. Cybersecurity plays a vital role in safeguarding this infrastructure and preventing any disruptions that could affect other industries.

Power systems are crucial, particularly in terms of national security, and any cybersecurity intrusion could result in power outages or monetary losses. The importance of electricity systems in vital infrastructure highlights the urgency of having strong cybersecurity safeguards. The continuous digitization of power systems, exemplified by the integration of smart grids, Internet of Things (IoT) devices, energy management, and advanced control systems, is expected to result in unparalleled efficiency and responsiveness. However, this digital transformation also introduces new risks, as the same mechanisms designed to enhance functionality can now be targeted by malicious actors. Cyberattacks on power systems exploit vulnerabilities, which can result in energy delivery interruptions, compromised confidential information, and even physical harm to critical infrastructure. Understanding this intricate relationship is crucial for developing effective cybersecurity plans that mitigate threats and ensure the stability of our electrical infrastructure.

9.10.1 Fundamentals of cybersecurity

A discussion about cybersecurity should begin with a basic understanding of its nature. Cybersecurity refers to the procedures, technologies, processes, and policies used to protect networks, computers, and data against damage, illegal access, and cyberattacks. Its main objective is to ensure the security and resilience of digital assets, and it includes a proactive strategy for identifying, preventing, and responding to potential threats. A set of three interrelated concepts that serve as the foundation of cybersecurity are defined as follows [77,78]:

Confidentiality: To ensure that sensitive information is only accessible to authorized individuals, safeguards such as encryption, access limits, and secure communication routes must be put in place.

Integrity: It is also important to ensure that data remains reliable, consistent, and accurate throughout its existence. Cybersecurity techniques such as digital signatures, hashing, and checksums are essential for identifying and preventing illegal data changes.

Availability: Ensuring authorized users have prompt and continuous access to necessary resources and services. The goal of cybersecurity strategies is to prevent and minimize interruptions caused by unexpected events, technical malfunctions, or cyberattacks.

Having a thorough comprehension of and adherence to the principles of availability, integrity, and confidentiality provides a framework for creating effective cybersecurity solutions. Understanding important phrases and concepts is essential to navigating the world of cybersecurity. This includes crucial components such as:

Firewall: A security system acts as a barrier between a secure internal network and potentially unsafe external networks by monitoring and managing incoming and outgoing network traffic.

Malware: Malware, which includes trojan horses, worms, viruses, and ransomware, is a type of destructive software that is designed to compromise and corrupt computer systems.
Authentication: The process used to verify the identity of a user, system, or device to ensure that only authorized parties are granted access.

Encryption: Data encryption is the process of converting plaintext into a coded format to prevent unauthorized access. It is a crucial tool for safeguarding private information during transmission and storage.

Phishing: Phishing is a type of cybercrime where perpetrators trick victims into revealing sensitive information, such as login credentials or bank account details, through fraudulent emails, texts, or websites.

A strong foundation for understanding cybersecurity and implementing measures to protect digital assets can be established by mastering these core concepts.

9.10.2 Cybersecurity threats

In the rapidly changing digital world, cybersecurity threats have become increasingly complex, varied, and pervasive. It is crucial to understand the nature of these threats to strengthen defenses against future cyberattacks. Common cybersecurity risks include malicious software that are:

(a) *Malware*: Aims to interfere with, harm, or gain unauthorized access to computer systems. This group includes spyware, trojan horses, worms, and viruses.
(b) *Phishing*: Phishing attacks are attempts by dishonest individuals or groups to impersonate reliable organizations in emails, chats, or webpages with the goal of obtaining sensitive information, such as financial information or login passwords.
(c) *Ransomware*: Ransomware is a type of virus that encrypts user data and demands payment, often in Bitcoin, to unlock it. These attacks pose a serious threat to businesses of all kinds as they can completely destroy systems.
(d) *Denial-of-Service (DoS)* and *Distributed Denial-of-Service (DDoS) Attacks*: A denial-of-service attack attempts to saturate a system, network, or website with traffic to prevent authorized users from accessing it. Such attacks cause disruptions to regular operations and may result in service outages.

Power systems, as vital infrastructure, are susceptible to various threats that could have severe consequences. Some notable dangers include:

(a) *SCADA and Industrial Control System (ICS) Attacks*: Cyberattacks targeting Industrial Control Systems (ICS) and Supervisory Control and Data Acquisition (SCADA) can potentially interfere with power system control processes, leading to operational failures.

(b) *Malware Targeting Power Systems*: Malicious programs designed to exploit vulnerabilities in power system components can compromise the integrity and performance of critical equipment in the electrical grid.
(c) *Insider Threats*: Dangers that come from within a company, when workers or outside contractors abuse their access rights to subvert power system security, whether on purpose or accidentally.

Studying real-world events provides invaluable insight into the potential impact of cybersecurity failures. Notable case studies include:

(a) *Stuxnet Worm (2010)*: An advanced worm created to attack Iran's nuclear facilities, which demonstrated the harm that can be caused to physical infrastructure by cyber assaults.
(b) *Ukraine Power Grid Cyberattack (2015 and 2016)*: Cyberattacks on Ukraine's power infrastructure resulted in widespread power outages, highlighting the vulnerability of power systems to well-planned and directed attacks.
(c) *NotPetya Ransomware (2017)*: The NotPetya ransomware outbreak caused major disruptions throughout the world, highlighting the potential collateral harm that can arise from cyber disasters. It is important to note that the ransomware did not deliberately target power systems.

The lessons learned from previous cybersecurity events involving power systems emphasize the need for frequent awareness and training campaigns, multi-layered defenses, and ongoing monitoring and incident response. As demonstrated by the 2015 and 2016 hack on the Ukraine power grid [79], regular training programs and heightened awareness campaigns are crucial in recognizing the human aspect as a possible risk. To build the human element of cybersecurity, it is essential to ensure that operators and staff possess the skills required to identify and respond to potential attacks. Additionally, it is crucial to recognize the importance of multiple layers of security, as a single line of defense is insufficient to combat the constantly evolving cyber threat landscape. Acquiring knowledge from events such as the 2010 Stuxnet worm [80], which exploited several vulnerabilities, highlights the need for a comprehensive defense plan that integrates technological measures, operational guidelines, and increased human awareness. Lessons learned also underscore the importance of incident response capabilities and ongoing monitoring, as demonstrated by events such as the hack on the Ukrainian power grid. Establishing strong procedures for continuing vigilance, rapid detection, and effective reaction ensures a proactive stance against cyberthreats [81]. This enhances the overall resilience of power systems in the face of a constantly changing cybersecurity landscape. Besides the Ukrainian power grid cyberattack, there are several other power system failures listed as follows [82–84].

- Northeast Blackout in the USA (2003) was caused by a software bug resulting in alarm system failure and affected more than 50 million customers.
- The Italy Blackout (2003) occurred due to cascading failures between power and communication infrastructures and left about 56 million people without electricity in Italy.

- Arizona Blackout (2007) affected 100,000 customers because of unexpected activation of the load shedding program.
- Florida Blackout (2008) left 1 million customers without electricity because of disabled relay protection during a diagnostic process.
- Southwest Blackout (2017) was triggered by a monitoring equipment failure at a substation and left around 2.7 million customers without electricity.
- Ukraine Blackout (2015) was initiated by remote cyber intrusions after a malware installation and caused about 225,000 customers to lose their electricity.

These examples show how the undisrupted operation of power systems is important. Although similar events might occur in microgrids, due to the islanding mode operating opportunity the effects of unwanted events can be kept locally without spreading the large areas. However, due to the extensive use of ICTs, microgrids are open to cyberattacks locally, as well, and precautions must be taken to detect and stop the attacks.

9.10.3 Cybersecurity for power systems

The increasing penetration of renewable energy sources such as wind and solar photovoltaics to advanced microgrids as distributed energy resources (DERs) brings new challenges to electrical power distribution networks. Microgrids become more complicated with new issues contrary to their reliability and resiliency operation expectations. Therefore, the inclusion of device control and energy management system (EMS) in microgrids using DERs becomes inevitable resulting in a network structure as given in Figures 9.4 and 9.18. As shown in these figures data is collected from every part of the grid and evaluated by control and management centers, which are also called supervisory control and data acquisition (SCADA) centers. After the evaluation of the data, the generated or developed control and management signals are sent back to the related devices or units for the smooth operation of the system. It is obvious that the increasing installations of renewable DERs in microgrids has gradually converted conventional power distribution systems from traditionally centralized electric power systems to localized power generation and consumption-type power systems by reducing inefficiencies and vulnerabilities included in the long-distance power delivery.

Microgrids with DERs might supply power to local loads during unexpected events such as earthquakes, floating, winter conditions, and cyberattacks by operating in islanding mode. The ability to be operated locally independent from the main power grid, microgrids can make important contributions to electric power system operations by strengthening resiliency.

Since microgrids include two-way data transmission either from each sub-unit to control and management centers, which are local sites or at a remote central site, the information and communication technologies (ICTs) are extensively used in microgrid-type electric power distribution systems. The data transmitted from individual units to control and management centers and vice versa is open to cyberattacks. Thus, the electric power networks equipped with ICTs can be considered as cyber-physical systems. The use of ICTs increases the ability of controllability and manageability of power systems. However, they may also cause unexpected cyberattacks such as remotely manipulating generators by affecting their control signals with a malware virus such as Trojan. The manipulated generators may be overloaded unintentionally and fail, which causes the collapse of the regional electric power system such as the one described in [85]. The attackers usually interfere with the power grids to gain economic benefit, to get profit from contingency in electricity markets, and to create/support terrorism [86].

The stable operation of the power system is one of the important issues in keeping power generation, transmission, distribution, and consumption at an equilibrium operating condition. The main condition of the stable operation of the power system is to make generated power equal to the sum of consumption and losses. Hence the power system is controlled by means of excitation and load–frequency adjustment as explained in Chapter 6. The load–frequency control (LFC) has two control loops, one is the primary control, which is based on the drop of the generator, and the other one is the secondary frequency control to maintain the frequency within the scheduled range and power flow on tie lines within the line capacity range. The attackers such as adversaries can perturb the scheduled frequency by preventing the correct signal from being transferred to the control center.

> The wrong or delayed frequency data may lead the power system to misoperate and collapse.

the objective of cybersecurity of LFC operation has been described in [86] as (1) to find the malicious measurements and prevent incorrect area control error (ACE) and (2) to maintain the balance between generation and demand powers during untrusted measurements. The cyberattacks on load–frequency control of power systems are classified in literature [86–88] as given below.

The Strategic Attack is classified into six sub-classes replay attack, denial of service (DoS) attack, data integrity attack, timing attack, covert attack, and zero dynamics attack. *The replay attack* is a sort of manipulation attack, in which the attacker monitors and records the measured data and replays it later at another time as if it belongs to the current case. *DoS attack* blocks the data transmission channels by pumping extensive data from the field. In *data integrity attacks*, the transmitted data is modified, and false data is sent to the controllers. If the attackers cause a delay in transmission time, the controllers or the decision-makers become late to generate the required signal. These types of attacks are called *timing attacks* and might have serious effects on system operation. Another strategic attack is the one called *covert attack*, in which the attack signal is hidden by replacing the measured output with a calculated system response. The attacker is able to access the communication channels and inject false data that might cause hazardous effects. The *Zero dynamic attack* uses decoupled outputs and the zeros of linear system transfer function to develop attack strategies. This attack requires a good knowledge of system dynamics and the use of related state equations.

Template Attacks affect the magnitude of the signal to be transmitted. In this attack, the magnitude is scaled by a constant or a bias signal. A constant, a pulse, or a random signal is fed continuously. The signal can also be modified by a resonance source in terms of frequency as a template attack.

Location Attack is related to the structure of cyberphysical systems (CBS). The attack can be injected at various load–frequency control system parts such as sensors, controllers, actuators, and loads. The attack on the load side is usually done by affecting the amount of change in load by false signals.

Example 9.6

A two-area load–frequency control system simulation was given in Examples 6.4 and 6.5. Assume that the measured data is passing through a communication network before being used by automatic generation control (AGC) units as shown in Figure E9.6(a). Analyze the effects of the data integrity cyberattacks on the FLC-controlled two-area load–frequency control system, which was simulated in Example 6.5.

Answer 9.6

Data integrity attack: The attackers modify the data by adding a constant or a variable to the measured signal such as described in (E9.6.1)

$$e_1(t) = AGC_1(t) + d \tag{E9.6.1}$$

where d is a disturbance added by the attacker. Without d, the LFC system operates normally resulting in the plots given in Figure E9.5(b). Let the change in demand power be $\Delta P_L = 0.188$ pu and inject a disturbance $d = -2$ at the 30th second of the operation, which lasts 50 s, and analyze the system operating without any controller as seen in Figure E9.6(b).

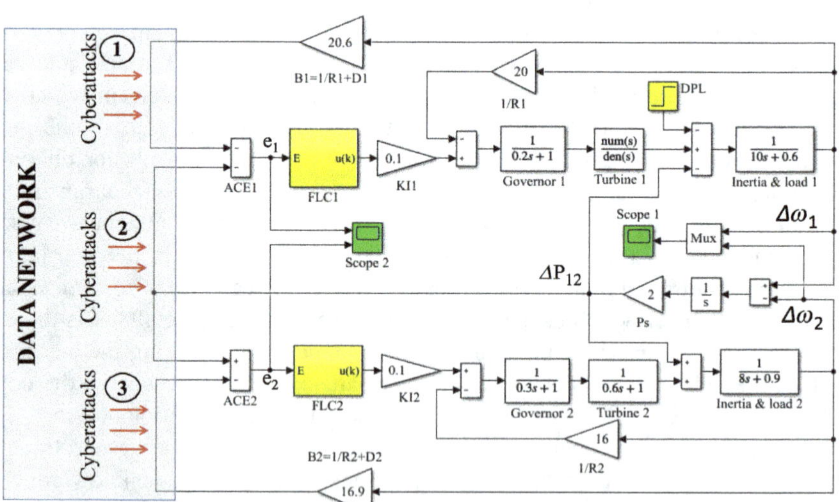

Figure E9.6(a) Block diagram of a two-area LFC open to cyberattacks

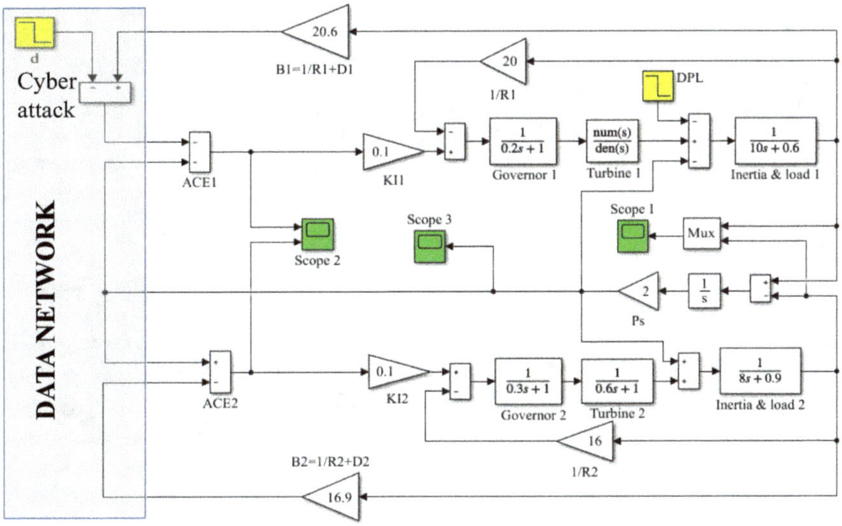

Figure E9.6(b) A data integrity attack on a two-area LFC without any controllers

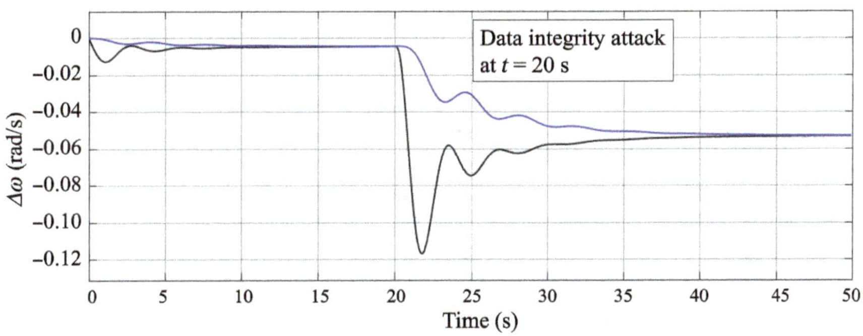

Figure E9.6(c) Response of an uncontrolled two-area LFC under integrity attack

The response of the LFC under attack when there is no controller is given in Figure E9.6(c). A data integrity attack, d, has been injected into ACE1 when data passes through the data network at the instant of t = 20 s. Before the attack, the steady-state values of the frequency changes are measured as -4.729×10^{-3} at t = 15.00 s. If a disturbance $d = -20$ is injected into the system at the cyberattack point at t = 20 s as given in Figure E9.6(c), the steady-state response drops down to -5.316×10^{-2} rad/s and remains at this value. Due to the injected disturbance, the LFC system keeps operating at stable conditions as if everything is normal. However, a large frequency drop will occur causing frequency-sensitive loads to operate under their nominal performances. The higher values of disturbance d cause more frequency drop in the system that the standard conditions might not be

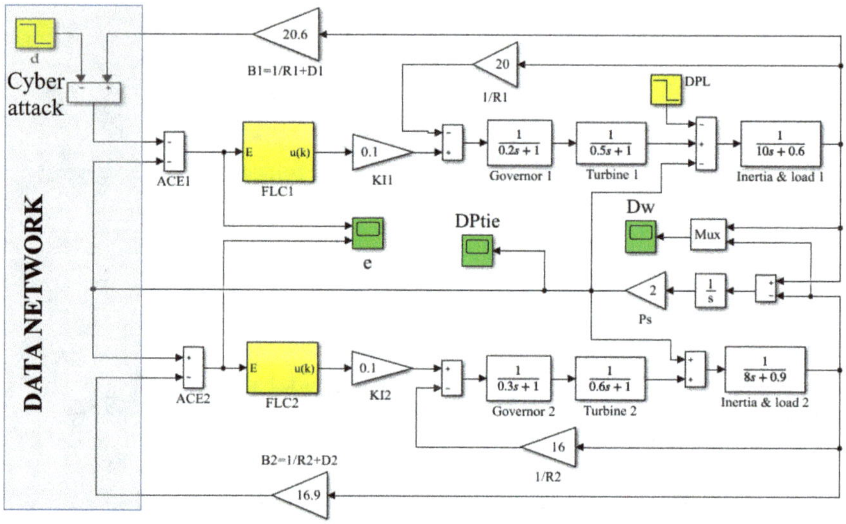

Figure E9.6(d) A data integrity attack on a two-area LFC system controlled by an FLC

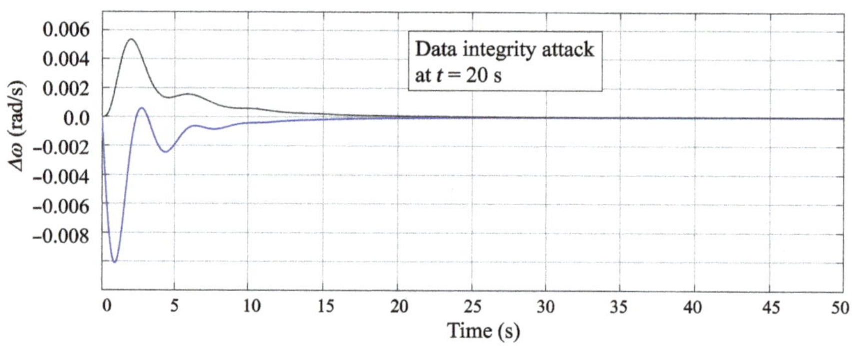

Figure E9.6(e) Response of an FL-controlled two-area LFC under integrity attack

satisfied. Under the attack conditions, the system operator tries to prevent any frequency changes by taking some precautions that may not needed normally.

Let us see what happens if this two-area LFC system is controlled by a fuzzy logic controller (FLC) when there is the same data integrity attack at the same point of the system. The Simulink simulation block diagram of the described system is given in Figure E9.6(d) where two FLCs are located in both areas in order to optimize the ACE from both areas.

The response of the FLC-controlled LFC under attack is given in Figure E9.6(e). A data integrity attack, d, has been injected into ACE1 when data passes through the data network at the instant of $t = 20$ s. Before the attack the steady-state values of the frequency changes are going toward zero, If a disturbance $d = -20$ is injected into the system at the cyberattack point at $t = 20$ s as given in Figure E9.6 (e), the steady-state response does not change and remains around zero. The FLC controls the two load–frequency control systems under data integration cyberattack and keeps the system operating as if there is no infection. The higher values of disturbance d may cause the controller to fail resulting in more frequency changes in the system that the standard conditions might not be satisfied. Under the attack conditions, the system operator tries to prevent any frequency changes by taking some precautions that may not needed normally.

In Example 9.6, the FLC was able to compensate for the effects of the infected data and keep the frequency changes around zero. However, the controller may not be able to stop or eliminate the effects of the cyberattacks. Therefore, the detection of cyber-attacks is an important topic in power systems and microgrids to be studied [87–89].

Design Example 9.7

Develop a timing attack scenario for the LFC system given in Examples 6.5 and 9.6. To develop a timing attack, assume that the area error signal or area error control signal is delayed by a time delay τ, where τ can be taken as equal to 1 or 2 sampling period.

Design Example 9.8

Develop a template attack scenario for the LFC system given in Examples 6.5 and 9.6. To develop a template attack, assume that the area error signal or area error control signal is multiplied by a constant c, where c can be taken between 1 and 100.

Design Example 9.9

Develop a location attack scenario for the LFC system given in Examples 6.5 and 9.6. To develop a location attack, assume that the change in demand power is infected and increased by an amount of z, where z can be taken between 0.1 and 1 pu.

9.10.4 Future trends in cybersecurity for power systems

Three major factors are expected to significantly alter the cybersecurity landscape for power systems in the future. Firstly, new developments in cybersecurity are poised to completely transform the defense systems used to safeguard vital infra-structure. Technological innovations such as decentralized cybersecurity archi-tectures and quantum-resistant cryptography offer stronger defenses against increasingly complex cyberattacks. Second, the cybersecurity paradigm will be influenced by anticipated opportunities and challenges as power systems increas-ingly adopt advanced technologies. The growing attack surface presented by con-nected devices may present challenges, but it also offers opportunities for creative solutions that utilize automation and real-time threat intelligence. Finally, the combination of machine learning (ML) and artificial intelligence (AI) is a crucial factor in enhancing power system cybersecurity. The use of AI and ML algorithms improves the proactive capabilities of cybersecurity measures by analyzing large datasets, identifying abnormalities, and predicting possible attacks with unparalleled accuracy. Power systems can strengthen their cybersecurity posture and move resiliently and adaptably across the changing digital landscape by adopting these upcoming trends.

9.11 Problems

P9.1 A residential site with 50 houses gets the electrical power from photovoltaic (PV) arrays installed on the roofs. The distribution system does not have any other sources except PV arrays and backup batteries. Each house has an installed power capacity of 5 kW and a demand power of 3 kW. Assume that the PV power is sufficient enough when they operate with 80% capacity during sunny days.

Develop a power management algorithm for this system so that the houses always get the required power for their priority loads.

P9.2 Assume that the residential site described in P.9.1 is also connected to a utility grid and gets its required power from the utility when the PV power is not sufficient. Develop a power management algorithm for this system so that the houses always get the required power for all their loads.

P9.3 A residential site with 100 houses gets the electrical power from wind, PV, and utility grid. There are no backup batteries installed. Develop a PMA for this distributed system for energy sustainability. Assume that the priority use of the sources are PV, wind, and utility.

P9.4 The installed loads in a house are listed as refrigerator, air-conditioner, the dishwashing machine, washing machine, drier, iron, vacuum cleaner, TV, lights, house heating, water heating, computer, and other low-power electronic devices. This house gets its electricity from PV and the utility grid.

(a) Categorize the loads in this house as shiftable, non-shiftable, and postponable.

(b) Develop a PMA so that the use of power from the utility grid is minimized while the user comfort is maintained.

(c) Make suggestions for the user to reduce the bill paid for electricity dissipation.

P9.5 Consider the distributed power system given in Figure 9.7. When this system is operated without any power management and controller, the power demand curves are obtained arbitrarily such that the difference between the lowest and highest demand points becomes large. As a result of this, the power transmitted in the distribution power lines is very light sometimes while it is very high sometimes. During the higher peak hours, the power line is overloaded and gets overheated causing high power losses and fast aging of the line.

In order to reduce the power losses and increase the life of the power lines, the company decided to reduce the gap between the highest and lowest demand points and operate the power lines with a power transmission around the planned rated value.

Design a power management algorithm to meet what the power company has planned.

P9.6 A distributed power network uses a centralized control approach because the wind and PV systems supplying power to the network are installed close to each other and generated power from both sources is collected in an AC power pool. The power network is also connected to the utility power grid. However, the utility grid is only activated when the power generation from wind and PV is not high enough.

Design a PMA and an FDM for this system and explain how FDM and PMA are used to manage this power system for power loss minimization and energy efficiency.

P9.7 Develop an energy management (EM) algorithm so that the sources are used according to the priority and management constraints, which are given below.

The priority of the sources:

1. Solar PV
2. Wind energy system
3. Storage
4. Combined heat and power
5. Utility grid

Constraints:

The state of charge (SOC) of the batteries should not be smaller than 20%.

The demand for power to be supplied is categorized as low, planned, and high.

Consider SOC, demand power, and the availability of the sources to develop the EM algorithm.

P9.8 The output voltage of a PV array, V_{DC}, varies between 0 V and 400 V. This PV array is being used to supply power to a DC load. In order to keep the load voltage V_0 constant at 200 V, The DC/DC converter (Chopper) given in Figure 9.9 is used for interfacing the PV array to the load.

The load voltage V_0 is controlled by the MOSFET's gate signals i_{g1} and i_{g2} which are in the form of pulses as shown in Figures 9.10 and 9.11. The gate signal pulses are obtained by comparing a reference sawtooth signal with the control signal u, which is generated by a controller. If the chopping frequency is 1000 Hz calculate the following:

(a) What would be the conducting period t_{ON} in order to have an output voltage $V_0 = 200$ V from the DC/DC converter when the PV array voltage is 400 V?
(b) What would be the conducting period t_{ON} in order to have an output voltage $V_0 = 200$ V from the DC/DC converter when the PV array voltage is 350 V?
(c) What would be the conducting period t_{ON} in order to have an output voltage $V_0 = 200$ V from the DC/DC converter when the PV array voltage is 300 V?
(d) What would be the output voltage V_0 if the conducting period t_{ON} is 0.00025 s when the PV array voltage is 400 V, 350 V, 300 V, or 250 V?

P9.9 Draw utility-connected connection diagrams of the following electrical sources. Show all required converters and filters in your diagrams.
   (a) Utility-connected wind energy systems (WES).
   (b) Utility-connected solar photovoltaic (PV) systems.
   (c) Stand-alone wind energy systems (WES).
   (d) Stand-alone solar photovoltaic (PV) systems.

P9.10 An e-car charging station is getting its power from a PV array and a wind energy system. The PV array can generate 10 kWp with an operating voltage of 500 V DC under the best environmental conditions. On the other hand, the wind energy system is able to generate 20 kWp with 500 V output voltage and 50 Hz frequency.

The charging station (a DC Power storage) is also being used to supply power to the utility grid during peak consumption hours besides its car charging duty.

Draw a detailed connection diagram of the system described above.

CAUTION: Your connection diagram must include a PV energy system, Wind energy system, charging station, and utility grid besides the required power electronic converters. You can represent the converters by blocks. However, you must draw the required three-phase, single-phase, and DC lines. Do not include control blocks in your drawing.

References

[1] Wood A.J. and Wollenberg B.F. *Power Generation, Operation and Control*. New York: John Wiley & Sons, 1984.

[2] Benini L. and De Micheli G. *Dynamic Power Management: Design Techniques and CAD Tools*. Dordrecht: Kluwer, 1998.

[3] Maghsoodlou F., Masiello R. and Ray T. 'Energy management systems', *IEEE Power Energy*. 2004, 2(5), pp. 49–57.

[4] Miller R.H. *Power System Operation*. New York: McGraw-Hill, 1983.

[5] Zhu J. *Optimization of Power System Operation*. Piscataway, NJ: Wiley-IEEE Press, 2015.

[6] U.S. Department of Energy Office of Energy Efficiency and Renewable Energy, 'Using distributed energy resources: a how-to guide for federal facility managers'—Distributed energy resources: a how-to guide, www.eren.doe.gov/femp/

[7] Lasseter R.H. 'Micro-grids', *IEEE Power Engineering Society Winter Meeting. Conference Proceedings* (Cat. No. 02CH37309), New York, NY, USA, 2002, Vol. 1, pp. 305–308, https://doi.org/10.1109/PESW.2002.985003.

[8] Moudud M., Meegahapola L., Vahidnia A. and Datta M. 'Analysis and mitigation of low-frequency oscillations in hybrid AC/DC microgrids with dynamic loads', *IET Generation, Transmission & Distribution*. 2019; 13 (9):1477–1488.

[9] Soni N., Doolla S. and Chandorkar M.C. 'Analysis of frequency transients in isolated microgrids', *IEEE Transactions on Industry Applications*. 2017; 53 (6):5940–5951, https://doi.org/10.1109/TIA.2017.2746619.

[10] Hamzeh M., Ghafouri M., Karimi H., Sheshyekani K. and Guerrero J.M. 'Power oscillations damping in DC microgrids', *IEEE Transactions on Energy Conversion*. 2016; 31(3):970–980, https://doi.org/10.1109/TEC.2016.2542266.

[11] Guo J., Chen T., Chaudhuri B. and Hui S.Y.R. 'Stability of isolated microgrids with renewable generation and smart loads', *IEEE Transactions on Sustainable Energy*. 2020; 11(4):2845–2854, https://doi.org/10.1109/TSTE.2020.2980276.

[12] Zhao Z., Xi J., Xu J., *et al.* 'Assessment and mitigation of multi-mode oscillations in wind-solar hybrid multi-microgrids', in *IEEE Transactions on Smart Grid*. 2024;15(2):1330–1345, https://doi.org/10.1109/TSG.2023.3307178.

[13] Saghafi H., Karshenas H. and Bakhshai A. 'Application of integrated series compensator in damping power oscillations in standalone microgrids', *Canadian Journal of Electrical and Computer Engineering*. 2015; 38(1):2–9.

[14] Ahmed M., Vahidnia A., Datta M. and Meegahapola L. 'Adaptive power oscillation damping controller for a hybrid AC/DC microgrid', *IEEE Access*. 2020; 8:69482–69495, https://doi.org/10.1109/ACCESS.2020.2985978.

[15] Colson C.M. and Nehrir M.H. 'A review of challenges to real-time power management of microgrids', *2009 IEEE Power & Energy Society General Meeting*, Calgary, AB, Canada, 2009, pp. 1–8, https://doi.org/10.1109/PES.2009.5275343.

[16] Sarker P.S., Venkataramanan V., Sebastian Cardenas D., *et al.* 'Cyber-physical security and resiliency analysis testbed for critical microgrids with IEEE 2030.5,' *2020 8th Workshop on Modeling and Simulation of Cyber-Physical Energy Systems*, Sydney, NSW, Australia, 2020, pp. 1–6, https://doi.org/10.1109/MSCPES49613.2020.9133689.

[17] Driesen J. and Belmans R. 'Distributed generation: challenges and possible solutions', *IEEE Power Engineering Society General Meeting*, Montreal, QC, Canada, 18–22 June 2006.

[18] Dondi P., Bayoumi D., Haederli C., Julian D. and Suter M. 'Network integration of distributed power generation', *Journal of Power Sources*. 2002; 106:1–9.

[19] Bacha, S., Picault D., Burger B., Etxeberria-Otadui I. and Martins J. 'Photovoltaics in microgrids: an overview of grid integration and energy management aspects,' *IEEE Industrial Electronics Magazine*. 2015; 9(1):33–46, https://doi.org/10.1109/MIE.2014.2366499.

[20] Zubieta L.E. 'Are microgrids the future of energy? DC microgrids from concept to demonstration to deployment,' *IEEE Electrification Magazine*. 2016; 4(2):37–44, https://doi.org/10.1109/MELE.2016.2544238.

[21] Bollen M. and Hassan F. *Integration of Distributed Generation in the Power System*. New York: John Wiley & Sons, 2011.

[22] Andleeb M., Lateef Khan K., Hussain S. and Javed Iqbal S. 'Non-linear modeling and control of DC-DC buck and boost converters for EV application,' *1st International Conference on Sustainable Technology for Power and Energy Systems (STPES)*, Srinagar, India, 2022, pp. 1–6, https://doi.org/10.1109/STPES54845.2022.10006486.

[23] Azer P. and Emadi A., 'Generalized state space average model for multi-phase interleaved buck, boost and buck-boost DC–DC converters: transient, steady-state and switching dynamics,' *IEEE Access*. 2020; 8:77735–77745, https://doi.org/10.1109/ACCESS.2020.2987277.

[24] Rao A.V. and Guruswamy K.P. 'Analysis, design and simulation of non-inverting buck-boost DC–DC converter for battery charging', *International Conference on Disruptive Technologies for Multi-Disciplinary Research and Applications (CENTCON)*, Bengaluru, India, 2021, pp. 79–84, https://doi.org/10.1109/CENTCON52345.2021.9688165.

[25] Kaye R. and Kalam A. 'Non-isolated DC/DC buck/boost converters for off-grid hybrid renewable system', *29th Australasian Universities Power Engineering Conference (AUPEC)*, Nadi, Fiji, 2019, pp. 1–5, https://doi.org/10.1109/AUPEC48547.2019.211787.

[26] Dewan S.B. and Straughen A. *Power Semiconductor Circuits*. New York: John Wiley & Sons, 1975.

[27] Altas I.H. 'A simplified dynamic model for the control of PV array-PMDC motor scheme', *International Conference on Electrical Machines—ICEM'98*, 2–4 September 1998, İstanbul, Turkey, pp. 1190–1195.

[28] Mengi O.O. 'An intelligent energy management system for sustainability in renewable energy systems', PhD Thesis, Printed in Turkish as 'Yenilenebilir Enerji Sistemlerinde Süreklilik İçin Akıllı Bir Enerji Yönetim Sistemi', Karadeniz Technical University, 2011.

[29] Trudnowski D.J., McReynolds W.L. and Johnson M. 'Real-time very short-term load prediction for power-system automatic generation control', *IEEE Transactions on Control Systems Technology*. 2001; 9(2):254–260.

[30] Sinha A.K. and Mondal J.K. 'Dynamic state estimator using ANN based bus load prediction', *IEEE Transactions on Power Systems*. 1999; 14(4):1219–1225.

[31] Matthewman P.D. and Nicholson H. 'Techniques for load prediction in the electricity-supply industry', *Proceedings of the Institution of Electrical Engineering*. 1968; 115(10):1451–1457.

[32] De Brabandere K., Vanthournout K., Driesen J., Deconinck G. and Belmans R. 'Control of microgrids', IEEE Power Engineering Society General Meeting, Tampa, FL, USA, 24–28 June 2007.

[33] Lopes J.A.P., Moreira C.L. and Madureira A.G. 'Defining control strategies for microgrids islanded operation', *IEEE Transactions on Power Systems*. 2006; 21(2):916–924.

[34] Xiang J., Hill D.J. and Ma J. 'Distributed power control for transient stability of multimachine power systems', *IEEE Journal on Emerging and Selected Topics in Circuits and Systems*. 2017; 7(3):383–392.

[35] Cakmak R. 'Intelligency power dissipation and energy management in utility integrated residential areas with PV installations', PhD Thesis, *Turkish 'Şebeke Bağlantılı Fv Güneş Enerji Sistemine Sahip Konutlarda Akıllı Güç Tüketimi Ve Enerji Yönetimi', under study*, Karadeniz Technical University, Trabzon, Turkey.

[36] Antonanzas J., Osorio R. Escobar, R., *et al.* 'Review of photovoltaic power forecasting', *Solar Energy*. 2016; 136:78–111. https://doi.org/10.1016/j.solener.2016.06.069.

[37] Ahmed R., Sreeram V., Mishra Y. and Arif M.D. 'A review and evaluation of the state-of-the-art in PV solar power forecasting: techniques and optimization', *Renewable and Sustainable Energy Reviews*. 2020; 124:109792.

[38] Wan C., Zhao J., Song Y., *et al.* 'Photovoltaic and solar power forecasting for smart grid energy management', *CSEE Journal of Power and Energy Systems*. 2015; 1(4):38–46.

[39] Guermoui M., Melgani F., Gairaa. K. and Mekhalfi M.L. 'A comprehensive review of hybrid models for solar radiation forecasting', *Journal of Cleaner Production*. 2020;258:120357, https://doi.org/10.1016/j.jclepro.2020.120357.

[40] Chang W.Y. 'A literature review of wind forecasting methods', *Journal of Power and Energy Engineering*. 2014; 2(4):161.

[41] Gao M., Li J., Hong F. and Long D. 'Day-ahead power forecasting in a large-scale photovoltaic plant based on weather classification using LSTM', *Energy*. 2019; 187:115838.

[42] Bitar E.Y., Rajagopal R., Khargonekar P.P., Poolla K. and Varaiya P. 'Bringing wind energy to market', *IEEE Transactions on Power Systems*. 2012; 27(3):1225–1235.

[43] Mellit A. Pavan A.M., Ogliari E., Leva S. and Lughi V. 'Advanced methods for photovoltaic output power forecasting: a review', *Applied Sciences*. 2020; 10(2):487.

[44] Rahimi N., Park S., Choi W., *et al.* 'A comprehensive review on ensemble solar power forecasting algorithms', *Journal of Electrical Engineering & Technology*. 2023; 18(2):719–733.

[45] Carrillo C., Obando Montaño A.F., Cidrás J., Díaz-Dorado E. 'Review of power curve modelling for wind turbines', *Renewable and Sustainable Energy Reviews*. 2013;21:572–581, https://doi.org/10.1016/j.rser.2013.01.012.

[46] Zang H., Cheng L., Ding T., *et al.* 'Day-ahead photovoltaic power forecasting approach based on deep convolutional neural networks and meta learning', *International Journal of Electrical Power & Energy Systems*. 2020; 118:105790.

[47] Lan H., Yin H., Hong Y.Y., *et al.* 'Day-ahead spatio-temporal forecasting of solar irradiation along a navigation route', *Applied Energy*. 2018; 211:15–27.

[48] Yan J., Zhang H., Liu Y., *et al.* 'Forecasting the high penetration of wind power on multiple scales using multi-to-multi mapping', *IEEE Transactions on Power Systems*. 2018; 33(3):3276–3284.

[49] Gigoni L., Betti A., Crisostom E., *et al.* 'Day-ahead hourly forecasting of power generation from photovoltaic plants', *IEEE Transactions on Sustainable Energy*. 2017; 9(2):831–842.

[50] Ogliari E., Dolara A., Manzolini G. and Leva S. 'Physical and hybrid methods comparison for the day ahead PV output power forecast', *Renewable Energy*. 2017; 113:11–21.

[51] Hong Y.Y., Martinez J.J.F. and Fajardo A.C. 'Day-ahead solar irradiation forecasting utilizing gramian angular field and convolutional long short-term memory', *IEEE Access*. 2020; 8:18741–18753.

[52] Zhang C.Y., Chen C.P., Gan M. and Chen L. 'Predictive deep Boltzmann machine for multiperiod wind speed forecasting', *IEEE Transactions on Sustainable Energy*. 2015; 6(4):1416–1425.

[53] Karakuş O., Kuruoğlu E.E. and Altınkaya M.A. 'One-day ahead wind speed/power prediction based on polynomial autoregressive model', *IET Renewable Power Generation*. 2017; 11(11):1430–1439.

[54] Khodayar M., Wang J. and Manthouri M. 'Interval deep generative neural network for wind speed forecasting', *IEEE Transactions on Smart Grid*. 2018; 10(4):3974–3989.

[55] Zhang Y., Sun H. and Guo Y. 'Wind power prediction based on PSO-SVR and grey combination model', *IEEE Access*. 2019; 7:136254–136267.

[56] Tong C., Li J., Lang C., *et al.* 'An efficient deep model for day-ahead electricity load forecasting with stacked denoising auto-encoders', *Journal of Parallel and Distributed Computing*. 2018; 117:267–273.

[57] Lahouar A. and Slama J.B.H. 'Day-ahead load forecast using random forest and expert input selection', *Energy Conversion and Management*. 2015; 103:1040–1051.

[58] Cai M., Pipattanasomporn M. and Rahman S. 'Day-ahead building-level load forecasts using deep learning vs. traditional time-series techniques', *Applied Energy*. 2019; 236:1078–1088.

[59] Chen H., Wang S., Wang S. and Li Y. 'Day-ahead aggregated load forecasting based on two-terminal sparse coding and deep neural network fusion', *Electric Power Systems Research*. 2019; 177:105987.

[60] Liu X., Zhang Z. and Song Z. 'A comparative study of the data-driven day-ahead hourly provincial load forecasting methods: from classical data mining to deep learning', *Renewable and Sustainable Energy Reviews*. 2020; 119:109632.

[61] Park K., Yoon S. and Hwang E. 'Hybrid load forecasting for mixed-use complex based on the characteristic load decomposition by pilot signals', *IEEE Access*. 2019; 7:12297–12306.

[62] Chang W.Y. 'A literature review of wind forecasting methods', *Journal of Power and Energy Engineering*. 2014; 2(4):161.

[63] Zang H., Cheng L., Ding T., *et al.* 'Hybrid method for short-term photovoltaic power forecasting based on deep convolutional neural network', *IET Generation, Transmission & Distribution*. 2018; 12(20):4557–4567.

[64] Qing X. and Niu Y. 'Hourly day-ahead solar irradiance prediction using weather forecasts by LSTM', *Energy*. 2018; 148:461–468.

[65] Srivastava S. and Lessmann S. 'A comparative study of LSTM neural networks in forecasting day-ahead global horizontal irradiance with satellite data', *Solar Energy*. 2018; 162:232–247.

[66] Zhou H., Zhang Y., Yang L., *et al.* 'Short-term photovoltaic power forecasting based on long short term memory neural network and attention mechanism', *IEEE Access*. 2019; 7:78063–78074.

[67] Shi H., Xu M. and Li R. 'Deep learning for household load forecasting—a novel pooling deep RNN', *IEEE Transactions on Smart Grid*. 2017; 9 (5):5271–5280.

[68] Raza M.Q., Mithulananthan N., Li J., Lee K.Y. and Gooi H.B. 'An ensemble framework for day-ahead forecast of PV output power in smart grids', *IEEE Transactions on Industrial Informatics*. 2018; 15(8):4624–4634.

[69] Liu H., Mi X.W. and Li Y.F. 'Wind speed forecasting method based on deep learning strategy using empirical wavelet transform, long short term memory neural network and Elman neural network', *Energy Conversion and Management*. 2018; 156:498–514.

[70] Shi X., Lei X., Huang Q., *et al.* 'Hourly day-ahead wind power prediction using the hybrid model of variational model decomposition and long short-term memory', *Energies*. 2018; 11(11):3227.

[71] He X., Nie Y., Guo H. and Wang J. 'Research on a novel combination system on the basis of deep learning and swarm intelligence optimization algorithm for wind speed forecasting', *IEEE Access*. 2020; 8:51482–51499.

[72] Öztemel E., *Yapay Sinir Ağları*. İstanbul: Papatya Yayıncılık, 2003.

[73] MERRA-2 Meteorological Re-Analysis. Available at http://www.soda-pro.com/web-services/meteo-data/merra.

[74] Copernicus Atmosphere Monitoring Service (CAMS) Radiation Service, Available at http://www.soda-pro.com/web-services/radiation/camsradiation-service.

[75] Çevik S., Çakmak R. and Altaş İ.H. 'A day ahead hourly solar radiation forecasting by artificial neural networks: a case study for Trabzon province', *2017 International Artificial Intelligence and Data Processing Symposium (IDAP)*, Malatya, Turkey, 2017, pp. 1–6, https://doi.org/10.1109/IDAP.2017.8090223

[76] Anderson J. 'Computer security planning study', Technical Report ESD-TR-73-51. Air Force Electronic System Division, 1972.

[77] Vanderham J. 'Toward a better understanding of cybersecurity', *Digital Threats: Research and Practice*. 2021; 2(3):Article 18.

[78] Styczynski J. and Beach-Westmoreland N. 'When the lights went out', Booz Allen Hamilton, 2019. C.07.031.19. Available at www.boozallen.com/ICS (accessed 28 January 2024).

[79] Karnouskos S. 'Stuxnet worm impact on industrial cyber-physical system security,' *IECON 2011—37th Annual Conference of the IEEE Industrial*

Electronics Society, Melbourne, VIC, Australia, 2011, pp. 4490–4494, https://doi.org/10.1109/IECON.2011.6120048.

[80] Sheikh Z.A. and Singh Y. 'A hybrid threat assessment model for security of cyber physical systems', *2022 Seventh International Conference on Parallel, Distributed and Grid Computing (PDGC)*, Solan, Himachal Pradesh, India, 2022, pp. 582–587, https://doi.org/10.1109/PDGC56933.2022.10053332.

[81] Fernandez E.B. 'Preventing and unifying threats in cyberphysical systems,' *2016 IEEE 17th International Symposium on High Assurance Systems Engineering (HASE)*, Orlando, FL, USA, 2016, pp. 292–293, https://doi.org/10.1109/HASE.2016.50.

[82] Bompard E., Huang T., Wu Y. and Cremenescu M. 'Classification and trend analysis of threats origins to the security of power systems,' *International Journal of Electrical Power & Energy Systems*. 2013; 50(1):50–64.

[83] Lloyd's, 'Business blackout: the insurance implications of a cyber attack on the US power grid,' 2015.

[84] Buldyrev S.V., Parshani R., Paul G., Eugene Stanley H. and Havlin S. 'Catastrophic cascade of failures in interdependent networks,' *Nature*. 2010; 464:1025–1028.

[85] Smith E., Corzine, D., Racey, D., *et al.* 'Going beyond cybersecurity compliance,' *IEEE Power and Energy Magazine*. 2016; 14(5):48–56.

[86] Saxena S., Bhatia S. and Gupta R. 'Cybersecurity analysis of load frequency control (LFC) in power systems: a survey', *Designs*. 2021; 5:52. https://doi.org/10.3390/designs5030052.

[87] Sarangan S., Singh V.K. and Govindarasu M. 'Cyber attack-defense analysis for automatic generation control with renewable energy sources', Proceedings of the 2018 North American Power Symposium (NAPS), Fargo, ND, USA, 9–11 September 2018; pp. 1–6.

[88] Mohan A.M., Meskin N., Mehrjerdi H. 'A comprehensive review of the cyber-attacks and cyber-security on load frequency control of power systems', *Energies*. 2020: 13:3860. https://doi.org/10.3390/en13153860.

[89] Wang Q., Cai X., Tang Y. and Ni M. 'Methods of cyber-attack identification for power systems based on bilateral cyber-physical information', *Electrical Power and Energy Systems*. 2021; 125:106515, https://doi.org/10.1016/j.ijepes.2020.106515.

Index

www.ingramcontent.com/pod-product-compliance
Lightning Source LLC
LaVergne TN
LVHW011538200926
848484LV00001B/281